Environmental Fluid Mechanics

ENVE 264

Cal Poly State

University

ENVIRONMENTAL ENGINEERING

create.mheducation.com

ISBN-13: 9781121845534

ISBN-10: 1121845533

Contents

Full Table of Contents   1
Introduction and Basic Concepts   25
Properties of Fluids   61
Pressure and Fluid Statics   99
Fluid Kinematics   157
Bernoulli and Energy Equations   209
Momentum Analysis of Flow Systems   267
Dimensional Analysis and Modeling   315
Internal Flow   371
Property Tables and Charts (SI Units)   461
Property Tables and Charts (English Units)   479
Glossary   493
Index   507
Credits   523

FLUID MECHANICS

■ ■ ■ ■ ■ ■ ■

FUNDAMENTALS AND APPLICATIONS

Third Edition

FLUID MECHANICS

FUNDAMENTALS AND APPLICATIONS

THIRD EDITION

YUNUS A. ÇENGEL
Department of
Mechanical
Engineering
University of Nevada,
Reno

JOHN M. CIMBALA
Department of
Mechanical and
Nuclear Engineering
The Pennsylvania
State University

FLUID MECHANICS: FUNDAMENTALS AND APPLICATIONS, THIRD EDITION

Published by McGraw-Hill, a business unit of The McGraw-Hill Companies, Inc., 1221 Avenue of the Americas, New York, NY 10020. Copyright © 2014 by The McGraw-Hill Companies, Inc. All rights reserved. Printed in the United States of America. Previous editions © 2006 and 2010. No part of this publication may be reproduced or distributed in any form or by any means, or stored in a database or retrieval system, without the prior written consent of The McGraw-Hill Companies, Inc., including, but not limited to, in any network or other electronic storage or transmission, or broadcast for distance learning.

Some ancillaries, including electronic and print components, may not be available to customers outside the United States.

This book is printed on acid-free paper.

1 2 3 4 5 6 7 8 9 0 DOW/DOW 1 0 9 8 7 6 5 4 3

ISBN 978-0-07-338032-2
MHID 0-07-338032-6

Senior Vice President, Products & Markets: *Kurt L. Strand*
Vice President, General Manager: *Marty Lange*
Vice President, Content Production & Technology Services: *Kimberly Meriwether David*
Managing Director: *Michael Lange*
Executive Editor: *Bill Stenquist*
Marketing Manager: *Curt Reynolds*
Development Editor: *Lorraine Buczek*
Director, Content Production: *Terri Schiesl*
Project Manager: *Melissa M. Leick*
Buyer: *Susan K. Culbertson*
Media Project Manager: *Prashanthi Nadipalli*
Cover Image: *Purestock/SuperStock.*
Cover Designer: *Studio Montage, St. Louis, MO*
Typeface: *10.5/12 Times Roman*
Compositor: *RPK Editorial Services*
Printer: *R. R. Donnelly—Willard*

Library of Congress Cataloging-in-Publication Data on File

www.mhhe.com

Dedication

To all students, with the hope of stimulating their desire to explore our marvelous world, of which fluid mechanics is a small but fascinating part. And to our wives Zehra and Suzy for their unending support.

ABOUT THE AUTHORS

Yunus A. Çengel is Professor Emeritus of Mechanical Engineering at the University of Nevada, Reno. He received his B.S. in mechanical engineering from Istanbul Technical University and his M.S. and Ph.D. in mechanical engineering from North Carolina State University. His research areas are renewable energy, desalination, exergy analysis, heat transfer enhancement, radiation heat transfer, and energy conservation. He served as the director of the Industrial Assessment Center (IAC) at the University of Nevada, Reno, from 1996 to 2000. He has led teams of engineering students to numerous manufacturing facilities in Northern Nevada and California to do industrial assessments, and has prepared energy conservation, waste minimization, and productivity enhancement reports for them.

Dr. Çengel is the coauthor of the widely adopted textbook *Thermodynamics: An Engineering Approach*, 7th edition (2011), published by McGraw-Hill. He is also the co-author of the textbook *Heat and Mass Transfer: Fundamentals & Applications*, 4th Edition (2011), and the coauthor of the textbook *Fundamentals of Thermal-Fluid Sciences*, 4th edition (2012), both published by McGraw-Hill. Some of his textbooks have been translated to Chinese, Japanese, Korean, Spanish, Turkish, Italian, and Greek.

Dr. Çengel is the recipient of several outstanding teacher awards, and he has received the ASEE Meriam/Wiley Distinguished Author Award for excellence in authorship in 1992 and again in 2000.

Dr. Çengel is a registered Professional Engineer in the State of Nevada, and is a member of the American Society of Mechanical Engineers (ASME) and the American Society for Engineering Education (ASEE).

John M. Cimbala is Professor of Mechanical Engineering at The Pennsylvania State University, University Park. He received his B.S. in Aerospace Engineering from Penn State and his M.S. in Aeronautics from the California Institute of Technology (CalTech). He received his Ph.D. in Aeronautics from CalTech in 1984 under the supervision of Professor Anatol Roshko, to whom he will be forever grateful. His research areas include experimental and computational fluid mechanics and heat transfer, turbulence, turbulence modeling, turbomachinery, indoor air quality, and air pollution control. Professor Cimbala completed sabbatical leaves at NASA Langley Research Center (1993-94), where he advanced his knowledge of computational fluid dynamics (CFD), and at Weir American Hydo (2010-11), where he performed CFD analyses to assist in the design of hydroturbines.

Dr. Cimbala is the coauthor of three other textbooks: *Indoor Air Quality Engineering: Environmental Health and Control of Indoor Pollutants* (2003), published by Marcel-Dekker, Inc.; *Essentials of Fluid Mechanics: Fundamentals and Applications* (2008); and *Fundamentals of Thermal-Fluid Sciences*, 4th edition (2012), both published by McGraw-Hill. He has also contributed to parts of other books, and is the author or co-author of dozens of journal and conference papers. More information can be found at www.mne.psu.edu/cimbala.

Professor Cimbala is the recipient of several outstanding teaching awards and views his book writing as an extension of his love of teaching. He is a member of the American Institute of Aeronautics and Astronautics (AIAA), the American Society of Mechanical Engineers (ASME), the American Society for Engineering Education (ASEE), and the American Physical Society (APS).

BRIEF CONTENTS

CHAPTER ONE
INTRODUCTION AND BASIC CONCEPTS   1

CHAPTER TWO
PROPERTIES OF FLUIDS   37

CHAPTER THREE
PRESSURE AND FLUID STATICS   75

CHAPTER FOUR
FLUID KINEMATICS   133

CHAPTER FIVE
BERNOULLI AND ENERGY EQUATIONS   185

CHAPTER SIX
MOMENTUM ANALYSIS OF FLOW SYSTEMS   243

CHAPTER SEVEN
DIMENSIONAL ANALYSIS AND MODELING   291

CHAPTER EIGHT
INTERNAL FLOW   347

CHAPTER NINE
DIFFERENTIAL ANALYSIS OF FLUID FLOW   437

CHAPTER TEN
APPROXIMATE SOLUTIONS OF THE NAVIER–STOKES EQUATION   515

CHAPTER ELEVEN
EXTERNAL FLOW: DRAG AND LIFT   607

CHAPTER TWELVE
COMPRESSIBLE FLOW   659

CHAPTER THIRTEEN
OPEN-CHANNEL FLOW   725

CHAPTER FOURTEEN
TURBOMACHINERY   787

CHAPTER FIFTEEN
INTRODUCTION TO COMPUTATIONAL FLUID DYNAMICS   879

CONTENTS

Preface   xv

CHAPTER ONE

INTRODUCTION AND BASIC CONCEPTS   1

1–1   Introduction   2
　　　What Is a Fluid?   2
　　　Application Areas of Fluid Mechanics   4

1–2   A Brief History of Fluid Mechanics   6

1–3   The No-Slip Condition   8

1–4   Classification of Fluid Flows   9
　　　Viscous versus Inviscid Regions of Flow   10
　　　Internal versus External Flow   10
　　　Compressible versus Incompressible Flow   10
　　　Laminar versus Turbulent Flow   11
　　　Natural (or Unforced) versus Forced Flow   11
　　　Steady versus Unsteady Flow   12
　　　One-, Two-, and Three-Dimensional Flows   13

1–5   System and Control Volume   14

1–6   Importance of Dimensions and Units   15
　　　Some SI and English Units   17
　　　Dimensional Homogeneity   19
　　　Unity Conversion Ratios   20

1–7   Modeling in Engineering   21

1–8   Problem-Solving Technique   23
　　　Step 1: Problem Statement   24
　　　Step 2: Schematic   24
　　　Step 3: Assumptions and Approximations   24
　　　Step 4: Physical Laws   24
　　　Step 5: Properties   24
　　　Step 6: Calculations   24
　　　Step 7: Reasoning, Verification, and Discussion   25

1–9   Engineering Software Packages   25
　　　Engineering Equation Solver (EES)   26
　　　CFD Software   27

1–10 Accuracy, Precision, and Significant Digits   28
　　　Summary   31
　　　References and Suggested Reading   31
　　　Application Spotlight: What Nuclear Blasts
　　　　and Raindrops Have in Common   32
　　　Problems   33

CHAPTER TWO

PROPERTIES OF FLUIDS   37

2–1   Introduction   38
　　　Continuum   38

2–2   Density and Specific Gravity   39
　　　Density of Ideal Gases   40

2–3   Vapor Pressure and Cavitation   41

2–4   Energy and Specific Heats   43

2–5   Compressibility and Speed of Sound   44
　　　Coefficient of Compressibility   44
　　　Coefficient of Volume Expansion   46
　　　Speed of Sound and Mach Number   48

2–6   Viscosity   50

2–7   Surface Tension and Capillary Effect   55
　　　Capillary Effect   58
　　　Summary   61
　　　Application Spotlight: Cavitation   62
　　　References and Suggested Reading   63
　　　Problems   63

CHAPTER THREE

PRESSURE AND FLUID STATICS   75

3–1   Pressure   76
　　　Pressure at a Point   77
　　　Variation of Pressure with Depth   78

3–2   Pressure Measurement Devices   81
　　　The Barometer   81
　　　The Manometer   84
　　　Other Pressure Measurement Devices   88

3–3   Introduction to Fluid Statics   89

3–4   Hydrostatic Forces on Submerged
　　　Plane Surfaces   89
　　　Special Case: Submerged Rectangular Plate   92

3–5   Hydrostatic Forces on Submerged
　　　Curved Surfaces   95

3–6 Buoyancy and Stability 98
Stability of Immersed and Floating Bodies 101

3–7 Fluids in Rigid-Body Motion 103
Special Case 1: Fluids at Rest 105
Special Case 2: Free Fall of a Fluid Body 105
Acceleration on a Straight Path 106
Rotation in a Cylindrical Container 107

Summary 111
References and Suggested Reading 112
Problems 112

CHAPTER FOUR
FLUID KINEMATICS 133

4–1 Lagrangian and Eulerian Descriptions 134
Acceleration Field 136
Material Derivative 139

4–2 Flow Patterns and Flow Visualization 141
Streamlines and Streamtubes 141
Pathlines 142
Streaklines 144
Timelines 146
Refractive Flow Visualization Techniques 147
Surface Flow Visualization Techniques 148

4–3 Plots of Fluid Flow Data 148
Profile Plots 149
Vector Plots 149
Contour Plots 150

4–4 Other Kinematic Descriptions 151
Types of Motion or Deformation of Fluid Elements 151

4–5 Vorticity and Rotationality 156
Comparison of Two Circular Flows 159

4–6 The Reynolds Transport Theorem 160
Alternate Derivation of the Reynolds Transport Theorem 165
Relationship between Material Derivative and RTT 167

Summary 168
Application Spotlight: Fluidic Actuators 169
References and Suggested Reading 170
Problems 170

CHAPTER FIVE
BERNOULLI AND ENERGY EQUATIONS 185

5–1 Introduction 186
Conservation of Mass 186

The Linear Momentum Equation 186
Conservation of Energy 186

5–2 Conservation of Mass 187
Mass and Volume Flow Rates 187
Conservation of Mass Principle 189
Moving or Deforming Control Volumes 191
Mass Balance for Steady-Flow Processes 191
Special Case: Incompressible Flow 192

5–3 Mechanical Energy and Efficiency 194

5–4 The Bernoulli Equation 199
Acceleration of a Fluid Particle 199
Derivation of the Bernoulli Equation 200
Force Balance across Streamlines 202
Unsteady, Compressible Flow 202
Static, Dynamic, and Stagnation Pressures 202
Limitations on the Use of the Bernoulli Equation 204
Hydraulic Grade Line (HGL) and Energy Grade Line (EGL) 205
Applications of the Bernoulli Equation 207

5–5 General Energy Equation 214
Energy Transfer by Heat, Q 215
Energy Transfer by Work, W 215

5–6 Energy Analysis of Steady Flows 219
Special Case: Incompressible Flow with No Mechanical Work Devices and Negligible Friction 221
Kinetic Energy Correction Factor, α 221

Summary 228
References and Suggested Reading 229
Problems 230

CHAPTER SIX
MOMENTUM ANALYSIS OF FLOW SYSTEMS 243

6–1 Newton's Laws 244
6–2 Choosing a Control Volume 245
6–3 Forces Acting on a Control Volume 246
6–4 The Linear Momentum Equation 249
Special Cases 251
Momentum-Flux Correction Factor, β 251
Steady Flow 253
Flow with No External Forces 254

6–5 Review of Rotational Motion and Angular Momentum 263

6–6   The Angular Momentum Equation   265

Special Cases   267
Flow with No External Moments   268
Radial-Flow Devices   269

Application Spotlight: Manta Ray
     Swimming   273

Summary   275
References and Suggested Reading   275
Problems   276

C H A P T E R   S E V E N

**DIMENSIONAL ANALYSIS
AND MODELING   291**

7–1   Dimensions and Units   292

7–2   Dimensional Homogeneity   293

Nondimensionalization of Equations   294

7–3   Dimensional Analysis and Similarity   299

7–4   The Method of Repeating Variables
and The Buckingham Pi Theorem   303

Historical Spotlight: Persons Honored
     by Nondimensional Parameters   311

7–5   Experimental Testing, Modeling,
and Incomplete Similarity   319

Setup of an Experiment and Correlation of
     Experimental Data   319
Incomplete Similarity   320
Wind Tunnel Testing   320
Flows with Free Surfaces   323

Application Spotlight: How a Fly Flies   326
Summary   327
References and Suggested Reading   327
Problems   327

C H A P T E R   E I G H T

INTERNAL FLOW   347

8–1   Introduction   348

8–2   Laminar and Turbulent Flows   349

Reynolds Number   350

8–3   The Entrance Region   351

Entry Lengths   352

8–4   Laminar Flow in Pipes   353

Pressure Drop and Head Loss   355
Effect of Gravity on Velocity and Flow Rate
     in Laminar Flow   357
Laminar Flow in Noncircular Pipes   358

8–5   Turbulent Flow in Pipes   361

Turbulent Shear Stress   363
Turbulent Velocity Profile   364
The Moody Chart and the Colebrook Equation   367
Types of Fluid Flow Problems   369

8–6   Minor Losses   374

8–7   Piping Networks and Pump Selection   381

Series and Parallel Pipes   381
Piping Systems with Pumps and Turbines   383

8–8   Flow Rate and Velocity Measurement   391

Pitot and Pitot-Static Probes   391
Obstruction Flowmeters: Orifice, Venturi,
     and Nozzle Meters   392
Positive Displacement Flowmeters   396
Turbine Flowmeters   397
Variable-Area Flowmeters (Rotameters)   398
Ultrasonic Flowmeters   399
Electromagnetic Flowmeters   401
Vortex Flowmeters   402
Thermal (Hot-Wire and Hot-Film) Anemometers   402
Laser Doppler Velocimetry   404
Particle Image Velocimetry   406
Introduction to Biofluid Mechanics   408

Application Spotlight: PIV Applied to Cardiac
     Flow   416

Summary   417
References and Suggested Reading   418
Problems   419

C H A P T E R   N I N E

DIFFERENTIAL ANALYSIS OF FLUID FLOW   437

9–1   Introduction   438

9–2   Conservation of Mass—The Continuity
Equation   438

Derivation Using the Divergence Theorem   439
Derivation Using an Infinitesimal Control Volume   440
Alternative Form of the Continuity Equation   443
Continuity Equation in Cylindrical Coordinates   444
Special Cases of the Continuity Equation   444

9–3   The Stream Function   450

The Stream Function in Cartesian Coordinates   450
The Stream Function in Cylindrical Coordinates   457
The Compressible Stream Function   458

12/20/12   10:30 AM

9–4 The Differential Linear Momentum Equation—
Cauchy's Equation   459

Derivation Using the Divergence Theorem   459
Derivation Using an Infinitesimal Control Volume   460
Alternative Form of Cauchy's Equation   463
Derivation Using Newton's Second Law   463

9–5 The Navier–Stokes Equation   464

Introduction   464
Newtonian versus Non-Newtonian Fluids   465
Derivation of the Navier–Stokes Equation for Incompressible,
   Isothermal Flow   466
Continuity and Navier–Stokes Equations in Cartesian
   Coordinates   468
Continuity and Navier–Stokes Equations in Cylindrical
   Coordinates   469

9–6 Differential Analysis of Fluid Flow
Problems   470

Calculation of the Pressure Field for a Known Velocity
   Field   470
Exact Solutions of the Continuity and Navier–Stokes
   Equations   475
Differential Analysis of Biofluid Mechanics Flows   493

Application Spotlight: The No-Slip Boundary
   Condition   498

Summary   499
References and Suggested Reading   499
Problems   499

CHAPTER TEN

**APPROXIMATE SOLUTIONS OF THE NAVIER–
STOKES EQUATION   515**

10–1 Introduction   516
10–2 Nondimensionalized Equations of Motion   517
10–3 The Creeping Flow Approximation   520

Drag on a Sphere in Creeping Flow   523

10–4 Approximation for Inviscid Regions of Flow   525

Derivation of the Bernoulli Equation in Inviscid
   Regions of Flow   526

10–5 The Irrotational Flow Approximation   529

Continuity Equation   529
Momentum Equation   531
Derivation of the Bernoulli Equation in Irrotational
   Regions of Flow   531
Two-Dimensional Irrotational Regions of Flow   534
Superposition in Irrotational Regions of Flow   538
Elementary Planar Irrotational Flows   538
Irrotational Flows Formed by Superposition   545

10–6 The Boundary Layer Approximation   554

The Boundary Layer Equations   559
The Boundary Layer Procedure   564
Displacement Thickness   568
Momentum Thickness   571
Turbulent Flat Plate Boundary Layer   572
Boundary Layers with Pressure Gradients   578
The Momentum Integral Technique for Boundary Layers   583

Summary   591
References and Suggested Reading   592

Application Spotlight: Droplet Formation   593

Problems   594

CHAPTER ELEVEN

EXTERNAL FLOW: DRAG AND LIFT   607

11–1 Introduction   608
11–2 Drag and Lift   610
11–3 Friction and Pressure Drag   614

Reducing Drag by Streamlining   615
Flow Separation   616

11–4 Drag Coefficients of Common Geometries   617

Biological Systems and Drag   618
Drag Coefficients of Vehicles   621
Superposition   623

11–5 Parallel Flow Over Flat Plates   625

Friction Coefficient   627

11–6 Flow Over Cylinders And Spheres   629

Effect of Surface Roughness   632

11–7 Lift   634

Finite-Span Wings and Induced Drag   638
Lift Generated by Spinning   639

Summary   643
References and Suggested Reading   644

Application Spotlight: Drag Reduction   645

Problems   646

CHAPTER TWELVE

COMPRESSIBLE FLOW   659

12–1 Stagnation Properties   660
12–2 One-Dimensional Isentropic Flow   663

Variation of Fluid Velocity with Flow Area   665
Property Relations for Isentropic Flow of Ideal Gases   667

12–3 Isentropic Flow Through Nozzles   669

Converging Nozzles   670
Converging–Diverging Nozzles   674

12–4 Shock Waves and Expansion Waves   678

Normal Shocks   678
Oblique Shocks   684
Prandtl–Meyer Expansion Waves   688

12–5 Duct Flow With Heat Transfer and Negligible Friction (Rayleigh Flow)   693

Property Relations for Rayleigh Flow   699
Choked Rayleigh Flow   700

12–6 Adiabatic Duct Flow With Friction (Fanno Flow)   702

Property Relations for Fanno Flow   705
Choked Fanno Flow   708

Application Spotlight: Shock-Wave/ Boundary-Layer Interactions   712

Summary   713
References and Suggested Reading   714
Problems   714

C H A P T E R   T H I R T E E N
OPEN-CHANNEL FLOW   725

13–1 Classification of Open-Channel Flows   726

Uniform and Varied Flows   726
Laminar and Turbulent Flows in Channels   727

13–2 Froude Number and Wave Speed   729

Speed of Surface Waves   731

13–3 Specific Energy   733

13–4 Conservation of Mass and Energy Equations   736

13–5 Uniform Flow in Channels   737

Critical Uniform Flow   739
Superposition Method for Nonuniform Perimeters   740

13–6 Best Hydraulic Cross Sections   743

Rectangular Channels   745
Trapezoidal Channels   745

13–7 Gradually Varied Flow   747

Liquid Surface Profiles in Open Channels, $y(x)$ 749
Some Representative Surface Profiles   752
Numerical Solution of Surface Profile   754

13–8 Rapidly Varied Flow and The Hydraulic Jump   757

13–9 Flow Control and Measurement   761

Underflow Gates   762
Overflow Gates   764

Application Spotlight: Bridge Scour   771

Summary   772
References and Suggested Reading   773
Problems   773

C H A P T E R   F O U R T E E N
TURBOMACHINERY   787

14–1 Classifications and Terminology   788

14–2 Pumps   790

Pump Performance Curves and Matching a Pump to a Piping System   791
Pump Cavitation and Net Positive Suction Head   797
Pumps in Series and Parallel   800
Positive-Displacement Pumps   803
Dynamic Pumps   806
Centrifugal Pumps   806
Axial Pumps   816

14–3 Pump Scaling Laws   824

Dimensional Analysis   824
Pump Specific Speed   827
Affinity Laws   829

14–4 Turbines   833

Positive-Displacement Turbines   834
Dynamic Turbines   834
Impulse Turbines   835
Reaction Turbines   837
Gas and Steam Turbines   847
Wind Turbines   847

14–5 Turbine Scaling Laws   855

Dimensionless Turbine Parameters   855
Turbine Specific Speed   857

Application Spotlight: Rotary Fuel Atomizers   861

Summary   862
References and Suggested Reading   862
Problems   863

C H A P T E R   F I F T E E N
INTRODUCTION TO COMPUTATIONAL FLUID DYNAMICS   879

15–1 Introduction and Fundamentals   880

Motivation   880
Equations of Motion   880

Solution Procedure   881
Additional Equations of Motion   883
Grid Generation and Grid Independence   883
Boundary Conditions   888
Practice Makes Perfect   893

15–2   Laminar CFD Calculations   893

Pipe Flow Entrance Region at Re = 500   893
Flow around a Circular Cylinder at Re = 150   897

15–3   Turbulent CFD Calculations   902

Flow around a Circular Cylinder at Re = 10,000   905
Flow around a Circular Cylinder at Re = 10^7 907
Design of the Stator for a Vane-Axial Flow Fan   907

15–4   CFD With Heat Transfer   915

Temperature Rise through a Cross-Flow Heat Exchanger   915
Cooling of an Array of Integrated Circuit Chips   917

15–5   Compressible Flow CFD Calculations   922

Compressible Flow through a Converging–Diverging
   Nozzle   923
Oblique Shocks over a Wedge   927

15–6   Open-Channel Flow CFD Calculations   928

Flow over a Bump on the Bottom of a Channel   929
Flow through a Sluice Gate (Hydraulic Jump)   930

Application Spotlight: A Virtual Stomach   931

Summary   932
References and Suggested Reading   932
Problems   933

APPENDIX 1
PROPERTY TABLES AND CHARTS (SI UNITS)   939

TABLE A–1   Molar Mass, Gas Constant, and
             Ideal-Gas Specfic Heats of Some
             Substances   940

TABLE A–2   Boiling and Freezing Point
             Properties   941

TABLE A–3   Properties of Saturated Water   942

TABLE A–4   Properties of Saturated
             Refrigerant-134a   943

TABLE A–5   Properties of Saturated Ammonia   944

TABLE A–6   Properties of Saturated Propane   945

TABLE A–7   Properties of Liquids   946

TABLE A–8   Properties of Liquid Metals   947

TABLE A–9   Properties of Air at 1 atm
             Pressure   948

TABLE A–10   Properties of Gases at 1 atm
             Pressure   949

TABLE A–11   Properties of the Atmosphere at High
             Altitude   951

FIGURE A–12 The Moody Chart for the Friction Factor
             for Fully Developed Flow in Circular
             Pipes   952

TABLE A–13   One-Dimensional Isentropic
             Compressible Flow Functions for an
             Ideal Gas with $k = 1.4$ 953

TABLE A–14   One-Dimensional Normal Shock
             Functions for an Ideal Gas with
             $k = 1.4$ 954

TABLE A–15   Rayleigh Flow Functions for an Ideal
             Gas with $k = 1.4$ 955

TABLE A–16   Fanno Flow Functions for an Ideal Gas
             with $k = 1.4$ 956

APPENDIX 2
PROPERTY TABLES AND CHARTS (ENGLISH UNITS)   957

TABLE A–1E   Molar Mass, Gas Constant, and
             Ideal-Gas Specific Heats of Some
             Substances   958

TABLE A–2E   Boiling and Freezing Point
             Properties   959

TABLE A–3E   Properties of Saturated Water   960

TABLE A–4E   Properties of Saturated
             Refrigerant-134a   961

TABLE A–5E   Properties of Saturated Ammonia   962

TABLE A–6E   Properties of Saturated Propane   963

TABLE A–7E   Properties of Liquids   964

TABLE A–8E   Properties of Liquid Metals   965

TABLE A–9E   Properties of Air at 1 atm
             Pressure   966

TABLE A–10E Properties of Gases at 1 atm
             Pressure   967

TABLE A–11E Properties of the Atmosphere at High
             Altitude   969

Glossary   971
Index   983

PREFACE

BACKGROUND

Fluid mechanics is an exciting and fascinating subject with unlimited practical applications ranging from microscopic biological systems to automobiles, airplanes, and spacecraft propulsion. Fluid mechanics has also historically been one of the most challenging subjects for undergraduate students because proper analysis of fluid mechanics problems requires not only knowledge of the concepts but also physical intuition and experience. Our hope is that this book, through its careful explanations of concepts and its use of numerous practical examples, sketches, figures, and photographs, bridges the gap between knowledge and the proper application of that knowledge.

Fluid mechanics is a mature subject; the basic equations and approximations are well established and can be found in any introductory textbook. Our book is distinguished from other introductory books because we present the subject in a *progressive order* from simple to more difficult, building each chapter upon foundations laid down in earlier chapters. We provide more diagrams and photographs that other books because fluid mechanics, is by its nature, a highly visual subject. Only by illustrating the concepts discussed, can students fully appreciate the mathematical significance of the material.

OBJECTIVES

This book has been written for the first fluid mechanics course for undergraduate engineering students. There is sufficient material for a two-course sequence, if desired. We assume that readers will have an adequate background in calculus, physics, engineering mechanics, and thermodynamics. The objectives of this text are

- To present the *basic principles and equations* of fluid mechanics.

- To show numerous and diverse real-world *engineering examples* to give the student the intuition necessary for correct application of fluid mechanics principles in engineering applications.

- To develop an *intuitive understanding* of fluid mechanics by emphasizing the physics, and reinforcing that understanding through illustrative figures and photographs.

The book contains enough material to allow considerable flexibility in teaching the course. Aeronautics and aerospace engineers might emphasize potential flow, drag and lift, compressible flow, turbomachinery, and CFD, while mechanical or civil engineering instructors might choose to emphasize pipe flows and open-channel flows, respectively.

NEW TO THE THIRD EDITION

In this edition, the overall content and order of presentation has not changed significantly except for the following: the visual impact of all figures and photographs has been enhanced by a full color treatment. We also added new photographs throughout the book, often replacing existing diagrams with photographs in order to convey the practical real-life applications of the material. Several new Application Spotlights have been added to the end of selected chapters. These introduce students to industrial applications and exciting research projects being conducted by leaders in the field about material presented in the chapter. We hope these motivate students to see the relevance and application of the materials they are studying. New sections on Biofluids have been added to Chapters 8 and 9, written by guest author Keefe Manning of The Pennsylvania State University, along with bio-related examples and homework problems in those chapters.

New solved example problems were added to some chapters and several new end-of-chapter problems or modifications to existing problems were made to make them more versatile and practical. Most significant is the addition of Fundamentals of Engineering (FE) exam-type problems to help students prepare to take their Professional Engineering exams. Finally, the end-of-chapter problems that require Computational Fluid Dynamics (CFD) have been moved to the text website (www.mhhe.com/cengel) where updates based on software or operating system changes can be better managed.

PHILOSOPHY AND GOAL

The Third Edition of *Fluid Mechanics: Fundamentals and Applications* has the same goals and philosophy as the other texts by lead author Yunus Çengel.

- Communicates directly with tomorrow's engineers in a *simple yet precise* manner
- Leads students toward a clear understanding and firm grasp of the *basic principles* of fluid mechanics
- Encourages creative thinking and development of a *deeper understanding* and *intuitive feel* for fluid mechanics
- Is read by students with *interest* and *enthusiasm* rather than merely as a guide to solve homework problems

The best way to learn is by practice. Special effort is made throughout the book to reinforce the material that was presented earlier (in each chapter as well as in material from previous chapters). Many of the illustrated example problems and end-of-chapter problems are comprehensive and encourage students to review and revisit concepts and intuitions gained previously.

Throughout the book, we show examples generated by computational fluid dynamics (CFD). We also provide an introductory chapter on the subject. Our goal is not to teach the details about numerical algorithms associated with CFD—this is more properly presented in a separate course. Rather, our intent is to introduce undergraduate students to the capabilities and limitations of CFD as an *engineering tool*. We use CFD solutions in much the same way as experimental results are used from wind tunnel tests (i.e., to reinforce understanding of the physics of fluid flows and to provide quality flow visualizations that help explain fluid behavior). With dozens of CFD end-of-chapter problems posted on the website, instructors have ample opportunity to introduce the basics of CFD throughout the course.

CONTENT AND ORGANIZATION

This book is organized into 15 chapters beginning with fundamental concepts of fluids, fluid properties, and fluid flows and ending with an introduction to computational fluid dynamics.

- Chapter 1 provides a basic introduction to fluids, classifications of fluid flow, control volume versus system formulations, dimensions, units, significant digits, and problem-solving techniques.

- Chapter 2 is devoted to fluid properties such as density, vapor pressure, specific heats, speed of sound, viscosity, and surface tension.

- Chapter 3 deals with fluid statics and pressure, including manometers and barometers, hydrostatic forces on submerged surfaces, buoyancy and stability, and fluids in rigid-body motion.

- Chapter 4 covers topics related to fluid kinematics, such as the differences between Lagrangian and Eulerian descriptions of fluid flows, flow patterns, flow visualization, vorticity and rotationality, and the Reynolds transport theorem.

- Chapter 5 introduces the fundamental conservation laws of mass, momentum, and energy, with emphasis on the proper use of the mass, Bernoulli, and energy equations and the engineering applications of these equations.

- Chapter 6 applies the Reynolds transport theorem to linear momentum and angular momentum and emphasizes practical engineering applications of finite control volume momentum analysis.

- Chapter 7 reinforces the concept of dimensional homogeneity and introduces the Buckingham Pi theorem of dimensional analysis, dynamic similarity, and the method of repeating variables—material that is useful throughout the rest of the book and in many disciplines in science and engineering.

- Chapter 8 is devoted to flow in pipes and ducts. We discuss the differences between laminar and turbulent flow, friction losses in pipes and ducts, and minor losses in piping networks. We also explain how to properly select a pump or fan to match a piping network. Finally, we discuss various experimental devices that are used to measure flow rate and velocity, and provide a brief introduction to biofluid mechanics.

- Chapter 9 deals with differential analysis of fluid flow and includes derivation and application of the continuity equation, the Cauchy equation, and the Navier-Stokes equation. We also introduce the stream function and describe its usefulness in analysis of fluid flows, and we provide a brief introduction to biofluids. Finally, we point out some of the unique aspects of differential analysis related to biofluid mechanics.

- Chapter 10 discusses several *approximations* of the Navier–Stokes equation and provides example solutions for each approximation, including creeping flow, inviscid flow, irrotational (potential) flow, and boundary layers.

- Chapter 11 covers forces on bodies (drag and lift), explaining the distinction between friction and pressure drag, and providing drag coefficients for many common geometries. This chapter emphasizes the practical application of wind tunnel measurements coupled with dynamic similarity and dimensional analysis concepts introduced earlier in Chapter 7.

- Chapter 12 extends fluid flow analysis to compressible flow, where the behavior of gases is greatly affected by the Mach number. In this chapter, the concepts of expansion waves, normal and oblique shock waves, and choked flow are introduced.

- Chapter 13 deals with open-channel flow and some of the unique features associated with the flow of liquids with a free surface, such as surface waves and hydraulic jumps.

- Chapter 14 examines turbomachinery in more detail, including pumps, fans, and turbines. An emphasis is placed on how pumps and turbines work, rather than on their detailed design. We also discuss overall pump and turbine design, based on dynamic similarity laws and simplified velocity vector analyses.

- Chapter 15 describes the fundamental concepts of computational fluid dyamics (CFD) and shows students how to use commercial CFD codes as tools to solve complex fluid mechanics problems. We emphasize the *application* of CFD rather than the algorithms used in CFD codes.

Each chapter contains a wealth of end-of-chapter homework problems. Most of the problems that require calculation use the SI system of units, however about 20 percent use English units. A comprehensive set of appendices is provided, giving the thermodynamic and fluid properties of several materials, in addition to air and water, along with some useful plots and tables. Many of the end-of-chapter problems require the use of material properties from the appendices to enhance the realism of the problems.

LEARNING TOOLS

EMPHASIS ON PHYSICS

A distinctive feature of this book is its emphasis on the physical aspects of the subject matter in addition to mathematical representations and manipulations. The authors believe that the emphasis in undergraduate education should remain on *developing a sense of underlying physical mechanisms* and a *mastery of solving practical problems* that an engineer is likely to face in the real world. Developing an intuitive understanding should also make the course a more motivating and worthwhile experience for the students.

EFFECTIVE USE OF ASSOCIATION

An observant mind should have no difficulty understanding engineering sciences. After all, the principles of engineering sciences are based on our *everyday experiences* and *experimental observations*. Therefore, a physical, intuitive approach is used throughout this text. Frequently, *parallels are drawn* between the subject matter and students' everyday experiences so that they can relate the subject matter to what they already know.

SELF-INSTRUCTING

The material in the text is introduced at a level that an average student can follow comfortably. It speaks *to* students, not *over* students. In fact, it is *self-instructive*. Noting that the principles of science are based on experimental observations, most of the derivations in this text are largely based on physical arguments, and thus they are easy to follow and understand.

EXTENSIVE USE OF ARTWORK AND PHOTOGRAPHS

Figures are important learning tools that help the students "get the picture," and the text makes effective use of graphics. It contains more figures, photographs, and illustrations than any other book in this category. Figures attract attention and stimulate curiosity and interest. Most of the figures in this text are intended to serve as a means of emphasizing some key concepts that would otherwise go unnoticed; some serve as page summaries.

CONSISTENT COLOR SCHEME FOR FIGURES

The figures have a consistent color scheme applied for all arrows.

- Blue: ($\longrightarrow$) motion related, like velocity vectors
- Green: ($\longrightarrow$) force and pressure related, and torque
- Black: ($\longrightarrow$) distance related arrows and dimensions
- Red: ($\longrightarrow$) energy related, like heat and work
- Purple: ($\longrightarrow$) acceleration and gravity vectors, vorticity, and miscellaneous

NUMEROUS WORKED-OUT EXAMPLES

All chapters contain numerous worked-out *examples* that both clarify the material and illustrate the use of basic principles in a context that helps develops the student's intuition. An *intuitive* and *systematic* approach is used in the solution of all example problems. The solution methodology starts with a statement of the problem, and all objectives are identified. The assumptions and approximations are then stated together with their justifications. Any properties needed to solve the problem are listed separately. Numerical values are used together with numbers to emphasize that without units, numbers are meaningless. The significance of each example's result is discussed following the solution. This methodical approach is also followed and provided in the solutions to the end-of-chapter problems, available to instructors.

A WEALTH OF REALISTIC END-OF-CHAPTER PROBLEMS

The end-of-chapter problems are grouped under specific topics to make problem selection easier for both instructors and students. Within each group of problems are *Concept Questions,* indicated by "C," to check the students' level of understanding of basic concepts. Problems under *Fundamentals of Engineering (FE) Exam Problems* are designed to help students prepare for the *Fundamentals of Engineering* exam, as they prepare for their Professional Engineering license. The problems under *Review Problems* are more comprehensive in nature and are not directly tied to any specific section of a chapter—in some cases they require review of material learned in previous chapters. Problems designated as *Design and Essay* are intended to encourage students to make engineering judgments, to conduct independent exploration of topics of interest, and to communicate their findings in a professional manner. Problems designated by an "E" are in English units, and SI users can ignore them. Problems with the ⊙ icon are solved using EES, and complete solutions together with parametric studies are included the text website. Problems with the ▣ icon are comprehensive in nature and are intended to be solved with a computer, preferably using the EES software. Several economics- and safety-related problems are incorporated throughout to enhance cost and safety awareness among engineering students. Answers to selected problems are listed immediately following the problem for convenience to students.

USE OF COMMON NOTATION

The use of different notation for the same quantities in different engineering courses has long been a source of discontent and confusion. A student taking both fluid mechanics and heat transfer, for example, has to use the notation Q for volume flow rate in one course, and for heat transfer in the other. The need to unify notation in engineering education has often been raised, even in some reports of conferences sponsored by the National Science Foundation through Foundation Coalitions, but little effort has been made to date in this regard. For example, refer to the final report of the *Mini-Conference on Energy Stem Innovations*, May 28 and 29, 2003, University of Wisconsin. In this text we made a conscious effort to minimize this conflict by adopting the familiar thermodynamic notation $\dot{V}$ for volume flow rate, thus reserving the notation Q for heat transfer. Also, we consistently use an overdot to denote time rate. We think that both students and instructors will appreciate this effort to promote a common notation.

A CHOICE OF SI ALONE OR SI/ENGLISH UNITS

In recognition of the fact that English units are still widely used in some industries, both SI and English units are used in this text, with an emphasis on SI. The material in this text can be covered using combined SI/English units or SI units alone, depending on the preference of the instructor. The property tables and charts in the appendices are presented in both units, except the ones that involve dimensionless quantities. Problems, tables, and charts in English units are designated by "E" after the number for easy recognition, and they can be ignored easily by the SI users.

COMBINED COVERAGE OF BERNOULLI AND ENERGY EQUATIONS

The Bernoulli equation is one of the most frequently used equations in fluid mechanics, but it is also one of the most misused. Therefore, it is important to emphasize the limitations on the use of this idealized equation and to show how to properly account for imperfections and irreversible losses. In Chapter 5, we do this by introducing the energy equation right after the Bernoulli equation and demonstrating how the solutions of many practical engineering problems differ from those obtained using the Bernoulli equation. This helps students develop a realistic view of the Bernoulli equation.

A SEPARATE CHAPTER ON CFD

Commercial *Computational Fluid Dynamics* (CFD) codes are widely used in engineering practice in the design and analysis of flow systems, and it has become exceedingly important for engineers to have a solid understanding of the fundamental aspects, capabilities, and limitations of CFD. Recognizing that most undergraduate engineering curriculums do not have room for a full course on CFD, a separate chapter is included here to make up for this deficiency and to equip students with an adequate background on the strengths and weaknesses of CFD.

APPLICATION SPOTLIGHTS

Throughout the book are highlighted examples called *Application Spotlights* where a real-world application of fluid mechanics is shown. A unique feature of these special examples is that they are written by *guest authors*. The Application Spotlights are designed to show students how fluid mechanics has diverse applications in a wide variety of fields. They also include eye-catching photographs from the guest authors' research.

GLOSSARY OF FLUID MECHANICS TERMS

Throughout the chapters, when an important key term or concept is introduced and defined, it appears in **black** boldface type. Fundamental fluid mechanics terms and concepts appear in **red** boldface type, and these fundamental terms also appear in a comprehensive end-of-book glossary developed by Professor James Brasseur of The Pennsylvania State University. This unique glossary is an excellent learning and review tool for students as they move forward in their study of fluid mechanics. In addition, students can test their knowledge of these fundamental terms by using the interactive flash cards and other resources located on our accompanying website (www.mhhe.com/cengel).

CONVERSION FACTORS

Frequently used conversion factors, physical constants, and properties of air and water at 20°C and atmospheric pressure are listed on the front inner cover pages of the text for easy reference.

NOMENCLATURE

A list of the major symbols, subscripts, and superscripts used in the text are listed on the inside back cover pages of the text for easy reference.

SUPPLEMENTS

These supplements are available to adopters of the book:

Text Website

Web support is provided for the book on the text specific website at www.mhhe.com/cengel. Visit this robust site for book and supplement information, errata, author information, and further resources for instructors and students.

Engineering Equation Solver (EES)

Developed by Sanford Klein and William Beckman from the University of Wisconsin–Madison, this software combines equation-solving capability and engineering property data. EES can do optimization, parametric analysis, and linear and nonlinear regression, and provides publication-quality plotting capabilities. Thermodynamics and transport properties for air, water, and many other fluids are built-in and EES allows the user to enter property data or functional relationships.

ACKNOWLEDGMENTS

The authors would like to acknowledge with appreciation the numerous and valuable comments, suggestions, constructive criticisms, and praise from the following evaluators and reviewers of the third edition:

Bass Abushakra
Milwaukee School of Engineering

John G. Cherng
University of Michigan—Dearborn

Peter Fox
Arizona State University

Sathya Gangadbaran
Embry Riddle Aeronautical University

Jonathan Istok
Oregon State University

Tim Lee
McGill University

Nagy Nosseir
San Diego State University

Robert Spall
Utah State University

We also thank those who were acknowledged in the first and second editions of this book, but are too numerous to mention again here. Special thanks go to Gary S. Settles and his associates at Penn State (Lori Dodson-Dreibelbis, J. D. Miller, and Gabrielle Tremblay) for creating the exciting narrated video clips that are found on the book's website. The authors also thank James Brasseur of Penn State for creating the precise glossary of fluid mechanics terms, Glenn Brown of Oklahoma State for providing many items of historical interest throughout the text, guest authors David F. Hill (parts of Chapter 13) and Keefe Manning (sections on biofluids), Mehmet Kanoglu of University of Gaziantep for preparing FE Exam problems and the solutions of EES problems, and Tahsin Engin of Sakarya University for contributing several end-of-chapter problems.

We also acknowledge the Korean translation team, who in the translation process, pointed out several errors and inconsistencies in the first and second editions that have now been corrected. The team includes Yun-ho Choi, Ajou University; Nae-Hyun Kim, University of Incheon; Woonjean Park, Korea University of Technology & Education; Wonnam Lee, Dankook University; Sang-Won Cha, Suwon University; Man Yeong Ha, Pusan National University; and Yeol Lee, Korea Aerospace University.

Finally, special thanks must go to our families, especially our wives, Zehra Çengel and Suzanne Cimbala, for their continued patience, understanding, and support throughout the preparation of this book, which involved many long hours when they had to handle family concerns on their own because their husbands' faces were glued to a computer screen.

Yunus A. Çengel
John M. Cimbala

Online Resources for Students and Instructors

Online Resources available at www.mhhe.com/cengel

Your home page for teaching and studying fluid mechanics, the *Fluid Mechanics: Fundamentals and Applications* text-specific website offers resources for both instructors and students.

For the student, this website offer various resources, including:

- **FE Exam Interactive Review Quizzes**—chapter-based self-quizzes provide hints for solutions and correct solution methods, and help students prepare for the NCEES Fundamentals of Engineering Examination.
- **Glossary of Key Terms in Fluid Mechanics**—full text and chapter-based glossaries.
- **Weblinks**—helpful weblinks to relevant fluid mechanics sites.

For the instructor, this password-protected website offers various resources, including:

- **Electronic Solutions Manual**—provides PDF files with detailed solutions to all text homework problems.
- **Image Library**—provide electronic files for text figures for easy integration into your course presentations, exams, and assignments.
- **Sample Syllabi**—make it easier for you to map out your course using this text for different course durations (one quarter, one semester, etc.) and for different disciplines (ME approach, Civil approach, etc.).
- **Transition Guides**—compare coverage to other popular introductory fluid mechanics books at the section level to aid transition to teaching from our text.
- **Links to ANSYS Workbench®, FLUENT FLOWLAB®, and EES (Engineering Equation Solver) download sites**—the academic versions of these powerful software programs are available free to departments of educational institutions who adopt this text.
- CFD homework problems and solutions designed for use with various CFD packages.

McGraw-Hill Connect® Engineering provides online presentation, assignment, and assessment solutions. It connects your students with the tools and resources they'll need to achieve success. With Connect Engineering, you can deliver assignments, quizzes, and tests online. A robust set of questions and activities are presented and aligned with the textbook's learning outcomes. As an instructor, you can edit existing questions and author entirely new problems. Track individual student performance—by question, assignment, or in relation to the class overall—with detailed grade reports. Integrate grade reports easily with Learning Management Systems (LMS), such as WebCT and Blackboard—and much more. ConnectPlus Engineering provides students with all the advantages of Connect Engineering, plus 24/7 online access to an eBook. This media-rich version of the book is available through the McGraw-Hill Connect platform and allows seamless integration of text, media, and assessments. To learn more, visit **www.mcgrawhillconnect.com.**

CHAPTER

1

INTRODUCTION AND BASIC CONCEPTS

In this introductory chapter, we present the basic concepts commonly used in the analysis of fluid flow. We start this chapter with a discussion of the phases of matter and the numerous ways of classification of fluid flow, such as *viscous versus inviscid regions of flow, internal versus external flow, compressible versus incompressible flow, laminar versus turbulent flow, natural versus forced flow,* and *steady versus unsteady flow.* We also discuss the *no-slip condition* at solid–fluid interfaces and present a brief history of the development of fluid mechanics.

After presenting the concepts of *system* and *control volume,* we review the *unit systems* that will be used. We then discuss how mathematical models for engineering problems are prepared and how to interpret the results obtained from the analysis of such models. This is followed by a presentation of an intuitive systematic *problem-solving technique* that can be used as a model in solving engineering problems. Finally, we discuss accuracy, precision, and significant digits in engineering measurements and calculations.

OBJECTIVES

When you finish reading this chapter, you should be able to

■ Understand the basic concepts of fluid mechanics

■ Recognize the various types of fluid flow problems encountered in practice

■ Model engineering problems and solve them in a systematic manner

■ Have a working knowledge of accuracy, precision, and significant digits, and recognize the importance of dimensional homogeneity in engineering calculations

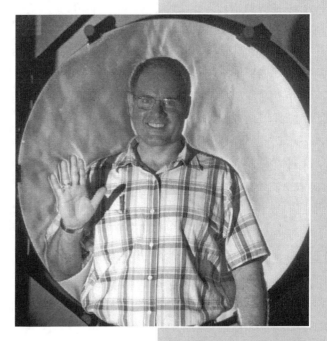

Schlieren image showing the thermal plume produced by Professor Cimbala as he welcomes you to the fascinating world of fluid mechanics.

Michael J. Hargather and Brent A. Craven, Penn State Gas Dynamics Lab. Used by Permission.

FIGURE 1–1
Fluid mechanics deals with liquids and gases in motion or at rest.
© McGraw-Hill Education

FIGURE 1–2
Deformation of a rubber block placed between two parallel plates under the influence of a shear force. The shear stress shown is that on the rubber—an equal but opposite shear stress acts on the upper plate.

1–1 · INTRODUCTION

Mechanics is the oldest physical science that deals with both stationary and moving bodies under the influence of forces. The branch of mechanics that deals with bodies at rest is called **statics,** while the branch that deals with bodies in motion is called **dynamics.** The subcategory **fluid mechanics** is defined as the science that deals with the behavior of fluids at rest (*fluid statics*) or in motion (*fluid dynamics*), and the interaction of fluids with solids or other fluids at the boundaries. Fluid mechanics is also referred to as **fluid dynamics** by considering fluids at rest as a special case of motion with zero velocity (Fig. 1–1).

Fluid mechanics itself is also divided into several categories. The study of the motion of fluids that can be approximated as incompressible (such as liquids, especially water, and gases at low speeds) is usually referred to as **hydrodynamics.** A subcategory of hydrodynamics is **hydraulics,** which deals with liquid flows in pipes and open channels. **Gas dynamics** deals with the flow of fluids that undergo significant density changes, such as the flow of gases through nozzles at high speeds. The category **aerodynamics** deals with the flow of gases (especially air) over bodies such as aircraft, rockets, and automobiles at high or low speeds. Some other specialized categories such as **meteorology, oceanography,** and **hydrology** deal with naturally occurring flows.

What Is a Fluid?

You will recall from physics that a substance exists in three primary phases: solid, liquid, and gas. (At very high temperatures, it also exists as plasma.) A substance in the liquid or gas phase is referred to as a **fluid.** Distinction between a solid and a fluid is made on the basis of the substance's ability to resist an applied shear (or tangential) stress that tends to change its shape. A solid can resist an applied shear stress by deforming, whereas *a fluid deforms continuously under the influence of a shear stress*, no matter how small. In solids, stress is proportional to *strain*, but in fluids, stress is proportional to *strain rate*. When a constant shear force is applied, a solid eventually stops deforming at some fixed strain angle, whereas a fluid never stops deforming and approaches a constant *rate* of strain.

Consider a rectangular rubber block tightly placed between two plates. As the upper plate is pulled with a force F while the lower plate is held fixed, the rubber block deforms, as shown in Fig. 1–2. The angle of deformation α (called the *shear strain* or *angular displacement*) increases in proportion to the applied force F. Assuming there is no slip between the rubber and the plates, the upper surface of the rubber is displaced by an amount equal to the displacement of the upper plate while the lower surface remains stationary. In equilibrium, the net force acting on the upper plate in the horizontal direction must be zero, and thus a force equal and opposite to F must be acting on the plate. This opposing force that develops at the plate–rubber interface due to friction is expressed as $F = \tau A$, where τ is the shear stress and A is the contact area between the upper plate and the rubber. When the force is removed, the rubber returns to its original position. This phenomenon would also be observed with other solids such as a steel block provided that the applied force does not exceed the elastic range. If this experiment were repeated with a fluid (with two large parallel plates placed in a large body of water, for example), the fluid layer in contact with the upper plate would move with the plate continuously at the velocity of the plate no matter how small the force F. The fluid velocity would decrease with depth because of friction between fluid layers, reaching zero at the lower plate.

You will recall from statics that **stress** is defined as force per unit area and is determined by dividing the force by the area upon which it acts. The normal component of a force acting on a surface per unit area is called the **normal stress,** and the tangential component of a force acting on a surface per unit area is called **shear stress** (Fig. 1–3). In a fluid at rest, the normal stress is called **pressure.** A fluid at rest is at a state of zero shear stress. When the walls are removed or a liquid container is tilted, a shear develops as the liquid moves to re-establish a horizontal free surface.

In a liquid, groups of molecules can move relative to each other, but the volume remains relatively constant because of the strong cohesive forces between the molecules. As a result, a liquid takes the shape of the container it is in, and it forms a free surface in a larger container in a gravitational field. A gas, on the other hand, expands until it encounters the walls of the container and fills the entire available space. This is because the gas molecules are widely spaced, and the cohesive forces between them are very small. Unlike liquids, a gas in an open container cannot form a free surface (Fig. 1–4).

Although solids and fluids are easily distinguished in most cases, this distinction is not so clear in some borderline cases. For example, *asphalt* appears and behaves as a solid since it resists shear stress for short periods of time. When these forces are exerted over extended periods of time, however, the asphalt deforms slowly, behaving as a fluid. Some plastics, lead, and slurry mixtures exhibit similar behavior. Such borderline cases are beyond the scope of this text. The fluids we deal with in this text will be clearly recognizable as fluids.

Intermolecular bonds are strongest in solids and weakest in gases. One reason is that molecules in solids are closely packed together, whereas in gases they are separated by relatively large distances (Fig. 1–5). The molecules in a solid are arranged in a pattern that is repeated throughout. Because of the small distances between molecules in a solid, the attractive forces of molecules on each other are large and keep the molecules at fixed positions. The molecular spacing in the liquid phase is not much different from that of

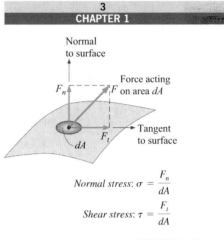

Normal to surface

Force acting on area dA

F_n

F

Tangent to surface

F_t

dA

Normal stress: $\sigma = \dfrac{F_n}{dA}$

Shear stress: $\tau = \dfrac{F_t}{dA}$

FIGURE 1–3

The normal stress and shear stress at the surface of a fluid element. For fluids at rest, the shear stress is zero and pressure is the only normal stress.

Free surface

Liquid

Gas

FIGURE 1–4

Unlike a liquid, a gas does not form a free surface, and it expands to fill the entire available space.

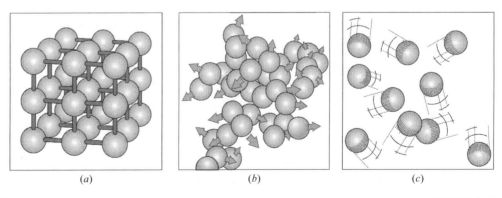

(a)   (b)   (c)

FIGURE 1–5

The arrangement of atoms in different phases: (a) molecules are at relatively fixed positions in a solid, (b) groups of molecules move about each other in the liquid phase, and (c) individual molecules move about at random in the gas phase.

the solid phase, except the molecules are no longer at fixed positions relative to each other and they can rotate and translate freely. In a liquid, the intermolecular forces are weaker relative to solids, but still strong compared with gases. The distances between molecules generally increase slightly as a solid turns liquid, with water being a notable exception.

In the gas phase, the molecules are far apart from each other, and molecular ordering is nonexistent. Gas molecules move about at random, continually colliding with each other and the walls of the container in which they are confined. Particularly at low densities, the intermolecular forces are very small, and collisions are the only mode of interaction between the molecules. Molecules in the gas phase are at a considerably higher energy level than they are in the liquid or solid phase. Therefore, the gas must release a large amount of its energy before it can condense or freeze.

Gas and *vapor* are often used as synonymous words. The vapor phase of a substance is customarily called a *gas* when it is above the critical temperature. *Vapor* usually implies that the current phase is not far from a state of condensation.

Any practical fluid system consists of a large number of molecules, and the properties of the system naturally depend on the behavior of these molecules. For example, the pressure of a gas in a container is the result of momentum transfer between the molecules and the walls of the container. However, one does not need to know the behavior of the gas molecules to determine the pressure in the container. It is sufficient to attach a pressure gage to the container (Fig. 1–6). This macroscopic or *classical* approach does not require a knowledge of the behavior of individual molecules and provides a direct and easy way to analyze engineering problems. The more elaborate microscopic or *statistical* approach, based on the average behavior of large groups of individual molecules, is rather involved and is used in this text only in a supporting role.

Application Areas of Fluid Mechanics

It is important to develop a good understanding of the basic principles of fluid mechanics, since fluid mechanics is widely used both in everyday activities and in the design of modern engineering systems from vacuum cleaners to supersonic aircraft. For example, fluid mechanics plays a vital role in the human body. The heart is constantly pumping blood to all parts of the human body through the arteries and veins, and the lungs are the sites of airflow in alternating directions. All artificial hearts, breathing machines, and dialysis systems are designed using fluid dynamics (Fig. 1–7).

An ordinary house is, in some respects, an exhibition hall filled with applications of fluid mechanics. The piping systems for water, natural gas, and sewage for an individual house and the entire city are designed primarily on the basis of fluid mechanics. The same is also true for the piping and ducting network of heating and air-conditioning systems. A refrigerator involves tubes through which the refrigerant flows, a compressor that pressurizes the refrigerant, and two heat exchangers where the refrigerant absorbs and rejects heat. Fluid mechanics plays a major role in the design of all these components. Even the operation of ordinary faucets is based on fluid mechanics.

We can also see numerous applications of fluid mechanics in an automobile. All components associated with the transportation of the fuel from the fuel tank to the cylinders—the fuel line, fuel pump, and fuel injectors or

FIGURE 1–6
On a microscopic scale, pressure is determined by the interaction of individual gas molecules. However, we can measure the pressure on a macroscopic scale with a pressure gage.

FIGURE 1–7
Fluid dynamics is used extensively in the design of artificial hearts. Shown here is the Penn State Electric Total Artificial Heart.

Photo courtesy of the Biomedical Photography Lab, Penn State Biomedical Engineering Institute. Used by Permission.

carburetors—as well as the mixing of the fuel and the air in the cylinders and the purging of combustion gases in exhaust pipes—are analyzed using fluid mechanics. Fluid mechanics is also used in the design of the heating and air-conditioning system, the hydraulic brakes, the power steering, the automatic transmission, the lubrication systems, the cooling system of the engine block including the radiator and the water pump, and even the tires. The sleek streamlined shape of recent model cars is the result of efforts to minimize drag by using extensive analysis of flow over surfaces.

On a broader scale, fluid mechanics plays a major part in the design and analysis of aircraft, boats, submarines, rockets, jet engines, wind turbines, biomedical devices, cooling systems for electronic components, and transportation systems for moving water, crude oil, and natural gas. It is also considered in the design of buildings, bridges, and even billboards to make sure that the structures can withstand wind loading. Numerous natural phenomena such as the rain cycle, weather patterns, the rise of ground water to the tops of trees, winds, ocean waves, and currents in large water bodies are also governed by the principles of fluid mechanics (Fig. 1–8).

Natural flows and weather
© Glen Allison/Betty RF

Boats
© Doug Menuez/Getty RF

Aircraft and spacecraft
© Photo Link/Getty RF

Power plants
© Malcom Fife/Getty RF

Human body
© Ryan McVay/Getty RF

Cars
© Mark Evans/Getty RF

Wind turbines
© F. Schussler/PhotoLink/Getty RF

Piping and plumbing systems
Photo by John M. Cimbala.

Industrial applications
Digital Vision/PunchStock

FIGURE 1–8
Some application areas of fluid mechanics.

FIGURE 1–9
Segment of Pergamon pipeline.
Each clay pipe section was
13 to 18 cm in diameter.

Courtesy Gunther Garbrecht.
Used by permission.

FIGURE 1–10
A mine hoist powered
by a reversible water wheel.

G. Agricola, De Re Metalica, *Basel, 1556.*

1–2 ▪ A BRIEF HISTORY OF FLUID MECHANICS[1]

One of the first engineering problems humankind faced as cities were developed was the supply of water for domestic use and irrigation of crops. Our urban lifestyles can be retained only with abundant water, and it is clear from archeology that every successful civilization of prehistory invested in the construction and maintenance of water systems. The Roman aqueducts, some of which are still in use, are the best known examples. However, perhaps the most impressive engineering from a technical viewpoint was done at the Hellenistic city of Pergamon in present-day Turkey. There, from 283 to 133 BC, they built a series of pressurized lead and clay pipelines (Fig. 1–9), up to 45 km long that operated at pressures exceeding 1.7 MPa (180 m of head). Unfortunately, the names of almost all these early builders are lost to history.

The earliest recognized contribution to fluid mechanics theory was made by the Greek mathematician Archimedes (285–212 BC). He formulated and applied the buoyancy principle in history's first nondestructive test to determine the gold content of the crown of King Hiero II. The Romans built great aqueducts and educated many conquered people on the benefits of clean water, but overall had a poor understanding of fluids theory. (Perhaps they shouldn't have killed Archimedes when they sacked Syracuse.)

During the Middle Ages, the application of fluid machinery slowly but steadily expanded. Elegant piston pumps were developed for dewatering mines, and the watermill and windmill were perfected to grind grain, forge metal, and for other tasks. For the first time in recorded human history, significant work was being done without the power of a muscle supplied by a person or animal, and these inventions are generally credited with enabling the later industrial revolution. Again the creators of most of the progress are unknown, but the devices themselves were well documented by several technical writers such as Georgius Agricola (Fig. 1–10).

The Renaissance brought continued development of fluid systems and machines, but more importantly, the scientific method was perfected and adopted throughout Europe. Simon Stevin (1548–1617), Galileo Galilei (1564–1642), Edme Mariotte (1620–1684), and Evangelista Torricelli (1608–1647) were among the first to apply the method to fluids as they investigated hydrostatic pressure distributions and vacuums. That work was integrated and refined by the brilliant mathematician and philosopher, Blaise Pascal (1623–1662). The Italian monk, Benedetto Castelli (1577–1644) was the first person to publish a statement of the continuity principle for fluids. Besides formulating his equations of motion for solids, Sir Isaac Newton (1643–1727) applied his laws to fluids and explored fluid inertia and resistance, free jets, and viscosity. That effort was built upon by Daniel Bernoulli (1700–1782), a Swiss, and his associate Leonard Euler (1707–1783). Together, their work defined the energy and momentum equations. Bernoulli's 1738 classic treatise *Hydrodynamica* may be considered the first fluid mechanics text. Finally, Jean d'Alembert (1717–1789) developed the idea of velocity and acceleration components, a differential expression of

[1] This section is contributed by Professor Glenn Brown of Oklahoma State University.

continuity, and his "paradox" of zero resistance to steady uniform motion over a body.

The development of fluid mechanics theory through the end of the eighteenth century had little impact on engineering since fluid properties and parameters were poorly quantified, and most theories were abstractions that could not be quantified for design purposes. That was to change with the development of the French school of engineering led by Riche de Prony (1755–1839). Prony (still known for his brake to measure shaft power) and his associates in Paris at the École Polytechnique and the École des Ponts et Chaussées were the first to integrate calculus and scientific theory into the engineering curriculum, which became the model for the rest of the world. (So now you know whom to blame for your painful freshman year.) Antonie Chezy (1718–1798), Louis Navier (1785–1836), Gaspard Coriolis (1792–1843), Henry Darcy (1803–1858), and many other contributors to fluid engineering and theory were students and/or instructors at the schools.

By the mid nineteenth century, fundamental advances were coming on several fronts. The physician Jean Poiseuille (1799–1869) had accurately measured flow in capillary tubes for multiple fluids, while in Germany Gotthilf Hagen (1797–1884) had differentiated between laminar and turbulent flow in pipes. In England, Lord Osborne Reynolds (1842–1912) continued that work (Fig. 1–11) and developed the dimensionless number that bears his name. Similarly, in parallel to the early work of Navier, George Stokes (1819–1903) completed the general equation of fluid motion (with friction) that takes their names. William Froude (1810–1879) almost single-handedly developed the procedures and proved the value of physical model testing. American expertise had become equal to the Europeans as demonstrated by James Francis' (1815–1892) and Lester Pelton's (1829–1908) pioneering work in turbines and Clemens Herschel's (1842–1930) invention of the Venturi meter.

In addition to Reynolds and Stokes, many notable contributions were made to fluid theory in the late nineteenth century by Irish and English scientists, including William Thomson, Lord Kelvin (1824–1907), William Strutt, Lord Rayleigh (1842–1919), and Sir Horace Lamb (1849–1934). These individuals investigated a large number of problems, including dimensional analysis, irrotational flow, vortex motion, cavitation, and waves. In a broader sense,

FIGURE 1–11
Osborne Reynolds' original apparatus for demonstrating the onset of turbulence in pipes, being operated by John Lienhard at the University of Manchester in 1975.

Photo courtesy of John Lienhard, University of Houston. Used by permission.

FIGURE 1–12
The Wright brothers take flight at Kitty Hawk.

Library of Congress Prints & Photographs Division [LC-DIG-ppprs-00626]

FIGURE 1–13
Old and new wind turbine technologies north of Woodward, OK. The modern turbines have 1.6 MW capacities.

Photo courtesy of the Oklahoma Wind Power Initiative. Used by permission.

their work also explored the links between fluid mechanics, thermodynamics, and heat transfer.

The dawn of the twentieth century brought two monumental developments. First, in 1903, the self-taught Wright brothers (Wilbur, 1867–1912; Orville, 1871–1948) invented the airplane through application of theory and determined experimentation. Their primitive invention was complete and contained all the major aspects of modern aircraft (Fig. 1–12). The Navier–Stokes equations were of little use up to this time because they were too difficult to solve. In a pioneering paper in 1904, the German Ludwig Prandtl (1875–1953) showed that fluid flows can be divided into a layer near the walls, the *boundary layer,* where the friction effects are significant, and an outer layer where such effects are negligible and the simplified Euler and Bernoulli equations are applicable. His students, Theodor von Kármán (1881–1963), Paul Blasius (1883–1970), Johann Nikuradse (1894–1979), and others, built on that theory in both hydraulic and aerodynamic applications. (During World War II, both sides benefited from the theory as Prandtl remained in Germany while his best student, the Hungarian-born von Kármán, worked in America.)

The mid twentieth century could be considered a golden age of fluid mechanics applications. Existing theories were adequate for the tasks at hand, and fluid properties and parameters were well defined. These supported a huge expansion of the aeronautical, chemical, industrial, and water resources sectors; each of which pushed fluid mechanics in new directions. Fluid mechanics research and work in the late twentieth century were dominated by the development of the digital computer in America. The ability to solve large complex problems, such as global climate modeling or the optimization of a turbine blade, has provided a benefit to our society that the eighteenth-century developers of fluid mechanics could never have imagined (Fig. 1–13). The principles presented in the following pages have been applied to flows ranging from a moment at the microscopic scale to 50 years of simulation for an entire river basin. It is truly mind-boggling.

Where will fluid mechanics go in the twenty-first century and beyond? Frankly, even a limited extrapolation beyond the present would be sheer folly. However, if history tells us anything, it is that engineers will be applying what they know to benefit society, researching what they don't know, and having a great time in the process.

1–3 ▪ THE NO-SLIP CONDITION

Fluid flow is often confined by solid surfaces, and it is important to understand how the presence of solid surfaces affects fluid flow. We know that water in a river cannot flow through large rocks, and must go around them. That is, the water velocity normal to the rock surface must be zero, and water approaching the surface normally comes to a complete stop at the surface. What is not as obvious is that water approaching the rock at any angle also comes to a complete stop at the rock surface, and thus the tangential velocity of water at the surface is also zero.

Consider the flow of a fluid in a stationary pipe or over a solid surface that is nonporous (i.e., impermeable to the fluid). All experimental observations indicate that a fluid in motion comes to a complete stop at the surface and assumes a zero velocity relative to the surface. That is, a fluid in direct contact with a solid "sticks" to the surface, and there is no slip. This is known as the **no-slip condition.** The fluid property responsible for the no-slip condition and the development of the boundary layer is *viscosity* and is discussed in Chap. 2.

The photograph in Fig. 1–14 clearly shows the evolution of a velocity gradient as a result of the fluid sticking to the surface of a blunt nose. The layer that sticks to the surface slows the adjacent fluid layer because of viscous forces between the fluid layers, which slows the next layer, and so on. A consequence of the no-slip condition is that all velocity profiles must have zero values with respect to the surface at the points of contact between a fluid and a solid surface (Fig. 1–15). Therefore, the no-slip condition is responsible for the development of the velocity profile. The flow region adjacent to the wall in which the viscous effects (and thus the velocity gradients) are significant is called the **boundary layer.** Another consequence of the no-slip condition is the *surface drag*, or *skin friction drag*, which is the force a fluid exerts on a surface in the flow direction.

When a fluid is forced to flow over a curved surface, such as the back side of a cylinder, the boundary layer may no longer remain attached to the surface and separates from the surface—a process called **flow separation** (Fig. 1–16). We emphasize that the no-slip condition applies *everywhere* along the surface, even downstream of the separation point. Flow separation is discussed in greater detail in Chap. 9.

A phenomenon similar to the no-slip condition occurs in heat transfer. When two bodies at different temperatures are brought into contact, heat transfer occurs such that both bodies assume the same temperature at the points of contact. Therefore, a fluid and a solid surface have the same temperature at the points of contact. This is known as **no-temperature-jump condition.**

1–4 ▪ CLASSIFICATION OF FLUID FLOWS

Earlier we defined *fluid mechanics* as the science that deals with the behavior of fluids at rest or in motion, and the interaction of fluids with solids or other fluids at the boundaries. There is a wide variety of fluid flow problems encountered in practice, and it is usually convenient to classify them on the basis of some common characteristics to make it feasible to study them in groups. There are many ways to classify fluid flow problems, and here we present some general categories.

FIGURE 1–14

The development of a velocity profile due to the no-slip condition as a fluid flows over a blunt nose.

"Hunter Rouse: Laminar and Turbulent Flow Film." Copyright IIHR-Hydroscience & Engineering, The University of Iowa. Used by permission.

FIGURE 1–15

A fluid flowing over a stationary surface comes to a complete stop at the surface because of the no-slip condition.

FIGURE 1–16

Flow separation during flow over a curved surface.

From G. M. Homsy et al, "Multi-Media Fluid Mechanics," Cambridge Univ. Press (2001). ISBN 0-521-78748-3. Reprinted by permission.

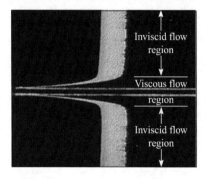

FIGURE 1–17

The flow of an originally uniform fluid stream over a flat plate, and the regions of viscous flow (next to the plate on both sides) and inviscid flow (away from the plate).

Fundamentals of Boundary Layers,
National Committee from Fluid Mechanics Films,
© *Education Development Center.*

FIGURE 1–18

External flow over a tennis ball, and the turbulent wake region behind.

Courtesy NASA and Cislunar Aerospace, Inc.

Viscous versus Inviscid Regions of Flow

When two fluid layers move relative to each other, a friction force develops between them and the slower layer tries to slow down the faster layer. This internal resistance to flow is quantified by the fluid property *viscosity*, which is a measure of internal stickiness of the fluid. Viscosity is caused by cohesive forces between the molecules in liquids and by molecular collisions in gases. There is no fluid with zero viscosity, and thus all fluid flows involve viscous effects to some degree. Flows in which the frictional effects are significant are called **viscous flows.** However, in many flows of practical interest, there are *regions* (typically regions not close to solid surfaces) where viscous forces are negligibly small compared to inertial or pressure forces. Neglecting the viscous terms in such **inviscid flow regions** greatly simplifies the analysis without much loss in accuracy.

The development of viscous and inviscid regions of flow as a result of inserting a flat plate parallel into a fluid stream of uniform velocity is shown in Fig. 1–17. The fluid sticks to the plate on both sides because of the no-slip condition, and the thin boundary layer in which the viscous effects are significant near the plate surface is the *viscous flow region.* The region of flow on both sides away from the plate and largely unaffected by the presence of the plate is the *inviscid flow region.*

Internal versus External Flow

A fluid flow is classified as being internal or external, depending on whether the fluid flows in a confined space or over a surface. The flow of an unbounded fluid over a surface such as a plate, a wire, or a pipe is **external flow.** The flow in a pipe or duct is **internal flow** if the fluid is completely bounded by solid surfaces. Water flow in a pipe, for example, is internal flow, and airflow over a ball or over an exposed pipe during a windy day is external flow (Fig. 1–18). The flow of liquids in a duct is called *open-channel flow* if the duct is only partially filled with the liquid and there is a free surface. The flows of water in rivers and irrigation ditches are examples of such flows.

Internal flows are dominated by the influence of viscosity throughout the flow field. In external flows the viscous effects are limited to boundary layers near solid surfaces and to wake regions downstream of bodies.

Compressible versus Incompressible Flow

A flow is classified as being *compressible* or *incompressible,* depending on the level of variation of density during flow. Incompressibility is an approximation, in which the flow is said to be **incompressible** if the density remains nearly constant throughout. Therefore, the volume of every portion of fluid remains unchanged over the course of its motion when the flow is approximated as incompressible.

The densities of liquids are essentially constant, and thus the flow of liquids is typically incompressible. Therefore, liquids are usually referred to as *incompressible substances.* A pressure of 210 atm, for example, causes the density of liquid water at 1 atm to change by just 1 percent. Gases, on the other hand, are highly compressible. A pressure change of just 0.01 atm, for example, causes a change of 1 percent in the density of atmospheric air.

When analyzing rockets, spacecraft, and other systems that involve high-speed gas flows (Fig. 1–19), the flow speed is often expressed in terms of the dimensionless **Mach number** defined as

$$Ma = \frac{V}{c} = \frac{\text{Speed of flow}}{\text{Speed of sound}}$$

where c is the **speed of sound** whose value is 346 m/s in air at room temperature at sea level. A flow is called **sonic** when Ma = 1, **subsonic** when Ma < 1, **supersonic** when Ma > 1, and **hypersonic** when Ma ≫ 1. Dimensionless parameters are discussed in detail in Chapter 7.

Liquid flows are incompressible to a high level of accuracy, but the level of variation of density in gas flows and the consequent level of approximation made when modeling gas flows as incompressible depends on the Mach number. Gas flows can often be approximated as incompressible if the density changes are under about 5 percent, which is usually the case when Ma < 0.3. Therefore, the compressibility effects of air at room temperature can be neglected at speeds under about 100 m/s.

Small density changes of liquids corresponding to large pressure changes can still have important consequences. The irritating "water hammer" in a water pipe, for example, is caused by the vibrations of the pipe generated by the reflection of pressure waves following the sudden closing of the valves.

Laminar versus Turbulent Flow

Some flows are smooth and orderly while others are rather chaotic. The highly ordered fluid motion characterized by smooth layers of fluid is called **laminar.** The word *laminar* comes from the movement of adjacent fluid particles together in "laminae." The flow of high-viscosity fluids such as oils at low velocities is typically laminar. The highly disordered fluid motion that typically occurs at high velocities and is characterized by velocity fluctuations is called **turbulent** (Fig. 1–20). The flow of low-viscosity fluids such as air at high velocities is typically turbulent. A flow that alternates between being laminar and turbulent is called **transitional.** The experiments conducted by Osborne Reynolds in the 1880s resulted in the establishment of the dimensionless **Reynolds number, Re,** as the key parameter for the determination of the flow regime in pipes (Chap. 8).

Natural (or Unforced) versus Forced Flow

A fluid flow is said to be natural or forced, depending on how the fluid motion is initiated. In **forced flow,** a fluid is forced to flow over a surface or in a pipe by external means such as a pump or a fan. In **natural flows,** fluid motion is due to natural means such as the buoyancy effect, which manifests itself as the rise of warmer (and thus lighter) fluid and the fall of cooler (and thus denser) fluid (Fig. 1–21). In solar hot-water systems, for example, the thermosiphoning effect is commonly used to replace pumps by placing the water tank sufficiently above the solar collectors.

FIGURE 1–19
Schlieren image of the spherical shock wave produced by a bursting balloon at the Penn State Gas Dynamics Lab. Several secondary shocks are seen in the air surrounding the ballon.
Photo by G. S. Settles, Penn State University. Used by permission.

Laminar

Transitional

Turbulent

FIGURE 1–20
Laminar, transitional, and turbulent flows over a flat plate.
Courtesy ONERA, photograph by Werlé.

FIGURE 1–21

In this schlieren image of a girl in a swimming suit, the rise of lighter, warmer air adjacent to her body indicates that humans and warm-blooded animals are surrounded by thermal plumes of rising warm air.

G. S. Settles, Gas Dynamics Lab,
Penn State University. Used by permission.

Steady versus Unsteady Flow

The terms *steady* and *uniform* are used frequently in engineering, and thus it is important to have a clear understanding of their meanings. The term **steady** implies *no change of properties, velocity, temperature, etc., at a point with time.* The opposite of steady is **unsteady.** The term **uniform** implies *no change with location* over a specified region. These meanings are consistent with their everyday use (steady girlfriend, uniform distribution, etc.).

The terms *unsteady* and *transient* are often used interchangeably, but these terms are not synonyms. In fluid mechanics, *unsteady* is the most general term that applies to any flow that is not steady, but **transient** is typically used for developing flows. When a rocket engine is fired up, for example, there are transient effects (the pressure builds up inside the rocket engine, the flow accelerates, etc.) until the engine settles down and operates steadily. The term **periodic** refers to the kind of unsteady flow in which the flow oscillates about a steady mean.

Many devices such as turbines, compressors, boilers, condensers, and heat exchangers operate for long periods of time under the same conditions, and they are classified as *steady-flow devices.* (Note that the flow field near the rotating blades of a turbomachine is of course unsteady, but we consider the overall flow field rather than the details at some localities when we classify devices.) During steady flow, the fluid properties can change from point to point within a device, but at any fixed point they remain constant. Therefore, the volume, the mass, and the total energy content of a steady-flow device or flow section remain constant in steady operation. A simple analogy is shown in Fig. 1–22.

Steady-flow conditions can be closely approximated by devices that are intended for continuous operation such as turbines, pumps, boilers, condensers, and heat exchangers of power plants or refrigeration systems. Some cyclic devices, such as reciprocating engines or compressors, do not satisfy the steady-flow conditions since the flow at the inlets and the exits is

FIGURE 1–22

Comparison of (*a*) instantaneous snapshot of an unsteady flow, and (*b*) long exposure picture of the same flow.

Photos by Eric G. Paterson. Used by permission.

(*a*)

(*b*)

pulsating and not steady. However, the fluid properties vary with time in a periodic manner, and the flow through these devices can still be analyzed as a steady-flow process by using time-averaged values for the properties.

Some fascinating visualizations of fluid flow are provided in the book *An Album of Fluid Motion* by Milton Van Dyke (1982). A nice illustration of an unsteady-flow field is shown in Fig. 1–23, taken from Van Dyke's book. Figure 1–23a is an instantaneous snapshot from a high-speed motion picture; it reveals large, alternating, swirling, turbulent eddies that are shed into the periodically oscillating wake from the blunt base of the object. The eddies produce shock waves that move upstream alternately over the top and bottom surfaces of the airfoil in an unsteady fashion. Figure 1–23b shows the *same* flow field, but the film is exposed for a longer time so that the image is time averaged over 12 cycles. The resulting time-averaged flow field appears "steady" since the details of the unsteady oscillations have been lost in the long exposure.

One of the most important jobs of an engineer is to determine whether it is sufficient to study only the time-averaged "steady" flow features of a problem, or whether a more detailed study of the unsteady features is required. If the engineer were interested only in the overall properties of the flow field (such as the time-averaged drag coefficient, the mean velocity, and pressure fields), a time-averaged description like that of Fig. 1–23b, time-averaged experimental measurements, or an analytical or numerical calculation of the time-averaged flow field would be sufficient. However, if the engineer were interested in details about the unsteady-flow field, such as flow-induced vibrations, unsteady pressure fluctuations, or the sound waves emitted from the turbulent eddies or the shock waves, a time-averaged description of the flow field would be insufficient.

Most of the analytical and computational examples provided in this textbook deal with steady or time-averaged flows, although we occasionally point out some relevant unsteady-flow features as well when appropriate.

One-, Two-, and Three-Dimensional Flows

A flow field is best characterized by its velocity distribution, and thus a flow is said to be one-, two-, or three-dimensional if the flow velocity varies in one, two, or three primary dimensions, respectively. A typical fluid flow involves a three-dimensional geometry, and the velocity may vary in all three dimensions, rendering the flow three-dimensional [$\vec{V}(x, y, z)$ in rectangular or $\vec{V}(r, \theta, z)$ in cylindrical coordinates]. However, the variation of velocity in certain directions can be small relative to the variation in other directions and can be ignored with negligible error. In such cases, the flow can be modeled conveniently as being one- or two-dimensional, which is easier to analyze.

Consider steady flow of a fluid entering from a large tank into a circular pipe. The fluid velocity everywhere on the pipe surface is zero because of the no-slip condition, and the flow is two-dimensional in the entrance region of the pipe since the velocity changes in both the r- and z-directions, but not in the θ-direction. The velocity profile develops fully and remains unchanged after some distance from the inlet (about 10 pipe diameters in turbulent flow, and less in laminar pipe flow, as in Fig. 1–24), and the flow in this region is said to be *fully developed*. The fully developed flow in a circular pipe is *one-dimensional* since the velocity varies in the radial r-direction but not in the angular θ- or axial z-directions, as shown in Fig. 1–24. That is, the velocity profile is the same at any axial z-location, and it is symmetric about the axis of the pipe.

(a)

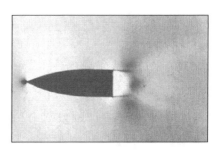

(b)

FIGURE 1–23
Oscillating wake of a blunt-based airfoil at Mach number 0.6. Photo (a) is an instantaneous image, while photo (b) is a long-exposure (time-averaged) image.

(a) Dyment, A., Flodrops, J. P. & Gryson, P. 1982 *in* Flow Visualization II, *W. Merzkirch, ed., 331–336. Washington: Hemisphere. Used by permission of Arthur Dyment.*

(b) Dyment, A. & Gryson, P. 1978 in Inst. Mèc. Fluides Lille, *No. 78-5. Used by permission of Arthur Dyment.*

FIGURE 1–24
The development of the velocity profile in a circular pipe. $V = V(r, z)$ and thus the flow is two-dimensional in the entrance region, and becomes one-dimensional downstream when the velocity profile fully develops and remains unchanged in the flow direction, $V = V(r)$.

Developing velocity profile, $V(r, z)$

Fully developed velocity profile, $V(r)$

FIGURE 1–25
Flow over a car antenna is approximately two-dimensional except near the top and bottom of the antenna.

Note that the dimensionality of the flow also depends on the choice of coordinate system and its orientation. The pipe flow discussed, for example, is one-dimensional in cylindrical coordinates, but two-dimensional in Cartesian coordinates—illustrating the importance of choosing the most appropriate coordinate system. Also note that even in this simple flow, the velocity cannot be uniform across the cross section of the pipe because of the no-slip condition. However, at a well-rounded entrance to the pipe, the velocity profile may be approximated as being nearly uniform across the pipe, since the velocity is nearly constant at all radii except very close to the pipe wall.

A flow may be approximated as *two-dimensional* when the aspect ratio is large and the flow does not change appreciably along the longer dimension. For example, the flow of air over a car antenna can be considered two-dimensional except near its ends since the antenna's length is much greater than its diameter, and the airflow hitting the antenna is fairly uniform (Fig. 1–25).

Axis of symmetry

FIGURE 1–26
Axisymmetric flow over a bullet.

EXAMPLE 1–1 **Axisymmetric Flow over a Bullet**

Consider a bullet piercing through calm air during a short time interval in which the bullet's speed is nearly constant. Determine if the time-averaged airflow over the bullet during its flight is one-, two-, or three-dimensional (Fig. 1–26).

SOLUTION It is to be determined whether airflow over a bullet is one-, two-, or three-dimensional.
Assumptions There are no significant winds and the bullet is not spinning.
Analysis The bullet possesses an axis of symmetry and is therefore an axisymmetric body. The airflow upstream of the bullet is parallel to this axis, and we expect the time-averaged airflow to be rotationally symmetric about the axis—such flows are said to be axisymmetric. The velocity in this case varies with axial distance z and radial distance r, but not with angle θ. Therefore, the time-averaged airflow over the bullet is **two-dimensional**.
Discussion While the time-averaged airflow is axisymmetric, the *instantaneous* airflow is not, as illustrated in Fig. 1–23. In Cartesian coordinates, the flow would be three-dimensional. Finally, many bullets also spin.

SURROUNDINGS

SYSTEM

BOUNDARY

FIGURE 1–27
System, surroundings, and boundary.

1–5 ▪ SYSTEM AND CONTROL VOLUME

A **system** is defined as a *quantity of matter or a region in space chosen for study*. The mass or region outside the system is called the **surroundings.** The real or imaginary surface that separates the system from its surroundings is called the **boundary** (Fig. 1–27). The boundary of a system can be *fixed* or *movable*. Note that the boundary is the contact surface shared by both the system and the surroundings. Mathematically speaking, the boundary has zero thickness, and thus it can neither contain any mass nor occupy any volume in space.

Systems may be considered to be *closed* or *open*, depending on whether a fixed mass or a volume in space is chosen for study. A **closed system** (also known as a **control mass** or simply a *system* when the context makes it clear) consists of a fixed amount of mass, and no mass can cross its boundary. But energy, in the form of heat or work, can cross the boundary, and the volume of a closed system does not have to be fixed. If, as a special case, even energy is not allowed to cross the boundary, that system is called an **isolated system.**

Consider the piston–cylinder device shown in Fig. 1–28. Let us say that we would like to find out what happens to the enclosed gas when it is heated. Since we are focusing our attention on the gas, it is our system. The inner surfaces of the piston and the cylinder form the boundary, and since no mass is crossing this boundary, it is a closed system. Notice that energy may cross the boundary, and part of the boundary (the inner surface of the piston, in this case) may move. Everything outside the gas, including the piston and the cylinder, is the surroundings.

An **open system,** or a **control volume,** as it is often called, is a *selected region in space.* It usually encloses a device that involves mass flow such as a compressor, turbine, or nozzle. Flow through these devices is best studied by selecting the region within the device as the control volume. Both mass and energy can cross the boundary (the *control surface*) of a control volume.

A large number of engineering problems involve mass flow in and out of an open system and, therefore, are modeled as *control volumes.* A water heater, a car radiator, a turbine, and a compressor all involve mass flow and should be analyzed as control volumes (open systems) instead of as control masses (closed systems). In general, *any arbitrary region in space* can be selected as a control volume. There are no concrete rules for the selection of control volumes, but a wise choice certainly makes the analysis much easier. If we were to analyze the flow of air through a nozzle, for example, a good choice for the control volume would be the region within the nozzle, or perhaps surrounding the entire nozzle.

A control volume can be fixed in size and shape, as in the case of a nozzle, or it may involve a moving boundary, as shown in Fig. 1–29. Most control volumes, however, have fixed boundaries and thus do not involve any moving boundaries. A control volume may also involve heat and work interactions just as a closed system, in addition to mass interaction.

1–6 · IMPORTANCE OF DIMENSIONS AND UNITS

Any physical quantity can be characterized by **dimensions.** The magnitudes assigned to the dimensions are called **units.** Some basic dimensions such as mass m, length L, time t, and temperature T are selected as **primary** or **fundamental dimensions,** while others such as velocity V, energy E, and volume V are expressed in terms of the primary dimensions and are called **secondary dimensions,** or **derived dimensions.**

FIGURE 1–28
A closed system with a moving boundary.

(*a*) A control volume (CV) with real and imaginary boundaries

(*b*) A control volume (CV) with fixed and moving boundaries as well as real and imaginary boundaries

FIGURE 1–29
A control volume may involve fixed, moving, real, and imaginary boundaries.

TABLE 1–1

The seven fundamental (or primary) dimensions and their units in SI

Dimension	Unit
Length	meter (m)
Mass	kilogram (kg)
Time	second (s)
Temperature	kelvin (K)
Electric current	ampere (A)
Amount of light	candela (cd)
Amount of matter	mole (mol)

TABLE 1–2

Standard prefixes in SI units

Multiple	Prefix
10^{24}	yotta, Y
10^{21}	zetta, Z
10^{18}	exa, E
10^{15}	peta, P
10^{12}	tera, T
10^{9}	giga, G
10^{6}	mega, M
10^{3}	kilo, k
10^{2}	hecto, h
10^{1}	deka, da
10^{-1}	deci, d
10^{-2}	centi, c
10^{-3}	milli, m
10^{-6}	micro, μ
10^{-9}	nano, n
10^{-12}	pico, p
10^{-15}	femto, f
10^{-18}	atto, a
10^{-21}	zepto, z
10^{-24}	yocto, y

A number of unit systems have been developed over the years. Despite strong efforts in the scientific and engineering community to unify the world with a single unit system, two sets of units are still in common use today: the **English system,** which is also known as the *United States Customary System* (USCS), and the metric **SI** (from *Le Système International d' Unités)*, which is also known as the *International System.* The SI is a simple and logical system based on a decimal relationship between the various units, and it is being used for scientific and engineering work in most of the industrialized nations, including England. The English system, however, has no apparent systematic numerical base, and various units in this system are related to each other rather arbitrarily (12 in = 1 ft, 1 mile = 5280 ft, 4 qt = 1 gal, etc.), which makes it confusing and difficult to learn. The United States is the only industrialized country that has not yet fully converted to the metric system.

The systematic efforts to develop a universally acceptable system of units dates back to 1790 when the French National Assembly charged the French Academy of Sciences to come up with such a unit system. An early version of the metric system was soon developed in France, but it did not find universal acceptance until 1875 when *The Metric Convention Treaty* was prepared and signed by 17 nations, including the United States. In this international treaty, meter and gram were established as the metric units for length and mass, respectively, and a *General Conference of Weights and Measures* (CGPM) was established that was to meet every six years. In 1960, the CGPM produced the SI, which was based on six fundamental quantities, and their units were adopted in 1954 at the Tenth General Conference of Weights and Measures: *meter* (m) for length, *kilogram* (kg) for mass, *second* (s) for time, *ampere* (A) for electric current, *degree Kelvin* (°K) for temperature, and *candela* (cd) for luminous intensity (amount of light). In 1971, the CGPM added a seventh fundamental quantity and unit: *mole* (mol) for the amount of matter.

Based on the notational scheme introduced in 1967, the degree symbol was officially dropped from the absolute temperature unit, and all unit names were to be written without capitalization even if they were derived from proper names (Table 1–1). However, the abbreviation of a unit was to be capitalized if the unit was derived from a proper name. For example, the SI unit of force, which is named after Sir Isaac Newton (1647–1723), is *newton* (not Newton), and it is abbreviated as N. Also, the full name of a unit may be pluralized, but its abbreviation cannot. For example, the length of an object can be 5 m or 5 meters, *not* 5 ms or 5 meter. Finally, no period is to be used in unit abbreviations unless they appear at the end of a sentence. For example, the proper abbreviation of meter is m (not m.).

The recent move toward the metric system in the United States seems to have started in 1968 when Congress, in response to what was happening in the rest of the world, passed a Metric Study Act. Congress continued to promote a voluntary switch to the metric system by passing the Metric Conversion Act in 1975. A trade bill passed by Congress in 1988 set a September 1992 deadline for all federal agencies to convert to the metric system. However, the deadlines were relaxed later with no clear plans for the future.

As pointed out, the SI is based on a decimal relationship between units. The prefixes used to express the multiples of the various units are listed in Table 1–2.

They are standard for all units, and the student is encouraged to memorize some of them because of their widespread use (Fig. 1–30).

Some SI and English Units

In SI, the units of mass, length, and time are the kilogram (kg), meter (m), and second (s), respectively. The respective units in the English system are the pound-mass (lbm), foot (ft), and second (s). The pound symbol *lb* is actually the abbreviation of *libra*, which was the ancient Roman unit of weight. The English retained this symbol even after the end of the Roman occupation of Britain in 410. The mass and length units in the two systems are related to each other by

$$1 \text{ lbm} = 0.45359 \text{ kg}$$

$$1 \text{ ft} = 0.3048 \text{ m}$$

In the English system, force is often considered to be one of the primary dimensions and is assigned a nonderived unit. This is a source of confusion and error that necessitates the use of a dimensional constant (g_c) in many formulas. To avoid this nuisance, we consider force to be a secondary dimension whose unit is derived from Newton's second law, i.e.,

$$\text{Force} = (\text{Mass})(\text{Acceleration})$$

or
$$F = ma \qquad (1\text{–}1)$$

In SI, the force unit is the newton (N), and it is defined as the *force required to accelerate a mass of 1 kg at a rate of 1 m/s²*. In the English system, the force unit is the **pound-force** (lbf) and is defined as the *force required to accelerate a mass of 32.174 lbm (1 slug) at a rate of 1 ft/s²* (Fig. 1–31). That is,

$$1 \text{ N} = 1 \text{ kg·m/s}^2$$

$$1 \text{ lbf} = 32.174 \text{ lbm·ft/s}^2$$

A force of 1 N is roughly equivalent to the weight of a small apple ($m = 102$ g), whereas a force of 1 lbf is roughly equivalent to the weight of four medium apples ($m_{\text{total}} = 454$ g), as shown in Fig. 1–32. Another force unit in common use in many European countries is the *kilogram-force* (kgf), which is the weight of 1 kg mass at sea level (1 kgf = 9.807 N).

The term **weight** is often incorrectly used to express mass, particularly by the "weight watchers." Unlike mass, weight W is a *force*. It is the gravitational force applied to a body, and its magnitude is determined from an equation based on Newton's second law,

$$W = mg \qquad (\text{N}) \qquad (1\text{–}2)$$

where m is the mass of the body, and g is the local gravitational acceleration (g is 9.807 m/s² or 32.174 ft/s² at sea level and 45° latitude). An ordinary bathroom scale measures the gravitational force acting on a body. The weight per unit volume of a substance is called the **specific weight** γ and is determined from $\gamma = \rho g$, where ρ is density.

FIGURE 1–30
The SI unit prefixes are used in all branches of engineering.

FIGURE 1–31
The definition of the force units.

FIGURE 1–32
The relative magnitudes of the force units newton (N), kilogram-force (kgf), and pound-force (lbf).

FIGURE 1–33
A body weighing 150 lbf on earth will weigh only 25 lbf on the moon.

FIGURE 1–34
The weight of a unit mass at sea level.

FIGURE 1–35
A typical match yields about one Btu (or one kJ) of energy if completely burned.

Photo by John M. Cimbala.

The mass of a body remains the same regardless of its location in the universe. Its weight, however, changes with a change in gravitational acceleration. A body weighs less on top of a mountain since g decreases (by a small amount) with altitude. On the surface of the moon, an astronaut weighs about one-sixth of what she or he normally weighs on earth (Fig. 1–33).

At sea level a mass of 1 kg weighs 9.807 N, as illustrated in Fig. 1–34. A mass of 1 lbm, however, weighs 1 lbf, which misleads people to believe that pound-mass and pound-force can be used interchangeably as pound (lb), which is a major source of error in the English system.

It should be noted that the *gravity force* acting on a mass is due to the *attraction* between the masses, and thus it is proportional to the magnitudes of the masses and inversely proportional to the square of the distance between them. Therefore, the gravitational acceleration g at a location depends on the *local density* of the earth's crust, the *distance* to the center of the earth, and to a lesser extent, the positions of the moon and the sun. The value of g varies with location from 9.8295 m/s^2 at 4500 m below sea level to 7.3218 m/s^2 at 100,000 m above sea level. However, at altitudes up to 30,000 m, the variation of g from the sea-level value of 9.807 m/s^2, is less than 1 percent. Therefore, for most practical purposes, the gravitational acceleration can be assumed to be *constant* at 9.807 m/s^2, often rounded to 9.81 m/s^2. It is interesting to note that the value of g increases with distance below sea level, reaches a maximum at about 4500 m below sea level, and then starts decreasing. (What do you think the value of g is at the center of the earth?)

The primary cause of confusion between mass and weight is that mass is usually measured *indirectly* by measuring the *gravity force* it exerts. This approach also assumes that the forces exerted by other effects such as air buoyancy and fluid motion are negligible. This is like measuring the distance to a star by measuring its red shift, or measuring the altitude of an airplane by measuring barometric pressure. Both of these are also indirect measurements. The correct *direct* way of measuring mass is to compare it to a known mass. This is cumbersome, however, and it is mostly used for calibration and measuring precious metals.

Work, which is a form of energy, can simply be defined as force times distance; therefore, it has the unit "newton-meter (N·m)," which is called a **joule** (J). That is,

$$1 \text{ J} = 1 \text{ N·m} \tag{1–3}$$

A more common unit for energy in SI is the kilojoule (1 kJ = 10^3 J). In the English system, the energy unit is the **Btu** (British thermal unit), which is defined as the energy required to raise the temperature of 1 lbm of water at 68°F by 1°F. In the metric system, the amount of energy needed to raise the temperature of 1 g of water at 14.5°C by 1°C is defined as 1 **calorie** (cal), and 1 cal = 4.1868 J. The magnitudes of the kilojoule and Btu are very nearly the same (1 Btu = 1.0551 kJ). Here is a good way to get a feel for these units: If you light a typical match and let it burn itself out, it yields approximately one Btu (or one kJ) of energy (Fig. 1–35).

The unit for time rate of energy is joule per second (J/s), which is called a **watt** (W). In the case of work, the time rate of energy is called *power*. A commonly used unit of power is horsepower (hp), which is equivalent to 745.7 W. Electrical energy typically is expressed in the unit kilowatt-hour (kWh), which is equivalent to 3600 kJ. An electric appliance with a rated power of 1 kW consumes 1 kWh of electricity when running continuously for one hour. When dealing with electric power generation, the units kW and kWh are often confused. Note that kW or kJ/s is a unit of power, whereas kWh is a unit of energy. Therefore, statements like "the new wind turbine will generate 50 kW of electricity per year" are meaningless and incorrect. A correct statement should be something like "the new wind turbine with a rated power of 50 kW will generate 120,000 kWh of electricity per year."

Dimensional Homogeneity

We all know that you cannot add apples and oranges. But we somehow manage to do it (by mistake, of course). In engineering, all equations must be *dimensionally homogeneous*. That is, every term in an equation must have the same dimensions. If, at some stage of an analysis, we find ourselves in a position to add two quantities that have different dimensions or units, it is a clear indication that we have made an error at an earlier stage. So checking dimensions (or units) can serve as a valuable tool to spot errors.

■ *EXAMPLE 1–2* **Electric Power Generation by a Wind Turbine**

A school is paying $0.09/kWh for electric power. To reduce its power bill, the school installs a wind turbine (Fig 1–36) with a rated power of 30 kW. If the turbine operates 2200 hours per year at the rated power, determine the amount of electric power generated by the wind turbine and the money saved by the school per year.

SOLUTION A wind turbine is installed to generate electricity. The amount of electric energy generated and the money saved per year are to be determined.
Analysis The wind turbine generates electric energy at a rate of 30 kW or 30 kJ/s. Then the total amount of electric energy generated per year becomes

$$\text{Total energy} = (\text{Energy per unit time})(\text{Time interval})$$
$$= (30 \text{ kW})(2200 \text{ h})$$
$$= \mathbf{66{,}000 \text{ kWh}}$$

The money saved per year is the monetary value of this energy determined as

$$\text{Money saved} = (\text{Total energy})(\text{Unit cost of energy})$$
$$= (66{,}000 \text{ kWh})(\$0.09/\text{kWh})$$
$$= \mathbf{\$5940}$$

Discussion The annual electric energy production also could be determined in kJ by unit manipulations as

$$\text{Total energy} = (30 \text{ kW})(2200 \text{ h})\left(\frac{3600 \text{ s}}{1 \text{ h}}\right)\left(\frac{1 \text{ kJ/s}}{1 \text{ kW}}\right) = 2.38 \times 10^8 \text{ kJ}$$

which is equivalent to 66,000 kWh (1 kWh = 3600 kJ).

FIGURE 1–36
A wind turbine, as discussed in
Example 1–2.
Photo by Andy Cimbala.

We all know from experience that units can give terrible headaches if they are not used carefully in solving a problem. However, with some attention and skill, units can be used to our advantage. They can be used to check formulas; sometimes they can even be used to *derive* formulas, as explained in the following example.

FIGURE 1–37
Schematic for Example 1–3.

EXAMPLE 1–3   Obtaining Formulas from Unit Considerations

A tank is filled with oil whose density is $\rho = 850$ kg/m³. If the volume of the tank is $V = 2$ m³, determine the amount of mass m in the tank.

SOLUTION   The volume of an oil tank is given. The mass of oil is to be determined.

Assumptions   Oil is a nearly incompressible substance and thus its density is constant.

Analysis   A sketch of the system just described is given in Fig. 1–37. Suppose we forgot the formula that relates mass to density and volume. However, we know that mass has the unit of kilograms. That is, whatever calculations we do, we should end up with the unit of kilograms. Putting the given information into perspective, we have

$$\rho = 850 \text{ kg/m}^3 \quad \text{and} \quad V = 2 \text{ m}^3$$

It is obvious that we can eliminate m³ and end up with kg by multiplying these two quantities. Therefore, the formula we are looking for should be

$$m = \rho V$$

Thus,

$$m = (850 \text{ kg/m}^3)(2 \text{ m}^3) = \mathbf{1700 \ kg}$$

Discussion   Note that this approach may not work for more complicated formulas. Nondimensional constants also may be present in the formulas, and these cannot be derived from unit considerations alone.

You should keep in mind that a formula that is not dimensionally homogeneous is definitely wrong (Fig. 1–38), but a dimensionally homogeneous formula is not necessarily right.

Unity Conversion Ratios

Just as all nonprimary dimensions can be formed by suitable combinations of primary dimensions, *all nonprimary units (**secondary units**) can be formed by combinations of primary units.* Force units, for example, can be expressed as

$$N = kg \frac{m}{s^2} \quad \text{and} \quad lbf = 32.174 \ lbm \frac{ft}{s^2}$$

They can also be expressed more conveniently as **unity conversion ratios** as

$$\frac{N}{kg \cdot m/s^2} = 1 \quad \text{and} \quad \frac{lbf}{32.174 \ lbm \cdot ft/s^2} = 1$$

FIGURE 1–38
Always check the units in your calculations.

Unity conversion ratios are identically equal to 1 and are unitless, and thus such ratios (or their inverses) can be inserted conveniently into any calculation to properly convert units (Fig 1–39). You are encouraged to always use unity conversion ratios such as those given here when converting units. Some textbooks insert the archaic gravitational constant g_c defined as $g_c = 32.174$ lbm·ft/lbf·s^2 = kg·m/N·s^2 = 1 into equations in order to force units to match. This practice leads to unnecessary confusion and is strongly discouraged by the present authors. We recommend that you instead use unity conversion ratios.

FIGURE 1–39

Every unity conversion ratio (as well as its inverse) is exactly equal to one. Shown here are a few commonly used unity conversion ratios.

■
■ *EXAMPLE 1–4*   **The Weight of One Pound-Mass**
■
■ Using unity conversion ratios, show that 1.00 lbm weighs 1.00 lbf on earth
■ (Fig. 1–40).

Solution   A mass of 1.00 lbm is subjected to standard earth gravity. Its weight in lbf is to be determined.
Assumptions   Standard sea-level conditions are assumed.
Properties   The gravitational constant is $g = 32.174$ ft/s^2.
Analysis   We apply Newton's second law to calculate the weight (force) that corresponds to the known mass and acceleration. The weight of any object is equal to its mass times the local value of gravitational acceleration. Thus,

$$W = mg = (1.00 \text{ lbm})(32.174 \text{ ft/s}^2)\left(\frac{1 \text{ lbf}}{32.174 \text{ lbm·ft/s}^2}\right) = 1.00 \text{ lbf}$$

Discussion   The quantity in large parentheses in this equation is a unity conversion ratio. Mass is the same regardless of its location. However, on some other planet with a different value of gravitational acceleration, the weight of 1 lbm would differ from that calculated here.

FIGURE 1–40
A mass of 1 lbm weighs 1 lbf on earth.

When you buy a box of breakfast cereal, the printing may say "Net weight: One pound (454 grams)." (See Fig. 1–41.) Technically, this means that the cereal inside the box weighs 1.00 lbf on earth and has a *mass* of 453.6 g (0.4536 kg). Using Newton's second law, the actual weight of the cereal on earth is

$$W = mg = (453.6 \text{ g})(9.81 \text{ m/s}^2)\left(\frac{1 \text{ N}}{1 \text{ kg·m/s}^2}\right)\left(\frac{1 \text{ kg}}{1000 \text{ g}}\right) = 4.49 \text{ N}$$

1–7 · MODELING IN ENGINEERING

An engineering device or process can be studied either *experimentally* (testing and taking measurements) or *analytically* (by analysis or calculations). The experimental approach has the advantage that we deal with the actual physical system, and the desired quantity is determined by measurement,

FIGURE 1–41
A quirk in the metric system of units.

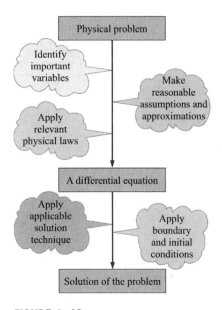

FIGURE 1–42
Mathematical modeling of physical problems.

within the limits of experimental error. However, this approach is expensive, time-consuming, and often impractical. Besides, the system we are studying may not even exist. For example, the entire heating and plumbing systems of a building must usually be sized *before* the building is actually built on the basis of the specifications given. The analytical approach (including the numerical approach) has the advantage that it is fast and inexpensive, but the results obtained are subject to the accuracy of the assumptions, approximations, and idealizations made in the analysis. In engineering studies, often a good compromise is reached by reducing the choices to just a few by analysis, and then verifying the findings experimentally.

The descriptions of most scientific problems involve equations that relate the changes in some key variables to each other. Usually the smaller the increment chosen in the changing variables, the more general and accurate the description. In the limiting case of infinitesimal or differential changes in variables, we obtain *differential equations* that provide precise mathematical formulations for the physical principles and laws by representing the rates of change as *derivatives*. Therefore, differential equations are used to investigate a wide variety of problems in sciences and engineering (Fig. 1–42). However, many problems encountered in practice can be solved without resorting to differential equations and the complications associated with them.

The study of physical phenomena involves two important steps. In the first step, all the variables that affect the phenomena are identified, reasonable assumptions and approximations are made, and the interdependence of these variables is studied. The relevant physical laws and principles are invoked, and the problem is formulated mathematically. The equation itself is very instructive as it shows the degree of dependence of some variables on others, and the relative importance of various terms. In the second step, the problem is solved using an appropriate approach, and the results are interpreted.

Many processes that seem to occur in nature randomly and without any order are, in fact, being governed by some visible or not-so-visible physical laws. Whether we notice them or not, these laws are there, governing consistently and predictably over what seem to be ordinary events. Most of these laws are well defined and well understood by scientists. This makes it possible to predict the course of an event before it actually occurs or to study various aspects of an event mathematically without actually running expensive and time-consuming experiments. This is where the power of analysis lies. Very accurate results to meaningful practical problems can be obtained with relatively little effort by using a suitable and realistic mathematical model. The preparation of such models requires an adequate knowledge of the natural phenomena involved and the relevant laws, as well as sound judgment. An unrealistic model will obviously give inaccurate and thus unacceptable results.

An analyst working on an engineering problem often finds himself or herself in a position to make a choice between a very accurate but complex model, and a simple but not-so-accurate model. The right choice depends on the situation at hand. The right choice is usually the simplest model that

(a) Actual engineering problem   (b) Minimum essential model of the engineering problem

FIGURE 1–43

Simplified models are often used in fluid mechanics to obtain approximate solutions to difficult engineering problems. Here, the helicopter's rotor is modeled by a disk, across which is imposed a sudden change in pressure. The helicopter's body is modeled by a simple ellipsoid. This simplified model yields the essential features of the overall air flow field in the vicinity of the ground.

Photo by John M. Cimbala.

yields satisfactory results (Fig 1–43). Also, it is important to consider the actual operating conditions when selecting equipment.

Preparing very accurate but complex models is usually not so difficult. But such models are not much use to an analyst if they are very difficult and time-consuming to solve. At the minimum, the model should reflect the essential features of the physical problem it represents. There are many significant real-world problems that can be analyzed with a simple model. But it should always be kept in mind that the results obtained from an analysis are at best as accurate as the assumptions made in simplifying the problem. Therefore, the solution obtained should not be applied to situations for which the original assumptions do not hold.

A solution that is not quite consistent with the observed nature of the problem indicates that the mathematical model used is too crude. In that case, a more realistic model should be prepared by eliminating one or more of the questionable assumptions. This will result in a more complex problem that, of course, is more difficult to solve. Thus any solution to a problem should be interpreted within the context of its formulation.

1–8 ▪ PROBLEM-SOLVING TECHNIQUE

The first step in learning any science is to grasp the fundamentals and to gain a sound knowledge of it. The next step is to master the fundamentals by testing this knowledge. This is done by solving significant real-world problems. Solving such problems, especially complicated ones, requires a systematic approach. By using a step-by-step approach, an engineer can reduce the

FIGURE 1–44
A step-by-step approach can greatly simplify problem solving.

Given: Air temperature in Denver
To be found: Density of air
Missing information: Atmospheric pressure
Assumption #1: Take $P = 1$ atm (Inappropriate. Ignores effect of altitude. Will cause more than 15% error.)
Assumption #2: Take $P = 0.83$ atm (Appropriate. Ignores only minor effects such as weather.)

FIGURE 1–45
The assumptions made while solving an engineering problem must be reasonable and justifiable.

---

solution of a complicated problem into the solution of a series of simple problems (Fig. 1–44). When you are solving a problem, we recommend that you use the following steps zealously as applicable. This will help you avoid some of the common pitfalls associated with problem solving.

Step 1: Problem Statement
In your own words, briefly state the problem, the key information given, and the quantities to be found. This is to make sure that you understand the problem and the objectives before you attempt to solve the problem.

Step 2: Schematic
Draw a realistic sketch of the physical system involved, and list the relevant information on the figure. The sketch does not have to be something elaborate, but it should resemble the actual system and show the key features. Indicate any energy and mass interactions with the surroundings. Listing the given information on the sketch helps one to see the entire problem at once. Also, check for properties that remain constant during a process (such as temperature during an isothermal process), and indicate them on the sketch.

Step 3: Assumptions and Approximations
State any appropriate assumptions and approximations made to simplify the problem to make it possible to obtain a solution. Justify the questionable assumptions. Assume reasonable values for missing quantities that are necessary. For example, in the absence of specific data for atmospheric pressure, it can be taken to be 1 atm. However, it should be noted in the analysis that the atmospheric pressure decreases with increasing elevation. For example, it drops to 0.83 atm in Denver (elevation 1610 m) (Fig. 1–45).

Step 4: Physical Laws
Apply all the relevant basic physical laws and principles (such as the conservation of mass), and reduce them to their simplest form by utilizing the assumptions made. However, the region to which a physical law is applied must be clearly identified first. For example, the increase in speed of water flowing through a nozzle is analyzed by applying conservation of mass between the inlet and outlet of the nozzle.

Step 5: Properties
Determine the unknown properties at known states necessary to solve the problem from property relations or tables. List the properties separately, and indicate their source, if applicable.

Step 6: Calculations
Substitute the known quantities into the simplified relations and perform the calculations to determine the unknowns. Pay particular attention to the units and unit cancellations, and remember that a dimensional quantity without a unit is meaningless. Also, don't give a false implication of high precision

by copying all the digits from the screen of the calculator—round the final results to an appropriate number of significant digits (Section 1–10).

Step 7: Reasoning, Verification, and Discussion

Check to make sure that the results obtained are reasonable and intuitive, and verify the validity of the questionable assumptions. Repeat the calculations that resulted in unreasonable values. For example, under the same test conditions the aerodynamic drag acting on a car should *not* increase after streamlining the shape of the car (Fig. 1–46).

Also, point out the significance of the results, and discuss their implications. State the conclusions that can be drawn from the results, and any recommendations that can be made from them. Emphasize the limitations under which the results are applicable, and caution against any possible misunderstandings and using the results in situations where the underlying assumptions do not apply. For example, if you determined that using a larger-diameter pipe in a proposed pipeline will cost an additional $5000 in materials, but it will reduce the annual pumping costs by $3000, indicate that the larger-diameter pipeline will pay for its cost differential from the electricity it saves in less than two years. However, also state that only additional material costs associated with the larger-diameter pipeline are considered in the analysis.

Keep in mind that the solutions you present to your instructors, and any engineering analysis presented to others, is a form of communication. Therefore neatness, organization, completeness, and visual appearance are of utmost importance for maximum effectiveness (Fig 1–47). Besides, neatness also serves as a great checking tool since it is very easy to spot errors and inconsistencies in neat work. Carelessness and skipping steps to save time often end up costing more time and unnecessary anxiety.

The approach described here is used in the solved example problems without explicitly stating each step, as well as in the Solutions Manual of this text. For some problems, some of the steps may not be applicable or necessary. For example, often it is not practical to list the properties separately. However, we cannot overemphasize the importance of a logical and orderly approach to problem solving. Most difficulties encountered while solving a problem are not due to a lack of knowledge; rather, they are due to a lack of organization. You are strongly encouraged to follow these steps in problem solving until you develop your own approach that works best for you.

1–9 · ENGINEERING SOFTWARE PACKAGES

You may be wondering why we are about to undertake an in-depth study of the fundamentals of another engineering science. After all, almost all such problems we are likely to encounter in practice can be solved using one of several sophisticated software packages readily available in the market today. These software packages not only give the desired numerical results, but also supply the outputs in colorful graphical form for impressive presentations. It is unthinkable to practice engineering today without using some of these packages. This tremendous computing power available to us at the touch of a button is both a blessing and a curse. It certainly enables engineers to solve problems easily and quickly, but it also opens the door for

FIGURE 1–46
The results obtained from an engineering analysis must be checked for reasonableness.

FIGURE 1–47
Neatness and organization are highly valued by employers.

abuses and misinformation. In the hands of poorly educated people, these software packages are as dangerous as sophisticated powerful weapons in the hands of poorly trained soldiers.

Thinking that a person who can use the engineering software packages without proper training in the fundamentals can practice engineering is like thinking that a person who can use a wrench can work as a car mechanic. If it were true that the engineering students do not need all these fundamental courses they are taking because practically everything can be done by computers quickly and easily, then it would also be true that the employers would no longer need high-salaried engineers since any person who knows how to use a word-processing program can also learn how to use those software packages. However, the statistics show that the need for engineers is on the rise, not on the decline, despite the availability of these powerful packages.

We should always remember that all the computing power and the engineering software packages available today are just *tools*, and tools have meaning only in the hands of masters. Having the best word-processing program does not make a person a good writer, but it certainly makes the job of a good writer much easier and makes the writer more productive (Fig. 1–48). Hand calculators did not eliminate the need to teach our children how to add or subtract, and sophisticated medical software packages did not take the place of medical school training. Neither will engineering software packages replace the traditional engineering education. They will simply cause a shift in emphasis in the courses from mathematics to physics. That is, more time will be spent in the classroom discussing the physical aspects of the problems in greater detail, and less time on the mechanics of solution procedures.

All these marvelous and powerful tools available today put an extra burden on today's engineers. They must still have a thorough understanding of the fundamentals, develop a "feel" of the physical phenomena, be able to put the data into proper perspective, and make sound engineering judgments, just like their predecessors. However, they must do it much better, and much faster, using more realistic models because of the powerful tools available today. The engineers in the past had to rely on hand calculations, slide rules, and later hand calculators and computers. Today they rely on software packages. The easy access to such power and the possibility of a simple misunderstanding or misinterpretation causing great damage make it more important today than ever to have solid training in the fundamentals of engineering. In this text we make an extra effort to put the emphasis on developing an intuitive and physical understanding of natural phenomena instead of on the mathematical details of solution procedures.

FIGURE 1–48
An excellent word-processing program does not make a person a good writer; it simply makes a good writer a more efficient writer.
© *Ingram Publishing RF*

Engineering Equation Solver (EES)

EES is a program that solves systems of linear or nonlinear algebraic or differential equations numerically. It has a large library of built-in thermodynamic property functions as well as mathematical functions, and allows the user to supply additional property data. Unlike some software packages, EES does not solve engineering problems; it only solves the equations supplied by the user. Therefore, the user must understand the problem and formulate it by applying any relevant physical laws and relations. EES saves the user considerable time and effort by simply solving the resulting mathematical equations. This makes it possible to attempt significant engineering problems not suitable for hand calculations and to conduct parametric studies quickly and conveniently. EES is a very powerful yet intuitive program that is very easy to use, as shown in Example 1–5. The use and capabilities of EES are explained in Appendix 3 on the text website.

■
■ *EXAMPLE 1–5*   **Solving a System of Equations with EES**
■
■ The difference of two numbers is 4, and the sum of the squares of these two
■ numbers is equal to the sum of the numbers plus 20. Determine these two
■ numbers.

SOLUTION   Relations are given for the difference and the sum of the squares of two numbers. The two numbers are to be determined.
Analysis   We start the EES program by double-clicking on its icon, open a new file, and type the following on the blank screen that appears:

$$x-y=4$$
$$x\,\hat{}\,2+y\,\hat{}\,2=x+y+20$$

which is an exact mathematical expression of the problem statement with *x* and *y* denoting the unknown numbers. The solution to this system of two nonlinear equations with two unknowns is obtained by a single click on the "calculator" icon on the taskbar. It gives (Fig. 1–49)

$$x=5 \quad \text{and} \quad y=1$$

Discussion   Note that all we did is formulate the problem as we would on paper; EES took care of all the mathematical details of solution. Also note that equations can be linear or nonlinear, and they can be entered in any order with unknowns on either side. Friendly equation solvers such as EES allow the user to concentrate on the physics of the problem without worrying about the mathematical complexities associated with the solution of the resulting system of equations.

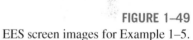

FIGURE 1–49
EES screen images for Example 1–5.

CFD Software

Computational fluid dynamics (CFD) is used extensively in engineering and research, and we discuss CFD in detail in Chapter 15. We also show example solutions from CFD throughout the textbook since CFD graphics are great for illustrating flow streamlines, velocity, and pressure distributions, etc.– beyond what we are able to visualize in the laboratory. However, because there are several different commercial CFD packages available for users, and student access to these codes is highly dependent on departmental licenses, we do not provide end-of-chapter CFD problems that are tied to any particular CFD package. Instead, we provide some general CFD problems in Chapter 15 , and we also maintain a website (see link at www.mhhe.com/cengel) containing CFD problems that can be solved with a number of different CFD programs. Students are encouraged to work through some of these problems to become familiar with CFD.

1–10 · ACCURACY, PRECISION, AND SIGNIFICANT DIGITS

In engineering calculations, the supplied information is not known to more than a certain number of significant digits, usually three digits. Consequently, the results obtained cannot possibly be precise to more significant digits. Reporting results in more significant digits implies greater precision than exists, and it should be avoided.

Regardless of the system of units employed, engineers must be aware of three principles that govern the proper use of numbers: accuracy, precision, and significant digits. For engineering measurements, they are defined as follows:

- **Accuracy error** (*inaccuracy*) is the value of one reading minus the true value. In general, accuracy of a set of measurements refers to the closeness of the average reading to the true value. Accuracy is generally associated with repeatable, fixed errors.

- **Precision error** is the value of one reading minus the average of readings. In general, precision of a set of measurements refers to the fineness of the resolution and the repeatability of the instrument. Precision is generally associated with unrepeatable, random errors.

- **Significant digits** are digits that are relevant and meaningful.

A measurement or calculation can be very precise without being very accurate, and vice versa. For example, suppose the true value of wind speed is 25.00 m/s. Two anemometers A and B take five wind speed readings each:

Anemometer A: 25.50, 25.69, 25.52, 25.58, and 25.61 m/s. Average of all readings = 25.58 m/s.

Anemometer B: 26.3, 24.5, 23.9, 26.8, and 23.6 m/s. Average of all readings = 25.02 m/s.

Clearly, anemometer A is more precise, since none of the readings differs by more than 0.11 m/s from the average. However, the average is 25.58 m/s, 0.58 m/s greater than the true wind speed; this indicates significant **bias error,** also called **constant error** or **systematic error.** On the other hand, anemometer B is not very precise, since its readings swing wildly from the average; but its overall average is much closer to the true value. Hence, anemometer B is more accurate than anemometer A, at least for this set of readings, even though it is less precise. The difference between accuracy and precision can be illustrated effectively by analogy to shooting arrows at a target, as sketched in Fig. 1–50. Shooter A is very precise, but not very accurate, while shooter B has better overall accuracy, but less precision.

Many engineers do not pay proper attention to the number of significant digits in their calculations. The least significant numeral in a number implies the precision of the measurement or calculation. For example, a result written as 1.23 (three significant digits) *implies* that the result is precise to within one digit in the second decimal place; i.e., the number is somewhere between 1.22 and 1.24. Expressing this number with any more digits would be misleading. The number of significant digits is most easily evaluated when the number is written in exponential notation; the number of significant digits can then simply be counted, including zeroes. Alternatively, the

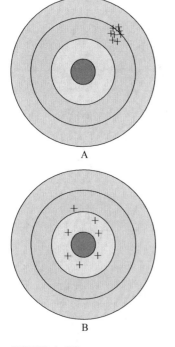

FIGURE 1–50
Illustration of accuracy versus precision. Shooter A is more precise, but less accurate, while shooter B is more accurate, but less precise.

least significant digit can be underlined to indicate the author's intent. Some examples are shown in Table 1–3.

When performing calculations or manipulations of several parameters, the final result is generally only as precise as the least precise parameter in the problem. For example, suppose A and B are multiplied to obtain C. If $A = 2.3601$ (five significant digits), and $B = 0.34$ (two significant digits), then $C = 0.80$ (only two digits are significant in the final result). Note that most students are tempted to write $C = 0.802434$, with six significant digits, since that is what is displayed on a calculator after multiplying these two numbers.

Let's analyze this simple example carefully. Suppose the exact value of B is 0.33501, which is read by the instrument as 0.34. Also suppose A is exactly 2.3601, as measured by a more accurate and precise instrument. In this case, $C = A \times B = 0.79066$ to five significant digits. Note that our first answer, $C = 0.80$ is off by one digit in the second decimal place. Likewise, if B is 0.34499, and is read by the instrument as 0.34, the product of A and B would be 0.81421 to five significant digits. Our original answer of 0.80 is again off by one digit in the second decimal place. The main point here is that 0.80 (to two significant digits) is the best one can expect from this multiplication since, to begin with, one of the values had only two significant digits. Another way of looking at this is to say that beyond the first two digits in the answer, the rest of the digits are meaningless or not significant. For example, if one reports what the calculator displays, 2.3601 times 0.34 equals 0.802434, the last four digits are *meaningless*. As shown, the final result may lie between 0.79 and 0.81—any digits beyond the two significant digits are not only meaningless, but *misleading*, since they imply to the reader more precision than is really there.

As another example, consider a 3.75-L container filled with gasoline whose density is 0.845 kg/L, and determine its mass. Probably the first thought that comes to your mind is to multiply the volume and density to obtain 3.16875 kg for the mass, which falsely implies that the mass so determined is precise to six significant digits. In reality, however, the mass cannot be more precise than three significant digits since both the volume and the density are precise to three significant digits only. Therefore, the result should be rounded to three significant digits, and the mass should be reported to be 3.17 kg instead of what the calculator displays (Fig. 1–51). The result 3.16875 kg would be correct only if the volume and density were given to be 3.75000 L and 0.845000 kg/L, respectively. The value 3.75 L implies that we are fairly confident that the volume is precise within ±0.01 L, and it cannot be 3.74 or 3.76 L. However, the volume can be 3.746, 3.750, 3.753, etc., since they all round to 3.75 L.

You should also be aware that sometimes we knowingly introduce small errors in order to avoid the trouble of searching for more accurate data. For example, when dealing with liquid water, we often use the value of 1000 kg/m^3 for density, which is the density value of pure water at 0°C. Using this value at 75°C will result in an error of 2.5 percent since the density at this temperature is 975 kg/m^3. The minerals and impurities in the water will introduce additional error. This being the case, you should have no reservation in rounding the final results to a reasonable number of significant digits. Besides, having a few percent uncertainty in the results of engineering analysis is usually the norm, not the exception.

TABLE 1–3

Significant digits

Number	Exponential Notation	Number of Significant Digits
12.3	1.23×10^1	3
123,000	1.23×10^5	3
0.00123	1.23×10^{-3}	3
40,300	4.03×10^4	3
40,300.	4.0300×10^4	5
0.005600	5.600×10^{-3}	4
0.0056	5.6×10^{-3}	2
0.006	$6. \times 10^{-3}$	1

Given: Volume: $V = 3.75$ L

Density: $\rho = 0.845$ kg/L

(3 significant digits

Also, $3.75 \times 0.845 = 3.16875$

Find: Mass: $m = \rho V = 3.16875$ kg

Rounding to 3 significant digits
$m = 3.17$ kg

FIGURE 1–51

A result with more significant digits than that of given data falsely implies more precision.

When writing intermediate results in a computation, it is advisable to keep several "extra" digits to avoid round-off errors; however, the final result should be written with the number of significant digits taken into consideration. You must also keep in mind that a certain number of significant digits of precision in the result does not necessarily imply the same number of digits of overall *accuracy*. Bias error in one of the readings may, for example, significantly reduce the overall accuracy of the result, perhaps even rendering the last significant digit meaningless, and reducing the overall number of reliable digits by one. Experimentally determined values are subject to measurement errors, and such errors are reflected in the results obtained. For example, if the density of a substance has an uncertainty of 2 percent, then the mass determined using this density value will also have an uncertainty of 2 percent.

Finally, when the number of significant digits is unknown, the accepted engineering standard is three significant digits. Therefore, if the length of a pipe is given to be 40 m, we will assume it to be 40.0 m in order to justify using three significant digits in the final results.

FIGURE 1–52
Photo for Example 1–6 for the measurement of volume flow rate.
Photo by John M. Cimbala.

EXAMPLE 1–6   Significant Digits and Volume Flow Rate

Jennifer is conducting an experiment that uses cooling water from a garden hose. In order to calculate the volume flow rate of water through the hose, she times how long it takes to fill a container (Fig. 1–52). The volume of water collected is $V = 1.1$ gal in time period $\Delta t = 45.62$ s, as measured with a stopwatch. Calculate the volume flow rate of water through the hose in units of cubic meters per minute.

SOLUTION Volume flow rate is to be determined from measurements of volume and time period.
Assumptions **1** Jennifer recorded her measurements properly, such that the volume measurement is precise to two significant digits while the time period is precise to four significant digits. **2** No water is lost due to splashing out of the container.
Analysis Volume flow rate $\dot{V}$ is volume displaced per unit time and is expressed as

Volume flow rate:
$$\dot{V} = \frac{\Delta V}{\Delta t}$$

Substituting the measured values, the volume flow rate is determined to be

$$\dot{V} = \frac{1.1 \text{ gal}}{45.62 \text{ s}} \left(\frac{3.7854 \times 10^{-3} \text{ m}^3}{1 \text{ gal}} \right) \left(\frac{60 \text{ s}}{1 \text{ min}} \right) = 5.5 \times 10^{-3} \text{m}^3/\text{min}$$

Discussion The final result is listed to two significant digits since we cannot be confident of any more precision than that. If this were an intermediate step in subsequent calculations, a few extra digits would be carried along to avoid accumulated round-off error. In such a case, the volume flow rate would be written as $\dot{V} = 5.4765 \times 10^{-3}$ m^3/min. Based on the given information, we cannot say anything about the *accuracy* of our result, since we have no information about systematic errors in either the volume measurement or the time measurement.

Exact time span = 45.623451 … s

(a)   (b)   (c)   (d)

FIGURE 1–53
An instrument with many digits of resolution (stopwatch *c*) may be less accurate than an instrument with few digits of resolution (stopwatch *a*). What can you say about stopwatches *b* and *d*?

Also keep in mind that good precision does not guarantee good accuracy. For example, if the batteries in the stopwatch were weak, its accuracy could be quite poor, yet the readout would still be displayed to four significant digits of precision.

In common practice, precision is often associated with *resolution*, which is a measure of how finely the instrument can report the measurement. For example, a digital voltmeter with five digits on its display is said to be more precise than a digital voltmeter with only three digits. However, the number of displayed digits has nothing to do with the overall *accuracy* of the measurement. An instrument can be very precise without being very accurate when there are significant bias errors. Likewise, an instrument with very few displayed digits can be more accurate than one with many digits (Fig. 1–53).

SUMMARY

In this chapter some basic concepts of fluid mechanics are introduced and discussed. A substance in the liquid or gas phase is referred to as a *fluid*. *Fluid mechanics* is the science that deals with the behavior of fluids at rest or in motion and the interaction of fluids with solids or other fluids at the boundaries.

The flow of an unbounded fluid over a surface is *external flow*, and the flow in a pipe or duct is *internal flow* if the fluid is completely bounded by solid surfaces. A fluid flow is classified as being *compressible* or *incompressible*, depending on the density variation of the fluid during flow. The densities of liquids are essentially constant, and thus the flow of liquids is typically incompressible. The term *steady* implies *no change with time*. The opposite of steady is *unsteady*. The term *uniform* implies *no change with location* over a specified region. A flow is said to be *one-dimensional* when the properties or variables change in one dimension only. A fluid in direct contact with a solid surface sticks to

the surface and there is no slip. This is known as the *no-slip condition*, which leads to the formation of *boundary layers* along solid surfaces. In this book we concentrate on steady incompressible viscous flows—both internal and external.

A system of fixed mass is called a *closed system*, and a system that involves mass transfer across its boundaries is called an *open system* or *control volume*. A large number of engineering problems involve mass flow in and out of a system and are therefore modeled as control volumes.

In engineering calculations, it is important to pay particular attention to the units of the quantities to avoid errors caused by inconsistent units, and to follow a systematic approach. It is also important to recognize that the information given is not known to more than a certain number of significant digits, and the results obtained cannot possibly be accurate to more significant digits. The information given on dimensions and units; problem-solving technique; and accuracy, precision, and significant digits will be used throughout the entire text.

REFERENCES AND SUGGESTED READING

1. American Society for Testing and Materials. *Standards for Metric Practice.* ASTM E 380-79, January 1980.

2. G. M. Homsy, H. Aref, K. S. Breuer, S. Hochgreb, J. R. Koseff, B. R. Munson, K. G. Powell, C. R. Robertson,

and S. T. Thoroddsen. *Multi-Media Fluid Mechanics* (CD). Cambridge: Cambridge University Press, 2000.

3. M. Van Dyke. *An Album of Fluid Motion.* Stanford, CA: The Parabolic Press, 1982.

APPLICATION SPOTLIGHT ■ What Nuclear Blasts and Raindrops Have in Common

Guest Author: Lorenz Sigurdson, Vortex Fluid Dynamics Lab, University of Alberta

(a)   (b)

FIGURE 1–54

Comparison of the vortex structure created by: (a) a water drop after impacting a pool of water (inverted) and (b) an above-ground nuclear test in Nevada in 1957 (U.S. Department of Energy). The spherical drop was illuminated at the time of impact with the clear pool of water. Details of a similar experimental procedure necessary to create a drop photograph are given by Peck and Sigurdson (1994) and Peck et al. (1995). The tracers added to the flow in the bomb case were primarily heat and dust. The heat is from the original fireball, which for this particular test (the "Priscilla" event of Operation Plumbob) was large enough to reach the ground from where the bomb was initially suspended. Therefore, the tracer's initial geometric condition was a sphere intersecting the ground.

(a) © Stepan Kapl/Shutterstock.com.

(b) United States Department of Energy. Photo from Lorenz Sigurdson.

Why do the two images in Fig. 1–54 look alike? Figure 1–54b shows an above-ground nuclear test performed by the U.S. Department of Energy in 1957. An atomic blast created a fireball on the order of 100 m in diameter. Expansion is so quick that a compressible flow feature occurs: an expanding spherical shock wave. The image shown in Fig. 1–54a is an everyday innocuous event: an *inverted* image of a water drop after it has fallen into a pool of water, looking from below the pool surface. It could have fallen from your spoon into a cup of coffee, or been a secondary splash after a raindrop hit a lake. Why is there such a strong similarity between these two vastly different events? The application of fundamental principles of fluid mechanics learned in this book will help you understand much of the answer, although one can go much deeper.

The water has higher *density* (Chap. 2) than air, so the drop has experienced negative *buoyancy* (Chap. 3) as it has fallen through the air before impact. The fireball of hot gas is less dense than the cool air surrounding it, so it has positive buoyancy and rises. The *shock wave* (Chap. 12) reflecting from the ground also imparts a positive upward force to the fireball. The primary structure at the top of each image is called a *vortex ring*. This ring is a mini-tornado of concentrated *vorticity* (Chap. 4) with the ends of the tornado looping around to close on itself. The laws of *kinematics* (Chap. 4) tell us that this vortex ring will carry the fluid in a direction toward the top of the page. This is expected in both cases from the forces applied and the law of conservation of momentum applied through a *control volume analysis* (Chap. 5). One could also analyze this problem with *differential analysis* (Chaps. 9 and 10) or with *computational fluid dynamics* (Chap. 15). But why does the *shape* of the tracer material look so similar? This occurs if there is approximate *geometric* and *kinematic similarity* (Chap. 7), and if the *flow visualization* (Chap. 4) technique is similar.

Further knowledge of kinematics and vortex dynamics can help explain the similarity of the vortex structure in the images to much greater detail, as discussed by Sigurdson (1997) and Peck and Sigurdson (1994). Look at the lobes dangling beneath the primary vortex ring, the striations in the "stalk," and the ring at the base of each structure. There is also topological similarity of this structure to other vortex structures occurring in turbulence. Comparison of the drop and bomb has given us a better understanding of how turbulent structures are created and evolve. What other secrets of fluid mechanics are left to be revealed in explaining the similarity between these two flows?

References

Peck, B., and Sigurdson, L.W., "The Three-Dimensional Vortex Structure of an Impacting Water Drop," *Phys. Fluids*, 6(2) (Part 1), p. 564, 1994.

Peck, B., Sigurdson, L.W., Faulkner, B., and Buttar, I., "An Apparatus to Study Drop-Formed Vortex Rings," *Meas. Sci. Tech.*, 6, p. 1538, 1995.

Sigurdson, L.W., "Flow Visualization in Turbulent Large-Scale Structure Research," Chapter 6 in *Atlas of Visualization*, Vol. III, Flow Visualization Society of Japan, eds., CRC Press, pp. 99–113, 1997.

PROBLEMS*

Introduction, Classification, and System

1–1C   What is a fluid? How does it differ from a solid? How does a gas differ from a liquid?

1–2C   Consider the flow of air over the wings of an aircraft. Is this flow internal or external? How about the flow of gases through a jet engine?

1–3C   Define incompressible flow and incompressible fluid. Must the flow of a compressible fluid necessarily be treated as compressible?

1–4C   Define internal, external, and open-channel flows.

1–5C   How is the Mach number of a flow defined? What does a Mach number of 2 indicate?

1–6C   When an airplane is flying at a constant speed relative to the ground, is it correct to say that the Mach number of this airplane is also constant?

1–7C   Consider the flow of air at a Mach number of 0.12. Should this flow be approximated as being incompressible?

1–8C   What is the no-slip condition? What causes it?

1–9C   What is forced flow? How does it differ from natural flow? Is flow caused by winds forced or natural flow?

1–10C   What is a boundary layer? What causes a boundary layer to develop?

1–11C   What is the difference between the classical and the statistical approaches?

1–12C   What is a steady-flow process?

1–13C   Define stress, normal stress, shear stress, and pressure.

1–14C   When analyzing the acceleration of gases as they flow through a nozzle, what would you choose as your system? What type of system is this?

1–15C   When is a system a closed system, and when is it a control volume?

1–16C   You are trying to understand how a reciprocating air compressor (a piston-cylinder device) works. What system would you use? What type of system is this?

1–17C   What are system, surroundings, and boundary?

---

* Problems designated by a "C" are concept questions, and students are encouraged to answer them all. Problems designated by an "E" are in English units, and the SI users can ignore them. Problems with the ⟨icon⟩ icon are solved using EES, and complete solutions together with parametric studies are included on the text website. Problems with the ⟨icon⟩ icon are comprehensive in nature and are intended to be solved with an equation solver such as EES.

Mass, Force, and Units

1–18C   Explain why the light-year has the dimension of length.

1–19C   What is the difference between kg-mass and kg-force?

1–20C   What is the difference between pound-mass and pound-force?

1–21C   In a news article, it is stated that a recently developed geared turbofan engine produces 15,000 pounds of thrust to propel the aircraft forward. Is "pound" mentioned here lbm or lbf? Explain.

1–22C   What is the net force acting on a car cruising at a constant velocity of 70 km/h (a) on a level road and (b) on an uphill road?

1–23   A 6-kg plastic tank that has a volume of 0.18 m^3 is filled with liquid water. Assuming the density of water is 1000 kg/m^3, determine the weight of the combined system.

1–24   What is the weight, in N, of an object with a mass of 200 kg at a location where $g = 9.6$ m/s^2?

1–25   What is the weight of a 1-kg substance in N, kN, $kg \cdot m/s^2$, kgf, $lbm \cdot ft/s^2$, and lbf?

1–26   Determine the mass and the weight of the air contained in a room whose dimensions are 6 m × 6 m × 8 m. Assume the density of the air is 1.16 kg/m^3.   *Answers: 334.1 kg, 3277 N*

1–27   While solving a problem, a person ends up with the equation $E = 16$ kJ + 7 kJ/kg at some stage. Here E is the total energy and has the unit of kilojoules. Determine how to correct the error and discuss what may have caused it.

1–28E   A 195-lbm astronaut took his bathroom scale (a spring scale) and a beam scale (compares masses) to the moon where the local gravity is $g = 5.48$ ft/s^2. Determine how much he will weigh (a) on the spring scale and (b) on the beam scale.   *Answers: (a) 33.2 lbf, (b) 195 lbf*

1–29   The acceleration of high-speed aircraft is sometimes expressed in g's (in multiples of the standard acceleration of gravity). Determine the net force, in N, that a 90-kg man would experience in an aircraft whose acceleration is 6 g's.

1–30   A 5-kg rock is thrown upward with a force of 150 N at a location where the local gravitational acceleration is 9.79 m/s^2. Determine the acceleration of the rock, in m/s^2.

1–31   Solve Prob. 1–30 using EES (or other) software. Print out the entire solution, including the numerical results with proper units.

1–32   The value of the gravitational acceleration g decreases with elevation from 9.807 m/s^2 at sea level to 9.767 m/s^2 at an altitude of 13,000 m, where large passenger planes cruise. Determine the percent reduction in the weight of an airplane cruising at 13,000 m relative to its weight at sea level.

1–33 At 45° latitude, the gravitational acceleration as a function of elevation z above sea level is given by $g = a - bz$, where $a = 9.807$ m/s^2 and $b = 3.32 \times 10^{-6}$ s^{-2}. Determine the height above sea level where the weight of an object will decrease by 1 percent. *Answer:* 29,500 m

1–34 A 4-kW resistance heater in a water heater runs for 2 hours to raise the water temperature to the desired level. Determine the amount of electric energy used in both kWh and kJ.

1–35 The gas tank of a car is filled with a nozzle that discharges gasoline at a constant flow rate. Based on unit considerations of quantities, obtain a relation for the filling time in terms of the volume V of the tank (in L) and the discharge rate of gasoline ($\dot{V}$, in L/s).

1–36 A pool of volume V (in m^3) is to be filled with water using a hose of diameter D (in m). If the average discharge velocity is V (in m/s) and the filling time is t (in s), obtain a relation for the volume of the pool based on unit considerations of quantities involved.

1–37 Based on unit considerations alone, show that the power needed to accelerate a car of mass m (in kg) from rest to velocity V (in m/s) in time interval t (in s) is proportional to mass and the square of the velocity of the car and inversely proportional to the time interval.

1–38 An airplane flies horizontally at 70 m/s. Its propeller delivers 1500 N of thrust (forward force) to overcome aerodynamic drag (backward force). Using dimensional reasoning and unity converstion ratios, calculate the useful power delivered by the propeller in units of kW and horsepower.

1–39 If the airplane of Problem 1–38 weighs 1450 lbf, estimate the lift force produced by the airplane's wings (in lbf and newtons) when flying at 70.0 m/s.

1–40E The boom of a fire truck raises a fireman (and his equipment—total weight 280 lbf) 40 ft into the air to fight a building fire. (*a*) Showing all your work and using unity conversion ratios, calculate the work done by the boom on the fireman in units of Btu. (*b*) If the useful power supplied by the boom to lift the fireman is 3.50 hp, estimate how long it takes to lift the fireman.

1–41 A man goes to a traditional market to buy a steak for dinner. He finds a 12-oz steak (1 lbm = 16 oz) for $3.15. He then goes to the adjacent international market and finds a 320-g steak of identical quality for $3.30. Which steak is the better buy?

1–42 Water at 20°C from a garden hose fills a 2.0 L container in 2.85 s. Using unity converstion ratios and showing all your work, calculate the volume flow rate in liters per minute (Lpm) and the mass flow rate in kg/s.

1–43 A forklift raises a 90.5 kg crate 1.80 m. (*a*) Showing all your work and using unity conversion ratios, calculate the work done by the forklift on the crane, in units of kJ. (*b*) If it takes 12.3 seconds to lift the crate, calculate the useful power supplied to the crate in kilowatts.

Modeling and Solving Engineering Problems

1–44C When modeling an engineering process, how is the right choice made between a simple but crude and a complex but accurate model? Is the complex model necessarily a better choice since it is more accurate?

1–45C What is the difference between the analytical and experimental approach to engineering problems? Discuss the advantages and disadvantages of each approach.

1–46C What is the importance of modeling in engineering? How are the mathematical models for engineering processes prepared?

1–47C What is the difference between precision and accuracy? Can a measurement be very precise but inaccurate? Explain.

1–48C How do the differential equations in the study of a physical problem arise?

1–49C What is the value of the engineering software packages in (*a*) engineering education and (*b*) engineering practice?

1–50 Solve this system of three equations with three unknowns using EES:

$$2x - y + z = 9$$
$$3x^2 + 2y = z + 2$$
$$xy + 2z = 14$$

1–51 Solve this system of two equations with two unknowns using EES:

$$x^3 - y^2 = 10.5$$
$$3xy + y = 4.6$$

1–52 Determine a positive real root of this equation using EES:

$$3.5x^3 - 10x^{0.5} - 3x = -4$$

1–53 Solve this system of three equations with three unknowns using EES:

$$x^2y - z = 1.5$$
$$x - 3y^{0.5} + xz = -2$$
$$x + y - z = 4.2$$

Review Problems

1–54 The reactive force developed by a jet engine to push an airplane forward is called thrust, and the thrust developed by the engine of a Boeing 777 is about 85,000 lbf. Express this thrust in N and kgf.

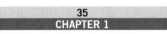

1–55 The weight of bodies may change somewhat from one location to another as a result of the variation of the gravitational acceleration g with elevation. Accounting for this variation using the relation in Prob. 1–33, determine the weight of an 80.0-kg person at sea level ($z = 0$), in Denver ($z = 1610$ m), and on the top of Mount Everest ($z = 8848$ m).

1–56E A student buys a 5000 Btu window air conditioner for his apartment bedroom. He monitors it for one hour on a hot day and determines that it operates approximately 60 percent of the time (duty cycle = 60 percent) to keep the room at nearly constant temperature. (a) Showing all your work and using unity conversion ratios, calculate the rate of heat transfer into the bedroom through the walls, windows, etc. in units of Btu/h and in units of kW. (b) If the energy efficiency ratio (EER) of the air conditioner is 9.0 and electricity costs 7.5 cents per kilowatt-hr, calculate how much it costs (in cents) for him to run the air conditioner for one hour.

1–57 For liquids, the dynamic viscosity μ, which is a measure of resistance against flow is approximated as $\mu = a10^{b/(T-c)}$, where T is the absolute temperature, and a, b and c are experimental constants. Using the data listed in Table A-7 for methanol at 20°C, 40°C and 60°C, determine the constant a, b and c.

1–58 An important design consideration in two-phase pipe flow of solid-liquid mixtures is the terminal settling velocity below, which the flow becomes unstable and eventually the pipe becomes clogged. On the basis of extended transportation tests, the terminal settling velocity of a solid particle in the rest water given by $V_L = F_L \sqrt{2gD(S - 1)}$, where F_L is an experimental coefficient, g the gravitational acceleration, D the pipe diameter, and S the specific gravity of solid particle. What is the dimension of F_L? Is this equation dimensionally homogeneous?

1–59 Consider the flow of air through a wind turbine whose blades sweep an area of diameter D (in m). The average air velocity through the swept area is V (in m/s). On the bases of the units of the quantities involved, show that the mass flow rate of air (in kg/s) through the swept area is proportional to air density, the wind velocity, and the square of the diameter of the swept area.

1–60 The drag force exerted on a car by air depends on a dimensionless drag coefficient, the density of air, the car velocity, and the frontal area of the car. That is, F_D = function ($C_{Drag,}$ $A_{front,}$ ρ, V). Based on unit considerations alone, obtain a relation for the drag force.

FIGURE P1–60

Fundamentals of Engineering (FE) Exam Problems

1–61 The speed of an aircraft is given to be 260 m/s in air. If the speed of sound at that location is 330 m/s, the flight of aircraft is
(a) Sonic   (b) Subsonic   (c) Supersonic   (d) Hypersonic

1–62 The speed of an aircraft is given to be 1250 km/h. If the speed of sound at that location is 315 m/s, the Mach number is
(a) 0.5   (b) 0.85   (c) 1.0   (d) 1.10   (e) 1.20

1–63 If mass, heat, and work are not allowed to cross the boundaries of a system, the system is called
(a) Isolated (b) Isothermal (c) Adiabatic (d) Control mass
(e) Control volume

1–64 The weight of a 10-kg mass at sea level is
(a) 9.81 N   (b) 32.2 kgf   (c) 98.1 N   (d) 10 N   (e) 100 N

1–65 The weight of a 1-lbm mass is
(a) 1 lbm·ft/s² (b) 9.81 lbf (c) 9.81 N (d) 32.2 lbf (e) 1 lbf

1–66 One kJ is NOT equal to
(a) 1 kPa·m³ (b) 1 kN·m (c) 0.001 MJ (d) 1000 J (e) 1 kg·m²/s²

1–67 Which is a unit for the amount of energy?
(a) Btu/h   (b) kWh   (c) kcal/h   (d) hp   (e) kW

1–68 A hydroelectric power plant operates at its rated power of 7 MW. If the plant has produced 26 million kWh of electricity in a specified year, the number of hours the plant has operated that year is
(a) 1125 h   (b) 2460 h   (c) 2893 h   (d) 3714 h   (e) 8760 h

Design and Essay Problems

1–69 Write an essay on the various mass- and volume-measurement devices used throughout history. Also, explain the development of the modern units for mass and volume.

1–70 Search the Internet to find out how to properly add or subtract numbers while taking into consideration the number of significant digits. Write a summary of the proper technique, then use the technique to solve the following cases: (a) 1.006 + 23.47, (b) 703,200 − 80.4, and (c) 4.6903 − 14.58. Be careful to express your final answer to the appropriate number of significant digits.

PROPERTIES OF FLUIDS

In this chapter, we discuss properties that are encountered in the analysis of fluid flow. First we discuss *intensive* and *extensive properties* and define *density* and *specific gravity*. This is followed by a discussion of the properties *vapor pressure*, *energy* and its various forms, the *specific heats* of ideal gases and incompressible substances, the *coefficient of compressibility*, and the *speed of sound*. Then we discuss the property *viscosity*, which plays a dominant role in most aspects of fluid flow. Finally, we present the property *surface tension* and determine the *capillary rise* from static equilibrium conditions. The property *pressure* is discussed in Chap. 3 together with fluid statics.

OBJECTIVES

When you finish reading this chapter, you should be able to

■ Have a working knowledge of the basic properties of fluids and understand the continuum approximation

■ Have a working knowledge of viscosity and the consequences of the frictional effects it causes in fluid flow

■ Calculate the capillary rise (or drop) in tubes due to the surface tension effect

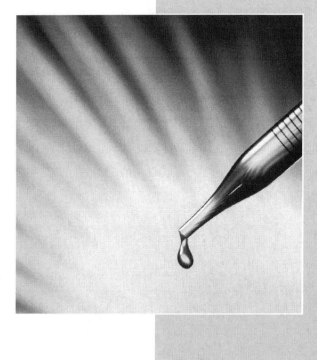

A drop forms when liquid is forced out of a small tube. The shape of the drop is determined by a balance of pressure, gravity, and surface tension forces.
Royalty-Free/CORBIS

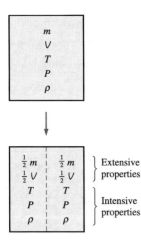

FIGURE 2–1
Criterion to differentiate intensive and extensive properties.

FIGURE 2–2
The length scale associated with most flows, such as seagulls in flight, is orders of magnitude larger than the mean free path of the air molecules. Therefore, here, and for all fluid flows considered in this book, the continuum idealization is appropriate.

PhotoLink/Getty RF

2–1 · INTRODUCTION

Any characteristic of a system is called a **property.** Some familiar properties are pressure P, temperature T, volume V, and mass m. The list can be extended to include less familiar ones such as viscosity, thermal conductivity, modulus of elasticity, thermal expansion coefficient, electric resistivity, and even velocity and elevation.

Properties are considered to be either *intensive* or *extensive*. **Intensive properties** are those that are independent of the mass of the system, such as temperature, pressure, and density. **Extensive properties** are those whose values depend on the size—or extent—of the system. Total mass, total volume V, and total momentum are some examples of extensive properties. An easy way to determine whether a property is intensive or extensive is to divide the system into two equal parts with an imaginary partition, as shown in Fig. 2–1. Each part will have the same value of intensive properties as the original system, but half the value of the extensive properties.

Generally, uppercase letters are used to denote extensive properties (with mass m being a major exception), and lowercase letters are used for intensive properties (with pressure P and temperature T being the obvious exceptions).

Extensive properties per unit mass are called **specific properties.** Some examples of specific properties are specific volume ($v = V/m$) and specific total energy ($e = E/m$).

The state of a system is described by its properties. But we know from experience that we do not need to specify all the properties in order to fix a state. Once the values of a sufficient number of properties are specified, the rest of the properties assume certain values. That is, specifying a certain number of properties is sufficient to fix a state. The number of properties required to fix the state of a system is given by the **state postulate:** *The state of a simple compressible system is completely specified by two independent, intensive properties.*

Two properties are independent if one property can be varied while the other one is held constant. Not all properties are independent, and some are defined in terms of others, as explained in Section 2–2.

Continuum

A fluid is composed of molecules which may be widely spaced apart, especially in the gas phase. Yet it is convenient to disregard the atomic nature of the fluid and view it as continuous, homogeneous matter with no holes, that is, a **continuum.** The continuum idealization allows us to treat properties as point functions and to assume that the properties vary continually in space with no jump discontinuities. This idealization is valid as long as the size of the system we deal with is large relative to the space between the molecules (Fig. 2–2). This is the case in practically all problems, except some specialized ones. The continuum idealization is implicit in many statements we make, such as "the density of water in a glass is the same at any point."

To have a sense of the distances involved at the molecular level, consider a container filled with oxygen at atmospheric conditions. The diameter of an oxygen molecule is about 3×10^{-10} m and its mass is 5.3×10^{-26} kg. Also, the *mean free path* of oxygen at 1 atm pressure and 20°C is 6.3×10^{-8} m. That is, an oxygen molecule travels, on average, a distance of 6.3×10^{-8} m (about 200 times its diameter) before it collides with another molecule.

Also, there are about 3×10^{16} molecules of oxygen in the tiny volume of 1 mm^3 at 1 atm pressure and 20°C (Fig. 2–3). The continuum model is applicable as long as the characteristic length of the system (such as its diameter) is much larger than the mean free path of the molecules. At very low pressure, e.g., at very high elevations, the mean free path may become large (for example, it is about 0.1 m for atmospheric air at an elevation of 100 km). For such cases the **rarefied gas flow theory** should be used, and the impact of individual molecules should be considered. In this text we limit our consideration to substances that can be modeled as a continuum.

2–2 · DENSITY AND SPECIFIC GRAVITY

Density is defined as *mass per unit volume* (Fig. 2–4). That is,

Density:
$$\rho = \frac{m}{V} \quad (\text{kg/m}^3) \tag{2-1}$$

The reciprocal of density is the **specific volume v,** which is defined as *volume per unit mass.* That is, $v = V/m = 1/\rho$. For a differential volume element of mass δm and volume δV, density can be expressed as $\rho = \delta m/\delta V$.

The density of a substance, in general, depends on temperature and pressure. The density of most gases is proportional to pressure and inversely proportional to temperature. Liquids and solids, on the other hand, are essentially incompressible substances, and the variation of their density with pressure is usually negligible. At 20°C, for example, the density of water changes from 998 kg/m^3 at 1 atm to 1003 kg/m^3 at 100 atm, a change of just 0.5 percent. The density of liquids and solids depends more strongly on temperature than it does on pressure. At 1 atm, for example, the density of water changes from 998 kg/m^3 at 20°C to 975 kg/m^3 at 75°C, a change of 2.3 percent, which can still be neglected in many engineering analyses.

Sometimes the density of a substance is given relative to the density of a well-known substance. Then it is called **specific gravity,** or **relative density,** and is defined as *the ratio of the density of a substance to the density of some standard substance at a specified temperature* (usually water at 4°C, for which $\rho_{H_2O} = 1000$ kg/m^3). That is,

Specific gravity:
$$SG = \frac{\rho}{\rho_{H_2O}} \tag{2-2}$$

Note that the specific gravity of a substance is a dimensionless quantity. However, in SI units, the numerical value of the specific gravity of a substance is exactly equal to its density in g/cm^3 or kg/L (or 0.001 times the density in kg/m^3) since the density of water at 4°C is 1 g/cm^3 = 1 kg/L = 1000 kg/m^3. The specific gravity of mercury at 20°C, for example, is 13.6. Therefore, its density at 20°C is 13.6 g/cm^3 = 13.6 kg/L = 13,600 kg/m^3. The specific gravities of some substances at 20°C are given in Table 2–1. Note that substances with specific gravities less than 1 are lighter than water, and thus they would float on water (if immiscible).

The weight of a unit volume of a substance is called **specific weight** or **weight density** and is expressed as

Specific weight:
$$\gamma_s = \rho g \quad (\text{N/m}^3) \tag{2-3}$$

where g is the gravitational acceleration.

FIGURE 2–3
Despite the relatively large gaps between molecules, a gas can usually be treated as a continuum because of the very large number of molecules even in an extremely small volume.

FIGURE 2–4
Density is mass per unit volume; specific volume is volume per unit mass.

TABLE 2–1	
The specific gravity of some substances at 20°C and 1 atm unless stated otherwise	
Substance	SG
Water	1.0
Blood (at 37°C)	1.06
Seawater	1.025
Gasoline	0.68
Ethyl alcohol	0.790
Mercury	13.6
Balsa wood	0.17
Dense oak wood	0.93
Gold	19.3
Bones	1.7–2.0
Ice (at 0°C)	0.916
Air	0.001204

Recall from Chap. 1 that the densities of liquids are essentially constant, and thus they can often be approximated as being incompressible substances during most processes without sacrificing much in accuracy.

Density of Ideal Gases

Property tables provide very accurate and precise information about the properties, but sometimes it is convenient to have some simple relations among the properties that are sufficiently general and reasonably accurate. Any equation that relates the pressure, temperature, and density (or specific volume) of a substance is called an **equation of state.** The simplest and best-known equation of state for substances in the gas phase is the **ideal-gas equation of state,** expressed as

$$P v = RT \quad \text{or} \quad P = \rho RT \tag{2–4}$$

where P is the absolute pressure, v is the specific volume, T is the thermodynamic (absolute) temperature, ρ is the density, and R is the gas constant. The gas constant R is different for each gas and is determined from $R = R_u/M$, where R_u is the **universal gas constant** whose value is $R_u = 8.314$ kJ/ kmol·K $= 1.986$ Btu/lbmol·R, and M is the *molar mass* (also called *molecular weight*) of the gas. The values of R and M for several substances are given in Table A–1.

The thermodynamic temperature scale in the SI is the **Kelvin scale,** and the temperature unit on this scale is the **kelvin,** designated by K. In the English system, it is the **Rankine scale,** and the temperature unit on this scale is the **rankine,** R. Various temperature scales are related to each other by

$$T(\text{K}) = T(^\circ\text{C}) + 273.15 = T(\text{R})/1.8 \tag{2–5}$$

$$T(\text{R}) = T(^\circ\text{F}) + 459.67 = 1.8\, T(\text{K}) \tag{2–6}$$

It is common practice to round the constants 273.15 and 459.67 to 273 and 460, respectively, but we do not encourage this practice.

Equation 2–4, the ideal-gas equation of state, is also called simply the **ideal-gas relation,** and a gas that obeys this relation is called an **ideal gas.** For an ideal gas of volume V, mass m, and number of moles $N = m/M$, the ideal-gas equation of state can also be written as $PV = mRT$ or $PV = NR_uT$. For a fixed mass m, writing the ideal-gas relation twice and simplifying, the properties of an ideal gas at two different states are related to each other by $P_1V_1/T_1 = P_2V_2/T_2$.

An ideal gas is a hypothetical substance that obeys the relation $P v = RT$. It has been experimentally observed that the ideal-gas relation closely approximates the P-v-T behavior of real gases at low densities. At low pressures and high temperatures, the density of a gas decreases and the gas behaves like an ideal gas (Fig. 2–5). In the range of practical interest, many familiar gases such as air, nitrogen, oxygen, hydrogen, helium, argon, neon, and krypton and even heavier gases such as carbon dioxide can be treated as ideal gases with negligible error (often less than 1 percent). Dense gases such as water vapor in steam power plants and refrigerant vapor in refrigerators, air conditioners, and heat pumps, however, should not be treated as ideal gases since they usually exist at a state near saturation.

FIGURE 2–5

Air behaves as an ideal gas, even at very high speeds. In this schlieren image, a bullet traveling at about the speed of sound bursts through both sides of a balloon, forming two expanding shock waves. The turbulent wake of the bullet is also visible.

Photograph by Gary S. Settles, Penn State Gas Dynamics Lab. Used by permission.

EXAMPLE 2–1   Density, Specific Gravity, and Mass of Air in a Room

Determine the density, specific gravity, and mass of the air in a room whose dimensions are 4 m × 5 m × 6 m at 100 kPa and 25°C (Fig. 2–6).

SOLUTION   The density, specific gravity, and mass of the air in a room are to be determined.
Assumptions   At specified conditions, air can be treated as an ideal gas.
Properties   The gas constant of air is $R = 0.287$ kPa·m³/kg·K.
Analysis   The density of the air is determined from the ideal-gas relation $P = \rho RT$ to be

$$\rho = \frac{P}{RT} = \frac{100 \text{ kPa}}{(0.287 \text{ kPa·m}^3/\text{kg·K})(25 + 273.15) \text{ K}} = \textbf{1.17 kg/m}^3$$

Then the specific gravity of the air becomes

$$\text{SG} = \frac{\rho}{\rho_{\text{H}_2\text{O}}} = \frac{1.17 \text{ kg/m}^3}{1000 \text{ kg/m}^3} = \textbf{0.00117}$$

Finally, the volume and the mass of the air in the room are

$$V = (4 \text{ m})(5 \text{ m})(6 \text{ m}) = 120 \text{ m}^3$$

$$m = \rho V = (1.17 \text{ kg/m}^3)(120 \text{ m}^3) = \textbf{140 kg}$$

Discussion   Note that we converted the temperature to (absolute) unit K from (relative) unit °C before using it in the ideal-gas relation.

FIGURE 2–6
Schematic for Example 2–1.

2–3 ▪ VAPOR PRESSURE AND CAVITATION

It is well-established that temperature and pressure are dependent properties for pure substances during phase-change processes, and there is one-to-one correspondence between temperature and pressure. At a given pressure, the temperature at which a pure substance changes phase is called the **saturation temperature** T_{sat}. Likewise, at a given temperature, the pressure at which a pure substance changes phase is called the **saturation pressure** P_{sat}. At an absolute pressure of 1 standard atmosphere (1 atm or 101.325 kPa), for example, the saturation temperature of water is 100°C. Conversely, at a temperature of 100°C, the saturation pressure of water is 1 atm.

The **vapor pressure** P_v of a pure substance is defined as *the pressure exerted by its vapor in phase equilibrium with its liquid at a given temperature* (Fig. 2–7). P_v is a property of the pure substance, and turns out to be identical to the saturation pressure P_{sat} of the liquid ($P_v = P_{\text{sat}}$). We must be careful not to confuse vapor pressure with *partial pressure*. **Partial pressure** is defined as *the pressure of a gas or vapor in a mixture with other gases.* For example, atmospheric air is a mixture of dry air and water vapor, and atmospheric pressure is the sum of the partial pressure of dry air and the partial pressure of water vapor. The partial pressure of water vapor constitutes a small fraction (usually under 3 percent) of the atmospheric pressure since air is mostly nitrogen and oxygen. The partial pressure of a vapor must be less than or equal to the vapor pressure if there is no liquid present. However, when both vapor and liquid are present and the system is in phase equilibrium, the partial pressure of the vapor must equal the vapor pressure, and the system is said to be *saturated*. The rate of evaporation from open water

Water molecules — vapor phase

Water molecules — liquid phase

FIGURE 2–7
The vapor pressure (saturation pressure) of a pure substance (e.g., water) is the pressure exerted by its vapor molecules when the system is in phase equilibrium with its liquid molecules at a given temperature.

TABLE 2–2

Saturation (or vapor) pressure of water at various temperatures

Temperature T, °C	Saturation Pressure P_{sat}, kPa
−10	0.260
−5	0.403
0	0.611
5	0.872
10	1.23
15	1.71
20	2.34
25	3.17
30	4.25
40	7.38
50	12.35
100	101.3 (1 atm)
150	475.8
200	1554
250	3973
300	8581

FIGURE 2–8

Cavitation damage on a 16-mm by 23-mm aluminum sample tested at 60 m/s for 2.5 hours. The sample was located at the cavity collapse region downstream of a cavity generator specifically designed to produce high damage potential.

Photo by David Stinebring, ARL/Pennsylvania State University. Used by permission.

bodies such as lakes is controlled by the difference between the vapor pressure and the partial pressure. For example, the vapor pressure of water at 20°C is 2.34 kPa. Therefore, a bucket of water at 20°C left in a room with dry air at 1 atm will continue evaporating until one of two things happens: the water evaporates away (there is not enough water to establish phase equilibrium in the room), or the evaporation stops when the partial pressure of the water vapor in the room rises to 2.34 kPa at which point phase equilibrium is established.

For phase-change processes between the liquid and vapor phases of a pure substance, the saturation pressure and the vapor pressure are equivalent since the vapor is pure. Note that the pressure value would be the same whether it is measured in the vapor or liquid phase (provided that it is measured at a location close to the liquid–vapor interface to avoid any hydrostatic effects). Vapor pressure increases with temperature. Thus, a substance at higher pressure boils at higher temperature. For example, water boils at 134°C in a pressure cooker operating at 3 atm absolute pressure, but it boils at 93°C in an ordinary pan at a 2000-m elevation, where the atmospheric pressure is 0.8 atm. The saturation (or vapor) pressures are given in Appendices 1 and 2 for various substances. An abridged table for water is given in Table 2–2 for easy reference.

The reason for our interest in vapor pressure is the possibility of the liquid pressure in liquid-flow systems dropping below the vapor pressure at some locations, and the resulting unplanned vaporization. For example, water at 10°C may vaporize and form bubbles at locations (such as the tip regions of impellers or suction sides of pumps) where the pressure drops below 1.23 kPa. The vapor bubbles (called **cavitation bubbles** since they form "cavities" in the liquid) collapse as they are swept away from the low-pressure regions, generating highly destructive, extremely high-pressure waves. This phenomenon, which is a common cause for drop in performance and even the erosion of impeller blades, is called **cavitation,** and it is an important consideration in the design of hydraulic turbines and pumps.

Cavitation must be avoided (or at least minimized) in most flow systems since it reduces performance, generates annoying vibrations and noise, and causes damage to equipment. We note that some flow systems use cavitation to their *advantage*, e.g., high-speed "supercavitating" torpedoes. The pressure spikes resulting from the large number of bubbles collapsing near a solid surface over a long period of time may cause erosion, surface pitting, fatigue failure, and the eventual destruction of the components or machinery (Fig. 2–8). The presence of cavitation in a flow system can be sensed by its characteristic tumbling sound.

EXAMPLE 2–2   Minimum Pressure to Avoid Cavitation

In a water distribution system, the temperature of water is observed to be as high as 30°C. Determine the minimum pressure allowed in the system to avoid cavitation.

SOLUTION   The minimum pressure in a water distribution system to avoid cavitation is to be determined.
Properties   The vapor pressure of water at 30°C is 4.25 kPa (Table 2–2).

Analysis To avoid cavitation, the pressure anywhere in the flow should not be allowed to drop below the vapor (or saturation) pressure at the given temperature. That is,

$$P_{min} = P_{sat@30°C} = 4.25 \text{ kPa}$$

Therefore, the pressure should be maintained above 4.25 kPa everywhere in the flow.

Discussion Note that the vapor pressure increases with increasing temperature, and thus the risk of cavitation is greater at higher fluid temperatures.

2–4 ▪ ENERGY AND SPECIFIC HEATS

Energy can exist in numerous forms such as thermal, mechanical, kinetic, potential, electrical, magnetic, chemical, and nuclear (Fig. 2–9) and their sum constitutes the **total energy** E (or e on a unit mass basis) of a system. The forms of energy related to the molecular structure of a system and the degree of the molecular activity are referred to as the *microscopic energy*. The sum of all microscopic forms of energy is called the **internal energy** of a system, and is denoted by U (or u on a unit mass basis).

The *macroscopic* energy of a system is related to motion and the influence of some external effects such as gravity, magnetism, electricity, and surface tension. The energy that a system possesses as a result of its motion is called **kinetic energy.** When all parts of a system move with the same velocity, the kinetic energy per unit mass is expressed as ke $= V^2/2$ where V denotes the velocity of the system relative to some fixed reference frame. The energy that a system possesses as a result of its elevation in a gravitational field is called **potential energy** and is expressed on a per-unit mass basis as pe $= gz$ where g is the gravitational acceleration and z is the elevation of the center of gravity of the system relative to some arbitrarily selected reference plane.

In daily life, we frequently refer to the sensible and latent forms of internal energy as **heat,** and we talk about the heat content of bodies. In engineering, however, those forms of energy are usually referred to as **thermal energy** to prevent any confusion with *heat transfer.*

The international unit of energy is the *joule* (J) or *kilojoule* (1 kJ $=$ 1000 J). A joule is 1 N times 1 m. In the English system, the unit of energy is the *British thermal unit* (Btu), which is defined as the energy needed to raise the temperature of 1 lbm of water at 68°F by 1°F. The magnitudes of kJ and Btu are almost identical (1 Btu $=$ 1.0551 kJ). Another well-known unit of energy is the *calorie* (1 cal $=$ 4.1868 J), which is defined as the energy needed to raise the temperature of 1 g of water at 14.5°C by 1°C.

In the analysis of systems that involve fluid flow, we frequently encounter the combination of properties u and Pv. For convenience, this combination is called **enthalpy** h. That is,

Enthalpy: $$h = u + Pv = u + \frac{P}{\rho} \qquad (2\text{–}7)$$

where P/ρ is the *flow energy*, also called the *flow work*, which is the energy per unit mass needed to move the fluid and maintain flow. In the energy analysis of flowing fluids, it is convenient to treat the flow energy as part of the energy of the fluid and to represent the microscopic energy of a fluid

(a)

(b)

FIGURE 2–9

At least six different forms of energy are encountered in bringing power from a nuclear plant to your home, nuclear, thermal, mechanical, kinetic, magnetic, and electrical.

(a) © Creatas/PunchStock RF
(b) Comstock Images/Jupiterimages RF

FIGURE 2–10

The *internal energy u* represents the microscopic energy of a nonflowing fluid per unit mass, whereas *enthalpy h* represents the microscopic energy of a flowing fluid per unit mass.

FIGURE 2–11

Fluids, like solids, compress when the applied pressure is increased from P_1 to P_2.

stream by enthalpy h (Fig. 2–10). Note that enthalpy is a quantity per unit mass, and thus it is a *specific* property.

In the absence of such effects as magnetic, electric, and surface tension, a system is called a simple compressible system. The total energy of a simple compressible system consists of three parts: internal, kinetic, and potential energies. On a unit-mass basis, it is expressed as $e = u + \text{ke} + \text{pe}$. The fluid entering or leaving a control volume possesses an additional form of energy—the *flow energy* P/ρ. Then the total energy of a **flowing fluid** on a unit-mass basis becomes

$$e_{\text{flowing}} = P/\rho + e = h + \text{ke} + \text{pe} = h + \frac{V^2}{2} + gz \qquad (\text{kJ/kg}) \qquad \textbf{(2–8)}$$

where $h = P/\rho + u$ is the enthalpy, V is the magnitude of velocity, and z is the elevation of the system relative to some external reference point.

By using the enthalpy instead of the internal energy to represent the energy of a flowing fluid, we do not need to be concerned about the flow work. The energy associated with pushing the fluid is automatically taken care of by enthalpy. In fact, this is the main reason for defining the property enthalpy.

The differential and finite changes in the internal energy and enthalpy of an *ideal gas* can be expressed in terms of the specific heats as

$$du = c_v\, dT \quad \text{and} \quad dh = c_p\, dT \qquad \textbf{(2–9)}$$

where c_v and c_p are the constant-volume and constant-pressure specific heats of the ideal gas. Using specific heat values at the average temperature, the finite changes in internal energy and enthalpy can be expressed approximately as

$$\Delta u \cong c_{v,\text{avg}}\, \Delta T \quad \text{and} \quad \Delta h \cong c_{p,\text{avg}}\, \Delta T \qquad \textbf{(2–10)}$$

For *incompressible substances*, the constant-volume and constant-pressure specific heats are identical. Therefore, $c_p \cong c_v \cong c$ for liquids, and the change in the internal energy of liquids can be expressed as $\Delta u \cong c_{\text{avg}}\, \Delta T$.

Noting that $\rho = $ constant for incompressible substances, the differentiation of enthalpy $h = u + P/\rho$ gives $dh = du + dP/\rho$. Integrating, the enthalpy change becomes

$$\Delta h = \Delta u + \Delta P/\rho \cong c_{\text{avg}}\, \Delta T + \Delta P/\rho \qquad \textbf{(2–11)}$$

Therefore, $\Delta h = \Delta u \cong c_{\text{avg}}\, \Delta T$ for constant-pressure processes, and $\Delta h = \Delta P/\rho$ for constant-temperature processes in liquids.

2–5 · COMPRESSIBILITY AND SPEED OF SOUND

Coefficient of Compressibility

We know from experience that the volume (or density) of a fluid changes with a change in its temperature or pressure. Fluids usually expand as they are heated or depressurized and contract as they are cooled or pressurized. But the amount of volume change is different for different fluids, and we need to define properties that relate volume changes to the changes in pressure and temperature. Two such properties are the bulk modulus of elasticity κ and the coefficient of volume expansion β.

It is a common observation that a fluid contracts when more pressure is applied on it and expands when the pressure acting on it is reduced (Fig. 2–11). That is, fluids act like elastic solids with respect to pressure. Therefore, in an analogous manner to Young's modulus of elasticity for solids, it is appropriate to define a **coefficient of compressibility** κ (also called the **bulk modulus of compressibility** or **bulk modulus of elasticity**) for fluids as

$$\kappa = -v\left(\frac{\partial P}{\partial v}\right)_T = \rho\left(\frac{\partial P}{\partial \rho}\right)_T \quad \text{(Pa)} \tag{2-12}$$

It can also be expressed approximately in terms of finite changes as

$$\kappa \cong -\frac{\Delta P}{\Delta v/v} \cong \frac{\Delta P}{\Delta \rho/\rho} \quad (T = \text{constant}) \tag{2-13}$$

Noting that $\Delta v/v$ or $\Delta \rho/\rho$ is dimensionless, κ must have the dimension of pressure (Pa or psi). Also, the coefficient of compressibility represents the change in pressure corresponding to a fractional change in volume or density of the fluid while the temperature remains constant. Then it follows that the coefficient of compressibility of a truly incompressible substance ($v = \text{constant}$) is infinity.

A large value of κ indicates that a large change in pressure is needed to cause a small fractional change in volume, and thus a fluid with a large κ is essentially incompressible. This is typical for liquids, and explains why liquids are usually considered to be *incompressible*. For example, the pressure of water at normal atmospheric conditions must be raised to 210 atm to compress it 1 percent, corresponding to a coefficient of compressibility value of $\kappa = 21{,}000$ atm.

Small density changes in liquids can still cause interesting phenomena in piping systems such as the *water hammer*—characterized by a sound that resembles the sound produced when a pipe is "hammered." This occurs when a liquid in a piping network encounters an abrupt flow restriction (such as a closing valve) and is locally compressed. The acoustic waves that are produced strike the pipe surfaces, bends, and valves as they propagate and reflect along the pipe, causing the pipe to vibrate and produce the familiar sound. In addition to the irritating sound, water hammering can be quite destructive, leading to leaks or even structural damage. The effect can be suppressed with a *water hammer arrestor* (Fig. 2–12), which is a volumetric chamber containing either a bellows or piston to absorb the shock. For large pipes, a vertical tube called a *surge tower* often is used. A surge tower has a free air surface at the top and is virtually maintenance free.

Note that volume and pressure are inversely proportional (volume decreases as pressure is increased and thus $\partial P/\partial v$ is a negative quantity), and the negative sign in the definition (Eq. 2–12) ensures that κ is a positive quantity. Also, differentiating $\rho = 1/v$ gives $d\rho = -dv/v^2$, which can be rearranged as

$$\frac{d\rho}{\rho} = -\frac{dv}{v} \tag{2-14}$$

That is, the fractional changes in the specific volume and the density of a fluid are equal in magnitude but opposite in sign.

For an ideal gas, $P = \rho RT$ and $(\partial P/\partial \rho)_T = RT = P/\rho$, and thus

$$\kappa_{\text{ideal gas}} = P \quad \text{(Pa)} \tag{2-15}$$

Therefore, the coefficient of compressibility of an ideal gas is equal to its absolute pressure, and the coefficient of compressibility of the gas increases

(a)

(b)

FIGURE 2–12
Water hammer arrestors:
(a) A large surge tower built to protect the pipeline against water hammer damage.
Photo by Arris S. Tijsseling, visitor of the University of Adelaide, Australia. Used by permission.
(b) Much smaller arrestors used for supplying water to a household washing machine.
Photo provided courtesy of Oatey Co.

FIGURE 2–13
Natural convection over a woman's hand.

Photograph by Gary S. Settles, Penn State Gas Dynamics Lab. Used by permission.

(a) A substance with a large β

(b) A substance with a small β

FIGURE 2–14
The coefficient of volume expansion is a measure of the change in volume of a substance with temperature at constant pressure.

with increasing pressure. Substituting $\kappa = P$ into the definition of the coefficient of compressibility and rearranging gives

Ideal gas: $$\frac{\Delta \rho}{\rho} = \frac{\Delta P}{P} \quad (T = \text{constant}) \tag{2–16}$$

Therefore, the percent increase of density of an ideal gas during isothermal compression is equal to the percent increase in pressure.

For air at 1 atm pressure, $\kappa = P = 1$ atm and a decrease of 1 percent in volume ($\Delta V/V = -0.01$) corresponds to an increase of $\Delta P = 0.01$ atm in pressure. But for air at 1000 atm, $\kappa = 1000$ atm and a decrease of 1 percent in volume corresponds to an increase of $\Delta P = 10$ atm in pressure. Therefore, a small fractional change in the volume of a gas can cause a large change in pressure at very high pressures.

The inverse of the coefficient of compressibility is called the **isothermal compressibility** α and is expressed as

$$\alpha = \frac{1}{\kappa} = -\frac{1}{V}\left(\frac{\partial V}{\partial P}\right)_T = \frac{1}{\rho}\left(\frac{\partial \rho}{\partial P}\right)_T \quad (1/\text{Pa}) \tag{2–17}$$

The isothermal compressibility of a fluid represents the fractional change in volume or density corresponding to a unit change in pressure.

Coefficient of Volume Expansion

The density of a fluid, in general, depends more strongly on temperature than it does on pressure, and the variation of density with temperature is responsible for numerous natural phenomena such as winds, currents in oceans, rise of plumes in chimneys, the operation of hot-air balloons, heat transfer by natural convection, and even the rise of hot air and thus the phrase "heat rises" (Fig. 2–13). To quantify these effects, we need a property that represents the *variation of the density of a fluid with temperature at constant pressure*.

The property that provides that information is the **coefficient of volume expansion** (or *volume expansivity*) β, defined as (Fig. 2–14)

$$\beta = \frac{1}{V}\left(\frac{\partial V}{\partial T}\right)_P = -\frac{1}{\rho}\left(\frac{\partial \rho}{\partial T}\right)_P \quad (1/\text{K}) \tag{2–18}$$

It can also be expressed approximately in terms of finite changes as

$$\beta \approx \frac{\Delta V/V}{\Delta T} = -\frac{\Delta \rho/\rho}{\Delta T} \quad (\text{at constant } P) \tag{2–19}$$

A large value of β for a fluid means a large change in density with temperature, and the product $\beta\,\Delta T$ represents the fraction of volume change of a fluid that corresponds to a temperature change of ΔT at constant pressure.

It can be shown that the volume expansion coefficient of an *ideal gas* ($P = \rho RT$) at a temperature T is equivalent to the inverse of the temperature:

$$\beta_{\text{ideal gas}} = \frac{1}{T} \quad (1/\text{K}) \tag{2–20}$$

where T is the *absolute* temperature.

In the study of natural convection currents, the condition of the main fluid body that surrounds the finite hot or cold regions is indicated by the subscript "infinity" to serve as a reminder that this is the value at a distance where the presence of the hot or cold region is not felt. In such cases, the volume expansion coefficient can be expressed approximately as

$$\beta \approx -\frac{(\rho_\infty - \rho)/\rho}{T_\infty - T} \quad \text{or} \quad \rho_\infty - \rho = \rho\beta(T - T_\infty) \qquad (2\text{–}21)$$

where ρ_∞ is the density and T_∞ is the temperature of the quiescent fluid away from the confined hot or cold fluid pocket.

We will see in Chap. 3 that natural convection currents are initiated by the *buoyancy force*, which is proportional to the *density difference*, which is in turn proportional to the *temperature difference* at constant pressure. Therefore, the larger the temperature difference between the hot or cold fluid pocket and the surrounding main fluid body, the *larger* the buoyancy force and thus the *stronger* the natural convection currents. A related phenomenon sometimes occurs when an aircraft flies near the speed of sound. The sudden drop in temperature produces condensation of water vapor on a visible vapor cloud (Fig. 2–15).

The combined effects of pressure and temperature changes on the volume change of a fluid can be determined by taking the specific volume to be a function of T and P. Differentiating $v = v(T, P)$ and using the definitions of the compression and expansion coefficients α and β give

$$dv = \left(\frac{\partial v}{\partial T}\right)_P dT + \left(\frac{\partial v}{\partial P}\right)_T dP = (\beta\, dT - \alpha\, dP)v \qquad (2\text{–}22)$$

Then the fractional change in volume (or density) due to changes in pressure and temperature can be expressed approximately as

$$\frac{\Delta v}{v} = -\frac{\Delta\rho}{\rho} \cong \beta\,\Delta T - \alpha\,\Delta P \qquad (2\text{–}23)$$

FIGURE 2–15
Vapor cloud around an F/A-18F Super Hornet as it flies near the speed of sound.
U.S. Navy photo by Photographer's Mate 3rd Class Jonathan Chandler.

EXAMPLE 2–3 Variation of Density with Temperature and Pressure

Consider water initially at 20°C and 1 atm. Determine the final density of the water (*a*) if it is heated to 50°C at a constant pressure of 1 atm, and (*b*) if it is compressed to 100-atm pressure at a constant temperature of 20°C. Take the isothermal compressibility of water to be $\alpha = 4.80 \times 10^{-5}$ atm^{-1}.

SOLUTION Water at a given temperature and pressure is considered. The densities of water after it is heated and after it is compressed are to be determined.
Assumptions **1** The coefficient of volume expansion and the isothermal compressibility of water are constant in the given temperature range. **2** An approximate analysis is performed by replacing differential changes in quantities by finite changes.
Properties The density of water at 20°C and 1 atm pressure is $\rho_1 = 998.0$ kg/m^3. The coefficient of volume expansion at the average temperature of (20 + 50)/2 = 35°C is $\beta = 0.337 \times 10^{-3}$ K^{-1}. The isothermal compressibility of water is given to be $\alpha = 4.80 \times 10^{-5}$ atm^{-1}.
Analysis When differential quantities are replaced by differences and the properties α and β are assumed to be constant, the change in density in

FIGURE 2–16

The variation of the coefficient of volume expansion β of water with temperature in the range of 20°C to 50°C.

Data were generated and plotted using EES.

terms of the changes in pressure and temperature is expressed approximately as (Eq. 2–23)

$$\Delta\rho = \alpha\rho\,\Delta P - \beta\rho\,\Delta T$$

(a) The change in density due to the change of temperature from 20°C to 50°C at constant pressure is

$$\Delta\rho = -\beta\rho\,\Delta T = -(0.337 \times 10^{-3}\,\mathrm{K}^{-1})(998\,\mathrm{kg/m^3})(50-20)\,\mathrm{K}$$
$$= -10.0\,\mathrm{kg/m^3}$$

Noting that $\Delta\rho = \rho_2 - \rho_1$, the density of water at 50°C and 1 atm is

$$\rho_2 = \rho_1 + \Delta\rho = 998.0 + (-10.0) = \mathbf{988.0\ kg/m^3}$$

which is almost identical to the listed value of 988.1 kg/m³ at 50°C in Table A–3. This is mostly due to β varying with temperature almost linearly, as shown in Fig. 2–16.

(b) The change in density due to a change of pressure from 1 atm to 100 atm at constant temperature is

$$\Delta\rho = \alpha\rho\,\Delta P = (4.80 \times 10^{-5}\,\mathrm{atm}^{-1})(998\,\mathrm{kg/m^3})(100-1)\,\mathrm{atm} = 4.7\,\mathrm{kg/m^3}$$

Then the density of water at 100 atm and 20°C becomes

$$\rho_2 = \rho_1 + \Delta\rho = 998.0 + 4.7 = \mathbf{1002.7\ kg/m^3}$$

Discussion Note that the density of water decreases while being heated and increases while being compressed, as expected. This problem can be solved more accurately using differential analysis when functional forms of properties are available.

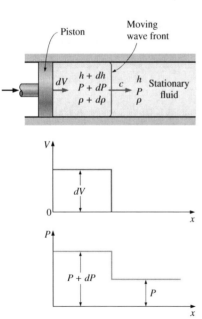

FIGURE 2–17

Propagation of a small pressure wave along a duct.

Speed of Sound and Mach Number

An important parameter in the study of compressible flow is the **speed of sound** (or the **sonic speed**), defined as the speed at which an infinitesimally small pressure wave travels through a medium. The pressure wave may be caused by a small disturbance, which creates a slight rise in local pressure.

To obtain a relation for the speed of sound in a medium, consider a duct that is filled with a fluid at rest, as shown in Fig. 2–17. A piston fitted in the duct is now moved to the right with a constant incremental velocity dV, creating a sonic wave. The wave front moves to the right through the fluid at the speed of sound c and separates the moving fluid adjacent to the piston from the fluid still at rest. The fluid to the left of the wave front experiences an incremental change in its thermodynamic properties, while the fluid on the right of the wave front maintains its original thermodynamic properties, as shown in Fig. 2–17.

To simplify the analysis, consider a control volume that encloses the wave front and moves with it, as shown in Fig. 2–18. To an observer traveling with the wave front, the fluid to the right appears to be moving toward the wave front with a speed of c and the fluid to the left to be moving away from the wave front with a speed of $c - dV$. Of course, the observer sees the control volume that encloses the wave front (and herself or himself) as stationary, and the observer is witnessing a steady-flow process. The mass balance for this single-stream, steady-flow process is expressed as

$$\dot{m}_{\mathrm{right}} = \dot{m}_{\mathrm{left}}$$

or

$$\rho A c = (\rho + d\rho)A(c - dV)$$

By canceling the cross-sectional (or flow) area A and neglecting the higher-order terms, this equation reduces to

$$c\,d\rho - \rho\,dV = 0$$

No heat or work crosses the boundaries of the control volume during this steady-flow process, and the potential energy change can be neglected. Then the steady-flow energy balance $e_{in} = e_{out}$ becomes

$$h + \frac{c^2}{2} = h + dh + \frac{(c - dV)^2}{2}$$

which yields

$$dh - c\,dV = 0$$

where we have neglected the second-order term dV^2. The amplitude of the ordinary sonic wave is very small and does not cause any appreciable change in the pressure and temperature of the fluid. Therefore, the propagation of a sonic wave is not only adiabatic but also very nearly isentropic. Then the thermodynamic relation $T\,ds = dh - dP/\rho$ (see Çengel and Boles, 2011) reduces to

$$T\,ds^{\,0} = dh - \frac{dP}{\rho}$$

or

$$dh = \frac{dP}{\rho}$$

Combining the above equations yields the desired expression for the speed of sound as

$$c^2 = \frac{dP}{d\rho} \quad \text{at } s = \text{constant}$$

or

$$c^2 = \left(\frac{\partial P}{\partial \rho}\right)_s \tag{2–24}$$

It is left as an exercise for the reader to show, by using thermodynamic property relations, that Eq. 2–24 can also be written as

$$c^2 = k\left(\frac{\partial P}{\partial \rho}\right)_T \tag{2–25}$$

where $k = c_p/c_v$ is the specific heat ratio of the fluid. Note that the speed of sound in a fluid is a function of the thermodynamic properties of that fluid Fig. 2–19.

When the fluid is an ideal gas ($P = \rho R T$), the differentiation in Eq. 2–25 can be performed to yield

$$c^2 = k\left(\frac{\partial P}{\partial \rho}\right)_T = k\left[\frac{\partial(\rho R T)}{\partial \rho}\right]_T = kRT$$

or

$$c = \sqrt{kRT} \tag{2–26}$$

49

Control volume traveling with the wave front

FIGURE 2–18
Control volume moving with the small pressure wave along a duct.

FIGURE 2–19
The speed of sound in air increases with temperature. At typical outside temperatures, c is about 340 m/s. In round numbers, therefore, the sound of thunder from a lightning strike travels about 1 km in 3 seconds. If you see the lightning and then hear the thunder less than 3 seconds later, you know that the lightning is close, and it is time to go indoors!

© Bear Dancer Studios/Mark Dierker

AIR   HELIUM

284 m/s   200 K   832 m/s

347 m/s   300 K   1019 m/s

1000 K

634 m/s   1861 m/s

FIGURE 2–20
The speed of sound changes with temperature and varies with the fluid.

AIR   V = 320 m/s
220 K   Ma = 1.08

AIR   V = 320 m/s
300 K   Ma = 0.92

FIGURE 2–21
The Mach number can be different at different temperatures even if the flight speed is the same.
© Alamy RF

AIR   Diffuser
V = 200 m/s
T = 30°C

FIGURE 2–22
Schematic for Example 12–4.

Noting that the gas constant R has a fixed value for a specified ideal gas and the specific heat ratio k of an ideal gas is, at most, a function of temperature, we see that the speed of sound in a specified ideal gas is a function of temperature alone (Fig. 2–20).

A second important parameter in the analysis of compressible fluid flow is the **Mach number Ma,** named after the Austrian physicist Ernst Mach (1838–1916). It is the ratio of the actual speed of the fluid (or an object in still fluid) to the speed of sound in the same fluid at the same state:

$$\text{Ma} = \frac{V}{c} \tag{2–27}$$

Note that the Mach number depends on the speed of sound, which depends on the state of the fluid. Therefore, the Mach number of an aircraft cruising at constant velocity in still air may be different at different locations (Fig. 2–21).

Fluid flow regimes are often described in terms of the flow Mach number. The flow is called **sonic** when Ma = 1, **subsonic** when Ma < 1, **supersonic** when Ma > 1, **hypersonic** when Ma ≫ 1, and **transonic** when Ma ≅ 1.

EXAMPLE 2–4   **Mach Number of Air Entering a Diffuser**

Air enters a diffuser shown in Fig. 2–22 with a speed of 200 m/s. Determine (*a*) the speed of sound and (*b*) the Mach number at the diffuser inlet when the air temperature is 30°C.

SOLUTION   Air enters a diffuser at high speed. The speed of sound and the Mach number are to be determined at the diffuser inlet.
Assumption   Air at the specified conditions behaves as an ideal gas.
Properties   The gas constant of air is $R = 0.287$ kJ/kg·K, and its specific heat ratio at 30°C is 1.4.
Analysis   We note that the speed of sound in a gas varies with temperature, which is given to be 30°C.

(*a*) The speed of sound in air at 30°C is determined from Eq. 2–26 to be

$$c = \sqrt{kRT} = \sqrt{(1.4)(0.287 \text{ kJ/kg·K})(303 \text{ K})\left(\frac{1000 \text{ m}^2/\text{s}^2}{1 \text{ kJ/kg}}\right)} = \textbf{349 m/s}$$

(*b*) Then the Mach number becomes

$$\text{Ma} = \frac{V}{c} = \frac{200 \text{ m/s}}{349 \text{ m/s}} = \textbf{0.573}$$

Discussion   The flow at the diffuser inlet is subsonic since Ma < 1.

2–6 · VISCOSITY

When two solid bodies in contact move relative to each other, a friction force develops at the contact surface in the direction opposite to motion. To move a table on the floor, for example, we have to apply a force to the table in the horizontal direction large enough to overcome the friction force.

The magnitude of the force needed to move the table depends on the *friction coefficient* between the table legs and the floor.

The situation is similar when a fluid moves relative to a solid or when two fluids move relative to each other. We move with relative ease in air, but not so in water. Moving in oil would be even more difficult, as can be observed by the slower downward motion of a glass ball dropped in a tube filled with oil. It appears that there is a property that represents the internal resistance of a fluid to motion or the "fluidity," and that property is the **viscosity.** The force a flowing fluid exerts on a body in the flow direction is called the **drag force,** and the magnitude of this force depends, in part, on viscosity (Fig. 2–23).

To obtain a relation for viscosity, consider a fluid layer between two very large parallel plates (or equivalently, two parallel plates immersed in a large body of a fluid) separated by a distance ℓ (Fig. 2–24). Now a constant parallel force F is applied to the upper plate while the lower plate is held fixed. After the initial transients, it is observed that the upper plate moves continuously under the influence of this force at a constant speed V. The fluid in contact with the upper plate sticks to the plate surface and moves with it at the same speed, and the shear stress τ acting on this fluid layer is

$$\tau = \frac{F}{A} \qquad (2\text{–}28)$$

where A is the contact area between the plate and the fluid. Note that the fluid layer deforms continuously under the influence of shear stress.

The fluid in contact with the lower plate assumes the velocity of that plate, which is zero (because of the no-slip condition—see Section 1–2). In steady laminar flow, the fluid velocity between the plates varies linearly between 0 and V, and thus the *velocity profile* and the *velocity gradient* are

$$u(y) = \frac{y}{\ell}V \quad \text{and} \quad \frac{du}{dy} = \frac{V}{\ell} \qquad (2\text{–}29)$$

where y is the vertical distance from the lower plate.

During a differential time interval dt, the sides of fluid particles along a vertical line MN rotate through a differential angle $d\beta$ while the upper plate moves a differential distance $da = V\,dt$. The angular displacement or deformation (or shear strain) can be expressed as

$$d\beta \approx \tan d\beta = \frac{da}{\ell} = \frac{V\,dt}{\ell} = \frac{du}{dy}dt \qquad (2\text{–}30)$$

Rearranging, the rate of deformation under the influence of shear stress τ becomes

$$\frac{d\beta}{dt} = \frac{du}{dy} \qquad (2\text{–}31)$$

Thus we conclude that the rate of deformation of a fluid element is equivalent to the velocity gradient du/dy. Further, it can be verified experimentally that for most fluids the rate of deformation (and thus the velocity gradient) is directly proportional to the shear stress τ,

$$\tau \propto \frac{d\beta}{dt} \quad \text{or} \quad \tau \propto \frac{du}{dy} \qquad (2\text{–}32)$$

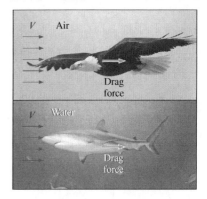

FIGURE 2–23
A fluid moving relative to a body exerts a drag force on the body, partly because of friction caused by viscosity.
© Digital Vision/Getty RF

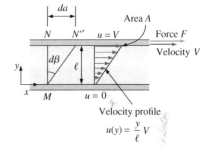

FIGURE 2–24
The behavior of a fluid in laminar flow between two parallel plates when the upper plate moves with a constant velocity.

Fluids for which the rate of deformation is linearly proportional to the shear stress are called **Newtonian fluids** after Sir Isaac Newton, who expressed it first in 1687. Most common fluids such as water, air, gasoline, and oils are Newtonian fluids. Blood and liquid plastics are examples of non-Newtonian fluids.

In one-dimensional shear flow of Newtonian fluids, shear stress can be expressed by the linear relationship

$$\text{Shear stress:} \qquad \tau = \mu \frac{du}{dy} \quad (\text{N/m}^2) \qquad (2\text{--}33)$$

where the constant of proportionality μ is called the **coefficient of viscosity** or the **dynamic** (or **absolute**) **viscosity** of the fluid, whose unit is kg/m·s, or equivalently, N·s/m² (or Pa·s where Pa is the pressure unit pascal). A common viscosity unit is **poise,** which is equivalent to 0.1 Pa·s (or *centipoise*, which is one-hundredth of a poise). The viscosity of water at 20°C is 1.002 centipoise, and thus the unit centipoise serves as a useful reference. A plot of shear stress versus the rate of deformation (velocity gradient) for a Newtonian fluid is a straight line whose slope is the viscosity of the fluid, as shown in Fig. 2–25. Note that viscosity is independent of the rate of deformation for Newtonian fluids. Since the rate of deformation is proportional to the strain rate, Fig. 2–25 reveals that viscosity is actually a coefficient in a stress–strain relationship.

The **shear force** acting on a Newtonian fluid layer (or, by Newton's third law, the force acting on the plate) is

$$\text{Shear force:} \qquad F = \tau A = \mu A \frac{du}{dy} \quad (\text{N}) \qquad (2\text{--}34)$$

where again A is the contact area between the plate and the fluid. Then the force F required to move the upper plate in Fig. 2–24 at a constant speed of V while the lower plate remains stationary is

$$F = \mu A \frac{V}{\ell} \quad (\text{N}) \qquad (2\text{--}35)$$

This relation can alternately be used to calculate μ when the force F is measured. Therefore, the experimental setup just described can be used to measure the viscosity of fluids. Note that under identical conditions, the force F would be very different for different fluids.

For non-Newtonian fluids, the relationship between shear stress and rate of deformation is not linear, as shown in Fig. 2–26. The slope of the curve on the τ versus du/dy chart is referred to as the *apparent viscosity* of the fluid. Fluids for which the apparent viscosity increases with the rate of deformation (such as solutions with suspended starch or sand) are referred to as *dilatant* or *shear thickening fluids*, and those that exhibit the opposite behavior (the fluid becoming less viscous as it is sheared harder, such as some paints, polymer solutions, and fluids with suspended particles) are referred to as *pseudoplastic* or *shear thinning fluids*. Some materials such as toothpaste can resist a finite shear stress and thus behave as a solid, but deform continuously when the shear stress exceeds the yield stress and behave as a fluid. Such materials are referred to as Bingham plastics after Eugene C. Bingham (1878–1945), who did pioneering work on fluid viscosity for the U.S. National Bureau of Standards in the early twentieth century.

FIGURE 2–25

The rate of deformation (velocity gradient) of a Newtonian fluid is proportional to shear stress, and the constant of proportionality is the viscosity.

FIGURE 2–26

Variation of shear stress with the rate of deformation for Newtonian and non-Newtonian fluids (the slope of a curve at a point is the apparent viscosity of the fluid at that point).

In fluid mechanics and heat transfer, the ratio of dynamic viscosity to density appears frequently. For convenience, this ratio is given the name **kinematic viscosity** ν and is expressed as $\nu = \mu/\rho$. Two common units of kinematic viscosity are m^2/s and **stoke** (1 stoke = 1 cm^2/s = 0.0001 m^2/s).

In general, the viscosity of a fluid depends on both temperature and pressure, although the dependence on pressure is rather weak. For *liquids*, both the dynamic and kinematic viscosities are practically independent of pressure, and any small variation with pressure is usually disregarded, except at extremely high pressures. For *gases*, this is also the case for dynamic viscosity (at low to moderate pressures), but not for kinematic viscosity since the density of a gas is proportional to its pressure (Fig. 2–27).

The viscosity of a fluid is a measure of its "resistance to deformation." Viscosity is due to the internal frictional force that develops between different layers of fluids as they are forced to move relative to each other.

The viscosity of a fluid is directly related to the pumping power needed to transport a fluid in a pipe or to move a body (such as a car in air or a submarine in the sea) through a fluid. Viscosity is caused by the cohesive forces between the molecules in liquids and by the molecular collisions in gases, and it varies greatly with temperature. The viscosity of liquids decreases with temperature, whereas the viscosity of gases increases with temperature (Fig. 2–28). This is because in a liquid the molecules possess more energy at higher temperatures, and they can oppose the large cohesive intermolecular forces more strongly. As a result, the energized liquid molecules can move more freely.

In a gas, on the other hand, the intermolecular forces are negligible, and the gas molecules at high temperatures move randomly at higher velocities. This results in more molecular collisions per unit volume per unit time and therefore in greater resistance to flow. The kinetic theory of gases predicts the viscosity of gases to be proportional to the square root of temperature. That is, $\mu_{gas} \propto \sqrt{T}$. This prediction is confirmed by practical observations, but deviations for different gases need to be accounted for by incorporating some correction factors. The viscosity of gases is expressed as a function of temperature by the Sutherland correlation (from The U.S. Standard Atmosphere) as

Gases:
$$\mu = \frac{aT^{1/2}}{1 + b/T} \qquad \text{(2–36)}$$

where T is absolute temperature and a and b are experimentally determined constants. Note that measuring viscosity at two different temperatures is sufficient to determine these constants. For air at atmospheric conditions, the values of these constants are $a = 1.458 \times 10^{-6}$ kg/(m·s·$K^{1/2}$) and $b = 110.4$ K. The viscosity of gases is independent of pressure at low to moderate pressures (from a few percent of 1 atm to several atm). But viscosity increases at high pressures due to the increase in density.

For liquids, the viscosity is approximated as

Liquids:
$$\mu = a10^{b/(T - c)} \qquad \text{(2–37)}$$

where again T is absolute temperature and a, b, and c are experimentally determined constants. For water, using the values $a = 2.414 \times 10^{-5}$ N·s/m^2, $b = 247.8$ K, and $c = 140$ K results in less than 2.5 percent error in viscosity in the temperature range of 0°C to 370°C (Touloukian et al., 1975).

FIGURE 2–27

Dynamic viscosity, in general, does not depend on pressure, but kinematic viscosity does.

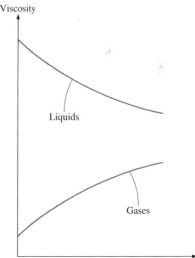

FIGURE 2–28

The viscosity of liquids decreases and the viscosity of gases increases with temperature.

54
PROPERTIES OF FLUIDS

TABLE 2–3

Dynamic viscosity of some fluids at 1 atm and 20°C (unless otherwise stated)

Fluid	Dynamic Viscosity μ, kg/m·s
Glycerin:	
−20°C	134.0
0°C	10.5
20°C	1.52
40°C	0.31
Engine oil:	
SAE 10W	0.10
SAE 10W30	0.17
SAE 30	0.29
SAE 50	0.86
Mercury	0.0015
Ethyl alcohol	0.0012
Water:	
0°C	0.0018
20°C	0.0010
100°C (liquid)	0.00028
100°C (vapor)	0.000012
Blood, 37°C	0.00040
Gasoline	0.00029
Ammonia	0.00015
Air	0.000018
Hydrogen, 0°C	0.0000088

The viscosities of some fluids at room temperature are listed in Table 2–3. They are plotted against temperature in Fig. 2–29. Note that the viscosities of different fluids differ by several orders of magnitude. Also note that it is more difficult to move an object in a higher-viscosity fluid such as engine oil than it is in a lower-viscosity fluid such as water. Liquids, in general, are much more viscous than gases.

Consider a fluid layer of thickness ℓ within a small gap between two concentric cylinders, such as the thin layer of oil in a journal bearing. The gap between the cylinders can be modeled as two parallel flat plates separated by the fluid. Noting that torque is T = FR (force times the moment arm, which is the radius R of the inner cylinder in this case), the tangential velocity is V = ωR (angular velocity times the radius), and taking the wetted surface area of the inner cylinder to be A = 2πRL by disregarding the shear stress acting on the two ends of the inner cylinder, torque can be expressed as

$$\text{T} = FR = \mu \, \frac{2\pi R^3 \omega L}{\ell} = \mu \frac{4\pi^2 R^3 \dot{n} L}{\ell} \tag{2–38}$$

where L is the length of the cylinder and $\dot{n}$ is the number of revolutions per unit time, which is usually expressed in rpm (revolutions per minute). Note that the angular distance traveled during one rotation is 2π rad, and thus the

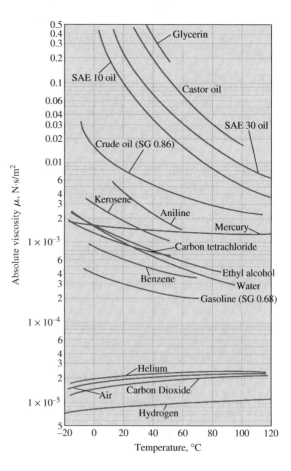

FIGURE 2–29

The variation of dynamic (absolute) viscosity of common fluids with temperature at 1 atm (1 N·s/m² = 1 kg/m·s = 0.020886 lbf·s/ft²).

Data from EES and F. M. White, Fluid Mechanics 7e. Copyright © 2011 The McGraw-Hill Companies, Inc. Used by permission.

relation between the angular velocity in rad/min and the rpm is $\omega = 2\pi\dot{n}$. Equation 2–38 can be used to calculate the viscosity of a fluid by measuring torque at a specified angular velocity. Therefore, two concentric cylinders can be used as a *viscometer*, a device that measures viscosity.

EXAMPLE 2–5   **Determining the Viscosity of a Fluid**

The viscosity of a fluid is to be measured by a viscometer constructed of two 40-cm-long concentric cylinders (Fig. 2–30). The outer diameter of the inner cylinder is 12 cm, and the gap between the two cylinders is 0.15 cm. The inner cylinder is rotated at 300 rpm, and the torque is measured to be 1.8 N·m. Determine the viscosity of the fluid.

SOLUTION The torque and the rpm of a double cylinder viscometer are given. The viscosity of the fluid is to be determined.
Assumptions **1** The inner cylinder is completely submerged in the fluid. **2** The viscous effects on the two ends of the inner cylinder are negligible.
Analysis The velocity profile is linear only when the curvature effects are negligible, and the profile can be approximated as being linear in this case since $\ell/R = 0.025 \ll 1$. Solving Eq. 2–38 for viscosity and substituting the given values, the viscosity of the fluid is determined to be

$$\mu = \frac{T\ell}{4\pi^2 R^3 \dot{n} L} = \frac{(1.8 \text{ N·m})(0.0015 \text{ m})}{4\pi^2 (0.06 \text{ m})^3 \left(300 \dfrac{1}{\text{min}}\right)\left(\dfrac{1 \text{ min}}{60 \text{ s}}\right)(0.4 \text{ m})} = 0.158 \text{ N·s/m}^2$$

Discussion Viscosity is a strong function of temperature, and a viscosity value without a corresponding temperature is of little usefulness. Therefore, the temperature of the fluid should have also been measured during this experiment, and reported with this calculation.

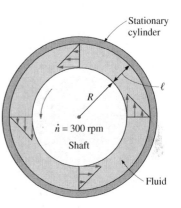

FIGURE 2–30
Schematic for Example 2–5
(not to scale).

2–7 ▪ SURFACE TENSION AND CAPILLARY EFFECT

It is often observed that a drop of blood forms a hump on a horizontal glass; a drop of mercury forms a near-perfect sphere and can be rolled just like a steel ball over a smooth surface; water droplets from rain or dew hang from branches or leaves of trees; a liquid fuel injected into an engine forms a mist of spherical droplets; water dripping from a leaky faucet falls as nearly spherical droplets; a soap bubble released into the air forms a nearly spherical shape; and water beads up into small drops on flower petals (Fig. 2–31*a*).

In these and other observances, liquid droplets behave like small balloons filled with the liquid, and the surface of the liquid acts like a stretched elastic membrane under tension. The pulling force that causes this tension acts parallel to the surface and is due to the attractive forces between the molecules of the liquid. The magnitude of this force per unit length is called **surface tension** or *coefficient of surface tension* σ_s and is usually expressed in the unit N/m (or lbf/ft in English units). This effect is also called *surface energy* (per unit area) and is expressed in the equivalent unit of N·m/m^2 or J/m^2. In this case, σ_s represents the stretching work that needs to be done to increase the surface area of the liquid by a unit amount.

(a)

(b)

(c)

FIGURE 2–31

Some consequences of surface tension: (a) drops of water beading up on a leaf, (b) a water strider sitting on top of the surface of water, and (c) a color schlieren image of the water strider revealing how the water surface dips down where its feet contact the water (it looks like two insects but the second one is just a shadow).

(a) © Don Paulson Photography/Purestock/ SuperStock RF
(b) NPS Photo by Rosalie LaRue.
(c) Photo courtesy of G. S. Settles, Gas Dynamics Lab, Penn State University, used by permission.

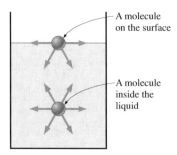

A molecule on the surface

A molecule inside the liquid

FIGURE 2–32

Attractive forces acting on a liquid molecule at the surface and deep inside the liquid.

To visualize how surface tension arises, we present a microscopic view in Fig. 2–32 by considering two liquid molecules, one at the surface and one deep within the liquid body. The attractive forces applied on the interior molecule by the surrounding molecules balance each other because of symmetry. But the attractive forces acting on the surface molecule are not symmetric, and the attractive forces applied by the gas molecules above are usually very small. Therefore, there is a net attractive force acting on the molecule at the surface of the liquid, which tends to pull the molecules on the surface toward the interior of the liquid. This force is balanced by the repulsive forces from the molecules below the surface that are trying to be compressed. The result is that the liquid minimizes its surface area. This is the reason for the tendency of liquid droplets to attain a spherical shape, which has the minimum surface area for a given volume.

You also may have observed, with amusement, that some insects can land on water or even walk on water (Fig. 2–31b) and that small steel needles can float on water. These phenomena are made possible by surface tension which balances the weights of these objects.

To understand the surface tension effect better, consider a liquid film (such as the film of a soap bubble) suspended on a U-shaped wire frame with a movable side (Fig. 2–33). Normally, the liquid film tends to pull the movable wire inward in order to minimize its surface area. A force F needs to be applied on the movable wire in the opposite direction to balance this pulling effect. Both sides of the thin film are surfaces exposed to air, and thus the length along which the surface tension acts in this case is $2b$. Then a force balance on the movable wire gives $F = 2b\sigma_s$, and thus the surface tension can be expressed as

$$\sigma_s = \frac{F}{2b} \tag{2–39}$$

Note that for $b = 0.5$ m, the measured force F (in N) is simply the surface tension in N/m. An apparatus of this kind with sufficient precision can be used to measure the surface tension of various liquids.

In the U-shaped wire frame apparatus, the movable wire is pulled to stretch the film and increase its surface area. When the movable wire is pulled a distance Δx, the surface area increases by $\Delta A = 2b\,\Delta x$, and the work W done during this stretching process is

$$W = \text{Force} \times \text{Distance} = F\,\Delta x = 2b\sigma_s\,\Delta x = \sigma_s\,\Delta A$$

where we have assumed that the force remains constant over the small distance. This result can also be interpreted as *the surface energy of the film is increased by an amount $\sigma_s\,\Delta A$ during this stretching process*, which is consistent with the alternative interpretation of σ_s as surface energy per unit area. This is similar to a rubber band having more potential (elastic) energy after it is stretched further. In the case of liquid film, the work is used to move liquid molecules from the interior parts to the surface against the attraction forces of other molecules. Therefore, surface tension also can be defined as *the work done per unit increase in the surface area of the liquid*.

The surface tension varies greatly from substance to substance, and with temperature for a given substance, as shown in Table 2–4. At 20°C, for example, the surface tension is 0.073 N/m for water and 0.440 N/m for mercury surrounded by atmospheric air. The surface tension of mercury is large enough that mercury droplets form nearly spherical balls that can be rolled like a solid ball on a smooth surface. The surface tension of a liquid, in general, decreases with temperature and becomes zero at the critical point (and thus there is no distinct liquid–vapor interface at temperatures above the critical point). The effect of pressure on surface tension is usually negligible.

The surface tension of a substance can be changed considerably by *impurities*. Therefore, certain chemicals, called *surfactants*, can be added to a liquid to decrease its surface tension. For example, soaps and detergents lower the surface tension of water and enable it to penetrate the small openings between fibers for more effective washing. But this also means that devices whose operation depends on surface tension (such as heat pipes) can be destroyed by the presence of impurities due to poor workmanship.

We speak of surface tension for liquids only at liquid–liquid or liquid–gas interfaces. Therefore, it is imperative that the adjacent liquid or gas be specified when specifying surface tension. Surface tension determines the size of the liquid droplets that form, and so a droplet that keeps growing by the addition of more mass breaks down when the surface tension can no longer hold it together. This is like a balloon that bursts while being inflated when the pressure inside rises above the strength of the balloon material.

A curved interface indicates a pressure difference (or "pressure jump") across the interface with pressure being higher on the concave side. Consider, for example, a droplet of liquid in air, an air (or other gas) bubble in water, or a soap bubble in air. The excess pressure ΔP above atmospheric pressure can be determined by considering a free-body diagram of half the droplet or bubble (Fig. 2–34). Noting that surface tension acts along the circumference and the pressure acts on the area, horizontal force balances for the droplet or air bubble and the soap bubble give

Droplet or air bubble: $(2\pi R)\sigma_s = (\pi R^2)\Delta P_{droplet} \rightarrow \Delta P_{droplet} = P_i - P_o = \dfrac{2\sigma_s}{R}$

$$(2\text{–}40)$$

Soap bubble: $2(2\pi R)\sigma_s = (\pi R^2)\Delta P_{bubble} \rightarrow \Delta P_{bubble} = P_i - P_o = \dfrac{4\sigma_s}{R}$ $(2\text{–}41)$

where P_i and P_o are the pressures inside and outside the droplet or bubble, respectively. When the droplet or bubble is in the atmosphere, P_o is simply atmospheric pressure. The extra factor of 2 in the force balance for the soap bubble is due to the existence of a soap film with *two* surfaces (inner and outer surfaces) and thus two circumferences in the cross section.

The excess pressure in a droplet of liquid in a gas (or a bubble of gas in a liquid) can also be determined by considering a differential increase in the radius of the droplet due to the addition of a differential amount of mass and interpreting the surface tension as the increase in the surface energy per unit area. Then the increase in the surface energy of the droplet during this differential expansion process becomes

$$\delta W_{surface} = \sigma_s\, dA = \sigma_s\, d(4\pi R^2) = 8\pi R\sigma_s\, dR$$

57
CHAPTER 2

FIGURE 2–33
Stretching a liquid film with a U-shaped wire, and the forces acting on the movable wire of length *b*.

TABLE 2–4

Surface tension of some fluids in air at 1 atm and 20°C (unless otherwise stated)

Fluid	Surface Tension σ_s, N/m*
†Water:	
0°C	0.076
20°C	0.073
100°C	0.059
300°C	0.014
Glycerin	0.063
SAE 30 oil	0.035
Mercury	0.440
Ethyl alcohol	0.023
Blood, 37°C	0.058
Gasoline	0.022
Ammonia	0.021
Soap solution	0.025
Kerosene	0.028

* Multiply by 0.06852 to convert to lbf/ft.
† See Appendices for more precise data for water.

(a) Half of a droplet or air bubble

(b) Half of a soap bubble

FIGURE 2–34
The free-body diagram of half of a droplet or air bubble and half of a soap bubble.

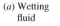

| (a) Wetting fluid | (b) Nonwetting fluid |

FIGURE 2–35
The contact angle for wetting and nonwetting fluids.

FIGURE 2–36
The meniscus of colored water in a 4-mm-inner-diameter glass tube. Note that the edge of the meniscus meets the wall of the capillary tube at a very small contact angle.

Photo by Gabrielle Tremblay. Used by permission.

The expansion work done during this differential process is determined by multiplying the force by distance to obtain

$$\delta W_{expansion} = \text{Force} \times \text{Distance} = F \, dR = (\Delta P A) \, dR = 4\pi R^2 \, \Delta P \, dR$$

Equating the two expressions above gives $\Delta P_{droplet} = 2\sigma_s/R$, which is the same relation obtained before and given in Eq. 2–40. Note that the excess pressure in a droplet or bubble is inversely proportional to the radius.

Capillary Effect

Another interesting consequence of surface tension is the **capillary effect,** which is the rise or fall of a liquid in a small-diameter tube inserted into the liquid. Such narrow tubes or confined flow channels are called **capillaries.** The rise of kerosene through a cotton wick inserted into the reservoir of a kerosene lamp is due to this effect. The capillary effect is also partially responsible for the rise of water to the top of tall trees. The curved free surface of a liquid in a capillary tube is called the **meniscus.**

It is commonly observed that water in a glass container curves up slightly at the edges where it touches the glass surface; but the opposite occurs for mercury: it curves down at the edges (Fig. 2–35). This effect is usually expressed by saying that water *wets* the glass (by sticking to it) while mercury does not. The strength of the capillary effect is quantified by the **contact** (or *wetting*) **angle** ϕ, defined as *the angle that the tangent to the liquid surface makes with the solid surface at the point of contact.* The surface tension force acts along this tangent line toward the solid surface. A liquid is said to wet the surface when $\phi < 90°$ and not to wet the surface when $\phi > 90°$. In atmospheric air, the contact angle of water (and most other organic liquids) with glass is nearly zero, $\phi \approx 0°$ (Fig. 2–36). Therefore, the surface tension force acts upward on water in a glass tube along the circumference, tending to pull the water up. As a result, water rises in the tube until the weight of the liquid in the tube above the liquid level of the reservoir balances the surface tension force. The contact angle is 130° for mercury–glass and 26° for kerosene–glass in air. Note that the contact angle, in general, is different in different environments (such as another gas or liquid in place of air).

The phenomenon of the capillary effect can be explained microscopically by considering *cohesive forces* (the forces between like molecules, such as water and water) and *adhesive forces* (the forces between unlike molecules, such as water and glass). The liquid molecules at the solid–liquid interface are subjected to both cohesive forces by other liquid molecules and adhesive forces by the molecules of the solid. The relative magnitudes of these forces determine whether a liquid wets a solid surface or not. Obviously, the water molecules are more strongly attracted to the glass molecules than they are to other water molecules, and thus water tends to rise along the glass surface. The opposite occurs for mercury, which causes the liquid surface near the glass wall to be suppressed (Fig. 2–37).

The magnitude of the capillary rise in a circular tube can be determined from a force balance on the cylindrical liquid column of height h in the tube (Fig. 2–38). The bottom of the liquid column is at the same level as the free surface of the reservoir, and thus the pressure there must be atmospheric pressure. This balances the atmospheric pressure acting at the top surface of the liquid column, and thus these two effects cancel each other. The weight of the liquid column is approximately

$$W = mg = \rho Vg = \rho g(\pi R^2 h)$$

Equating the vertical component of the surface tension force to the weight gives

$$W = F_{surface} \rightarrow \rho g(\pi R^2 h) = 2\pi R\sigma_s \cos \phi$$

Solving for h gives the capillary rise to be

Capillary rise: $\qquad h = \dfrac{2\sigma_s}{\rho g R} \cos \phi \qquad (R = \text{constant}) \qquad$ **(2–42)**

This relation is also valid for nonwetting liquids (such as mercury in glass) and gives the capillary drop. In this case $\phi > 90°$ and thus $\cos \phi < 0$, which makes h negative. Therefore, a negative value of capillary rise corresponds to a capillary drop (Fig. 2–37).

Note that the capillary rise is inversely proportional to the radius of the tube. Therefore, the thinner the tube is, the greater the rise (or fall) of the liquid in the tube. In practice, the capillary effect for water is usually negligible in tubes whose diameter is greater than 1 cm. When pressure measurements are made using manometers and barometers, it is important to use sufficiently large tubes to minimize the capillary effect. The capillary rise is also inversely proportional to the density of the liquid, as expected. Therefore, in general, lighter liquids experience greater capillary rises. Finally, it should be kept in mind that Eq. 2–42 is derived for constant-diameter tubes and should not be used for tubes of variable cross section.

FIGURE 2–37
The capillary rise of water and the capillary fall of mercury in a small-diameter glass tube.

FIGURE 2–38
The forces acting on a liquid column that has risen in a tube due to the capillary effect.

■
■ **EXAMPLE 2–6   The Capillary Rise of Water in a Tube**
■
■ A 0.6-mm-diameter glass tube is inserted into water at 20°C in a cup.
■ Determine the capillary rise of water in the tube (Fig. 2–39).

SOLUTION The rise of water in a slender tube as a result of the capillary effect is to be determined.
Assumptions **1** There are no impurities in the water and no contamination on the surfaces of the glass tube. **2** The experiment is conducted in atmospheric air.
Properties The surface tension of water at 20°C is 0.073 N/m (Table 2–4). The contact angle of water with glass is approximately 0° (from preceding text). We take the density of liquid water to be 1000 kg/m³.
Analysis The capillary rise is determined directly from Eq. 2–42 by substituting the given values, yielding

$$h = \frac{2\sigma_s}{\rho g R} \cos \phi = \frac{2(0.073 \text{ N/m})}{(1000 \text{ kg/m}^3)(9.81 \text{ m/s}^2)(0.3 \times 10^{-3}\text{m})}(\cos 0°)\left(\frac{1 \text{kg·m/s}^2}{1 \text{ N}}\right)$$

$$= 0.050 \text{ m} = \textbf{5.0 cm}$$

Therefore, water rises in the tube 5 cm above the liquid level in the cup.
Discussion Note that if the tube diameter were 1 cm, the capillary rise would be 0.3 mm, which is hardly noticeable to the eye. Actually, the capillary rise in a large-diameter tube occurs only at the rim. The center does not rise at all. Therefore, the capillary effect can be ignored for large-diameter tubes.

FIGURE 2–39
Schematic for Example 2–6.

FIGURE 2–40
Schematic for Example 2-7.

EXAMPLE 2–7   Using Capillary Rise to Generate Power in a Hydraulic Turbine

Reconsider Example 2–6. Realizing that water rises by 5 cm under the influence of surface tension without requiring any energy input from an external source, a person conceives the idea that power can be generated by drilling a hole in the tube just below the water level and feeding the water spilling out of the tube into a turbine (Fig. 2–40). The person takes this idea even further by suggesting that a series of tube banks can be used for this purpose and cascading can be incorporated to achieve practically feasible flow rates and elevation differences. Determine if this idea has any merit.

SOLUTION   Water that rises in tubes under the influence of the capillary effect is to be used to generate power by feeding it into a turbine. The validity of this suggestion is to be evaluated.

Analysis   The proposed system may appear like a stroke of genius, since the commonly used hydroelectric power plants generate electric power by simply capturing the potential energy of elevated water, and the capillary rise provides the mechanism to raise the water to any desired height without requiring any energy input.

When viewed from a thermodynamic point of view, the proposed system immediately can be labeled as a perpetual motion machine (PMM) since it continuously generates electric power without requiring any energy input. That is, the proposed system creates energy, which is a clear violation of the first law of thermodynamics or the conservation of energy principle, and it does not warrant any further consideration. But the fundamental principle of conservation of energy did not stop many from dreaming about being the first to prove nature wrong, and to come up with a trick to permanently solve the world's energy problems. Therefore, the impossibility of the proposed system should be demonstrated.

As you may recall from your physics courses (also to be discussed in the next chapter), the pressure in a static fluid varies in the vertical direction only and increases with increasing depth linearly. Then the pressure difference across the 5-cm-high water column in the tube becomes

$$\Delta P_{\text{water column in tube}} = P_2 - P_1 = \rho_{\text{water}}gh$$

$$= (1000 \text{ kg/m}^2)(9.81 \text{ m/s}^2)(0.05 \text{ m})\left(\frac{1 \text{ kN}}{1000 \text{ kg}\cdot\text{m/s}^2}\right)$$

$$= 0.49 \text{ kN/m}^2 \, (\approx 0.005 \text{ atm})$$

That is, the pressure at the top of the water column in the tube is 0.005 atm *less* than the pressure at the bottom. Noting that the pressure at the bottom of the water column is atmospheric pressure (since it is at the same horizontal line as the water surface in the cup) the pressure anywhere in the tube is below atmospheric pressure with the difference reaching 0.005 atm at the top. Therefore, if a hole is drilled in the tube, air will leak into the tube rather than water leaking out.

Discussion   The water column in the tube is motionless, and thus, there cannot be any unbalanced force acting on it (zero net force). The force due to the pressure difference across the meniscus between the atmospheric air and the water at the top of water column is balanced by the surface tension. If this surface-tension force were to disappear, the water in the tube would drop down under the influence of atmospheric pressure to the level of the free surface in the tube.

SUMMARY

In this chapter various properties commonly used in fluid mechanics are discussed. The mass-dependent properties of a system are called *extensive properties* and the others, *intensive properties*. Density is mass per unit volume, and *specific volume* is volume per unit mass. The *specific gravity* is defined as the ratio of the density of a substance to the density of water at 4°C,

$$SG = \frac{\rho}{\rho_{H_2O}}$$

The ideal-gas equation of state is expressed as

$$P = \rho RT$$

where P is the absolute pressure, T is the thermodynamic temperature, ρ is the density, and R is the gas constant.

At a given temperature, the pressure at which a pure substance changes phase is called the *saturation pressure*. For phase-change processes between the liquid and vapor phases of a pure substance, the saturation pressure is commonly called the *vapor pressure P_v*. Vapor bubbles that form in the low-pressure regions in a liquid (a phenomenon called *cavitation*) collapse as they are swept away from the low-pressure regions, generating highly destructive, extremely high-pressure waves.

Energy can exist in numerous forms, and their sum constitutes the *total energy E* (or e on a unit-mass basis) of a system. The sum of all microscopic forms of energy is called the *internal energy U* of a system. The energy that a system possesses as a result of its motion relative to some reference frame is called *kinetic energy* expressed per unit mass as $ke = V^2/2$, and the energy that a system possesses as a result of its elevation in a gravitational field is called *potential energy* expressed per unit mass as $pe = gz$.

The compressibility effects in a fluid are represented by the *coefficient of compressibility κ* (also called the *bulk modulus of elasticity*) defined as

$$\kappa = -v\left(\frac{\partial P}{\partial v}\right)_T = \rho\left(\frac{\partial P}{\partial \rho}\right)_T \cong -\frac{\Delta P}{\Delta v / v}$$

The property that represents the variation of the density of a fluid with temperature at constant pressure is the *volume expansion coefficient* (or volume expansivity) β, defined as

$$\beta = \frac{1}{v}\left(\frac{\partial v}{\partial T}\right)_P = -\frac{1}{\rho}\left(\frac{\partial \rho}{\partial T}\right)_P \cong -\frac{\Delta \rho / \rho}{\Delta T}$$

The velocity at which an infinitesimally small pressure wave travels through a medium is the *speed of sound*. For an ideal gas it is expressed as

$$c = \sqrt{\left(\frac{\partial P}{\partial \rho}\right)_s} = \sqrt{kRT}$$

The *Mach number* is the ratio of the actual speed of the fluid to the speed of sound at the same state:

$$Ma = \frac{V}{c}$$

The flow is called *sonic* when Ma $= 1$, *subsonic* when Ma < 1, *supersonic* when Ma > 1, *hypersonic* when Ma $\gg 1$, and *transonic* when Ma $\cong 1$.

The *viscosity* of a fluid is a measure of its resistance to deformation. The tangential force per unit area is called *shear stress* and is expressed for simple shear flow between plates (one-dimensional flow) as

$$\tau = \mu \frac{du}{dy}$$

where μ is the coefficient of viscosity or the *dynamic* (or *absolute*) *viscosity* of the fluid, u is the velocity component in the flow direction, and y is the direction normal to the flow direction. Fluids that obey this linear relationship are called *Newtonian fluids*. The ratio of dynamic viscosity to density is called the *kinematic viscosity ν*.

The pulling effect on the liquid molecules at an interface caused by the attractive forces of molecules per unit length is called *surface tension σ_s*. The excess pressure ΔP inside a spherical droplet or soap bubble, respectively, is given by

$$\Delta P_{droplet} = P_i - P_o = \frac{2\sigma_s}{R} \quad \text{and} \quad \Delta P_{soap\,bubble} = P_i - P_o = \frac{4\sigma_s}{R}$$

where P_i and P_o are the pressures inside and outside the droplet or soap bubble. The rise or fall of a liquid in a small-diameter tube inserted into the liquid due to surface tension is called the *capillary effect*. The capillary rise or drop is given by

$$h = \frac{2\sigma_s}{\rho g R} \cos \phi$$

where ϕ is the *contact angle*. The capillary rise is inversely proportional to the radius of the tube; for water, it is negligible for tubes whose diameter is larger than about 1 cm.

Density and viscosity are two of the most fundamental properties of fluids, and they are used extensively in the chapters that follow. In Chap. 3, the effect of density on the variation of pressure in a fluid is considered, and the hydrostatic forces acting on surfaces are determined. In Chap. 8, the pressure drop caused by viscous effects during flow is calculated and used in the determination of the pumping power requirements. Viscosity is also used as a key property in the formulation and solutions of the equations of fluid motion in Chaps. 9 and 10.

APPLICATION SPOTLIGHT ■ Cavitation

Guest Authors: G. C. Lauchle and M. L. Billet,
Penn State University

(a)

(b)

FIGURE 2–41

(a) *Vaporous cavitation* occurs in water that has very little entrained gas, such as that found very deep in a body of water. Cavitation bubbles are formed when the speed of the body—in this case the bulbous bow region of a surface ship sonar dome—increases to the point where the local static pressure falls below the vapor pressure of the water. The cavitation bubbles are filled essentially with water vapor. This type of cavitation is very violent and noisy. (b) On the other hand, in shallow water, there is much more entrained gas in the water to act as cavitation nuclei. That's because of the proximity of the dome to the atmosphere at the free surface. The cavitation bubbles first appear at a slower speed, and hence at a higher local static pressure. They are predominantly filled with the gases that are entrained in the water, so this is known as *gaseous cavitation*.

Reprinted by permission of G. C. Lauchle and M. L. Billet, Penn State University.

Cavitation is the rupture of a liquid, or of a fluid–solid interface, caused by a reduction of the local static pressure produced by the dynamic action of the fluid in the interior and/or boundaries of a liquid system. The rupture is the formation of a visible bubble. Liquids, such as water, contain many microscopic voids that act as *cavitation nuclei*. Cavitation occurs when these nuclei grow to a significant, visible size. Although boiling is also the formation of voids in a liquid, we usually separate this phenomenon from cavitation because it is caused by an increase in temperature, rather than by a reduction in pressure. Cavitation can be used in beneficial ways, such as in ultrasonic cleaners, etchers, and cutters. But more often than not, cavitation is to be avoided in fluid flow applications because it spoils hydrodynamic performance, it causes extremely loud noise and high vibration levels, and it damages (erodes) the surfaces that support it. When cavitation bubbles enter regions of high pressure and collapse, the underwater shock waves sometimes create minute amounts of light. This phenomenon is called *sonoluminescence*.

Body cavitation is illustrated in Fig. 2–41. The body is a model of the underwater bulbous bow region of a surface ship. It is shaped this way because located within it is a *so*und *na*vigation and *r*anging (sonar) system that is spherical in shape. This part of the surface ship is thus called a *sonar dome*. As ship speeds get faster and faster some of these domes start to cavitate and the noise created by the cavitation renders the sonar system useless. Naval architects and fluid dynamicists attempt to design these domes so that they will not cavitate. Model-scale testing allows the engineer to see first hand whether a given design provides improved cavitation performance. Because such tests are conducted in water tunnels, the conditions of the test water should have sufficient nuclei to model those conditions in which the prototype operates. This assures that the effect of liquid tension (nuclei distribution) is minimized. Important variables are the gas content level (nuclei distribution) of the water, the temperature, and the hydrostatic pressure at which the body operates. Cavitation first appears—as either the speed V is increased, or as the submergence depth h is decreased—at the minimum pressure point $C_{p_{min}}$ of the body. Thus, good hydrodynamic design requires $2(P_\infty - P_v)/\rho V^2 > C_{p_{min}}$, where ρ is density, $P_\infty = \rho g h$ is the reference to static pressure, C_p is the pressure coefficient (Chap. 7), and P_v is the vapor pressure of water.

References

Lauchle, G. C., Billet, M. L., and Deutsch, S., "High-Reynolds Number Liquid Flow Measurements," in *Lecture Notes in Engineering,* Vol. 46, *Frontiers in Experimental Fluid Mechanics,* Springer-Verlag, Berlin, edited by M. Gad-el-Hak, Chap. 3, pp. 95–158, 1989.

Ross, D., *Mechanics of Underwater Noise,* Peninsula Publ., Los Altos, CA, 1987.

Barber, B. P., Hiller, R. A., Löfstedt, R., Putterman, S. J., and Weninger, K. R., "Defining the Unknowns of Sonoluminescence," *Physics Reports,* Vol. 281, pp. 65–143, 1997.

REFERENCES AND SUGGESTED READING

1. J. D. Anderson *Modern Compressible Flow with Historical Perspective*, 3rd ed. New York: McGraw-Hill, 2003.

2. E. C. Bingham. "An Investigation of the Laws of Plastic Flow," *U.S. Bureau of Standards Bulletin*, 13, pp. 309–353, 1916.

3. Y. A. Cengel and M. A. Boles. *Thermodynamics: An Engineering Approach*, 7th ed. New York: McGraw-Hill, 2011.

4. D. C. Giancoli. *Physics*, 6th ed. Upper Saddle River, NJ: Pearson, 2004.

5. Y. S. Touloukian, S. C. Saxena, and P. Hestermans. *Thermophysical Properties of Matter, The TPRC Data Series*, Vol. 11, *Viscosity*. New York: Plenum, 1975.

6. L. Trefethen. "Surface Tension in Fluid Mechanics." In *Illustrated Experiments in Fluid Mechanics*. Cambridge, MA: MIT Press, 1972.

7. *The U.S. Standard Atmosphere*. Washington, DC: U.S. Government Printing Office, 1976.

8. M. Van Dyke. *An Album of Fluid Motion*. Stanford, CA: Parabolic Press, 1982.

9. C. L. Yaws, X. Lin, and L. Bu. "Calculate Viscosities for 355 Compounds. An Equation Can Be Used to Calculate Liquid Viscosity as a Function of Temperature," *Chemical Engineering*, 101, no. 4, pp. 1110–1128, April 1994.

10. C. L. Yaws. *Handbook of Viscosity*. 3 Vols. Houston, TX: Gulf Publishing, 1994.

PROBLEMS*

Density and Specific Gravity

2–1C For a substance, what is the difference between mass and molar mass? How are these two related?

2–2C What is the difference between intensive and extensive properties?

2–3C What is specific gravity? How is it related to density?

2–4C The specific weight of a system is defined as the weight per unit volume (note that this definition violates the normal specific property-naming convention). Is the specific weight an extensive or intensive property?

2–5C What is the state postulate?

2–6C Under what conditions is the ideal-gas assumption suitable for real gases?

2–7C What is the difference between R and R_u? How are these two related?

2–8 A fluid that occupies a volume of 24 L weighs 225 N at a location where the gravitational acceleration is 9.80 m/s². Determine the mass of this fluid and its density.

2–9 A 100-L container is filled with 1 kg of air at a temperature of 27°C. What is the pressure in the container?

2–10E A mass of 1-lbm of argon is maintained at 200 psia and 100°F in a tank. What is the volume of the tank?

2–11E What is the specific volume of oxygen at 40 psia and 80°F?

2–12E The air in an automobile tire with a volume of 2.60 ft³ is at 90°F and 20 psig. Determine the amount of air that must be added to raise the pressure to the recommended value of 30 psig. Assume the atmospheric pressure to be 14.6 psia and the temperature and the volume to remain constant.   *Answer:* 0.128 lbm

2–13 The pressure in an automobile tire depends on the temperature of the air in the tire. When the air temperature is 25°C, the pressure gage reads 210 kPa. If the volume of the

FIGURE P2–13
Stockbyte/GettyImages

---

* Problems designated by a "C" are concept questions, and students are encouraged to answer them all. Problems designated by an "E" are in English units, and the SI users can ignore them. Problems with the 🔵 icon are solved using EES, and complete solutions together with parametric studies are included on the text website. Problems with the 🔲 icon are comprehensive in nature and are intended to be solved with an equation solver such as EES.

tire is 0.025 m³, determine the pressure rise in the tire when the air temperature in the tire rises to 50°C. Also, determine the amount of air that must be bled off to restore pressure to its original value at this temperature. Assume the atmospheric pressure to be 100 kPa.

2–14 A spherical balloon with a diameter of 9 m is filled with helium at 20°C and 200 kPa. Determine the mole number and the mass of the helium in the balloon. *Answers:* 31.3 kmol, 125 kg

2–15 Reconsider Prob. 2–14. Using EES (or other) software, investigate the effect of the balloon diameter on the mass of helium contained in the balloon for the pressures of (*a*) 100 kPa and (*b*) 200 kPa. Let the diameter vary from 5 m to 15 m. Plot the mass of helium against the diameter for both cases.

2–16 A cylindrical tank of methanol has a mass of 40 kg and a volume of 51 L. Determine the methanol's weight, density, and specific gravity. Take the gravitational acceleration to be 9.81 m/s². Also, estimate how much force is needed to accelerate this tank linearly at 0.25 m/s².

2–17 The density of saturated liquid refrigerant–134a for $-20°C \leq T \leq 100°C$ is given in Table A–4. Using this value develop an expression in the form $\rho = aT^2 + bT + c$ for the density of refrigerant–134a as a function of absolute temperature, and determine relative error for each data set.

2–18E A rigid tank contains 40 lbm of air at 20 psia and 70°F. More air is added to the tank until the pressure and temperature rise to 35 psia and 90°F, respectively. Determine the amount of air added to the tank. *Answer:* 27.4 lbm

2–19 The density of atmospheric air varies with elevation, decreasing with increasing altitude. (*a*) Using the data given in the table, obtain a relation for the variation of density with elevation, and calculate the density at an elevation of 7000 m. (*b*) Calculate the mass of the atmosphere using the correlation you obtained. Assume the earth to be a perfect sphere with a radius of 6377 km, and take the thickness of the atmosphere to be 25 km.

r, km	ρ,kg/m³
6377	1.225
6378	1.112
6379	1.007
6380	0.9093
6381	0.8194
6382	0.7364
6383	0.6601
6385	0.5258
6387	0.4135
6392	0.1948
6397	0.08891
6402	0.04008

Vapor Pressure and Cavitation

2–20C What is cavitation? What causes it?

2–21C Does water boil at higher temperatures at higher pressures? Explain.

2–22C If the pressure of a substance is increased during a boiling process, will the temperature also increase or will it remain constant? Why?

2–23C What is vapor pressure? How is it related to saturation pressure?

2–24E The analysis of a propeller that operates in water at 70°F shows that the pressure at the tips of the propeller drops to 0.1 psia at high speeds. Determine if there is a danger of cavitation for this propeller.

2–25 A pump is used to transport water to a higher reservoir. If the water temperature is 20°C, determine the lowest pressure that can exist in the pump without cavitation.

2–26 In a piping system, the water temperature remains under 30°C. Determine the minimum pressure allowed in the system to avoid cavitation.

2–27 The analysis of a propeller that operates in water at 20°C shows that the pressure at the tips of the propeller drops to 2 kPa at high speeds. Determine if there is a danger of cavitation for this propeller.

Energy and Specific Heats

2–28C What is flow energy? Do fluids at rest possess any flow energy?

2–29C How do the energies of a flowing fluid and a fluid at rest compare? Name the specific forms of energy associated with each case.

2–30C What is the difference between the macroscopic and microscopic forms of energy?

2–31C What is total energy? Identify the different forms of energy that constitute the total energy.

2–32C List the forms of energy that contribute to the internal energy of a system.

2–33C How are heat, internal energy, and thermal energy related to each other?

2–34C Using average specific heats, explain how internal energy changes of ideal gases and incompressible substances can be determined.

2–35C Using average specific heats, explain how enthalpy changes of ideal gases and incompressible substances can be determined.

2–36 Saturated water vapor at 150°C (enthalpy $h = 2745.9$ kJ/kg) flows in a pipe at 50 m/s at an elevation of $z = 10$ m. Determine the total energy of vapor in J/kg relative to the ground level.

Compressibility

2–37C What does the coefficient of compressibility of a fluid represent? How does it differ from isothermal compressibility?

2–38C What does the coefficient of volume expansion of a fluid represent? How does it differ from the coefficient of compressibility?

2–39C Can the coefficient of compressibility of a fluid be negative? How about the coefficient of volume expansion?

2–40 Water at 15°C and 1 atm pressure is heated to 100°C at constant pressure. Using coefficient of volume expansion data, determine the change in the density of water.
Answer: −38.7 kg/m³

2–41 It is observed that the density of an ideal gas increases by 10 percent when compressed isothermally from 10 atm to 11 atm. Determine the percent increase in density of the gas if it is compressed isothermally from 1000 atm to 1001 atm.

2–42 Using the definition of the coefficient of volume expansion and the expression $\beta_{ideal\ gas} = 1/T$, show that the percent increase in the specific volume of an ideal gas during isobaric expansion is equal to the percent increase in absolute temperature.

2–43 Water at 1 atm pressure is compressed to 400 atm pressure isothermally. Determine the increase in the density of water. Take the isothermal compressibility of water to be 4.80×10^{-5} atm⁻¹.

2–44 The volume of an ideal gas is to be reduced by half by compressing it isothermally. Determine the required change in pressure.

2–45 Saturated refrigerant-134a liquid at 10°C is cooled to 0°C at constant pressure. Using coefficient of volume expansion data, determine the change in the density of the refrigerant.

2–46 A water tank is completely filled with liquid water at 20°C. The tank material is such that it can withstand tension caused by a volume expansion of 0.8 percent. Determine the maximum temperature rise allowed without jeopardizing safety. For simplicity, assume β = constant = β at 40°C.

2–47 Repeat Prob. 2–46 for a volume expansion of 1.5 percent for water.

2–48 The density of seawater at a free surface where the pressure is 98 kPa is approximately 1030 kg/m³. Taking the bulk modulus of elasticity of seawater to be 2.34×10^9 N/m² and expressing variation of pressure with depth z as $dP = \rho g\ dz$ determine the density and pressure at a depth of 2500 m. Disregard the effect of temperature.

2–49E Taking the coefficient of compressibility of water to be 7×10^5 psia, determine the pressure increase required to reduce the volume of water by (a) 1 percent and (b) 2 percent.

2–50E Ignoring any losses, estimate how much energy (in units of Btu) is required to raise the temperature of water in a 75-gallon hot-water tank from 60°F to 110°F.

2–51 Prove that the coefficient of volume expansion for an ideal gas is $\beta_{ideal\ gas} = 1/T$.

2–52 The ideal gas equation of state is very simple, but its range of applicability is limited. A more accurate but complicated equation is the Van der Waals equation of state given by

$$P = \frac{RT}{v - b} - \frac{a}{v^2}$$

where a and b are constants depending on critical pressure and temperatures of the gas. Predict the coefficient of compressibility of nitrogen gas at $T = 175$ K and $v = 0.00375$ m³/kg, assuming the nitrogen to obey the Van der Waals equation of state. Compare your result with the ideal gas value. Take $a = 0.175$ m⁶·kPa/kg² and $b = 0.00138$ m³/kg for the given conditions. The experimentally measured pressure of nitrogen is 10,000 kPa.

2–53 A frictionless piston-cylinder device contains 10 kg of water at 20°C at atmospheric pressure. An external force F is then applied on the piston until the pressure inside the cylinder increases to 100 atm. Assuming the coefficient of compressibility of water remains unchanged during the compression; estimate the energy needed to compress the water isothermally. *Answer:* 29.4 J

FIGURE P2–53

2–54 Reconsider Prob. 2–53. Assuming a linear pressure increase during the compression, estimate the energy needed to compress the water isothermally.

Speed of Sound

2–55C What is sound? How is it generated? How does it travel? Can sound waves travel in a vacuum?

2–56C In which medium does a sound wave travel faster: in cool air or in warm air?

2–57C In which medium will sound travel fastest for a given temperature: air, helium, or argon?

2–58C   In which medium does a sound wave travel faster: in air at 20°C and 1 atm or in air at 20°C and 5 atm?

2–59C   Does the Mach number of a gas flowing at a constant velocity remain constant? Explain.

2–60C   Is it realistic to approximate that the propagation of sound waves is an isentropic process? Explain.

2–61C   Is the sonic velocity in a specified medium a fixed quantity, or does it change as the properties of the medium change? Explain.

2–62   The Airbus A-340 passenger plane has a maximum takeoff weight of about 260,000 kg, a length of 64 m, a wing span of 60 m, a maximum cruising speed of 945 km/h, a seating capacity of 271 passengers, a maximum cruising altitude of 14,000 m, and a maximum range of 12,000 km. The air temperature at the crusing altitude is about −60°C. Determine the Mach number of this plane for the stated limiting conditions.

2–63   Carbon dioxide enters an adiabatic nozzle at 1200 K with a velocity of 50 m/s and leaves at 400 K. Assuming constant specific heats at room temperature, determine the Mach number (a) at the inlet and (b) at the exit of the nozzle. Assess the accuracy of the constant specific heat approximation.
Answers: (a) 0.0925, (b) 3.73

2–64   Nitrogen enters a steady-flow heat exchanger at 150 kPa, 10°C, and 100 m/s, and it receives heat in the amount of 120 kJ/kg as it flows through it. Nitrogen leaves the heat exchanger at 100 kPa with a velocity of 200 m/s. Determine the Mach number of the nitrogen at the inlet and the exit of the heat exchanger.

2–65   Assuming ideal gas behavior, determine the speed of sound in refrigerant-134a at 0.9 MPa and 60°C.

2–66   Determine the speed of sound in air at (a) 300 K and (b) 800 K. Also determine the Mach number of an aircraft moving in air at a velocity of 330 m/s for both cases.

2–67E   Steam flows through a device with a pressure of 120 psia, a temperature of 700°F, and a velocity of 900 ft/s. Determine the Mach number of the steam at this state by assuming ideal-gas behavior with $k = 1.3$.   *Answer:* 0.441

2–68E   [EES] Reconsider Prob. 2–67E. Using EES (or other) software, compare the Mach number of steam flow over the temperature range 350 to 700°F. Plot the Mach number as a function of temperature.

2–69E   Air expands isentropically from 170 psia and 200°F to 60 psia. Calculate the ratio of the initial to final speed of sound.   *Answer:* 1.16

2–70   Air expands isentropically from 2.2 MPa and 77°C to 0.4 MPa. Calculate the ratio of the initial to the final speed of sound.   *Answer:* 1.28

2–71   Repeat Prob. 2–70 for helium gas.

2–72   The isentropic process for an ideal gas is expressed as $Pv^k = $ constant. Using this process equation and the definition of the speed of sound (Eq. 2–24), obtain the expression for the speed of sound for an ideal gas (Eq. 2–26).

Viscosity

2–73C   What is viscosity? What is the cause of it in liquids and in gases? Do liquids or gases have higher dynamic viscosities?

2–74C   What is a Newtonian fluid? Is water a Newtonian fluid?

2–75C   How does the kinematic viscosity of (a) liquids and (b) gases vary with temperature?

2–76C   How does the dynamic viscosity of (a) liquids and (b) gases vary with temperature?

2–77C   Consider two identical small glass balls dropped into two identical containers, one filled with water and the other with oil. Which ball will reach the bottom of the container first? Why?

2–78E   The viscosity of a fluid is to be measured by a viscometer constructed of two 5-ft-long concentric cylinders. The inner diameter of the outer cylinder is 6 in, and the gap between the two cylinders is 0.035 in. The outer cylinder is rotated at 250 rpm, and the torque is measured to be 1.2 lbf·ft. Determine the viscosity of the fluid.   *Answer:* 0.000272 lbf·s/ft²

2–79   A 50-cm × 30-cm × 20-cm block weighing 150 N is to be moved at a constant velocity of 0.80 m/s on an inclined surface with a friction coefficient of 0.27. (a) Determine the force F that needs to be applied in the horizontal direction. (b) If a 0.40-mm-thick oil film with a dynamic viscosity of 0.012 Pa·s is applied between the block and inclined surface, determine the percent reduction in the required force.

FIGURE P2–79

2–80   Consider the flow of a fluid with viscosity μ through a circular pipe. The velocity profile in the pipe is given as $u(r) = u_{max}(1 - r^n/R^n)$, where u_{max} is the maximum flow velocity, which occurs at the centerline; r is the radial distance from the centerline; and $u(r)$ is the flow velocity at any position r. Develop a relation for the drag force exerted on the pipe wall by the fluid in the flow direction per unit length of the pipe.

$$u(r) = u_{max}(1 - r^n/R^n)$$

FIGURE P2–80

2–81 A thin 30-cm × 30-cm flat plate is pulled at 3 m/s horizontally through a 3.6-mm-thick oil layer sandwiched between two plates, one stationary and the other moving at a constant velocity of 0.3 m/s, as shown in Fig. P2–81. The dynamic viscosity of the oil is 0.027 Pa·s. Assuming the velocity in each oil layer to vary linearly, (a) plot the velocity profile and find the location where the oil velocity is zero and (b) determine the force that needs to be applied on the plate to maintain this motion.

FIGURE P2–81

2–82 A rotating viscometer consists of two concentric cylinders – an inner cylinder of radius R_i rotating at angular velocity (rotation rate) ω_i, and a stationary outer cylinder of inside radius R_o. In the tiny gap between the two cylinders is the fluid of viscosity μ. The length of the cylinders (into the page in Fig. P2–82) is L. L is large such that end effects

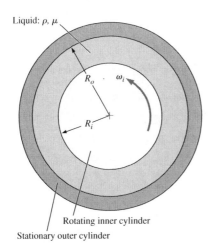

FIGURE P2-82

are negligible (we can treat this as a two-dimensional problem). Torque (T) is required to rotate the inner cylinder at constant speed. (a) Showing all of your work and algebra, generate an approximate expression for T as a function of the other variables. (b) Explain why your solution is only an *approximation*. In particular, do you expect the velocity profile in the gap to remain linear as the gap becomes larger and larger (i.e., if the outer radius R_o were to increase, all else staying the same)?

2–83 The clutch system shown in Fig. P2–83 is used to transmit torque through a 2-mm-thick oil film with $\mu = 0.38$ N·s/m² between two identical 30-cm-diameter disks. When the driving shaft rotates at a speed of 1450 rpm, the driven shaft is observed to rotate at 1398 rpm. Assuming a linear velocity profile for the oil film, determine the transmitted torque.

FIGURE P2–83

2–84 Reconsider Prob. 2–83. Using EES (or other) software, investigate the effect of oil film thickness on the torque transmitted. Let the film thickness vary from 0.1 mm to 10 mm. Plot your results, and state your conclusions.

2–85 The dynamic viscosity of carbon dioxide at 50°C and 200°C are 1.612×10^{-5} Pa·s and 2.276×10^{-5} Pa·s, respectively. Determine the constants a and b of Sutherland correlation for carbon dioxide at atmospheric pressure. Then predict the viscosity of carbon dioxide at 100°C and compare your result against the value given in Table A-10.

2–86 One of the widely used correlations to describe the variation of the viscosity of gases is the power-law equation given by $\mu/\mu_0 = (T/T_0)^n$, where μ_0 and T_0 are the reference viscosity and temperature, respectively. Using the power and Sutherland laws, examine the variation of the air viscosity for the temperature range 100°C (373 K) to 1000°C (1273 K). Plot your results to compare with values listed in Table A-9. Take the reference temperature as 0°C and $n = 0.666$ for the atmospheric air.

2–87 For flow over a plate, the variation of velocity with vertical distance y from the plate is given as $u(y) = ay - by^2$ where a and b are constants. Obtain a relation for the wall shear stress in terms of a, b, and μ.

2–88   In regions far from the entrance, fluid flow through a circular pipe is one dimensional, and the velocity profile for laminar flow is given by $u(r) = u_{max}(1 - r^2/R^2)$, where R is the radius of the pipe, r is the radial distance from the center of the pipe, and u_{max} is the maximum flow velocity, which occurs at the center. Obtain (a) a relation for the drag force applied by the fluid on a section of the pipe of length L and (b) the value of the drag force for water flow at 20°C with $R = 0.08$ m, $L = 30$ m, $u_{max} = 3$ m/s, and $\mu = 0.0010$ kg/m·s.

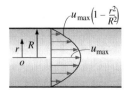

FIGURE P2–88

2–89   Repeat Prob. 2–88 for $u_{max} = 7$ m/s.   *Answer:* (b) 2.64 N

2–90   A frustum-shaped body is rotating at a constant angular speed of 200 rad/s in a container filled with SAE 10W oil at 20°C ($\mu = 0.100$ Pa·s), as shown in Fig. P2–90. If the thickness of the oil film on all sides is 1.2 mm, determine the power required to maintain this motion. Also determine the reduction in the required power input when the oil temperature rises to 80°C ($\mu = 0.0078$ Pa·s).

FIGURE P2–90

2–91   A rotating viscometer consists of two concentric cylinders—a stationary inner cyliner of radius R_i and an outer cylinder of inside radius R_o rotating at angular velocity (rotation rate) ω_o. In the tiny gap between the two cylinders is the fluid whose viscosity (μ) is to be measured. The length

of the cylinders (into the page in Fig. P2-91) is L. L is large such that end effects are negligible (we can treat this as a two-dimensional problem). Torque (T) is required to rotate the inner cylinder at constant speed. Showing all your work and algebra, generate an approximate expression of T as a function of the other variables.

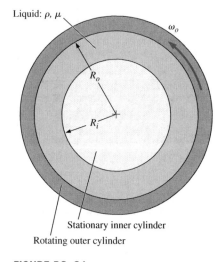

FIGURE P2–91

2–92   A large plate is pulled at a constant speed of $U = 4$ m/s over a fixed plate on 5-mm-thick engine oil film at 20°C. Assuming a half-parabolic velocity profile in the oil film, as sketched, determine the shear stress developed on the upper plate and its direction. What would happen if a linear velocity profile were assumed?

FIGURE P2–92

2–93   A cylinder of mass m slides down from rest in a vertical tube whose inner surface is covered by a viscous oil of film thickness h. If the diameter and height of the cylinder are D and L, respectively, derive an expression for the velocity of the cylinder as a function of time, t. Discuss what will happen as $t \to \infty$. Can this device serve as a viscometer?

FIGURE P2–93

2–94 A thin plate moves between two parallel, horizontal, stationary flat surfaces at a constant velocity of 5 m/s. The two stationary surfaces are spaced 4 cm apart, and the medium between them is filled with oil whose viscosity is 0.9 N·s/m². The part of the plate immersed in oil at any given time is 2-m long and 0.5-m wide. If the plate moves through the mid-plane between the surfaces, determine the force required to maintain this motion. What would your response be if the plate was 1 cm from the bottom surface (h_2) and 3 cm from the top surface (h_1)?

FIGURE P2–94

2–95 Reconsider Prob. 2–94. If the viscosity of the oil above the moving plate is 4 times that of the oil below the plate, determine the distance of the plate from the bottom surface (h_2) that will minimize the force needed to pull the plate between the two oils at constant velocity.

Surface Tension and Capillary Effect

2–96C What is surface tension? What is its cause? Why is the surface tension also called surface energy?

2–97C A small-diameter tube is inserted into a liquid whose contact angle is 110°. Will the level of liquid in the tube be higher or lower than the level of the rest of the liquid? Explain.

2–98C What is the capillary effect? What is its cause? How is it affected by the contact angle?

2–99C Consider a soap bubble. Is the pressure inside the bubble higher or lower than the pressure outside?

2–100C Is the capillary rise greater in small- or large-diameter tubes?

2–101 Consider a 0.15-mm diameter air bubble in a liquid. Determine the pressure difference between the inside and outside of the air bubble if the surface tension at the air-liquid interface is (a) 0.080 N/m and (b) 0.12 N/m.

2–102E A 2.4-in-diameter soap bubble is to be enlarged by blowing air into it. Taking the surface tension of soap solution to be 0.0027 lbf/ft, determine the work input required to inflate the bubble to a diameter of 2.7 in.

2–103 A 1.2-mm-diameter tube is inserted into an unknown liquid whose density is 960 kg/m³, and it is observed that the liquid rises 5 mm in the tube, making a contact angle of 15°. Determine the surface tension of the liquid.

2–104 Determine the gage pressure inside a soap bubble of diameter (a) 0.2 cm and (b) 5 cm at 20°C.

2–105E A 0.03-in-diameter glass tube is inserted into kerosene at 68°F. The contact angle of kerosene with a glass surface is 26°. Determine the capillary rise of kerosene in the tube. *Answer:* 0.65 in

FIGURE P2–105E

2–106 The surface tension of a liquid is to be measured using a liquid film suspended on a U-shaped wire frame with an 8-cm-long movable side. If the force needed to move the wire is 0.024 N, determine the surface tension of this liquid in air.

2–107 A capillary tube of 1.2 mm diameter is immersed vertically in water exposed to the atmosphere. Determine how high water will rise in the tube. Take the contact angle at the inner wall of the tube to be 6° and the surface tension to be 1.00 N/m. *Answer:* 0.338 m

2–108 A capillary tube is immersed vertically in a water container. Knowing that water starts to evaporate when the pressure drops below 2 kPa, determine the maximum capillary rise and tube diameter for this maximum-rise case. Take the contact angle at the inner wall of the tube to be 6° and the surface tension to be 1.00 N/m.

2–109 Contrary to what you might expect, a solid steel ball can float on water due to the surface tension effect. Determine the maximum diameter of a steel ball that would float on water at 20°C. What would your answer be for an aluminum ball? Take the densities of steel and aluminum balls to be 7800 kg/m³ and 2700 kg/m³, respectively.

2–110 Nutrients dissolved in water are carried to upper parts of plants by tiny tubes partly because of the capillary effect. Determine how high the water solution will rise in a tree in a 0.0026-mm-diameter tube as a result of the capillary effect. Treat the solution as water at 20°C with a contact angle of 15°. *Answer:* 11.1 m

Water solution — 0.0026 mm

FIGURE P2–110

Review Problems

2–111 Derive a relation for the capillary rise of a liquid between two large parallel plates a distance t apart inserted into the liquid vertically. Take the contact angle to be ϕ.

2–112 Consider a 55-cm-long journal bearing that is lubricated with oil whose viscosity is 0.1 kg/m·s at 20°C at the beginning of operation and 0.008 kg/m·s at the anticipated steady operating temperature of 80°C. The diameter of the shaft is 8 cm, and the average gap between the shaft and the journal is 0.08 cm. Determine the torque needed to overcome the bearing friction initially and during steady operation when the shaft is rotated at 1500 rpm.

2–113 The diameter of one arm of a U-tube is 5 mm while the other arm is large. If the U-tube contains some water, and both surfaces are exposed to atmospheric pressure, determine the difference between the water levels in the two arms.

2–114 The combustion in a gasoline engine may be approximated by a constant volume heat addition process, and the contents of the combustion chamber both before and after combustion as air. The conditions are 1.80 MPa and 450°C before the combustion and 1300°C after it. Determine the pressure at the end of the combustion process. *Answer:* 3916 kPa

Combustion chamber
1.80 MPa
450°C

FIGURE P2–114

2–115 A rigid tank contains an ideal gas at 300 kPa and 600 K. Half of the gas is withdrawn from the tank and the gas is at 100 kPa at the end of the process. Determine (*a*) the final temperature of the gas and (*b*) the final pressure if no mass were withdrawn from the tank and the same final temperature were reached at the end of the process.

2–116 The absolute pressure of an automobile tire is measured to be 320 kPa before a trip and 335 kPa after the trip. Assuming the volume of the tire remains constant at 0.022 m³, determine the percent increase in the absolute temperature of the air in the tire.

2–117E The pressure on the suction side of pumps is typically low, and the surfaces on that side of the pump are susceptible to cavitation, especially at high fluid temperatures. If the minimum pressure on the suction side of a water pump is 0.95 psia absolute, determine the maximum water temperature to avoid the danger of cavitation.

2–118 The composition of a liquid with suspended solid particles is generally characterized by the fraction of solid particles either by weight or mass, $C_{s,\,mass} = m_s/m_m$ or by volume, $C_{s,\,vol} = V_s/V_m$ where m is mass and V is volume. The subscripts s and m indicate solid and mixture, respectively. Develop an expression for the specific gravity of a water-based suspension in terms of $C_{s,\,mass}$ and $C_{s,\,vol}$.

2–119 The specific gravities of solids and carrier fluids of a slurry are usually known, but the specific gravity of the slurry depends on the concentration of the solid particles. Show that the specific gravity of a water-based slurry can be expressed in terms of the specific gravity of the solid SG_s and the mass concentration of the suspended solid particles $C_{s,\,mass}$ as

$$SG_m = \frac{1}{1 + C_{s,\,mass}(1/SG_s - 1)}$$

2–120 A 10-m³ tank contains nitrogen at 25°C and 800 kPa. Some nitrogen is allowed to escape until the pressure in the tank drops to 600 kPa. If the temperature at this point is 20°C, determine the amount of nitrogen that has escaped. *Answer:* 21.5 kg

2–121 A closed tank is partially filled with water at 60°C. If the air above the water is completely evacuated, determine the absolute pressure in the evacuated space. Assume the temperature to remain constant.

2–122 The variation of the dynamic viscosity of water with absolute temperature is given as

T, K	μ, Pa·s
273.15	1.787×10^{-3}
278.15	1.519×10^{-3}
283.15	1.307×10^{-3}
293.15	1.002×10^{-3}
303.15	7.975×10^{-4}
313.15	6.529×10^{-4}
333.15	4.665×10^{-4}
353.15	3.547×10^{-4}
373.15	2.828×10^{-4}

Using these tabulated data, develop a relation for viscosity in the form of $\mu = \mu(T) = A + BT + CT^2 + DT^3 + ET^4$. Using the relation developed, predict the dynamic viscosity of water at 50°C at which the reported value is 5.468×10^{-4} Pa·s. Compare your result with the results of Andrade's equation, which is given in the form of $\mu = D \cdot e^{B/T}$, where D and B are constants whose values are to be determined using the viscosity data given.

2–123 A newly produced pipe with diameter of 2 m and length 15 m is to be tested at 10 MPa using water at 15°C. After sealing both ends, the pipe is first filled with water and then the pressure is increased by pumping additional water into the test pipe until the test pressure is reached. Assuming no deformation in the pipe, determine how much additional water needs to be pumped into the pipe. Take the coefficient of compressibility to be 2.10×10^9 Pa. *Answer:* 224 kg

2–124 Although liquids, in general, are hard to compress, the compressibility effect (variation in the density) may become unavoidable at the great depths in the oceans due to enormous pressure increase. At a certain depth the pressure is reported to be 100 MPa and the average coefficient of compressibility is about 2350 MPa.

(*a*) Taking the liquid density at the free surface to be $\rho_0 = 1030$ kg/m³, obtain an analytical relation between density and pressure, and determine the density at the specified pressure. *Answer:* 1074 kg/m³

(*b*) Use Eq. 2–13 to estimate the density for the specified pressure and compare your result with that of part (*a*).

2–125 Consider laminar flow of a Newtonian fluid of viscosity μ between two parallel plates. The flow is one-dimensional, and the velocity profile is given as $u(y) = 4u_{max}[y/h - (y/h)^2]$, where y is the vertical coordinate from the bottom surface, h is the distance between the two plates,

and u_{max} is the maximum flow velocity that occurs at midplane. Develop a relation for the drag force exerted on both plates by the fluid in the flow direction per unit area of the plates.

$u(y) = 4u_{max}[y/h - (y/h)^2]$

FIGURE P2–125

2–126 Two immiscible Newtonian liquids flow steadily between two large parallel plates under the influence of an applied pressure gradient. The lower plate is fixed while the upper one is pulled with a constant velocity of $U = 10$ m/s. The thickness, h, of each layer of fluid is 0.5 m. The velocity profile for each layer is given by

$$V_1 = 6 + ay - 3y^2, \qquad -0.5 \leq y \leq 0$$
$$V_2 = b + cy - 9y^2, \qquad 0 \leq y \leq -0.5$$

where a, b, and c are constants.

(*a*) Determine the values of constants a, b, and c.

(*b*) Develop an expression for the viscosity ratio, e.g., $\mu_1/\mu_2 =$?

(*c*) Determine the forces and their directions exerted by the liquids on both plates if $\mu_1 = 10^{-3}$ Pa·s and each plate has a surface area of 4 m².

FIGURE P2–126

2–127 A shaft with a diameter of $D = 80$ mm and a length of $L = 400$ mm, shown in Fig. P2–127 is pulled with a constant velocity of $U = 5$ m/s through a bearing with variable diameter. The clearance between shaft and bearing, which varies from $h_1 = 1.2$ mm to $h_2 = 0.4$ mm, is filled with a Newtonian lubricant whose dynamic viscosity is 0.10 Pa·s. Determine the force required to maintain the axial movement of the shaft. *Answer:* 69 N

FIGURE P2–127

2–128 Reconsider Prob. 2–127. The shaft now rotates with a constant angular speed of $n = 1450$ rpm in a bearing with variable diameter. The clearance between shaft and bearing, which varies from $h_1 = 1.2$ mm to $h_2 = 0.4$ mm, is filled with a Newtonian lubricant whose dynamic viscosity is 0.1 Pa·s. Determine the torque required to maintain the motion.

2–129 A 10-cm-diameter cylindrical shaft rotates inside a 40-cm-long 10.3-cm diameter bearing. The space between the shaft and the bearing is completely filled with oil whose viscosity at anticipated operating temperature is 0.300 N·s/m². Determine the power required to overcome friction when the shaft rotates at a speed of (a) 600 rpm and (b) 1200 rpm.

2–130 Some rocks or bricks contain small air pockets in them and have a spongy structure. Assuming the air spaces form columns of an average diameter of 0.006 mm, determine how high water can rise in such a material. Take the surface tension of the air–water interface in that material to be 0.085 N/m.

Fundamentals of Engineering (FE) Exam Problems

2–131 The specific gravity of a fluid is specified to be 0.82. The specific volume of this fluid is
(a) 0.00100 m³/kg   (b) 0.00122 m³/kg   (c) 0.0082 m³/kg
(d) 82 m³/kg   (e) 820 m³/kg

2–132 The specific gravity of mercury is 13.6. The specific weight of mercury is
(a) 1.36 kN/m³ (b) 9.81 kN/m³ (c) 106 kN/m³ (d) 133 kN/m³
(e) 13,600 kN/m³

2–133 An ideal gas flows in a pipe at 20°C. The density of the gas is 1.9 kg/m³ and its molar mass is 44 kg/kmol. The pressure of the gas is
(a) 7 kPa (b) 72 kPa (c) 105 kPa (d) 460 kPa (e) 4630 kPa

2–134 A gas mixture consists of 3 kmol oxygen, 2 kmol nitrogen, and 0.5 kmol water vapor. The total pressure of the gas mixture is 100 kPa. The partial pressure of water vapor in this gas mixture is
(a) 5 kPa (b) 9.1 kPa (c) 10 kPa (d) 22.7 kPa (e) 100 kPa

2–135 Liquid water vaporizes into water vapor as it flows in the piping of a boiler. If the temperature of water in the pipe is 180°C, the vapor pressure of the water in the pipe is
(a) 1002 kPa (b) 180 kPa (c) 101.3 kPa (d) 18 kPa (e) 100 kPa

2–136 In a water distribution system, the pressure of water can be as low as 1.4 psia. The maximum temperature of water allowed in the piping to avoid cavitation is
(a) 50°F   (b) 77°F   (c) 100°F   (d) 113°F   (e) 140°F

2–137 The thermal energy of a system refers to
(a) Sensible energy (b) Latent energy
(c) Sensible + latent energies (d) Enthalpy (e) Internal energy

2–138 The difference between the energies of a flowing and stationary fluid per unit mass of the fluid is equal to
(a) Enthalpy   (b) Flow energy   (c) Sensible energy
(d) Kinetic energy   (e) Internal energy

2–139 The pressure of water is increased from 100 kPa to 1200 kPa by a pump. The temperature of water also increases by 0.15°C. The density of water is 1 kg/L and its specific heat is $c_p = 4.18$ kJ/kg·°C. The enthalpy change of the water during this process is
(a) 1100 kJ/kg (b) 0.63 kJ/kg (c) 1.1 kJ/kg (d) 1.73 kJ/kg
(e) 4.2 kJ/kg

2–140 The coefficient of compressibility of a truly incompressible substance is
(a) 0   (b) 0.5   (c) 1   (d) 100   (e) Infinity

2–141 The pressure of water at atmospheric pressure must be raised to 210 atm to compress it by 1 percent. Then, the coefficient of compressibility value of water is
(a) 209 atm   (b) 20,900 atm   (c) 21 atm   (d) 0.21 atm
(e) 210,000 atm

2–142 When a liquid in a piping network encounters an abrupt flow restriction (such as a closing valve), it is locally compressed. The resulting acoustic waves that are produced strike the pipe surfaces, bends, and valves as they propagate and reflect along the pipe, causing the pipe to vibrate and produce a familiar sound. This is known as
(a) Condensation   (b) Cavitation   (c) Water hammer
(d) Compression   (e) Water arrest

2–143 The density of a fluid decreases by 5 percent at constant pressure when its temperature increases by 10°C. The coefficient of volume expansion of this fluid is
(a) 0.01 K⁻¹ (b) 0.005 K⁻¹ (c) 0.1 K⁻¹ (d) 0.5 K⁻¹ (e) 5 K⁻¹

2–144 Water is compressed from 100 kPa to 5000 kPa at constant temperature. The initial density of water is 1000 kg/m³ and the isothermal compressibility of water is $\alpha = 4.8 \times 10^{-5}$ atm⁻¹. The final density of the water is
(a) 1000 kg/m³   (b) 1001.1 kg/m³   (c) 1002.3 kg/m³
(d) 1003.5 kg/m³   (e) 997.4 kg/m³

2–145 The speed of a spacecraft is given to be 1250 km/h in atmospheric air at −40°C. The Mach number of this flow is
(a) 35 .9   (b) 0.85   (c) 1.0   (d) 1.13   (e) 2.74

2–146 The dynamic viscosity of air at 20°C and 200 kPa is 1.83×10^{-5} kg/m·s. The kinematic viscosity of air at this state is
(a) 0.525×10^{-5} m²/s (b) 0.77×10^{-5} m²/s
(c) 1.47×10^{-5} m²/s   (d) 1.83×10^{-5} m²/s
(e) 0.380×10^{-5} m²/s

2–147 A viscometer constructed of two 30-cm-long concentric cylinders is used to measure the viscosity of a fluid. The outer diameter of the inner cylinder is 9 cm, and the gap between the two cylinders is 0.18 cm. The inner cylinder is rotated at 250 rpm, and the torque is measured to be 1.4 N·m. The viscosity of the fluid is
(a) 0.0084 N·s/m² (b) 0.017 N·s/m² (c) 0.062 N·s/m²
(d) 0.0049 N·s/m² (e) 0.56 N·s/m²

2–148 Which one is *not* a surface tension or surface energy (per unit area) unit?
(a) lbf/ft (b) N·m/m² (c) lbf/ft² (d) J/m² (e) Btu/ft²

2–149 The surface tension of soap water at 20°C is $\sigma_s = 0.025$ N/m. The gage pressure inside a soap bubble of diameter 2 cm at 20°C is
(a) 10 Pa (b) 5 Pa (c) 20 Pa (d) 40 Pa (e) 0.5 Pa

2–150 A 0.4-mm-diameter glass tube is inserted into water at 20°C in a cup. The surface tension of water at 20°C is $\sigma_s = 0.073$ N/m. The contact angle can be taken as zero degrees. The capillary rise of water in the tube is
(a) 2.9 cm (b) 7.4 cm (c) 5.1 cm
(d) 9.3 cm (e) 14.0 cm

Design and Essay Problems

2–151 Design an experiment to measure the viscosity of liquids using a vertical funnel with a cylindrical reservoir of height h and a narrow flow section of diameter D and length L. Making appropriate assumptions, obtain a relation for viscosity in terms of easily measurable quantities such as density and volume flow rate.

2–152 Write an essay on the rise of the fluid to the top of trees by capillary and other effects.

2–153 Write an essay on the oils used in car engines in different seasons and their viscosities.

2–154 Consider the flow of water through a clear tube. It is sometimes possible to observe cavitation in the throat created by pinching off the tube to a very small diameter as sketched. We assume incompressible flow with negligible gravitational effects and negligible irreversibilities. You will learn later (Chap. 5) that as the duct cross-sectional area decreases, the velocity increases and the pressure decreases according to

$$V_1 A_1 = V_2 A_2 \quad \text{and} \quad P_1 + \rho\frac{V_1^2}{2} = P_2 + \rho\frac{V_2^2}{2}$$

respectively, where V_1 and V_2 are the average velocities through cross-sectional areas A_1 and A_2. Thus, both the maximum

velocity and minimum pressure occur at the throat. (a) If the water is at 20°C, the inlet pressure is 20.803 kPa, and the throat diameter is one-twentieth of the inlet diameter, estimate the minimum average inlet velocity at which cavitation is likely to occur in the throat. (b) Repeat at a water temperature of 50°C. Explain why the required inlet velocity is higher or lower than that of part (a).

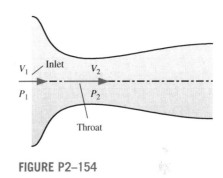

FIGURE P2–154

2–155 Even though steel is about 7 to 8 times denser than water, a steel paper clip or razor blade can be made to float on water! Explain and discuss. Predict what would happen if you mix some soap with the water.

FIGURE P2–155
Photo by John M. Cimbala.

PRESSURE AND FLUID STATICS

This chapter deals with forces applied by fluids at rest or in rigid-body motion. The fluid property responsible for those forces is *pressure,* which is a normal force exerted by a fluid per unit area. We start this chapter with a detailed discussion of pressure, including *absolute* and *gage pressures,* the pressure at a *point,* the *variation of pressure with depth* in a gravitational field, the *barometer,* the *manometer,* and other pressure measurement devices. This is followed by a discussion of the *hydrostatic forces* applied on submerged bodies with plane or curved surfaces. We then consider the *buoyant force* applied by fluids on submerged or floating bodies, and discuss the *stability* of such bodies. Finally, we apply Newton's second law of motion to a body of fluid in motion that acts as a rigid body and analyze the variation of pressure in fluids that undergo linear acceleration and in rotating containers. This chapter makes extensive use of force balances for bodies in static equilibrium, and it would be helpful if the relevant topics from statics are first reviewed.

OBJECTIVES

When you finish reading this chapter, you should be able to:

■ Determine the variation of pressure in a fluid at rest

■ Calculate pressure using various kinds of manometers

■ Calculate the forces and moments exerted by a fluid at rest on plane or curved submerged surfaces

■ Analyze the stability of floating and submerged bodies

■ Analyze the rigid-body motion of fluids in containers during linear acceleration or rotation

John Ninomiya flying a cluster of 72 helium-filled balloons over Temecula, California in April of 2003. The helium balloons displace approximately 230 m³ of air, providing the necessary buoyant force. Don't try this at home!

Photograph by Susan Dawson. Used by permission.

3–1 ▪ PRESSURE

Pressure is defined as *a normal force exerted by a fluid per unit area*. We speak of pressure only when we deal with a gas or a liquid. The counterpart of pressure in solids is *normal stress*. Since pressure is defined as force per unit area, it has the unit of newtons per square meter (N/m^2), which is called a **pascal** (Pa). That is,

$$1 Pa = 1 \ N/m^2$$

The pressure unit pascal is too small for most pressures encountered in practice. Therefore, its multiples *kilopascal* ($1 \ kPa = 10^3 \ Pa$) and *megapascal* ($1 \ MPa = 10^6 \ Pa$) are commonly used. Three other pressure units commonly used in practice, especially in Europe, are *bar, standard atmosphere,* and *kilogram-force per square centimeter:*

$$1 \ bar = 10^5 \ Pa = 0.1 \ MPa = 100 \ kPa$$

$$1 \ atm = 101{,}325 \ Pa = 101.325 \ kPa = 1.01325 \ bars$$

$$1 \ kgf/cm^2 = 9.807 \ N/cm^2 = 9.807 \times 10^4 \ N/m^2 = 9.807 \times 10^4 \ Pa$$

$$= 0.9807 \ bar$$

$$= 0.9679 \ atm$$

Note the pressure units bar, atm, and kgf/cm^2 are almost equivalent to each other. In the English system, the pressure unit is *pound-force per square inch* (lbf/in^2, or psi), and $1 \ atm = 14.696$ psi. The pressure units kgf/cm^2 and lbf/in^2 are also denoted by kg/cm^2 and lb/in^2, respectively, and they are commonly used in tire gages. It can be shown that $1 \ kgf/cm^2 = 14.223$ psi.

Pressure is also used on solid surfaces as synonymous to *normal stress,* which is the force acting perpendicular to the surface per unit area. For example, a 150-pound person with a total foot imprint area of 50 in^2 exerts a pressure of 150 $lbf/50 \ in^2 = 3.0$ psi on the floor (Fig. 3–1). If the person stands on one foot, the pressure doubles. If the person gains excessive weight, he or she is likely to encounter foot discomfort because of the increased pressure on the foot (the size of the bottom of the foot does not change with weight gain). This also explains how a person can walk on fresh snow without sinking by wearing large snowshoes, and how a person cuts with little effort when using a sharp knife.

The actual pressure at a given position is called the **absolute pressure,** and it is measured relative to absolute vacuum (i.e., absolute zero pressure). Most pressure-measuring devices, however, are calibrated to read zero in the atmosphere (Fig. 3–2), and so they indicate the difference between the absolute pressure and the local atmospheric pressure. This difference is called the **gage pressure.** P_{gage} can be positive or negative, but pressures below atmospheric pressure are sometimes called **vacuum pressures** and are measured by vacuum gages that indicate the difference between the atmospheric pressure and the absolute pressure. Absolute, gage, and vacuum pressures are related to each other by

$$P_{gage} = P_{abs} - P_{atm} \tag{3–1}$$

$$P_{vac} = P_{atm} - P_{abs} \tag{3–2}$$

This is illustrated in Fig. 3–3.

150 pounds   300 pounds

$A_{feet} = 50 \ in^2$

$P = 3$ psi $P = 6$ psi

$$P = \sigma_n = \frac{W}{A_{feet}} = \frac{150 \ lbf}{50 \ in^2} = 3 \ psi$$

FIGURE 3–1
The normal stress (or "pressure") on the feet of a chubby person is much greater than on the feet of a slim person.

FIGURE 3–2
Some basic pressure gages.
Dresser Instruments, Dresser, Inc. Used by permission.

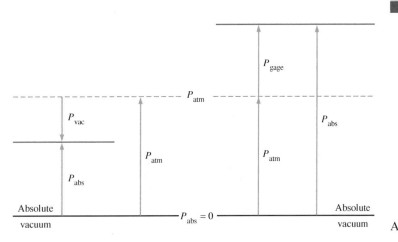

FIGURE 3–3
Absolute, gage, and vacuum pressures.

Like other pressure gages, the gage used to measure the air pressure in an automobile tire reads the gage pressure. Therefore, the common reading of 32.0 psi (2.25 kgf/cm^2) indicates a pressure of 32.0 psi above the atmospheric pressure. At a location where the atmospheric pressure is 14.3 psi, for example, the absolute pressure in the tire is 32.0 + 14.3 = 46.3 psi.

In thermodynamic relations and tables, absolute pressure is almost always used. Throughout this text, the pressure P will denote *absolute pressure* unless specified otherwise. Often the letters "a" (for absolute pressure) and "g" (for gage pressure) are added to pressure units (such as psia and psig) to clarify what is meant.

■ *EXAMPLE 3–1*   **Absolute Pressure of a Vacuum Chamber**

■ A vacuum gage connected to a chamber reads 5.8 psi at a location where
■ the atmospheric pressure is 14.5 psi. Determine the absolute pressure in the
■ chamber.

SOLUTION   The gage pressure of a vacuum chamber is given. The absolute pressure in the chamber is to be determined.
Analysis   The absolute pressure is easily determined from Eq. 3–2 to be

$$P_{abs} = P_{atm} - P_{vac} = 14.5 - 5.8 = \textbf{8.7 psi}$$

Discussion   Note that the *local* value of the atmospheric pressure is used when determining the absolute pressure.

Pressure at a Point

Pressure is the *compressive force* per unit area, and it gives the impression of being a vector. However, pressure at any point in a fluid is the same in all directions (Fig. 3–4). That is, it has magnitude but not a specific direction, and thus it is a scalar quantity. This can be demonstrated by considering a small wedge-shaped fluid element of unit length ($\Delta y = 1$ into the page) in equilibrium, as shown in Fig. 3–5. The mean pressures at the three surfaces are P_1, P_2, and P_3, and the force acting on a surface is the product of mean

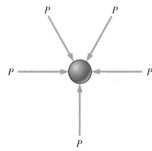

FIGURE 3–4
Pressure is a *scalar* quantity, not a vector; the pressure at a point in a fluid is the same in all directions.

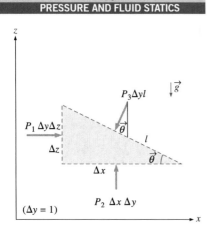

FIGURE 3–5
Forces acting on a wedge-shaped fluid element in equilibrium.

FIGURE 3–6
The pressure of a fluid at rest increases with depth (as a result of added weight).

pressure and the surface area. From Newton's second law, a force balance in the x- and z-directions gives

$$\sum F_x = ma_x = 0: \qquad P_1 \, \Delta y \Delta z - P_3 \, \Delta yl \sin \theta = 0 \qquad \text{(3–3a)}$$

$$\sum F_z = ma_z = 0: \qquad P_2 \, \Delta y \Delta x - P_3 \, \Delta yl \cos \theta - \frac{1}{2} \rho g \, \Delta x \, \Delta y \, \Delta z = 0 \quad \text{(3–3b)}$$

where ρ is the density and $W = mg = \rho g \, \Delta x \, \Delta y \, \Delta z/2$ is the weight of the fluid element. Noting that the wedge is a right triangle, we have $\Delta x = l \cos \theta$ and $\Delta z = l \sin \theta$. Substituting these geometric relations and dividing Eq. 3–3a by $\Delta y \, \Delta z$ and Eq. 3–3b by $\Delta x \, \Delta y$ gives

$$P_1 - P_3 = 0 \qquad \text{(3–4a)}$$

$$P_2 - P_3 - \frac{1}{2} \rho g \, \Delta z = 0 \qquad \text{(3–4b)}$$

The last term in Eq. 3–4b drops out as $\Delta z \to 0$ and the wedge becomes infinitesimal, and thus the fluid element shrinks to a point. Then combining the results of these two relations gives

$$P_1 = P_2 = P_3 = P \qquad \text{(3–5)}$$

regardless of the angle θ. We can repeat the analysis for an element in the yz-plane and obtain a similar result. Thus we conclude that *the pressure at a point in a fluid has the same magnitude in all directions*. This result is applicable to fluids in motion as well as fluids at rest since pressure is a *scalar*, not a vector.

Variation of Pressure with Depth

It will come as no surprise to you that pressure in a fluid at rest does not change in the horizontal direction. This can be shown easily by considering a thin horizontal layer of fluid and doing a force balance in any horizontal direction. However, this is not the case in the vertical direction in a gravity field. Pressure in a fluid increases with depth because more fluid rests on deeper layers, and the effect of this "extra weight" on a deeper layer is balanced by an increase in pressure (Fig. 3–6).

To obtain a relation for the variation of pressure with depth, consider a rectangular fluid element of height Δz, length Δx, and unit depth ($\Delta y = 1$ into the page) in equilibrium, as shown in Fig. 3–7. Assuming the density of the fluid ρ to be constant, a force balance in the vertical z-direction gives

$$\sum F_z = ma_z = 0: \qquad P_1 \, \Delta x \, \Delta y - P_2 \, \Delta x \, \Delta y - \rho g \, \Delta x \, \Delta y \, \Delta z = 0$$

where $W = mg = \rho g \, \Delta x \, \Delta y \, \Delta z$ is the weight of the fluid element and $\Delta z = z_2 - z_1$. Dividing by $\Delta x \, \Delta y$ and rearranging gives

$$\Delta P = P_2 - P_1 = -\rho g \, \Delta z = -\gamma_s \, \Delta z \qquad \text{(3–6)}$$

where $\gamma_s = \rho g$ is the *specific weight* of the fluid. Thus, we conclude that the pressure difference between two points in a constant density fluid is proportional to the vertical distance Δz between the points and the density ρ of the fluid. Noting the negative sign, *pressure in a static fluid increases linearly with depth*. This is what a diver experiences when diving deeper in a lake.

An easier equation to remember and apply between any two points in the same fluid under hydrostatic conditions is

$$P_{\text{below}} = P_{\text{above}} + \rho g |\Delta z| = P_{\text{above}} + \gamma_s |\Delta z| \qquad (3\text{-}7)$$

where "below" refers to the point at lower elevation (deeper in the fluid) and "above" refers to the point at higher elevation. If you use this equation consistently, you should avoid sign errors.

For a given fluid, the vertical distance Δz is sometimes used as a measure of pressure, and it is called the *pressure head*.

We also conclude from Eq. 3–6 that for small to moderate distances, the variation of pressure with height is negligible for gases because of their low density. The pressure in a tank containing a gas, for example, can be considered to be uniform since the weight of the gas is too small to make a significant difference. Also, the pressure in a room filled with air can be approximated as a constant (Fig. 3–8).

If we take the "above" point to be at the free surface of a liquid open to the atmosphere (Fig. 3–9), where the pressure is the atmospheric pressure P_{atm}, then from Eq. 3–7 the pressure at a depth h below the free surface becomes

$$P = P_{\text{atm}} + \rho g h \qquad \text{or} \qquad P_{\text{gage}} = \rho g h \qquad (3\text{-}8)$$

Liquids are essentially incompressible substances, and thus the variation of density with depth is negligible. This is also the case for gases when the elevation change is not very large. The variation of density of liquids or gases with temperature can be significant, however, and may need to be considered when high accuracy is desired. Also, at great depths such as those encountered in oceans, the change in the density of a liquid can be significant because of the compression by the tremendous amount of liquid weight above.

The gravitational acceleration g varies from 9.807 m/s² at sea level to 9.764 m/s² at an elevation of 14,000 m where large passenger planes cruise. This is a change of just 0.4 percent in this extreme case. Therefore, g can be approximated as a constant with negligible error.

For fluids whose density changes significantly with elevation, a relation for the variation of pressure with elevation can be obtained by dividing Eq. 3–6 by Δz, and taking the limit as $\Delta z \rightarrow 0$. This yields

$$\frac{dP}{dz} = -\rho g \qquad (3\text{-}9)$$

Note that dP is negative when dz is positive since pressure decreases in an upward direction. When the variation of density with elevation is known, the pressure difference between any two points 1 and 2 can be determined by integration to be

$$\Delta P = P_2 - P_1 = -\int_1^2 \rho g \, dz \qquad (3\text{-}10)$$

For constant density and constant gravitational acceleration, this relation reduces to Eq. 3–6, as expected.

Pressure in a fluid at rest is independent of the shape or cross section of the container. It changes with the vertical distance, but remains constant

FIGURE 3–7
Free-body diagram of a rectangular fluid element in equilibrium.

FIGURE 3–8
In a room filled with a gas, the variation of pressure with height is negligible.

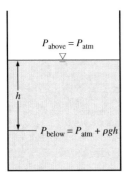

FIGURE 3–9
Pressure in a liquid at rest increases linearly with distance from the free surface.

in other directions. Therefore, the pressure is the same at all points on a horizontal plane in a given fluid. The Dutch mathematician Simon Stevin (1548–1620) published in 1586 the principle illustrated in Fig. 3–10. Note that the pressures at points A, B, C, D, E, F, and G are the same since they are at the same depth, and they are interconnected by the same static fluid. However, the pressures at points H and I are not the same since these two points cannot be interconnected by the same fluid (i.e., we cannot draw a curve from point I to point H while remaining in the same fluid at all times), although they are at the same depth. (Can you tell at which point the pressure is higher?) Also notice that the pressure force exerted by the fluid is always normal to the surface at the specified points.

A consequence of the pressure in a fluid remaining constant in the horizontal direction is that *the pressure applied to a confined fluid increases the pressure throughout by the same amount*. This is called **Pascal's law,** after Blaise Pascal (1623–1662). Pascal also knew that the force applied by a fluid is proportional to the surface area. He realized that two hydraulic cylinders of different areas could be connected, and the larger could be used to exert a proportionally greater force than that applied to the smaller. "Pascal's machine" has been the source of many inventions that are a part of our daily lives such as hydraulic brakes and lifts. This is what enables us to lift a car easily by one arm, as shown in Fig. 3–11. Noting that $P_1 = P_2$ since both pistons are at the same level (the effect of small height differences is negligible, especially at high pressures), the ratio of output force to input force is determined to be

$$P_1 = P_2 \quad \rightarrow \quad \frac{F_1}{A_1} = \frac{F_2}{A_2} \quad \rightarrow \quad \frac{F_2}{F_1} = \frac{A_2}{A_1} \qquad (3\text{–}11)$$

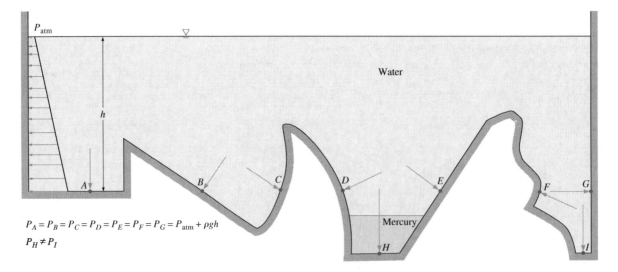

$P_A = P_B = P_C = P_D = P_E = P_F = P_G = P_{atm} + \rho g h$
$P_H \neq P_I$

FIGURE 3–10
Under hydrostatic conditions, the pressure is the same at all points on a horizontal plane in a given fluid regardless of geometry, provided that the points are interconnected by the same fluid.

The area ratio A_2/A_1 is called the *ideal mechanical advantage* of the hydraulic lift. Using a hydraulic car jack with a piston area ratio of $A_2/A_1 = 100$, for example, a person can lift a 1000-kg car by applying a force of just 10 kgf (= 90.8 N).

3–2 · PRESSURE MEASUREMENT DEVICES

The Barometer

Atmospheric pressure is measured by a device called a **barometer**; thus, the atmospheric pressure is often referred to as the *barometric pressure*.

The Italian Evangelista Torricelli (1608–1647) was the first to conclusively prove that the atmospheric pressure can be measured by inverting a mercury-filled tube into a mercury container that is open to the atmosphere, as shown in Fig. 3–12. The pressure at point B is equal to the atmospheric pressure, and the pressure at point C can be taken to be zero since there is only mercury vapor above point C and the pressure is very low relative to P_{atm} and can be neglected to an excellent approximation. Writing a force balance in the vertical direction gives

$$P_{atm} = \rho g h \qquad (3\text{–}12)$$

where ρ is the density of mercury, g is the local gravitational acceleration, and h is the height of the mercury column above the free surface. Note that the length and the cross-sectional area of the tube have no effect on the height of the fluid column of a barometer (Fig. 3–13).

A frequently used pressure unit is the *standard atmosphere*, which is defined as the pressure produced by a column of mercury 760 mm in height at 0°C ($\rho_{Hg} = 13{,}595$ kg/m³) under standard gravitational acceleration ($g = 9.807$ m/s²). If water instead of mercury were used to measure the standard atmospheric pressure, a water column of about 10.3 m would be needed. Pressure is sometimes expressed (especially by weather forecasters) in terms of the height of the mercury column. The standard atmospheric pressure, for example, is 760 mmHg (29.92 inHg) at 0°C. The unit mmHg is also called the **torr** in honor of Torricelli. Therefore, 1 atm = 760 torr and 1 torr = 133.3 Pa.

Atmospheric pressure P_{atm} changes from 101.325 kPa at sea level to 89.88, 79.50, 54.05, 26.5, and 5.53 kPa at altitudes of 1000, 2000, 5000, 10,000, and 20,000 meters, respectively. The typical atmospheric pressure in Denver (elevation = 1610 m), for example, is 83.4 kPa. Remember that the atmospheric pressure at a location is simply the weight of the air above that location per unit surface area. Therefore, it changes not only with elevation but also with weather conditions.

The decline of atmospheric pressure with elevation has far-reaching ramifications in daily life. For example, cooking takes longer at high altitudes since water boils at a lower temperature at lower atmospheric pressures. Nose bleeding is a common experience at high altitudes since the difference between the blood pressure and the atmospheric pressure is larger in this case, and the delicate walls of veins in the nose are often unable to withstand this extra stress.

For a given temperature, the density of air is lower at high altitudes, and thus a given volume contains less air and less oxygen. So it is no surprise

FIGURE 3–11
Lifting of a large weight by a small force by the application of Pascal's law. A common example is a hydraulic jack.
(Top) © Stockbyte/Getty RF

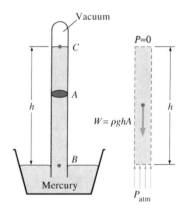

FIGURE 3–12
The basic barometer.

FIGURE 3–13
The length or the cross-sectional area of the tube has no effect on the height of the fluid column of a barometer, provided that the tube diameter is large enough to avoid surface tension (capillary) effects.

FIGURE 3–14
At high altitudes, a car engine generates less power and a person gets less oxygen because of the lower density of air.

that we tire more easily and experience breathing problems at high altitudes. To compensate for this effect, people living at higher altitudes develop more efficient lungs. Similarly, a 2.0-L car engine will act like a 1.7-L car engine at 1500 m altitude (unless it is turbocharged) because of the 15 percent drop in pressure and thus 15 percent drop in the density of air (Fig. 3–14). A fan or compressor will displace 15 percent less air at that altitude for the same volume displacement rate. Therefore, larger cooling fans may need to be selected for operation at high altitudes to ensure the specified mass flow rate. The lower pressure and thus lower density also affects lift and drag: airplanes need a longer runway at high altitudes to develop the required lift, and they climb to very high altitudes for cruising in order to reduce drag and thus achieve better fuel efficiency.

EXAMPLE 3–2   Measuring Atmospheric Pressure with a Barometer

Determine the atmospheric pressure at a location where the barometric reading is 740 mm Hg and the gravitational acceleration is $g = 9.805$ m/s². Assume the temperature of mercury to be 10°C, at which its density is 13,570 kg/m³.

SOLUTION The barometric reading at a location in height of mercury column is given. The atmospheric pressure is to be determined.
Assumptions The temperature of mercury is assumed to be 10°C.
Properties The density of mercury is given to be 13,570 kg/m³.
Analysis From Eq. 3–12, the atmospheric pressure is determined to be

$$P_{atm} = \rho g h$$

$$= (13{,}570 \text{ kg/m}^3)(9.805 \text{ m/s}^2)(0.740 \text{ m})\left(\frac{1 \text{ N}}{1 \text{ kg·m/s}^2}\right)\left(\frac{1 \text{ kPa}}{1000 \text{ N/m}^2}\right)$$

$$= 98.5 \text{ kPa}$$

Discussion Note that density changes with temperature, and thus this effect should be considered in calculations.

EXAMPLE 3–3   Gravity Driven Flow from an IV Bottle

Intravenous infusions usually are driven by gravity by hanging the fluid bottle at sufficient height to counteract the blood pressure in the vein and to force the fluid into the body (Fig. 3–15). The higher the bottle is raised, the higher the flow rate of the fluid will be. (a) If it is observed that the fluid and the blood pressures balance each other when the bottle is 1.2 m above the arm level, determine the gage pressure of the blood. (b) If the gage pressure of the fluid at the arm level needs to be 20 kPa for sufficient flow rate, determine how high the bottle must be placed. Take the density of the fluid to be 1020 kg/m³.

SOLUTION It is given that an IV fluid and the blood pressures balance each other when the bottle is at a certain height. The gage pressure of the blood and elevation of the bottle required to maintain flow at the desired rate are to be determined.

Assumptions **1** The IV fluid is incompressible. **2** The IV bottle is open to the atmosphere.

Properties The density of the IV fluid is given to be $\rho = 1020$ kg/m³.

Analysis (*a*) Noting that the IV fluid and the blood pressures balance each other when the bottle is 1.2 m above the arm level, the gage pressure of the blood in the arm is simply equal to the gage pressure of the IV fluid at a depth of 1.2 m,

$$P_{\text{gage, arm}} = P_{\text{abs}} - P_{\text{atm}} = \rho g h_{\text{arm-bottle}}$$

$$= (1020 \text{ kg/m}^3)(9.81 \text{ m/s}^2)(1.20 \text{ m})\left(\frac{1 \text{ kN}}{1000 \text{ kg·m/s}^2}\right)\left(\frac{1 \text{ kPa}}{1 \text{ kN/m}^2}\right)$$

$$= 12.0 \text{ kPa}$$

(*b*) To provide a gage pressure of 20 kPa at the arm level, the height of the surface of the IV fluid in the bottle from the arm level is again determined from $P_{\text{gage, arm}} = \rho g h_{\text{arm-bottle}}$ to be

$$h_{\text{arm-bottle}} = \frac{P_{\text{gage, arm}}}{\rho g}$$

$$= \frac{20 \text{ kPa}}{(1020 \text{ kg/m}^3)(9.81 \text{ m/s}^2)}\left(\frac{1000 \text{ kg·m/s}^2}{1 \text{ kN}}\right)\left(\frac{1 \text{ kN/m}^2}{1 \text{ kPa}}\right)$$

$$= 2.00 \text{ m}$$

FIGURE 3–15
Schematic for Example 3–3.

Discussion Note that the height of the reservoir can be used to control flow rates in gravity-driven flows. When there is flow, the pressure drop in the tube due to frictional effects also should be considered. For a specified flow rate, this requires raising the bottle a little higher to overcome the pressure drop.

EXAMPLE 3–4   Hydrostatic Pressure in a Solar Pond with Variable Density

Solar ponds are small artificial lakes of a few meters deep that are used to store solar energy. The rise of heated (and thus less dense) water to the surface is prevented by adding salt at the pond bottom. In a typical salt gradient solar pond, the density of water increases in the gradient zone, as shown in Fig. 3–16, and the density can be expressed as

$$\rho = \rho_0\sqrt{1 + \tan^2\left(\frac{\pi}{4}\frac{s}{H}\right)}$$

where ρ_0 is the density on the water surface, s is the vertical distance measured downward from the top of the gradient zone ($s = -z$), and H is the thickness of the gradient zone. For $H = 4$ m, $\rho_0 = 1040$ kg/m³, and a thickness of 0.8 m for the surface zone, calculate the gage pressure at the bottom of the gradient zone.

FIGURE 3–16
Schematic for Example 3–4.

SOLUTION The variation of density of saline water in the gradient zone of a solar pond with depth is given. The gage pressure at the bottom of the gradient zone is to be determined.

Assumptions The density in the surface zone of the pond is constant.

Properties The density of brine on the surface is given to be 1040 kg/m³.

Analysis We label the top and the bottom of the gradient zone as 1 and 2, respectively. Noting that the density of the surface zone is constant, the

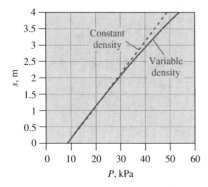

FIGURE 3–17
The variation of gage pressure with depth in the gradient zone of the solar pond.

FIGURE 3–18
A simple U-tube manometer, with high pressure applied to the right side.
Photo by John M. Cimbala.

gage pressure at the bottom of the surface zone (which is the top of the gradient zone) is

$$P_1 = \rho g h_1 = (1040 \text{ kg/m}^3)(9.81 \text{ m/s}^2)(0.8 \text{ m})\left(\frac{1 \text{ kN}}{1000 \text{ kg·m/s}^2}\right) = 8.16 \text{ kPa}$$

since $1 \text{ kN/m}^2 = 1 \text{ kPa}$. Since $s = -z$, the differential change in hydrostatic pressure across a vertical distance of ds is given by

$$dP = \rho g \, ds$$

Integrating from the top of the gradient zone (point 1 where $s = 0$) to any location s in the gradient zone (no subscript) gives

$$P - P_1 = \int_0^s \rho g \, ds \quad \rightarrow \quad P = P_1 + \int_0^s \rho_0 \sqrt{1 + \tan^2\left(\frac{\pi}{4}\frac{s}{H}\right)} g \, ds$$

Performing the integration gives the variation of gage pressure in the gradient zone to be

$$P = P_1 + \rho_0 g \frac{4H}{\pi} \sinh^{-1}\left(\tan \frac{\pi}{4}\frac{s}{H}\right)$$

Then the pressure at the bottom of the gradient zone ($s = H = 4$ m) becomes

$$P_2 = 8.16 \text{ kPa} + (1040 \text{ kg/m}^3)(9.81 \text{ m/s}^2)\frac{4(4 \text{ m})}{\pi} \sinh^{-1}\left(\tan \frac{\pi}{4}\frac{4}{4}\right)\left(\frac{1 \text{ kN}}{1000 \text{ kg·m/s}^2}\right)$$

$$= \mathbf{54.0 \text{ kPa (gage)}}$$

Discussion The variation of gage pressure in the gradient zone with depth is plotted in Fig. 3–17. The dashed line indicates the hydrostatic pressure for the case of constant density at 1040 kg/m³ and is given for reference. Note that the variation of pressure with depth is not linear when density varies with depth. That is why integration was required.

The Manometer

We notice from Eq. 3–6 that an elevation change of $-\Delta z$ in a fluid at rest corresponds to $\Delta P/\rho g$, which suggests that a fluid column can be used to measure pressure differences. A device based on this principle is called a **manometer,** and it is commonly used to measure small and moderate pressure differences. A manometer consists of a glass or plastic U-tube containing one or more fluids such as mercury, water, alcohol, or oil (Fig. 3–18). To keep the size of the manometer to a manageable level, heavy fluids such as mercury are used if large pressure differences are anticipated.

Consider the manometer shown in Fig. 3–19 that is used to measure the pressure in the tank. Since the gravitational effects of gases are negligible, the pressure anywhere in the tank and at position 1 has the same value. Furthermore, since pressure in a fluid does not vary in the horizontal direction within a fluid, the pressure at point 2 is the same as the pressure at point 1, $P_2 = P_1$.

The differential fluid column of height h is in static equilibrium, and it is open to the atmosphere. Then the pressure at point 2 is determined directly from Eq. 3–7 to be

$$P_2 = P_{\text{atm}} + \rho g h \qquad \text{(3–13)}$$

where ρ is the density of the manometer fluid in the tube. Note that the cross-sectional area of the tube has no effect on the differential height h, and thus the pressure exerted by the fluid. However, the diameter of the tube should be large enough (more than several millimeters) to ensure that the surface tension effect and thus the capillary rise is negligible.

FIGURE 3–19
The basic manometer.

■ EXAMPLE 3–5   Measuring Pressure with a Manometer

A manometer is used to measure the pressure of a gas in a tank. The fluid used has a specific gravity of 0.85, and the manometer column height is 55 cm, as shown in Fig. 3–20. If the local atmospheric pressure is 96 kPa, determine the absolute pressure within the tank.

SOLUTION The reading of a manometer attached to a tank and the atmospheric pressure are given. The absolute pressure in the tank is to be determined.

Assumptions The density of the gas in the tank is much lower than the density of the manometer fluid.

Properties The specific gravity of the manometer fluid is given to be 0.85. We take the standard density of water to be 1000 kg/m³.

Analysis The density of the fluid is obtained by multiplying its specific gravity by the density of water,

$$\rho = \text{SG} \,(\rho_{\text{H}_2\text{O}}) = (0.85)(1000 \text{ kg/m}^3) = 850 \text{ kg/m}^3$$

Then from Eq. 3–13,

$$P = P_{\text{atm}} + \rho g h$$

$$= 96 \text{ kPa} + (850 \text{ kg/m}^3)(9.81 \text{ m/s}^2)(0.55 \text{ m})\left(\frac{1 \text{ N}}{1 \text{ kg}\cdot\text{m/s}^2}\right)\left(\frac{1 \text{ kPa}}{1000 \text{ N/m}^2}\right)$$

$$= 100.6 \text{ kPa}$$

Discussion Note that the gage pressure in the tank is 4.6 kPa.

FIGURE 3–20
Schematic for Example 3–5.

Some manometers use a slanted or inclined tube in order to increase the resolution (precision) when reading the fluid height. Such devices are called **inclined manometers**.

Many engineering problems and some manometers involve multiple immiscible fluids of different densities stacked on top of each other. Such systems can be analyzed easily by remembering that (1) the pressure change across a fluid column of height h is $\Delta P = \rho g h$, (2) pressure increases downward in a given fluid and decreases upward (i.e., $P_{\text{bottom}} > P_{\text{top}}$), and (3) two points at the same elevation in a continuous fluid at rest are at the same pressure.

The last principle, which is a result of *Pascal's law,* allows us to "jump" from one fluid column to the next in manometers without worrying about pressure change as long as we stay in the same continuous fluid and the fluid is at rest. Then the pressure at any point can be determined by starting with a point of known pressure and adding or subtracting $\rho g h$ terms as we advance toward the point of interest. For example, the pressure at the bottom of the tank in Fig. 3–21 can be determined by starting at the free

FIGURE 3–21
In stacked-up fluid layers at rest, the pressure change across each fluid layer of density ρ and height h is $\rho g h$.

A flow section
or flow device

Fluid

FIGURE 3–22
Measuring the pressure drop across
a flow section or a flow device by a
differential manometer.

surface where the pressure is P_{atm}, moving downward until we reach point 1 at the bottom, and setting the result equal to P_1. It gives

$$P_{atm} + \rho_1 g h_1 + \rho_2 g h_2 + \rho_3 g h_3 = P_1$$

In the special case of all fluids having the same density, this relation reduces to $P_{atm} + \rho g(h_1 + h_2 + h_3) = P_1$.

Manometers are particularly well-suited to measure pressure drops across a horizontal flow section between two specified points due to the presence of a device such as a valve or heat exchanger or any resistance to flow. This is done by connecting the two legs of the manometer to these two points, as shown in Fig. 3–22. The working fluid can be either a gas or a liquid whose density is ρ_1. The density of the manometer fluid is ρ_2, and the differential fluid height is h. The two fluids must be immiscible, and ρ_2 must be greater than ρ_1.

A relation for the pressure difference $P_1 - P_2$ can be obtained by starting at point 1 with P_1, moving along the tube by adding or subtracting the $\rho g h$ terms until we reach point 2, and setting the result equal to P_2:

$$P_1 + \rho_1 g(a + h) - \rho_2 g h - \rho_1 g a = P_2 \qquad \text{(3–14)}$$

Note that we jumped from point A horizontally to point B and ignored the part underneath since the pressure at both points is the same. Simplifying,

$$P_1 - P_2 = (\rho_2 - \rho_1)g h \qquad \text{(3–15)}$$

Note that the distance a must be included in the analysis even though it has no effect on the result. Also, when the fluid flowing in the pipe is a gas, then $\rho_1 \ll \rho_2$ and the relation in Eq. 3–15 simplifies to $P_1 - P_2 \cong \rho_2 g h$.

FIGURE 3–23
Schematic for Example 3–3; drawing
not to scale.

EXAMPLE 3–6   Measuring Pressure with a Multifluid Manometer

The water in a tank is pressurized by air, and the pressure is measured by a multifluid manometer as shown in Fig. 3–23. The tank is located on a mountain at an altitude of 1400 m where the atmospheric pressure is 85.6 kPa. Determine the air pressure in the tank if $h_1 = 0.1$ m, $h_2 = 0.2$ m, and $h_3 = 0.35$ m. Take the densities of water, oil, and mercury to be 1000 kg/m^3, 850 kg/m^3, and 13,600 kg/m^3, respectively.

SOLUTION The pressure in a pressurized water tank is measured by a multifluid manometer. The air pressure in the tank is to be determined.
Assumption The air pressure in the tank is uniform (i.e., its variation with elevation is negligible due to its low density), and thus we can determine the pressure at the air–water interface.
Properties The densities of water, oil, and mercury are given to be 1000 kg/m^3, 850 kg/m^3, and 13,600 kg/m^3, respectively.
Analysis Starting with the pressure at point 1 at the air–water interface, moving along the tube by adding or subtracting the $\rho g h$ terms until we reach point 2, and setting the result equal to P_{atm} since the tube is open to the atmosphere gives

$$P_1 + \rho_{water} g h_1 + \rho_{oil} g h_2 - \rho_{mercury} g h_3 = P_2 = P_{atm}$$

Solving for P_1 and substituting,

$$P_1 = P_{atm} - \rho_{water}gh_1 - \rho_{oil}gh_2 + \rho_{mercury}gh_3$$
$$= P_{atm} + g(\rho_{mercury}h_3 - \rho_{water}h_1 - \rho_{oil}h_2)$$
$$= 85.6\ \text{kPa} + (9.81\ \text{m/s}^2)[(13,600\ \text{kg/m}^3)(0.35\ \text{m}) - (1000\ \text{kg/m}^3)(0.1\ \text{m})$$
$$- (850\ \text{kg/m}^3)(0.2\ \text{m})]\left(\frac{1\ \text{N}}{1\ \text{kg·m/s}^2}\right)\left(\frac{1\ \text{kPa}}{1000\ \text{N/m}^2}\right)$$
$$= \mathbf{130\ kPa}$$

Discussion Note that jumping horizontally from one tube to the next and realizing that pressure remains the same in the same fluid simplifies the analysis considerably. Also note that mercury is a toxic fluid, and mercury manometers and thermometers are being replaced by ones with safer fluids because of the risk of exposure to mercury vapor during an accident.

EXAMPLE 3–7 Analyzing a Multifluid Manometer with EES

Reconsider the multifluid manometer discussed in Example 3–6. Determine the air pressure in the tank using EES. Also determine what the differential fluid height h_3 would be for the same air pressure if the mercury in the last column were replaced by seawater with a density of 1030 kg/m^3.

SOLUTION The pressure in a water tank is measured by a multifluid manometer. The air pressure in the tank and the differential fluid height h_3 if mercury is replaced by seawater are to be determined using EES.

Analysis We start the EES program, open a new file, and type the following on the blank screen that appears (we express the atmospheric pressure in Pa for unit consistency):

g=9.81

Patm=85600

h1=0.1; h2=0.2; h3=0.35

rw=1000; roil=850; rm=13600

P1+rw*g*h1+roil*g*h2−rm*g*h3=Patm

Here P1 is the only unknown, and it is determined by EES to be

$$P_1 = 129647\ \text{Pa} \cong \mathbf{130\ kPa}$$

which is identical to the result obtained in Example 3–6. The height of the fluid column h_3 when mercury is replaced by seawater is determined easily by replacing "h3=0.35" by "P1=129647" and "rm=13600" by "rm=1030," and clicking on the calculator symbol. It gives

$$h_3 = \mathbf{4.62\ m}$$

Discussion Note that we used the screen like a paper pad and wrote down the relevant information together with the applicable relations in an organized manner. EES did the rest. Equations can be written on separate lines or on the same line by separating them by semicolons, and blank or comment lines can be inserted for readability. EES makes it very easy to ask "what if" questions and to perform parametric studies, as explained in Appendix 3 on the text website.

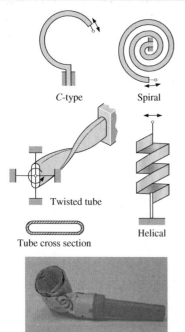

C-type   Spiral

Twisted tube

Tube cross section   Helical

FIGURE 3–24
Various types of Bourdon tubes used to measure pressure. They work on the same principle as party noise-makers (bottom photo) due to the flat tube cross section.

(Bottom) Photo by John M. Cimbala.

Weights

Oil
reservoir

Piston

Adjustable
plunger

F

A_e

Internal chamber   Oil

Reference pressure port   Crank

FIGURE 3–25
A deadweight tester is able to measure extremely high pressures (up to 10,000 psi in some applications).

Other Pressure Measurement Devices

Another type of commonly used mechanical pressure measurement device is the **Bourdon tube,** named after the French engineer and inventor Eugene Bourdon (1808–1884), which consists of a bent, coiled, or twisted hollow metal tube whose end is closed and connected to a dial indicator needle (Fig. 3–24). When the tube is open to the atmosphere, the tube is undeflected, and the needle on the dial at this state is calibrated to read zero (gage pressure). When the fluid inside the tube is pressurized, the tube stretches and moves the needle in proportion to the applied pressure.

Electronics have made their way into every aspect of life, including pressure measurement devices. Modern pressure sensors, called **pressure transducers,** use various techniques to convert the pressure effect to an electrical effect such as a change in voltage, resistance, or capacitance. Pressure transducers are smaller and faster, and they can be more sensitive, reliable, and precise than their mechanical counterparts. They can measure pressures from less than a millionth of 1 atm to several thousands of atm.

A wide variety of pressure transducers is available to measure gage, absolute, and differential pressures in a wide range of applications. *Gage pressure transducers* use the atmospheric pressure as a reference by venting the back side of the pressure-sensing diaphragm to the atmosphere, and they give a zero signal output at atmospheric pressure regardless of altitude. *Absolute pressure transducers* are calibrated to have a zero signal output at full vacuum. *Differential pressure transducers* measure the pressure difference between two locations directly instead of using two pressure transducers and taking their difference.

Strain-gage pressure transducers work by having a diaphragm deflect between two chambers open to the pressure inputs. As the diaphragm stretches in response to a change in pressure difference across it, the strain gage stretches and a Wheatstone bridge circuit amplifies the output. A capacitance transducer works similarly, but capacitance change is measured instead of resistance change as the diaphragm stretches.

Piezoelectric transducers, also called solid-state pressure transducers, work on the principle that an electric potential is generated in a crystalline substance when it is subjected to mechanical pressure. This phenomenon, first discovered by brothers Pierre and Jacques Curie in 1880, is called the piezoelectric (or press-electric) effect. Piezoelectric pressure transducers have a much faster frequency response compared to diaphragm units and are very suitable for high-pressure applications, but they are generally not as sensitive as diaphragm-type transducers, especially at low pressures.

Another type of mechanical pressure gage called a **deadweight tester** is used primarily for *calibration* and can measure extremely high pressures (Fig. 3–25). As its name implies, a deadweight tester measures pressure *directly* through application of a weight that provides a force per unit area—the fundamental definition of pressure. It is constructed with an internal chamber filled with a fluid (usually oil), along with a tight-fitting piston, cylinder, and plunger. Weights are applied to the top of the piston, which exerts a force on the oil in the chamber. The total force F acting on the oil at the piston–oil interface is the sum of the weight of the piston plus the applied weights. Since the piston cross-sectional area A_e is known, the pressure is calculated as $P = F/A_e$. The only significant source of error is that due to static friction along the interface between the piston and cylinder, but even this error is usually negligibly small. The reference pressure port is connected to either an unknown pressure that is to be measured or to a pressure sensor that is to be calibrated.

3–3 ▪ INTRODUCTION TO FLUID STATICS

Fluid statics deals with problems associated with fluids at rest. The fluid can be either gaseous or liquid. Fluid statics is generally referred to as *hydrostatics* when the fluid is a liquid and as *aerostatics* when the fluid is a gas. In fluid statics, there is no relative motion between adjacent fluid layers, and thus there are no shear (tangential) stresses in the fluid trying to deform it. The only stress we deal with in fluid statics is the *normal stress,* which is the pressure, and the variation of pressure is due only to the weight of the fluid. Therefore, the topic of fluid statics has significance only in gravity fields, and the force relations developed naturally involve the gravitational acceleration *g*. The force exerted on a surface by a fluid at rest is normal to the surface at the point of contact since there is no relative motion between the fluid and the solid surface, and thus there are no shear forces acting parallel to the surface.

Fluid statics is used to determine the forces acting on floating or submerged bodies and the forces developed by devices like hydraulic presses and car jacks. The design of many engineering systems such as water dams and liquid storage tanks requires the determination of the forces acting on their surfaces using fluid statics. The complete description of the resultant hydrostatic force acting on a submerged surface requires the determination of the magnitude, the direction, and the line of action of the force. In the following two sections, we consider the forces acting on both plane and curved surfaces of submerged bodies due to pressure.

FIGURE 3–26
Hoover Dam.
Courtesy United States Department of the Interior, Bureau of Reclamation-Lower Colorado Region.

3–4 ▪ HYDROSTATIC FORCES ON SUBMERGED PLANE SURFACES

A plate (such as a gate valve in a dam, the wall of a liquid storage tank, or the hull of a ship at rest) is subjected to fluid pressure distributed over its surface when exposed to a liquid (Fig. 3–26). On a *plane* surface, the hydrostatic forces form a system of parallel forces, and we often need to determine the *magnitude* of the force and its *point of application,* which is called the **center of pressure.** In most cases, the other side of the plate is open to the atmosphere (such as the dry side of a gate), and thus atmospheric pressure acts on both sides of the plate, yielding a zero resultant. In such cases, it is convenient to subtract atmospheric pressure and work with the gage pressure only (Fig. 3–27). For example, $P_{\text{gage}} = \rho g h$ at the bottom of the lake.

Consider the top surface of a flat plate of arbitrary shape completely submerged in a liquid, as shown in Fig. 3–28 together with its normal view. The plane of this surface (normal to the page) intersects the horizontal free surface at angle θ, and we take the line of intersection to be the *x*-axis (out of the page). The absolute pressure above the liquid is P_0, which is the local atmospheric pressure P_{atm} if the liquid is open to the atmosphere (but P_0

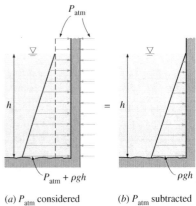

(a) P_{atm} considered   (b) P_{atm} subtracted

FIGURE 3–27
When analyzing hydrostatic forces on submerged surfaces, the atmospheric pressure can be subtracted for simplicity when it acts on both sides of the structure.

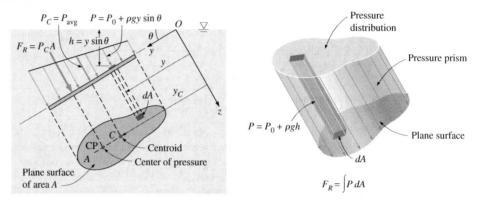

FIGURE 3–28
Hydrostatic force on an inclined plane surface completely submerged in a liquid.

may be different than P_{atm} if the space above the liquid is evacuated or pressurized). Then the absolute pressure at any point on the plate is

$$P = P_0 + \rho g h = P_0 + \rho g y \sin \theta \qquad (3\text{–}16)$$

where h is the vertical distance of the point from the free surface and y is the distance of the point from the x-axis (from point O in Fig. 3–28). The resultant hydrostatic force F_R acting on the surface is determined by integrating the force $P\, dA$ acting on a differential area dA over the entire surface area,

$$F_R = \int_A P\, dA = \int_A (P_0 + \rho g y \sin \theta)\, dA = P_0 A + \rho g \sin \theta \int_A y\, dA \qquad (3\text{–}17)$$

But the *first moment of area* $\int_A y\, dA$ is related to the y-coordinate of the centroid (or center) of the surface by

$$y_C = \frac{1}{A} \int_A y\, dA \qquad (3\text{–}18)$$

Substituting,

$$F_R = (P_0 + \rho g y_C \sin \theta)A = (P_0 + \rho g h_C)A = P_C A = P_{avg} A \qquad (3\text{–}19)$$

where $P_C = P_0 + \rho g h_C$ is the pressure at the centroid of the surface, which is equivalent to the *average* pressure P_{avg} on the surface, and $h_C = y_C \sin \theta$ is the *vertical distance* of the centroid from the free surface of the liquid (Fig. 3–29). Thus we conclude that:

> The magnitude of the resultant force acting on a plane surface of a completely submerged plate in a homogeneous (constant density) fluid is equal to the product of the pressure P_C at the centroid of the surface and the area A of the surface (Fig. 3–30).

The pressure P_0 is usually atmospheric pressure, which can be ignored in most force calculations since it acts on both sides of the plate. When this is not the case, a practical way of accounting for the contribution of P_0 to

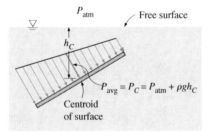

FIGURE 3–29
The pressure at the centroid of a plane surface is equivalent to the *average* pressure on the surface.

the resultant force is simply to add an equivalent depth $h_{equiv} = P_0/\rho g$ to h_C; that is, to assume the presence of an additional liquid layer of thickness h_{equiv} on top of the liquid with absolute vacuum above.

Next we need to determine the line of action of the resultant force F_R. Two parallel force systems are equivalent if they have the same magnitude and the same moment about any point. The line of action of the resultant hydrostatic force, in general, does not pass through the centroid of the surface—it lies underneath where the pressure is higher. The point of intersection of the line of action of the resultant force and the surface is the **center of pressure**. The vertical location of the line of action is determined by equating the moment of the resultant force to the moment of the distributed pressure force about the x-axis:

$$y_P F_R = \int_A yP\, dA = \int_A y(P_0 + \rho gy \sin\theta)\, dA = P_0 \int_A y\, dA + \rho g \sin\theta \int_A y^2\, dA$$

or

$$y_P F_R = P_0 y_C A + \rho g \sin\theta\, I_{xx,O} \qquad (3\text{-}20)$$

where y_P is the distance of the center of pressure from the x-axis (point O in Fig. 3–30) and $I_{xx,O} = \int_A y^2\, dA$ is the *second moment of area* (also called the *area moment of inertia*) about the x-axis. The second moments of area are widely available for common shapes in engineering handbooks, but they are usually given about the axes passing through the centroid of the area. Fortunately, the second moments of area about two parallel axes are related to each other by the *parallel axis theorem*, which in this case is expressed as

$$I_{xx,O} = I_{xx,C} + y_C^2 A \qquad (3\text{-}21)$$

where $I_{xx,C}$ is the second moment of area about the x-axis passing through the centroid of the area and y_C (the y-coordinate of the centroid) is the distance between the two parallel axes. Substituting the F_R relation from Eq. 3–19 and the $I_{xx,O}$ relation from Eq. 3–21 into Eq. 3–20 and solving for y_P yields

$$y_P = y_C + \frac{I_{xx,C}}{[y_C + P_0/(\rho g \sin\theta)]A} \qquad (3\text{-}22a)$$

For $P_0 = 0$, which is usually the case when the atmospheric pressure is ignored, it simplifies to

$$y_P = y_C + \frac{I_{xx,C}}{y_C A} \qquad (3\text{-}22b)$$

Knowing y_P, the vertical distance of the center of pressure from the free surface is determined from $h_P = y_P \sin\theta$.

The $I_{xx,C}$ values for some common areas are given in Fig. 3–31. For areas that possess symmetry about the y-axis, the center of pressure lies on the y-axis directly below the centroid. The location of the center of pressure in such cases is simply the point on the surface of the vertical plane of symmetry at a distance h_P from the free surface.

Pressure acts normal to the surface, and the hydrostatic forces acting on a flat plate of any shape form a volume whose base is the plate area and

FIGURE 3–30
The resultant force acting on a plane surface is equal to the product of the pressure at the centroid of the surface and the surface area, and its line of action passes through the center of pressure.

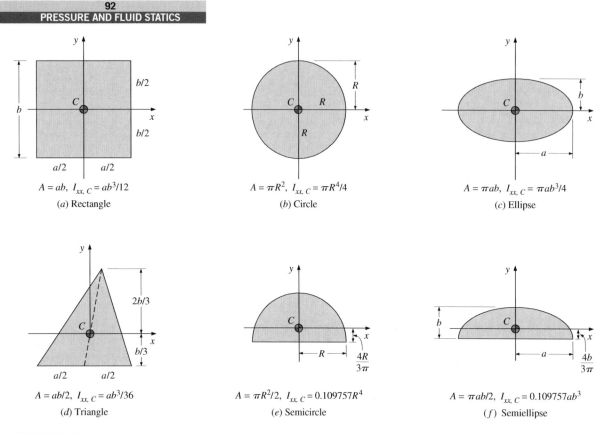

$A = ab, \; I_{xx,\,C} = ab^3/12$

(a) Rectangle

$A = \pi R^2, \; I_{xx,\,C} = \pi R^4/4$

(b) Circle

$A = \pi ab, \; I_{xx,\,C} = \pi ab^3/4$

(c) Ellipse

$A = ab/2, \; I_{xx,\,C} = ab^3/36$

(d) Triangle

$A = \pi R^2/2, \; I_{xx,\,C} = 0.109757 R^4$

(e) Semicircle

$A = \pi ab/2, \; I_{xx,\,C} = 0.109757 ab^3$

(f) Semiellipse

FIGURE 3–31
The centroid and the centroidal moments of inertia for some common geometries.

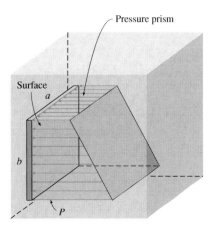

FIGURE 3–32
The hydrostatic forces acting on a plane surface form a pressure prism whose base (left face) is the surface and whose length is the pressure.

whose length is the linearly varying pressure, as shown in Fig. 3–32. This virtual **pressure prism** has an interesting physical interpretation: its *volume* is equal to the *magnitude* of the resultant hydrostatic force acting on the plate since $F_R = \int P \, dA$, and the line of action of this force passes through the *centroid* of this homogeneous prism. The projection of the centroid on the plate is the *pressure center*. Therefore, with the concept of pressure prism, the problem of describing the resultant hydrostatic force on a plane surface reduces to finding the volume and the two coordinates of the centroid of this pressure prism.

Special Case: Submerged Rectangular Plate

Consider a completely submerged rectangular flat plate of height b and width a tilted at an angle θ from the horizontal and whose top edge is horizontal and is at a distance s from the free surface along the plane of the plate, as shown in Fig. 3–33a. The resultant hydrostatic force on the upper surface is equal to the average pressure, which is the pressure at the midpoint of the surface, times the surface area A. That is,

Tilted rectangular plate: $\qquad F_R = P_C A = [P_0 + \rho g(s + b/2) \sin \theta] ab$ $\qquad$ **(3–23)**

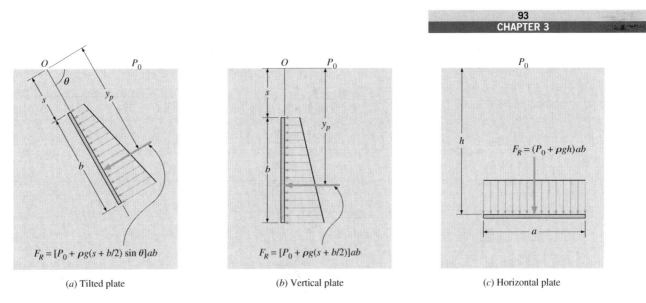

FIGURE 3–33
Hydrostatic force acting on the top surface of a submerged rectangular plate for tilted, vertical, and horizontal cases.

The force acts at a vertical distance of $h_P = y_P \sin \theta$ from the free surface directly beneath the centroid of the plate where, from Eq. 3–22a,

$$y_P = s + \frac{b}{2} + \frac{ab^3/12}{[s + b/2 + P_0/(\rho g \sin \theta)]ab}$$

$$= s + \frac{b}{2} + \frac{b^2}{12[s + b/2 + P_0/(\rho g \sin \theta)]} \qquad (3\text{--}24)$$

When the upper edge of the plate is at the free surface and thus $s = 0$, Eq. 3–23 reduces to

Tilted rectangular plate ($s = 0$): $F_R = [P_0 + \rho g(b \sin \theta)/2]ab$ (3–25)

For a completely submerged *vertical* plate ($\theta = 90°$) whose top edge is horizontal, the hydrostatic force can be obtained by setting $\sin \theta = 1$ (Fig. 3–33b)

Vertical rectangular plate: $F_R = [P_0 + \rho g(s + b/2)]ab$ (3–26)

Vertical rectangular plate ($s = 0$): $F_R = (P_0 + \rho g b/2)ab$ (3–27)

When the effect of P_0 is ignored since it acts on both sides of the plate, the hydrostatic force on a vertical rectangular surface of height b whose top edge is horizontal and at the free surface is $F_R = \rho g ab^2/2$ acting at a distance of $2b/3$ from the free surface directly beneath the centroid of the plate.

The pressure distribution on a submerged *horizontal* surface is uniform, and its magnitude is $P = P_0 + \rho g h$, where h is the distance of the surface from the free surface. Therefore, the hydrostatic force acting on a horizontal rectangular surface is

Horizontal rectangular plate: $F_R = (P_0 + \rho g h)ab$ (3–28)

and it acts through the midpoint of the plate (Fig. 3–32c).

Lake

8 m

1.2 m

1 m

FIGURE 3–34
Schematic for Example 3–8.

EXAMPLE 3–8   **Hydrostatic Force Acting on the Door of a Submerged Car**

A heavy car plunges into a lake during an accident and lands at the bottom of the lake on its wheels (Fig. 3–34). The door is 1.2 m high and 1 m wide, and the top edge of the door is 8 m below the free surface of the water. Determine the hydrostatic force on the door and the location of the pressure center, and discuss if the driver can open the door.

SOLUTION   A car is submerged in water. The hydrostatic force on the door is to be determined, and the likelihood of the driver opening the door is to be assessed.

Assumptions   **1** The bottom surface of the lake is horizontal. **2** The passenger cabin is well-sealed so that no water leaks inside. **3** The door can be approximated as a vertical rectangular plate. **4** The pressure in the passenger cabin remains at atmospheric value since there is no water leaking in, and thus no compression of the air inside. Therefore, atmospheric pressure cancels out in the calculations since it acts on both sides of the door. **5** The weight of the car is larger than the buoyant force acting on it.

Properties   We take the density of lake water to be 1000 kg/m³ throughout.

Analysis   The average (gage) pressure on the door is the pressure value at the centroid (midpoint) of the door and is determined to be

$$P_{avg} = P_C = \rho g h_C = \rho g(s + b/2)$$

$$= (1000 \text{ kg/m}^3)(9.81 \text{ m/s}^2)(8 + 1.2/2 \text{ m})\left(\frac{1 \text{ kN}}{1000 \text{ kg·m/s}^2}\right)$$

$$= 84.4 \text{ kN/m}^2$$

Then the resultant hydrostatic force on the door becomes

$$F_R = P_{avg}A = (84.4 \text{ kN/m}^2)(1 \text{ m} \times 1.2 \text{ m}) = 101.3 \text{ kN}$$

The pressure center is directly under the midpoint of the door, and its distance from the surface of the lake is determined from Eq. 3–24 by setting $P_0 = 0$, yielding

$$y_P = s + \frac{b}{2} + \frac{b^2}{12(s + b/2)} = 8 + \frac{1.2}{2} + \frac{1.2^2}{12(8 + 1.2/2)} = 8.61 \text{ m}$$

Discussion   A strong person can lift 100 kg, which is a weight of 981 N or about 1 kN. Also, the person can apply the force at a point farthest from the hinges (1 m farther) for maximum effect and generate a moment of 1 kN·m. The resultant hydrostatic force acts under the midpoint of the door, and thus a distance of 0.5 m from the hinges. This generates a moment of 50.6 kN·m, which is about 50 times the moment the driver can possibly generate. Therefore, it is impossible for the driver to open the door of the car. The driver's best bet is to let some water in (by rolling the window down a little, for example) and to keep his or her head close to the ceiling. The driver should be able to open the door shortly before the car is filled with water since at that point the pressures on both sides of the door are nearly the same and opening the door in water is almost as easy as opening it in air.

3–5 · HYDROSTATIC FORCES ON SUBMERGED CURVED SURFACES

In many practical applications, submerged surfaces are not flat (Fig. 3–35). For a submerged curved surface, the determination of the resultant hydrostatic force is more involved since it typically requires integration of the pressure forces that change direction along the curved surface. The concept of the pressure prism in this case is not much help either because of the complicated shapes involved.

The easiest way to determine the resultant hydrostatic force F_R acting on a two-dimensional curved surface is to determine the horizontal and vertical components F_H and F_V separately. This is done by considering the free-body diagram of the liquid block enclosed by the curved surface and the two plane surfaces (one horizontal and one vertical) passing through the two ends of the curved surface, as shown in Fig. 3–36. Note that the vertical surface of the liquid block considered is simply the projection of the curved surface on a *vertical plane,* and the horizontal surface is the projection of the curved surface on a *horizontal plane.* The resultant force acting on the curved solid surface is then equal and opposite to the force acting on the curved liquid surface (Newton's third law).

The force acting on the imaginary horizontal or vertical plane surface and its line of action can be determined as discussed in Section 3–4. The weight of the enclosed liquid block of volume V is simply $W = \rho g V$, and it acts downward through the centroid of this volume. Noting that the fluid block is in static equilibrium, the force balances in the horizontal and vertical directions give

Horizontal force component on curved surface: $$F_H = F_x \qquad (3\text{–}29)$$

Vertical force component on curved surface: $$F_V = F_y \pm W \qquad (3\text{–}30)$$

FIGURE 3–35
In many structures of practical application, the submerged surfaces are not flat, but curved as here at Glen Canyon Dam in Utah and Arizona.
© Corbis RF

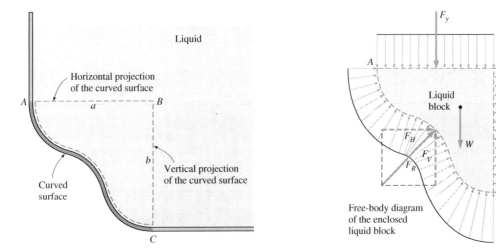

FIGURE 3–36
Determination of the hydrostatic force acting on a submerged curved surface.

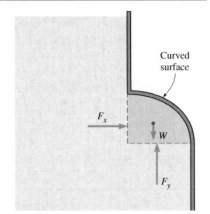

FIGURE 3–37
When a curved surface is above the liquid, the weight of the liquid and the vertical component of the hydrostatic force act in the opposite directions.

FIGURE 3–38
The hydrostatic force acting on a circular surface always passes through the center of the circle since the pressure forces are normal to the surface and they all pass through the center.

where the summation $F_y \pm W$ is a vector addition (i.e., add magnitudes if both act in the same direction and subtract if they act in opposite directions). Thus, we conclude that

1. The horizontal component of the hydrostatic force acting on a curved surface is equal (in both magnitude and the line of action) to the hydrostatic force acting on the vertical projection of the curved surface.
2. The vertical component of the hydrostatic force acting on a curved surface is equal to the hydrostatic force acting on the horizontal projection of the curved surface, plus (minus, if acting in the opposite direction) the weight of the fluid block.

The magnitude of the resultant hydrostatic force acting on the curved surface is $F_R = \sqrt{F_H^2 + F_V^2}$, and the tangent of the angle it makes with the horizontal is $\tan \alpha = F_V/F_H$. The exact location of the line of action of the resultant force (e.g., its distance from one of the end points of the curved surface) can be determined by taking a moment about an appropriate point. These discussions are valid for all curved surfaces regardless of whether they are above or below the liquid. Note that in the case of a *curved surface above a liquid*, the weight of the liquid is *subtracted* from the vertical component of the hydrostatic force since they act in opposite directions (Fig. 3–37).

When the curved surface is a *circular arc* (full circle or any part of it), the resultant hydrostatic force acting on the surface always passes through the center of the circle. This is because the pressure forces are normal to the surface, and all lines normal to the surface of a circle pass through the center of the circle. Thus, the pressure forces form a concurrent force system at the center, which can be reduced to a single equivalent force at that point (Fig. 3–38).

Finally, the hydrostatic force acting on a plane or curved surface submerged in a **multilayered fluid** of different densities can be determined by considering different parts of surfaces in different fluids as different surfaces, finding the force on each part, and then adding them using vector addition. For a plane surface, it can be expressed as (Fig. 3–39)

Plane surface in a multilayered fluid: $$F_R = \sum F_{R,i} = \sum P_{C,i} A_i \qquad (3\text{–}31)$$

where $P_{C,i} = P_0 + \rho_i g h_{C,i}$ is the pressure at the centroid of the portion of the surface in fluid i and A_i is the area of the plate in that fluid. The line of action of this equivalent force can be determined from the requirement that the moment of the equivalent force about any point is equal to the sum of the moments of the individual forces about the same point.

EXAMPLE 3–9   A Gravity-Controlled Cylindrical Gate

A long solid cylinder of radius 0.8 m hinged at point A is used as an automatic gate, as shown in Fig. 3–40. When the water level reaches 5 m, the gate opens by turning about the hinge at point A. Determine (a) the hydrostatic force acting on the cylinder and its line of action when the gate opens and (b) the weight of the cylinder per m length of the cylinder.

SOLUTION The height of a water reservoir is controlled by a cylindrical gate hinged to the reservoir. The hydrostatic force on the cylinder and the weight of the cylinder per m length are to be determined.

Assumptions **1** Friction at the hinge is negligible. **2** Atmospheric pressure acts on both sides of the gate, and thus it cancels out.

Properties We take the density of water to be 1000 kg/m^3 throughout.

Analysis (*a*) We consider the free-body diagram of the liquid block enclosed by the circular surface of the cylinder and its vertical and horizontal projections. The hydrostatic forces acting on the vertical and horizontal plane surfaces as well as the weight of the liquid block are determined as

Horizontal force on vertical surface:

$$F_H = F_x = P_{avg}A = \rho g h_C A = \rho g(s + R/2)A$$

$$= (1000 \text{ kg/m}^3)(9.81 \text{ m/s}^2)(4.2 + 0.8/2 \text{ m})(0.8 \text{ m} \times 1 \text{ m})\left(\frac{1 \text{ kN}}{1000 \text{ kg·m/s}^2}\right)$$

$$= 36.1 \text{ kN}$$

Vertical force on horizontal surface (upward):

$$F_y = P_{avg}A = \rho g h_C A = \rho g h_{bottom} A$$

$$= (1000 \text{ kg/m}^3)(9.81 \text{ m/s}^2)(5 \text{ m})(0.8 \text{ m} \times 1 \text{ m})\left(\frac{1 \text{ kN}}{1000 \text{ kg·m/s}^2}\right)$$

$$= 39.2 \text{ kN}$$

Weight (downward) of fluid block for one m width into the page:

$$W = mg = \rho g V = \rho g(R^2 - \pi R^2/4)(1 \text{ m})$$

$$= (1000 \text{ kg/m}^3)(9.81 \text{ m/s}^2)(0.8 \text{ m})^2(1 - \pi/4)(1 \text{ m})\left(\frac{1 \text{ kN}}{1000 \text{ kg·m/s}^2}\right)$$

$$= 1.3 \text{ kN}$$

Therefore, the net upward vertical force is

$$F_V = F_y - W = 39.2 - 1.3 = 37.9 \text{ kN}$$

Then the magnitude and direction of the hydrostatic force acting on the cylindrical surface become

$$F_R = \sqrt{F_H^2 + F_V^2} = \sqrt{36.1^2 + 37.9^2} = \mathbf{52.3 \text{ kN}}$$

$$\tan \theta = F_V/F_H = 37.9/36.1 = 1.05 \rightarrow \theta = 46.4°$$

Therefore, the magnitude of the hydrostatic force acting on the cylinder is 52.3 kN per m length of the cylinder, and its line of action passes through the center of the cylinder making an angle 46.4° with the horizontal.

(*b*) When the water level is 5 m high, the gate is about to open and thus the reaction force at the bottom of the cylinder is zero. Then the forces other than those at the hinge acting on the cylinder are its weight, acting through the center, and the hydrostatic force exerted by water. Taking a moment about point *A* at the location of the hinge and equating it to zero gives

$$F_R R \sin \theta - W_{cyl} R = 0 \rightarrow W_{cyl} = F_R \sin \theta = (52.3 \text{ kN}) \sin 46.4° = \mathbf{37.9 \text{ kN}}$$

Discussion The weight of the cylinder per m length is determined to be 37.9 kN. It can be shown that this corresponds to a mass of 3863 kg per m length and to a density of 1921 kg/m^3 for the material of the cylinder.

FIGURE 3–39

The hydrostatic force on a surface submerged in a multilayered fluid can be determined by considering parts of the surface in different fluids as different surfaces.

FIGURE 3–40

Schematic for Example 3–9 and the free-body diagram of the liquid underneath the cylinder.

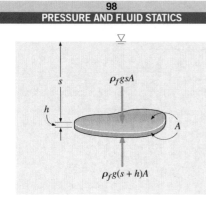

FIGURE 3–41

A flat plate of uniform thickness h submerged in a liquid parallel to the free surface.

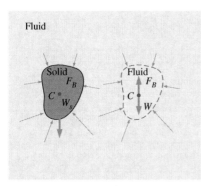

FIGURE 3–42

The buoyant forces acting on a solid body submerged in a fluid and on a fluid body of the same shape at the same depth are identical. The buoyant force F_B acts upward through the centroid C of the displaced volume and is equal in magnitude to the weight W of the displaced fluid, but is opposite in direction. For a solid of uniform density, its weight W_s also acts through the centroid, but its magnitude is not necessarily equal to that of the fluid it displaces. (Here $W_s > W$ and thus $W_s > F_B$; this solid body would sink.)

3–6 · BUOYANCY AND STABILITY

It is a common experience that an object feels lighter and weighs less in a liquid than it does in air. This can be demonstrated easily by weighing a heavy object in water by a waterproof spring scale. Also, objects made of wood or other light materials float on water. These and other observations suggest that a fluid exerts an upward force on a body immersed in it. This force that tends to lift the body is called the **buoyant force** and is denoted by F_B.

The buoyant force is caused by the increase of pressure with depth in a fluid. Consider, for example, a flat plate of thickness h submerged in a liquid of density ρ_f parallel to the free surface, as shown in Fig. 3–41. The area of the top (and also bottom) surface of the plate is A, and its distance to the free surface is s. The gage pressures at the top and bottom surfaces of the plate are $\rho_f g s$ and $\rho_f g(s + h)$, respectively. Then the hydrostatic force $F_{\text{top}} = \rho_f g s A$ acts downward on the top surface, and the larger force $F_{\text{bottom}} = \rho_f g(s + h)A$ acts upward on the bottom surface of the plate. The difference between these two forces is a net upward force, which is the *buoyant force,*

$$F_B = F_{\text{bottom}} - F_{\text{top}} = \rho_f g(s + h)A - \rho_f g s A = \rho_f g h A = \rho_f g V \quad (3\text{--}32)$$

where $V = hA$ is the volume of the plate. But the relation $\rho_f g V$ is simply the weight of the liquid whose volume is equal to the volume of the plate. Thus, we conclude that *the buoyant force acting on the plate is equal to the weight of the liquid displaced by the plate.* For a fluid with constant density, the buoyant force is independent of the distance of the body from the free surface. It is also independent of the density of the solid body.

The relation in Eq. 3–32 is developed for a simple geometry, but it is valid for any body regardless of its shape. This can be shown mathematically by a force balance, or simply by this argument: Consider an arbitrarily shaped solid body submerged in a fluid at rest and compare it to a body of fluid of the same shape indicated by dashed lines at the same vertical location (Fig. 3–42). The buoyant forces acting on these two bodies are the same since the pressure distributions, which depend only on elevation, are the same at the boundaries of both. The imaginary fluid body is in static equilibrium, and thus the net force and net moment acting on it are zero. Therefore, the upward buoyant force must be equal to the weight of the imaginary fluid body whose volume is equal to the volume of the solid body. Further, the weight and the buoyant force must have the same line of action to have a zero moment. This is known as **Archimedes' principle,** after the Greek mathematician Archimedes (287–212 BC), and is expressed as

The buoyant force acting on a body of uniform density immersed in a fluid is equal to the weight of the fluid displaced by the body, and it acts upward through the centroid of the displaced volume.

For *floating* bodies, the weight of the entire body must be equal to the buoyant force, which is the weight of the fluid whose volume is equal to the volume of the *submerged portion* of the floating body. That is,

$$F_B = W \rightarrow \rho_f g V_{\text{sub}} = \rho_{\text{avg, body}} g V_{\text{total}} \rightarrow \frac{V_{\text{sub}}}{V_{\text{total}}} = \frac{\rho_{\text{avg, body}}}{\rho_f} \quad (3\text{--}33)$$

Therefore, the submerged volume fraction of a floating body is equal to the ratio of the average density of the body to the density of the fluid. Note that when the density ratio is equal to or greater than one, the floating body becomes completely submerged.

It follows from these discussions that a body immersed in a fluid (1) remains at rest at any location in the fluid where its average density is equal to the density of the fluid, (2) sinks to the bottom when its average density is greater than the density of the fluid, and (3) rises to the surface of the fluid and floats when the average density of the body is less than the density of the fluid (Fig. 3–43).

The buoyant force is proportional to the density of the fluid, and thus we might think that the buoyant force exerted by gases such as air is negligible. This is certainly the case in general, but there are significant exceptions. For example, the volume of a person is about 0.1 m³, and taking the density of air to be 1.2 kg/m³, the buoyant force exerted by air on the person is

$$F_B = \rho_f g V = (1.2 \text{ kg/m}^3)(9.81 \text{ m/s}^2)(0.1 \text{ m}^3) \cong 1.2 \text{ N}$$

The weight of an 80-kg person is $80 \times 9.81 = 788$ N. Therefore, ignoring the buoyancy in this case results in an error in weight of just 0.15 percent, which is negligible. But the buoyancy effects in gases dominate some important natural phenomena such as the rise of warm air in a cooler environment and thus the onset of natural convection currents, the rise of hot-air or helium balloons, and air movements in the atmosphere. A helium balloon, for example, rises as a result of the buoyancy effect until it reaches an altitude where the density of air (which decreases with altitude) equals the density of helium in the balloon—assuming the balloon does not burst by then, and ignoring the weight of the balloon's skin. Hot air balloons (Fig. 3–44) work by similar principles.

Archimedes' principle is also used in geology by considering the continents to be floating on a sea of magma.

EXAMPLE 3–10   Measuring Specific Gravity by a Hydrometer

If you have a seawater aquarium, you have probably used a small cylindrical glass tube with a lead-weight at its bottom to measure the salinity of the water by simply watching how deep the tube sinks. Such a device that floats in a vertical position and is used to measure the specific gravity of a liquid is called a *hydrometer* (Fig. 3–45). The top part of the hydrometer extends above the liquid surface, and the divisions on it allow one to read the specific gravity directly. The hydrometer is calibrated such that in pure water it reads exactly 1.0 at the air–water interface. (*a*) Obtain a relation for the specific gravity of a liquid as a function of distance Δz from the mark corresponding to pure water and (*b*) determine the mass of lead that must be poured into a 1-cm-diameter, 20-cm-long hydrometer if it is to float halfway (the 10-cm mark) in pure water.

SOLUTION The specific gravity of a liquid is to be measured by a hydrometer. A relation between specific gravity and the vertical distance from the reference level is to be obtained, and the amount of lead that needs to be added into the tube for a certain hydrometer is to be determined.

Assumptions **1** The weight of the glass tube is negligible relative to the weight of the lead added. **2** The curvature of the tube bottom is disregarded.

FIGURE 3–43
A solid body dropped into a fluid will sink, float, or remain at rest at any point in the fluid, depending on its average density relative to the density of the fluid.

FIGURE 3–44
The altitude of a hot air balloon is controlled by the temperature difference between the air inside and outside the balloon, since warm air is less dense than cold air. When the balloon is neither rising nor falling, the upward buoyant force exactly balances the downward weight.
© PhotoLink/Getty RF

FIGURE 3–45
Schematic for Example 3–10.

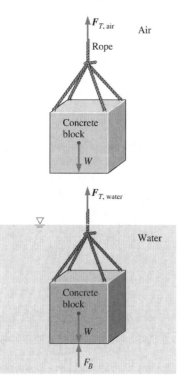

FIGURE 3–46
Schematic for Example 3–11.

Properties We take the density of pure water to be 1000 kg/m³.

Analysis (*a*) Noting that the hydrometer is in static equilibrium, the buoyant force F_B exerted by the liquid must always be equal to the weight W of the hydrometer. In pure water (subscript *w*), we let the vertical distance between the bottom of the hydrometer and the free surface of water be z_0. Setting $F_{B,\,w} = W$ in this case gives

$$W_{\text{hydro}} = F_{B,\,w} = \rho_w g V_{\text{sub}} = \rho_w g A z_0 \qquad (1)$$

where A is the cross-sectional area of the tube, and ρ_w is the density of pure water.

In a fluid lighter than water ($\rho_f < \rho_w$), the hydrometer will sink deeper, and the liquid level will be a distance of Δz above z_0. Again setting $F_B = W$ gives

$$W_{\text{hydro}} = F_{B,\,f} = \rho_f g V_{\text{sub}} = \rho_f g A(z_0 + \Delta z) \qquad (2)$$

This relation is also valid for fluids heavier than water by taking Δz to be a negative quantity. Setting Eqs. (1) and (2) here equal to each other since the weight of the hydrometer is constant and rearranging gives

$$\rho_w g A z_0 = \rho_f g A(z_0 + \Delta z) \quad \rightarrow \quad SG_f = \frac{\rho_f}{\rho_w} = \frac{z_0}{z_0 + \Delta z}$$

which is the relation between the specific gravity of the fluid and Δz. Note that z_0 is constant for a given hydrometer and Δz is negative for fluids heavier than pure water.

(*b*) Disregarding the weight of the glass tube, the amount of lead that needs to be added to the tube is determined from the requirement that the weight of the lead be equal to the buoyant force. When the hydrometer is floating with half of it submerged in water, the buoyant force acting on it is

$$F_B = \rho_w g V_{\text{sub}}$$

Equating F_B to the weight of lead gives

$$W = mg = \rho_w g V_{\text{sub}}$$

Solving for m and substituting, the mass of lead is determined to be

$$m = \rho_w V_{\text{sub}} = \rho_w(\pi R^2 h_{\text{sub}}) = (1000 \text{ kg/m}^3)[\pi(0.005 \text{ m})^2(0.1 \text{ m})] = \mathbf{0.00785 \text{ kg}}$$

Discussion Note that if the hydrometer were required to sink only 5 cm in water, the required mass of lead would be one-half of this amount. Also, the assumption that the weight of the glass tube is negligible is questionable since the mass of lead is only 7.85 g.

EXAMPLE 3–11   **Weight Loss of an Object in Seawater**

A crane is used to lower weights into the sea (density = 1025 kg/m³) for an underwater construction project (Fig. 3–46). Determine the tension in the rope of the crane due to a rectangular 0.4-m × 0.4-m × 3-m concrete block (density = 2300 kg/m³) when it is (*a*) suspended in the air and (*b*) completely immersed in water.

SOLUTION A concrete block is lowered into the sea. The tension in the rope is to be determined before and after the block is in water.

Assumptions **1** The buoyant force in air is negligible. **2** The weight of the ropes is negligible.

Properties The densities are given to be 1025 kg/m³ for seawater and 2300 kg/m³ for concrete.

Analysis (*a*) Consider a free-body diagram of the concrete block. The forces acting on the concrete block in air are its weight and the upward pull action (tension) by the rope. These two forces must balance each other, and thus the tension in the rope must be equal to the weight of the block:

$$V = (0.4 \text{ m})(0.4 \text{ m})(3 \text{ m}) = 0.48 \text{ m}^3$$

$$F_{T,\text{air}} = W = \rho_{\text{concrete}} g V$$

$$= (2300 \text{ kg/m}^3)(9.81 \text{ m/s}^2)(0.48 \text{ m}^3)\left(\frac{1 \text{ kN}}{1000 \text{ kg}\cdot\text{m/s}^2}\right) = \textbf{10.8 kN}$$

(*b*) When the block is immersed in water, there is the additional force of buoyancy acting upward. The force balance in this case gives

$$F_B = \rho_f g V = (1025 \text{ kg/m}^3)(9.81 \text{ m/s}^2)(0.48 \text{ m}^3)\left(\frac{1 \text{ kN}}{1000 \text{ kg}\cdot\text{m/s}^2}\right) = 4.8 \text{ kN}$$

$$F_{T,\text{water}} = W - F_B = 10.8 - 4.8 = \textbf{6.0 kN}$$

Discussion Note that the weight of the concrete block, and thus the tension of the rope, decreases by (10.8 − 6.0)/10.8 = 55 percent in water.

FIGURE 3–47
For floating bodies such as ships, stability is an important consideration for safety.
© *Corbis RF*

Stability of Immersed and Floating Bodies

An important application of the buoyancy concept is the assessment of the stability of immersed and floating bodies with no external attachments. This topic is of great importance in the design of ships and submarines (Fig. 3–47). Here we provide some general qualitative discussions on vertical and rotational stability.

We use the classic "ball on the floor" analogy to explain the fundamental concepts of stability and instability. Shown in Fig. 3–48 are three balls at rest on the floor. Case (*a*) is **stable** since any small disturbance (someone moves the ball to the right or left) generates a restoring force (due to gravity) that returns it to its initial position. Case (*b*) is **neutrally stable** because if someone moves the ball to the right or left, it would stay put at its new location. It has no tendency to move back to its original location, nor does it continue to move away. Case (*c*) is a situation in which the ball may be at rest at the moment, but any disturbance, even an infinitesimal one, causes the ball to roll off the hill—it does not return to its original position; rather it *diverges* from it. This situation is **unstable.** What about a case where the ball is on an *inclined* floor? It is not appropriate to discuss stability for this case since the ball is not in a state of equilibrium. In other words, it cannot be at rest and would roll down the hill even without any disturbance.

For an immersed or floating body in static equilibrium, the weight and the buoyant force acting on the body balance each other, and such bodies are

(*a*) Stable

(*b*) Neutrally stable

(*c*) Unstable

FIGURE 3–48
Stability is easily understood by analyzing a ball on the floor.

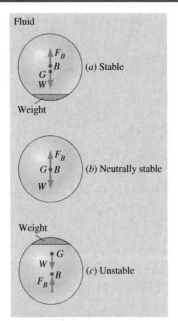

FIGURE 3–49
An immersed neutrally buoyant body is (*a*) stable if the center of gravity *G* is directly below the center of buoyancy *B* of the body, (*b*) neutrally stable if *G* and *B* are coincident, and (*c*) unstable if *G* is directly above *B*.

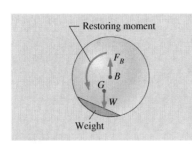

FIGURE 3–50
When the center of gravity *G* of an immersed neutrally buoyant body is not vertically aligned with the center of buoyancy *B* of the body, it is not in an equilibrium state and would rotate to its stable state, even without any disturbance.

inherently stable in the *vertical direction.* If an immersed neutrally buoyant body is raised or lowered to a different depth in an incompressible fluid, the body will remain in equilibrium at that location. If a floating body is raised or lowered somewhat by a vertical force, the body will return to its original position as soon as the external effect is removed. Therefore, a floating body possesses vertical stability, while an immersed neutrally buoyant body is neutrally stable since it does not return to its original position after a disturbance.

The *rotational stability* of an *immersed body* depends on the relative locations of the *center of gravity G* of the body and the *center of buoyancy B,* which is the centroid of the displaced volume. An immersed body is *stable* if the body is bottom-heavy and thus point *G* is directly below point *B* (Fig. 3–49*a*). A rotational disturbance of the body in such cases produces a *restoring moment* to return the body to its original stable position. Thus, a stable design for a submarine calls for the engines and the cabins for the crew to be located at the lower half in order to shift the weight to the bottom as much as possible. Hot-air or helium balloons (which can be viewed as being immersed in air) are also stable since the heavy cage that carries the load is at the bottom. An immersed body whose center of gravity *G* is directly above point *B* is *unstable,* and any disturbance will cause this body to turn upside down (Fig 3–49*c*). A body for which *G* and *B* coincide is *neutrally stable* (Fig 3–49*b*). This is the case for bodies whose density is constant throughout. For such bodies, there is no tendency to overturn or right themselves.

What about a case where the center of gravity is not vertically aligned with the center of buoyancy, as in Fig. 3–50? It is not appropriate to discuss stability for this case since the body is not in a state of equilibrium. In other words, it cannot be at rest and would rotate toward its stable state even without any disturbance. The restoring moment in the case shown in Fig. 3–50 is counterclockwise and causes the body to rotate counterclockwise so as to align point *G* vertically with point *B*. Note that there may be some oscillation, but eventually the body settles down at its stable equilibrium state [case (*a*) of Fig. 3–49]. The initial stability of the body of Fig. 3–50 is analogous to that of the ball on an inclined floor. Can you predict what would happen if the weight in the body of Fig. 3–50 were on the opposite side of the body?

The rotational stability criteria are similar for *floating bodies.* Again, if the floating body is bottom-heavy and thus the center of gravity *G* is directly below the center of buoyancy *B,* the body is always stable. But unlike immersed bodies, a floating body may still be stable when *G* is directly above *B* (Fig. 3–51). This is because the centroid of the displaced volume shifts to the side to a point *B′* during a rotational disturbance while the center of gravity *G* of the body remains unchanged. If point *B′* is sufficiently far, these two forces create a restoring moment and return the body to the original position. A measure of stability for floating bodies is the **metacentric height *GM*,** which is the distance between the center of gravity *G* and the metacenter *M*—the intersection point of the lines of action of the buoyant force through the body before and after rotation. The metacenter may be considered to be a fixed point for most hull shapes for small rolling angles up to about 20°. A floating body is stable if point *M* is above point *G,* and thus *GM* is positive, and unstable if point *M* is below point *G,* and thus *GM* is negative. In the

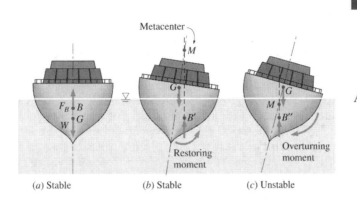

FIGURE 3–51
A floating body is *stable* if the body is (*a*) bottom-heavy and thus the center of gravity *G* is below the centroid *B* of the body, or (*b*) if the metacenter *M* is above point *G*. However, the body is (*c*) *unstable* if point *M* is below point *G*.

latter case, the weight and the buoyant force acting on the tilted body generate an overturning moment instead of a restoring moment, causing the body to capsize. The length of the metacentric height *GM* above *G* is a measure of the stability: the larger it is, the more stable is the floating body.

As already discussed, a boat can tilt to some maximum angle without capsizing, but beyond that angle it overturns (and sinks). We make a final analogy between the stability of floating objects and the stability of a ball rolling along the floor. Namely, imagine the ball in a trough between two hills (Fig. 3–52). The ball returns to its stable equilibrium position after being perturbed—up to a limit. If the perturbation amplitude is too great, the ball rolls down the opposite side of the hill and does not return to its equilibrium position. This situation is described as stable up to some limiting level of disturbance, but unstable beyond.

FIGURE 3–52
A ball in a trough between two hills is stable for small disturbances, but unstable for large disturbances.

3–7 · FLUIDS IN RIGID-BODY MOTION

We showed in Section 3–1 that pressure at a given point has the same magnitude in all directions, and thus it is a *scalar* function. In this section we obtain relations for the variation of pressure in fluids moving like a solid body with or without acceleration in the absence of any shear stresses (i.e., no motion between fluid layers relative to each other).

Many fluids such as milk and gasoline are transported in tankers. In an accelerating tanker, the fluid rushes to the back, and some initial splashing occurs. But then a new free surface (usually nonhorizontal) is formed, each fluid particle assumes the same acceleration, and the entire fluid moves like a rigid body. No shear stresses exist within the fluid body since there is no deformation and thus no change in shape. Rigid-body motion of a fluid also occurs when the fluid is contained in a tank that rotates about an axis.

Consider a differential rectangular fluid element of side lengths dx, dy, and dz in the x-, y-, and z-directions, respectively, with the z-axis being upward in the vertical direction (Fig. 3–53). Noting that the differential fluid element behaves like a *rigid body, Newton's second law of motion* for this element can be expressed as

$$\delta \vec{F} = \delta m \cdot \vec{a} \tag{3–34}$$

where $\delta m = \rho \, dV = \rho \, dx \, dy \, dz$ is the mass of the fluid element, $\vec{a}$ is the acceleration, and $\delta \vec{F}$ is the net force acting on the element.

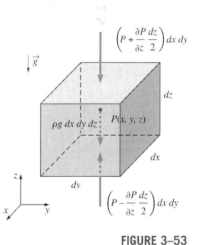

FIGURE 3–53
The surface and body forces acting on a differential fluid element in the vertical direction.

The forces acting on the fluid element consist of *body forces* such as gravity that act throughout the entire body of the element and are proportional to the volume of the body (and also electrical and magnetic forces, which will not be considered in this text), and *surface forces* such as the pressure forces that act on the surface of the element and are proportional to the surface area (shear stresses are also surface forces, but they do not apply in this case since the relative positions of fluid elements remain unchanged). The surface forces appear as the fluid element is isolated from its surroundings for analysis, and the effect of the detached body is replaced by a force at that location. Note that pressure represents the compressive force applied on the fluid element by the surrounding fluid and is always normal to the surface and inward toward the surface.

Taking the pressure at the center of the element to be P, the pressures at the top and bottom surfaces of the element can be expressed as $P + (\partial P/\partial z) \, dz/2$ and $P - (\partial P/\partial z) \, dz/2$, respectively. Noting that the pressure force acting on a surface is equal to the average pressure multiplied by the surface area, the net surface force acting on the element in the z-direction is the difference between the pressure forces acting on the bottom and top faces,

$$\delta F_{S,z} = \left(P - \frac{\partial P}{\partial z}\frac{dz}{2} \right) dx\,dy - \left(P + \frac{\partial P}{\partial z}\frac{dz}{2} \right) dx\,dy = -\frac{\partial P}{\partial z}\,dx\,dy\,dz \qquad \textbf{(3–35)}$$

Similarly, the net surface forces in the x- and y-directions are

$$\delta F_{S,x} = -\frac{\partial P}{\partial x}\,dx\,dy\,dz \quad \text{and} \quad \delta F_{S,y} = -\frac{\partial P}{\partial y}\,dx\,dy\,dz \qquad \textbf{(3–36)}$$

Then the surface force (which is simply the pressure force) acting on the entire element can be expressed in vector form as

$$\delta \vec{F}_S = \delta F_{S,x}\,\vec{i} + \delta F_{S,y}\,\vec{j} + \delta F_{S,z}\,\vec{k}$$

$$= -\left(\frac{\partial P}{\partial x}\vec{i} + \frac{\partial P}{\partial y}\vec{j} + \frac{\partial P}{\partial z}\vec{k} \right) dx\,dy\,dz = -\vec{\nabla}P\,dx\,dy\,dz \qquad \textbf{(3–37)}$$

where $\vec{i}$, $\vec{j}$, and $\vec{k}$ are the unit vectors in the x-, y-, and z-directions, respectively, and

$$\vec{\nabla}P = \frac{\partial P}{\partial x}\vec{i} + \frac{\partial P}{\partial y}\vec{j} + \frac{\partial P}{\partial z}\vec{k} \qquad \textbf{(3–38)}$$

is the *pressure gradient*. Note that $\vec{\nabla}$ or "del" is a vector operator that is used to express the gradients of a scalar function compactly in vector form. Also, the *gradient* of a scalar function is expressed in a given *direction* and thus it is a *vector* quantity.

The only body force acting on the fluid element is the weight of the element acting in the negative z-direction, and it is expressed as $\delta F_{B,z} = -g\delta m = -\rho g\,dx\,dy\,dz$ or in vector form as

$$\delta \vec{F}_{B,z} = -g\delta m\vec{k} = -\rho g\,dx\,dy\,dz\vec{k} \qquad \textbf{(3–39)}$$

Then the total force acting on the element becomes

$$\delta \vec{F} = \delta \vec{F}_S + \delta \vec{F}_B = -(\vec{\nabla}P + \rho g\vec{k})\,dx\,dy\,dz \qquad \textbf{(3–40)}$$

Substituting into Newton's second law of motion $\delta\vec{F} = \delta m \cdot \vec{a} = \rho \, dx \, dy \, dz \cdot \vec{a}$ and canceling $dx \, dy \, dz$, the general **equation of motion** for a fluid that acts as a rigid body (no shear stresses) is determined to be

Rigid-body motion of fluids: $\qquad \vec{\nabla}P + \rho g \vec{k} = -\rho \vec{a}$ $\qquad$ (3-41)

Resolving the vectors into their components, this relation can be expressed more explicitly as

$$\frac{\partial P}{\partial x}\vec{i} + \frac{\partial P}{\partial y}\vec{j} + \frac{\partial P}{\partial z}\vec{k} + \rho g \vec{k} = -\rho(a_x\vec{i} + a_y\vec{j} + a_z\vec{k}) \qquad (3\text{-}42)$$

or, in scalar form in the three orthogonal directions as

Accelerating fluids: $\qquad \dfrac{\partial P}{\partial x} = -\rho a_x, \quad \dfrac{\partial P}{\partial y} = -\rho a_y, \quad \text{and} \quad \dfrac{\partial P}{\partial z} = -\rho(g + a_z)$ $\quad$ (3-43)

where a_x, a_y, and a_z are accelerations in the x-, y-, and z-directions, respectively.

Special Case 1: Fluids at Rest

For fluids at rest or moving on a straight path at constant velocity, all components of acceleration are zero, and the relations in Eqs. 3–43 reduce to

Fluids at rest: $\qquad \dfrac{\partial P}{\partial x} = 0, \quad \dfrac{\partial P}{\partial y} = 0, \quad \text{and} \quad \dfrac{dP}{dz} = -\rho g$ $\qquad$ (3-44)

which confirm that, in fluids at rest, the pressure remains constant in any horizontal direction (P is independent of x and y) and varies only in the vertical direction as a result of gravity [and thus $P = P(z)$]. These relations are applicable for both compressible and incompressible fluids (Fig. 3–54).

Special Case 2: Free Fall of a Fluid Body

A freely falling body accelerates under the influence of gravity. When the air resistance is negligible, the acceleration of the body equals the gravitational acceleration, and acceleration in any horizontal direction is zero. Therefore, $a_x = a_y = 0$ and $a_z = -g$. Then the equations of motion for accelerating fluids (Eqs. 3–43) reduce to

Free-falling fluids: $\qquad \dfrac{\partial P}{\partial x} = \dfrac{\partial P}{\partial y} = \dfrac{\partial P}{\partial z} = 0 \quad \rightarrow \quad P = \text{constant}$ $\quad$ (3-45)

Therefore, in a frame of reference moving with the fluid, it behaves like it is in an environment with zero gravity. (This is the situation in an orbiting spacecraft, by the way. Gravity is *not* zero up there, despite what many people think!) Also, the gage pressure in a drop of liquid in free fall is zero throughout. (Actually, the gage pressure is slightly above zero due to surface tension, which holds the drop intact.)

When the direction of motion is reversed and the fluid is forced to accelerate vertically with $a_z = +g$ by placing the fluid container in an elevator or a space vehicle propelled upward by a rocket engine, the pressure gradient in the z-direction is $\partial P/\partial z = -2\rho g$. Therefore, the pressure difference across a fluid layer now doubles relative to the stationary fluid case (Fig. 3–55).

FIGURE 3–54
A glass of water at rest is a special case of a fluid in rigid-body motion. If the glass of water were moving at constant velocity in any direction, the hydrostatic equations would still apply.

© Imagestate Media (John Foxx)/Imagestate RF

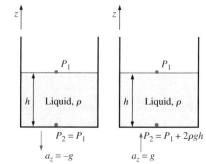

(*a*) Free fall of a liquid

(*b*) Upward acceleration of a liquid with $a_z = +g$

FIGURE 3–55
The effect of acceleration on the pressure of a liquid during free fall and upward acceleration.

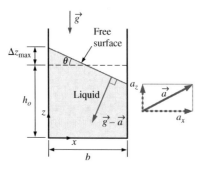

FIGURE 3–56
Rigid-body motion of a liquid in a linearly accelerating tank. The system behaves like a fluid at rest except that $\vec{g} - \vec{a}$ replaces $\vec{g}$ in the hydrostatic equations.

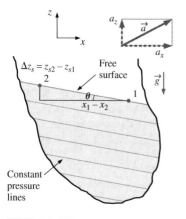

FIGURE 3–57
Lines of constant pressure (which are the projections of the surfaces of constant pressure on the xz-plane) in a linearly accelerating liquid. Also shown is the vertical rise.

Acceleration on a Straight Path

Consider a container partially filled with a liquid. The container is moving on a straight path with a constant acceleration. We take the projection of the path of motion on the horizontal plane to be the x-axis, and the projection on the vertical plane to be the z-axis, as shown in Fig. 3–56. The x- and z-components of acceleration are a_x and a_z. There is no movement in the y-direction, and thus the acceleration in that direction is zero, $a_y = 0$. Then the equations of motion for accelerating fluids (Eqs. 3–43) reduce to

$$\frac{\partial P}{\partial x} = -\rho a_x, \quad \frac{\partial P}{\partial y} = 0, \quad \text{and} \quad \frac{\partial P}{\partial z} = -\rho(g + a_z) \tag{3–46}$$

Therefore, pressure is independent of y. Then the total differential of $P = P(x, z)$, which is $(\partial P/\partial x)dx + (\partial P/\partial z)\, dz$, becomes

$$dP = -\rho a_x\, dx - \rho(g + a_z)\, dz \tag{3–47}$$

For $\rho = $ constant, the pressure difference between two points 1 and 2 in the fluid is determined by integration to be

$$P_2 - P_1 = -\rho a_x(x_2 - x_1) - \rho(g + a_z)(z_2 - z_1) \tag{3–48}$$

Taking point 1 to be the origin ($x = 0$, $z = 0$) where the pressure is P_0 and point 2 to be any point in the fluid (no subscript), the pressure distribution is expressed as

Pressure variation: $\qquad P = P_0 - \rho a_x x - \rho(g + a_z)z \tag{3–49}$

The vertical rise (or drop) of the free surface at point 2 relative to point 1 is determined by choosing both 1 and 2 on the free surface (so that $P_1 = P_2$), and solving Eq. 3–48 for $z_2 - z_1$ (Fig. 3–57),

Vertical rise of surface: $\qquad \Delta z_s = z_{s2} - z_{s1} = -\frac{a_x}{g + a_z}(x_2 - x_1) \tag{3–50}$

where z_s is the z-coordinate of the liquid's free surface. The equation for surfaces of constant pressure, called **isobars,** is obtained from Eq. 3–47 by setting $dP = 0$ and replacing z by z_{isobar}, which is the z-coordinate (the vertical distance) of the surface as a function of x. It gives

Surfaces of constant pressure: $\qquad \dfrac{dz_{\text{isobar}}}{dx} = -\dfrac{a_x}{g + a_z} = \text{constant} \tag{3–51}$

Thus we conclude that the isobars (including the free surface) in an incompressible fluid with constant acceleration in linear motion are parallel surfaces whose slope in the xz-plane is

Slope of isobars: $\qquad \text{Slope} = \dfrac{dz_{\text{isobar}}}{dx} = -\dfrac{a_x}{g + a_z} = -\tan\theta \tag{3–52}$

Obviously, the free surface of such a fluid is a *plane* surface, and it is inclined unless $a_x = 0$ (the acceleration is in the vertical direction only). Also, conservation of mass, together with the assumption of incompressibility ($\rho = constant$), requires that the volume of the fluid remain constant before and during acceleration. Therefore, the rise of fluid level on one side must be balanced by a drop of fluid level on the other side.

■ **EXAMPLE 3–12   Overflow from a Water Tank During Acceleration**

An 80-cm-high fish tank of cross section 2 m × 0.6 m that is partially filled with water is to be transported on the back of a truck (Fig. 3–58). The truck accelerates from 0 to 90 km/h in 10 s. If it is desired that no water spills during acceleration, determine the allowable initial water height in the tank. Would you recommend the tank to be aligned with the long or short side parallel to the direction of motion?

SOLUTION   A fish tank is to be transported on a truck. The allowable water height to avoid spill of water during acceleration and the proper orientation are to be determined.

Assumptions   **1** The road is horizontal during acceleration so that acceleration has no vertical component ($a_z = 0$). **2** Effects of splashing, braking, shifting gears, driving over bumps, climbing hills, etc., are assumed to be secondary and are not considered. **3** The acceleration remains constant.

Analysis   We take the x-axis to be the direction of motion, the z-axis to be the upward vertical direction, and the origin to be the lower left corner of the tank. Noting that the truck goes from 0 to 90 km/h in 10 s, the acceleration of the truck is

$$a_x = \frac{\Delta V}{\Delta t} = \frac{(90 - 0)\text{ km/h}}{10\text{ s}}\left(\frac{1\text{ m/s}}{3.6\text{ km/h}}\right) = 2.5\text{ m/s}^2$$

The tangent of the angle the free surface makes with the horizontal is

$$\tan\theta = \frac{a_x}{g + a_z} = \frac{2.5}{9.81 + 0} = 0.255 \qquad (\text{and thus } \theta = 14.3°)$$

The maximum vertical rise of the free surface occurs at the back of the tank, and the vertical midplane experiences no rise or drop during acceleration since it is a plane of symmetry. Then the vertical rise at the back of the tank relative to the midplane for the two possible orientations becomes

Case 1: The long side is parallel to the direction of motion:

$$\Delta z_{s1} = (b_1/2)\tan\theta = [(2\text{ m})/2] \times 0.255 = 0.255\text{ m} = \textbf{25.5 cm}$$

Case 2: The short side is parallel to the direction of motion:

$$\Delta z_{s2} = (b_2/2)\tan\theta = [(0.6\text{ m})/2] \times 0.255 = 0.076\text{ m} = \textbf{7.6 cm}$$

Therefore, assuming tipping is not a problem, **the tank should definitely be oriented such that its short side is parallel to the direction of motion.** Emptying the tank such that its free surface level drops just 7.6 cm in this case will be adequate to avoid spilling during acceleration.

Discussion   Note that the orientation of the tank is important in controlling the vertical rise. Also, the analysis is valid for any fluid with constant density, not just water, since we used no information that pertains to water in the solution.

FIGURE 3–58
Schematic for Example 3–12.

Rotation in a Cylindrical Container

We know from experience that when a glass filled with water is rotated about its axis, the fluid is forced outward as a result of the so-called centrifugal

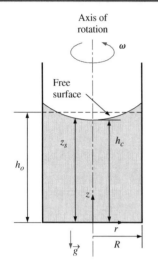

FIGURE 3–59
Rigid-body motion of a liquid in a rotating vertical cylindrical container.

FIGURE 3–60
Surfaces of constant pressure in a rotating liquid.

force (but more properly explained in terms of centripetal acceleration), and the free surface of the liquid becomes concave. This is known as the *forced vortex motion*.

Consider a vertical cylindrical container partially filled with a liquid. The container is now rotated about its axis at a constant angular velocity of ω, as shown in Fig. 3–59. After initial transients, the liquid will move as a rigid body together with the container. There is no deformation, and thus there can be no shear stress, and every fluid particle in the container moves with the same angular velocity.

This problem is best analyzed in cylindrical coordinates (r, θ, z), with z taken along the centerline of the container directed from the bottom toward the free surface, since the shape of the container is a cylinder, and the fluid particles undergo a circular motion. The centripetal acceleration of a fluid particle rotating with a constant angular velocity of ω at a distance r from the axis of rotation is $r\omega^2$ and is directed radially toward the axis of rotation (negative r-direction). That is, $a_r = -r\omega^2$. There is symmetry about the z-axis, which is the axis of rotation, and thus there is no θ dependence. Then $P = P(r, z)$ and $a_\theta = 0$. Also, $a_z = 0$ since there is no motion in the z-direction.

Then the equation of motion for accelerating fluids (Eq. 3–41) reduces to

$$\frac{\partial P}{\partial r} = \rho r\omega^2, \quad \frac{\partial P}{\partial \theta} = 0, \quad \text{and} \quad \frac{\partial P}{\partial z} = -\rho g \tag{3-53}$$

Then the total differential of $P = P(r, z)$, which is $dP = (\partial P/\partial r)dr + (\partial P/\partial z)dz$, becomes

$$dP = \rho r\omega^2\, dr - \rho g\, dz \tag{3-54}$$

The equation for surfaces of constant pressure is obtained by setting $dP = 0$ and replacing z by z_{isobar}, which is the z-value (the vertical distance) of the surface as a function of r. It gives

$$\frac{dz_{\text{isobar}}}{dr} = \frac{r\omega^2}{g} \tag{3-55}$$

Integrating, the equation for the surfaces of constant pressure is determined to be

Surfaces of constant pressure: $\qquad z_{\text{isobar}} = \dfrac{\omega^2}{2g} r^2 + C_1 \tag{3-56}$

which is the equation of a *parabola*. Thus we conclude that the surfaces of constant pressure, including the free surface, are *paraboloids of revolution* (Fig. 3–60).

The value of the integration constant C_1 is different for different paraboloids of constant pressure (i.e., for different isobars). For the free surface, setting $r = 0$ in Eq. 3–56 gives $z_{\text{isobar}}(0) = C_1 = h_c$, where h_c is the distance of the free surface from the bottom of the container along the axis of rotation (Fig. 3–59). Then the equation for the free surface becomes

$$z_s = \frac{\omega^2}{2g} r^2 + h_c \tag{3-57}$$

where z_s is the distance of the free surface from the bottom of the container at radius r. The underlying assumption in this analysis is that there is sufficient liquid in the container so that the entire bottom surface remains covered with liquid.

The volume of a cylindrical shell element of radius r, height z_s, and thickness dr is $dV = 2\pi r z_s\, dr$. Then the volume of the paraboloid formed by the free surface is

$$V = \int_{r=0}^{R} 2\pi z_s r\, dr = 2\pi \int_{r=0}^{R} \left(\frac{\omega^2}{2g} r^2 + h_c \right) r\, dr = \pi R^2 \left(\frac{\omega^2 R^2}{4g} + h_c \right) \qquad (3\text{--}58)$$

Since mass is conserved and density is constant, this volume must be equal to the original volume of the fluid in the container, which is

$$V = \pi R^2 h_0 \qquad (3\text{--}59)$$

where h_0 is the original height of the fluid in the container with no rotation. Setting these two volumes equal to each other, the height of the fluid along the centerline of the cylindrical container becomes

$$h_c = h_0 - \frac{\omega^2 R^2}{4g} \qquad (3\text{--}60)$$

Then the equation of the free surface becomes

Free surface: $\qquad\qquad z_s = h_0 - \frac{\omega^2}{4g}(R^2 - 2r^2) \qquad (3\text{--}61)$

The paraboloid shape is independent of fluid properties, so the same free surface equation applies to *any* liquid. For example, spinning liquid mercury forms a parabolic mirror that is useful in astronomy (Fig. 3–61).

The maximum vertical height occurs at the edge where $r = R$, and the *maximum height difference* between the edge and the center of the free surface is determined by evaluating z_s at $r = R$ and also at $r = 0$, and taking their difference,

Maximum height difference: $\qquad \Delta z_{s,\,max} = z_s(R) - z_s(0) = \frac{\omega^2}{2g} R^2 \qquad (3\text{--}62)$

When $\rho = $ constant, the pressure difference between two points 1 and 2 in the fluid is determined by integrating $dP = \rho r \omega^2\, dr - \rho g\, dz$. This yields

$$P_2 - P_1 = \frac{\rho \omega^2}{2}(r_2^2 - r_1^2) - \rho g(z_2 - z_1) \qquad (3\text{--}63)$$

Taking point 1 to be the origin ($r = 0$, $z = 0$) where the pressure is P_0 and point 2 to be any point in the fluid (no subscript), the pressure distribution is expressed as

Pressure variation: $\qquad\qquad P = P_0 + \frac{\rho \omega^2}{2} r^2 - \rho g z \qquad (3\text{--}64)$

Note that at a fixed radius, the pressure varies hydrostatically in the vertical direction, as in a fluid at rest. For a fixed vertical distance z, the pressure varies with the square of the radial distance r, increasing from the centerline toward the outer edge. In any horizontal plane, the pressure difference between the center and edge of the container of radius R is $\Delta P = \rho \omega^2 R^2 / 2$.

FIGURE 3–61
The 6-meter spinning liquid-mercury mirror of the Large Zenith Telescope located near Vancouver, British Columbia.

Photo courtesy of Paul Hickson, The University of British Columbia. Used by permission.

FIGURE 3–62
Schematic for Example 3–13.

EXAMPLE 3–13   **Rising of a Liquid During Rotation**

A 20-cm-diameter, 60-cm-high vertical cylindrical container, shown in Fig. 3–62, is partially filled with 50-cm-high liquid whose density is 850 kg/m³. Now the cylinder is rotated at a constant speed. Determine the rotational speed at which the liquid will start spilling from the edges of the container.

SOLUTION A vertical cylindrical container partially filled with a liquid is rotated. The angular speed at which the liquid will start spilling is to be determined.

Assumptions **1** The increase in the rotational speed is very slow so that the liquid in the container always acts as a rigid body. **2** The bottom surface of the container remains covered with liquid during rotation (no dry spots).

Analysis Taking the center of the bottom surface of the rotating vertical cylinder as the origin ($r = 0$, $z = 0$), the equation for the free surface of the liquid is given as

$$z_s = h_0 - \frac{\omega^2}{4g}(R^2 - 2r^2)$$

Then the vertical height of the liquid at the edge of the container where $r = R$ becomes

$$z_s(R) = h_0 + \frac{\omega^2 R^2}{4g}$$

where $h_0 = 0.5$ m is the original height of the liquid before rotation. Just before the liquid starts spilling, the height of the liquid at the edge of the container equals the height of the container, and thus $z_s(R) = H = 0.6$ m. Solving the last equation for ω and substituting, the maximum rotational speed of the container is determined to be

$$\omega = \sqrt{\frac{4g(H - h_0)}{R^2}} = \sqrt{\frac{4(9.81 \text{ m/s}^2)[(0.6 - 0.5) \text{ m}]}{(0.1 \text{ m})^2}} = 19.8 \text{ rad/s}$$

Noting that one complete revolution corresponds to 2π rad, the rotational speed of the container can also be expressed in terms of revolutions per minute (rpm) as

$$\dot{n} = \frac{\omega}{2\pi} = \frac{19.8 \text{ rad/s}}{2\pi \text{ rad/rev}}\left(\frac{60 \text{ s}}{1 \text{ min}}\right) = 189 \text{ rpm}$$

Therefore, the rotational speed of this container should be limited to 189 rpm to avoid any spill of liquid as a result of the centrifugal effect.

Discussion Note that the analysis is valid for any liquid since the result is independent of density or any other fluid property. We should also verify that our assumption of no dry spots is valid. The liquid height at the center is

$$z_s(0) = h_0 - \frac{\omega^2 R^2}{4g} = 0.4 \text{ m}$$

Since $z_s(0)$ is positive, our assumption is validated.

SUMMARY

The normal force exerted by a fluid per unit area is called *pressure*, and its SI unit is the *pascal*, 1 Pa $\equiv$ 1 N/m^2. The pressure relative to absolute vacuum is called the *absolute pressure*, and the difference between the absolute pressure and the local atmospheric pressure is called the *gage pressure*. Pressures below atmospheric pressure are sometimes called *vacuum pressures*. The absolute, gage, and vacuum pressures are related by

$$P_{\text{gage}} = P_{\text{abs}} - P_{\text{atm}}$$

$$P_{\text{vac}} = P_{\text{atm}} - P_{\text{abs}} = -P_{\text{gage}}$$

The pressure at a point in a fluid has the same magnitude in all directions. The variation of pressure with elevation in a fluid at rest is given by

$$\frac{dP}{dz} = -\rho g$$

where the positive z-direction is taken to be upward by convention. When the density of the fluid is constant, the pressure difference across a fluid layer of thickness Δz is

$$P_{\text{below}} = P_{\text{above}} + \rho g|\Delta z| = P_{\text{above}} + \gamma_s|\Delta z|$$

The absolute and gage pressures in a static liquid open to the atmosphere at a depth h from the free surface are

$$P = P_{\text{atm}} + \rho g h \quad \text{and} \quad P_{\text{gage}} = \rho g h$$

The pressure in a fluid at rest does not vary in the horizontal direction. *Pascal's law* states that the pressure applied to a confined fluid increases the pressure throughout by the same amount. The atmospheric pressure can be measured by a *barometer* and is given by

$$P_{\text{atm}} = \rho g h$$

where h is the height of the liquid column.

Fluid statics deals with problems associated with fluids at rest, and it is called *hydrostatics* when the fluid is a liquid. The magnitude of the resultant force acting on a plane surface of a completely submerged plate in a homogeneous fluid is equal to the product of the pressure P_C at the centroid of the surface and the area A of the surface and is expressed as

$$F_R = (P_0 + \rho g h_C)A = P_C A = P_{\text{avg}}A$$

where $h_C = y_C \sin \theta$ is the *vertical distance* of the centroid from the free surface of the liquid. The pressure P_0 is usually atmospheric pressure, which cancels out in most cases since it acts on both sides of the plate. The point of intersection of the line of action of the resultant force and the surface is the *center of pressure*. The vertical location of the line of action of the resultant force is given by

$$y_P = y_C + \frac{I_{xx, C}}{[y_C + P_0/(\rho g \sin \theta)]A}$$

where $I_{xx, C}$ is the second moment of area about the x-axis passing through the centroid of the area.

A fluid exerts an upward force on a body immersed in it. This force is called the *buoyant force* and is expressed as

$$F_B = \rho_f g V$$

where V is the volume of the body. This is known as *Archimedes' principle* and is expressed as: the buoyant force acting on a body immersed in a fluid is equal to the weight of the fluid displaced by the body; it acts upward through the centroid of the displaced volume. In a fluid with constant density, the buoyant force is independent of the distance of the body from the free surface. For *floating* bodies, the submerged volume fraction of the body is equal to the ratio of the average density of the body to the density of the fluid.

The general *equation of motion* for a fluid that acts as a rigid body is

$$\vec{\nabla}P + \rho g\vec{k} = -\rho\vec{a}$$

When gravity is aligned in the $-z$-direction, it is expressed in scalar form as

$$\frac{\partial P}{\partial x} = -\rho a_x, \quad \frac{\partial P}{\partial y} = -\rho a_y, \quad \text{and} \quad \frac{\partial P}{\partial z} = -\rho(g + a_z)$$

where a_x, a_y, and a_z are accelerations in the x-, y-, and z-directions, respectively. During *linearly accelerating motion* in the xz-plane, the pressure distribution is expressed as

$$P = P_0 - \rho a_x x - \rho(g + a_z)z$$

The surfaces of constant pressure (including the free surface) in a liquid with constant acceleration in linear motion are parallel surfaces whose slope in some xz-plane is

$$\text{Slope} = \frac{dz_{\text{isobar}}}{dx} = -\frac{a_x}{g + a_z} = -\tan \theta$$

During rigid-body motion of a liquid in a *rotating cylinder*, the surfaces of constant pressure are *paraboloids of revolution*. The equation for the free surface is

$$z_s = h_0 - \frac{\omega^2}{4g}(R^2 - 2r^2)$$

where z_s is the distance of the free surface from the bottom of the container at radius r and h_0 is the original height of the fluid in the container with no rotation. The variation of pressure in the liquid is expressed as

$$P = P_0 + \frac{\rho\omega^2}{2}r^2 - \rho g z$$

where P_0 is the pressure at the origin ($r = 0$, $z = 0$).

Pressure is a fundamental property, and it is hard to imagine a significant fluid flow problem that does not involve pressure. Therefore, you will see this property in all chapters in the rest of this book. The consideration of hydrostatic forces acting on plane or curved surfaces, however, is mostly limited to this chapter.

REFERENCES AND SUGGESTED READING

1. F. P. Beer, E. R. Johnston, Jr., E. R. Eisenberg, and G. H. Staab. *Vector Mechanics for Engineers, Statics,* 10th ed. New York: McGraw-Hill, 2012.

2. D. C. Giancoli. *Physics,* 6th ed. Upper Saddle River, NJ: Prentice Hall, 2012.

PROBLEMS*

Pressure, Manometer, and Barometer

3–1C Someone claims that the absolute pressure in a liquid of constant density doubles when the depth is doubled. Do you agree? Explain.

3–2C A tiny steel cube is suspended in water by a string. If the lengths of the sides of the cube are very small, how would you compare the magnitudes of the pressures on the top, bottom, and side surfaces of the cube?

3–3C Express Pascal's law, and give a real-world example of it.

3–4C Consider two identical fans, one at sea level and the other on top of a high mountain, running at identical speeds. How would you compare (a) the volume flow rates and (b) the mass flow rates of these two fans?

3–5C What is the difference between gage pressure and absolute pressure?

3–6C Explain why some people experience nose bleeding and some others experience shortness of breath at high elevations.

3–7 The piston of a vertical piston-cylinder device containing a gas has a mass of 40 kg and a cross-sectional area of 0.012 m² (Fig P3–7). The local atmospheric pressure is 95 kPa, and the gravitational acceleration is 9.81 m/s². (a) Determine the pressure inside the cylinder. (b) If some heat is transferred to the gas and its volume is doubled, do you expect the pressure inside the cylinder to change?

$P_{atm} = 95$ kPa
$m = 40$ kg

$A = 0.012$ m²

FIGURE P3–7

* Problems designated by a "C" are concept questions, and students are encouraged to answer them all. Problems designated by an "E" are in English units, and the SI users can ignore them. Problems with the ⚙ icon are solved using EES, and complete solutions together with parametric studies are included on the text website. Problems with the 🔧 icon are comprehensive in nature and are intended to be solved with an equation solver such as EES.

3–8 A vacuum gage connected to a chamber reads 36 kPa at a location where the atmospheric pressure is 92 kPa. Determine the absolute pressure in the chamber.

3–9E The pressure at the exit of an air compressor is 150 psia. What is this pressure in kPa?

3–10E The pressure in a water line is 1500 kPa. What is the line pressure in (a) lbf/ft² units and (b) lbf/in² (psi) units?

3–11E A manometer is used to measure the air pressure in a tank. The fluid used has a specific gravity of 1.25, and the differential height between the two arms of the manometer is 28 in. If the local atmospheric pressure is 12.7 psia, determine the absolute pressure in the tank for the cases of the manometer arm with the (a) higher and (b) lower fluid level being attached to the tank.

3–12 The water in a tank is pressurized by air, and the pressure is measured by a multifluid manometer as shown in Fig. P3–12. Determine the gage pressure of air in the tank if $h_1 = 0.4$ m, $h_2 = 0.6$ m, and $h_3 = 0.8$ m. Take the densities of water, oil, and mercury to be 1000 kg/m³, 850 kg/m³, and 13,600 kg/m³, respectively.

Air
1
Water
Oil
h_1
h_2
2
h_3
Mercury

FIGURE P3–12

3–13 Determine the atmospheric pressure at a location where the barometric reading is 735 mmHg. Take the density of mercury to be 13,600 kg/m³.

3–14 The gage pressure in a liquid at a depth of 3 m is read to be 28 kPa. Determine the gage pressure in the same liquid at a depth of 12 m.

3–15 The absolute pressure in water at a depth of 8 m is read to be 175 kPa. Determine (*a*) the local atmospheric pressure, and (*b*) the absolute pressure at a depth of 8 m in a liquid whose specific gravity is 0.78 at the same location.

3–16E Show that 1 kgf/cm^2 = 14.223 psi.

3–17E A 200-lb man has a total foot imprint area of 72 in^2. Determine the pressure this man exerts on the ground if (*a*) he stands on both feet and (*b*) he stands on one foot.

3–18 Consider a 55-kg woman who has a total foot imprint area of 400 cm^2. She wishes to walk on the snow, but the snow cannot withstand pressures greater than 0.5 kPa. Determine the minimum size of the snowshoes needed (imprint area per shoe) to enable her to walk on the snow without sinking.

3–19 A vacuum gage connected to a tank reads 45 kPa at a location where the barometric reading is 755 mmHg. Determine the absolute pressure in the tank. Take ρ_{Hg} = 13,590 kg/m^3.
Answer: 55.6 kPa

3–20E A pressure gage connected to a tank reads 50 psi at a location where the barometric reading is 29.1 inHg. Determine the absolute pressure in the tank. Take ρ_{Hg} = 848.4 lbm/ft^3.
Answer: 64.3 psia

3–21 A pressure gage connected to a tank reads 500 kPa at a location where the atmospheric pressure is 94 kPa. Determine the absolute pressure in the tank.

3–22 If the pressure inside a rubber balloon is 1500 mmHg, what is this pressure in pounds-force per square inch (psi)?
Answer: 29.0 psi

3–23 The vacuum pressure of a condenser is given to be 80 kPa. If the atmospheric pressure is 98 kPa, what is the gage pressure and absolute pressure in kPa, kN/m^2, lbf/in^2, psi, and mmHg.

3–24 Water from a reservoir is raised in a vertical tube of internal diameter D = 30 cm under the influence of the pulling force F of a piston. Determine the force needed to raise the water to a height of h = 1.5 m above the free surface. What would your response be for h = 3 m? Also, taking the atmospheric pressure to be 96 kPa, plot the absolute water pressure at the piston face as h varies from 0 to 3 m.

FIGURE P3–24

3–25 The barometer of a mountain hiker reads 980 mbars at the beginning of a hiking trip and 790 mbars at the end.

Neglecting the effect of altitude on local gravitational acceleration, determine the vertical distance climbed. Assume an average air density of 1.20 kg/m^3.   *Answer:* 1614 m

3–26 The basic barometer can be used to measure the height of a building. If the barometric readings at the top and at the bottom of a building are 730 and 755 mmHg, respectively, determine the height of the building. Assume an average air density of 1.18 kg/m^3.

FIGURE P3–26

3–27 [EES] Solve Prob. 3–26 using EES (or other) software. Print out the entire solution, including the numerical results with proper units, and take the density of mercury to be 13,600 kg/m^3.

3–28 Determine the pressure exerted on a diver at 20 m below the free surface of the sea. Assume a barometric pressure of 101 kPa and a specific gravity of 1.03 for seawater.   *Answer:* 303 kPa

3–29E Determine the pressure exerted on the surface of a submarine cruising 225 ft below the free surface of the sea. Assume that the barometric pressure is 14.7 psia and the specific gravity of seawater is 1.03.

3–30 A gas is contained in a vertical, frictionless piston–cylinder device. The piston has a mass of 4 kg and a cross-sectional area of 35 cm^2. A compressed spring above the piston exerts a force of 60 N on the piston. If the atmospheric pressure is 95 kPa, determine the pressure inside the cylinder.
Answer: 123.4 kPa

FIGURE P3–30

3–31 Reconsider Prob. 3–30. Using EES (or other) software, investigate the effect of the spring force in the range of 0 to 500 N on the pressure inside the cylinder. Plot the pressure against the spring force, and discuss the results.

3–32 Both a gage and a manometer are attached to a gas tank to measure its pressure. If the reading on the pressure gage is 65 kPa, determine the distance between the two fluid levels of the manometer if the fluid is (a) mercury ($\rho = 13{,}600$ kg/m^3) or (b) water ($\rho = 1000$ kg/m^3).

$P_g = 65$ kPa

Gas

$h = ?$

FIGURE P3–32

3–33 Reconsider Prob. 3–32. Using EES (or other) software, investigate the effect of the manometer fluid density in the range of 800 to 13,000 kg/m^3 on the differential fluid height of the manometer. Plot the differential fluid height against the density, and discuss the results.

3–34 The variation of pressure P in a gas with density ρ is is given by $P = C\rho^n$ where C and n and are constants with $P = P_0$ and $\rho = \rho_0$ at elevation $z = 0$. Obtain a relation for the variaton of P with elevation in terms of z, g, n, P_0 and ρ_0.

3–35 The system shown in the figure is used to accurately measure the pressure changes when the pressure is increased

Pipe

Water

Δh

Glycerin, SG = 1.26

$D = 30$ mm

$d = 3$ mm

FIGURE P3–35

by ΔP in the water pipe. When $\Delta h = 70$ mm, what is the change in the pipe pressure?

3–36 The manometer shown in the figure is designed to measure pressures of up to a maximum of 100 Pa. If the reading error is estimated to be ± 0.5 mm, what should the ratio of d/D be in order for the error associated with pressure measurement not to exceed 2.5% of the full scale.

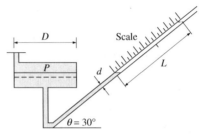

D Scale

P d L

$\theta = 30°$

FIGURE P3–36

3–37 A manometer containing oil ($\rho = 850$ kg/m^3) is attached to a tank filled with air. If the oil-level difference between the two columns is 150 cm and the atmospheric pressure is 98 kPa, determine the absolute pressure of the air in the tank.   *Answer:* 111 kPa

3–38 A mercury manometer ($\rho = 13{,}600$ kg/m^3) is connected to an air duct to measure the pressure inside. The difference in the manometer levels is 10 mm, and the atmospheric pressure is 100 kPa. (a) Judging from Fig. P3–38, determine if the pressure in the duct is above or below the atmospheric pressure. (b) Determine the absolute pressure in the duct.

AIR

$P = ?$

$h = 10$ mm

FIGURE P3–38

3–39 Repeat Prob. 3–38 for a differential mercury height of 30 mm.

3–40 Blood pressure is usually measured by wrapping a closed air-filled jacket equipped with a pressure gage around the upper arm of a person at the level of the heart. Using a mercury manometer and a stethoscope, the systolic pressure (the maximum pressure when the heart is pumping) and the diastolic pressure (the minimum pressure when the heart is resting) are measured in mmHg. The systolic and diastolic pressures of a healthy person are about 120 mmHg and 80 mmHg, respectively, and are indicated as 120/80. Express both of these gage pressures in kPa, psi, and meter water column.

3–41   The maximum blood pressure in the upper arm of a healthy person is about 120 mmHg. If a vertical tube open to the atmosphere is connected to the vein in the arm of the person, determine how high the blood will rise in the tube. Take the density of the blood to be 1040 kg/m^3.

FIGURE P3–41

3–42   Consider a 1.73-m-tall man standing vertically in water and completely submerged in a pool. Determine the difference between the pressures acting at the head and at the toes of this man, in kPa.

3–43   Consider a U-tube whose arms are open to the atmosphere. Now water is poured into the U-tube from one arm, and light oil ($\rho = 790$ kg/m^3) from the other. One arm contains 70-cm-high water, while the other arm contains both fluids with an oil-to-water height ratio of 6. Determine the height of each fluid in that arm.

FIGURE P3–43

3–44   The hydraulic lift in a car repair shop has an output diameter of 40 cm and is to lift cars up to 1800 kg. Determine the fluid gage pressure that must be maintained in the reservoir.

3–45   Freshwater and seawater flowing in parallel horizontal pipelines are connected to each other by a double U-tube manometer, as shown in Fig. P3–45. Determine the pressure difference between the two pipelines. Take the density of seawater at that location to be $\rho = 1035$ kg/m^3. Can the air column be ignored in the analysis?

FIGURE P3–45

3–46   Repeat Prob. 3–45 by replacing the air with oil whose specific gravity is 0.72.

3–47E   The pressure in a natural gas pipeline is measured by the manometer shown in Fig. P3–47E with one of the arms open to the atmosphere where the local atmospheric pressure is 14.2 psia. Determine the absolute pressure in the pipeline.

FIGURE P3–47E

3–48E   Repeat Prob. 3–47E by replacing air by oil with a specific gravity of 0.69.

3–49   The gage pressure of the air in the tank shown in Fig. P3–49 is measured to be 65 kPa. Determine the differential height h of the mercury column.

FIGURE P3–49

3–50 Repeat Prob. 3–49 for a gage pressure of 45 kPa.

3–51 The 500-kg load on the hydraulic lift shown in Fig. P3–51 is to be raised by pouring oil ($\rho = 780$ kg/m³) into a thin tube. Determine how high h should be in order to begin to raise the weight.

FIGURE P3–51

3–52E Two oil tanks are connected to each other through a manometer. If the difference between the mercury levels in the two arms is 32 in, determine the pressure difference between the two tanks. The densities of oil and mercury are 45 lbm/ft³ and 848 lbm/ft³, respectively.

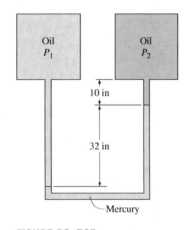

FIGURE P3–52E

3–53 Pressure is often given in terms of a liquid column and is expressed as "pressure head." Express the standard atmospheric pressure in terms of (*a*) mercury (SG = 13.6), (*b*) water (SG = 1.0), and (*c*) glycerin (SG = 1.26) columns. Explain why we usually use mercury in manometers.

3–54 Two chambers with the same fluid at their base are separated by a 30-cm-diameter piston whose weight is 25 N, as shown in Fig. P3–54. Calculate the gage pressures in chambers *A* and *B*.

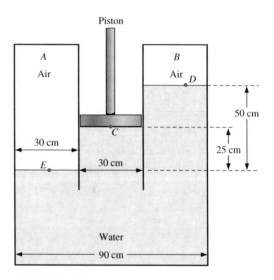

FIGURE P3–54

3–55 Consider a double-fluid manometer attached to an air pipe shown in Fig. P3–55. If the specific gravity of one fluid is 13.55, determine the specific gravity of the other fluid for the indicated absolute pressure of air. Take the atmospheric pressure to be 100 kPa. *Answer:* 1.34

FIGURE P3–55

3–56 The pressure difference between an oil pipe and water pipe is measured by a double-fluid manometer, as shown in

Fig. P3–56. For the given fluid heights and specific gravities, calculate the pressure difference $\Delta P = P_B - P_A$.

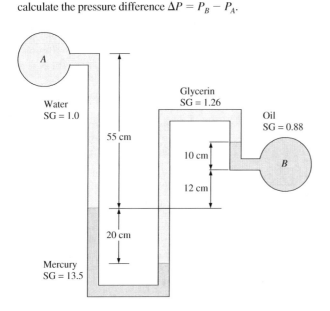

FIGURE P3–56

3–57 Consider the system shown in Fig. P3–57. If a change of 0.9 kPa in the pressure of air causes the brine-mercury interface in the right column to drop by 5 mm in the brine level in the right column while the pressure in the brine pipe remains constant, determine the ratio of A_2/A_1.

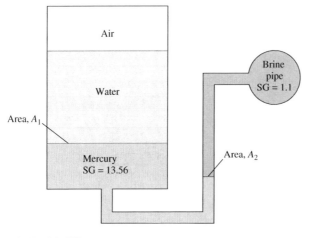

FIGURE P3–57

3–58 Two water tanks are connected to each other through a mercury manometer with inclined tubes, as shown in Fig. P3–58. If the pressure difference between the two tanks is 20 kPa, calculate a and θ.

FIGURE P3–58

3–59 Consider a hydraulic jack being used in a car repair shop, as in Fig. P3–59. The pistons have an area of $A_1 = 0.8 \text{ cm}^2$ and $A_2 = 0.04 \text{ m}^2$. Hydraulic oil with a specific gravity of 0.870 is pumped in as the small piston on the left side is pushed up and down, slowly raising the larger piston on the right side. A car that weighs 13,000 N is to be jacked up. (*a*) At the beginning, when both pistons are at the same elevation ($h = 0$), calculate the force F_1 in newtons required to hold the weight of the car. (*b*) Repeat the calculation after the car has been lifted two meters ($h = 2$ m). Compare and discuss.

FIGURE P3–59

Fluid Statics: Hydrostatic Forces on Plane and Curved Surfaces

3–60C Define the resultant hydrostatic force acting on a submerged surface, and the center of pressure.

3–61C Someone claims that she can determine the magnitude of the hydrostatic force acting on a plane surface submerged in water regardless of its shape and orientation if she knew the vertical distance of the centroid of the surface from the free surface and the area of the surface. Is this a valid claim? Explain.

3–62C A submerged horizontal flat plate is suspended in water by a string attached at the centroid of its upper surface. Now the plate is rotated 45° about an axis that passes through its centroid. Discuss the change in the hydrostatic force acting on the top surface of this plate as a result of this rotation. Assume the plate remains submerged at all times.

3–63C You may have noticed that dams are much thicker at the bottom. Explain why dams are built that way.

3–64C Consider a submerged curved surface. Explain how you would determine the horizontal component of the hydrostatic force acting on this surface.

3–65C Consider a submerged curved surface. Explain how you would determine the vertical component of the hydrostatic force acting on this surface.

3–66C Consider a circular surface subjected to hydrostatic forces by a constant density liquid. If the magnitudes of the horizontal and vertical components of the resultant hydrostatic force are determined, explain how you would find the line of action of this force.

3–67 Consider a heavy car submerged in water in a lake with a flat bottom. The driver's side door of the car is 1.1 m high and 0.9 m wide, and the top edge of the door is 10 m below the water surface. Determine the net force acting on the door (normal to its surface) and the location of the pressure center if (*a*) the car is well-sealed and it contains air at atmospheric pressure and (*b*) the car is filled with water.

3–68E A long, solid cylinder of radius 2 ft hinged at point *A* is used as an automatic gate, as shown in Fig. P3–68E. When the water level reaches 15 ft, the cylindrical gate opens by turning about the hinge at point *A*. Determine (*a*) the hydrostatic force acting on the cylinder and its line of action when the gate opens and (*b*) the weight of the cylinder per ft length of the cylinder.

FIGURE P3–68E

3–69 Consider a 8-m-long, 8-m-wide, and 2-m-high aboveground swimming pool that is filled with water to the rim. (*a*) Determine the hydrostatic force on each wall and the distance of the line of action of this force from the ground. (*b*) If the height of the walls of the pool is doubled and the pool is filled, will the hydrostatic force on each wall double or quadruple? Why? *Answer:* (*a*) 157 kN

3–70E Consider a 200-ft-high, 1200-ft-wide dam filled to capacity. Determine (*a*) the hydrostatic force on the dam and

(*b*) the force per unit area of the dam near the top and near the bottom.

3–71 A room in the lower level of a cruise ship has a 30-cm-diameter circular window. If the midpoint of the window is 4 m below the water surface, determine the hydrostatic force acting on the window, and the pressure center. Take the specific gravity of seawater to be 1.025. *Answers:* 2840 N, 4.001 m

FIGURE P3–71

3–72 The water side of the wall of a 70-m-long dam is a quarter circle with a radius of 7 m. Determine the hydrostatic force on the dam and its line of action when the dam is filled to the rim.

3–73 For a gate width of 2 m into the paper (Fig. P3–73), determine the force required to hold the gate *ABC* at its location. *Answer:* 17.8 kN

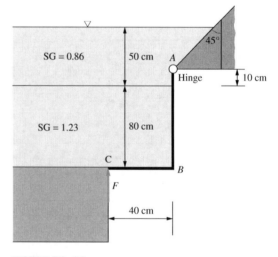

FIGURE P3–73

3–74 Determine the resultant force acting on the 0.7-m-high and 0.7-m-wide triangular gate shown in Fig. P3–74 and its line of action.

FIGURE P3–74

3–75 A 6-m-high, 5-m-wide rectangular plate blocks the end of a 5-m-deep freshwater channel, as shown in Fig. P3–75. The plate is hinged about a horizontal axis along its upper edge through a point A and is restrained from opening by a fixed ridge at point B. Determine the force exerted on the plate by the ridge.

FIGURE P3–75

3–76 Reconsider Prob. 3–75. Using EES (or other) software, investigate the effect of water depth on the force exerted on the plate by the ridge. Let the water depth vary from 0 to 5 m in increments of 0.5 m. Tabulate and plot your results.

3–77E The flow of water from a reservoir is controlled by a 5-ft-wide L-shaped gate hinged at point A, as shown

FIGURE P3–77E

in Fig. P3–77E. If it is desired that the gate open when the water height is 12 ft, determine the mass of the required weight W. *Answer:* 30,900 lbm

3–78E Repeat Prob. 3–77E for a water height of 8 ft.

3–79 A water trough of semicircular cross section of radius 0.6 m consists of two symmetric parts hinged to each other at the bottom, as shown in Fig. P3–79. The two parts are held together by a cable and turnbuckle placed every 3 m along the length of the trough. Calculate the tension in each cable when the trough is filled to the rim.

FIGURE P3–79

3–80 A cylindrical tank is fully filled with water (Fig. P3–80). In order to increase the flow from the tank, an additional pressure is applied to the water surface by a compressor. For $P_0 = 0$, $P_0 = 3$ bar, and $P_0 = 10$ bar, calculate the hydrostatic force on the surface A exerted by water.

FIGURE P3–80

3–81 An open settling tank shown in the figure contains a liquid suspension. Determine the resultant force acting on the gate and its line of action if the liquid density is 850 kg/m^3. *Answers:* 140 kN, 1.64 m from bottom

FIGURE P3–81

3–82 From Prob. 3-81, knowing that the density of the suspension depends on liquid depth and changes linearly from 800 kg/m³ to 900 kg/m³ in the vertical direction, determine the resultant force acting on the gate *ABC*, and its line of action.

3–83 The 2.5 m × 8.1 m × 6 m tank shown below is filled by oil of SG = 0.88. Determine (*a*) the magnitude and the location of the line of action of the resultant force acting on surface *AB* and (*b*) the pressure force acting on surface *BD*. Will the force acting on surface *BD* equal the weight of the oil in the tank? Explain.

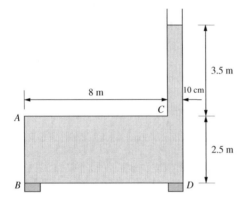

FIGURE P3–83

3–84 The two sides of a V-shaped water trough are hinged to each other at the bottom where they meet, as shown in Fig. P3–84, making an angle of 45° with the ground from both sides. Each side is 0.75 m wide, and the two parts are held together by a cable and turnbuckle placed every 6 m along the length of the trough. Calculate the tension in each cable when the trough is filled to the rim. *Answer: 5510 N*

FIGURE P3–84

3–85 Repeat Prob. 3–84 for the case of a partially filled trough with a water height of 0.4 m directly above the hinge.

3–86 A retaining wall against a mud slide is to be constructed by placing 1.2-m-high and 0.25-m-wide rectangular concrete blocks (ρ = 2700 kg/m³) side by side, as shown in Fig. P3–86. The friction coefficient between the ground and the concrete blocks is f = 0.4, and the density of the mud is about 1400 kg/m³. There is concern that the concrete blocks may slide or tip over the lower left edge as the mud level rises.

Determine the mud height at which (*a*) the blocks will overcome friction and start sliding and (*b*) the blocks will tip over.

FIGURE P3–86

3–87 Repeat Prob. 3–86 for 0.4-m-wide concrete blocks.

3–88 A 4-m-long quarter-circular gate of radius 3 m and of negligible weight is hinged about its upper edge *A*, as shown in Fig. P3–88. The gate controls the flow of water over the ledge at *B*, where the gate is pressed by a spring. Determine the minimum spring force required to keep the gate closed when the water level rises to *A* at the upper edge of the gate.

FIGURE P3–88

3–89 Repeat Prob. 3–88 for a radius of 4 m for the gate. *Answer: 314 kN*

3–90 Consider a flat plate of thickness *t*, width *w* into the page, and length *b* submerged in water, as in Fig. P3–90. The depth of water from the surface to the center of the plate is *H*,

FIGURE P3–90

and angle θ is defined relative to the center of the plate. (a) Generate an equation for the force F on the upper face of the plate as a function of (at most) H, b, t, w, g, ρ, and θ. Ignore atmospheric pressure. In other words, calculate the force that is *in addition to* the force due to atmospheric pressure. (b) As a test of your equation, let $H = 1.25$ m, $b = 1$ m, $t = 0.2$ m, $w = 1$ m, $g = 9.807$ m/s^2, $\rho = 998.3$ kg/m^3, and $\theta = 30°$. If your equation is correct, you should get a force of 11.4 kN.

3–91 The weight of the gate separating the two fluids is such that the system shown in Fig. P3–91 is at static equilibrium. If it is known that $F_1/F_2 = 1.70$, determine h/H.

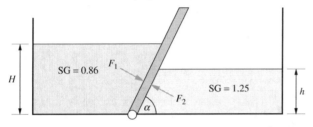

FIGURE P3–91

3–92 Consider a 1-m wide inclined gate of negligible weight that separates water from another fluid. What would be the volume of the concrete block (SG = 2.4) immersed in water to keep the gate at the position shown? Disregard any frictional effects.

FIGURE P3–92

3–93 The parabolic shaped gate with a width of 2 m shown in Fig. P3–93 is hinged at point B. Determine the force F needed to keep the gate stationary.

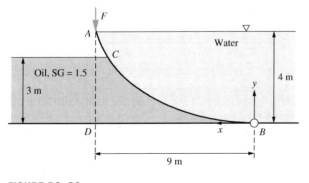

FIGURE P3–93

Buoyancy

3–94C What is buoyant force? What causes it? What is the magnitude of the buoyant force acting on a submerged body whose volume is V? What are the direction and the line of action of the buoyant force?

3–95C Consider two identical spherical balls submerged in water at different depths. Will the buoyant forces acting on these two balls be the same or different? Explain.

3–96C Consider two 5-cm-diameter spherical balls—one made of aluminum, the other of iron—submerged in water. Will the buoyant forces acting on these two balls be the same or different? Explain.

3–97C Consider a 3-kg copper cube and a 3-kg copper ball submerged in a liquid. Will the buoyant forces acting on these two bodies be the same or different? Explain.

3–98C Discuss the stability of (a) a submerged and (b) a floating body whose center of gravity is above the center of buoyancy.

3–99 The density of a liquid is to be determined by an old 1-cm-diameter cylindrical hydrometer whose division marks are completely wiped out. The hydrometer is first dropped in water, and the water level is marked. The hydrometer is then dropped into the other liquid, and it is observed that the mark for water has risen 0.3 cm above the liquid–air interface (Fig. P3–99). If the height of the original water mark is 12.3 cm, determine the density of the liquid.

FIGURE P3–99

3–100E A crane is used to lower weights into a lake for an underwater construction project. Determine the tension in the rope of the crane due to a 3-ft-diameter spherical steel block (density = 494 lbm/ft^3) when it is (a) suspended in the air and (b) completely immersed in water.

3–101 The volume and the average density of an irregularly shaped body are to be determined by using a spring scale. The body weighs 7200 N in air and 4790 N in water. Determine the volume and the density of the body. State your assumptions.

3–102   Consider a large cubic ice block floating in seawater. The specific gravities of ice and seawater are 0.92 and 1.025, respectively. If a 25-cm-high portion of the ice block extends above the surface of the water, determine the height of the ice block below the surface.   *Answer:* 2.19 m

FIGURE P3–102

3–103   A spherical shell made of a material with a density of 1600 kg/m³ is placed in water. If the inner and outer radii of the shell are $R_1 = 5$ cm, $R_2 = 6$ cm, determine the percentage of the shell's total volume that would be submerged.

3–104   An inverted cone is placed in a water tank as shown. If the weight of the cone is 16.5 N, what is the tensile force in the cord connecting the cone to the bottom of the tank?

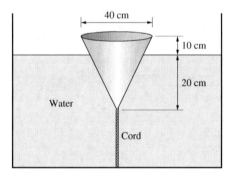

FIGURE P3–104

3–105   The weight of a body is usually measured by disregarding buoyancy force applied by the air. Consider a 20-cm-diameter spherical body of density 7800 kg/m³. What is the percentage error associated with the neglecting of air buoyancy?

3–106   A 170-kg granite rock ($\rho = 2700$ kg/m³) is dropped into a lake. A man dives in and tries to lift the rock. Determine how much force the man needs to apply to lift it from the bottom of the lake. Do you think he can do it?

3–107   It is said that Archimedes discovered his principle during a bath while thinking about how he could determine if King Hiero's crown was actually made of pure gold. While in the bathtub, he conceived the idea that he could determine

the average density of an irregularly shaped object by weighing it in air and also in water. If the crown weighed 3.55 kgf (= 34.8 N) in air and 3.25 kgf (= 31.9 N) in water, determine if the crown is made of pure gold. The density of gold is 19,300 kg/m³. Discuss how you can solve this problem without weighing the crown in water but by using an ordinary bucket with no calibration for volume. You may weigh anything in air.

3–108   The hull of a boat has a volume of 180 m³, and the total mass of the boat when empty is 8560 kg. Determine how much load this boat can carry without sinking (*a*) in a lake and (*b*) in seawater with a specific gravity of 1.03.

Fluids in Rigid-Body Motion

3–109C   Under what conditions can a moving body of fluid be treated as a rigid body?

3–110C   Consider a glass of water. Compare the water pressures at the bottom surface for the following cases: the glass is (*a*) stationary, (*b*) moving up at constant velocity, (*c*) moving down at constant velocity, and (*d*) moving horizontally at constant velocity.

3–111C   Consider two identical glasses of water, one stationary and the other moving on a horizontal plane with constant acceleration. Assuming no splashing or spilling occurs, which glass will have a higher pressure at the (*a*) front, (*b*) midpoint, and (*c*) back of the bottom surface?

3–112C   Consider a vertical cylindrical container partially filled with water. Now the cylinder is rotated about its axis at a specified angular velocity, and rigid-body motion is established. Discuss how the pressure will be affected at the midpoint and at the edges of the bottom surface due to rotation.

3–113   A water tank is being towed by a truck on a level road, and the angle the free surface makes with the horizontal is measured to be 12°. Determine the acceleration of the truck.

3–114   Consider two water tanks filled with water. The first tank is 8 m high and is stationary, while the second tank is 2 m high and is moving upward with an acceleration of 5 m/s². Which tank will have a higher pressure at the bottom?

3–115   A water tank is being towed on an uphill road that makes 14° with the horizontal with a constant acceleration of 3.5 m/s² in the direction of motion. Determine the angle the free surface of water makes with the horizontal. What would your answer be if the direction of motion were downward on the same road with the same acceleration?

3–116E   A 3-ft-diameter vertical cylindrical tank open to the atmosphere contains 1-ft-high water. The tank is now rotated about the centerline, and the water level drops at the center while it rises at the edges. Determine the angular velocity at which the bottom of the tank will first be exposed. Also determine the maximum water height at this moment.

FIGURE P3–116E

3–117 A 60-cm-high, 40-cm-diameter cylindrical water tank is being transported on a level road. The highest acceleration anticipated is 4 m/s². Determine the allowable initial water height in the tank if no water is to spill out during acceleration.   *Answer:* 51.8 cm

3–118 A 30-cm-diameter, 90-cm-high vertical cylindrical container is partially filled with 60-cm-high water. Now the cylinder is rotated at a constant angular speed of 180 rpm. Determine how much the liquid level at the center of the cylinder will drop as a result of this rotational motion.

3–119 A fish tank that contains 60-cm-high water is moved in the cabin of an elevator. Determine the pressure at the bottom of the tank when the elevator is (*a*) stationary, (*b*) moving up with an upward acceleration of 3 m/s², and (*c*) moving down with a downward acceleration of 3 m/s².

3–120 A 3-m-diameter vertical cylindrical milk tank rotates at a constant rate of 12 rpm. If the pressure at the center of the bottom surface is 130 kPa, determine the pressure at the edge of the bottom surface of the tank. Take the density of the milk to be 1030 kg/m³.

3–121 Consider a tank of rectangular cross-section partially filled with a liquid placed on an inclined surface, as shown in the figure. When frictional effects are negligible, show that the slope of the liquid surface will be the same as the slope of the inclined surface when the tank is released. What can you say about the slope of the free surface when the friction is significant?

FIGURE P3–121

3–122 The bottom quarter of a vertical cylindrical tank of total height 0.4 m and diameter 0.3 m is filled with a liquid

(SG > 1, like glycerin) and the rest with water, as shown in the figure. The tank is now rotated about its vertical axis at a constant angular speed of ω. Determine (*a*) the value of the angular speed when the point *P* on the axis at the liquid-liquid interface touches the bottom of the tank and (*b*) the amount of water that would be spilled out at this angular speed.

FIGURE P3–122

3–123 Milk with a density of 1020 kg/m³ is transported on a level road in a 9-m-long, 3-m-diameter cylindrical tanker. The tanker is completely filled with milk (no air space), and it accelerates at 4 m/s². If the minimum pressure in the tanker is 100 kPa, determine the maximum pressure difference and the location of the maximum pressure.   *Answer:* 66.7 kPa

FIGURE P3–123

3–124 Repeat Prob. 3–123 for a deceleration of 2.5 m/s².

3–125 The distance between the centers of the two arms of a U-tube open to the atmosphere is 30 cm, and the U-tube contains 20-cm-high alcohol in both arms. Now the U-tube is

FIGURE P3–125

rotated about the left arm at 4.2 rad/s. Determine the elevation difference between the fluid surfaces in the two arms.

3–126 A 1.2-m-diameter, 3-m-high sealed vertical cylinder is completely filled with gasoline whose density is 740 kg/m^3. The tank is now rotated about its vertical axis at a rate of 70 rpm. Determine (a) the difference between the pressures at the centers of the bottom and top surfaces and (b) the difference between the pressures at the center and the edge of the bottom surface.

FIGURE P3–126

3–127 Reconsider Prob. 3–126. Using EES (or other) software, investigate the effect of rotational speed on the pressure difference between the center and the edge of the bottom surface of the cylinder. Let the rotational speed vary from 0 rpm to 500 rpm in increments of 50 rpm. Tabulate and plot your results.

3–128E A 15-ft-long, 6-ft-high rectangular tank open to the atmosphere is towed by a truck on a level road. The tank is filled with water to a depth of 5 ft. Determine the maximum acceleration or deceleration allowed if no water is to spill during towing.

3–129E An 8-ft-long tank open to the atmosphere initially contains 3-ft-high water. It is being towed by a truck on a

FIGURE P3–129E

level road. The truck driver applies the brakes and the water level at the front rises 0.5 ft above the initial level. Determine the deceleration of the truck.   *Answer:* 4.03 ft/s^2

3–130 A 3-m-diameter, 7-m-long cylindrical tank is completely filled with water. The tank is pulled by a truck on a level road with the 7-m-long axis being horizontal. Determine the pressure difference between the front and back ends of the tank along a horizontal line when the truck (a) accelerates at 3 m/s^2 and (b) decelerates at 4 m/s^2.

3–131 The rectangular tank is filled with heavy oil (like glycerin) at the bottom and water at the top, as shown in the figure. The tank is now moved to the right horizontally with a constant acceleration and ¼ of water is spilled out as a result from the back. Using geometrical considerations, determine how high the point A at the back of the tank on the oil-water interface will rise under this acceleration. *Answer:* 0.25 m

FIGURE P3–131

3–132 A sealed box filled with a liquid shown in the figure can be used to measure the acceleration of vehicles by measuring the pressure at top point A at back of the box while point B is kept at atmospheric pressure. Obtain a relation between the pressure P_A and the acceleration a.

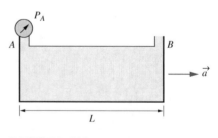

FIGURE P3–132

3–133 A centrifugal pump consists simply of a shaft and a few blades attached normally to the shaft. If the shaft is rotated at a constant rate of 2400 rpm, what would the theoretical pump head due to this rotation be? Take the impeller diameter to be 35 cm and neglect the blade tip effects. *Answer:* 98.5 m

3–134 A U-tube is rotating at a constant angular velocity of ω. The liquid (glycerin) rises to the levels shown in Fig. P3–134. Obtain a relation for ω in terms of g, h, and L.

FIGURE P3–134

FIGURE P3–137

Review Problems

3–135 An air-conditioning system requires a 34-m-long section of 12-cm-diameter ductwork to be laid underwater. Determine the upward force the water will exert on the duct. Take the densities of air and water to be 1.3 kg/m³ and 1000 kg/m³, respectively.

3–136 The 0.5-m-radius semi-circular gate shown in the figure is hinged through the top edge AB. Find the required force to be applied at the center of gravity to keep the gate closed.   *Answer:* 11.3 kN

3–138 A 30-cm-diameter vertical cylindrical vessel is rotated about its vertical axis at a constant angular velocity of 100 rad/s. If the pressure at the midpoint of the inner top surface is atmospheric pressure like the outer surface, determine the total upward force acting upon the entire top surface inside the cylinder.

3–139 Balloons are often filled with helium gas because it weighs only about one-seventh of what air weighs under identical conditions. The buoyancy force, which can be expressed as $F_b = \rho_{air} g V_{balloon}$, will push the balloon upward. If the balloon has a diameter of 12 m and carries two people, 70 kg each, determine the acceleration of the balloon when it is first released. Assume the density of air is $\rho = 1.16$ kg/m³, and neglect the weight of the ropes and the cage.   *Answer:* 25.7 m/s²

FIGURE P3–136

FIGURE P3–139

3–137 If the rate of rotational speed of the 3-tube system shown in Fig. P3–137 is $\omega = 10$ rad/s, determine the water heights in each tube leg. At what rotational speed will the middle tube be completely empty?

3–140 Reconsider Prob. 3–139. Using EES (or other) software, investigate the effect of the number of people carried in the balloon on acceleration. Plot the acceleration against the number of people, and discuss the results.

3–141   Determine the maximum amount of load, in kg, the balloon described in Prob. 3–139 can carry.   *Answer:* 521 kg

3–142E   The pressure in a steam boiler is given to be 90 kgf/cm². Express this pressure in psi, kPa, atm, and bars.

3–143   The basic barometer can be used as an altitude-measuring device in airplanes. The ground control reports a barometric reading of 760 mmHg while the pilot's reading is 420 mmHg. Estimate the altitude of the plane from ground level if the average air density is 1.20 kg/m³.   *Answer:* 3853 m

3–144   The lower half of a 12-m-high cylindrical container is filled with water (ρ = 1000 kg/m³) and the upper half with oil that has a specific gravity of 0.85. Determine the pressure difference between the top and bottom of the cylinder.   *Answer:* 109 kPa

FIGURE P3–144

3–145   A vertical, frictionless piston–cylinder device contains a gas at 500 kPa. The atmospheric pressure outside is 100 kPa, and the piston area is 30 cm². Determine the mass of the piston.

3–146   A pressure cooker cooks a lot faster than an ordinary pan by maintaining a higher pressure and temperature inside. The lid of a pressure cooker is well sealed, and steam can escape only through an opening in the middle of the lid. A sep-

arate metal piece, the petcock, sits on top of this opening and prevents steam from escaping until the pressure force overcomes the weight of the petcock. The periodic escape of the steam in this manner prevents any potentially dangerous pressure buildup and keeps the pressure inside at a constant value. Determine the mass of the petcock of a pressure cooker whose operation pressure is 120 kPa gage and has an opening cross-sectional area of 3 mm². Assume an atmospheric pressure of 101 kPa, and draw the free-body diagram of the petcock.   *Answer:* 36.7 g

3–147   A glass tube is attached to a water pipe, as shown in Fig. P3–147. If the water pressure at the bottom of the tube is 115 kPa and the local atmospheric pressure is 98 kPa, determine how high the water will rise in the tube, in m. Assume g = 9.8 m/s² at that location and take the density of water to be 1000 kg/m³.

FIGURE P3–147

3–148   The average atmospheric pressure on earth is approximated as a function of altitude by the relation P_{atm} = 101.325 $(1 - 0.02256z)^{5.256}$, where P_{atm} is the atmospheric pressure in kPa and z is the altitude in km with z = 0 at sea level. Determine the approximate atmospheric pressures at Atlanta (z = 306 m), Denver (z = 1610 m), Mexico City (z = 2309 m), and the top of Mount Everest (z = 8848 m).

3–149   When measuring small pressure differences with a manometer, often one arm of the manometer is inclined to improve the accuracy of the reading. (The pressure difference is still proportional to the *vertical* distance and not the actual length of the fluid along the tube.) The air pressure in

FIGURE P3–146

FIGURE P3–149

a circular duct is to be measured using a manometer whose open arm is inclined 25° from the horizontal, as shown in Fig. P3–149. The density of the liquid in the manometer is 0.81 kg/L, and the vertical distance between the fluid levels in the two arms of the manometer is 8 cm. Determine the gage pressure of air in the duct and the length of the fluid column in the inclined arm above the fluid level in the vertical arm.

3–150E Consider a U-tube whose arms are open to the atmosphere. Now equal volumes of water and light oil ($\rho = 49.3$ lbm/ft^3) are poured from different arms. A person blows from the oil side of the U-tube until the contact surface of the two fluids moves to the bottom of the U-tube, and thus the liquid levels in the two arms are the same. If the fluid height in each arm is 40 in, determine the gage pressure the person exerts on the oil by blowing.

FIGURE P3–150E

3–151 An elastic air balloon having a diameter of 30 cm is attached to the base of a container partially filled with water at +4°C, as shown in Fig. P3–151. If the pressure of the air above the water is gradually increased from 100 kPa to 1.6 MPa, will the force on the cable change? If so, what is the percent change in the force? Assume the pressure on the

free surface and the diameter of the balloon are related by $P = CD^n$, where C is a constant and $n = -2$. The weight of the balloon and the air in it is negligible. *Answer:* 98.4 percent

3–152 [EES] Reconsider Prob. 3–151. Using EES (or other) software, investigate the effect of air pressure above water on the cable force. Let this pressure vary from 0.5 MPa to 15 MPa. Plot the cable force versus the air pressure.

3–153 A gasoline line is connected to a pressure gage through a double-U manometer, as shown in Fig. P3–153. If the reading of the pressure gage is 260 kPa, determine the gage pressure of the gasoline line.

FIGURE P3–153

3–154 Repeat Prob. 3–153 for a pressure gage reading of 330 kPa.

3–155E A water pipe is connected to a double-U manometer as shown in Fig. P3–155E at a location where the local atmospheric pressure is 14.2 psia. Determine the absolute pressure at the center of the pipe.

FIGURE P3–155E

3–156 The pressure of water flowing through a pipe is measured by the arrangement shown in Fig. P3–156. For the values given, calculate the pressure in the pipe.

FIGURE P3–151

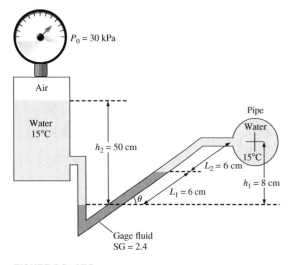

$P_0 = 30$ kPa

Air

Water
15°C

Pipe

Water

15°C

$h_2 = 50$ cm

$L_2 = 6$ cm

$h_1 = 8$ cm

$L_1 = 6$ cm

θ

Gage fluid
SG = 2.4

FIGURE P3–156

3–157 Consider a U-tube filled with mercury as shown in Fig. P3–157. The diameter of the right arm of the U-tube is $D = 1.5$ cm, and the diameter of the left arm is twice that. Heavy oil with a specific gravity of 2.72 is poured into the left arm, forcing some mercury from the left arm into the right one. Determine the maximum amount of oil that can be added into the left arm. *Answer:* 0.0884 L

Oil poured in here
SG = 2.72

12 cm

$2D$

$D = 1.5$ cm

Mercury
SG = 13.6

FIGURE P3–157

3–158 It is well known that the temperature of the atmosphere varies with altitude. In the troposphere, which extends to an altitude of 11 km, for example, the variation of temperature can be approximated by $T = T_0 - \beta z$, where T_0 is the temperature at sea level, which can be taken to be 288.15 K, and $\beta = 0.0065$ K/m. The gravitational acceleration also changes with altitude as $g(z) = g_0/(1 + z/6,370,320)^2$ where $g_0 = 9.807$ m/s^2

and z is the elevation from sea level in m. Obtain a relation for the variation of pressure in the troposphere (*a*) by ignoring and (*b*) by considering the variation of g with altitude.

3–159 The variation of pressure with density in a thick gas layer is given by $P = C\rho^n$, where C and n are constants. Noting that the pressure change across a differential fluid layer of thickness dz in the vertical z-direction is given as $dP = -\rho g\, dz$, obtain a relation for pressure as a function of elevation z. Take the pressure and density at $z = 0$ to be P_0 and ρ_0, respectively.

3–160 A 3-m-high, 6-m-wide rectangular gate is hinged at the top edge at A and is restrained by a fixed ridge at B. Determine the hydrostatic force exerted on the gate by the 5-m-high water and the location of the pressure center.

2 m

Water

3 m

A

Gate

B

FIGURE P3–160

3–161 Repeat Prob. 3–160 for a total water height of 2 m.

3–162E A semicircular 40-ft-diameter tunnel is to be built under a 150-ft-deep, 800-ft-long lake, as shown in Fig. P3–162E. Determine the total hydrostatic force acting on the roof of the tunnel.

Water

150 ft

Tunnel

40 ft

FIGURE P3–162E

3–163 A 30-ton, 4-m-diameter hemispherical dome on a level surface is filled with water, as shown in Fig. P3–163. Someone claims that he can lift this dome by making use of Pascal's law by attaching a long tube to the top and filling it with water. Determine the required height of water in the tube to lift the dome. Disregard the weight of the tube and the water in it. *Answer:* 0.72 m

FIGURE P3–163

3–164 The water in a 25-m-deep reservoir is kept inside by a 150-m-wide wall whose cross section is an equilateral triangle, as shown in Fig. P3–164. Determine (*a*) the total force (hydrostatic + atmospheric) acting on the inner surface of the wall and its line of action and (*b*) the magnitude of the horizontal component of this force. Take $P_{atm} = 100$ kPa.

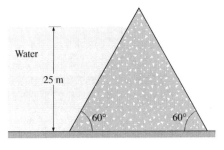

FIGURE P3–164

3–165 A U-tube contains water in the right arm, and another liquid in the left arm. It is observed that when the U-tube rotates at 50 rpm about an axis that is 15 cm from the right arm and 5 cm from the left arm, the liquid levels in

FIGURE P3–165

both arms become the same, and the fluids meet at the axis of rotation. Determine the density of the fluid in the left arm.

3–166 A 1-m-diameter, 2-m-high vertical cylinder is completely filled with gasoline whose density is 740 kg/m³. The tank is now rotated about its vertical axis at a rate of 130 rpm, while being accelerated upward at 5 m/s². Determine (*a*) the difference between the pressures at the centers of the bottom and top surfaces and (*b*) the difference between the pressures at the center and the edge of the bottom surface.

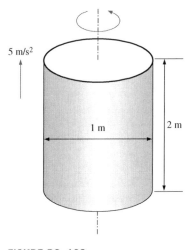

FIGURE P3–166

3–167 A 5-m-long, 4-m-high tank contains 2.5-m-deep water when not in motion and is open to the atmosphere through a vent in the middle. The tank is now accelerated to the right on a level surface at 2 m/s². Determine the maximum pressure in the tank relative to the atmospheric pressure. *Answer:* 29.5 kPa

FIGURE P3–167

3–168 Reconsider Prob. 3–167. Using EES (or other) software, investigate the effect of acceleration on the slope of the free surface of water in the tank. Let the acceleration vary from 0 m/s² to 15 m/s² in increments of 1 m/s². Tabulate and plot your results.

3–169 A cylindrical container whose weight is 65 N is inverted and pressed into the water, as shown in Fig. P3–169. Determine the differential height h of the manometer and the force F needed to hold the container at the position shown.

FIGURE P3–169

3–170 The average density of icebergs is about 917 kg/m^3. (a) Determine the percentage of the total volume of an iceberg submerged in seawater of density 1042 kg/m^3. (b) Although icebergs are mostly submerged, they are observed to turn over. Explain how this can happen. (*Hint*: Consider the temperatures of icebergs and seawater.)

3–171 The density of a floating body can be determined by tying weights to the body until both the body and the weights are completely submerged, and then weighing them separately in air. Consider a wood log that weighs 1540 N in air. If it takes 34 kg of lead ($\rho = 11,300$ kg/m^3) to completely sink the log and the lead in water, determine the average density of the log.   *Answer:* 835 kg/m^3

3–172   The 280-kg, 6-m-wide rectangular gate shown in Fig. P3–172 is hinged at B and leans against the floor at A making an angle of 45° with the horizontal.

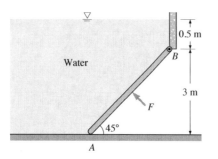

FIGURE P3–172

The gate is to be opened from its lower edge by applying a normal force at its center. Determine the minimum force F required to open the water gate.   *Answer:* 626 kN

3–173 Repeat Prob. 3–172 for a water height of 0.8 m above the hinge at B.

Fundamentals of Engineering (FE) Exam Problems

3–174 The absolute pressure in a tank is measured to be 35 kPa. If the atmospheric pressure is 100 kPa, the vacuum pressure in the tank is
(*a*) 35 kPa   (*b*) 100 kPa   (*c*) 135 psi
(*d*) 0 kPa   (*e*) 65 kPa

3–175 The pressure difference between the top and bottom of a water body with a depth of 10 m is (Take the density of water to be 1000 kg/m^3.)
(*a*) 98,100 kPa   (*b*) 98.1 kPa   (*c*) 100 kPa
(*d*) 10 kPa   (*e*) 1.9 kPa

3–176 The gage pressure in a pipe is measured by a manometer containing mercury ($\rho = 13,600$ kg/m^3). The top of the mercury is open to the atmosphere and the atmospheric pressure is 100 kPa. If the mercury column height is 24 cm, the gage pressure in the pipe is
(*a*) 32 kPa   (*b*) 24 kPa   (*c*) 76 kPa
(*d*) 124 kPa   (*e*) 68 kPa

3–177 Consider a hydraulic car jack with a piston diameter ratio of 9. A person can lift a 2000-kg car by applying a force of
(*a*) 2000 N   (*b*) 200 N   (*c*) 19,620 N
(*d*) 19.6 N   (*e*) 18,000 N

3–178 The atmospheric pressure in a location is measured by a mercury ($\rho = 13,600$ kg/m^3) barometer. If the height of the mercury column is 715 mm, the atmospheric pressure at that location is
(*a*) 85.6 kPa   (*b*) 93.7 kPa   (*c*) 95.4 kPa
(*d*) 100 kPa   (*e*) 101 kPa

3–179 A manometer is used to measure the pressure of a gas in a tank. The manometer fluid is water ($\rho = 1000$ kg/m^3) and the manometer column height is 1.8 m. If the local atmospheric pressure is 100 kPa, the absolute pressure within the tank is
(*a*) 17,760 kPa   (*b*) 100 kPa   (*c*) 180 kPa
(*d*) 101 kPa   (*e*) 118 kPa

3–180 Consider the vertical rectangular wall of a water tank with a width of 5 m and a height of 8 m. The other side of the wall is open to the atmosphere. The resultant hydrostatic force on this wall is
(*a*) 1570 kN   (*b*) 2380 kN   (*c*) 2505 kN
(*d*) 1410 kN   (*e*) 404 kN

3–181 A vertical rectangular wall with a width of 20 m and a height of 12 m is holding a 7-m-deep water body. The resultant hydrostatic force acting on this wall is
(*a*) 1370 kN (*b*) 4807 kN (*c*) 8240 kN
(*d*) 9740 kN (*e*) 11,670 kN

3–182 A vertical rectangular wall with a width of 20 m and a height of 12 m is holding a 7-m-deep water body. The line of action y_p for the resultant hydrostatic force on this wall is (disregard the atmospheric pressure)
(*a*) 5 m (*b*) 4.0 m (*c*) 4.67 m (*d*) 9.67 m (*e*) 2.33 m

3–183 A rectangular plate with a width of 16 m and a height of 12 m is located 4 m below a water surface. The plate is tilted and makes a 35° angle with the horizontal. The resultant hydrostatic force acting on the top surface of this plate is
(*a*) 10,800 kN (*b*) 9745 kN (*c*) 8470 kN
(*d*) 6400 kN (*e*) 5190 kN

3–184 A 2-m-long and 3-m-wide horizontal rectangular plate is submerged in water. The distance of the top surface from the free surface is 5 m. The atmospheric pressure is 95 kPa. Considering atmospheric pressure, the hydrostatic force acting on the top surface of this plate is
(*a*) 307 kN (*b*) 688 kN (*c*) 747 kN
(*d*) 864 kN (*e*) 2950 kN

3–185 A 1.8-m-diameter and 3.6-m-long cylindrical container contains a fluid with a specific gravity of 0.73. The container is positioned vertically and is full of the fluid. Disregarding atmospheric pressure, the hydrostatic force acting on the top and bottom surfaces of this container, respectively, are
(*a*) 0 kN, 65.6 kN (*b*) 65.6 kN, 0 kN (*c*) 65.6 kN, 65.6 kN
(*d*) 25.5 kN, 0 kN (*e*) 0 kN, 25.5 kN

3–186 Consider a 6-m-diameter spherical gate holding a body of water whose height is equal to the diameter of the gate. Atmospheric pressure acts on both sides of the gate. The horizontal component of the hydrostatic force acting on this curved surface is
(*a*) 709 kN (*b*) 832 kN (*c*) 848 kN
(*d*) 972 kN (*e*) 1124 kN

3–187 Consider a 6-m-diameter spherical gate holding a body of water whose height is equal to the diameter of the gate. Atmospheric pressure acts on both sides of the gate. The vertical component of the hydrostatic force acting on this curved surface is
(*a*) 89 kN (*b*) 270 kN (*c*) 327 kN
(*d*) 416 kN (*e*) 505 kN

3–188 A 0.75-cm-diameter spherical object is completely submerged in water. The buoyant force acting on this object is
(*a*) 13,000 N (*b*) 9835 N (*c*) 5460 N
(*d*) 2167 N (*e*) 1267 N

3–189 A 3-kg object with a density of 7500 kg/m³ is placed in water. The weight of this object in water is
(*a*) 29.4 N (*b*) 25.5 N (*c*) 14.7 N (*d*) 30 N (*e*) 3 N

3–190 A 7-m-diameter hot air balloon is neither rising nor falling. The density of atmospheric air is 1.3 kg/m³. The total mass of the balloon including the people on board is
(*a*) 234 kg (*b*) 207 kg (*c*) 180 kg (*d*) 163 kg (*e*) 134 kg

3–191 A 10-kg object with a density of 900 kg/m³ is placed in a fluid with a density of 1100 kg/m³. The fraction of the volume of the object submerged in water is
(*a*) 0.637 (*b*) 0.716 (*c*) 0.818 (*d*) 0.90 (*e*) 1

3–192 Consider a cubical water tank with a side length of 3 m. The tank is half filled with water, and is open to the atmosphere with a pressure of 100 kPa. Now, a truck carrying this tank is accelerated at a rate of 5 m/s². The maximum pressure in the water is
(*a*) 115 kPa (*b*) 122 kPa (*c*) 129 kPa
(*d*) 137 kPa (*e*) 153 kPa

3–193 A 15-cm-diameter, 40-cm-high vertical cylindrical container is partially filled with 25-cm-high water. Now the cylinder is rotated at a constant speed of 20 rad/s. The maximum height difference between the edge and the center of the free surface is
(*a*) 15 cm (*b*) 7.2 cm (*c*) 5.4 cm (*d*) 9.5 cm (*e*) 11 .5 cm

3–194 A 20-cm-diameter, 40-cm-high vertical cylindrical container is partially filled with 25-cm-high water. Now the cylinder is rotated at a constant speed of 15 rad/s. The height of water at the center of the cylinder is
(*a*) 25 cm (*b*) 19.5 cm (*c*) 22.7 cm
(*d*) 17.7 cm (*e*) 15 cm

3–195 A 15-cm-diameter, 50-cm-high vertical cylindrical container is partially filled with 30-cm-high water. Now the cylinder is rotated at a constant speed of 20 rad/s. The pressure difference between the center and edge of the container at the base surface is
(*a*) 7327 Pa (*b*) 8750 Pa (*c*) 9930 Pa
(*d*) 1045 Pa (*e*) 1125 Pa

Design and Essay Problems

3–196 Shoes are to be designed to enable people of up to 80 kg to walk on freshwater or seawater. The shoes are to be made of blown plastic in the shape of a sphere, a (American) football, or a loaf of French bread. Determine the equivalent diameter of each shoe and comment on the proposed shapes from the stability point of view. What is your assessment of the marketability of these shoes?

3–197 The volume of a rock is to be determined without using any volume measurement devices. Explain how you would do this with a waterproof spring scale.

3–198   The density of stainless steel is about 8000 kg/m^3 (eight times denser than water), but a razor blade can float on water, even with some added weights. The water is at 20°C. The blade shown in the photograph is 4.3 cm long and 2.2 cm wide. For simplicity, the center cut-out area of the razor blade has been taped so that only the outer edges of the blade contribute to surface tension effects. Because the razor blade has sharp corners, the contact angle is not relevant. Rather, the limiting case is when the water contacts the blade vertically as sketched (effective contact angle along the edge of the blade is 180°). (a) Considering surface tension *alone*, estimate (in grams) how much total mass (razor blade + weights placed on top of it) can be supported. (b) Refine your analysis by considering that the razor blade pushes the water down, and thus hydrostatic pressure effects are also present. *Hint*: You will also need to know that due to the curvature of the meniscus, the maximum possible

depth is $h = \sqrt{\dfrac{2\sigma_s}{\rho g}}$.

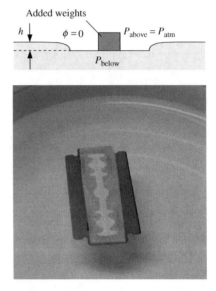

FIGURE P3–198

(Bottom) Photo by John M. Cimbala.

FLUID KINEMATICS

Fluid kinematics deals with describing the motion of fluids without necessarily considering the forces and moments that *cause* the motion. In this chapter, we introduce several kinematic concepts related to flowing fluids. We discuss the *material derivative* and its role in transforming the conservation equations from the *Lagrangian description of fluid flow* (following a *fluid particle*) to the *Eulerian description of fluid flow* (pertaining to a *flow field*). We then discuss various ways to visualize flow fields—*streamlines, streaklines, pathlines, timelines,* optical methods *schlieren* and *shadowgraph,* and *surface* methods; and we describe three ways to plot flow data—*profile plots, vector plots,* and *contour plots.* We explain the four fundamental kinematic properties of fluid motion and deformation—*rate of translation, rate of rotation, linear strain rate,* and *shear strain rate.* The concepts of *vorticity, rotationality,* and *irrotationality* in fluid flows are then discussed. Finally, we discuss the *Reynolds transport theorem (RTT),* emphasizing its role in transforming the equations of motion from those following a *system* to those pertaining to fluid flow into and out of a *control volume.* The analogy between material derivative for infinitesimal fluid elements and RTT for finite control volumes is explained.

OBJECTIVES

When you finish reading this chapter, you should be able to

- Understand the role of the material derivative in transforming between Lagrangian and Eulerian descriptions

- Distinguish between various types of flow visualizations and methods of plotting the characteristics of a fluid flow

- Appreciate the many ways that fluids move and deform

- Distinguish between rotational and irrotational regions of flow based on the flow property vorticity

- Understand the usefulness of the Reynolds transport theorem

Satellite image of a hurricane near the Florida coast; water droplets move with the air, enabling us to visualize the counterclockwise swirling motion. However, the major portion of the hurricane is actually *irrotational,* while only the core (the eye of the storm) is *rotational.*
© StockTrek/Getty RF

FIGURE 4–1
With a small number of objects,
such as billiard balls on a pool table,
individual objects can be tracked.

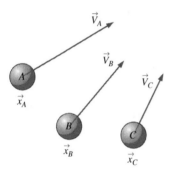

FIGURE 4–2
In the Lagrangian description, we
must keep track of the position and
velocity of individual particles.

4–1 · LAGRANGIAN AND EULERIAN DESCRIPTIONS

The subject called **kinematics** concerns the study of *motion*. In fluid dynamics, *fluid kinematics* is the study of how fluids flow and how to describe fluid motion. From a fundamental point of view, there are two distinct ways to describe motion. The first and most familiar method is the one you learned in high school physics—to follow the path of individual objects. For example, we have all seen physics experiments in which a ball on a pool table or a puck on an air hockey table collides with another ball or puck or with the wall (Fig. 4–1). Newton's laws are used to describe the motion of such objects, and we can accurately predict where they go and how momentum and kinetic energy are exchanged from one object to another. The kinematics of such experiments involves keeping track of the **position vector** of each object, $\vec{x}_A$, $\vec{x}_B$, . . . , and the **velocity vector** of each object, $\vec{V}_A$, $\vec{V}_B$, . . . , as functions of time (Fig. 4–2). When this method is applied to a flowing fluid, we call it the **Lagrangian description** of fluid motion after the Italian mathematician Joseph Louis Lagrange (1736–1813). Lagrangian analysis is analogous to the (closed) **system analysis** that you learned in thermodynamics; namely, we follow a mass of fixed identity. The Lagrangian description requires us to track the position and velocity of each individual fluid parcel, which we refer to as a **fluid particle,** and take to be a parcel of fixed identity.

As you can imagine, this method of describing motion is much more difficult for fluids than for billiard balls! First of all we cannot easily define and identify fluid particles as they move around. Secondly, a fluid is a **continuum** (from a macroscopic point of view), so interactions between fluid particles are not as easy to describe as are interactions between distinct objects like billiard balls or air hockey pucks. Furthermore, the fluid particles continually *deform* as they move in the flow.

From a *microscopic* point of view, a fluid is composed of *billions* of molecules that are continuously banging into one another, somewhat like billiard balls; but the task of following even a subset of these molecules is quite difficult, even for our fastest and largest computers. Nevertheless, there are many practical applications of the Lagrangian description, such as the tracking of passive scalars in a flow to model contaminant transport, rarefied gas dynamics calculations concerning reentry of a spaceship into the earth's atmosphere, and the development of flow visualization and measurement systems based on particle tracking (as discussed in Section 4–2).

A more common method of describing fluid flow is the **Eulerian description** of fluid motion, named after the Swiss mathematician Leonhard Euler (1707–1783). In the Eulerian description of fluid flow, a finite volume called a **flow domain** or **control volume** is defined, through which fluid flows in and out. Instead of tracking individual fluid particles, we define **field variables,** functions of space and time, within the control volume. The field variable at a particular location at a particular time is the value of the variable for whichever fluid particle happens to occupy that location at that time. For example, the **pressure field** is a **scalar field variable;** for general unsteady three-dimensional fluid flow in Cartesian coordinates,

Pressure field: $P = P(x, y, z, t)$ **(4–1)**

We define the **velocity field** as a **vector field variable** in similar fashion,

Velocity field: $\vec{V} = \vec{V}(x, y, z, t)$ **(4–2)**

Likewise, the **acceleration field** is also a vector field variable,

Acceleration field: $\qquad\qquad \vec{a} = \vec{a}(x, y, z, t)$ **(4–3)**

Collectively, these (and other) field variables define the **flow field**. The velocity field of Eq. 4–2 is expanded in Cartesian coordinates (x, y, z), $(\vec{i}, \vec{j}, \vec{k})$ as

$$\vec{V} = (u, v, w) = u(x, y, z, t)\vec{i} + v(x, y, z, t)\vec{j} + w(x, y, z, t)\vec{k} \qquad \textbf{(4–4)}$$

A similar expansion can be performed for the acceleration field of Eq. 4–3. In the Eulerian description, all such field variables are defined at any location (x, y, z) in the control volume and at any instant in time t (Fig. 4–3). In the Eulerian description we don't really care what happens to individual fluid particles; rather we are concerned with the pressure, velocity, acceleration, etc., of whichever fluid particle happens to be at the location of interest at the time of interest.

The difference between these two descriptions is made clearer by imagining a person standing beside a river, measuring its properties. In the Lagrangian approach, he throws in a probe that moves downstream with the water. In the Eulerian approach, he anchors the probe at a fixed location in the water.

While there are many occasions in which the Lagrangian description is useful, the Eulerian description is often more convenient for fluid mechanics applications. Furthermore, experimental measurements are generally more suited to the Eulerian description. In a wind tunnel, for example, velocity or pressure probes are usually placed at a fixed location in the flow, measuring $\vec{V}(x, y, z, t)$ or $P(x, y, z, t)$. However, whereas the equations of motion in the Lagrangian description following individual fluid particles are well known (e.g., Newton's second law), the equations of motion of fluid flow are not so readily apparent in the Eulerian description and must be carefully derived. We do this for control volume (integral) analysis via the Reynolds transport theorem at the end of this chapter. We derive the differential equations of motion in Chap. 9.

(a)

(b)

FIGURE 4–3

(*a*) In the Eulerian description, we define field variables, such as the pressure field and the velocity field, at any location and instant in time. (*b*) For example, the air speed probe mounted under the wing of an airplane measures the air speed at that location.

(Bottom) Photo by John M. Cimbala.

■ *EXAMPLE 4–1*   **A Steady Two-Dimensional Velocity Field**

A steady, incompressible, two-dimensional velocity field is given by

$$\vec{V} = (u, v) = (0.5 + 0.8x)\,\vec{i} + (1.5 - 0.8y)\,\vec{j} \qquad \textbf{(1)}$$

where the *x*- and *y*-coordinates are in meters and the magnitude of velocity is in m/s. A **stagnation point** is defined as *a point in the flow field where the velocity is zero*. (*a*) Determine if there are any stagnation points in this flow field and, if so, where? (*b*) Sketch velocity vectors at several locations in the domain between $x = -2$ m to 2 m and $y = 0$ m to 5 m; qualitatively describe the flow field.

SOLUTION For the given velocity field, the location(s) of stagnation point(s) are to be determined. Several velocity vectors are to be sketched and the velocity field is to be described.
Assumptions **1** The flow is steady and incompressible. **2** The flow is two-dimensional, implying no *z*-component of velocity and no variation of *u* or *v* with *z*.
Analysis (*a*) Since $\vec{V}$ is a vector, *all* its components must equal zero in order for $\vec{V}$ itself to be zero. Using Eq. 4–4 and setting Eq. 1 equal to zero,

Stagnation point: $\qquad \begin{aligned} u &= 0.5 + 0.8x = 0 \quad \rightarrow \quad x = -0.625 \text{ m} \\ v &= 1.5 - 0.8y = 0 \quad \rightarrow \quad y = 1.875 \text{ m} \end{aligned}$

Yes. There is one stagnation point located at $x = -0.625$ **m**, $y = 1.875$ **m**.

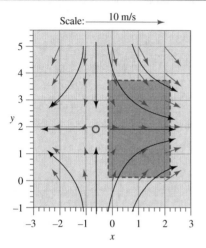

FIGURE 4–4
Velocity vectors (blue arrows) for the velocity field of Example 4–1. The scale is shown by the top arrow, and the solid black curves represent the approximate shapes of some streamlines, based on the calculated velocity vectors. The stagnation point is indicated by the blue circle. The shaded region represents a portion of the flow field that can approximate flow into an inlet (Fig. 4–5).

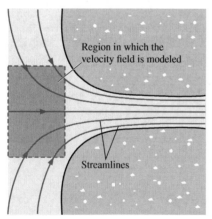

FIGURE 4–5
Flow field near the bell mouth inlet of a hydroelectric dam; a portion of the velocity field of Example 4–1 may be used as a first-order approximation of this physical flow field. The shaded region corresponds to that of Fig. 4–4.

(*b*) The *x*- and *y*-components of velocity are calculated from Eq. 1 for several (x, y) locations in the specified range. For example, at the point ($x = 2$ m, $y = 3$ m), $u = 2.10$ m/s and $v = -0.900$ m/s. The magnitude of velocity (the *speed*) at that point is 2.28 m/s. At this and at an array of other locations, the velocity vector is constructed from its two components, the results of which are shown in Fig. 4–4. The flow can be described as stagnation point flow in which flow enters from the top and bottom and spreads out to the right and left about a horizontal line of symmetry at $y = 1.875$ m. The stagnation point of part (*a*) is indicated by the blue circle in Fig. 4–4.

If we look only at the shaded portion of Fig. 4–4, this flow field models a converging, accelerating flow from the left to the right. Such a flow might be encountered, for example, near the submerged bell mouth inlet of a hydro-electric dam (Fig. 4–5). The useful portion of the given velocity field may be thought of as a first-order approximation of the shaded portion of the physical flow field of Fig. 4–5.

Discussion It can be verified from the material in Chap. 9 that this flow field is physically valid because it satisfies the differential equation for conservation of mass.

Acceleration Field

As you should recall from your study of thermodynamics, the fundamental conservation laws (such as conservation of mass and the first law of thermodynamics) are expressed for a *system* of fixed identity (also called a *closed system*). In cases where analysis of a *control volume* (also called an *open system*) is more convenient than system analysis, it is necessary to rewrite these fundamental laws into forms applicable to the control volume. The same principle applies here. In fact, there is a direct analogy between systems versus control volumes in thermodynamics and Lagrangian versus Eulerian descriptions in fluid dynamics. The equations of motion for fluid flow (such as Newton's second law) are written for a fluid particle, which we also call a **material particle.** If we were to follow a particular fluid particle as it moves around in the flow, we would be employing the Lagrangian description, and the equations of motion would be directly applicable. For example, we would define the particle's location in space in terms of a **material position vector** ($x_{particle}(t)$, $y_{particle}(t)$, $z_{particle}(t)$). However, some mathematical manipulation is then necessary to convert the equations of motion into forms applicable to the Eulerian description.

Consider, for example, Newton's second law applied to our fluid particle,

Newton's second law: $$\vec{F}_{particle} = m_{particle}\vec{a}_{particle} \qquad \text{(4–5)}$$

where $\vec{F}_{particle}$ is the net force acting on the fluid particle, $m_{particle}$ is its mass, and $\vec{a}_{particle}$ is its acceleration (Fig. 4–6). By definition, the acceleration of the fluid particle is the time derivative of the particle's velocity,

Acceleration of a fluid particle: $$\vec{a}_{particle} = \frac{d\vec{V}_{particle}}{dt} \qquad \text{(4–6)}$$

However, at any instant in time *t*, the velocity of the particle is the same as the local value of the velocity *field* at the location ($x_{particle}(t)$, $y_{particle}(t)$, $z_{particle}(t)$) of the particle, since the fluid particle moves with the fluid by definition. In other words, $\vec{V}_{\text{particle}}(t) \equiv \vec{V}(x_{\text{particle}}(t), y_{\text{particle}}(t), z_{\text{particle}}(t), t)$. To take the time derivative in Eq. 4–6, we must therefore use the *chain rule*, since the dependent variable $(\vec{V})$ is a function of *four* independent variables $(x_{\text{particle}}, y_{\text{particle}}, z_{\text{particle}}, \text{and } t)$,

$$\vec{a}_{\text{particle}} = \frac{d\vec{V}_{\text{particle}}}{dt} = \frac{d\vec{V}}{dt} = \frac{d\vec{V}(x_{\text{particle}}, y_{\text{particle}}, z_{\text{particle}}, t)}{dt}$$

$$= \frac{\partial \vec{V}}{\partial t}\frac{dt}{dt} + \frac{\partial \vec{V}}{\partial x_{\text{particle}}}\frac{dx_{\text{particle}}}{dt} + \frac{\partial \vec{V}}{\partial y_{\text{particle}}}\frac{dy_{\text{particle}}}{dt} + \frac{\partial \vec{V}}{\partial z_{\text{particle}}}\frac{dz_{\text{particle}}}{dt}$$ (4–7)

In Eq. 4–7, ∂ is the **partial derivative operator** and d is the **total derivative operator**. Consider the second term on the right-hand side of Eq. 4–7. Since the acceleration is defined as that *following a fluid particle* (Lagrangian description), the rate of change of the particle's x-position with respect to time is $dx_{\text{particle}}/dt = u$ (Fig. 4–7), where u is the x-component of the velocity vector defined by Eq. 4–4. Similarly, $dy_{\text{particle}}/dt = v$ and $dz_{\text{particle}}/dt = w$. Furthermore, at any instant in time under consideration, the material position vector $(x_{\text{particle}}, y_{\text{particle}}, z_{\text{particle}})$ of the fluid particle in the Lagrangian frame is equal to the position vector (x, y, z) in the Eulerian frame. Equation 4–7 thus becomes

$$\vec{a}_{\text{particle}}(x, y, z, t) = \frac{d\vec{V}}{dt} = \frac{\partial \vec{V}}{\partial t} + u\frac{\partial \vec{V}}{\partial x} + v\frac{\partial \vec{V}}{\partial y} + w\frac{\partial \vec{V}}{\partial z}$$ (4–8)

where we have also used the (obvious) fact that $dt/dt = 1$. Finally, at any instant in time t, the acceleration field of Eq. 4–3 must equal the acceleration of the fluid particle that happens to occupy the location (x, y, z) at that time t. Why? Because the fluid particle is by definition accelerating with the fluid flow. Hence, *we may replace $\vec{a}_{\text{particle}}$ with $\vec{a}(x, y, z, t)$ in Eqs. 4–7 and 4–8 to transform from the Lagrangian to the Eulerian frame of reference.* In vector form, Eq. 4–8 is written as

Acceleration of a fluid particle expressed as a field variable:

$$\vec{a}(x, y, z, t) = \frac{d\vec{V}}{dt} = \frac{\partial \vec{V}}{\partial t} + (\vec{V} \cdot \vec{\nabla})\vec{V}$$ (4–9)

where $\vec{\nabla}$ is the **gradient operator** or **del operator**, a vector operator that is defined in Cartesian coordinates as

Gradient or del operator: $\quad \vec{\nabla} = \left(\frac{\partial}{\partial x}, \frac{\partial}{\partial y}, \frac{\partial}{\partial z}\right) = \vec{i}\frac{\partial}{\partial x} + \vec{j}\frac{\partial}{\partial y} + \vec{k}\frac{\partial}{\partial z}$ (4–10)

In Cartesian coordinates then, the components of the acceleration vector are

Cartesian coordinates:

$$a_x = \frac{\partial u}{\partial t} + u\frac{\partial u}{\partial x} + v\frac{\partial u}{\partial y} + w\frac{\partial u}{\partial z}$$

$$a_y = \frac{\partial v}{\partial t} + u\frac{\partial v}{\partial x} + v\frac{\partial v}{\partial y} + w\frac{\partial v}{\partial z}$$ (4–11)

$$a_z = \frac{\partial w}{\partial t} + u\frac{\partial w}{\partial x} + v\frac{\partial w}{\partial y} + w\frac{\partial w}{\partial z}$$

The first term on the right-hand side of Eq. 4–9, $\partial\vec{V}/\partial t$, is called the **local acceleration** and is nonzero only for unsteady flows. The second term, $(\vec{V}\cdot\vec{\nabla})\vec{V}$, is called the **advective acceleration** (sometimes the **convective**

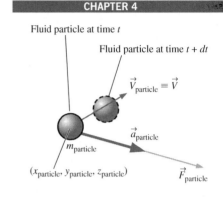

FIGURE 4–6

Newton's second law applied to a fluid particle; the acceleration vector (purple arrow) is in the same direction as the force vector (green arrow), but the velocity vector (blue arrow) may act in a different direction.

FIGURE 4–7

When following a fluid particle, the x-component of velocity, u, is defined as dx_{particle}/dt. Similarly, $v = dy_{\text{particle}}/dt$ and $w = dz_{\text{particle}}/dt$. Movement is shown here only in two dimensions for simplicity.

FIGURE 4–8

Flow of water through the nozzle of a garden hose illustrates that fluid particles may accelerate, even in a steady flow. In this example, the exit speed of the water is much higher than the water speed in the hose, implying that fluid particles have accelerated even though the flow is steady.

acceleration); *this term can be nonzero even for steady flows.* It accounts for the effect of the fluid particle moving (advecting or convecting) to a new location in the flow, where the velocity field is different. For example, consider steady flow of water through a garden hose nozzle (Fig. 4–8). We define *steady* in the Eulerian frame of reference to be when properties at any point in the flow field do not change with respect to time. Since the velocity at the exit of the nozzle is larger than that at the nozzle entrance, fluid particles clearly accelerate, even though the flow is steady. The acceleration is nonzero because of the advective acceleration terms in Eq. 4–9. Note that while the flow is steady from the point of view of a fixed observer in the Eulerian reference frame, it is *not* steady from the Lagrangian reference frame moving with a fluid particle that enters the nozzle and accelerates as it passes through the nozzle.

EXAMPLE 4–2   Acceleration of a Fluid Particle through a Nozzle

Nadeen is washing her car, using a nozzle similar to the one sketched in Fig. 4–8. The nozzle is 3.90 in (0.325 ft) long, with an inlet diameter of 0.420 in (0.0350 ft) and an outlet diameter of 0.182 in (see Fig. 4–9). The volume flow rate through the garden hose (and through the nozzle) is $\dot{V} = 0.841$ gal/min (0.00187 ft³/s), and the flow is steady. Estimate the magnitude of the acceleration of a fluid particle moving down the centerline of the nozzle.

SOLUTION The acceleration following a fluid particle down the center of a nozzle is to be estimated.
Assumptions **1** The flow is steady and incompressible. **2** The x-direction is taken along the centerline of the nozzle. **3** By symmetry, $v = w = 0$ along the centerline, but u increases through the nozzle.
Analysis The flow is steady, so you may be tempted to say that the acceleration is zero. However, even though the local acceleration $\partial \vec{V}/\partial t$ is identically zero for this steady flow field, the advective acceleration $(\vec{V}\cdot\vec{\nabla})\vec{V}$ is *not* zero. We first calculate the average x-component of velocity at the inlet and outlet of the nozzle by dividing volume flow rate by cross-sectional area:

Inlet speed:

$$u_{inlet} \cong \frac{\dot{V}}{A_{inlet}} = \frac{4\dot{V}}{\pi D_{inlet}^2} = \frac{4(0.00187 \text{ ft}^3/\text{s})}{\pi(0.0350 \text{ ft})^2} = 1.95 \text{ ft/s}$$

Similarly, the average outlet speed is $u_{outlet} = 10.4$ ft/s. We now calculate the acceleration two ways, with equivalent results. First, a simple average value of acceleration in the x-direction is calculated based on the change in speed divided by an estimate of the **residence time** of a fluid particle in the nozzle, $\Delta t = \Delta x/u_{avg}$ (Fig. 4–10). By the fundamental definition of acceleration as the rate of change of velocity,

$$\textit{Method A:} \quad a_x \cong \frac{\Delta u}{\Delta t} = \frac{u_{outlet} - u_{inlet}}{\Delta x/u_{avg}} = \frac{u_{outlet} - u_{inlet}}{2\,\Delta x/(u_{outlet} + u_{inlet})} = \frac{u_{outlet}^2 - u_{inlet}^2}{2\,\Delta x}$$

The second method uses the equation for acceleration field components in Cartesian coordinates, Eq. 4–11,

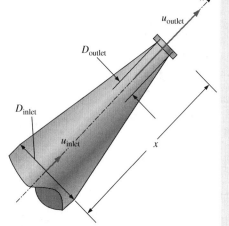

FIGURE 4–9

Flow of water through the nozzle of Example 4–2.

Method B: $a_x = \cancel{\dfrac{\partial u}{\partial t}}_{\underbrace{}_{\text{Steady}} 0} + u\dfrac{\partial u}{\partial x} + \cancel{v\dfrac{\partial u}{\partial y}}_{\underbrace{}_{\substack{0 \\ v\,=\,0 \text{ along centerline}}}} + \cancel{w\dfrac{\partial u}{\partial z}}_{\underbrace{}_{\substack{0 \\ w\,=\,0 \text{ along centerline}}}} \cong u_{\text{avg}}\dfrac{\Delta u}{\Delta x}$

Here we see that only one advective term is nonzero. We approximate the average speed through the nozzle as the average of the inlet and outlet speeds, and we use a **first-order finite difference approximation** (Fig. 4–11) for the average value of derivative $\partial u/\partial x$ through the centerline of the nozzle:

$$a_x \cong \frac{u_{\text{outlet}} + u_{\text{inlet}}}{2}\frac{u_{\text{outlet}} - u_{\text{inlet}}}{\Delta x} = \frac{u_{\text{outlet}}^2 - u_{\text{inlet}}^2}{2\,\Delta x}$$

The result of method B is identical to that of method A. Substitution of the given values yields

Axial acceleration:

$$a_x \cong \frac{u_{\text{outlet}}^2 - u_{\text{inlet}}^2}{2\,\Delta x} = \frac{(10.4 \text{ ft/s})^2 - (1.95 \text{ ft/s})^2}{2(0.325 \text{ ft})} = \mathbf{160 \text{ ft/s}^2}$$

Discussion Fluid particles are accelerated through the nozzle at nearly five times the acceleration of gravity (almost five *g*'s)! This simple example clearly illustrates that the acceleration of a fluid particle can be nonzero, even in steady flow. Note that the acceleration is actually a **point function**, whereas we have estimated a simple average acceleration through the entire nozzle.

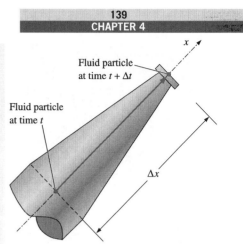

FIGURE 4–10
Residence time Δt is defined as the time it takes for a fluid particle to travel through the nozzle from inlet to outlet (distance Δx).

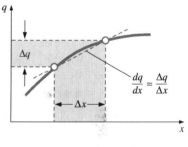

FIGURE 4–11
A *first-order finite difference approximation* for derivative dq/dx is simply the change in dependent variable (q) divided by the change in independent variable (x).

Material Derivative

The total derivative operator d/dt in Eq. 4–9 is given a special name, the **material derivative**; it is assigned a special notation, D/Dt, in order to emphasize that it is formed by *following a fluid particle as it moves through the flow field* (Fig. 4–12). Other names for the material derivative include **total, particle, Lagrangian, Eulerian,** and **substantial derivative.**

Material derivative: $\qquad \dfrac{D}{Dt} = \dfrac{d}{dt} = \dfrac{\partial}{\partial t} + (\vec{V}\cdot\vec{\nabla})$ (4–12)

When we apply the material derivative of Eq. 4–12 to the velocity field, the result is the acceleration field as expressed by Eq. 4–9, which is thus sometimes called the **material acceleration,**

Material acceleration: $\qquad \vec{a}(x, y, z, t) = \dfrac{D\vec{V}}{Dt} = \dfrac{d\vec{V}}{dt} = \dfrac{\partial\vec{V}}{\partial t} + (\vec{V}\cdot\vec{\nabla})\vec{V}$ (4–13)

Equation 4–12 can also be applied to other fluid properties besides velocity, both scalars and vectors. For example, the material derivative of pressure is written as

Material derivative of pressure: $\qquad \dfrac{DP}{Dt} = \dfrac{dP}{dt} = \dfrac{\partial P}{\partial t} + (\vec{V}\cdot\vec{\nabla})P$ (4–14)

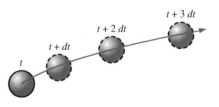

FIGURE 4–12
The material derivative D/Dt is defined by following a fluid particle as it moves throughout the flow field. In this illustration, the fluid particle is accelerating to the right as it moves up and to the right.

FIGURE 4–13
The material derivative D/Dt is composed of a *local* or *unsteady* part and a *convective* or *advective* part.

Equation 4–14 represents the time rate of change of pressure following a fluid particle as it moves through the flow and contains both local (unsteady) and advective components (Fig. 4–13).

EXAMPLE 4–3   Material Acceleration of a Steady Velocity Field

Consider the steady, incompressible, two-dimensional velocity field of Example 4–1. (*a*) Calculate the material acceleration at the point (x = 2 m, y = 3 m). (*b*) Sketch the material acceleration vectors at the same array of x- and y-values as in Example 4–1.

SOLUTION For the given velocity field, the material acceleration vector is to be calculated at a particular point and plotted at an array of locations in the flow field.
Assumptions **1** The flow is steady and incompressible. **2** The flow is two-dimensional, implying no z-component of velocity and no variation of u or v with z.
Analysis (*a*) Using the velocity field of Eq. 1 of Example 4–1 and the equation for material acceleration components in Cartesian coordinates (Eq. 4–11), we write expressions for the two nonzero components of the acceleration vector:

$$a_x = \frac{\partial u}{\partial t} + u\frac{\partial u}{\partial x} + v\frac{\partial u}{\partial y} + w\frac{\partial u}{\partial z}$$

$$= 0 + (0.5 + 0.8x)(0.8) + (1.5 - 0.8y)(0) + 0 = (0.4 + 0.64x) \text{ m/s}^2$$

and

$$a_y = \frac{\partial v}{\partial t} + u\frac{\partial v}{\partial x} + v\frac{\partial v}{\partial y} + w\frac{\partial v}{\partial z}$$

$$= 0 + (0.5 + 0.8x)(0) + (1.5 - 0.8y)(-0.8) + 0 = (-1.2 + 0.64y) \text{ m/s}^2$$

At the point (x = 2 m, y = 3 m), a_x = 1.68 m/s² and a_y = 0.720 m/s².
(*b*) The equations in part (*a*) are applied to an array of x- and y-values in the flow domain within the given limits, and the acceleration vectors are plotted in Fig. 4–14.
Discussion The acceleration field is nonzero, even though the flow is *steady*. Above the stagnation point (above y = 1.875 m), the acceleration vectors plotted in Fig. 4–14 point upward, increasing in magnitude away from the stagnation point. To the right of the stagnation point (to the right of x = −0.625 m), the acceleration vectors point to the right, again increasing in magnitude away from the stagnation point. This agrees qualitatively with the velocity vectors of Fig. 4–4 and the streamlines sketched in Fig. 4–14; namely, in the upper-right portion of the flow field, fluid particles are accelerated in the upper-right direction and therefore veer in the counterclockwise direction due to **centripetal acceleration** toward the upper right. The flow below y = 1.875 m is a mirror image of the flow above this symmetry line, and the flow to the left of x = −0.625 m is a mirror image of the flow to the right of this symmetry line.

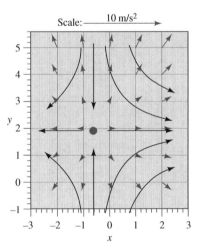

FIGURE 4–14
Acceleration vectors (purple arrows) for the velocity field of Examples 4–1 and 4–3. The scale is shown by the top arrow, and the solid black curves represent the approximate shapes of some streamlines, based on the calculated velocity vectors (see Fig. 4–4). The stagnation point is indicated by the red circle.

4–2 • FLOW PATTERNS AND FLOW VISUALIZATION

While quantitative study of fluid dynamics requires advanced mathematics, much can be learned from **flow visualization**—the visual examination of flow field features. Flow visualization is useful not only in physical experiments (Fig. 4–15), but in *numerical* solutions as well [**computational fluid dynamics (CFD)**]. In fact, the very first thing an engineer using CFD does after obtaining a numerical solution is simulate some form of flow visualization, so that he or she can see the "whole picture" rather than merely a list of numbers and quantitative data. Why? Because the human mind is designed to rapidly process an incredible amount of visual information; as they say, a picture is worth a thousand words. There are many types of flow patterns that can be visualized, both physically (experimentally) and/or computationally.

141
CHAPTER 4

FIGURE 4–15
Spinning baseball.
The late F. N. M. Brown devoted many years to developing and using smoke visualization in wind tunnels at the University of Notre Dame. Here the flow speed is about 77 ft/s and the ball is rotated at 630 rpm.
Photograph courtesy of T. J. Mueller.

Streamlines and Streamtubes

A **streamline** is a curve that is everywhere tangent to the instantaneous local velocity vector.

Streamlines are useful as indicators of the instantaneous direction of fluid motion throughout the flow field. For example, regions of recirculating flow and separation of a fluid off of a solid wall are easily identified by the streamline pattern. Streamlines cannot be directly observed experimentally except in steady flow fields, in which they are coincident with pathlines and streaklines, to be discussed next. Mathematically, however, we can write a simple expression for a streamline based on its definition.

Consider an infinitesimal arc length $d\vec{r} = dx\vec{i} + dy\vec{j} + dz\vec{k}$ along a streamline; $d\vec{r}$ must be parallel to the local velocity vector $\vec{V} = u\vec{i} + v\vec{j} + w\vec{k}$ by definition of the streamline. By simple geometric arguments using similar triangles, we know that the components of $d\vec{r}$ must be proportional to those of $\vec{V}$ (Fig. 4–16). Hence,

Equation for a streamline: $\dfrac{dr}{V} = \dfrac{dx}{u} = \dfrac{dy}{v} = \dfrac{dz}{w}$ **(4–15)**

where dr is the magnitude of $d\vec{r}$ and V is the speed, the magnitude of velocity vector $\vec{V}$. Equation 4–15 is illustrated in two dimensions for simplicity in Fig. 4–16. For a known velocity field, we integrate Eq. 4–15 to obtain equations for the streamlines. In two dimensions, (x, y), (u, v), the following differential equation is obtained:

Streamline in the xy-plane: $\left(\dfrac{dy}{dx}\right)_{\text{along a streamline}} = \dfrac{v}{u}$ **(4–16)**

In some simple cases, Eq. 4–16 may be solvable analytically; in the general case, it must be solved numerically. In either case, an arbitrary constant of integration appears. Each chosen value of the constant represents a different streamline. The *family* of curves that satisfy Eq. 4–16 therefore represents streamlines of the flow field.

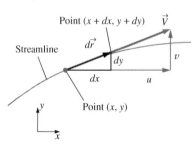

FIGURE 4–16
For two-dimensional flow in the *xy*-plane, arc length $d\vec{r} = (dx, dy)$ along a *streamline* is everywhere tangent to the local instantaneous velocity vector $\vec{V} = (u, v)$.

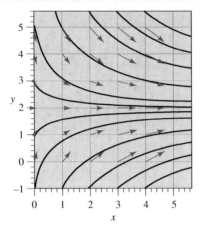

FIGURE 4–17
Streamlines (solid black curves) for the velocity field of Example 4–4; velocity vectors of Fig. 4–4 (blue arrows) are superimposed for comparison.

EXAMPLE 4–4   Streamlines in the *xy*-Plane—An Analytical Solution

For the steady, incompressible, two-dimensional velocity field of Example 4–1, plot several streamlines in the right half of the flow ($x > 0$) and compare to the velocity vectors plotted in Fig. 4–4.

SOLUTION   An analytical expression for streamlines is to be generated and plotted in the upper-right quadrant.

Assumptions   **1** The flow is steady and incompressible. **2** The flow is two-dimensional, implying no *z*-component of velocity and no variation of *u* or *v* with *z*.

Analysis   Equation 4–16 is applicable here; thus, along a streamline,

$$\frac{dy}{dx} = \frac{v}{u} = \frac{1.5 - 0.8y}{0.5 + 0.8x}$$

We solve this differential equation by separation of variables:

$$\frac{dy}{1.5 - 0.8y} = \frac{dx}{0.5 + 0.8x} \quad \rightarrow \quad \int \frac{dy}{1.5 - 0.8y} = \int \frac{dx}{0.5 + 0.8x}$$

After some algebra, we solve for *y* as a function of *x* along a streamline,

$$y = \frac{C}{0.8(0.5 + 0.8x)} + 1.875$$

where *C* is a constant of integration that can be set to various values in order to plot the streamlines. Several streamlines of the given flow field are shown in Fig. 4–17.

Discussion   The velocity vectors of Fig. 4–4 are superimposed on the streamlines of Fig. 4–17; the agreement is excellent in the sense that the velocity vectors point everywhere tangent to the streamlines. Note that speed cannot be determined directly from the streamlines alone.

A **streamtube** consists of a bundle of streamlines (Fig. 4–18), much like a communications cable consists of a bundle of fiber-optic cables. Since streamlines are everywhere parallel to the local velocity, fluid cannot cross a streamline by definition. By extension, *fluid within a streamtube must remain there and cannot cross the boundary of the streamtube.* You must keep in mind that both streamlines and streamtubes are instantaneous quantities, defined at a particular instant in time according to the velocity field at that instant. In an *unsteady* flow, the streamline pattern may change significantly with time. Nevertheless, at any instant in time, the mass flow rate passing through any cross-sectional slice of a given streamtube must remain the same. For example, in a converging portion of an incompressible flow field, the diameter of the streamtube must decrease as the velocity increases in order to conserve mass (Fig. 4–19*a*). Likewise, the streamtube diameter increases in diverging portions of an incompressible flow (Fig. 4–19*b*).

Pathlines

A **pathline** is the actual path traveled by an individual fluid particle over some time period.

Streamlines

Streamtube

FIGURE 4–18
A *streamtube* consists of a bundle of individual streamlines.

FIGURE 4–19
In an incompressible flow field, a streamtube (*a*) decreases in diameter as the flow accelerates or converges and (*b*) increases in diameter as the flow decelerates or diverges.

Pathlines are the easiest of the flow patterns to understand. A pathline is a Lagrangian concept in that we simply follow the path of an individual fluid particle as it moves around in the flow field (Fig. 4–20). Thus, a pathline is the same as the fluid particle's material position vector $(x_{particle}(t), y_{particle}(t), z_{particle}(t))$, discussed in Section 4–1, traced out over some finite time interval. In a physical experiment, you can imagine a tracer fluid particle that is marked somehow—either by color or brightness—such that it is easily distinguishable from surrounding fluid particles. Now imagine a camera with the shutter open for a certain time period, $t_{start} < t < t_{end}$, in which the particle's path is recorded; the resulting curve is called a pathline. An intriguing example is shown in Fig. 4–21 for the case of waves moving along the surface of water in a tank. Neutrally buoyant white **tracer particles** are suspended in the water, and a time-exposure photograph is taken for one complete wave period. The result is pathlines that are elliptical in shape, showing that fluid particles bob up and down and forward and backward, but return to their original position upon completion of one wave period; there is no net forward motion. You may have experienced something similar while bobbing up and down on ocean waves at the beach.

FIGURE 4–20
A *pathline* is formed by following the actual path of a fluid particle.

FIGURE 4–21
Pathlines produced by white tracer particles suspended in water and captured by time-exposure photography; as waves pass horizontally, each particle moves in an elliptical path during one wave period.

Wallet, A. & Ruellan, F. 1950, La Houille Blanche 5:483–489. Used by permission.

A modern experimental technique called **particle image velocimetry** (**PIV**) utilizes short segments of particle pathlines to measure the velocity field over an entire plane in a flow (Adrian, 1991). (Recent advances also extend the technique to three dimensions.) In PIV, tiny tracer particles are suspended in the fluid, much like in Fig. 4–21. However, the flow is illuminated by two flashes of light (usually a light sheet from a laser as in Fig. 4–22) to produce two bright spots (recorded by a camera) for each moving particle. Then, both the magnitude and direction of the velocity vector at each particle location can be inferred, assuming that the tracer particles are small enough that they move with the fluid. Modern digital photography and fast computers have enabled PIV to be performed rapidly enough so that *unsteady* features of a flow field can also be measured. PIV is discussed in more detail in Chap. 8.

FIGURE 4–22

Stereo PIV measurements of the wing tip vortex in the wake of a NACA-66 airfoil at angle of attack. Color contours denote the local vorticity, normalized by the minimum value, as indicated in the color map. Vectors denote fluid motion in the plane of measurement. The black line denotes the location of the upstream wing trailing edge. Coordinates are normalized by the airfoil chord, and the origin is the wing root.

Photo by Michael H. Krane, ARL-Penn State.

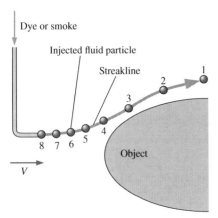

FIGURE 4–23

A *streakline* is formed by continuous introduction of dye or smoke from a point in the flow. Labeled tracer particles (1 through 8) were introduced sequentially.

Pathlines can also be calculated numerically for a known velocity field. Specifically, the location of the tracer particle is integrated over time from some starting location $\vec{x}_{start}$ and starting time t_{start} to some later time t.

Tracer particle location at time t:
$$\vec{x} = \vec{x}_{start} + \int_{t_{start}}^{t} \vec{V}\, dt \qquad (4-17)$$

When Eq. 4–17 is calculated for t between t_{start} and t_{end}, a plot of $\vec{x}(t)$ is the pathline of the fluid particle during that time interval, as illustrated in Fig. 4–20. For some simple flow fields, Eq. 4–17 can be integrated analytically. For more complex flows, we must perform a numerical integration.

If the velocity field is steady, individual fluid particles follow streamlines. Thus, *for steady flow, pathlines are identical to streamlines.*

Streaklines

A **streakline** is the locus of fluid particles that have passed sequentially through a prescribed point in the flow.

Streaklines are the most common flow pattern generated in a physical experiment. If you insert a small tube into a flow and introduce a continuous stream of tracer fluid (dye in a water flow or smoke in an airflow), the observed pattern is a streakline. Figure 4–23 shows a tracer being injected into a free-stream flow containing an object, such as a wing. The circles represent individual injected tracer fluid particles, released at a uniform time interval. As the particles are forced out of the way by the object, they accelerate around the shoulder of the object, as indicated by the increased distance between individual tracer particles in that region. The streakline is formed by connecting all the circles into a smooth curve. In physical experiments in a wind or water tunnel, the smoke or dye is injected *continuously*, not as individual particles, and the resulting flow pattern is by definition a streakline. In Fig. 4–23, tracer particle 1 was released at an earlier time than tracer particle 2, and so on. The location of an individual tracer particle is determined by the surrounding velocity field from the moment of its injection into the flow until the present time. If the flow is unsteady, the surrounding velocity field changes, and we cannot expect the resulting streakline to resemble a streamline or pathline at any given instant in time. However, *if the flow is steady, streamlines, pathlines, and streaklines are identical* (Fig. 4–24).

Streaklines are often confused with streamlines or pathlines. While the three flow patterns are identical in steady flow, they can be quite different in unsteady flow. The main difference is that a streamline represents an *instantaneous* flow pattern at a given instant in time, while a streakline and a pathline are flow patterns that have some *age* and thus a *time history* associated with them. A streakline is an instantaneous snapshot of a *time-integrated* flow pattern. A pathline, on the other hand, is the *time-exposed* flow path of an individual particle over some time period.

The time-integrative property of streaklines is vividly illustrated in an experiment by Cimbala et al. (1988), reproduced here as Fig. 4–25. The authors used a **smoke wire** for flow visualization in a wind tunnel. In operation, the smoke wire is a thin vertical wire that is coated with mineral oil. The oil breaks up into beads along the length of the wire due to surface tension effects. When an electric current heats the wire, each little bead of oil produces a streakline of smoke. In Fig. 4–25a, streaklines are introduced from a smoke wire located just downstream of a circular cylinder of diameter D aligned normal to the plane of view. (When multiple streaklines are introduced along a line, as in Fig. 4–25, we refer to this as a **rake** of streaklines.) The Reynolds number of the flow is Re $= \rho VD/\mu = 93$. Because of unsteady **vortices** shed in an alternating pattern from the cylinder, the smoke collects into a clearly defined periodic pattern called a **Kármán vortex street.** A similar pattern can be seen at much larger scale in the air flow in the wake of an island (Fig. 4–26).

From Fig. 4–25a alone, you may think that the shed vortices continue to exist to several hundred diameters downstream of the cylinder. However, the streakline pattern of this figure is misleading! In Fig. 4–25b, the smoke wire is placed 150 diameters downstream of the cylinder. The resulting streaklines are straight, indicating that the shed vortices have in reality disappeared by this downstream distance. The flow is steady and parallel at this location, and there are no more vortices; viscous diffusion has caused adjacent vortices of opposite sign to cancel each other out by around 100 cylinder diameters. The patterns of Fig. 4–25a near $x/D = 150$ are merely *remnants* of the vortex street that existed upstream. The streaklines of Fig. 4–25b, however, show the correct features of the flow at that location. The streaklines generated at $x/D = 150$ are identical to streamlines or pathlines in that region of the flow—straight, nearly horizontal lines—since the flow is steady there.

For a known velocity field, a streakline can be generated numerically. We need to follow the paths of a continuous stream of tracer particles from the time of their injection into the flow until the present time, using Eq. 4–17. Mathematically, the location of a tracer particle is integrated over time from the time of its injection t_{inject} to the present time t_{present}. Equation 4–17 becomes

Integrated tracer particle location: $\vec{x} = \vec{x}_{\text{injection}} + \int_{t_{\text{inject}}}^{t_{\text{present}}} \vec{V} \, dt$ (4–18)

FIGURE 4–24
Streaklines produced by colored fluid introduced upstream; since the flow is steady, these streaklines are the same as streamlines and pathlines.
Courtesy ONERA. Photograph by Werlé.

FIGURE 4–25
Smoke streaklines introduced by a smoke wire at two different locations in the wake of a circular cylinder: (a) smoke wire just downstream of the cylinder and (b) smoke wire located at $x/D = 150$. The time-integrative nature of streaklines is clearly seen by comparing the two photographs.
Photos by John M. Cimbala.

FIGURE 4–26
Kármán vortices visible in the clouds in the wake of Alexander Selkirk Island in the southern Pacific Ocean.
Photo from Landsat 7 WRS Path 6 Row 83, center: -33.18, -79.99, 9/15/1999, earthobservatory.nasa.gov. Courtesy of NASA.

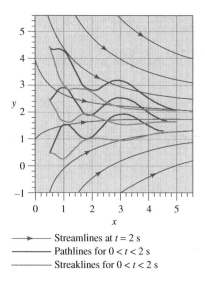

Streamlines at $t = 2$ s
Pathlines for $0 < t < 2$ s
Streaklines for $0 < t < 2$ s

FIGURE 4–27

Streamlines, pathlines, and streaklines for the oscillating velocity field of Example 4–5. The streaklines and pathlines are wavy because of their integrated time history, but the streamlines are not wavy since they represent an instantaneous snapshot of the velocity field.

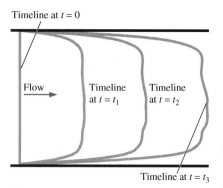

FIGURE 4–28

Timelines are formed by marking a line of fluid particles, and then watching that line move (and deform) through the flow field; timelines are shown at $t = 0$, t_1, t_2, and t_3.

In a complex unsteady flow, the time integration must be performed numerically as the velocity field changes with time. When the locus of tracer particle locations at $t = t_{present}$ is connected by a smooth curve, the result is the desired streakline.

EXAMPLE 4–5   Comparison of Flow Patterns in an Unsteady Flow

An *unsteady*, incompressible, two-dimensional velocity field is given by

$$\vec{V} = (u, v) = (0.5 + 0.8x)\vec{i} + (1.5 + 2.5\sin(\omega t) - 0.8y)\vec{j} \qquad (1)$$

where the angular frequency ω is equal to 2π rad/s (a physical frequency of 1 Hz). This velocity field is identical to that of Eq. 1 of Example 4–1 except for the additional periodic term in the v-component of velocity. In fact, since the period of oscillation is 1 s, when time t is any integral multiple of $\frac{1}{2}$ s ($t = 0, \frac{1}{2}, 1, \frac{3}{2}, 2, \ldots$ s), the sine term in Eq. 1 is zero and the velocity field is instantaneously identical to that of Example 4–1. Physically, we imagine flow into a large bell mouth inlet that is oscillating up and down at a frequency of 1 Hz. Consider two complete cycles of flow from $t = 0$ s to $t = 2$ s. Compare instantaneous streamlines at $t = 2$ s to pathlines and streaklines generated during the time period from $t = 0$ s to $t = 2$ s.

SOLUTION   Streamlines, pathlines, and streaklines are to be generated and compared for the given unsteady velocity field.

Assumptions   **1** The flow is incompressible. **2** The flow is two-dimensional, implying no z-component of velocity and no variation of u or v with z.

Analysis   The instantaneous streamlines at $t = 2$ s are identical to those of Fig. 4–17, and several of them are replotted in Fig. 4–27. To simulate pathlines, we use the Runge–Kutta numerical integration technique to march in time from $t = 0$ s to $t = 2$ s, tracing the path of fluid particles released at three locations: ($x = 0.5$ m, $y = 0.5$ m), ($x = 0.5$ m, $y = 2.5$ m), and ($x = 0.5$ m, $y = 4.5$ m). These pathlines are shown in Fig. 4–27, along with the streamlines. Finally, streaklines are simulated by following the paths of *many* fluid tracer particles released at the given three locations at times between $t = 0$ s and $t = 2$ s, and connecting the locus of their positions at $t = 2$ s. These streaklines are also plotted in Fig. 4–27.

Discussion   Since the flow is unsteady, the streamlines, pathlines, and streaklines are *not* coincident. In fact, they differ significantly from each other. Note that the streaklines and pathlines are wavy due to the undulating v-component of velocity. Two complete periods of oscillation have occurred between $t = 0$ s and $t = 2$ s, as verified by a careful look at the pathlines and streaklines. The streamlines have no such waviness since they have no time history; they represent an instantaneous snapshot of the velocity field at $t = 2$ s.

Timelines

A **timeline** is a set of adjacent fluid particles that were marked at the same (earlier) instant in time.

Timelines are particularly useful in situations where the uniformity of a flow (or lack thereof) is to be examined. Figure 4–28 illustrates timelines in a channel flow between two parallel walls. Because of friction at the walls, the fluid velocity there is zero (the no-slip condition), and the top and bottom of the timeline are anchored at their starting locations. In regions of the flow away from the walls, the marked fluid particles move at the local fluid velocity, deforming the timeline. In the example of Fig. 4–28, the speed near the center of the channel is fairly uniform, but small deviations tend to amplify with time as the timeline stretches. Timelines can be generated experimentally in a water channel through use of a **hydrogen bubble wire.** When a short burst of electric current is sent through the cathode wire, electrolysis of the water occurs and tiny hydrogen gas bubbles form at the wire. Since the bubbles are so small, their buoyancy is nearly negligible, and the bubbles follow the water flow nicely (Fig. 4–29).

Refractive Flow Visualization Techniques

Another category of flow visualization is based on the **refractive property** of light waves. As you recall from your study of physics, the speed of light through one material may differ somewhat from that in another material, or even in the *same* material if its density changes. As light travels through one fluid into a fluid with a different index of refraction, the light rays bend (they are **refracted**).

There are two primary flow visualization techniques that utilize the fact that the index of refraction in air (or other gases) varies with density. They are the **shadowgraph technique** and the **schlieren technique** (Settles, 2001). **Interferometry** is a visualization technique that utilizes the related *phase change* of light as it passes through air of varying densities as the basis for flow visualization and is not discussed here (see Merzkirch, 1987). All these techniques are useful for flow visualization in flow fields where density changes from one location in the flow to another, such as natural convection flows (temperature differences cause the density variations), mixing flows (fluid species cause the density variations), and supersonic flows (shock waves and expansion waves cause the density variations).

Unlike flow visualizations involving streaklines, pathlines, and timelines, the shadowgraph and schlieren methods do not require injection of a visible

FIGURE 4–29
Timelines produced by a hydrogen bubble wire are used to visualize the boundary layer velocity profile shape along a flat plate. Flow is from left to right, and the hydrogen bubble wire is located to the left of the field of view. Bubbles near the wall reveal a flow instability that leads to turbulence.

Bippes, H. 1972 Sitzungsber, Heidelb. Akad. Wiss. Math. Naturwiss. Kl., no. 3, 103–180; NASA TM-75243, 1978.

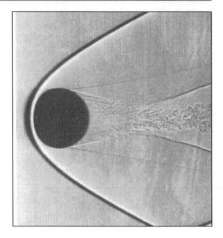

FIGURE 4–30

Shadowgram of a 14.3 mm sphere in free flight through air at Ma = 3.0. A shock wave is clearly visible in the shadow as a dark band that curves around the sphere and is called a *bow wave* (see Chap. 12).

A. C. Charters, Air Flow Branch, U.S. Army Ballistic Research Laboratory.

FIGURE 4–31

Schlieren image of natural convection due to a barbeque grill.

G. S. Settles, Gas Dynamics Lab, Penn State University. Used by permission.

tracer (smoke or dye). Rather, density differences and the refractive property of light provide the necessary means for visualizing regions of activity in the flow field, allowing us to "see the invisible." The image (a **shadowgram**) produced by the shadowgraph method is formed when the refracted rays of light rearrange the shadow cast onto a viewing screen or camera focal plane, causing bright or dark patterns to appear in the shadow. The dark patterns indicate the location where the refracted rays *originate*, while the bright patterns mark where these rays *end up*, and can be misleading. As a result, the dark regions are less distorted than the bright regions and are more useful in the interpretation of the shadowgram. In the shadowgram of Fig. 4–30, for example, we can be confident of the shape and position of the bow shock wave (the dark band), but the refracted bright light has distorted the front of the sphere's shadow.

A shadowgram is not a true optical image; it is, after all, merely a shadow. A **schlieren image,** however, involves lenses (or mirrors) and a knife edge or other cutoff device to block the refracted light and is a true focused optical image. Schlieren imaging is more complicated to set up than is shadowgraphy (see Settles, 2001 for details) but has a number of advantages. For example, a schlieren image does not suffer from optical distortion by the refracted light rays. Schlieren imaging is also more sensitive to weak density gradients such as those caused by natural convection (Fig. 4–31) or by gradual phenomena like expansion fans in supersonic flow. Color schlieren imaging techniques have also been developed. Finally, one can adjust more components in a schlieren setup, such as the location, orientation, and type of the cutoff device, in order to produce an image that is most useful for the problem at hand.

Surface Flow Visualization Techniques

Finally, we briefly mention some flow visualization techniques that are useful along solid surfaces. The direction of fluid flow immediately above a solid surface can be visualized with **tufts**—short, flexible strings glued to the surface at one end that point in the flow direction. Tufts are especially useful for locating regions of flow separation, where the flow direction reverses.

A technique called **surface oil visualization** can be used for the same purpose—oil placed on the surface forms streaks called **friction lines** that indicate the direction of flow. If it rains lightly when your car is dirty (especially in the winter when salt is on the roads), you may have noticed streaks along the hood and sides of the car, or even on the windshield. This is similar to what is observed with surface oil visualization.

Lastly, there are pressure-sensitive and temperature-sensitive paints that enable researchers to observe the pressure or temperature distribution along solid surfaces.

4–3 ▪ PLOTS OF FLUID FLOW DATA

Regardless of how the results are obtained (analytically, experimentally, or computationally), it is usually necessary to *plot* flow data in ways that enable the reader to get a feel for how the flow properties vary in time and/or space. You are already familiar with *time plots*, which are especially useful in turbulent flows (e.g., a velocity component plotted as a function of time), and *xy*-plots (e.g., pressure as a function of radius). In this section, we discuss three additional types of plots that are useful in fluid mechanics—profile plots, vector plots, and contour plots.

Profile Plots

A **profile plot** indicates how the value of a scalar property varies along some desired direction in the flow field.

Profile plots are the simplest of the three to understand because they are like the common *xy*-plots that you have generated since grade school. Namely, you plot how one variable *y* varies as a function of a second variable *x*. In fluid mechanics, profile plots of *any* scalar variable (pressure, temperature, density, etc.) can be created, but the most common one used in this book is the *velocity profile plot*. We note that since velocity is a vector quantity, we usually plot either the magnitude of velocity or one of the components of the velocity vector as a function of distance in some desired direction.

For example, one of the timelines in the boundary layer flow of Fig. 4–29 is converted into a velocity profile plot by recognizing that at a given instant in time, the horizontal distance traveled by a hydrogen bubble at vertical location *y* is proportional to the local *x*-component of velocity *u*. We plot *u* as a function of *y* in Fig. 4–32. The values of *u* for the plot can also be obtained analytically (see Chaps. 9 and 10), experimentally using PIV or some kind of local velocity measurement device (see Chap. 8), or computationally (see Chap. 15). Note that it is more physically meaningful in this example to plot *u* on the *abscissa* (horizontal axis) rather than on the *ordinate* (vertical axis) even though it is the dependent variable, since position *y* is then in its proper orientation (up) rather than across.

Finally, it is common to add arrows to velocity profile plots to make them more visually appealing, although no additional information is provided by the arrows. If more than one component of velocity is plotted by the arrow, the *direction* of the local velocity vector is indicated and the velocity profile plot becomes a velocity *vector* plot.

Vector Plots

A **vector plot** is an array of arrows indicating the magnitude and direction of a vector property at an instant in time.

While streamlines indicate the *direction* of the instantaneous velocity field, they do not directly indicate the *magnitude* of the velocity (i.e., the speed). A useful flow pattern for both experimental and computational fluid flows is thus the vector plot, which consists of an array of arrows that indicate both magnitude *and* direction of an instantaneous vector property. We have already seen an example of a velocity vector plot in Fig. 4–4 and an acceleration vector plot in Fig. 4–14. These were generated analytically. Vector plots can also be generated from experimentally obtained data (e.g., from PIV measurements) or numerically from CFD calculations.

To further illustrate vector plots, we generate a two-dimensional flow field consisting of free-stream flow impinging on a block of rectangular cross section. We perform CFD calculations, and the results are shown in

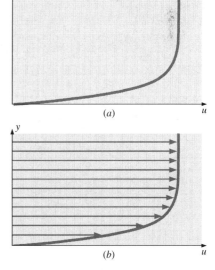

FIGURE 4–32

Profile plots of the horizontal component of velocity as a function of vertical distance; flow in the boundary layer growing along a horizontal flat plate: (*a*) standard profile plot and (*b*) profile plot with arrows.

150
FLUID KINEMATICS

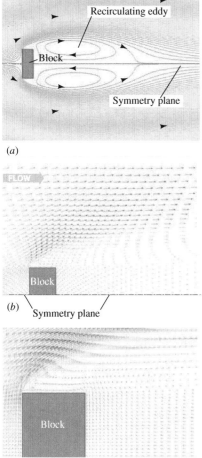

(a)

(b)

(c)

FIGURE 4–33
Results of CFD calculations of flow
impinging on a block; (a) streamlines,
(b) velocity vector plot of the upper
half of the flow, and (c) velocity vector
plot, close-up view revealing more
details in the separated flow region.

Fig. 4–33. Note that this flow is by nature turbulent and unsteady, but only the long-time averaged results are calculated and displayed here. Streamlines are plotted in Fig. 4–33a; a view of the entire block and a large portion of its wake is shown. The closed streamlines above and below the symmetry plane indicate large recirculating eddies, one above and one below the line of symmetry. A velocity vector plot is shown in Fig. 4–33b. (Only the upper half of the flow is shown because of symmetry.) It is clear from this plot that the flow accelerates around the upstream corner of the block, so much so in fact that the boundary layer cannot negotiate the sharp corner and separates off the block, producing the large recirculating eddies downstream of the block. (Note that these velocity vectors are time-averaged values; the instantaneous vectors change in both magnitude and direction with time as vortices are shed from the body, similar to those of Fig. 4–25a.) A close-up view of the separated flow region is plotted in Fig. 4–33c, where we verify the reverse flow in the lower half of the large recirculating eddy.

The vectors of Fig. 4–33 are colored by velocity magnitude, but with modern CFD codes and postprocessors, the vectors can be colored according to some other flow property such as pressure (red for high pressure and blue for low pressure) or temperature (red for hot and blue for cold). In this manner, one can easily visualize not only the magnitude and direction of the flow, but other properties as well, simultaneously.

Contour Plots

> A **contour plot** shows curves of constant values of a scalar property (or magnitude of a vector property) at an instant in time.

If you do any hiking, you are familiar with contour maps of mountain trails. The maps consist of a series of closed curves, each indicating a constant elevation or altitude. Near the center of a group of such curves is the mountain peak or valley; the actual peak or valley is a *point* on the map showing the highest or lowest elevation. Such maps are useful in that not only do you get a bird's-eye view of the streams and trails, etc., but you can also easily see your elevation and where the trail is flat or steep. In fluid mechanics, the same principle is applied to various scalar flow properties; contour plots (also called **isocontour plots**) are generated of pressure, temperature, velocity magnitude, species concentration, properties of turbulence, etc. A contour plot can quickly reveal regions of high (or low) values of the flow property being studied.

A contour plot may consist simply of curves indicating various levels of the property; this is called a **contour line plot**. Alternatively, the contours can be filled in with either colors or shades of gray; this is called a **filled contour plot**. An example of pressure contours is shown in Fig. 4–34 for the same flow as in Fig. 4–33. In Fig. 4–34a, filled contours are shown using color to identify regions of different pressure levels—blue regions indicate low pressure and red regions indicate high pressure. It is clear from this figure that the pressure is highest at the front face of the block and lowest along the top of the block in the separated zone. The pressure is also low in the wake of the block, as expected. In Fig. 4–34b, the same pressure contours are shown, but as a contour line plot with labeled levels of gage pressure in units of pascal.

FLOW

Symmetry plane

(a)

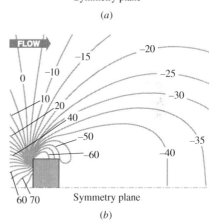

FLOW

0   −10   −15   −20
         −25
   10   −30
     20
      40
   −50   −35
        −60   −40

60 70   Symmetry plane

(b)

FIGURE 4–34

Contour plots of the pressure field due to flow impinging on a block, as produced by CFD calculations; only the upper half is shown due to symmetry; (*a*) filled color contour plot and (*b*) contour line plot where pressure values are displayed in units of Pa (pascals) gage pressure.

In CFD, contour plots are often displayed in vivid colors with red usually indicating the highest value of the scalar and blue the lowest. A healthy human eye can easily spot a red or blue region and thus locate regions of high or low value of the flow property. Because of the pretty pictures produced by CFD, computational fluid dynamics is sometimes given the nickname "colorful fluid dynamics."

4–4 ▪ OTHER KINEMATIC DESCRIPTIONS

Types of Motion or Deformation of Fluid Elements

In fluid mechanics, as in solid mechanics, an element may undergo four fundamental types of motion or deformation, as illustrated in two dimensions in Fig. 4–35: (*a*) **translation,** (*b*) **rotation,** (*c*) **linear strain** (sometimes called **extensional strain**), and (*d*) **shear strain.** The study of fluid dynamics is further complicated by the fact that all four types of motion or deformation usually occur simultaneously. Because fluid elements may be in constant motion, it is preferable in fluid dynamics to describe the motion and deformation of fluid elements in terms of *rates*. In particular, we discuss *velocity* (rate of translation), *angular velocity* (rate of rotation), *linear strain rate* (rate of linear strain), and *shear strain rate* (rate of shear strain). In order for these **deformation rates** to be useful in the calculation of fluid flows, we must express them in terms of velocity and derivatives of velocity.

Translation and rotation are easily understood since they are commonly observed in the motion of solid particles such as billiard balls (Fig. 4–1). A vector is required in order to fully describe the rate of translation in three dimensions. The **rate of translation vector** is described mathematically as the **velocity vector.** In Cartesian coordinates,

Rate of translation vector in Cartesian coordinates:

$$\vec{V} = u\vec{i} + v\vec{j} + w\vec{k} \tag{4–19}$$

In Fig. 4–35*a*, the fluid element has moved in the positive horizontal (*x*) direction; thus *u* is positive, while *v* (and *w*) are zero.

Rate of rotation (angular velocity) at a point is defined as *the average rotation rate of two initially perpendicular lines that intersect at that point.* In Fig. 4–35*b*, for example, consider the point at the bottom-left corner of the initially square fluid element. The left edge and the bottom edge of the element intersect at that point and are initially perpendicular. Both of these lines rotate counterclockwise, which is the mathematically positive direction. The angle between these two lines (or between *any* two initially perpendicular lines on this fluid element) remains at 90° since solid body rotation is illustrated in the figure. Therefore, both lines rotate at the same rate, and the rate of rotation in the plane is simply the component of angular velocity in that plane.

In the more general, but still two-dimensional case (Fig. 4–36), the fluid particle translates and deforms as it rotates, and the rate of rotation is calculated according to the definition given in the previous paragraph. Namely, we begin at time t_1 with two initially perpendicular lines (lines *a* and *b* in Fig. 4–36) that intersect at point *P* in the *xy*-plane. We follow these lines as they move and rotate in an infinitesimal increment of time $dt = t_2 - t_1$.

(a)

(b)

(c)

(d)

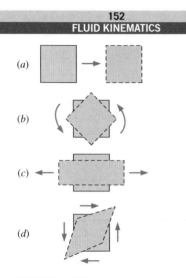

FIGURE 4–35
Fundamental types of fluid element motion or deformation: (*a*) translation, (*b*) rotation, (*c*) linear strain, and (*d*) shear strain.

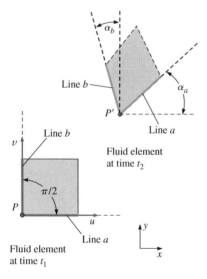

FIGURE 4–36
For a fluid element that translates and deforms as sketched, the *rate of rotation* at point *P* is defined as the average rotation rate of two initially perpendicular lines (lines *a* and *b*).

At time t_2, line *a* has rotated by angle α_a, and line *b* has rotated by angle α_b, and both lines have moved with the flow as sketched (both angle values are given in radians and are shown mathematically positive in the sketch). The average rotation angle is thus $(\alpha_a + \alpha_b)/2$, and the *rate of rotation* or angular velocity in the *xy*-plane is equal to the time derivative of this average rotation angle,

Rate of rotation of fluid element about point P in Fig. 4–36:

$$\omega = \frac{d}{dt}\left(\frac{\alpha_a + \alpha_b}{2}\right) = \frac{1}{2}\left(\frac{\partial v}{\partial x} - \frac{\partial u}{\partial y}\right) \tag{4–20}$$

It is left as an exercise to prove the right side of Eq. 4–20 where we have written ω in terms of velocity components u and v in place of angles α_a and α_b.

In three dimensions, we must define a *vector* for the rate of rotation at a point in the flow since its magnitude may differ in each of the three dimensions. Derivation of the rate of rotation vector in three dimensions can be found in many fluid mechanics books such as Kundu and Cohen (2011) and White (2005). The **rate of rotation vector** is equal to the **angular velocity vector** and is expressed in Cartesian coordinates as

Rate of rotation vector in Cartesian coordinates:

$$\vec{\omega} = \frac{1}{2}\left(\frac{\partial w}{\partial y} - \frac{\partial v}{\partial z}\right)\vec{i} + \frac{1}{2}\left(\frac{\partial u}{\partial z} - \frac{\partial w}{\partial x}\right)\vec{j} + \frac{1}{2}\left(\frac{\partial v}{\partial x} - \frac{\partial u}{\partial y}\right)\vec{k} \tag{4–21}$$

Linear strain rate is defined as *the rate of increase in length per unit length*. Mathematically, the linear strain rate of a fluid element depends on the initial orientation or direction of the line segment upon which we measure the linear strain. Thus, it cannot be expressed as a scalar or vector quantity. Instead, we define linear strain rate in some arbitrary direction, which we denote as the x_α-direction. For example, line segment *PQ* in Fig. 4–37 has an initial length of dx_α, and it grows to line segment *P'Q'* as shown. From the given definition and using the lengths marked in Fig. 4–37, the linear strain rate in the x_α-direction is

$$\varepsilon_{\alpha\alpha} = \frac{1}{dt}\left(\frac{P'Q' - PQ}{PQ}\right)$$

$$\cong \frac{1}{dt}\left(\frac{\overbrace{\left(u_\alpha + \frac{\partial u_\alpha}{\partial x_\alpha}dx_\alpha\right)dt + dx_\alpha - u_\alpha\,dt}^{\text{Length of } P'Q' \text{ in the } x_\alpha\text{-direction}} - \overbrace{dx_\alpha}^{\text{Length of } PQ \text{ in the } x_\alpha\text{-direction}}}{\underbrace{dx_\alpha}_{\text{Length of } PQ \text{ in the } x_\alpha\text{-direction}}}\right) = \frac{\partial u_\alpha}{\partial x_\alpha} \tag{4–22}$$

In Cartesian coordinates, we normally take the x_α-direction as that of each of the three coordinate axes, although we are not restricted to these directions.

Linear strain rate in Cartesian coordinates:

$$\varepsilon_{xx} = \frac{\partial u}{\partial x} \qquad \varepsilon_{yy} = \frac{\partial v}{\partial y} \qquad \varepsilon_{zz} = \frac{\partial w}{\partial z} \tag{4–23}$$

For the more general case, the fluid element moves and deforms as sketched in Fig. 4–36. It is left as an exercise to show that Eq. 4–23 is still valid for the general case.

Solid objects such as wires, rods, and beams stretch when pulled. You should recall from your study of engineering mechanics that when such an object stretches in one direction, it usually shrinks in direction(s) normal to that direction. The same is true of fluid elements. In Fig. 4–35c, the originally square fluid element stretches in the horizontal direction and shrinks in the vertical direction. The linear strain rate is thus positive horizontally and negative vertically.

If the flow is *incompressible*, the net volume of the fluid element must remain constant; thus if the element stretches in one direction, it must shrink by an appropriate amount in other direction(s) to compensate. The volume of a *compressible* fluid element, however, may increase or decrease as its density decreases or increases, respectively. (The mass of a fluid element must remain constant, but since $\rho = m/V$, density and volume are inversely proportional.) Consider for example a parcel of air in a cylinder being compressed by a piston (Fig. 4–38); the volume of the fluid element decreases while its density increases such that the fluid element's mass is conserved. The rate of increase of volume of a fluid element per unit volume is called its **volumetric strain rate** or **bulk strain rate.** This kinematic property is defined as *positive* when the volume *increases*. Another synonym of volumetric strain rate is **rate of volumetric dilatation,** which is easy to remember if you think about how the iris of your eye dilates (enlarges) when exposed to dim light. It turns out that the volumetric strain rate is the sum of the linear strain rates in three mutually orthogonal directions. In Cartesian coordinates (Eq. 4–23), the volumetric strain rate is thus

Volumetric strain rate in Cartesian coordinates:

$$\frac{1}{V}\frac{DV}{Dt} = \frac{1}{V}\frac{dV}{dt} = \varepsilon_{xx} + \varepsilon_{yy} + \varepsilon_{zz} = \frac{\partial u}{\partial x} + \frac{\partial v}{\partial y} + \frac{\partial w}{\partial z} \qquad (4\text{–}24)$$

In Eq. 4–24, the uppercase D notation is used to stress that we are talking about the volume *following a fluid element*, that is to say, the *material volume* of the fluid element, as in Eq. 4–12.

The volumetric strain rate is zero in an incompressible flow.

Shear strain rate is a more difficult deformation rate to describe and understand. **Shear strain rate** at a point is defined as *half of the rate of decrease of the angle between two initially perpendicular lines that intersect at the point.* (The reason for the half will become clear later when we combine shear strain rate and linear strain rate into one tensor.) In Fig. 4–35d, for example, the initially 90° angles at the lower-left corner and upper-right corner of the square fluid element decrease; this is by definition a *positive* shear strain. However, the angles at the upper-left and lower-right corners of the square fluid element increase as the initially square fluid element deforms; this is a *negative* shear strain. Obviously we cannot describe the shear strain rate in terms of only one scalar quantity or even in terms of one *vector* quantity for that matter. Rather, a full mathematical description of shear strain rate requires its specification in any *two mutually perpendicular directions*. In Cartesian coordinates, the axes themselves are the most obvious choice, although we are not restricted to these. Consider a fluid element in two dimensions in the *xy*-plane. The element translates and deforms with time as sketched in Fig. 4–39. Two initially mutually perpendicular lines

FIGURE 4–37

Linear strain rate in some arbitrary direction x_α is defined as the rate of increase in length per unit length in that direction. Linear strain rate would be *negative* if the line segment length were to *decrease*. Here we follow the increase in length of line segment PQ into line segment $P'Q'$, which yields a positive linear strain rate. Velocity components and distances are truncated to first-order since dx_α and dt are infinitesimally small.

FIGURE 4–38

Air being compressed by a piston in a cylinder; the volume of a fluid element in the cylinder decreases, corresponding to a negative rate of volumetric dilatation.

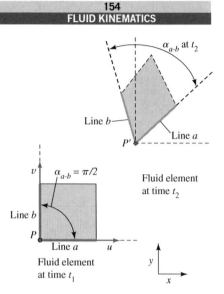

FIGURE 4–39

For a fluid element that translates and deforms as sketched, the *shear strain rate* at point P is defined as half of the rate of decrease of the angle between two initially perpendicular lines (lines a and b).

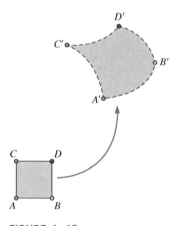

FIGURE 4–40

A fluid element illustrating translation, rotation, linear strain, shear strain, and volumetric strain.

(lines a and b in the x- and y-directions, respectively) are followed. The angle between these two lines decreases from $\pi/2$ (90°) to the angle marked $\alpha_{a\text{-}b}$ at t_2 in the sketch. It is left as an exercise to show that the shear strain rate at point P for initially perpendicular lines in the x- and y-directions is given by

Shear strain rate, initially perpendicular lines in the x- and y-directions:

$$\varepsilon_{xy} = -\frac{1}{2}\frac{d}{dt}\alpha_{a\text{-}b} = \frac{1}{2}\left(\frac{\partial u}{\partial y} + \frac{\partial v}{\partial x}\right) \tag{4–25}$$

Equation 4–25 can be easily extended to three dimensions. The shear strain rate is thus

Shear strain rate in Cartesian coordinates:

$$\varepsilon_{xy} = \frac{1}{2}\left(\frac{\partial u}{\partial y} + \frac{\partial v}{\partial x}\right) \quad \varepsilon_{zx} = \frac{1}{2}\left(\frac{\partial w}{\partial x} + \frac{\partial u}{\partial z}\right) \quad \varepsilon_{yz} = \frac{1}{2}\left(\frac{\partial v}{\partial z} + \frac{\partial w}{\partial y}\right) \tag{4–26}$$

Finally, it turns out that we can mathematically combine linear strain rate and shear strain rate into one symmetric second-order tensor called the **strain rate tensor,** which is a combination of Eqs. 4–23 and 4–26:

Strain rate tensor in Cartesian coordinates:

$$\varepsilon_{ij} = \begin{pmatrix} \varepsilon_{xx} & \varepsilon_{xy} & \varepsilon_{xz} \\ \varepsilon_{yx} & \varepsilon_{yy} & \varepsilon_{yz} \\ \varepsilon_{zx} & \varepsilon_{zy} & \varepsilon_{zz} \end{pmatrix} = \begin{pmatrix} \dfrac{\partial u}{\partial x} & \dfrac{1}{2}\left(\dfrac{\partial u}{\partial y} + \dfrac{\partial v}{\partial x}\right) & \dfrac{1}{2}\left(\dfrac{\partial u}{\partial z} + \dfrac{\partial w}{\partial x}\right) \\ \dfrac{1}{2}\left(\dfrac{\partial v}{\partial x} + \dfrac{\partial u}{\partial y}\right) & \dfrac{\partial v}{\partial y} & \dfrac{1}{2}\left(\dfrac{\partial v}{\partial z} + \dfrac{\partial w}{\partial y}\right) \\ \dfrac{1}{2}\left(\dfrac{\partial w}{\partial x} + \dfrac{\partial u}{\partial z}\right) & \dfrac{1}{2}\left(\dfrac{\partial w}{\partial y} + \dfrac{\partial v}{\partial z}\right) & \dfrac{\partial w}{\partial z} \end{pmatrix} \tag{4–27}$$

The strain rate tensor obeys all the laws of mathematical tensors, such as tensor invariants, transformation laws, and principal axes.

Figure 4–40 shows a general (although two-dimensional) situation in a compressible fluid flow in which all possible motions and deformations are present simultaneously. In particular, there is translation, rotation, linear strain, and shear strain. Because of the compressible nature of the fluid flow, there is also volumetric strain (dilatation). You should now have a better appreciation of the inherent complexity of fluid dynamics, and the mathematical sophistication required to fully describe fluid motion.

EXAMPLE 4–6   **Calculation of Kinematic Properties in a Two-Dimensional Flow**

Consider the steady, two-dimensional velocity field of Example 4–1:

$$\vec{V} = (u, v) = (0.5 + 0.8x)\vec{i} + (1.5 - 0.8y)\vec{j} \tag{1}$$

where lengths are in units of m, time in s, and velocities in m/s. There is a stagnation point at $(-0.625, 1.875)$ as shown in Fig. 4–41. Streamlines of the flow are also plotted in Fig. 4–41. Calculate the various kinematic properties, namely, the rate of translation, rate of rotation, linear strain rate, shear strain rate, and volumetric strain rate. Verify that this flow is incompressible.

SOLUTION We are to calculate several kinematic properties of a given velocity field and verify that the flow is incompressible.

Assumptions **1** The flow is steady. **2** The flow is two-dimensional, implying no z-component of velocity and no variation of u or v with z.

Analysis By Eq. 4–19, the rate of translation is simply the velocity vector itself, given by Eq. 1. Thus,

Rate of translation: $u = 0.5 + 0.8x$ $v = 1.5 - 0.8y$ $w = 0$ **(2)**

The rate of rotation is found from Eq. 4–21. In this case, since $w = 0$ everywhere, and since neither u nor v vary with z, the only nonzero component of rotation rate is in the z-direction. Thus,

Rate of rotation: $\vec{\omega} = \frac{1}{2}\left(\frac{\partial v}{\partial x} - \frac{\partial u}{\partial y}\right)\vec{k} = \frac{1}{2}(0 - 0)\vec{k} = 0$ **(3)**

In this case, we see that there is no net rotation of fluid particles as they move about. (This is a significant piece of information, to be discussed in more detail later in this chapter and also in Chap. 10.)

Linear strain rates can be calculated in any arbitrary direction using Eq. 4–23. In the x-, y-, and z-directions, the linear strain rates are

$$\varepsilon_{xx} = \frac{\partial u}{\partial x} = 0.8\ \text{s}^{-1} \quad \varepsilon_{yy} = \frac{\partial v}{\partial y} = -0.8\ \text{s}^{-1} \quad \varepsilon_{zz} = 0 \quad \textbf{(4)}$$

Thus, we predict that fluid particles *stretch* in the x-direction (positive linear strain rate) and *shrink* in the y-direction (negative linear strain rate). This is illustrated in Fig. 4–42, where we have marked an initially square parcel of fluid centered at (0.25, 4.25). By integrating Eqs. 2 with time, we calculate the location of the four corners of the marked fluid after an elapsed time of 1.5 s. Indeed this fluid parcel has stretched in the x-direction and has shrunk in the y-direction as predicted.

Shear strain rate is determined from Eq. 4–26. Because of the two-dimensionality, nonzero shear strain rates can occur only in the xy-plane. Using lines parallel to the x- and y-axes as our initially perpendicular lines, we calculate ε_{xy},

$$\varepsilon_{xy} = \frac{1}{2}\left(\frac{\partial u}{\partial y} + \frac{\partial v}{\partial x}\right) = \frac{1}{2}(0 + 0) = 0 \quad \textbf{(5)}$$

Thus, there is no shear strain in this flow, as also indicated by Fig. 4–42. Although the sample fluid particle deforms, it remains rectangular; its initially 90° corner angles remain at 90° throughout the time period of the calculation.

Finally, the volumetric strain rate is calculated from Eq. 4–24:

$$\frac{1}{V}\frac{DV}{Dt} = \varepsilon_{xx} + \varepsilon_{yy} + \varepsilon_{zz} = (0.8 - 0.8 + 0)\ \text{s}^{-1} = 0 \quad \textbf{(6)}$$

Since the volumetric strain rate is zero everywhere, we can say definitively that fluid particles are neither dilating (expanding) nor shrinking (compressing) in volume. Thus, **we verify that this flow is indeed incompressible.** In Fig. 4–42, the area of the shaded fluid particle (and thus its volume since it is a 2-D flow) remains constant as it moves and deforms in the flow field.

Discussion In this example it turns out that the linear strain rates (ε_{xx} and ε_{yy}) are nonzero, while the shear strain rates (ε_{xy} and its symmetric partner ε_{yx})

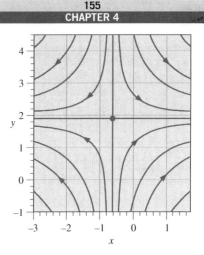

FIGURE 4–41
Streamlines for the velocity field of Example 4–6. The stagnation point is indicated by the red circle at $x = -0.625$ m and $y = 1.875$ m.

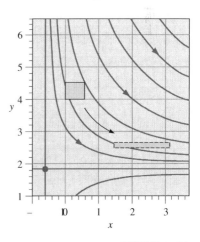

FIGURE 4–42
Deformation of an initially square parcel of marked fluid subjected to the velocity field of Example 4–6 for a time period of 1.5 s. The stagnation point is indicated by the red circle at $x = -0.625$ m and $y = 1.875$ m, and several streamlines are plotted.

are zero. This means that *the x- and y-axes of this flow field are the principal axes*. The (two-dimensional) strain rate tensor in this orientation is thus

$$\varepsilon_{ij} = \begin{pmatrix} \varepsilon_{xx} & \varepsilon_{xy} \\ \varepsilon_{yx} & \varepsilon_{yy} \end{pmatrix} = \begin{pmatrix} 0.8 & 0 \\ 0 & -0.8 \end{pmatrix} \text{s}^{-1} \qquad (7)$$

If we were to rotate the axes by some arbitrary angle, the new axes would *not* be principal axes, and all four elements of the strain rate tensor would be nonzero. You may recall rotating axes in your engineering mechanics classes through use of Mohr's circles to determine principal axes, maximum shear strains, etc. Similar analyses are performed in fluid mechanics.

4–5 · VORTICITY AND ROTATIONALITY

We have already defined the rate of rotation vector of a fluid element (see Eq. 4–21). A closely related kinematic property of great importance to the analysis of fluid flows is the **vorticity vector,** defined mathematically as the curl of the velocity vector $\vec{V}$,

Vorticity vector: $$\qquad \vec{\zeta} = \vec{\nabla} \times \vec{V} = \text{curl}(\vec{V}) \qquad (4\text{–}28)$$

Physically, you can tell the direction of the vorticity vector by using the right-hand rule for cross product (Fig. 4–43). The symbol ζ used for vorticity is the Greek letter *zeta*. You should note that this symbol for vorticity is *not* universal among fluid mechanics textbooks; some authors use the Greek letter *omega* (ω) while still others use uppercase *omega* (Ω). In this book, $\vec{\omega}$ is used to denote the rate of rotation vector (angular velocity vector) of a fluid element. It turns out that the rate of rotation vector is equal to half of the vorticity vector,

Rate of rotation vector: $$\qquad \vec{\omega} = \frac{1}{2} \vec{\nabla} \times \vec{V} = \frac{1}{2}\text{curl}(\vec{V}) = \frac{\vec{\zeta}}{2} \qquad (4\text{–}29)$$

Thus, *vorticity is a measure of rotation of a fluid particle*. Specifically,

Vorticity is equal to twice the angular velocity of a fluid particle (Fig. 4–44).

If the vorticity at a point in a flow field is nonzero, the fluid particle that happens to occupy that point in space is rotating; the flow in that region is called **rotational.** Likewise, if the vorticity in a region of the flow is zero (or negligibly small), fluid particles there are not rotating; the flow in that region is called **irrotational.** Physically, fluid particles in a rotational region of flow rotate end over end as they move along in the flow. For example, fluid particles within the viscous boundary layer near a solid wall are rotational (and thus have nonzero vorticity), while fluid particles outside the boundary layer are irrotational (and their vorticity is zero). Both of these cases are illustrated in Fig. 4–45.

Rotation of fluid elements is associated with wakes, boundary layers, flow through turbomachinery (fans, turbines, compressors, etc.), and flow with heat transfer. The vorticity of a fluid element cannot change except through the action of viscosity, nonuniform heating (temperature gradients), or other nonuniform phenomena. Thus if a flow originates in an irrotational region, it remains irrotational until some nonuniform process alters it. For example,

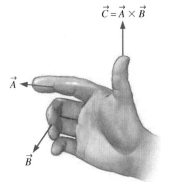

FIGURE 4–43
The direction of a vector cross product is determined by the right-hand rule.

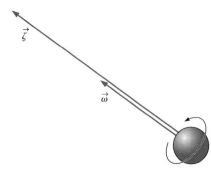

FIGURE 4–44
The *vorticity vector* is equal to twice the angular velocity vector of a rotating fluid particle.

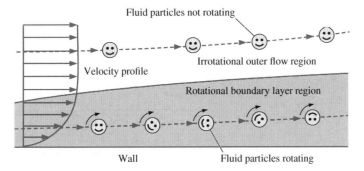

Fluid particles not rotating

Irrotational outer flow region

Velocity profile

Rotational boundary layer region

Wall   Fluid particles rotating

FIGURE 4–45
The difference between rotational and irrotational flow: fluid elements in a rotational region of the flow rotate, but those in an irrotational region of the flow do not.

air entering an inlet from quiescent (still) surroundings is irrotational and remains so unless it encounters an object in its path or is subjected to non-uniform heating. If a region of flow can be approximated as irrotational, the equations of motion are greatly simplified, as you will see in Chap. 10.

In Cartesian coordinates, $(\vec{i}, \vec{j}, \vec{k})$, (x, y, z), and (u, v, w), Eq. 4–28 is expanded as follows:

Vorticity vector in Cartesian coordinates:

$$\vec{\zeta} = \left(\frac{\partial w}{\partial y} - \frac{\partial v}{\partial z}\right)\vec{i} + \left(\frac{\partial u}{\partial z} - \frac{\partial w}{\partial x}\right)\vec{j} + \left(\frac{\partial v}{\partial x} - \frac{\partial u}{\partial y}\right)\vec{k} \tag{4–30}$$

If the flow is two-dimensional in the xy-plane, the z-component of velocity (w) is zero and neither u nor v varies with z. Thus the first two components of Eq. 4–30 are identically zero and the vorticity reduces to

Two-dimensional flow in Cartesian coordinates:

$$\vec{\zeta} = \left(\frac{\partial v}{\partial x} - \frac{\partial u}{\partial y}\right)\vec{k} \tag{4–31}$$

Note that if a flow is two-dimensional in the xy-plane, the vorticity vector must point in either the z- or $-z$-direction (Fig. 4–46).

FIGURE 4–46
For two-dimensional flow in the xy-plane, the vorticity vector always points in the z- or $-z$-direction. In this illustration, the flag-shaped fluid particle rotates in the counterclockwise direction as it moves in the xy-plane; its vorticity points in the positive z-direction as shown.

■ *EXAMPLE 4–7*   **Vorticity Contours in a Two-Dimensional Flow**

Consider the CFD calculation of two-dimensional free-stream flow impinging on a block of rectangular cross section, as shown in Figs. 4–33 and 4–34. Plot vorticity contours and discuss.

SOLUTION We are to calculate the vorticity field for a given velocity field produced by CFD and then generate a contour plot of vorticity.
Analysis Since the flow is two-dimensional, the only nonzero component of vorticity is in the z-direction, normal to the page in Figs. 4–33 and 4–34. A contour plot of the z-component of vorticity for this flow field is shown in Fig. 4–47. The blue region near the upper-left corner of the block indicates large negative values of vorticity, implying *clockwise* rotation of fluid particles in that region. This is due to the large velocity gradients encountered in this portion of the flow field; the boundary layer separates off the wall at the corner

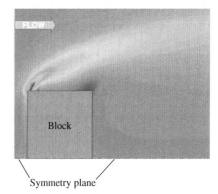

FLOW

Block

Symmetry plane

FIGURE 4–47
Contour plot of the vorticity field ζ_z due to flow impinging on a block, as produced by CFD calculations; only the upper half is shown due to symmetry. Blue regions represent large negative vorticity, and red regions represent large positive vorticity.

of the body and forms a thin **shear layer** across which the velocity changes rapidly. The concentration of vorticity in the shear layer diminishes as vorticity diffuses downstream. The small red region near the top right corner of the block represents a region of *positive* vorticity (counterclockwise rotation)—a secondary flow pattern caused by the flow separation.

Discussion We expect the magnitude of vorticity to be highest in regions where spatial derivatives of velocity are high (see Eq. 4–30). Close examination reveals that the blue region in Fig. 4–47 does indeed correspond to large velocity gradients in Fig. 4–33. Keep in mind that the vorticity field of Fig. 4–47 is time-averaged. The instantaneous flow field is in reality turbulent and unsteady, and vortices are shed from the bluff body.

EXAMPLE 4–8   **Determination of Rotationality in a Two-Dimensional Flow**

Consider the following steady, incompressible, two-dimensional velocity field:

$$\vec{V} = (u, v) = x^2 \vec{i} + (-2xy - 1)\vec{j} \tag{1}$$

Is this flow rotational or irrotational? Sketch some streamlines in the first quadrant and discuss.

SOLUTION We are to determine whether a flow with a given velocity field is rotational or irrotational, and we are to draw some streamlines in the first quadrant.

Analysis Since the flow is two-dimensional, Eq. 4–31 is applicable. Thus,

Vorticity: $$\vec{\zeta} = \left(\frac{\partial v}{\partial x} - \frac{\partial u}{\partial y}\right)\vec{k} = (-2y - 0)\vec{k} = -2y\vec{k} \tag{2}$$

Since the vorticity is nonzero, this flow is **rotational**. In Fig. 4–48 we plot several streamlines of the flow in the first quadrant; we see that fluid moves downward and to the right. The translation and deformation of a fluid parcel is also shown: at $\Delta t = 0$, the fluid parcel is square, at $\Delta t = 0.25$ s, it has moved and deformed, and at $\Delta t = 0.50$ s, the parcel has moved farther and is further deformed. In particular, the right-most portion of the fluid parcel moves faster to the right and faster downward compared to the left-most portion, stretching the parcel in the x-direction and squashing it in the vertical direction. It is clear that there is also a net *clockwise* rotation of the fluid parcel, which agrees with the result of Eq. 2.

Discussion From Eq. 4–29, individual fluid particles rotate at an angular velocity equal to $\vec{\omega} = -y\vec{k}$, half of the vorticity vector. Since $\vec{\omega}$ is not constant, this flow is *not* solid-body rotation. Rather, $\vec{\omega}$ is a linear function of y. Further analysis reveals that this flow field is incompressible; the area (and volume) of the shaded regions representing the fluid parcel in Fig. 4–48 remains constant at all three instants in time.

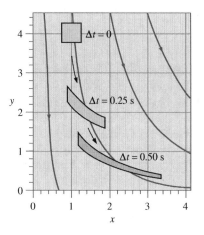

FIGURE 4–48

Deformation of an initially square fluid parcel subjected to the velocity field of Example 4–8 for a time period of 0.25 s and 0.50 s. Several streamlines are also plotted in the first quadrant. It is clear that this flow is *rotational*.

In cylindrical coordinates, $(\vec{e}_r, \vec{e}_\theta, \vec{e}_z)$, (r, θ, z), and (u_r, u_θ, u_z), Eq. 4–28 is expanded as

Vorticity vector in cylindrical coordinates:

$$\vec{\zeta} = \left(\frac{1}{r}\frac{\partial u_z}{\partial \theta} - \frac{\partial u_\theta}{\partial z}\right)\vec{e}_r + \left(\frac{\partial u_r}{\partial z} - \frac{\partial u_z}{\partial r}\right)\vec{e}_\theta + \frac{1}{r}\left(\frac{\partial(r u_\theta)}{\partial r} - \frac{\partial u_r}{\partial \theta}\right)\vec{e}_z \tag{4–32}$$

For two-dimensional flow in the $r\theta$-plane, Eq. 4–32 reduces to

Two-dimensional flow in cylindrical coordinates:

$$\vec{\zeta} = \frac{1}{r}\left(\frac{\partial(ru_\theta)}{\partial r} - \frac{\partial u_r}{\partial \theta}\right)\vec{k} \tag{4–33}$$

where $\vec{k}$ is used as the unit vector in the z-direction in place of $\vec{e}_z$. Note that if a flow is two-dimensional in the $r\theta$-plane, the vorticity vector must point in either the z- or $-z$-direction (Fig. 4–49).

Comparison of Two Circular Flows

Not all flows with circular streamlines are rotational. To illustrate this point, we consider two incompressible, steady, two-dimensional flows, both of which have circular streamlines in the $r\theta$-plane:

Flow A—solid-body rotation: $\qquad u_r = 0 \quad$ and $\quad u_\theta = \omega r \tag{4–34}$

Flow B—line vortex: $\qquad u_r = 0 \quad$ and $\quad u_\theta = \dfrac{K}{r} \tag{4–35}$

where ω and K are constants. (Alert readers will note that u_θ in Eq. 4–35 is infinite at $r = 0$, which is of course physically impossible; we ignore the region close to the origin to avoid this problem.) Since the radial component of velocity is zero in both cases, the streamlines are circles about the origin. The velocity profiles for the two flows, along with their streamlines, are sketched in Fig. 4–50. We now calculate and compare the vorticity field for each of these flows, using Eq. 4–33.

Flow A—solid-body rotation: $\qquad \vec{\zeta} = \dfrac{1}{r}\left(\dfrac{\partial(\omega r^2)}{\partial r} - 0\right)\vec{k} = 2\omega\vec{k} \tag{4–36}$

Flow B—line vortex: $\qquad \vec{\zeta} = \dfrac{1}{r}\left(\dfrac{\partial(K)}{\partial r} - 0\right)\vec{k} = 0 \tag{4–37}$

Not surprisingly, the vorticity for solid-body rotation is nonzero. In fact, it is a constant of magnitude twice the angular velocity and pointing in the same direction. (This agrees with Eq. 4–29.) *Flow A is rotational*. Physically, this means that individual fluid particles rotate as they revolve around the origin (Fig. 4–50*a*). By contrast, the vorticity of the line vortex is zero everywhere (except right at the origin, which is a mathematical singularity). *Flow B is irrotational*. Physically, fluid particles do *not* rotate as they revolve in circles about the origin (Fig. 4–50*b*).

A simple analogy can be made between flow A and a merry-go-round or roundabout, and flow B and a Ferris wheel (Fig. 4–51). As children revolve around a roundabout, they also rotate at the same angular velocity as that of the ride itself. This is analogous to a rotational flow. In contrast, children on a Ferris wheel always remain oriented in an upright position as they trace out their circular path. This is analogous to an irrotational flow.

FIGURE 4–49
For a two-dimensional flow in the $r\theta$-plane, the vorticity vector always points in the z (or $-z$) direction. In this illustration, the flag-shaped fluid particle rotates in the clockwise direction as it moves in the $r\theta$-plane; its vorticity points in the $-z$-direction as shown.

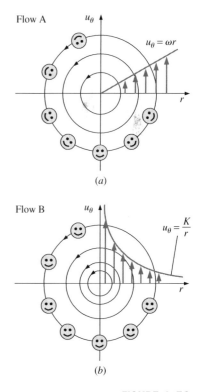

FIGURE 4–50
Streamlines and velocity profiles for (*a*) flow A, solid-body rotation and (*b*) flow B, a line vortex. Flow A is rotational, but flow B is irrotational everywhere except at the origin.

(a)   (b)

FIGURE 4–51

A simple analogy: (a) *rotational* circular flow is analogous to a roundabout, while (b) *irrotational* circular flow is analogous to a Ferris wheel.

(a) Mc Graw-Hill Companies, Inc. Mark Dierker, photographer (b) © DAJ/Getty RF

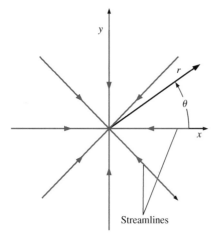

FIGURE 4–52

Streamlines in the $r\theta$-plane for the case of a line sink.

> ### EXAMPLE 4–9   Determination of Rotationality of a Line Sink
>
> A simple two-dimensional velocity field called a **line sink** is often used to simulate fluid being sucked into a line along the z-axis. Suppose the volume flow rate per unit length along the z-axis, $\dot{V}/L$, is known, where $\dot{V}$ is a negative quantity. In two dimensions in the $r\theta$-plane,
>
> *Line sink:*
> $$u_r = \frac{\dot{V}}{2\pi L}\frac{1}{r} \quad \text{and} \quad u_\theta = 0 \tag{1}$$
>
> Draw several streamlines of the flow and calculate the vorticity. Is this flow rotational or irrotational?
>
> **SOLUTION** Streamlines of the given flow field are to be sketched and the rotationality of the flow is to be determined.
>
> *Analysis* Since there is only radial flow and no tangential flow, we know immediately that all streamlines must be rays into the origin. Several streamlines are sketched in Fig. 4–52. The vorticity is calculated from Eq. 4–33:
>
> $$\vec{\zeta} = \frac{1}{r}\left(\frac{\partial(ru_\theta)}{\partial r} - \frac{\partial}{\partial\theta}u_r\right)\vec{k} = \frac{1}{r}\left(0 - \frac{\partial}{\partial\theta}\left(\frac{\dot{V}}{2\pi L}\frac{1}{r}\right)\right)\vec{k} = 0 \tag{2}$$
>
> Since the vorticity vector is everywhere zero, this flow field is irrotational.
>
> *Discussion* Many practical flow fields involving suction, such as flow into inlets and hoods, can be approximated quite accurately by assuming irrotational flow (Heinsohn and Cimbala, 2003).

4–6 · THE REYNOLDS TRANSPORT THEOREM

In thermodynamics and solid mechanics we often work with a *system* (also called a *closed system*), defined as a *quantity of matter of fixed identity*. In fluid dynamics, it is more common to work with a *control volume* (also called an *open system*), defined as a *region in space chosen for study*. The size and shape of a system may change during a process, but no mass crosses its boundaries. A control volume, on the other hand, allows mass to flow in or out across its boundaries, which are called the **control surface.** A control volume may also move and deform during a process, but many real-world applications involve fixed, nondeformable control volumes.

Figure 4–53 illustrates both a system and a control volume for the case of deodorant being sprayed from a spray can. When analyzing the spraying process, a natural choice for our analysis is either the moving, deforming fluid (a system) or the volume bounded by the inner surfaces of the can (a control volume). These two choices are identical before the deodorant is sprayed. When some contents of the can are discharged, the system approach considers the discharged mass as part of the system and tracks it (a difficult job indeed); thus the mass of the system remains constant. Conceptually, this is equivalent to attaching a flat balloon to the nozzle of the can and letting the spray inflate the balloon. The inner surface of the balloon now becomes part of the boundary of the system. The control volume approach, however, is not concerned at all with the deodorant that has escaped the can (other than its properties at the exit), and thus the mass of the control volume decreases during this process while its volume remains constant. Therefore, the system approach treats the spraying process as an expansion of the system's volume, whereas the control volume approach considers it as a fluid discharge through the control surface of the fixed control volume.

Most principles of fluid mechanics are adopted from solid mechanics, where the physical laws dealing with the time rates of change of extensive properties are expressed for systems. In fluid mechanics, it is usually more convenient to work with control volumes, and thus there is a need to relate the changes in a control volume to the changes in a system. The relationship between the time rates of change of an extensive property for a system and for a control volume is expressed by the **Reynolds transport theorem (RTT),** which provides the link between the system and control volume approaches (Fig. 4–54). RTT is named after the English engineer, Osborne Reynolds (1842–1912), who did much to advance its application in fluid mechanics.

The general form of the Reynolds transport theorem can be derived by considering a system with an arbitrary shape and arbitrary interactions, but the derivation is rather involved. To help you grasp the fundamental meaning of the theorem, we derive it first in a straightforward manner using a simple geometry and then generalize the results.

Consider flow from left to right through a diverging (expanding) portion of a flow field as sketched in Fig. 4–55. The upper and lower bounds of the fluid under consideration are *streamlines* of the flow, and we assume uniform flow through any cross section between these two streamlines. We choose the control volume to be fixed between sections (1) and (2) of the flow field. Both (1) and (2) are normal to the direction of flow. At some initial time t, the system coincides with the control volume, and thus the system and control volume are identical (the greenish-shaded region in Fig. 4–55). During time interval Δt, the system moves in the flow direction at uniform speeds V_1 at section (1) and V_2 at section (2). The system at this later time is indicated by the hatched region. The region uncovered by the system during this motion is designated as section I (part of the CV), and the new region

FIGURE 4–53
Two methods of analyzing the spraying of deodorant from a spray can: (*a*) We follow the fluid as it moves and deforms. This is the *system approach*—no mass crosses the boundary, and the total mass of the system remains fixed. (*b*) We consider a fixed interior volume of the can. This is the *control volume approach*—mass crosses the boundary.

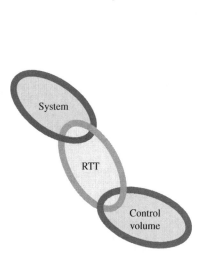

FIGURE 4–54
The *Reynolds transport theorem* (RTT) provides a link between the system approach and the control volume approach.

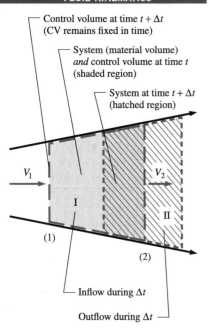

Control volume at time $t + \Delta t$
(CV remains fixed in time)

System (material volume) *and* control volume at time t (shaded region)

System at time $t + \Delta t$ (hatched region)

V_1

V_2

I

II

(1)

(2)

Inflow during Δt

Outflow during Δt

At time t: Sys = CV
At time $t + \Delta t$: Sys = CV − I + II

FIGURE 4–55

A moving *system* (hatched region) and a fixed *control volume* (shaded region) in a diverging portion of a flow field at times t and $t + \Delta t$. The upper and lower bounds are streamlines of the flow.

covered by the system is designated as section II (not part of the CV). Therefore, at time $t + \Delta t$, the system consists of the same fluid, but it occupies the region CV − I + II. The control volume is fixed in space, and thus it remains as the shaded region marked CV at all times.

Let B represent any **extensive property** (such as mass, energy, or momentum), and let $b = B/m$ represent the corresponding **intensive property.** Noting that extensive properties are additive, the extensive property B of the system at times t and $t + \Delta t$ is expressed as

$$B_{\text{sys}, t} = B_{\text{CV}, t} \quad \text{(the system and CV coincide at time } t\text{)}$$

$$B_{\text{sys}, t+\Delta t} = B_{\text{CV}, t+\Delta t} - B_{\text{I}, t+\Delta t} + B_{\text{II}, t+\Delta t}$$

Subtracting the first equation from the second one and dividing by Δt gives

$$\frac{B_{\text{sys}, t+\Delta t} - B_{\text{sys}, t}}{\Delta t} = \frac{B_{\text{CV}, t+\Delta t} - B_{\text{CV}, t}}{\Delta t} - \frac{B_{\text{I}, t+\Delta t}}{\Delta t} + \frac{B_{\text{II}, t+\Delta t}}{\Delta t}$$

Taking the limit as $\Delta t \to 0$, and using the definition of derivative, we get

$$\frac{dB_{\text{sys}}}{dt} = \frac{dB_{\text{CV}}}{dt} - \dot{B}_{\text{in}} + \dot{B}_{\text{out}} \tag{4–38}$$

or

$$\frac{dB_{\text{sys}}}{dt} = \frac{dB_{\text{CV}}}{dt} - b_1\rho_1 V_1 A_1 + b_2\rho_2 V_2 A_2$$

since

$$B_{\text{I}, t+\Delta t} = b_1 m_{\text{I}, t+\Delta t} = b_1 \rho_1 V_{\text{I}, t+\Delta t} = b_1 \rho_1 V_1 \Delta t\, A_1$$

$$B_{\text{II}, t+\Delta t} = b_2 m_{\text{II}, t+\Delta t} = b_2 \rho_2 V_{\text{II}, t+\Delta t} = b_2 \rho_2 V_2 \Delta t\, A_2$$

and

$$\dot{B}_{\text{in}} = \dot{B}_{\text{I}} = \lim_{\Delta t \to 0} \frac{B_{\text{I}, t+\Delta t}}{\Delta t} = \lim_{\Delta t \to 0} \frac{b_1 \rho_1 V_1 \Delta t\, A_1}{\Delta t} = b_1 \rho_1 V_1 A_1$$

$$\dot{B}_{\text{out}} = \dot{B}_{\text{II}} = \lim_{\Delta t \to 0} \frac{B_{\text{II}, t+\Delta t}}{\Delta t} = \lim_{\Delta t \to 0} \frac{b_2 \rho_2 V_2 \Delta t\, A_2}{\Delta t} = b_2 \rho_2 V_2 A_2$$

where A_1 and A_2 are the cross-sectional areas at locations 1 and 2. Equation 4–38 states that *the time rate of change of the property B of the system is equal to the time rate of change of B of the control volume plus the net flux of B out of the control volume by mass crossing the control surface.* This is the desired relation since it relates the change of a property of a system to the change of that property for a control volume. Note that Eq. 4–38 applies at any instant in time, where it is assumed that the system and the control volume occupy the same space at that particular instant in time.

The influx $\dot{B}_{\text{in}}$ and outflux $\dot{B}_{\text{out}}$ of the property B in this case are easy to determine since there is only one inlet and one outlet, and the velocities are approximately normal to the surfaces at sections (1) and (2). In general, however, we may have several inlet and outlet ports, and the velocity may not be normal to the control surface at the point of entry. Also, the velocity may not be uniform. To generalize the process, we consider a differential surface area dA on the control surface and denote its **unit outer normal** by $\vec{n}$. The flow rate of property b through dA is $\rho b \vec{V} \cdot \vec{n}\, dA$ since the dot product $\vec{V} \cdot \vec{n}$ gives the normal component of the velocity. Then the net rate of outflow through the entire control surface is determined by integration to be (Fig. 4–56)

$$\dot{B}_{net} = \dot{B}_{out} - \dot{B}_{in} = \int_{CS} \rho b \vec{V} \cdot \vec{n} \, dA \qquad \text{(inflow if negative)} \qquad \textbf{(4-39)}$$

An important aspect of this relation is that it automatically subtracts the inflow from the outflow, as explained next. The dot product of the velocity vector at a point on the control surface and the outer normal at that point is $\vec{V} \cdot \vec{n} = |\vec{V}||\vec{n}| \cos\theta = |\vec{V}| \cos\theta$, where θ is the angle between the velocity vector and the outer normal, as shown in Fig. 4–57. For $\theta < 90°$, $\cos\theta > 0$ and thus $\vec{V} \cdot \vec{n} > 0$ for outflow of mass from the control volume, and for $\theta > 90°$, $\cos\theta < 0$ and thus $\vec{V} \cdot \vec{n} < 0$ for inflow of mass into the control volume. Therefore, the differential quantity $\rho b \vec{V} \cdot \vec{n} \, dA$ is positive for mass flowing out of the control volume, and negative for mass flowing into the control volume, and its integral over the entire control surface gives the rate of net outflow of the property B by mass.

The properties within the control volume may vary with position, in general. In such a case, the total amount of property B within the control volume must be determined by integration:

$$B_{CV} = \int_{CV} \rho b \, d\mathcal{V} \qquad \textbf{(4-40)}$$

The term dB_{CV}/dt in Eq. 4–38 is thus equal to $\dfrac{d}{dt}\displaystyle\int_{CV} \rho b \, d\mathcal{V}$, and represents the time rate of change of the property B content of the control volume. A positive value for dB_{CV}/dt indicates an increase in the B content, and a negative value indicates a decrease. Substituting Eqs. 4–39 and 4–40 into Eq. 4–38 yields the Reynolds transport theorem, also known as the *system-to-control-volume transformation* for a fixed control volume:

RTT, fixed CV:
$$\frac{dB_{sys}}{dt} = \frac{d}{dt}\int_{CV} \rho b \, d\mathcal{V} + \int_{CS} \rho b \vec{V} \cdot \vec{n} \, dA \qquad \textbf{(4-41)}$$

Since the control volume is not moving or deforming with time, the time derivative on the right-hand side can be moved inside the integral, since the domain of integration does not change with time. (In other words, it is irrelevant whether we differentiate or integrate first.) But the time derivative in that case must be expressed as a *partial* derivative ($\partial/\partial t$) since density and the quantity b may depend not only on time, but also on the position within the control volume. Thus, an alternate form of the Reynolds transport theorem for a fixed control volume is

Alternate RTT, fixed CV:
$$\frac{dB_{sys}}{dt} = \int_{CV} \frac{\partial}{\partial t}(\rho b) \, d\mathcal{V} + \int_{CS} \rho b \vec{V} \cdot \vec{n} \, dA \qquad \textbf{(4-42)}$$

It turns out that Eq. 4–42 is also valid for the most general case of a moving and/or deforming control volume, provided that velocity vector $\vec{V}$ is an *absolute* velocity (as viewed from a fixed reference frame).

Next we consider yet *another* alternative form of the RTT. Equation 4–41 was derived for a *fixed* control volume. However, many practical systems such as turbine and propeller blades involve nonfixed control volumes. Fortunately, Eq. 4–41 is also valid for *moving* and/or *deforming* control volumes provided that the absolute fluid velocity $\vec{V}$ in the last term is replaced by the **relative velocity** $\vec{V}_r$,

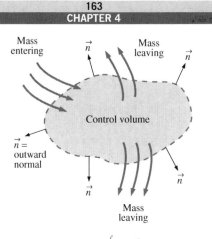

$$\dot{B}_{net} = \dot{B}_{out} - \dot{B}_{in} = \int_{CS} \rho b \vec{V} \cdot \vec{n} \, dA$$

FIGURE 4–56
The integral of $b\rho\vec{V}\cdot\vec{n}\,dA$ over the control surface gives the net amount of the property B flowing out of the control volume (into the control volume if it is negative) per unit time.

$\vec{V} \cdot \vec{n} = |\vec{V}||\vec{n}| \cos\theta = V \cos\theta$
If $\theta < 90°$, then $\cos\theta > 0$ (outflow).
If $\theta > 90°$, then $\cos\theta < 0$ (inflow).
If $\theta = 90°$, then $\cos\theta = 0$ (no flow).

FIGURE 4–57
Outflow and inflow of mass across the differential area of a control surface.

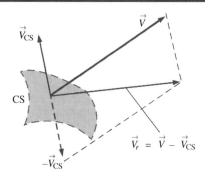

FIGURE 4–58

Relative velocity crossing a control surface is found by vector addition of the absolute velocity of the fluid and the negative of the local velocity of the control surface.

Relative velocity:
$$\vec{V}_r = \vec{V} - \vec{V}_{CS} \tag{4–43}$$

where $\vec{V}_{CS}$ is the local velocity of the control surface (Fig. 4–58). The most general form of the Reynolds transport theorem is thus

RTT, nonfixed CV:
$$\frac{dB_{sys}}{dt} = \frac{d}{dt}\int_{CV} \rho b \, dV + \int_{CS} \rho b \vec{V}_r \cdot \vec{n} \, dA \tag{4–44}$$

Note that for a control volume that moves and/or deforms with time, the time derivative is applied *after* integration in Eq. 4–44. As a simple example of a moving control volume, consider a toy car moving at a constant absolute velocity $\vec{V}_{car} = 10$ km/h to the right. A high-speed jet of water (absolute velocity $= \vec{V}_{jet} = 25$ km/h to the right) strikes the back of the car and propels it (Fig. 4–59). If we draw a control volume around the car, the relative velocity is $\vec{V}_r = 25 - 10 = 15$ km/h to the right. This represents the velocity at which an observer moving with the control volume (moving with the car) would observe the fluid crossing the control surface. In other words, $\vec{V}_r$ is the fluid velocity expressed relative to a coordinate system moving *with* the control volume.

Finally, by application of the Leibniz theorem, it can be shown that the Reynolds transport theorem for a general moving and/or deforming control volume (Eq. 4–44) is equivalent to the form given by Eq. 4–42, which is repeated here:

Alternate RTT, nonfixed CV:
$$\frac{dB_{sys}}{dt} = \int_{CV} \frac{\partial}{\partial t}(\rho b) \, dV + \int_{CS} \rho b \vec{V} \cdot \vec{n} \, dA \tag{4–45}$$

In contrast to Eq. 4–44, the velocity vector $\vec{V}$ in Eq. 4–45 must be taken as the *absolute* velocity (as viewed from a fixed reference frame) in order to apply to a nonfixed control volume.

During steady flow, the amount of the property B within the control volume remains constant in time, and thus the time derivative in Eq. 4–44 becomes zero. Then the Reynolds transport theorem reduces to

Absolute reference frame:

Control volume

RTT, steady flow:
$$\frac{dB_{sys}}{dt} = \int_{CS} \rho b \vec{V}_r \cdot \vec{n} \, dA \tag{4–46}$$

Note that unlike the control volume, the property B content of the system may still change with time during a steady process. But in this case the change must be equal to the net property transported by mass across the control surface (an advective rather than an unsteady effect).

Relative reference frame:

Control volume

In most practical engineering applications of the RTT, fluid crosses the boundary of the control volume at a finite number of well-defined inlets and outlets (Fig. 4–60). In such cases, it is convenient to cut the control surface directly across each inlet and outlet and replace the surface integral in Eq. 4–44 with approximate algebraic expressions at each inlet and outlet based on the *average* values of fluid properties crossing the boundary. We define ρ_{avg}, b_{avg}, and $V_{r,\,avg}$ as the average values of ρ, b, and V_r, respectively,

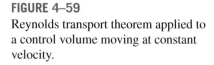

across an inlet or outlet of cross-sectional area A [e.g., $b_{avg} = \dfrac{1}{A}\displaystyle\int_A b\, dA$]. The surface integrals in the RTT (Eq. 4–44), when applied over an inlet or outlet

FIGURE 4–59

Reynolds transport theorem applied to a control volume moving at constant velocity.

of cross-sectional area A, are then *approximated* by pulling property b out of the surface integral and replacing it with its average. This yields

$$\int_A \rho b \vec{V}_r \cdot \vec{n} \, dA \cong b_{\text{avg}} \int_A \rho \vec{V}_r \cdot \vec{n} \, dA = b_{\text{avg}} \dot{m}_r$$

where $\dot{m}_r$ is the mass flow rate through the inlet or outlet relative to the (moving) control surface. The approximation in this equation is exact when property b is uniform over cross-sectional area A. Equation 4–44 thus becomes

$$\frac{dB_{\text{sys}}}{dt} = \frac{d}{dt} \int_{\text{CV}} \rho b \, dV + \sum_{\substack{\text{out} \\ \text{for each outlet}}} \dot{m}_r b_{\text{avg}} - \sum_{\substack{\text{in} \\ \text{for each inlet}}} \dot{m}_r b_{\text{avg}} \tag{4–47}$$

In some applications, we may wish to rewrite Eq. 4–47 in terms of volume (rather than mass) flow rate. In such cases, we make a further approximation that $\dot{m}_r \approx \rho_{\text{avg}} \dot{V}_r = \rho_{\text{avg}} V_{r,\text{avg}} A$. This approximation is exact when fluid density ρ is uniform over A. Equation 4–47 then reduces to

Approximate RTT for well-defined inlets and outlets:

$$\frac{dB_{\text{sys}}}{dt} = \frac{d}{dt} \int_{\text{CV}} \rho b \, dV + \sum_{\substack{\text{out} \\ \text{for each outlet}}} \rho_{\text{avg}} b_{\text{avg}} V_{r,\text{avg}} A - \sum_{\substack{\text{in} \\ \text{for each inlet}}} \rho_{\text{avg}} b_{\text{avg}} V_{r,\text{avg}} A \tag{4–48}$$

Note that these approximations simplify the analysis greatly but may not always be accurate, especially in cases where the velocity distribution across the inlet or outlet is not very uniform (e.g., pipe flows; Fig. 4–60). In particular, the control surface integral of Eq. 4–45 becomes *nonlinear* when property b contains a velocity term (e.g., when applying RTT to the linear momentum equation, $b = \vec{V}$), and the approximation of Eq. 4–48 leads to errors. Fortunately we can eliminate the errors by including *correction factors* in Eq. 4–48, as discussed in Chaps. 5 and 6.

Equations 4–47 and 4–48 apply to fixed *or* moving control volumes, but as discussed previously, the *relative velocity* must be used for the case of a nonfixed control volume. In Eq. 4–47 for example, the mass flow rate $\dot{m}_r$ is relative to the (moving) control surface, hence the r subscript.

*Alternate Derivation of the Reynolds Transport Theorem

A more elegant mathematical derivation of the Reynolds transport theorem is possible through use of the **Leibniz theorem** (see Kundu and Cohen, 2011). You may be familiar with the one-dimensional version of this theorem, which allows you to differentiate an integral whose limits of integration are functions of the variable with which you need to differentiate (Fig. 4–61):

One-dimensional Leibniz theorem:

$$\frac{d}{dt} \int_{x=a(t)}^{x=b(t)} G(x,t) \, dx = \int_a^b \frac{\partial G}{\partial t} \, dx + \frac{db}{dt} G(b,t) - \frac{da}{dt} G(a,t) \tag{4–49}$$

---

* This section may be omitted without loss of continuity.

FIGURE 4–60
An example control volume in which there is one well-defined inlet (1) and two well-defined outlets (2 and 3). In such cases, the control surface integral in the RTT can be more conveniently written in terms of the average values of fluid properties crossing each inlet and outlet.

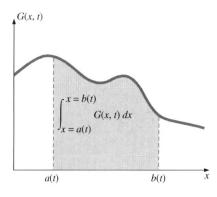

FIGURE 4–61
The *one-dimensional Leibniz theorem* is required when calculating the time derivative of an integral (with respect to x) for which the limits of the integral are functions of time.

The Leibniz theorem takes into account the change of limits $a(t)$ and $b(t)$ with respect to time, as well as the unsteady changes of integrand $G(x, t)$ with time.

EXAMPLE 4–10   One-Dimensional Leibniz Integration

Reduce the following expression as far as possible:

$$F(t) = \frac{d}{dt} \int_{x=0}^{x=Ct} e^{-x^2} \, dx \qquad (1)$$

SOLUTION $F(t)$ is to be evaluated from the given expression.

Analysis   We could try integrating first and then differentiating, but since Eq. 1 is of the form of Eq. 4–49, we use the one-dimensional Leibniz theorem. Here, $G(x, t) = e^{-x^2}$ (G is not a function of time in this simple example). The limits of integration are $a(t) = 0$ and $b(t) = Ct$. Thus,

$$F(t) = \int_a^b \underbrace{\frac{\partial G}{\partial t}}_{0} dx + \underbrace{\frac{db}{dt}}_{C} \underbrace{G(b, t)}_{e^{-b^2}} - \underbrace{\frac{da}{dt}}_{0} G(a, t) \quad \rightarrow \quad F(t) = Ce^{-C^2 t^2} \qquad (2)$$

Discussion   You are welcome to try to obtain the same solution without using the Leibniz theorem.

In three dimensions, the Leibniz theorem for a *volume* integral is

Three-dimensional Leibniz theorem:

$$\frac{d}{dt} \int_{V(t)} G(x, y, z, t) \, dV = \int_{V(t)} \frac{\partial G}{\partial t} \, dV + \int_{A(t)} G\vec{V}_A \cdot \vec{n} \, dA \qquad (4\text{–}50)$$

where $V(t)$ is a moving and/or deforming volume (a function of time), $A(t)$ is its surface (boundary), and $\vec{V}_A$ is the absolute velocity of this (moving) surface (Fig. 4–62). Equation 4–50 is valid for *any* volume, moving and/or deforming arbitrarily in space and time. For consistency with the previous analyses, we set integrand G to ρb for application to fluid flow,

Three-dimensional Leibniz theorem applied to fluid flow:

$$\frac{d}{dt} \int_{V(t)} \rho b \, dV = \int_{V(t)} \frac{\partial}{\partial t} (\rho b) \, dV + \int_{A(t)} \rho b \vec{V}_A \cdot \vec{n} \, dA \qquad (4\text{–}51)$$

If we apply the Leibniz theorem to the special case of a **material volume** (a system of fixed identity moving with the fluid flow), then $\vec{V}_A = \vec{V}$ everywhere on the material surface since it moves *with* the fluid. Here $\vec{V}$ is the local fluid velocity, and Eq. 4–51 becomes

Leibniz theorem applied to a material volume:

$$\frac{d}{dt} \int_{V(t)} \rho b \, dV = \frac{dB_{sys}}{dt} = \int_{V(t)} \frac{\partial}{\partial t} (\rho b) \, dV + \int_{A(t)} \rho b \vec{V} \cdot \vec{n} \, dA \qquad (4\text{–}52)$$

Equation 4–52 is valid at any instant in time t. We define our control volume such that at this time t, the control volume and the system occupy the same space; in other words, they are *coincident*. At some later time $t + \Delta t$, the system has moved and deformed with the flow, but the control volume

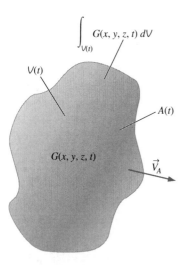

FIGURE 4–62

The *three-dimensional Leibniz theorem* is required when calculating the time derivative of a volume integral for which the volume itself moves and/or deforms with time. It turns out that the three-dimensional form of the Leibniz theorem can be used in an alternative derivation of the Reynolds transport theorem.

may have moved and deformed differently (Fig. 4–63). The key, however, is that *at time t, the system (material volume) and control volume are one and the same.* Thus, the volume integral on the right-hand side of Eq. 4–52 can be evaluated over the *control volume* at time *t*, and the surface integral can be evaluated over the *control surface* at time *t*. Hence,

General RTT, nonfixed CV:
$$\frac{dB_{\text{sys}}}{dt} = \int_{\text{CV}} \frac{\partial}{\partial t}(\rho b)\, dV + \int_{\text{CS}} \rho b \vec{V} \cdot \vec{n}\, dA \qquad (4\text{–}53)$$

This expression is identical to that of Eq. 4–42 and is valid for an arbitrarily shaped, moving, and/or deforming control volume at time *t*. Keep in mind that $\vec{V}$ in Eq. 4–53 is the *absolute* fluid velocity.

■
■ **EXAMPLE 4–11**   **Reynolds Transport Theorem**
■   **in Terms of Relative Velocity**
■
■ Beginning with the Leibniz theorem and the general Reynolds transport theorem for an arbitrarily moving and deforming control volume, Eq. 4–53, prove that Eq. 4–44 is valid.

SOLUTION   Equation 4–44 is to be proven.
Analysis The general three-dimensional version of the Leibniz theorem, Eq. 4–50, applies to *any* volume. We choose to apply it to the control volume of interest, which can be moving and/or deforming differently than the material volume (Fig. 4–63). Setting G to ρb, Eq. 4–50 becomes

$$\frac{d}{dt} \int_{\text{CV}} \rho b\, dV = \int_{\text{CV}} \frac{\partial}{\partial t}(\rho b)\, dV + \int_{\text{CS}} \rho b \vec{V}_{\text{CS}} \cdot \vec{n}\, dA \qquad (1)$$

We solve Eq. 4–53 for the control volume integral,

$$\int_{\text{CV}} \frac{\partial}{\partial t}(\rho b)\, dV = \frac{dB_{\text{sys}}}{dt} - \int_{\text{CS}} \rho b \vec{V} \cdot \vec{n}\, dA \qquad (2)$$

Substituting Eq. 2 into Eq. 1, we get

$$\frac{d}{dt} \int_{\text{CV}} \rho b\, dV = \frac{dB_{\text{sys}}}{dt} - \int_{\text{CS}} \rho b \vec{V} \cdot \vec{n}\, dA + \int_{\text{CS}} \rho b \vec{V}_{\text{CS}} \cdot \vec{n}\, dA \qquad (3)$$

Combining the last two terms and rearranging,

$$\frac{dB_{\text{sys}}}{dt} = \frac{d}{dt} \int_{\text{CV}} \rho b\, dV + \int_{\text{CS}} \rho b (\vec{V} - \vec{V}_{\text{CS}}) \cdot \vec{n}\, dA \qquad (4)$$

But recall that the relative velocity is defined by Eq. 4–43. Thus,

RTT in terms of relative velocity:
$$\frac{dB_{\text{sys}}}{dt} = \frac{d}{dt} \int_{\text{CV}} \rho b\, dV + \int_{\text{CS}} \rho b \vec{V}_r \cdot \vec{n}\, dA \qquad (5)$$

Discussion Equation 5 is indeed identical to Eq. 4–44, and the power and elegance of the Leibniz theorem are demonstrated.

Relationship between Material Derivative and RTT

You may have noticed a similarity or analogy between the material derivative discussed in Section 4–1 and the Reynolds transport theorem discussed here. In fact, both analyses represent methods to transform from fundamentally

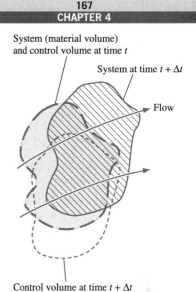

System (material volume) and control volume at time *t*

System at time *t* + Δ*t*

Flow

Control volume at time *t* + Δ*t*

FIGURE 4–63
The material volume (system) and control volume occupy the same space at time *t* (the greenish shaded area), but move and deform differently. At a later time they are *not* coincident.

FIGURE 4–64

The Reynolds transport theorem for finite volumes (integral analysis) is analogous to the material derivative for infinitesimal volumes (differential analysis). In both cases, we transform from a Lagrangian or system viewpoint to an Eulerian or control volume viewpoint.

Lagrangian concepts to Eulerian interpretations of those concepts. While the Reynolds transport theorem deals with finite-size control volumes and the material derivative deals with infinitesimal fluid particles, the same fundamental physical interpretation applies to both (Fig. 4–64). In fact, the Reynolds transport theorem can be thought of as the integral counterpart of the material derivative. In either case, the total rate of change of some property following an identified portion of fluid consists of two parts: There is a local or unsteady part that accounts for changes in the flow field with time (compare the first term on the right-hand side of Eq. 4–12 to that of Eq. 4–45). There is also an advective part that accounts for the movement of fluid from one region of the flow to another (compare the second term on the right-hand sides of Eqs. 4–12 and 4–45).

Just as the material derivative can be applied to any fluid property, scalar or vector, the Reynolds transport theorem can be applied to any scalar or vector property as well. In Chaps. 5 and 6, we apply the Reynolds transport theorem to conservation of mass, energy, momentum, and angular momentum by choosing parameter B to be mass, energy, momentum, and angular momentum, respectively. In this fashion we can easily convert from the fundamental system conservation laws (Lagrangian viewpoint) to forms that are valid and useful in a control volume analysis (Eulerian viewpoint).

SUMMARY

Fluid kinematics is concerned with describing fluid motion, without necessarily analyzing the forces responsible for such motion. There are two fundamental descriptions of fluid motion—*Lagrangian* and *Eulerian*. In a Lagrangian description, we follow individual fluid particles or collections of fluid particles, while in the Eulerian description, we define a *control volume* through which fluid flows in and out. We transform equations of motion from Lagrangian to Eulerian through use of the *material derivative* for infinitesimal fluid particles and through use of the *Reynolds transport theorem (RTT)* for systems of finite volume. For some extensive property B or its corresponding intensive property b,

Material derivative:
$$\frac{Db}{Dt} = \frac{\partial b}{\partial t} + (\vec{V} \cdot \vec{\nabla})b$$

General RTT, nonfixed CV:
$$\frac{dB_{sys}}{dt} = \int_{CV} \frac{\partial}{\partial t}(\rho b)\, d\mathcal{V} + \int_{CS} \rho b \vec{V} \cdot \vec{n}\, dA$$

In both equations, the total change of the property following a fluid particle or following a system is composed of two parts: a *local* (unsteady) part and an *advective* (movement) part.

There are various ways to visualize and analyze flow fields—*streamlines, streaklines, pathlines, timelines, surface imaging, shadowgraphy, schlieren imaging, profile plots,* vector plots, and *contour plots*. We define each of these and provide examples in this chapter. In general unsteady flow, streamlines, streaklines, and pathlines differ, but *in steady flow, streamlines, streaklines, and pathlines are coincident.*

Four fundamental rates of motion (*deformation rates*) are required to fully describe the kinematics of a fluid flow: *velocity* (rate of translation), *angular velocity* (rate of rotation), *linear strain rate*, and *shear strain rate*. *Vorticity* is a property of fluid flows that indicates the *rotationality* of fluid particles.

Vorticity vector: $\vec{\zeta} = \vec{\nabla} \times \vec{V} = \text{curl}(\vec{V}) = 2\vec{\omega}$

A region of flow is *irrotational* if the vorticity is zero in that region.

The concepts learned in this chapter are used repeatedly throughout the rest of the book. We use the RTT to transform the conservation laws from closed systems to control volumes in Chaps. 5 and 6, and again in Chap. 9 in the derivation of the differential equations of fluid motion. The role of vorticity and irrotationality is revisited in greater detail in Chap. 10 where we show that the irrotationality approximation leads to greatly reduced complexity in the solution of fluid flows. Finally, we use various types of flow visualization and data plots to describe the kinematics of example flow fields in nearly every chapter of this book.

APPLICATION SPOTLIGHT ■ Fluidic Actuators

Guest Author: Ganesh Raman,
Illinois Institute of Technology

Fluidic actuators are devices that use fluid logic circuits to produce oscillatory velocity or pressure perturbations in jets and shear layers for delaying separation, enhancing mixing, and suppressing noise. Fluidic actuators are potentially useful for shear flow control applications for many reasons: they have no moving parts; they can produce perturbations that are controllable in frequency, amplitude, and phase; they can operate in harsh thermal environments and are not susceptible to electromagnetic interference; and they are easy to integrate into a functioning device. Although fluidics technology has been around for many years, recent advances in miniaturization and microfabrication have made them very attractive candidates for practical use. The fluidic actuator produces a self-sustaining oscillatory flow using the principles of wall attachment and backflow that occur within miniature passages of the device.

Figure 4–65 demonstrates the application of a fluidic actuator for jet thrust vectoring. Fluidic thrust vectoring is important for future aircraft designs, since they can improve maneuverability without the complexity of additional surfaces near the nozzle exhaust. In the three images of Fig. 4–65, the primary jet exhausts from right to left and a single fluidic actuator is located at the top. Figure 4–65*a* shows the unperturbed jet. Figures 4–65*b* and *c* show the vectoring effect at two fluidic actuation levels. Changes to the primary jet are characterized using particle image velocimetry (PIV). A simplified explanation is as follows: In this technique tracer particles are introduced into the flow and illuminated by a thin laser light sheet that is pulsed to freeze particle motion. Laser light scattered by the particles is recorded at two instances in time using a digital camera. Using a spatial cross correlation, the local displacement vector is obtained. The results indicate that there exists the potential for integrating multiple fluidic sub-elements into aircraft components for improved performance.

Figure 4–65 is actually a combination vector plot and contour plot. Velocity vectors are superimposed on contour plots of velocity magnitude (speed). The red regions represent high speeds, and the blue regions represent low speeds.

References

Raman, G., Packiarajan, S., Papadopoulos, G., Weissman, C., and Raghu, S., "Jet Thrust Vectoring Using a Miniature Fluidic Oscillator," ASME FEDSM 2001-18057, 2001.

Raman, G., Raghu, S., and Bencic, T. J., "Cavity Resonance Suppression Using Miniature Fluidic Oscillators," AIAA Paper 99-1900, 1999.

(*a*)

(*b*)

(*c*)

FIGURE 4–65

Time-averaged mean velocity field of a fluidic actuator jet. Results are from 150 PIV realizations, overlaid on an image of the seeded flow. Every seventh and second velocity vector is shown in the horizontal and vertical directions, respectively. The color levels denote the magnitude of the velocity field. (*a*) No actuation; (*b*) single actuator operating at 3 psig; (*c*) single actuator operating at 9 psig.

Courtesy Ganesh Raman, Illinois Institute of Technology. Used by permission.

REFERENCES AND SUGGESTED READING

1. R. J. Adrian. "Particle-Imaging Technique for Experimental Fluid Mechanics," *Annual Reviews in Fluid Mechanics*, 23, pp. 261–304, 1991.

2. J. M. Cimbala, H. Nagib, and A. Roshko. "Large Structure in the Far Wakes of Two-Dimensional Bluff Bodies," *Journal of Fluid Mechanics,* 190, pp. 265–298, 1988.

3. R. J. Heinsohn and J. M. Cimbala. *Indoor Air Quality Engineering.* New York: Marcel-Dekker, 2003.

4. P. K. Kundu and I. M. Cohen *Fluid Mechanics.* Ed. 5, London, England: Elsevier Inc. 2011.

5. W. Merzkirch. *Flow Visualization*, 2nd ed. Orlando, FL: Academic Press, 1987.

6. G. S. Settles. *Schlieren and Shadowgraph Techniques: Visualizing Phenomena in Transparent Media.* Heidelberg: Springer-Verlag, 2001.

7. M. Van Dyke. *An Album of Fluid Motion.* Stanford, CA: The Parabolic Press, 1982.

8. F. M. White. *Viscous Fluid Flow*, 3rd ed. New York: McGraw-Hill, 2005.

PROBLEMS*

Introductory Problems

4–1C What does the word *kinematics* mean? Explain what the study of *fluid kinematics* involves.

4–2C Briefly discuss the difference between derivative operators d and ∂. If the derivative $\partial u / \partial x$ appears in an equation, what does this imply about variable u?

4–3 Consider steady flow of water through an axisymmetric garden hose nozzle (Fig. P4–3). Along the centerline of the nozzle, the water speed increases from $u_{entrance}$ to u_{exit} as sketched. Measurements reveal that the centerline water speed increases parabolically through the nozzle. Write an equation for centerline speed $u(x)$, based on the parameters given here, from $x = 0$ to $x = L$.

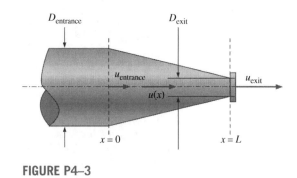

FIGURE P4–3

---

* Problems designated by a "C" are concept questions, and students are encouraged to answer them all. Problems designated by an "E" are in English units, and the SI users can ignore them. Problems with the ⊙ icon are solved using EES, and complete solutions together with parametric studies are included on the text website. Problems with the ⊡ icon are comprehensive in nature and are intended to be solved with an equation solver such as EES.

4–4 Consider the following steady, two-dimensional velocity field:

$$\vec{V} = (u, v) = (a^2 - (b - cx)^2)\vec{i} + (-2cby + 2c^2xy)\vec{j}$$

Is there a stagnation point in this flow field? If so, where is it?

4–5 A steady, two-dimensional velocity field is given by

$$\vec{V} = (u, v) = (-0.781 - 4.67x)\vec{i} + (-3.54 + 4.67y)\vec{j}$$

Calculate the location of the stagnation point.

4–6 Consider the following steady, two-dimensional velocity field:

$$\vec{V} = (u, v) = (0.66 + 2.1x)\vec{i} + (-2.7 - 2.1y)\vec{j}$$

Is there a stagnation point in this flow field? If so, where is it?
Answer: Yes; $x = -0.314$, $y = -1.29$

Lagrangian and Eulerian Descriptions

4–7C What is the *Eulerian description* of fluid motion? How does it differ from the Lagrangian description?

4–8C Is the Lagrangian method of fluid flow analysis more similar to study of a system or a control volume? Explain.

4–9C What is the *Lagrangian description* of fluid motion?

4–10C A stationary probe is placed in a fluid flow and measures pressure and temperature as functions of time at

FIGURE P4–10C

one location in the flow (Fig. P4–10C). Is this a Lagrangian or an Eulerian measurement? Explain.

4–11C A tiny neutrally buoyant electronic pressure probe is released into the inlet pipe of a water pump and transmits 2000 pressure readings per second as it passes through the pump. Is this a Lagrangian or an Eulerian measurement? Explain.

4–12C Define a *steady flow field* in the Eulerian reference frame. In such a steady flow, is it possible for a fluid particle to experience a nonzero acceleration?

4–13C List at least three other names for the material derivative, and write a brief explanation about why each name is appropriate.

4–14C A weather balloon is launched into the atmosphere by meteorologists. When the balloon reaches an altitude where it is neutrally buoyant, it transmits information about weather conditions to monitoring stations on the ground (Fig. P4–14C). Is this a Lagrangian or an Eulerian measurement? Explain.

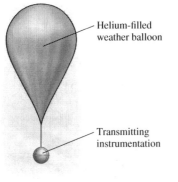

Helium-filled weather balloon

Transmitting instrumentation

FIGURE P4–14C

4–15C A Pitot-static probe can often be seen protruding from the underside of an airplane (Fig. P4–15C). As the

Probe

FIGURE P4–15C

airplane flies, the probe measures relative wind speed. Is this a Lagrangian or an Eulerian measurement? Explain.

4–16C Is the Eulerian method of fluid flow analysis more similar to study of a system or a control volume? Explain.

4–17 Consider steady, incompressible, two-dimensional flow through a converging duct (Fig. P4–17). A simple approximate velocity field for this flow is

$$\vec{V} = (u, v) = (U_0 + bx)\vec{i} - by\vec{j}$$

where U_0 is the horizontal speed at $x = 0$. Note that this equation ignores viscous effects along the walls but is a reasonable approximation throughout the majority of the flow field. Calculate the material acceleration for fluid particles passing through this duct. Give your answer in two ways: (1) as acceleration components a_x and a_y and (2) as acceleration vector $\vec{a}$.

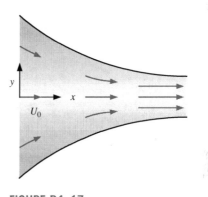

y

x

U_0

FIGURE P4–17

4–18 Converging duct flow is modeled by the steady, two-dimensional velocity field of Prob. 4–17. The pressure field is given by

$$P = P_0 - \frac{\rho}{2}\left[2U_0 bx + b^2(x^2 + y^2)\right]$$

where P_0 is the pressure at $x = 0$. Generate an expression for the rate of change of pressure *following a fluid particle*.

4–19 A steady, incompressible, two-dimensional velocity field is given by the following components in the xy-plane:

$$u = 1.85 + 2.33x + 0.656y$$

$$v = 0.754 - 2.18x - 2.33y$$

Calculate the acceleration field (find expressions for acceleration components a_x and a_y), and calculate the acceleration at the point $(x, y) = (-1, 2)$. *Answers: $a_x = 0.806$, $a_y = 2.21$*

4–20 A steady, incompressible, two-dimensional velocity field is given by the following components in the xy-plane:

$$u = 0.205 + 0.97x + 0.851y$$

$$v = -0.509 + 0.953x - 0.97y$$

Calculate the acceleration field (find expressions for acceleration components a_x and a_y) and calculate the acceleration at the point $(x, y) = (2, 1.5)$.

4–21 The velocity field for a flow is given by $\vec{V} = u\vec{i} + v\vec{j} + w\vec{k}$ where $u = 3x$, $v = -2y$, $w = 2z$. Find the streamline that will pass through the point $(1, 1, 0)$.

4–22 Consider steady flow of air through the diffuser portion of a wind tunnel (Fig. P4–22). Along the centerline of the diffuser, the air speed decreases from $u_{entrance}$ to u_{exit} as sketched. Measurements reveal that the centerline air speed decreases parabolically through the diffuser. Write an equation for centerline speed $u(x)$, based on the parameters given here, from $x = 0$ to $x = L$.

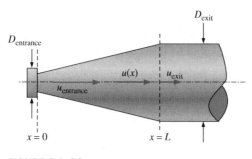

FIGURE P4–22

4–23 For the velocity field of Prob. 4–22, calculate the fluid acceleration along the diffuser centerline as a function of x and the given parameters. For $L = 1.56$ m, $u_{entrance} = 24.3$ m/s, and $u_{exit} = 16.8$ m/s, calculate the acceleration at $x = 0$ and $x = 1.0$ m. *Answers:* 0, -131 m/s^2

4–24 A steady, incompressible, two-dimensional (in the xy-plane) velocity field is given by

$$\vec{V} = (0.523 - 1.88x + 3.94y)\vec{i} + (-2.44 + 1.26x + 1.88y)\vec{j}$$

Calculate the acceleration at the point $(x, y) = (-1.55, 2.07)$.

4–25 For the velocity field of Prob. 4–3, calculate the fluid acceleration along the nozzle centerline as a function of x and the given parameters.

Flow Patterns and Flow Visualization

4–26C What is the definition of a *pathline*? What do pathlines indicate?

4–27C Consider the visualization of flow over a 12° cone in Fig. P4–27C. Are we seeing streamlines, streaklines, pathlines, or timelines? Explain.

FIGURE P4–27C

Visualization of flow over a 12° cone at a 16° angle of attack at a Reynolds number of 15,000. The visualization is produced by colored fluid injected into water from ports in the body.

Courtesy ONERA. Photograph by Werlé.

4–28C What is the definition of a *streamline*? What do streamlines indicate?

4–29C What is the definition of a *streakline*? How do streaklines differ from streamlines?

4–30C Consider the visualization of flow over a 15° delta wing in Fig. P4–30C. Are we seeing streamlines, streaklines, pathlines, or timelines? Explain.

FIGURE P4–30C

Visualization of flow over a 15° delta wing at a 20° angle of attack at a Reynolds number of 20,000. The visualization is produced by colored fluid injected into water from ports on the underside of the wing.

Courtesy ONERA. Photograph by Werlé.

4–31C Consider the visualization of ground vortex flow in Fig. P4–31C. Are we seeing streamlines, streaklines, pathlines, or timelines? Explain.

FIGURE P4–31C

Visualization of ground vortex flow. A high-speed round air jet impinges on the ground in the presence of a free-stream flow of air from left to right. (The ground is at the bottom of the picture.) The portion of the jet that travels upstream forms a recirculating flow known as a **ground vortex.** The visualization is produced by a smoke wire mounted vertically to the left of the field of view.

Photo by John M. Cimbala.

4–32C Consider the visualization of flow over a sphere in Fig. P4–32C. Are we seeing streamlines, streaklines, pathlines, or timelines? Explain.

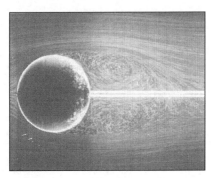

FIGURE P4–32C

Visualization of flow over a sphere at a Reynolds number of 15,000. The visualization is produced by a time exposure of air bubbles in water.

Courtesy ONERA. Photograph by Werlé.

4–33C What is the definition of a *timeline*? How can timelines be produced in a water channel? Name an application where timelines are more useful than streaklines.

4–34C Consider a cross-sectional slice through an array of heat exchanger tubes (Fig. P4–34C). For each desired piece of information, choose which kind of flow visualization plot (vector plot or contour plot) would be most appropriate, and explain why.

(*a*) The location of maximum fluid speed is to be visualized.

(*b*) Flow separation at the rear of the tubes is to be visualized.

(*c*) The temperature field throughout the plane is to be visualized.

(*d*) The distribution of the vorticity component normal to the plane is to be visualized.

FIGURE P4–34C

4–35 Converging duct flow (Fig. P4–17) is modeled by the steady, two-dimensional velocity field of Prob. 4–17. Generate an analytical expression for the flow streamlines.
Answer: $y = C/(U_0 + bx)$

4–36 The velocity field of a flow is described by $\vec{V} = (4x)\vec{i} + (5y + 3)\vec{j} + (3t^2)\vec{k}$. What is the pathline of a particle at a location (1 m, 2 m, 4 m) at time $t = 1$ s?

4–37 Consider the following steady, incompressible, two-dimensional velocity field:

$$\vec{V} = (u, v) = (4.35 + 0.656x)\vec{i} + (-1.22 - 0.656y)\vec{j}$$

Generate an analytical expression for the flow streamlines and draw several streamlines in the upper-right quadrant from $x = 0$ to 5 and $y = 0$ to 6.

4–38 Consider the steady, incompressible, two-dimensional velocity field of Prob. 4–37. Generate a velocity vector plot in the upper-right quadrant from $x = 0$ to 5 and $y = 0$ to 6.

4–39 Consider the steady, incompressible, two-dimensional velocity field of Prob. 4–37. Generate a vector plot of the acceleration field in the upper-right quadrant from $x = 0$ to 5 and $y = 0$ to 6.

4–40 A steady, incompressible, two-dimensional velocity field is given by

$$\vec{V} = (u, v) = (1 + 2.5x + y)\vec{i} + (-0.5 - 3x - 2.5y)\vec{j}$$

where the *x*- and *y*-coordinates are in m and the magnitude of velocity is in m/s.

(*a*) Determine if there are any stagnation points in this flow field, and if so, where they are.

(*b*) Sketch velocity vectors at several locations in the upper-right quadrant for $x = 0$ m to 4 m and $y = 0$ m to 4 m; qualitatively describe the flow field.

4–41 Consider the steady, incompressible, two-dimensional velocity field of Prob. 4–40.

(*a*) Calculate the material acceleration at the point ($x = 2$ m, $y = 3$ m). *Answers:* $a_x = 8.50$ m/s^2, $a_y = 8.00$ m/s^2

(*b*) Sketch the material acceleration vectors at the same array of *x*- and *y*-values as in Prob. 4–40.

4–42 The velocity field for *solid-body rotation* in the $r\theta$-plane (Fig. P4–42) is given by

$$u_r = 0 \qquad u_\theta = \omega r$$

where ω is the magnitude of the angular velocity ($\vec{\omega}$ points in the *z*-direction). For the case with $\omega = 1.5$ s^{-1}, plot a contour plot of velocity magnitude (speed). Specifically, draw curves of constant speed $V = 0.5$, 1.0, 1.5, 2.0, and 2.5 m/s. Be sure to label these speeds on your plot.

FIGURE P4–42

4–43 The velocity field for a *line vortex* in the $r\theta$-plane (Fig. P4–43) is given by

$$u_r = 0 \qquad u_\theta = \frac{K}{r}$$

where K is the *line vortex strength*. For the case with $K = 1.5$ m/s^2, plot a contour plot of velocity magnitude (speed).

FIGURE P4–43

Specifically, draw curves of constant speed $V = 0.5$, 1.0, 1.5, 2.0, and 2.5 m/s. Be sure to label these speeds on your plot.

4–44 The velocity field for a *line source* in the $r\theta$-plane (Fig. P4–44) is given by

$$u_r = \frac{m}{2\pi r} \qquad u_\theta = 0$$

where m is the line source strength. For the case with $m/(2\pi) = 1.5$ m^2/s, plot a contour plot of velocity magnitude (speed). Specifically, draw curves of constant speed $V = 0.5$, 1.0, 1.5, 2.0, and 2.5 m/s. Be sure to label these speeds on your plot.

FIGURE P4–44

4–45 A very small circular cylinder of radius R_i is rotating at angular velocity ω_i inside a much larger concentric cylinder of radius R_o that is rotating at angular velocity ω_o. A liquid of density ρ and viscosity μ is confined between the two cylinders, as in Fig. P4–45. Gravitational and end effects can be neglected (the flow is two-dimensional into the page).

FIGURE P4–45

If $\omega_i = \omega_o$ and a long time has passed, generate an expression for the tangential velocity profile, u_θ as a function of (at most) r, ω, R_i, R_o, ρ, and μ, where $\omega = \omega_i = \omega_o$. Also, calculate the torque exerted by the fluid on the inner cylinder and on the outer cylinder.

4–46 Consider the same two concentric cylinders of Prob. 4–45. This time, however, the inner cylinder is rotating, but the outer cylinder is stationary. In the limit, as the outer cylinder is very large compared to the inner cylinder (imagine the inner cylinder spinning very fast while its radius gets very small), what kind of flow does this approximate? Explain. After a long time has passed, generate an expression for the tangential velocity profile, namely u_θ as a function of (at most) r, ω_i, R_i, R_o, ρ, and μ. *Hint*: Your answer may contain an (unknown) constant, which can be obtained by specifying a boundary condition at the inner cylinder surface.

4–47E Converging duct flow is modeled by the steady, two-dimensional velocity field of Prob. 4–17. For the case in which $U_0 = 3.56$ ft/s and $b = 7.66$ s^{-1}, plot several streamlines from $x = 0$ ft to 5 ft and $y = -2$ ft to 2 ft. Be sure to show the *direction* of the streamlines.

Motion and Deformation of Fluid Elements; Vorticity and Rotationality

4–48C Explain the relationship between vorticity and rotationality.

4–49C Name and briefly describe the four fundamental types of motion or deformation of fluid particles.

4–50 Converging duct flow (Fig. P4–17) is modeled by the steady, two-dimensional velocity field of Prob. 4–17. Is this flow field rotational or irrotational? Show all your work.   *Answer: irrotational*

4–51 Converging duct flow is modeled by the steady, two-dimensional velocity field of Prob. 4–17. A fluid particle (*A*) is located on the *x*-axis at $x = x_A$ at time $t = 0$ (Fig. P4–51).

At some later time t, the fluid particle has moved downstream with the flow to some new location $x = x_{A'}$, as shown in the figure. Since the flow is symmetric about the *x*-axis, the fluid particle remains on the *x*-axis at all times. Generate an analytical expression for the *x*-location of the fluid particle at some arbitrary time t in terms of its initial location x_A and constants U_0 and b. In other words, develop an expression for $x_{A'}$. (*Hint*: We know that $u = dx_{particle}/dt$ following a fluid particle. Plug in u, separate variables, and integrate.)

4–52 Converging duct flow is modeled by the steady, two-dimensional velocity field of Prob. 4–17. Since the flow is symmetric about the *x*-axis, line segment *AB* along the *x*-axis remains on the axis, but stretches from length ξ to length $\xi + \Delta\xi$ as it flows along the channel centerline (Fig. P4–52). Generate an analytical expression for the change in length of the line segment, $\Delta\xi$. (*Hint*: Use the result of Prob. 4–51.)   *Answer: $(x_B - x_A)(e^{bt} - 1)$*

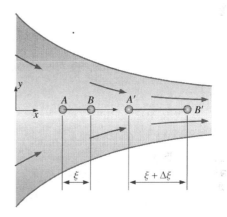

FIGURE P4–52

4–53 Using the results from Prob. 4–52 and the fundamental definition of linear strain rate (the rate of increase in length per unit length), develop an expression for the linear strain rate in the *x*-direction (ε_{xx}) of fluid particles located on the centerline of the channel. Compare your result to the general expression for ε_{xx} in terms of the velocity field, i.e., $\varepsilon_{xx} = \partial u/\partial x$. (*Hint*: Take the limit as time $t \to 0$. You may need to apply a truncated series expansion for e^{bt}.)   *Answer: b*

4–54 Converging duct flow is modeled by the steady, two-dimensional velocity field of Prob. 4–17. A fluid particle (*A*) is located at $x = x_A$ and $y = y_A$ at time $t = 0$ (Fig. P4–54). At some later time t, the fluid particle has moved downstream with the flow to some new location $x = x_{A'}$, $y = y_{A'}$, as shown in the figure. Generate an analytical expression for the *y*-location of the fluid particle at arbitrary time t in terms of its initial *y*-location y_A and constant b. In other words, develop an expression for $y_{A'}$. (*Hint*: We know that $v = dy_{particle}/dt$ following a fluid particle. Substitute the equation for v, separate variables, and integrate.)   *Answer: $y_A e^{-bt}$*

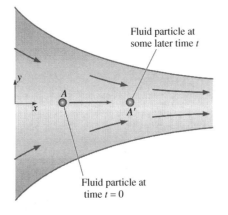

Fluid particle at some later time t

Fluid particle at time $t = 0$

FIGURE P4–51

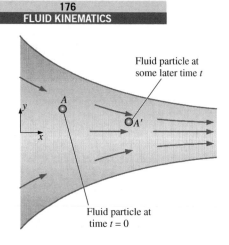

Fluid particle at
some later time t

Fluid particle at
time t = 0

FIGURE P4–54

4–55 Converging duct flow is modeled by the steady, two-dimensional velocity field of Prob. 4–17. As vertical line segment AB moves downstream it shrinks from length η to length $\eta + \Delta\eta$ as sketched in Fig. P4–55. Generate an analytical expression for the change in length of the line segment, $\Delta\eta$. Note that the change in length, $\Delta\eta$, is *negative*. (*Hint*: Use the result of Prob. 4–54.)

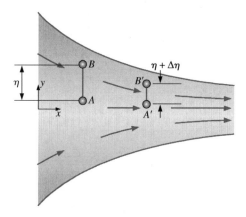

$\eta + \Delta\eta$

FIGURE P4–55

4–56 Using the results of Prob. 4–55 and the fundamental definition of linear strain rate (the rate of increase in length per unit length), develop an expression for the linear strain rate in the y-direction (ε_{yy}) of fluid particles moving down the channel. Compare your result to the general expression for ε_{yy} in terms of the velocity field, i.e., $\varepsilon_{yy} = \partial v/\partial y$. (*Hint*: Take the limit as time $t \to 0$. You may need to apply a truncated series expansion for e^{-bt}.)

4–57 Converging duct flow is modeled by the steady, two-dimensional velocity field of Prob. 4–17. Use the equation for volumetric strain rate to verify that this flow field is incompressible.

4–58 A general equation for a steady, two-dimensional velocity field that is linear in both spatial directions (x and y) is

$$\vec{V} = (u, v) = (U + a_1 x + b_1 y)\vec{i} + (V + a_2 x + b_2 y)\vec{j}$$

where U and V and the coefficients are constants. Their dimensions are assumed to be appropriately defined. Calculate the x- and y-components of the acceleration field.

4–59 For the velocity field of Prob. 4–58, what relationship must exist between the coefficients to ensure that the flow field is incompressible? *Answer*: $a_1 + b_2 = 0$

4–60 For the velocity field of Prob. 4–58, calculate the linear strain rates in the x- and y-directions. *Answers*: a_1, b_2

4–61 For the velocity field of Prob. 4–58, calculate the shear strain rate in the xy-plane.

4–62 Combine your results from Probs. 4–60 and 4–61 to form the two-dimensional strain rate tensor ε_{ij} in the xy-plane,

$$\varepsilon_{ij} = \begin{pmatrix} \varepsilon_{xx} & \varepsilon_{xy} \\ \varepsilon_{yx} & \varepsilon_{yy} \end{pmatrix}$$

Under what conditions would the x- and y-axes be principal axes? *Answer*: $b_1 + a_2 = 0$

4–63 For the velocity field of Prob. 4–58, calculate the vorticity vector. In which direction does the vorticity vector point? *Answer*: $(a_2 - b_1)\vec{k}$ in z − direction

4–64 Consider steady, incompressible, two-dimensional **shear flow** for which the velocity field is

$$\vec{V} = (u, v) = (a + by)\vec{i} + 0\vec{j}$$

where a and b are constants. Sketched in Fig. P4–64 is a small rectangular fluid particle of dimensions dx and dy at time t. The fluid particle moves and deforms with the flow such that at a later time (t + dt), the particle is no longer rectangular, as also shown in the figure. The initial location of each corner of the fluid particle is labeled in Fig. P4–64. The lower-left corner is at (x, y) at time t, where the x-component

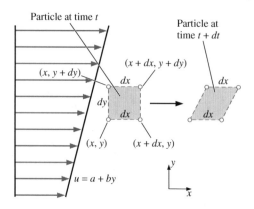

Particle at time t

Particle at
time t + dt

$(x, y + dy)$

$(x + dx, y + dy)$

dx

dx

dy

dx

dx

(x, y)

$(x + dx, y)$

$u = a + by$

FIGURE P4–64

of velocity is $u = a + by$. At the later time, this corner moves to $(x + u\,dt, y)$, or

$$(x + (a + by)\,dt, y)$$

(a) In similar fashion, calculate the location of each of the other three corners of the fluid particle at time $t + dt$.

(b) From the fundamental definition of *linear strain rate* (the rate of increase in length per unit length), calculate linear strain rates ε_{xx} and ε_{yy}.   *Answers: 0, 0*

(c) Compare your results with those obtained from the equations for ε_{xx} and ε_{yy} in Cartesian coordinates, i.e.,

$$\varepsilon_{xx} = \frac{\partial u}{\partial x} \qquad \varepsilon_{yy} = \frac{\partial v}{\partial y}$$

4–65 Use two methods to verify that the flow of Prob. 4–64 is incompressible: (a) by calculating the volume of the fluid particle at both times, and (b) by calculating the volumetric strain rate. Note that Prob. 4–64 should be completed before this problem.

4–66 Consider the steady, incompressible, two-dimensional flow field of Prob. 4–64. Using the results of Prob. 4–64(a), do the following:

(a) From the fundamental definition of *shear strain rate* (half of the rate of decrease of the angle between two initially perpendicular lines that intersect at a point), calculate shear strain rate ε_{xy} in the xy-plane. (*Hint*: Use the lower edge and the left edge of the fluid particle, which intersect at 90° at the lower-left corner of the particle at the initial time.)

(b) Compare your results with those obtained from the equation for ε_{xy} in Cartesian coordinates, i.e.,

$$\varepsilon_{xy} = \frac{1}{2}\left(\frac{\partial u}{\partial y} + \frac{\partial v}{\partial x}\right)$$

Answers: (a) b/2, (b) b/2

4–67 Consider the steady, incompressible, two-dimensional flow field of Prob. 4–64. Using the results of Prob. 4–64(a), do the following:

(a) From the fundamental definition of the *rate of rotation* (average rotation rate of two initially perpendicular lines that intersect at a point), calculate the rate of rotation of the fluid particle in the xy-plane, ω_z. (*Hint*: Use the lower edge and the left edge of the fluid particle, which intersect at 90° at the lower-left corner of the particle at the initial time.)

(b) Compare your results with those obtained from the equation for ω_z in Cartesian coordinates, i.e.,

$$\omega_z = \frac{1}{2}\left(\frac{\partial v}{\partial x} - \frac{\partial u}{\partial y}\right)$$

Answers: (a) −b/2, (b) −b/2

4–68 From the results of Prob. 4–67,

(a) Is this flow rotational or irrotational?

(b) Calculate the z-component of vorticity for this flow field.

4–69 A two-dimensional fluid element of dimensions dx and dy translates and distorts as shown in Fig. P4–69 during the infinitesimal time period $dt = t_2 - t_1$. The velocity components at point P at the initial time are u and v in the x- and y-directions, respectively. Show that the magnitude of the rate of rotation (angular velocity) about point P in the xy-plane is

$$\omega_z = \frac{1}{2}\left(\frac{\partial v}{\partial x} - \frac{\partial u}{\partial y}\right)$$

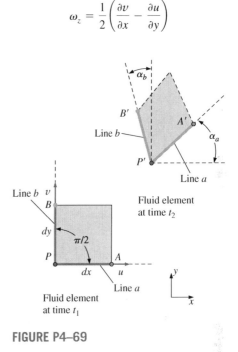

FIGURE P4–69

4–70 A two-dimensional fluid element of dimensions dx and dy translates and distorts as shown in Fig. P4–69 during the infinitesimal time period $dt = t_2 - t_1$. The velocity components at point P at the initial time are u and v in the x- and y-directions, respectively. Consider the line segment PA in Fig. P4–69, and show that the magnitude of the linear strain rate in the x-direction is

$$\varepsilon_{xx} = \frac{\partial u}{\partial x}$$

4–71 A two-dimensional fluid element of dimensions dx and dy translates and distorts as shown in Fig. P4–69 during the infinitesimal time period $dt = t_2 - t_1$. The velocity components at point P at the initial time are u and v in the x- and y-directions, respectively. Show that the magnitude of the shear strain rate about point P in the xy-plane is

$$\varepsilon_{xy} = \frac{1}{2}\left(\frac{\partial u}{\partial y} + \frac{\partial v}{\partial x}\right)$$

4–72 Consider a steady, two-dimensional, incompressible flow field in the xy-plane. The linear strain rate in the x-direction is 2.5 s^{-1}. Calculate the linear strain rate in the y-direction.

4–73 A cylindrical tank of water rotates in solid-body rotation, counterclockwise about its vertical axis (Fig. P4–73) at angular speed $\dot{n} = 175$ rpm. Calculate the vorticity of fluid particles in the tank. *Answer:* $36.7 \, \vec{k}$ rad/s

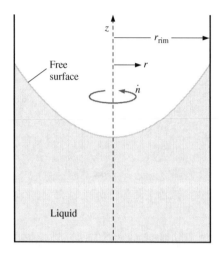

FIGURE P4–73

4–74 A cylindrical tank of water rotates about its vertical axis (Fig. P4–73). A PIV system is used to measure the vorticity field of the flow. The measured value of vorticity in the z-direction is -45.4 rad/s and is constant to within ± 0.5 percent everywhere that it is measured. Calculate the angular speed of rotation of the tank in rpm. Is the tank rotating clockwise or counterclockwise about the vertical axis?

4–75 A cylindrical tank of radius $r_{\text{rim}} = 0.354$ m rotates about its vertical axis (Fig. P4–73). The tank is partially filled with oil. The speed of the rim is 3.61 m/s in the counterclockwise direction (looking from the top), and the tank has been spinning long enough to be in solid-body rotation. For any fluid particle in the tank, calculate the magnitude of the component of vorticity in the vertical z-direction. *Answer:* 20.4 rad/s

4–76 Consider a two-dimensional, incompressible flow field in which an initially square fluid particle moves and deforms. The fluid particle dimension is a at time t and is aligned with the x- and y-axes as sketched in Fig. P4–76. At some later time, the particle is still aligned with the x- and

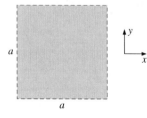

FIGURE P4–76

y-axes, but has deformed into a rectangle of horizontal length $2a$. What is the vertical length of the rectangular fluid particle at this later time?

4–77 Consider a two-dimensional, *compressible* flow field in which an initially square fluid particle moves and deforms. The fluid particle dimension is a at time t and is aligned with the x- and y-axes as sketched in Fig. P4–76. At some later time, the particle is still aligned with the x- and y-axes but has deformed into a rectangle of horizontal length $1.08a$ and vertical length $0.903a$. (The particle's dimension in the z-direction does not change since the flow is two-dimensional.) By what percentage has the density of the fluid particle increased or decreased?

4–78 Consider the following steady, three-dimensional velocity field:

$$\vec{V} = (u, v, w)$$
$$= (3.0 + 2.0x - y)\vec{i} + (2.0x - 2.0y)\vec{j} + (0.5xy)\vec{k}$$

Calculate the vorticity vector as a function of space (x, y, z).

4–79 Consider fully developed **Couette flow**—flow between two infinite parallel plates separated by distance h, with the top plate moving and the bottom plate stationary as illustrated in Fig. P4–79. The flow is steady, incompressible, and two-dimensional in the xy-plane. The velocity field is given by

$$\vec{V} = (u, v) = V\frac{y}{h}\vec{i} + 0\vec{j}$$

Is this flow rotational or irrotational? If it is rotational, calculate the vorticity component in the z-direction. Do fluid particles in this flow rotate clockwise or counterclockwise? *Answers:* yes, $-V/h$, clockwise

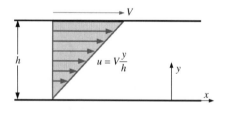

FIGURE P4–79

4–80 For the Couette flow of Fig. P4–79, calculate the linear strain rates in the x- and y-directions, and calculate the shear strain rate ε_{xy}.

4–81 Combine your results from Prob. 4–80 to form the two-dimensional strain rate tensor ε_{ij},

$$\varepsilon_{ij} = \begin{pmatrix} \varepsilon_{xx} & \varepsilon_{xy} \\ \varepsilon_{yx} & \varepsilon_{yy} \end{pmatrix}$$

Are the x- and y-axes principal axes?

4–82 A steady, three-dimensional velocity field is given by

$$\vec{V} = (u, v, w)$$
$$= (2.49 + 1.36x - 0.867y)\vec{i}$$
$$+ (1.95x - 1.36y)\vec{j} + (-0.458xy)\vec{k}$$

Calculate the vorticity vector as a function of space variables (x, y, z).

4–83 A steady, two-dimensional velocity field is given by

$$\vec{V} = (u, v)$$
$$= (2.85 + 1.26x - 0.896y)\vec{i}$$
$$+ (3.45x + cx - 1.26y)\vec{j}$$

Calculate constant c such that the flow field is irrotational.

4–84 A steady, three-dimensional velocity field is given by

$$\vec{V} = (1.35 + 2.78x + 0.754y + 4.21z)\vec{i}$$
$$+ (3.45 + cx - 2.78y + bz)\vec{j}$$
$$+ (-4.21x - 1.89y)\vec{k}$$

Calculate constants b and c such that the flow field is irrotational.

4–85 A steady, three-dimensional velocity field is given by

$$\vec{V} = (0.657 + 1.73x + 0.948y + az)\vec{i}$$
$$+ (2.61 + cx + 1.91y + bz)\vec{j}$$
$$+ (-2.73x - 3.66y - 3.64z)\vec{k}$$

Calculate constants a, b, and c such that the flow field is irrotational.

4–86E Converging duct flow is modeled by the steady, two-dimensional velocity field of Prob. 4–17. For the case in which $U_0 = 5.0$ ft/s and $b = 4.6$ s^{-1}, consider

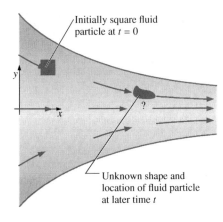

FIGURE P4–86E

an initially square fluid particle of edge dimension 0.5 ft, centered at $x = 0.5$ ft and $y = 1.0$ ft at $t = 0$ (Fig. P4–86E). Carefully calculate and plot where the fluid particle will be and what it will look like at time $t = 0.2$ s later. Comment on the fluid particle's distortion. (*Hint*: Use the results of Probs. 4–51 and 4–54.)

4–87E Based on the results of Prob. 4–86E, verify that this converging duct flow field is indeed incompressible.

Reynolds Transport Theorem

4–88C Briefly explain the similarities and differences between the material derivative and the Reynolds transport theorem.

4–89C Briefly explain the purpose of the Reynolds transport theorem (RTT). Write the RTT for extensive property B as a "word equation," explaining each term in your own words.

4–90C True or false: For each statement, choose whether the statement is true or false and discuss your answer briefly.

(*a*) The Reynolds transport theorem is useful for transforming conservation equations from their naturally occurring control volume forms to their system forms.

(*b*) The Reynolds transport theorem is applicable only to nondeforming control volumes.

(*c*) The Reynolds transport theorem can be applied to both steady and unsteady flow fields.

(*d*) The Reynolds transport theorem can be applied to both scalar and vector quantities.

4–91 Consider the integral $\dfrac{d}{dt}\displaystyle\int_{t}^{2t} x^{-2}dx$. Solve it two ways:

(*a*) Take the integral first and then the time derivative.
(*b*) Use Leibniz theorem. Compare your results.

4–92 Solve the integral $\dfrac{d}{dt}\displaystyle\int_{t}^{2t} x^{x}dx$ as far as you are able.

4–93 Consider the general form of the Reynolds transport theorem (RTT) given by

$$\frac{dB_{sys}}{dt} = \frac{d}{dt}\int_{CV} \rho b \, dV + \int_{CS} \rho b \vec{V}_r \cdot \vec{n} \, dA$$

where $\vec{V}_r$ is the velocity of the fluid relative to the control surface. Let B_{sys} be the mass m of a closed system of fluid particles. We know that for a system, $dm/dt = 0$ since no mass can enter or leave the system by definition. Use the given equation to derive the equation of conservation of mass for a control volume.

4–94 Consider the general form of the Reynolds transport theorem (RTT) as stated in Prob. 4–93. Let B_{sys} be the linear momentum $m\vec{V}$ of a system of fluid particles. We know that for a system, Newton's second law is

$$\sum \vec{F} = m\vec{a} = m\frac{d\vec{V}}{dt} = \frac{d}{dt}(m\vec{V})_{sys}$$

Use the RTT and Newton's second law to derive the linear momentum equation for a control volume.

4–95   Consider the general form of the Reynolds transport theorem (RTT) as stated in Prob. 4–93. Let B_{sys} be the angular momentum $\vec{H} = \vec{r} \times m\vec{V}$ of a system of fluid particles, where $\vec{r}$ is the moment arm. We know that for a system, conservation of angular momentum is

$$\sum \vec{M} = \frac{d}{dt} \vec{H}_{sys}$$

where $\sum \vec{M}$ is the net moment applied to the system. Use the RTT and the above equation to derive the equation of conservation of angular momentum for a control volume.

4–96   Reduce the following expression as far as possible:

$$F(t) = \frac{d}{dt} \int_{x=At}^{x=Bt} e^{-2x^2} \, dx$$

(*Hint*: Use the one-dimensional Leibniz theorem.)   *Answer:*
$Be^{-B^2 t^2} - Ae^{-A^2 t^2}$

Review Problems

4–97   Consider a steady, two-dimensional flow field in the xy-plane whose x-component of velocity is given by

$$u = a + b(x - c)^2$$

where a, b, and c are constants with appropriate dimensions. Of what form does the y-component of velocity need to be in order for the flow field to be incompressible? In other words, generate an expression for v as a function of x, y, and the constants of the given equation such that the flow is incompressible.   *Answer:* $-2b(x - c)y + f(x)$

4–98   In a steady, two-dimensional flow field in the xy-plane, the x-component of velocity is

$$u = ax + by + cx^2$$

where a, b, and c are constants with appropriate dimensions. Generate a general expression for velocity component v such that the flow field is incompressible.

4–99   Consider fully developed two-dimensional **Poiseuille flow**—flow between two infinite parallel plates separated by distance h, with both the top plate and bottom plate stationary, and a forced pressure gradient dP/dx driving the flow as illustrated in Fig. P4–99. (dP/dx is constant and negative.)

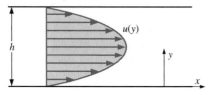

FIGURE P4–99

The flow is steady, incompressible, and two-dimensional in the xy-plane. The velocity components are given by

$$u = \frac{1}{2\mu} \frac{dP}{dx} (y^2 - hy) \quad v = 0$$

where μ is the fluid's viscosity. Is this flow rotational or irrotational? If it is rotational, calculate the vorticity component in the z-direction. Do fluid particles in this flow rotate clockwise or counterclockwise?

4–100   For the two-dimensional Poiseuille flow of Prob. 4–99, calculate the linear strain rates in the x- and y-directions, and calculate the shear strain rate ε_{xy}.

4–101   Combine your results from Prob. 4–100 to form the two-dimensional strain rate tensor ε_{ij} in the xy-plane,

$$\varepsilon_{ij} = \begin{pmatrix} \varepsilon_{xx} & \varepsilon_{xy} \\ \varepsilon_{yx} & \varepsilon_{yy} \end{pmatrix}$$

Are the x- and y-axes principal axes?

4–102   Consider the two-dimensional Poiseuille flow of Prob. 4–99. The fluid between the plates is water at 40°C. Let the gap height $h = 1.6$ mm and the pressure gradient $dP/dx = -230$ N/m^3. Calculate and plot seven *pathlines* from $t = 0$ to $t = 10$ s. The fluid particles are released at $x = 0$ and at $y = 0.2, 0.4, 0.6, 0.8, 1.0, 1.2,$ and 1.4 mm.

4–103   Consider the two-dimensional Poiseuille flow of Prob. 4–99. The fluid between the plates is water at 40°C. Let the gap height $h = 1.6$ mm and the pressure gradient $dP/dx = -230$ N/m^3. Calculate and plot seven *streaklines* generated from a dye rake that introduces dye streaks at $x = 0$ and at $y = 0.2, 0.4, 0.6, 0.8, 1.0, 1.2,$ and 1.4 mm (Fig. P4–103). The dye is introduced from $t = 0$ to $t = 10$ s, and the streaklines are to be plotted at $t = 10$ s.

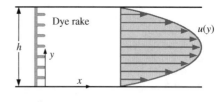

FIGURE P4–103

4–104   Repeat Prob. 4–103 except that the dye is introduced from $t = 0$ to $t = 10$ s, and the streaklines are to be plotted at $t = 12$ s instead of 10 s.

4–105   Compare the results of Probs. 4–103 and 4–104 and comment about the linear strain rate in the x-direction.

4–106   Consider the two-dimensional Poiseuille flow of Prob. 4–99. The fluid between the plates is water at 40°C. Let the gap height $h = 1.6$ mm and the pressure gradient $dP/dx = -230$ N/m³. Imagine a hydrogen bubble wire stretched vertically through the channel at $x = 0$ (Fig. P4–106). The wire is pulsed on and off such that bubbles are produced periodically to create *timelines*. Five distinct timelines are generated at $t = 0, 2.5, 5.0, 7.5,$ and 10.0 s. Calculate and plot what these five timelines look like at time $t = 12.5$ s.

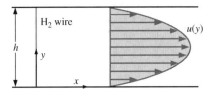

FIGURE P4–106

4–107 The velocity field of a flow is given by $\vec{V} = k(x^2 - y^2)\vec{i} - 2kxy\vec{j}$ where k is a constant. If the radius of curvature of a streamline is $R = [1 + y'^2]^{3/2}/|y''|$, determine the normal acceleration of a particle (which is normal to the streamline) passing through the position $x = 1, y = 2$.

4–108 The velocity field for an incompressible flow is given as $\vec{V} = 5x^2\vec{i} - 20xy\vec{j} + 100t\vec{k}$. Determine if this flow is steady. Also determine the velocity and acceleration of a particle at (1, 3, 3) at $t = 0.2$ s.

4–109 Consider fully developed axisymmetric Poiseuille flow—flow in a round pipe of radius R (diameter $D = 2R$), with a forced pressure gradient dP/dx driving the flow as illustrated in Fig. P4–109. (dP/dx is constant and negative.) The flow is steady, incompressible, and axisymmetric about the x-axis. The velocity components are given by

$$u = \frac{1}{4\mu}\frac{dP}{dx}(r^2 - R^2) \quad u_r = 0 \quad u_\theta = 0$$

where μ is the fluid's viscosity. Is this flow rotational or irrotational? If it is rotational, calculate the vorticity component in the circumferential (θ) direction and discuss the sign of the rotation.

FIGURE P4–109

4–110 For the axisymmetric Poiseuille flow of Prob. 4–109, calculate the linear strain rates in the x- and r-directions, and

calculate the shear strain rate ε_{xr}. The strain rate tensor in cylindrical coordinates (r, θ, x) and (u_r, u_θ, u_x), is

$$\varepsilon_{ij} = \begin{pmatrix} \varepsilon_{rr} & \varepsilon_{r\theta} & \varepsilon_{rx} \\ \varepsilon_{\theta r} & \varepsilon_{\theta\theta} & \varepsilon_{\theta x} \\ \varepsilon_{xr} & \varepsilon_{x\theta} & \varepsilon_{xx} \end{pmatrix}$$

$$= \begin{pmatrix} \dfrac{\partial u_r}{\partial r} & \dfrac{1}{2}\left(r\dfrac{\partial}{\partial r}\left(\dfrac{u_\theta}{r}\right) + \dfrac{1}{r}\dfrac{\partial u_r}{\partial \theta}\right) & \dfrac{1}{2}\left(\dfrac{\partial u_r}{\partial x} + \dfrac{\partial u_x}{\partial r}\right) \\ \dfrac{1}{2}\left(r\dfrac{\partial}{\partial r}\left(\dfrac{u_\theta}{r}\right) + \dfrac{1}{r}\dfrac{\partial u_r}{\partial \theta}\right) & \dfrac{1}{r}\dfrac{\partial u_\theta}{\partial \theta} + \dfrac{u_r}{r} & \dfrac{1}{2}\left(\dfrac{1}{r}\dfrac{\partial u_x}{\partial \theta} + \dfrac{\partial u_\theta}{\partial x}\right) \\ \dfrac{1}{2}\left(\dfrac{\partial u_r}{\partial x} + \dfrac{\partial u_x}{\partial r}\right) & \dfrac{1}{2}\left(\dfrac{1}{r}\dfrac{\partial u_x}{\partial \theta} + \dfrac{\partial u_\theta}{\partial x}\right) & \dfrac{\partial u_x}{\partial x} \end{pmatrix}$$

4–111 Combine your results from Prob. 4–110 to form the axisymmetric strain rate tensor ε_{ij},

$$\varepsilon_{ij} = \begin{pmatrix} \varepsilon_{rr} & \varepsilon_{rx} \\ \varepsilon_{xr} & \varepsilon_{xx} \end{pmatrix}$$

Are the x- and r-axes principal axes?

4–112 We approximate the flow of air into a vacuum cleaner attachment by the following velocity components in the centerplane (the xy-plane):

$$u = \frac{-\dot{V}x}{\pi L}\frac{x^2 + y^2 + b^2}{x^4 + 2x^2y^2 + 2x^2b^2 + y^4 - 2y^2b^2 + b^4}$$

and

$$v = \frac{-\dot{V}y}{\pi L}\frac{x^2 + y^2 - b^2}{x^4 + 2x^2y^2 + 2x^2b^2 + y^4 - 2y^2b^2 + b^4}$$

where b is the distance of the attachment above the floor, L is the length of the attachment, and $\dot{V}$ is the volume flow rate of air being sucked up into the hose (Fig. P4–112). Determine the location of any stagnation point(s) in this flow field. *Answer:* at the origin

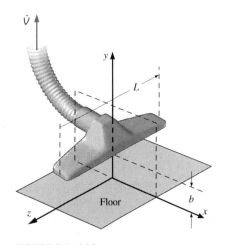

FIGURE P4–112

4–113 Consider the vacuum cleaner of Prob. 4–112. For the case where $b = 2.0$ cm, $L = 35$ cm, and $\dot{V} = 0.1098$ m^3/s, create a velocity vector plot in the upper half of the xy-plane from $x = -3$ to 3 cm and from $y = 0$ cm to 2.5 cm. Draw as many vectors as you need to get a good feel of the flow field. *Note*: The velocity is infinite at the point $(x, y) = (0, 2.0$ cm), so do not attempt to draw a velocity vector at that point.

4–114 Consider the approximate velocity field given for the vacuum cleaner of Prob. 4–112. Calculate the flow speed along the floor. Dust particles on the floor are most likely to be sucked up by the vacuum cleaner at the location of maximum speed. Where is that location? Do you think the vacuum cleaner will do a good job at sucking up dust directly below the inlet (at the origin)? Why or why not?

4–115 In a steady, two-dimensional flow field in the xy-plane, the x-component of velocity is

$$u = ax + by + cx^2 - dxy$$

where a, b, c, and d are constants with appropriate dimensions. Generate a general expression for velocity component v such that the flow field is incompressible.

4–116 There are numerous occasions in which a fairly uniform free-stream flow encounters a long circular cylinder aligned normal to the flow (Fig. P4–116). Examples include air flowing around a car antenna, wind blowing against a flag pole or telephone pole, wind hitting electrical wires, and ocean currents impinging on the submerged round beams that support oil platforms. In all these cases, the flow at the rear of the cylinder is separated and unsteady, and usually turbulent. However, the flow in the front half of the cylinder is much more steady and predictable. In fact, except for a very thin boundary layer near the cylinder surface, the flow field may be approximated by the following steady, two-dimensional velocity components in the xy- or $r\theta$-plane:

$$u_r = V \cos\theta\left(1 - \frac{a^2}{r^2}\right) \quad u_\theta = -V \sin\theta\left(1 + \frac{a^2}{r^2}\right)$$

Is this flow field rotational or irrotational? Explain.

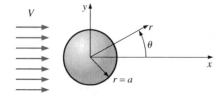

FIGURE P4–116

4–117 Consider the flow field of Prob. 4–116 (flow over a circular cylinder). Consider only the front half of the flow ($x < 0$). There is one stagnation point in the front half of the flow field. Where is it? Give your answer in both cylindrical (r, θ) coordinates and Cartesian (x, y) coordinates.

4–118 Consider the upstream half ($x < 0$) of the flow field of Prob. 4–116 (flow over a circular cylinder). We introduce a parameter called the **stream function** ψ, which is *constant along streamlines* in two-dimensional flows such as the one being considered here (Fig. P4–118). The velocity field of Prob. 4–116 corresponds to a stream function given by

$$\psi = V \sin\theta\left(r - \frac{a^2}{r}\right)$$

(*a*) Setting ψ to a constant, generate an equation for a streamline. (*Hint*: Use the quadratic rule to solve for r as a function of θ.)

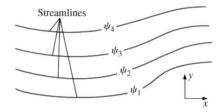

FIGURE P4–118

(*b*) For the particular case in which $V = 1.00$ m/s and cylinder radius $a = 10.0$ cm, plot several streamlines in the upstream half of the flow ($90° < \theta < 270°$). For consistency, plot in the range -0.4 m $< x < 0$ m, -0.2 m $< y < 0.2$ m, with stream function values evenly spaced between -0.16 m^2/s and 0.16 m^2/s.

4–119 Consider the flow field of Prob. 4–116 (flow over a circular cylinder). Calculate the two linear strain rates in the $r\theta$-plane; i.e., calculate ε_{rr} and $\varepsilon_{\theta\theta}$. Discuss whether fluid line segments stretch (or shrink) in this flow field. (*Hint*: The strain rate tensor in cylindrical coordinates is given in Prob. 4–110.)

4–120 Based on your results of Prob. 4–119, discuss the compressibility (or incompressibility) of this flow. *Answer:* flow is incompressible

4–121 Consider the flow field of Prob. 4–116 (flow over a circular cylinder). Calculate $\varepsilon_{r\theta}$, the shear strain rate in the $r\theta$-plane. Discuss whether fluid particles in this flow deform with shear or not. (*Hint*: The strain rate tensor in cylindrical coordinates is given in Prob. 4–110.)

Fundamentals of Engineering (FE) Exam Problems

4–122 A steady, incompressible, two-dimensional velocity field is given by

$$\vec{V} = (u, v) = (2.5 - 1.6x)\vec{i} + (0.7 + 1.6y)\vec{j}$$

where the x- and y-coordinates are in meters and the magnitude of velocity is in m/s. The values of x and y at the stagnation point, respectively, are

(*a*) 0.9375 m, 0.375 m   (*b*) 1.563 m, -0.4375 m
(*c*) 2.5 m, 0.7 m   (*d*) 0.731 m, 1.236 m   (*e*) -1.6 m, 0.8 m

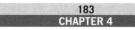

4–123 Water is flowing in a 3-cm-diameter garden hose at a rate of 30 L/min. A 20-cm nozzle is attached to the hose which decreases the diameter to 1.2 cm. The magnitude of the acceleration of a fluid particle moving down the center-line of the nozzle is
(*a*) 9.81 m/s^2 (*b*) 14.5 m/s^2 (*c*) 25.4 m/s^2 (*d*) 39.1 m/s^2
(*e*) 47.6 m/s^2

4–124 A steady, incompressible, two-dimensional velocity field is given by

$$\vec{V} = (u, v) = (2.5 - 1.6x)\vec{i} + (0.7 + 1.6y)\vec{j}$$

where the *x*- and *y*-coordinates are in meters and the magnitude of velocity is in m/s. The *x*-component of the acceleration vector a_x is
(*a*) 0.8*y* (*b*) $-1.6x$ (*c*) 2.5*x* − 1.6 (*d*) 2.56*x* − 4
(*e*) 2.56*x* + 0.8*y*

4–125 A steady, incompressible, two-dimensional velocity field is given by

$$\vec{V} = (u, v) = (2.5 - 1.6x)\vec{i} + (0.7 + 1.6y)\vec{j}$$

where the *x*- and *y*-coordinates are in meters and the magnitude of velocity is in m/s. The *x*- and *y*-component of material acceleration a_x and a_y at the point (*x* = 1 m, *y* = 1 m), respectively, in m/s^2, are
(*a*) −1.44, 3.68 (*b*) −1.6, 1.5 (*c*) 3.1, −1.32
(*d*) 2.56, −4 (*e*) −0.8, 1.6

4–126 A steady, incompressible, two-dimensional velocity field is given by

$$\vec{V} = (u, v) = (0.65 + 1.7x)\vec{i} + (1.3 - 1.7y)\vec{j}$$

where the *x*- and *y*-coordinates are in meters and the magnitude of velocity is in m/s. The *y*-component of the acceleration vector a_y is
(*a*) 1.7*y* (*b*) −1.7*y* (*c*) 2.89*y* − 2.21 (*d*) 3.0*x* − 2.73
(*e*) 0.84*y* + 1.42

4–127 A steady, incompressible, two-dimensional velocity field is given by

$$\vec{V} = (u, v) = (0.65 + 1.7x)\vec{i} + (1.3 - 1.7y)\vec{j}$$

where the *x*- and *y*-coordinates are in meters and the magnitude of velocity is in m/s. The *x*- and *y*-component of material acceleration a_x and a_y at the point (*x* = 0 m, *y* = 0 m), respectively, in m/s^2, are
(*a*) 0.37, −1.85 (*b*) −1.7, 1.7 (*c*) 1.105, −2.21
(*d*) 1.7, −1.7 (*e*) 0.65, 1.3

4–128 A steady, incompressible, two-dimensional velocity field is given by

$$\vec{V} = (u, v) = (0.65 + 1.7x)\vec{i} + (1.3 - 1.7y)\vec{j}$$

where the *x*- and *y*-coordinates are in meters and the magnitude of velocity is in m/s. The *x*- and *y*-component of velocity *u* and *v* at the point (*x* = 1 m, *y* = 2 m), respectively, in m/s, are
(*a*) 0.54, −2.31 (*b*) −1.9, 0.75 (*c*) 0.598, −2.21
(*d*) 2.35, −2.1 (*e*) 0.65, 1.3

4–129 The actual path traveled by an individual fluid particle over some period is called a
(*a*) Pathline (*b*) Streamtube (*c*) Streamline
(*d*) Streakline (*e*) Timeline

4–130 The locus of fluid particles that have passed sequentially through a prescribed point in the flow is called a
(*a*) Pathline (*b*) Streamtube (*c*) Streamline
(*d*) Streakline (*e*) Timeline

4–131 A curve that is everywhere tangent to the instantaneous local velocity vector is called a
(*a*) Pathline (*b*) Streamtube (*c*) Streamline
(*d*) Streakline (*e*) Timeline

4–132 An array of arrows indicating the magnitude and direction of a vector property at an instant in time is called a
(*a*) Profiler plot (*b*) Vector plot (*c*) Contour plot
(*d*) Velocity plot (*e*) Time plot

4–133 The CFD stands for
(*a*) Compressible fluid dynamics
(*b*) Compressed flow domain
(*c*) Circular flow dynamics
(*d*) Convective fluid dynamics
(*e*) Computational fluid dynamics

4–134 Which one is not a fundamental type of motion or deformation an element may undergo in fluid mechanics?
(*a*) Rotation (*b*) Converging (*c*) Translation
(*d*) Linear strain (*e*) Shear strain

4–135 A steady, incompressible, two-dimensional velocity field is given by

$$\vec{V} = (u, v) = (2.5 - 1.6x)\vec{i} + (0.7 + 1.6y)\vec{j}$$

where the *x*- and *y*-coordinates are in meters and the magnitude of velocity is in m/s. The linear strain rate in the *x*-direction in s^{-1} is
(*a*) −1.6 (*b*) 0.8 (*c*) 1.6 (*d*) 2.5 (*e*) −0.875

4-136 A steady, incompressible, two-dimensional velocity field is given by

$$\vec{V} = (u, v) = (2.5 - 1.6x)\vec{i} + (0.7 + 1.6y)\vec{j}$$

where the *x*- and *y*-coordinates are in meters and the magnitude of velocity is in m/s. The shear strain rate in s^{-1} is
(*a*) −1.6 (*b*) 1.6 (*c*) 2.5 (*d*) 0.7 (*e*) 0

4–137 A steady, two-dimensional velocity field is given by

$$\vec{V} = (u, v) = (2.5 - 1.6x)\vec{i} + (0.7 + 0.8y)\vec{j}$$

where the *x*- and *y*-coordinates are in meters and the magnitude of velocity is in m/s. The volumetric strain rate in s^{-1} is
(*a*) 0 (*b*) 3.2 (*c*) −0.8 (*d*) 0.8 (*e*) −1.6

4–138 If the vorticity in a region of the flow is zero, the flow is
(*a*) Motionless (*b*) Incompressible (*c*) Compressible
(*d*) Irrotational (*e*) Rotational

4–139 The angular velocity of a fluid particle is 20 rad/s. The vorticity of this fluid particle is
(*a*) 20 rad/s   (*b*) 40 rad/s   (*c*) 80 rad/s   (*d*) 10 rad/s
(*e*) 5 rad/s

4–140 A steady, incompressible, two-dimensional velocity field is given by

$$\vec{V} = (u, v) = (0.75 + 1.2x)\vec{i} + (2.25 - 1.2y)\vec{j}$$

where the *x*- and *y*-coordinates are in meters and the magnitude of velocity is in m/s. The vorticity of this flow is
(*a*) 0   (*b*) $1.2y\vec{k}$ (*c*) $-1.2y\vec{k}$ (*d*) $y\vec{k}$ (*e*) $-1.2xy\vec{k}$

4–141 A steady, incompressible, two-dimensional velocity field is given by

$$\vec{V} = (u, v) = (2xy + 1)\vec{i} + (-y^2 - 0.6)\vec{j}$$

where the *x*- and *y*-coordinates are in meters and the magnitude of velocity is in m/s. The angular velocity of this flow is
(*a*) 0   (*b*) $-2y\vec{k}$ (*c*) $2y\vec{k}$ (*d*) $-2x\vec{k}$ (*e*) $-x\vec{k}$

4–142 A cart is moving at a constant absolute velocity $\vec{V}_{cart} = 5$ km/h to the right. A high-speed jet of water at an absolute velocity of $\vec{V}_{jet} = 15$ km/h to the right strikes the back of the car. The relative velocity of the water is
(*a*) 0 km/h   (*b*) 5 km/h   (*c*) 10 km/h   (*d*) 15 km/h   (*e*) 20 km/h

BERNOULLI AND ENERGY EQUATIONS

This chapter deals with three equations commonly used in fluid mechanics: the mass, Bernoulli, and energy equations. The *mass equation* is an expression of the conservation of mass principle. The *Bernoulli equation* is concerned with the conservation of kinetic, potential, and flow energies of a fluid stream and their conversion to each other in regions of flow where net viscous forces are negligible and where other restrictive conditions apply. The *energy equation* is a statement of the conservation of energy principle. In fluid mechanics, it is convenient to separate *mechanical energy* from *thermal energy* and to consider the conversion of mechanical energy to thermal energy as a result of frictional effects as *mechanical energy loss*. Then the energy equation becomes the *mechanical energy balance*.

We start this chapter with an overview of conservation principles and the conservation of mass relation. This is followed by a discussion of various forms of mechanical energy and the efficiency of mechanical work devices such as pumps and turbines. Then we derive the Bernoulli equation by applying Newton's second law to a fluid element along a streamline and demonstrate its use in a variety of applications. We continue with the development of the energy equation in a form suitable for use in fluid mechanics and introduce the concept of *head loss*. Finally, we apply the energy equation to various engineering systems.

OBJECTIVES

When you finish reading this chapter, you should be able to

- Apply the conservation of mass equation to balance the incoming and outgoing flow rates in a flow system

- Recognize various forms of mechanical energy, and work with energy conversion efficiencies

- Understand the use and limitations of the Bernoulli equation, and apply it to solve a variety of fluid flow problems

- Work with the energy equation expressed in terms of heads, and use it to determine turbine power output and pumping power requirements

Wind turbine "farms" are being constructed all over the world to extract kinetic energy from the wind and convert it to electrical energy. The mass, energy, momentum, and angular momentum balances are utilized in the design of a wind turbine. The Bernoulli equation is also useful in the preliminary design stage.

© J. Luke/PhotoLink/Getty RF

FIGURE 5–1

Many fluid flow devices such as this Pelton wheel hydraulic turbine are analyzed by applying the conservation of mass and energy principles, along with the linear momentum equation.

Courtesy of Hydro Tasmania, www.hydro.com.au. Used by permission.

5–1 · INTRODUCTION

You are already familiar with numerous **conservation laws** such as the laws of conservation of mass, conservation of energy, and conservation of momentum. Historically, the conservation laws are first applied to a fixed quantity of matter called a *closed system* or just a *system*, and then extended to regions in space called *control volumes*. The conservation relations are also called *balance equations* since any conserved quantity must balance during a process. We now give a brief description of the conservation of mass and energy relations, and the linear momentum equation (Fig. 5–1).

Conservation of Mass

The conservation of mass relation for a closed system undergoing a change is expressed as $m_{\text{sys}} = $ constant or $dm_{\text{sys}}/dt = 0$, which is the statement that the mass of the system remains constant during a process. For a control volume (CV), mass balance is expressed in rate form as

Conservation of mass: $$\dot{m}_{\text{in}} - \dot{m}_{\text{out}} = \frac{dm_{\text{CV}}}{dt} \tag{5–1}$$

where $\dot{m}_{\text{in}}$ and $\dot{m}_{\text{out}}$ are the total rates of mass flow into and out of the control volume, respectively, and dm_{CV}/dt is the rate of change of mass within the control volume boundaries. In fluid mechanics, the conservation of mass relation written for a differential control volume is usually called the *continuity equation*. Conservation of mass is discussed in Section 5–2.

The Linear Momentum Equation

The product of the mass and the velocity of a body is called the *linear momentum* or just the *momentum* of the body, and the momentum of a rigid body of mass m moving with a velocity $\vec{V}$ is $m\vec{V}$. Newton's second law states that the acceleration of a body is proportional to the net force acting on it and is inversely proportional to its mass, and that the rate of change of the momentum of a body is equal to the net force acting on the body. Therefore, the momentum of a system remains constant only when the net force acting on it is zero, and thus the momentum of such systems is conserved. This is known as the *conservation of momentum principle*. In fluid mechanics, Newton's second law is usually referred to as the *linear momentum equation*, which is discussed in Chap. 6 together with the *angular momentum equation*.

Conservation of Energy

Energy can be transferred to or from a closed system by heat or work, and the conservation of energy principle requires that the net energy transfer to or from a system during a process be equal to the change in the energy content of the system. Control volumes involve energy transfer via mass flow also, and the *conservation of energy principle,* also called the *energy balance,* is expressed as

Conservation of energy: $$\dot{E}_{\text{in}} - \dot{E}_{\text{out}} = \frac{dE_{\text{CV}}}{dt} \tag{5–2}$$

where $\dot{E}_{\text{in}}$ and $\dot{E}_{\text{out}}$ are the total rates of energy transfer into and out of the control volume, respectively, and dE_{CV}/dt is the rate of change of energy within the control volume boundaries. In fluid mechanics, we usually limit our consideration to mechanical forms of energy only. Conservation of energy is discussed in Section 5–6.

5–2 ▪ CONSERVATION OF MASS

The conservation of mass principle is one of the most fundamental principles in nature. We are all familiar with this principle, and it is not difficult to understand. A person does not have to be a rocket scientist to figure out how much vinegar-and-oil dressing will be obtained by mixing 100 g of oil with 25 g of vinegar. Even chemical equations are balanced on the basis of the conservation of mass principle. When 16 kg of oxygen reacts with 2 kg of hydrogen, 18 kg of water is formed (Fig. 5–2). In an electrolysis process, the water separates back to 2 kg of hydrogen and 16 kg of oxygen.

Technically, mass is not exactly conserved. It turns out that mass m and energy E can be converted to each other according to the well-known formula proposed by Albert Einstein (1879–1955):

$$E = mc^2 \qquad (5–3)$$

where c is the speed of light in a vacuum, which is $c = 2.9979 \times 10^8$ m/s. This equation suggests that there is equivalence between mass and energy. All physical and chemical systems exhibit energy interactions with their surroundings, but the amount of energy involved is equivalent to an extremely small mass compared to the system's total mass. For example, when 1 kg of liquid water is formed from oxygen and hydrogen at normal atmospheric conditions, the amount of energy released is 15.8 MJ, which corresponds to a mass of only 1.76×10^{-10} kg. However, in nuclear reactions, the mass equivalence of the amount of energy interacted is a significant fraction of the total mass involved. Therefore, in most engineering analyses, we consider both mass and energy as conserved quantities.

For *closed systems*, the conservation of mass principle is implicitly used by requiring that the mass of the system remain constant during a process. For *control volumes*, however, mass can cross the boundaries, and so we must keep track of the amount of mass entering and leaving the control volume.

Mass and Volume Flow Rates

The amount of mass flowing through a cross section per unit time is called the **mass flow rate** and is denoted by $\dot{m}$. The dot over a symbol is used to indicate *time rate of change*.

A fluid flows into or out of a control volume, usually through pipes or ducts. The differential mass flow rate of fluid flowing across a small area element dA_c in a cross section of a pipe is proportional to dA_c itself, the fluid density ρ, and the component of the flow velocity normal to dA_c, which we denote as V_n, and is expressed as (Fig. 5–3)

$$\delta\dot{m} = \rho V_n \, dA_c \qquad (5–4)$$

Note that both δ and d are used to indicate differential quantities, but δ is typically used for quantities (such as heat, work, and mass transfer) that are *path functions* and have *inexact differentials*, while d is used for quantities

FIGURE 5–2

Mass is conserved even during chemical reactions.

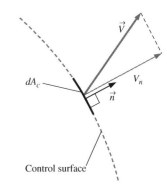

FIGURE 5–3

The normal velocity V_n for a surface is the component of velocity perpendicular to the surface.

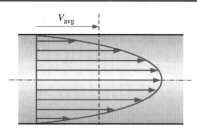

FIGURE 5–4
Average velocity V_{avg} is defined as the average speed through a cross section.

Cross section

FIGURE 5–5
The volume flow rate is the volume of fluid flowing through a cross section per unit time.

(such as properties) that are *point functions* and have *exact differentials*. For flow through an annulus of inner radius r_1 and outer radius r_2, for example,

$$\int_1^2 dA_c = A_{c2} - A_{c1} = \pi(r_2^2 - r_1^2) \text{ but } \int_1^2 \delta\dot{m} = \dot{m}_{total} \text{ (total mass flow rate}$$

through the annulus), not $\dot{m}_2 - \dot{m}_1$. For specified values of r_1 and r_2, the value of the integral of dA_c is fixed (thus the names point function and exact differential), but this is not the case for the integral of $\delta\dot{m}$ (thus the names path function and inexact differential).

The mass flow rate through the entire cross-sectional area of a pipe or duct is obtained by integration:

$$\dot{m} = \int_{A_c} \delta\dot{m} = \int_{A_c} \rho V_n \, dA_c \quad \text{(kg/s)} \tag{5–5}$$

While Eq. 5–5 is always valid (in fact it is *exact*), it is not always practical for engineering analyses because of the integral. We would like instead to express mass flow rate in terms of average values over a cross section of the pipe. In a general compressible flow, both ρ and V_n vary across the pipe. In many practical applications, however, the density is essentially uniform over the pipe cross section, and we can take ρ outside the integral of Eq. 5–5. Velocity, however, is *never* uniform over a cross section of a pipe because of the no-slip condition at the walls. Rather, the velocity varies from zero at the walls to some maximum value at or near the centerline of the pipe. We define the **average velocity** V_{avg} as the average value of V_n across the entire cross section of the pipe (Fig. 5–4),

Average velocity: $$V_{avg} = \frac{1}{A_c} \int_{A_c} V_n \, dA_c \tag{5–6}$$

where A_c is the area of the cross section normal to the flow direction. Note that if the speed were V_{avg} all through the cross section, the mass flow rate would be identical to that obtained by integrating the actual velocity profile. Thus for incompressible flow or even for compressible flow where ρ is approximated as uniform across A_c, Eq. 5–5 becomes

$$\dot{m} = \rho V_{avg} A_c \quad \text{(kg/s)} \tag{5–7}$$

For compressible flow, we can think of ρ as the bulk average density over the cross section, and then Eq. 5–7 can be used as a reasonable approximation. For simplicity, we drop the subscript on the average velocity. Unless otherwise stated, V denotes the average velocity in the flow direction. Also, A_c denotes the cross-sectional area normal to the flow direction.

The volume of the fluid flowing through a cross section per unit time is called the **volume flow rate** $\dot{V}$ (Fig. 5–5) and is given by

$$\dot{V} = \int_{A_c} V_n \, dA_c = V_{avg} A_c = V A_c \quad \text{(m}^3\text{/s)} \tag{5–8}$$

An early form of Eq. 5–8 was published in 1628 by the Italian monk Benedetto Castelli (circa 1577–1644). Note that many fluid mechanics textbooks use Q instead of $\dot{V}$ for volume flow rate. We use $\dot{V}$ to avoid confusion with heat transfer.

The mass and volume flow rates are related by

$$\dot{m} = \rho\dot{V} = \frac{\dot{V}}{v} \tag{5–9}$$

where v is the specific volume. This relation is analogous to $m = \rho V = V/v$, which is the relation between the mass and the volume of a fluid in a container.

Conservation of Mass Principle

The **conservation of mass principle** for a control volume can be expressed as: *The net mass transfer to or from a control volume during a time interval Δt is equal to the net change (increase or decrease) of the total mass within the control volume during Δt.* That is,

$$\begin{pmatrix} \text{Total mass entering} \\ \text{the CV during } \Delta t \end{pmatrix} - \begin{pmatrix} \text{Total mass leaving} \\ \text{the CV during } \Delta t \end{pmatrix} = \begin{pmatrix} \text{Net change of mass} \\ \text{within the CV during } \Delta t \end{pmatrix}$$

or

$$m_{\text{in}} - m_{\text{out}} = \Delta m_{\text{CV}} \quad (\text{kg}) \tag{5-10}$$

where $\Delta m_{\text{CV}} = m_{\text{final}} - m_{\text{initial}}$ is the change in the mass of the control volume during the process (Fig. 5–6). It can also be expressed in *rate form* as

$$\dot{m}_{\text{in}} - \dot{m}_{\text{out}} = dm_{\text{CV}}/dt \quad (\text{kg/s}) \tag{5-11}$$

where $\dot{m}_{\text{in}}$ and $\dot{m}_{\text{out}}$ are the total rates of mass flow into and out of the control volume, and dm_{CV}/dt is the rate of change of mass within the control volume boundaries. Equations 5–10 and 5–11 are often referred to as the **mass balance** and are applicable to any control volume undergoing any kind of process.

Consider a control volume of arbitrary shape, as shown in Fig. 5–7. The mass of a differential volume dV within the control volume is $dm = \rho \, dV$. The total mass within the control volume at any instant in time t is determined by integration to be

Total mass within the CV: $\qquad m_{\text{CV}} = \displaystyle\int_{\text{CV}} \rho \, dV \tag{5-12}$

Then the time rate of change of the amount of mass within the control volume is expressed as

Rate of change of mass within the CV: $\qquad \dfrac{dm_{\text{CV}}}{dt} = \dfrac{d}{dt}\displaystyle\int_{\text{CV}} \rho \, dV \tag{5-13}$

For the special case of no mass crossing the control surface (i.e., the control volume is a closed system), the conservation of mass principle reduces to $dm_{\text{CV}}/dt = 0$. This relation is valid whether the control volume is fixed, moving, or deforming.

Now consider mass flow into or out of the control volume through a differential area dA on the control surface of a fixed control volume. Let $\vec{n}$ be the outward unit vector of dA normal to dA and $\vec{V}$ be the flow velocity at dA relative to a fixed coordinate system, as shown in Fig. 5–7. In general, the velocity may cross dA at an angle θ off the normal of dA, and the mass flow rate is proportional to the normal component of velocity $\vec{V}_n = \vec{V} \cos \theta$ ranging from a maximum outflow of $\vec{V}$ for $\theta = 0$ (flow is normal to dA) to a minimum of zero for $\theta = 90°$ (flow is tangent to dA) to a maximum *inflow* of $\vec{V}$ for $\theta = 180°$ (flow is normal to dA but in the opposite direction).

FIGURE 5–6
Conservation of mass principle for an ordinary bathtub.

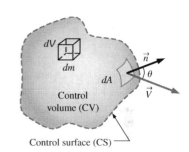

FIGURE 5–7
The differential control volume dV and the differential control surface dA used in the derivation of the conservation of mass relation.

Making use of the concept of dot product of two vectors, the magnitude of the normal component of velocity is

Normal component of velocity: $V_n = V \cos \theta = \vec{V} \cdot \vec{n}$ (5–14)

The mass flow rate through dA is proportional to the fluid density ρ, normal velocity V_n, and the flow area dA, and is expressed as

Differential mass flow rate: $\delta \dot{m} = \rho V_n \, dA = \rho(V \cos \theta) \, dA = \rho(\vec{V} \cdot \vec{n}) \, dA$ (5–15)

The net flow rate into or out of the control volume through the entire control surface is obtained by integrating $\delta \dot{m}$ over the entire control surface,

Net mass flow rate: $\dot{m}_{\text{net}} = \int_{\text{CS}} \delta \dot{m} = \int_{\text{CS}} \rho V_n \, dA = \int_{\text{CS}} \rho(\vec{V} \cdot \vec{n}) \, dA$ (5–16)

Note that $V_n = \vec{V} \cdot \vec{n} = V \cos \theta$ is positive for $\theta < 90°$ (outflow) and negative for $\theta > 90°$ (inflow). Therefore, the direction of flow is automatically accounted for, and the surface integral in Eq. 5–16 directly gives the *net* mass flow rate. A positive value for $\dot{m}_{\text{net}}$ indicates a net outflow of mass and a negative value indicates a net inflow of mass.

Rearranging Eq. 5–11 as $dm_{\text{CV}}/dt + \dot{m}_{\text{out}} - \dot{m}_{\text{in}} = 0$, the conservation of mass relation for a fixed control volume is then expressed as

General conservation of mass: $\dfrac{d}{dt} \int_{\text{CV}} \rho \, dV + \int_{\text{CS}} \rho(\vec{V} \cdot \vec{n}) \, dA = 0$ (5–17)

It states that *the time rate of change of mass within the control volume plus the net mass flow rate through the control surface is equal to zero.*

The general conservation of mass relation for a control volume can also be derived using the Reynolds transport theorem (RTT) by taking the property B to be the mass m (Chap. 4). Then we have $b = 1$ since dividing mass by mass to get the property per unit mass gives unity. Also, the mass of a closed system is constant, and thus its time derivative is zero. That is, $dm_{\text{sys}}/dt = 0$. Then the Reynolds transport equation reduces immediately to Eq. 5–17, as shown in Fig. 5–8, and thus illustrates that the Reynolds transport theorem is a very powerful tool indeed.

Splitting the surface integral in Eq. 5–17 into two parts—one for the outgoing flow streams (positive) and one for the incoming flow streams (negative)—the general conservation of mass relation can also be expressed as

$$\dfrac{d}{dt} \int_{\text{CV}} \rho \, dV + \sum_{\text{out}} \rho |V_n| A - \sum_{\text{in}} \rho |V_n| A = 0 \qquad (5\text{–}18)$$

where A represents the area for an inlet or outlet, and the summation signs are used to emphasize that *all* the inlets and outlets are to be considered. Using the definition of mass flow rate, Eq. 5–18 can also be expressed as

$$\dfrac{d}{dt} \int_{\text{CV}} \rho \, dV = \sum_{\text{in}} \dot{m} - \sum_{\text{out}} \dot{m} \quad \text{or} \quad \dfrac{dm_{\text{CV}}}{dt} = \sum_{\text{in}} \dot{m} - \sum_{\text{out}} \dot{m} \qquad (5\text{–}19)$$

There is considerable flexibility in the selection of a control volume when solving a problem. Many control volume choices are available, but some are more convenient to work with. A control volume should not introduce any unnecessary complications. A wise choice of a control volume can make the solution of a seemingly complicated problem rather easy. A simple rule in selecting a control volume is to make the control surface *normal to the flow*

FIGURE 5–8

The conservation of mass equation is obtained by replacing B in the Reynolds transport theorem by mass m, and b by 1 (m per unit mass $= m/m = 1$).

at all locations where it crosses the fluid flow, whenever possible. This way the dot product $\vec{V} \cdot \vec{n}$ simply becomes the magnitude of the velocity, and the integral $\int_A \rho(\vec{V} \cdot \vec{n}) \, dA$ becomes simply ρVA (Fig. 5–9).

Moving or Deforming Control Volumes

Equations 5–17 and 5–19 are also valid for moving control volumes provided that the *absolute velocity* $\vec{V}$ is replaced by the *relative velocity* $\vec{V}_r$, which is the fluid velocity relative to the control surface (Chap. 4). In the case of a moving but nondeforming control volume, relative velocity is the fluid velocity observed by a person moving with the control volume and is expressed as $\vec{V}_r = \vec{V} - \vec{V}_{CS}$, where $\vec{V}$ is the fluid velocity and $\vec{V}_{CS}$ is the velocity of the control surface, both relative to a fixed point outside. Note that this is a *vector* subtraction.

Some practical problems (such as the injection of medication through the needle of a syringe by the forced motion of the plunger) involve *deforming* control volumes. The conservation of mass relations developed can still be used for such deforming control volumes provided that the velocity of the fluid crossing a deforming part of the control surface is expressed relative to the control surface (that is, the fluid velocity should be expressed relative to a reference frame attached to the deforming part of the control surface). The relative velocity in this case at any point on the control surface is expressed again as $\vec{V}_r = \vec{V} - \vec{V}_{CS}$, where $\vec{V}_{CS}$ is the local velocity of the control surface at that point relative to a fixed point outside the control volume.

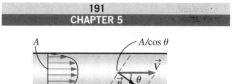

$V_n = V \cos \theta$

$\dot{m} = \rho(V \cos \theta)(A/\cos \theta) = \rho VA$

(a) Control surface *at an angle* to the flow

$\dot{m} = \rho VA$

(b) Control surface *normal* to the flow

FIGURE 5–9
A control surface should always be selected *normal to the flow* at all locations where it crosses the fluid flow to avoid complications, even though the result is the same.

Mass Balance for Steady-Flow Processes

During a steady-flow process, the total amount of mass contained within a control volume does not change with time (m_{CV} = constant). Then the conservation of mass principle requires that the total amount of mass entering a control volume equal the total amount of mass leaving it. For a garden hose nozzle in steady operation, for example, the amount of water entering the nozzle per unit time is equal to the amount of water leaving it per unit time.

When dealing with steady-flow processes, we are not interested in the amount of mass that flows in or out of a device over time; instead, we are interested in the amount of mass flowing per unit time, that is, *the mass flow rate* $\dot{m}$. *The conservation of mass principle* for a general steady-flow system with multiple inlets and outlets is expressed in rate form as (Fig. 5–10)

$\dot{m}_1 = 2$ kg/s $\dot{m}_2 = 3$ kg/s

CV

$\dot{m}_3 = \dot{m}_1 + \dot{m}_2 = 5$ kg/s

FIGURE 5–10
Conservation of mass principle for a two-inlet–one-outlet steady-flow system.

Steady flow: $$\sum_{\text{in}} \dot{m} = \sum_{\text{out}} \dot{m} \quad \text{(kg/s)} \tag{5–20}$$

It states that *the total rate of mass entering a control volume is equal to the total rate of mass leaving it.*

Many engineering devices such as nozzles, diffusers, turbines, compressors, and pumps involve a single stream (only one inlet and one outlet). For these cases, we typically denote the inlet state by the subscript 1 and the outlet state by the subscript 2, and drop the summation signs. Then Eq. 5–20 reduces, for *single-stream steady-flow systems*, to

Steady flow (single stream): $$\dot{m}_1 = \dot{m}_2 \quad \rightarrow \quad \rho_1 V_1 A_1 = \rho_2 V_2 A_2 \tag{5–21}$$

$\dot{m}_2 = 2$ kg/s
$\dot{V}_2 = 0.8$ m³/s

Air compressor

$\dot{m}_1 = 2$ kg/s
$\dot{V}_1 = 1.4$ m³/s

FIGURE 5–11
During a steady-flow process, volume flow rates are not necessarily conserved although mass flow rates are.

Special Case: Incompressible Flow

The conservation of mass relations can be simplified even further when the fluid is incompressible, which is usually the case for liquids. Canceling the density from both sides of the general steady-flow relation gives

Steady, incompressible flow: $\sum_{in} \dot{V} = \sum_{out} \dot{V}$ (m³/s)   (5–22)

For single-stream steady-flow systems Eq. 5–22 becomes

Steady, incompressible flow (single stream): $\dot{V}_1 = \dot{V}_2 \rightarrow V_1 A_1 = V_2 A_2$ (5–23)

It should always be kept in mind that there is no such thing as a "conservation of volume" principle. Therefore, the volume flow rates into and out of a steady-flow device may be different. The volume flow rate at the outlet of an air compressor is much less than that at the inlet even though the mass flow rate of air through the compressor is constant (Fig. 5–11). This is due to the higher density of air at the compressor exit. For steady flow of liquids, however, the volume flow rates remain nearly constant since liquids are essentially incompressible (constant-density) substances. Water flow through the nozzle of a garden hose is an example of the latter case.

The conservation of mass principle requires every bit of mass to be accounted for during a process. If you can balance your checkbook (by keeping track of deposits and withdrawals, or by simply observing the "conservation of money" principle), you should have no difficulty applying the conservation of mass principle to engineering systems.

FIGURE 5–12
Schematic for Example 5–1.
Photo by John M. Cimbala.

EXAMPLE 5–1   **Water Flow through a Garden Hose Nozzle**

A garden hose attached with a nozzle is used to fill a 10-gal bucket. The inner diameter of the hose is 2 cm, and it reduces to 0.8 cm at the nozzle exit (Fig. 5–12). If it takes 50 s to fill the bucket with water, determine (*a*) the volume and mass flow rates of water through the hose, and (*b*) the average velocity of water at the nozzle exit.

SOLUTION A garden hose is used to fill a water bucket. The volume and mass flow rates of water and the exit velocity are to be determined.
Assumptions **1** Water is a nearly incompressible substance. **2** Flow through the hose is steady. **3** There is no waste of water by splashing.
Properties We take the density of water to be 1000 kg/m³ = 1 kg/L.
Analysis (*a*) Noting that 10 gal of water are discharged in 50 s, the volume and mass flow rates of water are

$$\dot{V} = \frac{V}{\Delta t} = \frac{10 \text{ gal}}{50 \text{ s}} \left(\frac{3.7854 \text{ L}}{1 \text{ gal}} \right) = 0.757 \text{ L/s}$$

$$\dot{m} = \rho \dot{V} = (1 \text{ kg/L})(0.757 \text{ L/s}) = 0.757 \text{ kg/s}$$

(*b*) The cross-sectional area of the nozzle exit is

$$A_e = \pi r_e^2 = \pi (0.4 \text{ cm})^2 = 0.5027 \text{ cm}^2 = 0.5027 \times 10^{-4} \text{ m}^2$$

The volume flow rate through the hose and the nozzle is constant. Then the average velocity of water at the nozzle exit becomes

$$V_e = \frac{\dot{V}}{A_e} = \frac{0.757 \text{ L/s}}{0.5027 \times 10^{-4} \text{ m}^2} \left(\frac{1 \text{ m}^3}{1000 \text{ L}} \right) = 15.1 \text{ m/s}$$

Discussion It can be shown that the average velocity in the hose is 2.4 m/s. Therefore, the nozzle increases the water velocity by over six times.

EXAMPLE 5–2   Discharge of Water from a Tank

A 4-ft-high, 3-ft-diameter cylindrical water tank whose top is open to the atmosphere is initially filled with water. Now the discharge plug near the bottom of the tank is pulled out, and a water jet whose diameter is 0.5 in streams out (Fig. 5–13). The average velocity of the jet is approximated as $V = \sqrt{2gh}$, where h is the height of water in the tank measured from the center of the hole (a variable) and g is the gravitational acceleration. Determine how long it takes for the water level in the tank to drop to 2 ft from the bottom.

SOLUTION The plug near the bottom of a water tank is pulled out. The time it takes for half of the water in the tank to empty is to be determined. *Assumptions* **1** Water is a nearly incompressible substance. **2** The distance between the bottom of the tank and the center of the hole is negligible compared to the total water height. **3** The gravitational acceleration is 32.2 ft/s². *Analysis* We take the volume occupied by water as the control volume. The size of the control volume decreases in this case as the water level drops, and thus this is a variable control volume. (We could also treat this as a fixed control volume that consists of the interior volume of the tank by disregarding the air that replaces the space vacated by the water.) This is obviously an unsteady-flow problem since the properties (such as the amount of mass) within the control volume change with time.

The conservation of mass relation for a control volume undergoing any process is given in rate form as

$$\dot{m}_{\text{in}} - \dot{m}_{\text{out}} = \frac{dm_{\text{CV}}}{dt} \tag{1}$$

During this process no mass enters the control volume ($\dot{m}_{\text{in}} = 0$), and the mass flow rate of discharged water is

$$\dot{m}_{\text{out}} = (\rho V A)_{\text{out}} = \rho \sqrt{2gh} A_{\text{jet}} \tag{2}$$

where $A_{\text{jet}} = \pi D_{\text{jet}}^2 / 4$ is the cross-sectional area of the jet, which is constant. Noting that the density of water is constant, the mass of water in the tank at any time is

$$m_{\text{CV}} = \rho V = \rho A_{\text{tank}} h \tag{3}$$

where $A_{\text{tank}} = \pi D_{\text{tank}}^2 / 4$ is the base area of the cylindrical tank. Substituting Eqs. 2 and 3 into the mass balance relation (Eq. 1) gives

$$-\rho \sqrt{2gh} A_{\text{jet}} = \frac{d(\rho A_{\text{tank}} h)}{dt} \rightarrow -\rho \sqrt{2gh} (\pi D_{\text{jet}}^2/4) = \frac{\rho (\pi D_{\text{tank}}^2/4) dh}{dt}$$

FIGURE 5–13
Schematic for Example 5–2.

Canceling the densities and other common terms and separating the variables give

$$dt = -\frac{D_{tank}^2}{D_{jet}^2}\frac{dh}{\sqrt{2gh}}$$

Integrating from $t = 0$ at which $h = h_0$ to $t = t$ at which $h = h_2$ gives

$$\int_0^t dt = -\frac{D_{tank}^2}{D_{jet}^2\sqrt{2g}}\int_{h_0}^{h_2}\frac{dh}{\sqrt{h}} \rightarrow t = \frac{\sqrt{h_0}-\sqrt{h_2}}{\sqrt{g/2}}\left(\frac{D_{tank}}{D_{jet}}\right)^2$$

Substituting, the time of discharge is determined to be

$$t = \frac{\sqrt{4\text{ ft}}-\sqrt{2\text{ ft}}}{\sqrt{32.2/2\text{ ft/s}^2}}\left(\frac{3\times12\text{ in}}{0.5\text{ in}}\right)^2 = 757\text{ s} = \mathbf{12.6\text{ min}}$$

Therefore, it takes 12.6 min after the discharge hole is unplugged for half of the tank to be emptied.

Discussion   Using the same relation with $h_2 = 0$ gives $t = 43.1$ min for the discharge of the entire amount of water in the tank. Therefore, emptying the bottom half of the tank takes much longer than emptying the top half. This is due to the decrease in the average discharge velocity of water with decreasing h.

5–3 · MECHANICAL ENERGY AND EFFICIENCY

Many fluid systems are designed to transport a fluid from one location to another at a specified flow rate, velocity, and elevation difference, and the system may generate mechanical work in a turbine or it may consume mechanical work in a pump or fan during this process (Fig. 5–14). These systems do not involve the conversion of nuclear, chemical, or thermal energy to mechanical energy. Also, they do not involve heat transfer in any significant amount, and they operate essentially at constant temperature. Such systems can be analyzed conveniently by considering only the *mechanical forms of energy* and the frictional effects that cause the mechanical energy to be lost (i.e., to be converted to thermal energy that usually cannot be used for any useful purpose).

The **mechanical energy** is defined as *the form of energy that can be converted to mechanical work completely and directly by an ideal mechanical device such as an ideal turbine.* Kinetic and potential energies are the familiar forms of mechanical energy. Thermal energy is not mechanical energy, however, since it cannot be converted to work directly and completely (the second law of thermodynamics).

A pump transfers mechanical energy to a fluid by raising its pressure, and a turbine extracts mechanical energy from a fluid by dropping its pressure. Therefore, the pressure of a flowing fluid is also associated with its mechanical energy. In fact, the pressure unit Pa is equivalent to Pa = N/m² = N·m/m³ = J/m³, which is energy per unit volume, and the product Pv or its equivalent P/ρ has the unit J/kg, which is energy per unit mass. Note that pressure itself is not a form of energy. But a pressure force acting on a fluid through a distance produces work, called *flow work*, in the amount of P/ρ per unit mass. Flow work is expressed in terms of fluid properties, and it is convenient to view it as part of the energy of a flowing fluid and call

FIGURE 5–14

Mechanical energy is a useful concept for flows that do not involve significant heat transfer or energy conversion, such as the flow of gasoline from an underground tank into a car.

Royalty-Free/CORBIS

it *flow energy*. Therefore, the mechanical energy of a flowing fluid can be expressed on a unit-mass basis as

$$e_{\text{mech}} = \frac{P}{\rho} + \frac{V^2}{2} + gz$$

where P/ρ is the *flow energy*, $V^2/2$ is the *kinetic energy*, and gz is the *potential energy* of the fluid, all per unit mass. Then the mechanical energy change of a fluid during incompressible flow becomes

$$\Delta e_{\text{mech}} = \frac{P_2 - P_1}{\rho} + \frac{V_2^2 - V_1^2}{2} + g(z_2 - z_1) \quad \text{(kJ/kg)} \qquad \textbf{(5–24)}$$

Therefore, the mechanical energy of a fluid does not change during flow if its pressure, density, velocity, and elevation remain constant. In the absence of any irreversible losses, the mechanical energy change represents the mechanical work supplied to the fluid (if $\Delta e_{\text{mech}} > 0$) or extracted from the fluid (if $\Delta e_{\text{mech}} < 0$). The maximum (ideal) power generated by a turbine, for example, is $\dot{W}_{\text{max}} = \dot{m}\Delta e_{\text{mech}}$, as shown in Fig. 5–15.

Consider a container of height h filled with water, as shown in Fig. 5–16, with the reference level selected at the bottom surface. The gage pressure and the potential energy per unit mass are, respectively, $P_{\text{gage},\, A} = 0$ and $\text{pe}_A = gh$ at point A at the free surface, and $P_{\text{gage},\, B} = \rho gh$ and $\text{pe}_B = 0$ at point B at the bottom of the container. An ideal hydraulic turbine at the bottom elevation would produce the same work per unit mass $w_{\text{turbine}} = gh$ whether it receives water (or any other fluid with constant density) from the top or from the bottom of the container. Note that we are assuming ideal flow (no irreversible losses) through the pipe leading from the tank to the turbine and negligible kinetic energy at the turbine outlet. Therefore, the total available mechanical energy of water at the bottom is equivalent to that at the top.

The transfer of mechanical energy is usually accomplished by a rotating shaft, and thus mechanical work is often referred to as *shaft work*. A pump or a fan receives shaft work (usually from an electric motor) and transfers it to the fluid as mechanical energy (less frictional losses). A turbine, on the other hand, converts the mechanical energy of a fluid to shaft work. Because of irreversibilities such as friction, mechanical energy cannot be converted entirely from one mechanical form to another, and the **mechanical efficiency** of a device or process is defined as

$$\eta_{\text{mech}} = \frac{\text{Mechanical energy output}}{\text{Mechanical energy input}} = \frac{E_{\text{mech, out}}}{E_{\text{mech, in}}} = 1 - \frac{E_{\text{mech, loss}}}{E_{\text{mech, in}}} \qquad \textbf{(5–25)}$$

A conversion efficiency of less than 100 percent indicates that conversion is less than perfect and some losses have occurred during conversion. A mechanical efficiency of 74 percent indicates that 26 percent of the mechanical energy input is converted to thermal energy as a result of frictional heating (Fig 5–17), and this manifests itself as a slight rise in the temperature of the fluid.

In fluid systems, we are usually interested in increasing the pressure, velocity, and/or elevation of a fluid. This is done by *supplying mechanical energy* to the fluid by a pump, a fan, or a compressor (we refer to all of them as pumps). Or we are interested in the reverse process of *extracting mechanical*

$$\dot{W}_{\text{max}} = \dot{m}\Delta e_{\text{mech}} = \dot{m}g(z_1 - z_4) = \dot{m}gh$$

since $P_1 \approx P_4 = P_{\text{atm}}$ and $V_1 = V_4 \approx 0$

(a)

$$\dot{W}_{\text{max}} = \dot{m}\Delta e_{\text{mech}} = \dot{m}\frac{P_2 - P_3}{\rho} = \dot{m}\frac{\Delta P}{\rho}$$

since $V_2 \approx V_3$ and $z_2 \approx z_3$

(b)

FIGURE 5–15

Mechanical energy is illustrated by an ideal hydraulic turbine coupled with an ideal generator. In the absence of irreversible losses, the maximum produced power is proportional to (*a*) the change in water surface elevation from the upstream to the downstream reservoir or (*b*) (close-up view) the drop in water pressure from just upstream to just downstream of the turbine.

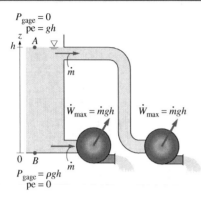

FIGURE 5–16
The available mechanical energy of water at the bottom of a container is equal to the available mechanical energy at any depth including the free surface of the container.

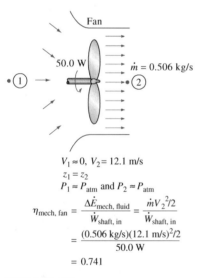

$V_1 \approx 0,\ V_2 = 12.1$ m/s
$z_1 = z_2$
$P_1 \approx P_{atm}$ and $P_2 \approx P_{atm}$

$$\eta_{mech,\ fan} = \frac{\Delta \dot{E}_{mech,\ fluid}}{\dot{W}_{shaft,\ in}} = \frac{\dot{m}V_2^2/2}{\dot{W}_{shaft,\ in}}$$

$$= \frac{(0.506\ \text{kg/s})(12.1\ \text{m/s})^2/2}{50.0\ \text{W}}$$

$$= 0.741$$

FIGURE 5–17
The mechanical efficiency of a fan is the ratio of the rate of increase of the mechanical energy of the air to the mechanical power input.

energy from a fluid by a turbine and producing mechanical power in the form of a rotating shaft that can drive a generator or any other rotary device. The degree of perfection of the conversion process between the mechanical work supplied or extracted and the mechanical energy of the fluid is expressed by the **pump efficiency** and **turbine efficiency.** In rate form, these are defined as

$$\eta_{pump} = \frac{\text{Mechanical power increase of the fluid}}{\text{Mechanical power input}} = \frac{\Delta \dot{E}_{mech,\ fluid}}{\dot{W}_{shaft,\ in}} = \frac{\dot{W}_{pump,\ u}}{\dot{W}_{pump}} \quad \text{(5–26)}$$

where $\Delta \dot{E}_{mech,\ fluid} = \dot{E}_{mech,\ out} - \dot{E}_{mech,\ in}$ is the rate of increase in the mechanical energy of the fluid, which is equivalent to the **useful pumping power** $\dot{W}_{pump,\ u}$ supplied to the fluid, and

$$\eta_{turbine} = \frac{\text{Mechanical power output}}{\text{Mechanical power decrease of the fluid}} = \frac{\dot{W}_{shaft,\ out}}{|\Delta \dot{E}_{mech,\ fluid}|} = \frac{\dot{W}_{turbine}}{\dot{W}_{turbine,e}} \quad \text{(5–27)}$$

where $|\Delta \dot{E}_{mech,\ fluid}| = \dot{E}_{mech,\ in} - \dot{E}_{mech,\ out}$ is the rate of decrease in the mechanical energy of the fluid, which is equivalent to the mechanical power extracted from the fluid by the turbine $\dot{W}_{turbine,\ e}$, and we use the absolute value sign to avoid negative values for efficiencies. A pump or turbine efficiency of 100 percent indicates perfect conversion between the shaft work and the mechanical energy of the fluid, and this value can be approached (but never attained) as the frictional effects are minimized.

The mechanical efficiency should not be confused with the **motor efficiency** and the **generator efficiency**, which are defined as

Motor:
$$\eta_{motor} = \frac{\text{Mechanical power output}}{\text{Electric power input}} = \frac{\dot{W}_{shaft,\ out}}{\dot{W}_{elect,\ in}} \quad \text{(5–28)}$$

and

Generator:
$$\eta_{generator} = \frac{\text{Electric power output}}{\text{Mechanical power input}} = \frac{\dot{W}_{elect,\ out}}{\dot{W}_{shaft,\ in}} \quad \text{(5–29)}$$

A pump is usually packaged together with its motor, and a turbine with its generator. Therefore, we are usually interested in the **combined** or **overall efficiency** of pump–motor and turbine–generator combinations (Fig. 5–18), which are defined as

$$\eta_{pump-motor} = \eta_{pump}\ \eta_{motor} = \frac{\dot{W}_{pump,\ u}}{\dot{W}_{elect,\ in}} = \frac{\Delta \dot{E}_{mech,\ fluid}}{\dot{W}_{elect,\ in}} \quad \text{(5–30)}$$

and

$$\eta_{turbine-gen} = \eta_{turbine}\ \eta_{generator} = \frac{\dot{W}_{elect,\ out}}{\dot{W}_{turbine,\ e}} = \frac{\dot{W}_{elect,\ out}}{|\Delta \dot{E}_{mech,\ fluid}|} \quad \text{(5–31)}$$

All the efficiencies just defined range between 0 and 100 percent. The lower limit of 0 percent corresponds to the conversion of the entire mechanical or electric energy input to thermal energy, and the device in this case functions like a resistance heater. The upper limit of 100 percent corresponds to the case of perfect conversion with no friction or other irreversibilities, and thus no conversion of mechanical or electric energy to thermal energy (no losses).

EXAMPLE 5–3   Performance of a Hydraulic Turbine–Generator

The water in a large lake is to be used to generate electricity by the installation of a hydraulic turbine–generator. The elevation difference between the
free surfaces upstream and downstream of the dam is 50 m (Fig. 5–19).
Water is to be supplied at a rate of 5000 kg/s. If the electric power generated is measured to be 1862 kW and the generator efficiency is 95 percent,
determine (a) the overall efficiency of the turbine–generator, (b) the mechanical efficiency of the turbine, and (c) the shaft power supplied by the turbine
to the generator.

SOLUTION A hydraulic turbine–generator is to generate electricity from the
water of a lake. The overall efficiency, the turbine efficiency, and the shaft
power are to be determined.
Assumptions **1** The elevation of the lake and that of the discharge site
remain constant. **2** Irreversible losses in the pipes are negligible.
Properties The density of water is taken to be $\rho = 1000$ kg/m^3.
Analysis (a) We perform our analysis from inlet (1) at the free surface of
the lake to outlet (2) at the free surface of the downstream discharge site. At
both free surfaces the pressure is atmospheric and the velocity is negligibly
small. The change in the water's mechanical energy per unit mass is then

$$e_{\text{mech, in}} - e_{\text{mech, out}} = \underbrace{\frac{P_{\text{in}} - P_{\text{out}}}{\rho}}_{0} + \underbrace{\frac{V_{\text{in}}^2 - V_{\text{out}}^2}{2}}_{0} + g(z_{\text{in}} - z_{\text{out}})$$

$$= gh$$

$$= (9.81 \text{ m/s}^2)(50 \text{ m})\left(\frac{1 \text{ kJ/kg}}{1000 \text{ m}^2/\text{s}^2}\right) = 0.491 \frac{\text{kJ}}{\text{kg}}$$

Then the rate at which mechanical energy is supplied to the turbine by the
fluid and the overall efficiency become

$$\left|\Delta \dot{E}_{\text{mech, fluid}}\right| = \dot{m}(e_{\text{mech, in}} - e_{\text{mech, out}}) = (5000 \text{ kg/s})(0.491 \text{ kJ/kg}) = 2455 \text{ kW}$$

$$\eta_{\text{overall}} = \eta_{\text{turbine-gen}} = \frac{\dot{W}_{\text{elect, out}}}{\left|\Delta \dot{E}_{\text{mech, fluid}}\right|} = \frac{1862 \text{ kW}}{2455 \text{ kW}} = \mathbf{0.760}$$

(b) Knowing the overall and generator efficiencies, the mechanical efficiency
of the turbine is determined from

$$\eta_{\text{turbine-gen}} = \eta_{\text{turbine}}\,\eta_{\text{generator}} \rightarrow \eta_{\text{turbine}} = \frac{\eta_{\text{turbine-gen}}}{\eta_{\text{generator}}} = \frac{0.76}{0.95} = \mathbf{0.800}$$

(c) The shaft power output is determined from the definition of mechanical
efficiency,

$$\dot{W}_{\text{shaft, out}} = \eta_{\text{turbine}}\left|\Delta \dot{E}_{\text{mech, fluid}}\right| = (0.800)(2455 \text{ kW}) = 1964 \text{ kW} \approx \mathbf{1960 \text{ kW}}$$

Discussion Note that the lake supplies 2455 kW of mechanical power to
the turbine, which converts 1964 kW of it to shaft power that drives the
generator, which generates 1862 kW of electric power. There are irreversible
losses through each component. Irreversible losses in the pipes are ignored
here; you will learn how to account for these in Chap. 8.

$\eta_{\text{turbine}} = 0.75$ $\eta_{\text{generator}} = 0.97$

$\eta_{\text{turbine-gen}} = \eta_{\text{turbine}}\eta_{\text{generator}}$
$= 0.75 \times 0.97$
$= 0.73$

FIGURE 5–18
The overall efficiency of a turbine–
generator is the product of the
efficiency of the turbine and the
efficiency of the generator, and
represents the fraction of the
mechanical power of the fluid
converted to electrical power.

FIGURE 5–19
Schematic for Example 5–3.

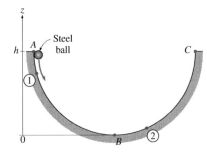

FIGURE 5–20
Schematic for Example 5–4.

EXAMPLE 5–4   **Conservation of Energy for an Oscillating Steel Ball**

The motion of a steel ball in a hemispherical bowl of radius h shown in Fig. 5–20 is to be analyzed. The ball is initially held at the highest location at point A, and then it is released. Obtain relations for the conservation of energy of the ball for the cases of frictionless and actual motions.

SOLUTION   A steel ball is released in a bowl. Relations for the energy balance are to be obtained.
Assumptions   For the frictionless case, friction between the ball, the bowl, and the air is negligible.
Analysis   When the ball is released, it accelerates under the influence of gravity, reaches a maximum velocity (and minimum elevation) at point B at the bottom of the bowl, and moves up toward point C on the opposite side. In the ideal case of frictionless motion, the ball will oscillate between points A and C. The actual motion involves the conversion of the kinetic and potential energies of the ball to each other, together with overcoming resistance to motion due to friction (doing frictional work). The general energy balance for any system undergoing any process is

$$\underbrace{E_{in} - E_{out}}_{\substack{\text{Net energy transfer} \\ \text{by heat, work, and mass}}} = \underbrace{\Delta E_{system}}_{\substack{\text{Change in internal, kinetic,} \\ \text{potential, etc., energies}}}$$

Then the energy balance (per unit mass) for the ball for a process from point 1 to point 2 becomes

$$-w_{friction} = (ke_2 + pe_2) - (ke_1 + pe_1)$$

or

$$\frac{V_1^2}{2} + gz_1 = \frac{V_2^2}{2} + gz_2 + w_{friction}$$

since there is no energy transfer by heat or mass and no change in the internal energy of the ball (the heat generated by frictional heating is dissipated to the surrounding air). The frictional work term $w_{friction}$ is often expressed as e_{loss} to represent the loss (conversion) of mechanical energy into thermal energy.
   For the idealized case of frictionless motion, the last relation reduces to

$$\frac{V_1^2}{2} + gz_1 = \frac{V_2^2}{2} + gz_2 \quad \text{or} \quad \frac{V^2}{2} + gz = C = \text{constant}$$

where the value of the constant is $C = gh$. That is, *when the frictional effects are negligible, the sum of the kinetic and potential energies of the ball remains constant.*
Discussion   This is certainly a more intuitive and convenient form of the conservation of energy equation for this and other similar processes such as the swinging motion of a pendulum. The relation obtained is analogous to the Bernoulli equation derived in Section 5–4.

Most processes encountered in practice involve only certain forms of energy, and in such cases it is more convenient to work with the simplified versions of the energy balance. For systems that involve only *mechanical forms of energy* and its transfer as *shaft work*, the conservation of energy principle can be expressed conveniently as

$$E_{mech, in} - E_{mech, out} = \Delta E_{mech, system} + E_{mech, loss} \qquad \text{(5–32)}$$

where $E_{mech, loss}$ represents the conversion of mechanical energy to thermal energy due to irreversibilities such as friction. For a system in steady operation, the rate of mechanical energy balance becomes $\dot{E}_{mech, in} = \dot{E}_{mech, out} + \dot{E}_{mech, loss}$ (Fig. 5–21).

5–4 · THE BERNOULLI EQUATION

The **Bernoulli equation** is *an approximate relation between pressure, velocity, and elevation,* and is valid in *regions of steady, incompressible flow where net frictional forces are negligible* (Fig. 5–22). Despite its simplicity, it has proven to be a very powerful tool in fluid mechanics. In this section, we derive the Bernoulli equation by applying the *conservation of linear momentum principle,* and we demonstrate both its usefulness and its limitations.

The key approximation in the derivation of the Bernoulli equation is that *viscous effects are negligibly small compared to inertial, gravitational, and pressure effects.* Since all fluids have viscosity (there is no such thing as an "inviscid fluid"), this approximation cannot be valid for an entire flow field of practical interest. In other words, we cannot apply the Bernoulli equation *everywhere* in a flow, no matter how small the fluid's viscosity. However, it turns out that the approximation *is* reasonable in certain *regions* of many practical flows. We refer to such regions as *inviscid regions of flow,* and we stress that they are *not* regions where the fluid itself is inviscid or frictionless, but rather they are regions where net viscous or frictional forces are negligibly small compared to other forces acting on fluid particles.

Care must be exercised when applying the Bernoulli equation since it is an approximation that applies only to inviscid regions of flow. In general, frictional effects are always important very close to solid walls (*boundary layers*) and directly downstream of bodies (*wakes*). Thus, the Bernoulli approximation is typically useful in flow regions outside of boundary layers and wakes, where the fluid motion is governed by the combined effects of pressure and gravity forces.

Acceleration of a Fluid Particle

The motion of a particle and the path it follows are described by the *velocity vector* as a function of time and space coordinates and the initial position of the particle. When the flow is *steady* (no change with time at a specified location), all particles that pass through the same point follow the same path (which is the *streamline*), and the velocity vectors remain tangent to the path at every point.

Often it is convenient to describe the motion of a particle in terms of its distance s along a streamline together with the radius of curvature along the streamline. The speed of the particle is related to the distance by $V = ds/dt$, which may vary along the streamline. In two-dimensional flow, the acceleration can be decomposed into two components: *streamwise acceleration* a_s along the streamline and *normal acceleration* a_n in the direction normal to the streamline, which is given as $a_n = V^2/R$. Note that streamwise

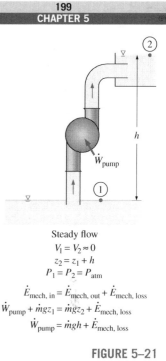

Steady flow
$V_1 = V_2 \approx 0$
$z_2 = z_1 + h$
$P_1 = P_2 = P_{atm}$

$\dot{E}_{mech, in} = \dot{E}_{mech, out} + \dot{E}_{mech, loss}$
$\dot{W}_{pump} + \dot{m}gz_1 = \dot{m}gz_2 + \dot{E}_{mech, loss}$
$\dot{W}_{pump} = \dot{m}gh + \dot{E}_{mech, loss}$

FIGURE 5–21
Many fluid flow problems involve mechanical forms of energy only, and such problems are conveniently solved by using a rate of *mechanical energy* balance.

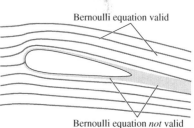

Bernoulli equation valid

Bernoulli equation *not* valid

FIGURE 5–22
The *Bernoulli equation* is an *approximate* equation that is valid only in *inviscid regions of flow* where net viscous forces are negligibly small compared to inertial, gravitational, or pressure forces. Such regions occur outside of *boundary layers* and *wakes.*

FIGURE 5–23
During steady flow, a fluid may not accelerate in time at a fixed point, but it may accelerate in space.

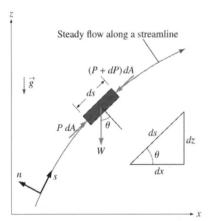

FIGURE 5–24
The forces acting on a fluid particle along a streamline.

acceleration is due to a change in speed along a streamline, and normal acceleration is due to a change in direction. For particles that move along a *straight path,* $a_n = 0$ since the radius of curvature is infinity and thus there is no change in direction. The Bernoulli equation results from a force balance along a streamline.

One may be tempted to think that acceleration is zero in steady flow since acceleration is the rate of change of velocity with time, and in steady flow there is no change with time. Well, a garden hose nozzle tells us that this understanding is not correct. Even in steady flow and thus constant mass flow rate, water accelerates through the nozzle (Fig. 5–23 as discussed in Chap. 4). *Steady* simply means *no change with time at a specified location,* but the value of a quantity may change from one location to another. In the case of a nozzle, the velocity of water remains constant at a specified point, but it changes from the inlet to the exit (water accelerates along the nozzle).

Mathematically, this can be expressed as follows: We take the velocity V of a fluid particle to be a function of s and t. Taking the total differential of $V(s, t)$ and dividing both sides by dt yield

$$dV = \frac{\partial V}{\partial s} ds + \frac{\partial V}{\partial t} dt \quad \text{and} \quad \frac{dV}{dt} = \frac{\partial V}{\partial s}\frac{ds}{dt} + \frac{\partial V}{\partial t} \tag{5–33}$$

In steady flow $\partial V/\partial t = 0$ and thus $V = V(s)$, and the acceleration in the s-direction becomes

$$a_s = \frac{dV}{dt} = \frac{\partial V}{\partial s}\frac{ds}{dt} = \frac{\partial V}{\partial s} V = V\frac{dV}{ds} \tag{5–34}$$

where $V = ds/dt$ if we are following a fluid particle as it moves along a streamline. Therefore, acceleration in steady flow is due to the change of velocity with position.

Derivation of the Bernoulli Equation

Consider the motion of a fluid particle in a flow field in steady flow. Applying Newton's second law (which is referred to as the *linear momentum equation* in fluid mechanics) in the s-direction on a particle moving along a streamline gives

$$\sum F_s = ma_s \tag{5–35}$$

In regions of flow where net frictional forces are negligible, there is no pump or turbine, and there is no heat transfer along the streamline, the significant forces acting in the s-direction are the pressure (acting on both sides) and the component of the weight of the particle in the s-direction (Fig. 5–24). Therefore, Eq. 5–35 becomes

$$P \, dA - (P + dP) \, dA - W \sin \theta = mV\frac{dV}{ds} \tag{5–36}$$

where θ is the angle between the normal of the streamline and the vertical z-axis at that point, $m = \rho V = \rho \, dA \, ds$ is the mass, $W = mg = \rho g \, dA \, ds$ is the weight of the fluid particle, and $\sin \theta = dz/ds$. Substituting,

$$-dP \, dA - \rho g \, dA \, ds \frac{dz}{ds} = \rho \, dA \, ds \, V\frac{dV}{ds} \tag{5–37}$$

Canceling dA from each term and simplifying,

$$-dP - \rho g \, dz = \rho V \, dV \qquad (5\text{–}38)$$

Noting that $V \, dV = \frac{1}{2} d(V^2)$ and dividing each term by ρ gives

$$\frac{dP}{\rho} + \frac{1}{2} d(V^2) + g \, dz = 0 \qquad (5\text{–}39)$$

Integrating,

Steady flow: $\displaystyle \int \frac{dP}{\rho} + \frac{V^2}{2} + gz = \text{constant (along a streamline)}$ (5–40)

since the last two terms are exact differentials. In the case of incompressible flow, the first term also becomes an exact differential, and integration gives

Steady, incompressible flow: $\displaystyle \frac{P}{\rho} + \frac{V^2}{2} + gz = \text{constant (along a streamline)}$ (5–41)

This is the famous **Bernoulli equation** (Fig. 5–25), which is commonly used in fluid mechanics for steady, incompressible flow along a streamline in inviscid regions of flow. The Bernoulli equation was first stated in words by the Swiss mathematician Daniel Bernoulli (1700–1782) in a text written in 1738 when he was working in St. Petersburg, Russia. It was later derived in equation form by his associate Leonhard Euler (1707–1783) in 1755.

The value of the constant in Eq. 5–41 can be evaluated at any point on the streamline where the pressure, density, velocity, and elevation are known. The Bernoulli equation can also be written between any two points on the same streamline as

Steady, incompressible flow: $\displaystyle \frac{P_1}{\rho} + \frac{V_1^2}{2} + gz_1 = \frac{P_2}{\rho} + \frac{V_2^2}{2} + gz_2$ (5–42)

We recognize $V^2/2$ as *kinetic energy,* gz as *potential energy,* and P/ρ as *flow energy,* all per unit mass. Therefore, the Bernoulli equation can be viewed as an expression of *mechanical energy balance* and can be stated as follows (Fig. 5–26):

> The sum of the kinetic, potential, and flow energies of a fluid particle is constant along a streamline during steady flow when compressibility and frictional effects are negligible.

The kinetic, potential, and flow energies are the mechanical forms of energy, as discussed in Section 5–3, and the Bernoulli equation can be viewed as the "conservation of mechanical energy principle." This is equivalent to the general conservation of energy principle for systems that do not involve any conversion of mechanical energy and thermal energy to each other, and thus the mechanical energy and thermal energy are conserved separately. The Bernoulli equation states that during steady, incompressible flow with negligible friction, the various forms of mechanical energy are converted to each other, but their sum remains constant. In other words, there is no dissipation of mechanical energy during such flows since there is no friction that converts mechanical energy to sensible thermal (internal) energy.

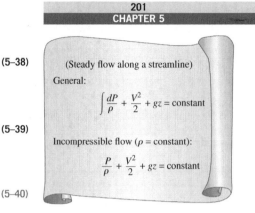

FIGURE 5–25
The incompressible Bernoulli equation is derived assuming incompressible flow, and thus it should not be used for flows with significant compressibility effects.

FIGURE 5–26
The Bernoulli equation states that the sum of the kinetic, potential, and flow energies (all per unit mass) of a fluid particle is constant along a streamline during steady flow.

Recall that energy is transferred to a system as work when a force is applied to the system through a distance. In the light of Newton's second law of motion, the Bernoulli equation can also be viewed as: *The work done by the pressure and gravity forces on the fluid particle is equal to the increase in the kinetic energy of the particle.*

The Bernoulli equation is obtained from Newton's second law for a fluid particle moving along a streamline. It can also be obtained from the *first law of thermodynamics* applied to a steady-flow system, as shown in Section 5–6.

Despite the highly restrictive approximations used in its derivation, the Bernoulli equation is commonly used in practice since a variety of practical fluid flow problems can be analyzed to reasonable accuracy with it. This is because many flows of practical engineering interest are steady (or at least steady in the mean), compressibility effects are relatively small, and net frictional forces are negligible in some regions of interest in the flow.

Force Balance across Streamlines

It is left as an exercise to show that a force balance in the direction n normal to the streamline yields the following relation applicable *across* the streamlines for steady, incompressible flow:

$$\frac{P}{\rho} + \int \frac{V^2}{R} \, dn + gz = \text{constant} \quad \text{(across streamlines)} \quad (5\text{–}43)$$

where R is the local radius of curvature of the streamline. For flow along curved streamlines (Fig 5–27a), the pressure *decreases* towards the center of curvature, and fluid particles experience a corresponding centripetal force and centripetal acceleration due to this pressure gradient.

For flow along a straight line, $R \to \infty$ and Eq. 5–43 reduces to $P/\rho + gz = $ constant or $P = -\rho g z + $ constant, which is an expression for the variation of hydrostatic pressure with vertical distance for a stationary fluid body. Therefore, the variation of pressure with elevation in steady, incompressible flow along a straight line in an inviscid region of flow is the same as that in the stationary fluid (Fig. 5–27b).

FIGURE 5–27
Pressure decreases towards the center of curvature when streamlines are curved (*a*), but the variation of pressure with elevation in steady, incompressible flow along a straight line (*b*) is the same as that in stationary fluid.

Unsteady, Compressible Flow

Similarly, using both terms in the acceleration expression (Eq. 5–33), it can be shown that the Bernoulli equation for *unsteady, compressible flow* is

$$\textit{Unsteady, compressible flow:} \quad \int \frac{dP}{\rho} + \int \frac{\partial V}{\partial t} \, ds + \frac{V^2}{2} + gz = \text{constant} \quad (5\text{–}44)$$

Static, Dynamic, and Stagnation Pressures

The Bernoulli equation states that the sum of the flow, kinetic, and potential energies of a fluid particle along a streamline is constant. Therefore, the kinetic and potential energies of the fluid can be converted to flow energy (and vice versa) during flow, causing the pressure to change. This phenomenon can be made more visible by multiplying the Bernoulli equation by the density ρ,

$$P + \rho \frac{V^2}{2} + \rho g z = \text{constant (along a streamline)} \quad (5\text{–}45)$$

Each term in this equation has pressure units, and thus each term represents some kind of pressure:

- P is the **static pressure** (it does not incorporate any dynamic effects); it represents the actual thermodynamic pressure of the fluid. This is the same as the pressure used in thermodynamics and property tables.
- $\rho V^2/2$ is the **dynamic pressure**; it represents the pressure rise when the fluid in motion is brought to a stop isentropically.
- ρgz is the **hydrostatic pressure** term, which is not pressure in a real sense since its value depends on the reference level selected; it accounts for the elevation effects, i.e., fluid weight on pressure. (Be careful of the sign—unlike hydrostatic pressure ρgh which *increases* with fluid depth h, the hydrostatic pressure term ρgz *decreases* with fluid depth.)

The sum of the static, dynamic, and hydrostatic pressures is called the **total pressure**. Therefore, the Bernoulli equation states that *the total pressure along a streamline is constant.*

The sum of the static and dynamic pressures is called the **stagnation pressure**, and it is expressed as

$$P_{\text{stag}} = P + \rho \frac{V^2}{2} \quad \text{(kPa)} \tag{5-46}$$

The stagnation pressure represents the pressure at a point where the fluid is brought to a complete stop isentropically. The static, dynamic, and stagnation pressures are shown in Fig. 5–28. When static and stagnation pressures are measured at a specified location, the fluid velocity at that location is calculated from

$$V = \sqrt{\frac{2(P_{\text{stag}} - P)}{\rho}} \tag{5-47}$$

Equation 5–47 is useful in the measurement of flow velocity when a combination of a static pressure tap and a Pitot tube is used, as illustrated in Fig. 5–28. A **static pressure tap** is simply a small hole drilled into a wall such that the plane of the hole is parallel to the flow direction. It measures the static pressure. A **Pitot tube** is a small tube with its open end aligned *into* the flow so as to sense the full impact pressure of the flowing fluid. It measures the stagnation pressure. In situations in which the static and stagnation pressure of a flowing *liquid* are greater than atmospheric pressure, a vertical transparent tube called a **piezometer tube** (or simply a **piezometer**) can be attached to the pressure tap and to the Pitot tube, as sketched in Fig. 5–28. The liquid rises in the piezometer tube to a column height (*head*) that is proportional to the pressure being measured. If the pressures to be measured are below atmospheric, or if measuring pressures in *gases*, piezometer tubes do not work. However, the static pressure tap and Pitot tube can still be used, but they must be connected to some other kind of pressure measurement device such as a U-tube manometer or a pressure transducer (Chap. 3). Sometimes it is convenient to integrate static pressure holes on a Pitot probe. The result is a **Pitot-static probe** (also called a **Pitot-Darcy probe**), as shown in Fig. 5–29 and discussed in more detail in Chap. 8. A Pitot-static probe connected to a pressure transducer or a manometer measures the dynamic pressure (and thus infers the fluid velocity) directly.

When the static pressure is measured by drilling a hole in the tube wall, care must be exercised to ensure that the opening of the hole is flush with the wall surface, with no extrusions before or after the hole (Fig. 5–30). Otherwise the reading would incorporate some dynamic effects, and thus it would be in error.

FIGURE 5–28
The static, dynamic, and stagnation pressures measured using piezometer tubes.

FIGURE 5–29
Close-up of a Pitot-static probe, showing the stagnation pressure hole and two of the five static circumferential pressure holes.
Photo by Po-Ya Abel Chuang. Used by permission.

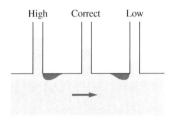

FIGURE 5–30
Careless drilling of the static pressure tap may result in an erroneous reading of the static pressure head.

FIGURE 5–31
Streaklines produced by colored fluid introduced upstream of an airfoil; since the flow is steady, the streaklines are the same as streamlines and pathlines. The stagnation streamline is marked.

Courtesy ONERA. Photograph by Werlé.

When a stationary body is immersed in a flowing stream, the fluid is brought to a stop at the nose of the body (the **stagnation point**). The flow streamline that extends from far upstream to the stagnation point is called the **stagnation streamline** (Fig. 5–31). For a two-dimensional flow in the *xy*-plane, the stagnation point is actually a *line* parallel to the *z*-axis, and the stagnation streamline is actually a *surface* that separates fluid that flows *over* the body from fluid that flows *under* the body. In an incompressible flow, the fluid decelerates nearly isentropically from its free-stream velocity to zero at the stagnation point, and the pressure at the stagnation point is thus the stagnation pressure.

Limitations on the Use of the Bernoulli Equation

The Bernoulli equation (Eq. 5–41) is one of the most frequently used and *misused* equations in fluid mechanics. Its versatility, simplicity, and ease of use make it a very valuable tool for use in analysis, but the same attributes also make it very tempting to misuse. Therefore, it is important to understand the restrictions on its applicability and observe the limitations on its use, as explained here:

1. **Steady flow**   The first limitation on the Bernoulli equation is that it is applicable to *steady flow*. Therefore, it should not be used during the transient start-up and shut-down periods, or during periods of change in the flow conditions. Note that there is an unsteady form of the Bernoulli equation (Eq. 5–44), discussion of which is beyond the scope of the present text (see Panton, 2005).

2. **Negligible viscous effects**   Every flow involves some friction, no matter how small, and *frictional effects* may or may not be negligible. The situation is complicated even more by the amount of error that can be tolerated. In general, frictional effects are negligible for short flow sections with large cross sections, especially at low flow velocities. Frictional effects are usually significant in long and narrow flow passages, in the wake region downstream of an object, and in *diverging flow sections* such as diffusers because of the increased possibility of the fluid separating from the walls in such geometries. Frictional effects are also significant near solid surfaces, and thus the Bernoulli equation is usually applicable along a streamline in the core region of the flow, but not along a streamline close to the surface (Fig. 5–32).
   A component that disturbs the streamlined structure of flow and thus causes considerable mixing and backflow such as a sharp entrance of a tube or a partially closed valve in a flow section can make the Bernoulli equation inapplicable.

3. **No shaft work**   The Bernoulli equation was derived from a force balance on a particle moving along a streamline. Therefore, the Bernoulli equation is not applicable in a flow section that involves a pump, turbine, fan, or any other machine or impeller since such devices disrupt the streamlines and carry out energy interactions with the fluid particles. When the flow section considered involves any of these devices, the energy equation should be used instead to account for the shaft work input or output. However, the Bernoulli equation can still be applied to a flow section prior to or past a machine (assuming, of course, that the other

restrictions on its use are satisfied). In such cases, the Bernoulli constant changes from upstream to downstream of the device.

4. **Incompressible flow** One of the approximations used in the derivation of the Bernoulli equation is that ρ = constant and thus the flow is incompressible. This condition is satisfied by liquids and also by gases at Mach numbers less than about 0.3 since compressibility effects and thus density variations of gases are negligible at such relatively low velocities. Note that there is a compressible form of the Bernoulli equation (Eqs. 5–40 and 5–44).

5. **Negligible heat transfer** The density of a gas is inversely proportional to temperature, and thus the Bernoulli equation should not be used for flow sections that involve significant temperature change such as heating or cooling sections.

6. **Flow along a streamline** Strictly speaking, the Bernoulli equation $P/\rho + V^2/2 + gz = C$ is applicable along a streamline, and the value of the constant C is generally different for different streamlines. However, when a region of the flow is *irrotational* and there is no *vorticity* in the flow field, the value of the constant C remains the same for all streamlines, and the Bernoulli equation becomes applicable *across* streamlines as well (Fig. 5–33). Therefore, we do not need to be concerned about the streamlines when the flow is irrotational, and we can apply the Bernoulli equation between any two points in the irrotational region of the flow (Chap. 10).

We derived the Bernoulli equation by considering two-dimensional flow in the *xz*-plane for simplicity, but the equation is valid for general three-dimensional flow as well, as long as it is applied along the same streamline. We should always keep in mind the approximations used in the derivation of the Bernoulli equation and make sure that they are valid before applying it.

Hydraulic Grade Line (HGL) and Energy Grade Line (EGL)

It is often convenient to represent the level of mechanical energy graphically using *heights* to facilitate visualization of the various terms of the Bernoulli equation. This is done by dividing each term of the Bernoulli equation by g to give

$$\frac{P}{\rho g} + \frac{V^2}{2g} + z = H = \text{constant} \quad \text{(along a streamline)} \quad (5\text{–}48)$$

Each term in this equation has the dimension of length and represents some kind of "head" of a flowing fluid as follows:

- $P/\rho g$ is the **pressure head**; it represents the height of a fluid column that produces the static pressure P.
- $V^2/2g$ is the **velocity head**; it represents the elevation needed for a fluid to reach the velocity V during frictionless free fall.
- z is the **elevation head**; it represents the potential energy of the fluid.

Also, H is the **total head** for the flow. Therefore, the Bernoulli equation is expressed in terms of heads as: *The sum of the pressure, velocity, and elevation heads along a streamline is constant during steady flow when compressibility and frictional effects are negligible* (Fig. 5–34).

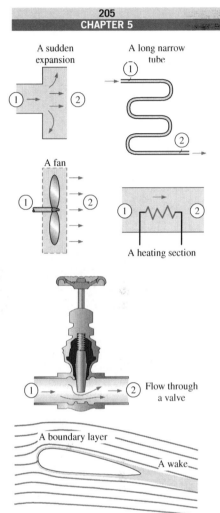

A sudden expansion

A long narrow tube

A fan

A heating section

Flow through a valve

A boundary layer

A wake

FIGURE 5–32
Frictional effects, heat transfer, and components that disturb the streamlined structure of flow make the Bernoulli equation invalid. It should *not* be used in any of the flows shown here.

Streamlines

$$\frac{P_1}{\rho} + \frac{V_1^2}{2} + gz_1 = \frac{P_2}{\rho} + \frac{V_2^2}{2} + gz_2$$

FIGURE 5–33

When the flow is irrotational, the
Bernoulli equation becomes applicable
between any two points along the flow
(not just on the same streamline).

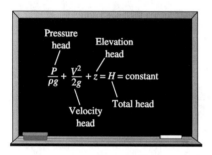

FIGURE 5–34

An alternative form of the Bernoulli
equation is expressed in terms of
heads as: *The sum of the pressure,
velocity, and elevation heads is
constant along a streamline.*

If a piezometer (which measures static pressure) is tapped into a pressurized pipe, as shown in Fig. 5–35, the liquid would rise to a height of $P/\rho g$
above the pipe center. The *hydraulic grade line* (HGL) is obtained by doing
this at several locations along the pipe and drawing a curve through the liquid levels in the piezometers. The vertical distance above the pipe center is
a measure of pressure within the pipe. Similarly, if a Pitot tube (measures
static + dynamic pressure) is tapped into a pipe, the liquid would rise to a
height of $P/\rho g + V^2/2g$ above the pipe center, or a distance of $V^2/2g$ above
the HGL. The *energy grade line* (EGL) is obtained by doing this at several
locations along the pipe and drawing a curve through the liquid levels in the
Pitot tubes.

Noting that the fluid also has elevation head z (unless the reference level is
taken to be the centerline of the pipe), the HGL and EGL are defined as follows: The line that represents the sum of the static pressure and the elevation
heads, $P/\rho g + z$, is called the **hydraulic grade line**. The line that represents
the total head of the fluid, $P/\rho g + V^2/2g + z$, is called the **energy grade
line**. The difference between the heights of EGL and HGL is equal to the
dynamic head, $V^2/2g$. We note the following about the HGL and EGL:

- For *stationary bodies* such as reservoirs or lakes, the EGL and HGL coincide with the free surface of the liquid. The elevation of the free surface z
  in such cases represents both the EGL and the HGL since the velocity is
  zero and the static (gage) pressure is zero.

- The EGL is always a distance $V^2/2g$ above the HGL. These two curves
  approach each other as the velocity decreases, and they diverge as the
  velocity increases. The height of the HGL decreases as the velocity increases, and vice versa.

- In an *idealized Bernoulli-type flow*, EGL is horizontal and its height
  remains constant. This would also be the case for HGL when the flow
  velocity is constant (Fig. 5–36).

- For *open-channel flow*, the HGL coincides with the free surface of the
  liquid, and the EGL is a distance $V^2/2g$ above the free surface.

- At a *pipe exit*, the pressure head is zero (atmospheric pressure) and thus
  the HGL coincides with the pipe outlet (location 3 on Fig. 5–35).

- The *mechanical energy loss* due to frictional effects (conversion to
  thermal energy) causes the EGL and HGL to slope downward in the
  direction of flow. The slope is a measure of the head loss in the pipe

FIGURE 5–35

The *hydraulic grade line* (HGL) and
the *energy grade line* (EGL) for free
discharge from a reservoir through a
horizontal pipe with a diffuser.

(discussed in detail in Chap. 8). A component that generates significant frictional effects such as a valve causes a sudden drop in both EGL and HGL at that location.

- A *steep jump* occurs in EGL and HGL whenever mechanical energy is added to the fluid (by a pump, for example). Likewise, a *steep drop* occurs in EGL and HGL whenever mechanical energy is removed from the fluid (by a turbine, for example), as shown in Fig. 5–37.

- The gage pressure of a fluid is zero at locations where the HGL *intersects* the fluid. The pressure in a flow section that lies above the HGL is negative, and the pressure in a section that lies below the HGL is positive (Fig. 5–38). Therefore, an accurate drawing of a piping system overlaid with the HGL can be used to determine the regions where the gage pressure in the pipe is negative (below atmospheric pressure).

The last remark enables us to avoid situations in which the pressure drops below the vapor pressure of the liquid (which may cause *cavitation*, as discussed in Chap. 2). Proper consideration is necessary in the placement of a liquid pump to ensure that the suction side pressure does not fall too low, especially at elevated temperatures where vapor pressure is higher than it is at low temperatures.

Now we examine Fig. 5–35 more closely. At point 0 (at the liquid surface), EGL and HGL are even with the liquid surface since there is no flow there. HGL decreases rapidly as the liquid accelerates into the pipe; however, EGL decreases very slowly through the well-rounded pipe inlet. EGL declines continually along the flow direction due to friction and other irreversible losses in the flow. EGL cannot increase in the flow direction unless energy is supplied to the fluid. HGL can rise or fall in the flow direction, but can never exceed EGL. HGL rises in the diffuser section as the velocity decreases, and the static pressure recovers somewhat; the total pressure does *not* recover, however, and EGL decreases through the diffuser. The difference between EGL and HGL is $V_1^2/2g$ at point 1, and $V_2^2/2g$ at point 2. Since $V_1 > V_2$, the difference between the two grade lines is larger at point 1 than at point 2. The downward slope of both grade lines is larger for the smaller diameter section of pipe since the frictional head loss is greater. Finally, HGL decays to the liquid surface at the outlet since the pressure there is atmospheric. However, EGL is still higher than HGL by the amount $V_2^2/2g$ since $V_3 = V_2$ at the outlet.

Applications of the Bernoulli Equation

So far, we have discussed the fundamental aspects of the Bernoulli equation. Now, we demonstrate its use in a wide range of applications through examples.

EXAMPLE 5–5   Spraying Water into the Air

Water is flowing from a garden hose (Fig. 5–39). A child places his thumb to cover most of the hose outlet, causing a thin jet of high-speed water to emerge. The pressure in the hose just upstream of his thumb is 400 kPa. If the hose is held upward, what is the maximum height that the jet could achieve?

SOLUTION   Water from a hose attached to the water main is sprayed into the air. The maximum height the water jet can rise is to be determined.

FIGURE 5–36
In an idealized Bernoulli-type flow, EGL is horizontal and its height remains constant. But this is not the case for HGL when the flow velocity varies along the flow.

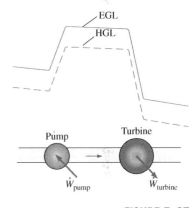

FIGURE 5–37
A *steep jump* occurs in EGL and HGL whenever mechanical energy is added to the fluid by a pump, and a *steep drop* occurs whenever mechanical energy is removed from the fluid by a turbine.

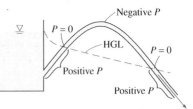

FIGURE 5–38
The gage pressure of a fluid is zero at locations where the HGL *intersects* the fluid, and the gage pressure is negative (vacuum) in a flow section that lies above the HGL.

FIGURE 5–39
Schematic for Example 5–5. Inset shows a magnified view of the hose outlet region.

Assumptions **1** The flow exiting into the air is steady, incompressible, and irrotational (so that the Bernoulli equation is applicable). **2** The surface tension effects are negligible. **3** The friction between the water and air is negligible. **4** The irreversibilities that occur at the outlet of the hose due to abrupt contraction are not taken into account.

Properties We take the density of water to be 1000 kg/m³.

Analysis This problem involves the conversion of flow, kinetic, and potential energies to each other without involving any pumps, turbines, and wasteful components with large frictional losses, and thus it is suitable for the use of the Bernoulli equation. The water height will be maximum under the stated assumptions. The velocity inside the hose is relatively low ($V_1^2 << V_j^2$, and thus $V_1 \cong 0$ compared to V_j) and we take the elevation just below the hose outlet as the reference level ($z_1 = 0$). At the top of the water trajectory $V_2 = 0$, and atmospheric pressure pertains. Then the Bernoulli equation along a streamline from 1 to 2 simplifies to

$$\frac{P_1}{\rho g} + \frac{\overset{\approx 0}{V_1^2}}{2g} + \overset{0}{z_1} = \frac{P_2}{\rho g} + \frac{\overset{0}{V_2^2}}{2g} + z_2 \quad \rightarrow \quad \frac{P_1}{\rho g} = \frac{P_{atm}}{\rho g} + z_2$$

Solving for z_2 and substituting,

$$z_2 = \frac{P_1 - P_{atm}}{\rho g} = \frac{P_{1,\,gage}}{\rho g} = \frac{400 \text{ kPa}}{(1000 \text{ kg/m}^3)(9.81 \text{ m/s}^2)}\left(\frac{1000 \text{ N/m}^2}{1 \text{ kPa}}\right)\left(\frac{1 \text{ kg·m/s}^2}{1 \text{ N}}\right)$$

$$= \mathbf{40.8 \text{ m}}$$

Therefore, the water jet can rise as high as 40.8 m into the sky in this case.

Discussion The result obtained by the Bernoulli equation represents the upper limit and should be interpreted accordingly. It tells us that the water cannot possibly rise more than 40.8 m, and, in all likelihood, the rise will be much less than 40.8 m due to irreversible losses that we neglected.

FIGURE 5–40
Schematic for Example 5–6.

EXAMPLE 5–6   Water Discharge from a Large Tank

A large tank open to the atmosphere is filled with water to a height of 5 m from the outlet tap (Fig. 5–40). A tap near the bottom of the tank is now opened, and water flows out from the smooth and rounded outlet. Determine the maximum water velocity at the outlet.

SOLUTION A tap near the bottom of a tank is opened. The maximum exit velocity of water from the tank is to be determined.

Assumptions **1** The flow is incompressible and irrotational (except very close to the walls). **2** The water drains slowly enough that the flow can be approximated as steady (actually quasi-steady when the tank begins to drain). **3** Irreversible losses in the tap region are neglected.

Analysis This problem involves the conversion of flow, kinetic, and potential energies to each other without involving any pumps, turbines, and wasteful components with large frictional losses, and thus it is suitable for the use of the Bernoulli equation. We take point 1 to be at the free surface of water so that $P_1 = P_{atm}$ (open to the atmosphere), $V_1^2 << V_2^2$ and thus $V_1 \cong 0$ compared to V_2 (the tank is very large relative to the outlet), $z_1 = 5$ m and $z_2 = 0$ (we take the reference level at the center of the outlet). Also, $P_2 = P_{atm}$ (water discharges into the atmosphere). For flow along a streamline from 1 to 2, the Bernoulli equation simplifies to

$$\frac{\cancel{P_1}}{\rho g} + \frac{\cancel{V_1^2}^{\approx 0}}{2g} + z_1 = \frac{\cancel{P_2}}{\rho g} + \frac{V_2^2}{2g} + \cancel{z_2}^{0} \quad \rightarrow \quad z_1 = \frac{V_2^2}{2g}$$

Solving for V_2 and substituting,

$$V_2 = \sqrt{2gz_1} = \sqrt{2(9.81 \ \text{m/s}^2)(5 \ \text{m})} = \textbf{9.9 m/s}$$

The relation $V = \sqrt{2gz}$ is called the **Torricelli equation.**

Therefore, the water leaves the tank with an initial maximum velocity of 9.9 m/s. This is the same velocity that would manifest if a solid were dropped a distance of 5 m in the absence of air friction drag. (What would the velocity be if the tap were at the bottom of the tank instead of on the side?)

Discussion If the orifice were sharp-edged instead of rounded, then the flow would be disturbed, and the average exit velocity would be less than 9.9 m/s. Care must be exercised when attempting to apply the Bernoulli equation to situations where abrupt expansions or contractions occur since the friction and flow disturbance in such cases may not be negligible. From conversion of mass, $(V_1/V_2)^2 = (D_2/D_1)^4$. So, for example, if $D_2/D_1 = 0.1$, then $(V_1/V_2)^2 = 0.0001$, and our approximation that $V_1^2 \ll V_2^2$ is justified.

EXAMPLE 5–7   **Siphoning Out Gasoline from a Fuel Tank**

During a trip to the beach ($P_{\text{atm}} = 1 \ \text{atm} = 101.3 \ \text{kPa}$), a car runs out of gasoline, and it becomes necessary to siphon gas out of the car of a Good Samaritan (Fig. 5–41). The siphon is a small-diameter hose, and to start the siphon it is necessary to insert one siphon end in the full gas tank, fill the hose with gasoline via suction, and then place the other end in a gas can below the level of the gas tank. The difference in pressure between point 1 (at the free surface of the gasoline in the tank) and point 2 (at the outlet of the tube) causes the liquid to flow from the higher to the lower elevation. Point 2 is located 0.75 m below point 1 in this case, and point 3 is located 2 m above point 1. The siphon diameter is 5 mm, and frictional losses in the siphon are to be disregarded. Determine (*a*) the minimum time to withdraw 4 L of gasoline from the tank to the can and (*b*) the pressure at point 3. The density of gasoline is 750 kg/m³.

SOLUTION Gasoline is to be siphoned from a tank. The minimum time it takes to withdraw 4 L of gasoline and the pressure at the highest point in the system are to be determined.

Assumptions **1** The flow is steady and incompressible. **2** Even though the Bernoulli equation is not valid through the pipe because of frictional losses, we employ the Bernoulli equation anyway in order to obtain a *best-case estimate.* **3** The change in the gasoline surface level inside the tank is negligible compared to elevations z_1 and z_2 during the siphoning period.

Properties The density of gasoline is given to be 750 kg/m³.

Analysis (*a*) We take point 1 to be at the free surface of gasoline in the tank so that $P_1 = P_{\text{atm}}$ (open to the atmosphere), $V_1 \cong 0$ (the tank is large relative to the tube diameter), and $z_2 = 0$ (point 2 is taken as the reference level). Also, $P_2 = P_{\text{atm}}$ (gasoline discharges into the atmosphere). Then the Bernoulli equation simplifies to

$$\frac{\cancel{P_1}}{\rho g} + \frac{\cancel{V_1^2}^{\approx 0}}{2g} + z_1 = \frac{\cancel{P_2}}{\rho g} + \frac{V_2^2}{2g} + \cancel{z_2}^{0} \quad \rightarrow \quad z_1 = \frac{V_2^2}{2g}$$

FIGURE 5–41
Schematic for Example 5–7.

Solving for V_2 and substituting,

$$V_2 = \sqrt{2gz_1} = \sqrt{2(9.81 \text{ m/s}^2)(0.75 \text{ m})} = 3.84 \text{ m/s}$$

The cross-sectional area of the tube and the flow rate of gasoline are

$$A = \pi D^2/4 = \pi(5 \times 10^{-3} \text{ m})^2/4 = 1.96 \times 10^{-5} \text{ m}^2$$

$$\dot{V} = V_2 A = (3.84 \text{ m/s})(1.96 \times 10^{-5} \text{ m}^2) = 7.53 \times 10^{-5} \text{ m}^3/\text{s} = 0.0753 \text{ L/s}$$

Then the time needed to siphon 4 L of gasoline becomes

$$\Delta t = \frac{V}{\dot{V}} = \frac{4 \text{ L}}{0.0753 \text{ L/s}} = 53.1 \text{ s}$$

(b) The pressure at point 3 is determined by writing the Bernoulli equation along a streamline between points 3 and 2. Noting that $V_2 = V_3$ (conservation of mass), $z_2 = 0$, and $P_2 = P_{atm}$,

$$\frac{P_2}{\rho g} + \frac{V_2^2}{2g} + \cancel{z_2}^{0} = \frac{P_3}{\rho g} + \frac{V_3^2}{2g} + z_3 \rightarrow \frac{P_{atm}}{\rho g} = \frac{P_3}{\rho g} + z_3$$

Solving for P_3 and substituting,

$$P_3 = P_{atm} - \rho g z_3$$

$$= 101.3 \text{ kPa} - (750 \text{ kg/m}^3)(9.81 \text{ m/s}^2)(2.75 \text{ m})\left(\frac{1 \text{ N}}{1 \text{ kg·m/s}^2}\right)\left(\frac{1 \text{ kPa}}{1000 \text{ N/m}^2}\right)$$

$$= 81.1 \text{ kPa}$$

Discussion The siphoning time is determined by neglecting frictional effects, and thus this is the *minimum time* required. In reality, the time will be longer than 53.1 s because of friction between the gasoline and the tube surface, along with other irreversible losses, as discussed in Chap. 8. Also, the pressure at point 3 is below the atmospheric pressure. If the elevation difference between points 1 and 3 is too high, the pressure at point 3 may drop below the vapor pressure of gasoline at the gasoline temperature, and some gasoline may evaporate (cavitate). The vapor then may form a pocket at the top and halt the flow of gasoline.

FIGURE 5–42
Schematic for Example 5–8.

EXAMPLE 5–8   Velocity Measurement by a Pitot Tube

A piezometer and a Pitot tube are tapped into a horizontal water pipe, as shown in Fig. 5–42, to measure static and stagnation (static + dynamic) pressures. For the indicated water column heights, determine the velocity at the center of the pipe.

SOLUTION The static and stagnation pressures in a horizontal pipe are measured. The velocity at the center of the pipe is to be determined.
Assumptions **1** The flow is steady and incompressible. **2** Points 1 and 2 are close enough together that the irreversible energy loss between these two points is negligible, and thus we can use the Bernoulli equation.
Analysis We take points 1 and 2 along the streamline at the centerline of the pipe, with point 1 directly under the piezometer and point 2 at the tip of

the Pitot tube. This is a steady flow with straight and parallel streamlines, and the gage pressures at points 1 and 2 can be expressed as

$$P_1 = \rho g(h_1 + h_2)$$

$$P_2 = \rho g(h_1 + h_2 + h_3)$$

Noting that $z_1 = z_2$, and point 2 is a stagnation point and thus $V_2 = 0$, the application of the Bernoulli equation between points 1 and 2 gives

$$\frac{P_1}{\rho g} + \frac{V_1^2}{2g} + z_1 = \frac{P_2}{\rho g} + \frac{V_2^2}{2g}^0 + z_2 \rightarrow \frac{V_1^2}{2g} = \frac{P_2 - P_1}{\rho g}$$

Substituting the P_1 and P_2 expressions gives

$$\frac{V_1^2}{2g} = \frac{P_2 - P_1}{\rho g} = \frac{\rho g(h_1 + h_2 + h_3) - \rho g(h_1 + h_2)}{\rho g} = h_3$$

Solving for V_1 and substituting,

$$V_1 = \sqrt{2gh_3} = \sqrt{2(9.81 \text{ m/s}^2)(0.12 \text{ m})} = \textbf{1.53 m/s}$$

Discussion Note that to determine the flow velocity, all we need is to measure the height of the excess fluid column in the Pitot tube compared to that in the piezometer tube.

EXAMPLE 5–9   The Rise of the Ocean Due to a Hurricane

A hurricane is a tropical storm formed over the ocean by low atmospheric pressures. As a hurricane approaches land, inordinate ocean swells (very high tides) accompany the hurricane. A Class-5 hurricane features winds in excess of 155 mph, although the wind velocity at the center "eye" is very low.

Figure 5–43 depicts a hurricane hovering over the ocean swell below. The atmospheric pressure 200 mi from the eye is 30.0 in Hg (at point 1, generally normal for the ocean) and the winds are calm. The atmospheric pressure at the eye of the storm is 22.0 in Hg. Estimate the ocean swell at (a) the eye of the hurricane at point 3 and (b) point 2, where the wind velocity is 155 mph. Take the density of seawater and mercury to be 64 lbm/ft³ and 848 lbm/ft³, respectively, and the density of air at normal sea-level temperature and pressure to be 0.076 lbm/ft³.

SOLUTION A hurricane is moving over the ocean. The amount of ocean swell at the eye and at active regions of the hurricane are to be determined.
Assumptions 1 The airflow within the hurricane is steady, incompressible, and irrotational (so that the Bernoulli equation is applicable). (This is certainly a very questionable assumption for a highly turbulent flow, but it is justified in the discussion.) 2 The effect of water sucked into the air is negligible.
Properties The densities of air at normal conditions, seawater, and mercury are given to be 0.076 lbm/ft³, 64.0 lbm/ft³, and 848 lbm/ft³, respectively.
Analysis (a) Reduced atmospheric pressure over the water causes the water to rise. Thus, decreased pressure at point 2 relative to point 1 causes the ocean water to rise at point 2. The same is true at point 3, where the storm air

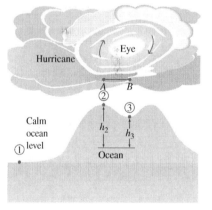

FIGURE 5–43
Schematic for Example 5–9. The vertical scale is greatly exaggerated.

FIGURE 5–44

The eye of hurricane Linda (1997 in the Pacific Ocean near Baja California) is clearly visible in this satellite photo.

© Brand X Pictures/PunchStock RF

velocity is negligible. The pressure difference given in terms of the mercury column height is expressed in terms of the seawater column height by

$$\Delta P = (\rho g h)_{Hg} = (\rho g h)_{sw} \rightarrow h_{sw} = \frac{\rho_{Hg}}{\rho_{sw}} h_{Hg}$$

Then the pressure difference between points 1 and 3 in terms of the seawater column height becomes

$$h_3 = \frac{\rho_{Hg}}{\rho_{sw}} h_{Hg} = \left(\frac{848 \text{ lbm/ft}^3}{64.0 \text{ lbm/ft}^3} \right)[(30 - 22) \text{ in Hg}]\left(\frac{1 \text{ ft}}{12 \text{ in}} \right) = \mathbf{8.83 \text{ ft}}$$

which is equivalent to the storm surge at the *eye of the hurricane* (Fig. 5–44) since the wind velocity there is negligible and there are no dynamic effects.

(b) To determine the additional rise of ocean water at point 2 due to the high winds at that point, we write the Bernoulli equation between points A and B, which are on top of points 2 and 3, respectively. Noting that $V_B \cong 0$ (the eye region of the hurricane is relatively calm) and $z_A = z_B$ (both points are on the same horizontal line), the Bernoulli equation simplifies to

$$\frac{P_A}{\rho g} + \frac{V_A^2}{2g} + \cancel{z_A} = \frac{P_B}{\rho g} + \overbrace{\cancel{\frac{V_B^2}{2g}}}^{0} + \cancel{z_B} \quad \rightarrow \quad \frac{P_B - P_A}{\rho g} = \frac{V_A^2}{2g}$$

Substituting,

$$\frac{P_B - P_A}{\rho g} = \frac{V_A^2}{2g} = \frac{(155 \text{ mph})^2}{2(32.2 \text{ ft/s}^2)}\left(\frac{1.4667 \text{ ft/s}}{1 \text{ mph}} \right)^2 = 803 \text{ ft}$$

where ρ is the density of air in the hurricane. Noting that the density of an ideal gas at constant temperature is proportional to absolute pressure and the density of air at the normal atmospheric pressure of 14.7 psia $\cong$ 30 in Hg is 0.076 lbm/ft³, the density of air in the hurricane is

$$\rho_{air} = \frac{P_{air}}{P_{atm \, air}} \rho_{atm \, air} = \left(\frac{22 \text{ in Hg}}{30 \text{ in Hg}} \right)(0.076 \text{ lbm/ft}^3) = 0.056 \text{ lbm/ft}^3$$

Using the relation developed above in part (a), the seawater column height equivalent to 803 ft of air column height is determined to be

$$h_{dynamic} = \frac{\rho_{air}}{\rho_{sw}} h_{air} = \left(\frac{0.056 \text{ lbm/ft}^3}{64 \text{ lbm/ft}^3} \right)(803 \text{ ft}) = 0.70 \text{ ft}$$

Therefore, the pressure at point 2 is 0.70 ft seawater column lower than the pressure at point 3 due to the high wind velocities, causing the ocean to rise an additional 0.70 ft. Then the total storm surge at point 2 becomes

$$h_2 = h_3 + h_{dynamic} = 8.83 + 0.70 = \mathbf{9.53 \text{ ft}}$$

Discussion This problem involves highly turbulent flow and the intense breakdown of the streamlines, and thus the applicability of the Bernoulli equation in part (b) is questionable. Furthermore, the flow in the eye of the storm is *not* irrotational, and the Bernoulli equation constant changes across streamlines (see Chap. 10). The Bernoulli analysis can be thought of as the limiting, ideal case, and shows that the rise of seawater due to high-velocity winds cannot be more than 0.70 ft.

The wind power of hurricanes is not the only cause of damage to coastal areas. Ocean flooding and erosion from excessive tides is just as serious, as are high waves generated by the storm turbulence and energy.

EXAMPLE 5–10   **Bernoulli Equation for Compressible Flow**

Derive the Bernoulli equation when the compressibility effects are not negligible for an ideal gas undergoing (a) an isothermal process and (b) an isentropic process.

SOLUTION The Bernoulli equation for compressible flow is to be obtained for an ideal gas for isothermal and isentropic processes.

Assumptions **1** The flow is steady and frictional effects are negligible. **2** The fluid is an ideal gas, so the relation $P = \rho RT$ is applicable. **3** The specific heats are constant so that P/ρ^k = constant during an isentropic process.

Analysis (a) When the compressibility effects are significant and the flow cannot be assumed to be incompressible, the Bernoulli equation is given by Eq. 5–40 as

$$\int \frac{dP}{\rho} + \frac{V^2}{2} + gz = \text{constant} \quad \text{(along a streamline)} \tag{1}$$

The compressibility effects can be properly accounted for by performing the integration $\int dP/\rho$ in Eq. 1. But this requires a relation between P and ρ for the process. For the *isothermal* expansion or compression of an ideal gas, the integral in Eq. 1 is performed easily by noting that T = constant and substituting $\rho = P/RT$,

$$\int \frac{dP}{\rho} = \int \frac{dP}{P/RT} = RT \ln P$$

Substituting into Eq. 1 gives the desired relation,

Isothermal process: $$RT \ln P + \frac{V^2}{2} + gz = \text{constant} \tag{2}$$

(b) A more practical case of compressible flow is the *isentropic flow of ideal gases* through equipment that involves high-speed fluid flow such as nozzles, diffusers, and the passages between turbine blades (Fig. 5–45). Isentropic (i.e., reversible and adiabatic) flow is closely approximated by these devices, and it is characterized by the relation $P/\rho^k = C$ = constant, where k is the specific heat ratio of the gas. Solving for ρ from $P/\rho^k = C$ gives $\rho = C^{-1/k}P^{1/k}$. Performing the integration,

$$\int \frac{dP}{\rho} = \int C^{1/k} P^{-1/k} \, dP = C^{1/k} \frac{P^{-1/k+1}}{-1/k+1} = \frac{P^{1/k}}{\rho} \frac{P^{-1/k+1}}{-1/k+1} = \left(\frac{k}{k-1}\right)\frac{P}{\rho} \tag{3}$$

Substituting, the Bernoulli equation for steady, isentropic, compressible flow of an ideal gas becomes

Isentropic flow: $$\left(\frac{k}{k-1}\right)\frac{P}{\rho} + \frac{V^2}{2} + gz = \text{constant} \tag{4a}$$

FIGURE 5–45
Compressible flow of a gas through turbine blades is often modeled as isentropic, and the compressible form of the Bernoulli equation is a reasonable approximation.
Royalty-Free/CORBIS

or

$$\left(\frac{k}{k-1}\right)\frac{P_1}{\rho_1} + \frac{V_1^2}{2} + gz_1 = \left(\frac{k}{k-1}\right)\frac{P_2}{\rho_2} + \frac{V_2^2}{2} + gz_2 \qquad \text{(4b)}$$

A common practical situation involves the acceleration of a gas from rest (stagnation conditions at state 1) with negligible change in elevation. In that case we have $z_1 = z_2$ and $V_1 = 0$. Noting that $\rho = P/RT$ for ideal gases, P/ρ^k = constant for isentropic flow, and the Mach number is defined as Ma = V/c where $c = \sqrt{kRT}$ is the local speed of sound for ideal gases, Eq. 4b simplifies to

$$\frac{P_1}{P_2} = \left[1 + \left(\frac{k-1}{2}\right)\text{Ma}_2^2\right]^{k/(k-1)} \qquad \text{(4c)}$$

where state 1 is the stagnation state and state 2 is any state along the flow. *Discussion* It can be shown that the results obtained using the compressible and incompressible equations deviate no more than 2 percent when the Mach number is less than 0.3. Therefore, the flow of an ideal gas can be considered to be incompressible when Ma $\leqslant$ 0.3. For atmospheric air at normal conditions, this corresponds to a flow speed of about 100 m/s or 360 km/h.

5–5 · GENERAL ENERGY EQUATION

One of the most fundamental laws in nature is the **first law of thermodynamics**, also known as the **conservation of energy principle**, which provides a sound basis for studying the relationships among the various forms of energy and energy interactions. It states that *energy can be neither created nor destroyed during a process; it can only change forms.* Therefore, every bit of energy must be accounted for during a process.

A rock falling off a cliff, for example, picks up speed as a result of its potential energy being converted to kinetic energy (Fig. 5–46). Experimental data show that the decrease in potential energy equals the increase in kinetic energy when the air resistance is negligible, thus confirming the conservation of energy principle. The conservation of energy principle also forms the backbone of the diet industry: a person who has a greater energy input (food) than energy output (exercise) will gain weight (store energy in the form of fat), and a person who has a smaller energy input than output will lose weight. The change in the energy content of a system is equal to the difference between the energy input and the energy output, and the conservation of energy principle for any system can be expressed simply as $E_{in} - E_{out} = \Delta E$.

The transfer of any quantity (such as mass, momentum, and energy) is recognized *at the boundary* as the quantity *crosses the boundary*. A quantity is said to *enter* a system (or control volume) if it crosses the boundary from the outside to the inside, and to *exit* the system if it moves in the reverse direction. A quantity that moves from one location to another within a system is not considered as a transferred quantity in an analysis since it does not enter or exit the system. Therefore, it is important to specify the system and thus clearly identify its boundaries before an engineering analysis is performed.

FIGURE 5–46
Energy cannot be created or destroyed during a process; it can only change forms.

The energy content of a fixed quantity of mass (a closed system) can be changed by two mechanisms: *heat transfer Q* and *work transfer W*. Then the conservation of energy for a fixed quantity of mass can be expressed in rate form as (Fig. 5–47)

$$\dot{Q}_{net\ in} + \dot{W}_{net\ in} = \frac{dE_{sys}}{dt} \quad \text{or} \quad \dot{Q}_{net\ in} + \dot{W}_{net\ in} = \frac{d}{dt} \int_{sys} \rho e\, dV \quad \text{(5–49)}$$

where the overdot stands for time rate of change, and $\dot{Q}_{net\ in} = \dot{Q}_{in} - \dot{Q}_{out}$ is the net rate of heat transfer to the system (negative, if from the system), $\dot{W}_{net\ in} = \dot{W}_{in} - \dot{W}_{out}$ is the net power input to the system in all forms (negative, if power output), and dE_{sys}/dt is the rate of change of the total energy content of the system. For simple compressible systems, total energy consists of internal, kinetic, and potential energies, and it is expressed on a unit-mass basis as (see Chap. 2)

$$e = u + \text{ke} + \text{pe} = u + \frac{V^2}{2} + gz \quad \text{(5–50)}$$

Note that total energy is a property, and its value does not change unless the state of the system changes.

FIGURE 5–47
The energy change of a system during a process is equal to the *net* work and heat transfer between the system and its surroundings.

Energy Transfer by Heat, Q

In daily life, we frequently refer to the sensible and latent forms of internal energy as *heat*, and talk about the heat content of bodies. Scientifically the more correct name for these forms of energy is *thermal energy*. For single-phase substances, a change in the thermal energy of a given mass results in a change in temperature, and thus temperature is a good representative of thermal energy. Thermal energy tends to move naturally in the direction of decreasing temperature. The transfer of energy from one system to another as a result of a temperature difference is called **heat transfer**. The warming up of a canned drink in a warmer room, for example, is due to heat transfer (Fig. 5–48). The time rate of heat transfer is called **heat transfer rate** and is denoted by $\dot{Q}$.

The direction of heat transfer is always from the higher-temperature body to the lower-temperature one. Once temperature equality is established, heat transfer stops. There cannot be any net heat transfer between two systems (or a system and its surroundings) that are at the same temperature.

A process during which there is no heat transfer is called an **adiabatic process**. There are two ways a process can be adiabatic: Either the system is well insulated so that only a negligible amount of heat can pass through the system boundary, or both the system and the surroundings are at the same temperature and therefore there is no driving force (temperature difference) for net heat transfer. An adiabatic process should not be confused with an isothermal process. Even though there is no net heat transfer during an adiabatic process, the energy content and thus the temperature of a system can still be changed by other means such as work transfer.

FIGURE 5–48
Temperature difference is the driving force for heat transfer. The larger the temperature difference, the higher is the rate of heat transfer.

Energy Transfer by Work, W

An energy interaction is **work** if it is associated with a force acting through a distance. A rising piston, a rotating shaft, and an electric wire crossing the system boundary are all associated with work interactions. The time rate of doing work is called **power** and is denoted by $\dot{W}$. Car engines and hydraulic,

FIGURE 5–49
The pressure force acting on (a) the moving boundary of a system in a piston-cylinder device, and (b) the differential surface area of a system of arbitrary shape.

steam, and gas turbines produce power ($\dot{W}_{shaft,\,in} < 0$); compressors, pumps, fans, and mixers consume power ($\dot{W}_{shaft,\,in} > 0$).

Work-consuming devices transfer energy to the fluid, and thus increase the energy of the fluid. A fan in a room, for example, mobilizes the air and increases its kinetic energy. The electric energy a fan consumes is first converted to mechanical energy by its motor that forces the shaft of the blades to rotate. This mechanical energy is then transferred to the air, as evidenced by the increase in air velocity. This energy transfer to air has nothing to do with a temperature difference, so it cannot be heat transfer. Therefore, it must be work. Air discharged by the fan eventually comes to a stop and thus loses its mechanical energy as a result of friction between air particles of different velocities. But this is not a "loss" in the real sense; it is simply the conversion of mechanical energy to an equivalent amount of thermal energy (which is of limited value, and thus the term *loss*) in accordance with the conservation of energy principle. If a fan runs a long time in a sealed room, we can sense the buildup of this thermal energy by a rise in air temperature.

A system may involve numerous forms of work, and the total work can be expressed as

$$W_{total} = W_{shaft} + W_{pressure} + W_{viscous} + W_{other} \qquad (5\text{–}51)$$

where W_{shaft} is the work transmitted by a rotating shaft, $W_{pressure}$ is the work done by the pressure forces on the control surface, $W_{viscous}$ is the work done by the normal and shear components of viscous forces on the control surface, and W_{other} is the work done by other forces such as electric, magnetic, and surface tension, which are insignificant for simple compressible systems and are not considered in this text. We do not consider $W_{viscous}$ either, since moving walls (such as fan blades or turbine runners) are usually *inside* the control volume and are not part of the control surface. But it should be kept in mind that the work done by shear forces as the blades shear through the fluid may need to be considered in a refined analysis of turbomachinery.

Shaft Work

Many flow systems involve a machine such as a pump, a turbine, a fan, or a compressor whose shaft protrudes through the control surface, and the work transfer associated with all such devices is simply referred to as *shaft work* W_{shaft}. The power transmitted via a rotating shaft is proportional to the shaft torque T_{shaft} and is expressed as

$$\dot{W}_{shaft} = \omega T_{shaft} = 2\pi \dot{n}\, T_{shaft} \qquad (5\text{–}52)$$

where ω is the angular speed of the shaft in rad/s and $\dot{n}$ is the number of revolutions of the shaft per unit time, often expressed in rev/min or rpm.

Work Done by Pressure Forces

Consider a gas being compressed in the piston-cylinder device shown in Fig. 5–49a. When the piston moves down a differential distance ds under the influence of the pressure force PA, where A is the cross-sectional area of the piston, the boundary work done *on* the system is $\delta W_{boundary} = PA\, ds$. Dividing both sides of this relation by the differential time interval dt gives the time rate of boundary work (i.e., *power*),

$$\delta \dot{W}_{pressure} = \delta \dot{W}_{boundary} = PA V_{piston}$$

where $V_{\text{piston}} = ds/dt$ is the piston speed, which is the speed of the moving boundary at the piston face.

Now consider a material chunk of fluid (a system) of arbitrary shape that moves with the flow and is free to deform under the influence of pressure, as shown in Fig. 5–49b. Pressure always acts inward and normal to the surface, and the pressure force acting on a differential area dA is $P\,dA$. Again noting that work is force times distance and distance traveled per unit time is velocity, the time rate at which work is done by pressure forces on this differential part of the system is

$$\delta\dot{W}_{\text{pressure}} = -P\,dA\,V_n = -P\,dA(\vec{V}\cdot\vec{n}) \tag{5–53}$$

since the normal component of velocity through the differential area dA is $V_n = V\cos\theta = \vec{V}\cdot\vec{n}$. Note that $\vec{n}$ is the outward normal of dA, and thus the quantity $\vec{V}\cdot\vec{n}$ is positive for expansion and negative for compression. The negative sign in Eq. 5–53 ensures that work done by pressure forces is positive when it is done *on* the system, and negative when it is done *by* the system, which agrees with our sign convention. The total rate of work done by pressure forces is obtained by integrating $\delta\dot{W}_{\text{pressure}}$ over the entire surface A,

$$\dot{W}_{\text{pressure, net in}} = -\int_A P(\vec{V}\cdot\vec{n})\,dA = -\int_A \frac{P}{\rho}\rho(\vec{V}\cdot\vec{n})\,dA \tag{5–54}$$

In light of these discussions, the net power transfer can be expressed as

$$\dot{W}_{\text{net in}} = \dot{W}_{\text{shaft, net in}} + \dot{W}_{\text{pressure, net in}} = \dot{W}_{\text{shaft, net in}} - \int_A P(\vec{V}\cdot\vec{n})\,dA \tag{5–55}$$

Then the rate form of the conservation of energy relation for a closed system becomes

$$\dot{Q}_{\text{net in}} + \dot{W}_{\text{shaft, net in}} + \dot{W}_{\text{pressure, net in}} = \frac{dE_{\text{sys}}}{dt} \tag{5–56}$$

To obtain a relation for the conservation of energy for a *control volume*, we apply the Reynolds transport theorem by replacing B with total energy E, and b with total energy per unit mass e, which is $e = u + \text{ke} + \text{pe} = u + V^2/2 + gz$ (Fig. 5–50). This yields

$$\frac{dE_{\text{sys}}}{dt} = \frac{d}{dt}\int_{\text{CV}} e\rho\,dV + \int_{\text{CS}} e\rho(\vec{V}_r\cdot\vec{n})A \tag{5–57}$$

Substituting the left-hand side of Eq. 5–56 into Eq. 5–57, the general form of the energy equation that applies to fixed, moving, or deforming control volumes becomes

$$\dot{Q}_{\text{net in}} + \dot{W}_{\text{shaft, net in}} + \dot{W}_{\text{pressure, net in}} = \frac{d}{dt}\int_{\text{CV}} e\rho\,dV + \int_{\text{CS}} e\rho(\vec{V}_r\cdot\vec{n})\,dA \tag{5–58}$$

which is stated in words as

$$\begin{pmatrix}\text{The net rate of energy}\\ \text{transfer into a CV by}\\ \text{heat and work transfer}\end{pmatrix} = \begin{pmatrix}\text{The time rate of}\\ \text{change of the energy}\\ \text{content of the CV}\end{pmatrix} + \begin{pmatrix}\text{The net flow rate of}\\ \text{energy out of the control}\\ \text{surface by mass flow}\end{pmatrix}$$

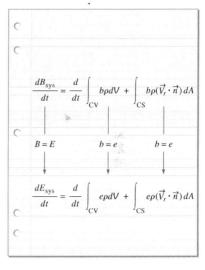

$$\frac{dB_{\text{sys}}}{dt} = \frac{d}{dt}\int_{\text{CV}} b\rho\,dV + \int_{\text{CS}} b\rho(\vec{V}_r\cdot\vec{n})\,dA$$

$B = E \qquad b = e \qquad b = e$

$$\frac{dE_{\text{sys}}}{dt} = \frac{d}{dt}\int_{\text{CV}} e\rho\,dV + \int_{\text{CS}} e\rho(\vec{V}_r\cdot\vec{n})\,dA$$

FIGURE 5–50

The conservation of energy equation is obtained by replacing B in the Reynolds transport theorem by energy E and b by e.

Here $\vec{V}_r = \vec{V} - \vec{V}_{CS}$ is the fluid velocity relative to the control surface, and the product $\rho(\vec{V}_r \cdot \vec{n})\, dA$ represents the mass flow rate through area element dA into or out of the control volume. Again noting that $\vec{n}$ is the outward normal of dA, the quantity $\vec{V}_r \cdot \vec{n}$ and thus mass flow is positive for outflow and negative for inflow.

Substituting the surface integral for the rate of pressure work from Eq. 5–54 into Eq. 5–58 and combining it with the surface integral on the right give

$$\dot{Q}_{\text{net in}} + \dot{W}_{\text{shaft, net in}} = \frac{d}{dt}\int_{CV} e\rho\, dV + \int_{CS}\left(\frac{P}{\rho} + e\right)\rho\,(\vec{V}_r \cdot \vec{n})dA \qquad \text{(5–59)}$$

This is a convenient form for the energy equation since pressure work is now combined with the energy of the fluid crossing the control surface and we no longer have to deal with pressure work.

The term $P/\rho = Pv = w_{\text{flow}}$ is the **flow work,** which is the work per unit mass associated with pushing a fluid into or out of a control volume. Note that the fluid velocity at a solid surface is equal to the velocity of the solid surface because of the no-slip condition. As a result, the pressure work along the portions of the control surface that coincide with nonmoving solid surfaces is zero. Therefore, pressure work for fixed control volumes can exist only along the imaginary part of the control surface where the fluid enters and leaves the control volume, i.e., inlets and outlets.

For a fixed control volume (no motion or deformation of the control volume), $\vec{V}_r = \vec{V}$ and the energy equation Eq. 5–59 becomes

Fixed CV: $\quad \dot{Q}_{\text{net in}} + \dot{W}_{\text{shaft, net in}} = \dfrac{d}{dt}\displaystyle\int_{CV} e\rho\, dV + \int_{CS}\left(\dfrac{P}{\rho} + e\right)\rho\,(\vec{V} \cdot \vec{n})\, dA \quad$ **(5–60)**

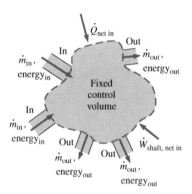

FIGURE 5–51

In a typical engineering problem, the control volume may contain many inlets and outlets; energy flows in at each inlet, and energy flows out at each outlet. Energy also enters the control volume through net heat transfer and net shaft work.

This equation is not in a convenient form for solving practical engineering problems because of the integrals, and thus it is desirable to rewrite it in terms of average velocities and mass flow rates through inlets and outlets. If $P/\rho + e$ is nearly uniform across an inlet or outlet, we can simply take it outside the integral. Noting that $\dot{m} = \displaystyle\int_{A_c} \rho(\vec{V}\cdot\vec{n})\, dA_c$ is the mass flow rate across an inlet or outlet, the rate of inflow or outflow of energy through the inlet or outlet can be approximated as $\dot{m}(P/\rho + e)$. Then the energy equation becomes (Fig. 5–51)

$$\dot{Q}_{\text{net in}} + \dot{W}_{\text{shaft, net in}} = \frac{d}{dt}\int_{CV} e\rho\, dV + \sum_{\text{out}} \dot{m}\left(\frac{P}{\rho} + e\right) - \sum_{\text{in}} \dot{m}\left(\frac{P}{\rho} + e\right) \qquad \text{(5–61)}$$

where $e = u + V^2/2 + gz$ (Eq. 5–50) is the total energy per unit mass for both the control volume and flow streams. Then,

$$\dot{Q}_{\text{net in}} + \dot{W}_{\text{shaft, net in}} = \frac{d}{dt}\int_{CV} e\rho\, dV + \sum_{\text{out}} \dot{m}\left(\frac{P}{\rho} + u + \frac{V^2}{2} + gz\right) - \sum_{\text{in}} \dot{m}\left(\frac{P}{\rho} + u + \frac{V^2}{2} + gz\right)$$

$$\text{(5–62)}$$

or

$$\dot{Q}_{\text{net in}} + \dot{W}_{\text{shaft, net in}} = \frac{d}{dt}\int_{CV} e\rho\, dV + \sum_{\text{out}} \dot{m}\left(h + \frac{V^2}{2} + gz\right) - \sum_{\text{in}} \dot{m}\left(h + \frac{V^2}{2} + gz\right)$$

$$\text{(5–63)}$$

where we used the definition of specific enthalpy $h = u + P\vee = u + P/\rho$. The last two equations are fairly general expressions of conservation of energy, but their use is still limited to fixed control volumes, uniform flow at inlets and outlets, and negligible work due to viscous forces and other effects. Also, the subscript "net in" stands for "net input," and thus any heat or work transfer is positive if *to* the system and negative if *from* the system.

5–6 · ENERGY ANALYSIS OF STEADY FLOWS

For steady flows, the time rate of change of the energy content of the control volume is zero, and Eq. 5–63 simplifies to

$$\dot{Q}_{\text{net in}} + \dot{W}_{\text{shaft, net in}} = \sum_{\text{out}} \dot{m}\left(h + \frac{V^2}{2} + gz\right) - \sum_{\text{in}} \dot{m}\left(h + \frac{V^2}{2} + gz\right) \quad \text{(5–64)}$$

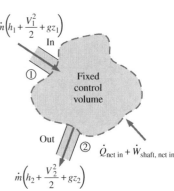

FIGURE 5–52

A control volume with only one inlet and one outlet and energy interactions.

It states that *during steady flow the net rate of energy transfer to a control volume by heat and work transfers is equal to the difference between the rates of outgoing and incoming energy flows by mass flow.*

Many practical problems involve just one inlet and one outlet (Fig. 5–52). The mass flow rate for such **single-stream devices** is the same at the inlet and outlet, and Eq. 5–64 reduces to

$$\dot{Q}_{\text{net in}} + \dot{W}_{\text{shaft, net in}} = \dot{m}\left(h_2 - h_1 + \frac{V_2^2 - V_1^2}{2} + g(z_2 - z_1)\right) \quad \text{(5–65)}$$

where subscripts 1 and 2 refer to the inlet and outlet, respectively. The steady-flow energy equation on a unit-mass basis is obtained by dividing Eq. 5–65 by the mass flow rate $\dot{m}$,

$$q_{\text{net in}} + w_{\text{shaft, net in}} = h_2 - h_1 + \frac{V_2^2 - V_1^2}{2} + g(z_2 - z_1) \quad \text{(5–66)}$$

where $q_{\text{net in}} = \dot{Q}_{\text{net in}}/\dot{m}$ is the net heat transfer to the fluid per unit mass and $w_{\text{shaft, net in}} = \dot{W}_{\text{shaft, net in}}/\dot{m}$ is the net shaft work input to the fluid per unit mass. Using the definition of enthalpy $h = u + P/\rho$ and rearranging, the steady-flow energy equation can also be expressed as

$$w_{\text{shaft, net in}} + \frac{P_1}{\rho_1} + \frac{V_1^2}{2} + gz_1 = \frac{P_2}{\rho_2} + \frac{V_2^2}{2} + gz_2 + (u_2 - u_1 - q_{\text{net in}}) \quad \text{(5–67)}$$

where u is the *internal energy*, P/ρ is the *flow energy*, $V^2/2$ is the *kinetic energy*, and gz is the *potential energy* of the fluid, all per unit mass. These relations are valid for both compressible and incompressible flows.

The left side of Eq. 5–67 represents the mechanical energy input, while the first three terms on the right side represent the mechanical energy output. If the flow is ideal with no irreversibilities such as friction, the total mechanical energy must be conserved, and the term in parentheses $(u_2 - u_1 - q_{\text{net in}})$ must equal zero. That is,

Ideal flow (no mechanical energy loss): $\qquad q_{\text{net in}} = u_2 - u_1 \quad \text{(5–68)}$

FIGURE 5–53

The lost mechanical energy in a fluid flow system results in an increase in the internal energy of the fluid and thus in a rise of fluid temperature.

Any increase in $u_2 - u_1$ above $q_{\text{net in}}$ is due to the irreversible conversion of mechanical energy to thermal energy, and thus $u_2 - u_1 - q_{\text{net in}}$ represents the mechanical energy loss per unit mass (Fig. 5–53). That is,

Real flow (with mechanical energy loss): $\qquad e_{\text{mech, loss}} = u_2 - u_1 - q_{\text{net in}} \quad \text{(5–69)}$

For single-phase fluids (a gas or a liquid), $u_2 - u_1 = c_v(T_2 - T_1)$ where c_v is the constant-volume specific heat.

The steady-flow energy equation on a unit-mass basis can be written conveniently as a **mechanical energy** balance,

$$e_{mech, in} = e_{mech, out} + e_{mech, loss} \tag{5–70}$$

or

$$w_{shaft, net in} + \frac{P_1}{\rho_1} + \frac{V_1^2}{2} + gz_1 = \frac{P_2}{\rho_2} + \frac{V_2^2}{2} + gz_2 + e_{mech, loss} \tag{5–71}$$

Noting that $w_{shaft, net in} = w_{pump} - w_{turbine}$, the mechanical energy balance can be written more explicitly as

$$\frac{P_1}{\rho_1} + \frac{V_1^2}{2} + gz_1 + w_{pump} = \frac{P_2}{\rho_2} + \frac{V_2^2}{2} + gz_2 + w_{turbine} + e_{mech, loss} \tag{5–72}$$

where w_{pump} is the mechanical work input (due to the presence of a pump, fan, compressor, etc.) and $w_{turbine}$ is the mechanical work output (due to a turbine). When the flow is incompressible, either absolute or gage pressure can be used for P since P_{atm}/ρ would appear on both sides and would cancel out.

Multiplying Eq. 5–72 by the mass flow rate $\dot{m}$ gives

$$\dot{m}\left(\frac{P_1}{\rho_1} + \frac{V_1^2}{2} + gz_1\right) + \dot{W}_{pump} = \dot{m}\left(\frac{P_2}{\rho_2} + \frac{V_2^2}{2} + gz_2\right) + \dot{W}_{turbine} + \dot{E}_{mech, loss} \tag{5–73}$$

where $\dot{W}_{pump}$ is the shaft power input through the pump's shaft, $\dot{W}_{turbine}$ is the shaft power output through the turbine's shaft, and $\dot{E}_{mech, loss}$ is the *total* mechanical power loss, which consists of pump and turbine losses as well as the frictional losses in the piping network. That is,

$$\dot{E}_{mech, loss} = \dot{E}_{mech loss, pump} + \dot{E}_{mech loss, turbine} + \dot{E}_{mech loss, piping}$$

By convention, irreversible pump and turbine losses are treated separately from irreversible losses due to other components of the piping system (Fig. 5–54). Thus, the energy equation is expressed in its most common form in terms of *heads* by dividing each term in Eq. 5–73 by $\dot{m}g$. The result is

$$\frac{P_1}{\rho_1 g} + \frac{V_1^2}{2g} + z_1 + h_{pump, u} = \frac{P_2}{\rho_2 g} + \frac{V_2^2}{2g} + z_2 + h_{turbine, e} + h_L \tag{5–74}$$

where

FIGURE 5–54

A typical power plant has numerous pipes, elbows, valves, pumps, and turbines, all of which have irreversible losses.

© Brand X Pictures PunchStock RF

- $h_{pump, u} = \dfrac{w_{pump, u}}{g} = \dfrac{\dot{W}_{pump, u}}{\dot{m}g} = \dfrac{\eta_{pump}\dot{W}_{pump}}{\dot{m}g}$ is the *useful head delivered to the fluid by the pump*. Because of irreversible losses in the pump, $h_{pump, u}$ is *less* than $\dot{W}_{pump}/\dot{m}g$ by the factor η_{pump}.

- $h_{turbine, e} = \dfrac{w_{turbine, e}}{g} = \dfrac{\dot{W}_{turbine, e}}{\dot{m}g} = \dfrac{\dot{W}_{turbine}}{\eta_{turbine}\dot{m}g}$ is the *extracted head removed from the fluid by the turbine*. Because of irreversible losses in the turbine, $h_{turbine, e}$ is *greater* than $\dot{W}_{turbine}/\dot{m}g$ by the factor $\eta_{turbine}$.

- $h_L = \dfrac{e_{mech loss, piping}}{g} = \dfrac{\dot{E}_{mech loss, piping}}{\dot{m}g}$ is the *irreversible head loss* between 1 and 2 due to all components of the piping system other than the pump or turbine.

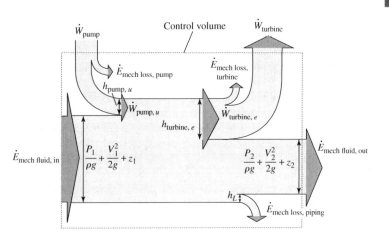

FIGURE 5–55
Mechanical energy flow chart for a fluid flow system that involves a pump and a turbine. Vertical dimensions show each energy term expressed as an equivalent column height of fluid, i.e., *head*, corresponding to each term of Eq. 5–74.

Note that the head loss h_L represents the frictional losses associated with fluid flow in piping, and it does not include the losses that occur within the pump or turbine due to the inefficiencies of these devices—these losses are taken into account by η_{pump} and η_{turbine}. Equation 5–74 is illustrated schematically in Fig. 5–55.

The *pump head* is zero if the piping system does not involve a pump, a fan, or a compressor, and the *turbine head* is zero if the system does not involve a turbine.

Special Case: Incompressible Flow with No Mechanical Work Devices and Negligible Friction

When piping losses are negligible, there is negligible dissipation of mechanical energy into thermal energy, and thus $h_L = e_{\text{mech loss, piping}}/g \cong 0$, as shown later in Example 5–11. Also, $h_{\text{pump, }u} = h_{\text{turbine, }e} = 0$ when there are no mechanical work devices such as fans, pumps, or turbines. Then Eq. 5–74 reduces to

$$\frac{P_1}{\rho g} + \frac{V_1^2}{2g} + z_1 = \frac{P_2}{\rho g} + \frac{V_2^2}{2g} + z_2 \quad \text{or} \quad \frac{P}{\rho g} + \frac{V^2}{2g} + z = \text{constant} \quad \textbf{(5–75)}$$

which is the **Bernoulli equation** derived earlier using Newton's second law of motion. Thus, the Bernoulli equation can be thought of as a degenerate form of the energy equation.

Kinetic Energy Correction Factor, α

The average flow velocity V_{avg} was defined such that the relation $\rho V_{\text{avg}} A$ gives the actual mass flow rate. Therefore, there is no such thing as a correction factor for mass flow rate. However, as Gaspard Coriolis (1792–1843) showed, the kinetic energy of a fluid stream obtained from $V^2/2$ is not the same as the actual kinetic energy of the fluid stream since the square of a sum is not equal to the sum of the squares of its components (Fig. 5–56). This error can be corrected by replacing the kinetic energy terms $V^2/2$ in the energy equation by $\alpha V_{\text{avg}}^2/2$, where α is the **kinetic energy correction factor.** By using equations for the variation of velocity with the radial distance, it can be shown that the correction factor is 2.0 for fully developed laminar pipe flow, and it ranges between 1.04 and 1.11 for fully developed turbulent flow in a round pipe.

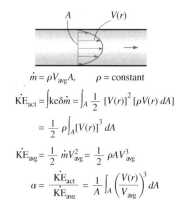

FIGURE 5–56
The determination of the *kinetic energy correction factor* using the actual velocity distribution $V(r)$ and the average velocity V_{avg} at a cross section.

The kinetic energy correction factors are often ignored (i.e., α is set equal to 1) in an elementary analysis since (1) most flows encountered in practice are turbulent, for which the correction factor is near unity, and (2) the kinetic energy terms are often small relative to the other terms in the energy equation, and multiplying them by a factor less than 2.0 does not make much difference. When the velocity and thus the kinetic energy are high, the flow turns turbulent, and a unity correction factor is more appropriate. However, you should keep in mind that you may encounter some situations for which these factors *are* significant, especially when the flow is laminar. Therefore, we recommend that you always include the kinetic energy correction factor when analyzing fluid flow problems. When the kinetic energy correction factors are included, the energy equations for *steady incompressible flow* (Eqs. 5–73 and 5–74) become

$$\dot{m}\left(\frac{P_1}{\rho} + \alpha_1 \frac{V_1^2}{2} + gz_1\right) + \dot{W}_{pump} = \dot{m}\left(\frac{P_2}{\rho} + \alpha_2 \frac{V_2^2}{2} + gz_2\right) + \dot{W}_{turbine} + \dot{E}_{mech, loss}$$

(5–76)

$$\frac{P_1}{\rho g} + \alpha_1 \frac{V_1^2}{2g} + z_1 + h_{pump, u} = \frac{P_2}{\rho g} + \alpha_2 \frac{V_2^2}{2g} + z_2 + h_{turbine, e} + h_L \qquad (5\text{–}77)$$

If the flow at an inlet or outlet is fully developed turbulent pipe flow, we recommend using $\alpha = 1.05$ as a reasonable estimate of the correction factor. This leads to a more conservative estimate of head loss, and it does not take much additional effort to include α in the equations.

FIGURE 5–57
Schematic for Example 5–11.

EXAMPLE 5–11   Effect of Friction on Fluid Temperature and Head Loss

Show that during steady and incompressible flow of a fluid in an adiabatic flow section (*a*) the temperature remains constant and there is no head loss when friction is ignored and (*b*) the temperature increases and some head loss occurs when frictional effects are considered. Discuss if it is possible for the fluid temperature to decrease during such flow (Fig. 5–57).

SOLUTION   Steady and incompressible flow through an adiabatic section is considered. The effects of friction on the temperature and the heat loss are to be determined.
Assumptions   **1** The flow is steady and incompressible. **2** The flow section is adiabatic and thus there is no heat transfer, $q_{net\ in} = 0$.
Analysis   The density of a fluid remains constant during incompressible flow and the entropy change is

$$\Delta s = c_v \ln \frac{T_2}{T_1}$$

This relation represents the entropy change of the fluid per unit mass as it flows through the flow section from state 1 at the inlet to state 2 at the outlet. Entropy change is caused by two effects: (1) heat transfer and (2) irreversibilities. Therefore, in the absence of heat transfer, entropy change is due to irreversibilities only, whose effect is always to increase entropy.

(*a*) The entropy change of the fluid in an adiabatic flow section ($q_{\text{net in}} = 0$) is zero when the process does not involve any irreversibilities such as friction and swirling, and thus for *reversible flow* we have

Temperature change:
$$\Delta s = c_v \ln \frac{T_2}{T_1} = 0 \quad \rightarrow \quad T_2 = T_1$$

Mechanical energy loss:

$$e_{\text{mech loss, piping}} = u_2 - u_1 - q_{\text{net in}} = c_v(T_2 - T_1) - q_{\text{net in}} = 0$$

Head loss:
$$h_L = e_{\text{mech loss, piping}}/g = 0$$

Thus we conclude that when heat transfer and frictional effects are negligible, (1) the temperature of the fluid remains constant, (2) no mechanical energy is converted to thermal energy, and (3) there is no irreversible head loss.

(*b*) When irreversibilities such as friction are taken into account, the entropy change is positive and thus we have:

Temperature change:
$$\Delta s = c_v \ln \frac{T_2}{T_1} > 0 \rightarrow T_2 > T_1$$

Mechanical energy loss: $e_{\text{mech loss, piping}} = u_2 - u_1 - q_{\text{net in}} = c_v(T_2 - T_1) > 0$

Head loss:
$$h_L = e_{\text{mech loss, piping}}/g > 0$$

Thus we conclude that when the flow is adiabatic and irreversible, (1) the temperature of the fluid increases, (2) some mechanical energy is converted to thermal energy, and (3) some irreversible head loss occurs.

Discussion It is impossible for the fluid temperature to decrease during steady, incompressible, adiabatic flow since this would require the entropy of an adiabatic system to decrease, which would be a violation of the second law of thermodynamics.

■
■
■
■ *EXAMPLE 5–12*   **Pumping Power and Frictional Heating in a Pump**

The pump of a water distribution system is powered by a 15-kW electric motor whose efficiency is 90 percent (Fig. 5–58). The water flow rate through the pump is 50 L/s. The diameters of the inlet and outlet pipes are the same, and the elevation difference across the pump is negligible. If the absolute pressures at the inlet and outlet of the pump are measured to be 100 kPa and 300 kPa, respectively, determine (*a*) the mechanical efficiency of the pump and (*b*) the temperature rise of water as it flows through the pump due to mechanical inefficiencies.

SOLUTION The pressures across a pump are measured. The mechanical efficiency of the pump and the temperature rise of water are to be determined.

Assumptions **1** The flow is steady and incompressible. **2** The pump is driven by an external motor so that the heat generated by the motor is dissipated to the atmosphere. **3** The elevation difference between the inlet and outlet of the pump is negligible, $z_1 \cong z_2$. **4** The inlet and outlet diameters are the

FIGURE 5–58
Schematic for Example 5–12.

same and thus the average inlet and outlet velocities are equal, $V_1 = V_2$.
5 The kinetic energy correction factors are equal, $\alpha_1 = \alpha_2$.
Properties We take the density of water to be 1 kg/L = 1000 kg/m³ and its
specific heat to be 4.18 kJ/kg·°C.
Analysis (*a*) The mass flow rate of water through the pump is

$$\dot{m} = \rho\dot{V} = (1\text{ kg/L})(50\text{ L/s}) = 50\text{ kg/s}$$

The motor draws 15 kW of power and is 90 percent efficient. Thus the
mechanical (shaft) power it delivers to the pump is

$$\dot{W}_{\text{pump, shaft}} = \eta_{\text{motor}}\dot{W}_{\text{electric}} = (0.90)(15\text{ kW}) = 13.5\text{ kW}$$

To determine the mechanical efficiency of the pump, we need to know the
increase in the mechanical energy of the fluid as it flows through the pump,
which is

$$\Delta\dot{E}_{\text{mech, fluid}} = \dot{E}_{\text{mech, out}} - \dot{E}_{\text{mech, in}} = \dot{m}\left(\frac{P_2}{\rho} + \alpha_2\frac{V_2^2}{2} + gz_2\right) - \dot{m}\left(\frac{P_1}{\rho} + \alpha_1\frac{V_1^2}{2} + gz_1\right)$$

Simplifying it for this case and substituting the given values,

$$\Delta\dot{E}_{\text{mech, fluid}} = \dot{m}\left(\frac{P_2 - P_1}{\rho}\right) = (50\text{ kg/s})\left(\frac{(300-100)\text{ kPa}}{1000\text{ kg/m}^3}\right)\left(\frac{1\text{ kJ}}{1\text{ kPa}\cdot\text{m}^3}\right) = 10.0\text{ kW}$$

Then the mechanical efficiency of the pump becomes

$$\eta_{\text{pump}} = \frac{\dot{W}_{\text{pump, }u}}{\dot{W}_{\text{pump, shaft}}} = \frac{\Delta\dot{E}_{\text{mech, fluid}}}{\dot{W}_{\text{pump, shaft}}} = \frac{10.0\text{ kW}}{13.5\text{ kW}} = 0.741 \quad\text{or}\quad 74.1\%$$

(*b*) Of the 13.5-kW mechanical power supplied by the pump, only 10.0 kW
is imparted to the fluid as mechanical energy. The remaining 3.5 kW is converted to thermal energy due to frictional effects, and this "lost" mechanical
energy manifests itself as a heating effect in the fluid,

$$\dot{E}_{\text{mech, loss}} = \dot{W}_{\text{pump,shaft}} - \Delta\dot{E}_{\text{mech, fluid}} = 13.5 - 10.0 = 3.5\text{kW}$$

The temperature rise of water due to this mechanical inefficiency is deter-
mined from the thermal energy balance, $\dot{E}_{\text{mech, loss}} = \dot{m}(u_2 - u_1) = \dot{m}c\Delta T$.
Solving for ΔT,

$$\Delta T = \frac{\dot{E}_{\text{mech, loss}}}{\dot{m}c} = \frac{3.5\text{ kW}}{(50\text{ kg/s})(4.18\text{ kJ/kg·°C})} = 0.017°C$$

Therefore, the water experiences a temperature rise of 0.017°C which is very
small, due to mechanical inefficiency, as it flows through the pump.
Discussion In an actual application, the temperature rise of water would
probably be less since part of the heat generated would be transferred to
the casing of the pump and from the casing to the surrounding air. If the entire
pump and motor were submerged in water, then the 1.5 kW dissipated due to
motor inefficiency would also be transferred to the surrounding water as heat.

FIGURE 5–59
Schematic for Example 5–13.

EXAMPLE 5–13   Hydroelectric Power Generation from a Dam

In a hydroelectric power plant, 100 m³/s of water flows from an elevation of
120 m to a turbine, where electric power is generated (Fig. 5–59). The total
irreversible head loss in the piping system from point 1 to point 2 (excluding

the turbine unit) is determined to be 35 m. If the overall efficiency of the turbine–generator is 80 percent, estimate the electric power output.

SOLUTION The available head, flow rate, head loss, and efficiency of a hydroelectric turbine are given. The electric power output is to be determined.
Assumptions **1** The flow is steady and incompressible. **2** Water levels at the reservoir and the discharge site remain constant.
Properties We take the density of water to be 1000 kg/m³.
Analysis The mass flow rate of water through the turbine is

$$\dot{m} = \rho \dot{V} = (1000 \text{ kg/m}^3)(100 \text{ m}^3/\text{s}) = 10^5 \text{ kg/s}$$

We take point 2 as the reference level, and thus $z_2 = 0$. Also, both points 1 and 2 are open to the atmosphere ($P_1 = P_2 = P_{atm}$) and the flow velocities are negligible at both points ($V_1 = V_2 = 0$). Then the energy equation for steady, incompressible flow reduces to

$$\frac{\cancel{P_1}}{\rho g} + \alpha_1 \frac{\cancel{V_1^2}}{2g} + z_1 + \cancel{h_{pump, u}}^{\,0} = \frac{\cancel{P_2}}{\rho g} + \alpha_2 \frac{\cancel{V_2^2}}{2g} + \cancel{z_2}^{\,0} + h_{turbine, e} + h_L$$

or

$$h_{turbine, e} = z_1 - h_L$$

Substituting, the extracted turbine head and the corresponding turbine power are

$$h_{turbine, e} = z_1 - h_L = 120 - 35 = 85 \text{ m}$$

$$\dot{W}_{turbine, e} = \dot{m} g h_{turbine, e} = (10^5 \text{ kg/s})(9.81 \text{ m/s}^2)(85 \text{ m})\left(\frac{1 \text{ kJ/kg}}{1000 \text{ m}^2/\text{s}^2}\right) = 83,400 \text{ kW}$$

Therefore, a perfect turbine–generator would generate 83,400 kW of electricity from this resource. The electric power generated by the actual unit is

$$\dot{W}_{electric} = \eta_{turbine-gen} \dot{W}_{turbine, e} = (0.80)(83.4 \text{ MW}) = \textbf{66.7 MW}$$

Discussion Note that the power generation would increase by almost 1 MW for each percentage point improvement in the efficiency of the turbine–generator unit. You will learn how to determine h_L in Chap. 8.

■ EXAMPLE 5–14   Fan Selection for Air Cooling of a Computer

A fan is to be selected to cool a computer case whose dimensions are 12 cm × 40 cm × 40 cm (Fig. 5–60). Half of the volume in the case is expected to be filled with components and the other half to be air space. A 5-cm-diameter hole is available at the back of the case for the installation of the fan that is to replace the air in the void spaces of the case once every second. Small low-power fan–motor combined units are available in the market and their efficiency is estimated to be 30 percent. Determine (*a*) the wattage of the fan–motor unit to be purchased and (*b*) the pressure difference across the fan. Take the air density to be 1.20 kg/m³.

Solution A fan is to cool a computer case by completely replacing the air inside once every second. The power of the fan and the pressure difference across it are to be determined.
Assumptions **1** The flow is steady and incompressible. **2** Losses other than those due to the inefficiency of the fan–motor unit are negligible. **3** The flow

FIGURE 5–60
Schematic for Example 5–14.

at the outlet is fairly uniform except near the center (due to the wake of the fan motor), and the kinetic energy correction factor at the outlet is 1.10.

Properties   The density of air is given to be 1.20 kg/m^3.

Analysis   (a) Noting that half of the volume of the case is occupied by the components, the air volume in the computer case is

$$V = (\text{Void fraction})(\text{Total case volume})$$

$$= 0.5(12 \text{ cm} \times 40 \text{ cm} \times 40 \text{ cm}) = 9600 \text{ cm}^3$$

Therefore, the volume and mass flow rates of air through the case are

$$\dot{V} = \frac{V}{\Delta t} = \frac{9600 \text{ cm}^3}{1\text{s}} = 9600 \text{ cm}^3/\text{s} = 9.6 \times 10^{-3} \text{m}^3/\text{s}$$

$$\dot{m} = \rho \dot{V} = (1.20 \text{ kg/m}^3)(9.6 \times 10^{-3} \text{m}^3/\text{s}) = 0.0115 \text{ kg/s}$$

The cross-sectional area of the opening in the case and the average air velocity through the outlet are

$$A = \frac{\pi D^2}{4} = \frac{\pi (0.05 \text{ m})^2}{4} = 1.96 \times 10^{-3} \text{m}^2$$

$$V = \frac{\dot{V}}{A} = \frac{9.6 \times 10^{-3} \text{m}^3/\text{s}}{1.96 \times 10^{-3} \text{m}^2} = 4.90 \text{ m/s}$$

We draw the control volume around the fan such that both the inlet and the outlet are at atmospheric pressure ($P_1 = P_2 = P_{\text{atm}}$), as shown in Fig. 5–60, where the inlet section 1 is large and far from the fan so that the flow velocity at the inlet section is negligible ($V_1 \cong 0$). Noting that $z_1 = z_2$ and frictional losses in the flow are disregarded, the mechanical losses consist of fan losses only and the energy equation (Eq. 5–76) simplifies to

$$\dot{m}\left(\cancel{\frac{P_1}{\rho}} + \alpha_1 \cancel{\frac{V_1^2}{2}}^{0} + \cancel{g z_1}\right) + \dot{W}_{\text{fan}} = \dot{m}\left(\cancel{\frac{P_2}{\rho}} + \alpha_2 \frac{V_2^2}{2} + \cancel{g z_2}\right) + \cancel{\dot{W}_{\text{turbine}}}^{0} + \dot{E}_{\text{mech loss, fan}}$$

Solving for $\dot{W}_{\text{fan}} - \dot{E}_{\text{mech loss, fan}} = \dot{W}_{\text{fan, }u}$ and substituting,

$$\dot{W}_{\text{fan, }u} = \dot{m}\alpha_2 \frac{V_2^2}{2} = (0.0115 \text{ kg/s})(1.10)\frac{(4.90 \text{ m/s})^2}{2}\left(\frac{1 \text{ N}}{1 \text{kg} \cdot \text{m/s}^2}\right) = 0.152 \text{ W}$$

Then the required electric power input to the fan is determined to be

$$\dot{W}_{\text{elect}} = \frac{\dot{W}_{\text{fan, }u}}{\eta_{\text{fan−motor}}} = \frac{0.152 \text{ W}}{0.3} = 0.506 \text{ W}$$

Therefore, a fan–motor rated at about a half watt is adequate for this job (Fig. 5–61). (b) To determine the pressure difference across the fan unit, we take points 3 and 4 to be on the two sides of the fan on a horizontal line. This time $z_3 = z_4$ again and $V_3 = V_4$ since the fan is a narrow cross section, and the energy equation reduces to

$$\dot{m}\frac{P_3}{\rho} + \dot{W}_{\text{fan}} = \dot{m}\frac{P_4}{\rho} + \dot{E}_{\text{mech loss, fan}} \quad \rightarrow \quad \dot{W}_{\text{fan, }u} = \dot{m}\frac{P_4 - P_3}{\rho}$$

FIGURE 5–61
The cooling fans used in computers and computer power supplies are typically small and consume only a few watts of electrical power.

© *PhotoDisc/Getty RF*

Solving for $P_4 - P_3$ and substituting,

$$P_4 - P_3 = \frac{\rho \dot{W}_{\text{fan},u}}{\dot{m}} = \frac{(1.2 \text{ kg/m}^3)(0.152 \text{ W})}{0.0115 \text{ kg/s}} \left(\frac{1 \text{Pa·m}^3}{1 \text{ Ws}} \right) = \textbf{15.8 Pa}$$

Therefore, the pressure rise across the fan is 15.8 Pa.

Discussion The efficiency of the fan–motor unit is given to be 30 percent, which means 30 percent of the electric power $\dot{W}_{\text{electric}}$ consumed by the unit is converted to useful mechanical energy while the rest (70 percent) is "lost" and converted to thermal energy. Also, a more powerful fan is required in an actual system to overcome frictional losses inside the computer case. Note that if we had ignored the kinetic energy correction factor at the outlet, the required electrical power and pressure rise would have been 10 percent lower in this case (0.460 W and 14.4 Pa, respectively).

■ EXAMPLE 5–15   Pumping Water from a Lake to a Pool

A submersible pump with a shaft power of 5 kW and an efficiency of 72 percent is used to pump water from a lake to a pool through a constant diameter pipe (Fig. 5–62). The free surface of the pool is 25 m above the free surface of the lake. If the irreversible head loss in the piping system is 4 m, determine the discharge rate of water and the pressure difference across the pump.

SOLUTION Water from a lake is pumped to a pool at a given elevation. For a given head loss, the flow rate and the pressure difference across the pump are to be determined.

Assumptions **1** The flow is steady and incompressible. **2** Both the lake and pool are large enough that their surface elevations remain fixed.

Properties We take the density of water to be 1 kg/L = 1000 kg/m³.

Analysis The pump delivers 5 kW of shaft power and is 72 percent efficient. The useful mechanical power it imparts to the water is

$$\dot{W}_{\text{pump }u} = \eta_{\text{pump}} \dot{W}_{\text{shaft}} = (0.72)(5 \text{ kW}) = 3.6 \text{ kW}$$

We take point 1 at the free surface of the lake, which is also taken as the reference level ($z_1 = 0$), and point 2 at the free surface of the pool. Also, both points 1 and 2 are open to the atmosphere ($P_1 = P_2 = P_{\text{atm}}$), and the velocities are negligible there ($V_1 \cong V_2 \cong 0$). Then the energy equation for steady, incompressible flow through a control volume between these two surfaces that includes the pump is expressed as

$$\dot{m}\left(\frac{P_1}{\rho} + \alpha_1 \frac{V_1^2}{2} + gz_1 \right) + \dot{W}_{\text{pump},u} = \dot{m}\left(\frac{P_2}{\rho} + \alpha_2 \frac{V_2^2}{2} + gz_2 \right)$$

$$+ \dot{W}_{\text{turbine},e} + \dot{E}_{\text{mech loss, piping}}$$

Under the stated assumptions, the energy equation reduces to

$$\dot{W}_{\text{pump},u} = \dot{m}gz_2 + \dot{E}_{\text{mech loss, piping}}$$

FIGURE 5–62
Schematic for Example 5–15.

Noting that $\dot{E}_{\text{mech loss, piping}} = \dot{m}gh_L$, the mass and volume flow rates of water become

$$\dot{m} = \frac{\dot{W}_{\text{pump},u}}{gz_2 + gh_L} = \frac{\dot{W}_{\text{pump},u}}{g(z_2 + h_L)} = \frac{3.6 \text{ kJ/s}}{(9.81 \text{ m/s}^2)(25 + 4 \text{ m})} \left(\frac{1000 \text{ m}^2/\text{s}^2}{1 \text{ kJ}}\right) = 12.7 \text{ kg/s}$$

$$\dot{V} = \frac{\dot{m}}{\rho} = \frac{12.7 \text{ kg/s}}{1000 \text{ kg/m}^3} = 12.7 \times 10^{-3} \text{ m}^3/\text{s} = 12.7 \text{ L/s}$$

We now take the pump as the control volume. Assuming that the elevation difference and the kinetic energy change across the pump are negligible, the energy equation for this control volume yields

$$\Delta P = P_{\text{out}} - P_{\text{in}} = \frac{\dot{W}_{\text{pump},u}}{\dot{V}} = \frac{3.6 \text{ kJ/s}}{12.7 \times 10^{-3} \text{ m}^3/\text{s}} \left(\frac{1 \text{ kN·m}}{1 \text{ kJ}}\right)\left(\frac{1 \text{ kPa}}{1 \text{ kN/m}^2}\right)$$

$$= 283 \text{ kPa}$$

Discussion   It can be shown that in the absence of head loss ($h_L = 0$) the flow rate of water would be 14.7 L/s, which is an increase of 16 percent. Therefore, frictional losses in pipes should be minimized since they always cause the flow rate to decrease.

SUMMARY

This chapter deals with the mass, Bernoulli, and energy equations and their applications. The amount of mass flowing through a cross section per unit time is called the *mass flow rate* and is expressed as

$$\dot{m} = \rho V A_c = \rho \dot{V}$$

where ρ is the density, V is the average velocity, $\dot{V}$ is the volume flow rate of the fluid, and A_c is the cross-sectional area normal to the flow direction. The conservation of mass relation for a control volume is expressed as

$$\frac{d}{dt}\int_{\text{CV}} \rho \, dV + \int_{\text{CS}} \rho(\vec{V}\cdot\vec{n}) \, dA = 0$$

It states that *the time rate of change of the mass within the control volume plus the net mass flow rate out of the control surface is equal to zero.*

In simpler terms,

$$\frac{dm_{\text{CV}}}{dt} = \sum_{\text{in}} \dot{m} - \sum_{\text{out}} \dot{m}$$

For steady-flow devices, the conservation of mass principle is expressed as

Steady flow: $$\sum_{\text{in}} \dot{m} = \sum_{\text{out}} \dot{m}$$

Steady flow (single stream):

$$\dot{m}_1 = \dot{m}_2 \quad \rightarrow \quad \rho_1 V_1 A_1 = \rho_2 V_2 A_2$$

Steady, incompressible flow: $$\sum_{\text{in}} \dot{V} = \sum_{\text{out}} \dot{V}$$

Steady, incompressible flow (single stream):

$$\dot{V}_1 = \dot{V}_2 \rightarrow V_1 A_1 = V_2 A_2$$

Mechanical energy is the form of energy associated with the velocity, elevation, and pressure of the fluid, and it can be converted to mechanical work completely and directly by an ideal mechanical device. The efficiencies of various *real* devices are defined as

$$\eta_{\text{pump}} = \frac{\Delta \dot{E}_{\text{mech, fluid}}}{\dot{W}_{\text{shaft, in}}} = \frac{\dot{W}_{\text{pump},u}}{\dot{W}_{\text{pump}}}$$

$$\eta_{\text{turbine}} = \frac{\dot{W}_{\text{shaft, out}}}{|\Delta \dot{E}_{\text{mech, fluid}}|} = \frac{\dot{W}_{\text{turbine}}}{\dot{W}_{\text{turbine},e}}$$

$$\eta_{\text{motor}} = \frac{\text{Mechanical power output}}{\text{Electric power input}} = \frac{\dot{W}_{\text{shaft, out}}}{\dot{W}_{\text{elect, in}}}$$

$$\eta_{\text{generator}} = \frac{\text{Electric power output}}{\text{Mechanical power input}} = \frac{\dot{W}_{\text{elect, out}}}{\dot{W}_{\text{shaft, in}}}$$

$$\eta_{\text{pump-motor}} = \eta_{\text{pump}}\eta_{\text{motor}} = \frac{\Delta \dot{E}_{\text{mech, fluid}}}{\dot{W}_{\text{elect, in}}} = \frac{\dot{W}_{\text{pump, }u}}{\dot{W}_{\text{elect, in}}}$$

$$\eta_{\text{turbine–gen}} = \eta_{\text{turbine}}\eta_{\text{generator}} = \frac{\dot{W}_{\text{elect, out}}}{|\Delta \dot{E}_{\text{mech, fluid}}|} = \frac{\dot{W}_{\text{elect, out}}}{\dot{W}_{\text{turbine, }e}}$$

The *Bernoulli equation* is a relation between pressure, velocity, and elevation in steady, incompressible flow, and is expressed along a streamline and in regions where net viscous forces are negligible as

$$\frac{P}{\rho} + \frac{V^2}{2} + gz = \text{constant}$$

It can also be expressed between any two points on a streamline as

$$\frac{P_1}{\rho} + \frac{V_1^2}{2} + gz_1 = \frac{P_2}{\rho} + \frac{V_2^2}{2} + gz_2$$

The Bernoulli equation is an expression of mechanical energy balance and can be stated as: *The sum of the kinetic, potential, and flow energies of a fluid particle is constant along a streamline during steady flow when the compressibility and frictional effects are negligible.* Multiplying the Bernoulli equation by density gives

$$P + \rho \frac{V^2}{2} + \rho gz = \text{constant}$$

where P is the *static pressure,* which represents the actual pressure of the fluid; $\rho V^2/2$ is the *dynamic pressure,* which represents the pressure rise when the fluid in motion is brought to a stop; and ρgz is the *hydrostatic pressure,* which accounts for the effects of fluid weight on pressure. The sum of the static, dynamic, and hydrostatic pressures is called the *total pressure.* The Bernoulli equation states that *the total pressure along a streamline is constant.* The sum of the static and dynamic pressures is called the *stagnation pressure,* which represents the pressure at a point where the fluid is brought to a complete stop in an isentropic manner.

The Bernoulli equation can also be represented in terms of "heads" by dividing each term by g,

$$\frac{P}{\rho g} + \frac{V^2}{2g} + z = H = \text{constant}$$

where $P/\rho g$ is the *pressure head,* which represents the height of a fluid column that produces the static pressure P; $V^2/2g$ is the *velocity head,* which represents the elevation needed for a fluid to reach the velocity V during frictionless free fall; and z is the *elevation head,* which represents the potential energy of the fluid. Also, H is the *total head* for the flow. The curve that represents the sum of the static pressure and the elevation heads, $P/\rho g + z$, is called the *hydraulic grade line* (HGL), and the curve that represents the total head of the fluid, $P/\rho g + V^2/2g + z$, is called the *energy grade line* (EGL).

The *energy equation* for steady, incompressible flow is

$$\frac{P_1}{\rho g} + \alpha_1 \frac{V_1^2}{2g} + z_1 + h_{\text{pump, }u}$$
$$= \frac{P_2}{\rho g} + \alpha_2 \frac{V_2^2}{2g} + z_2 + h_{\text{turbine, }e} + h_L$$

where

$$h_{\text{pump, }u} = \frac{w_{\text{pump, }u}}{g} = \frac{\dot{W}_{\text{pump, }u}}{\dot{m}g} = \frac{\eta_{\text{pump}}\dot{W}_{\text{pump}}}{\dot{m}g}$$

$$h_{\text{turbine, }e} = \frac{w_{\text{turbine, }e}}{g} = \frac{\dot{W}_{\text{turbine, }e}}{\dot{m}g} = \frac{\dot{W}_{\text{turbine}}}{\eta_{\text{turbine}}\dot{m}g}$$

$$h_L = \frac{e_{\text{mech loss, piping}}}{g} = \frac{\dot{E}_{\text{mech loss, piping}}}{\dot{m}g}$$

$$e_{\text{mech, loss}} = u_2 - u_1 - q_{\text{net in}}$$

The mass, Bernoulli, and energy equations are three of the most fundamental relations in fluid mechanics, and they are used extensively in the chapters that follow. In Chap. 6, either the Bernoulli equation or the energy equation is used together with the mass and momentum equations to determine the forces and torques acting on fluid systems. In Chaps. 8 and 14, the mass and energy equations are used to determine the pumping power requirements in fluid systems and in the design and analysis of turbomachinery. In Chaps. 12 and 13, the energy equation is also used to some extent in the analysis of compressible flow and open-channel flow.

REFERENCES AND SUGGESTED READING

1. R. C. Dorf, ed. in chief. *The Engineering Handbook,* 2nd ed. Boca Raton, FL: CRC Press, 2004.

2. R. L. Panton. *Incompressible Flow*, 3rd ed. New York: Wiley, 2005.

3. M. Van Dyke. *An Album of Fluid Motion*. Stanford, CA: The Parabolic Press, 1982.

PROBLEMS*

Conservation of Mass

5–1C Name four physical quantities that are conserved and two quantities that are not conserved during a process.

5–2C Define mass and volume flow rates. How are they related to each other?

5–3C Does the amount of mass entering a control volume have to be equal to the amount of mass leaving during an unsteady-flow process?

5–4C When is the flow through a control volume steady?

5–5C Consider a device with one inlet and one outlet. If the volume flow rates at the inlet and at the outlet are the same, is the flow through this device necessarily steady? Why?

5–6 In climates with low night-time temperatures, an energy-efficient way of cooling a house is to install a fan in the ceiling that draws air from the interior of the house and discharges it to a ventilated attic space. Consider a house whose interior air volume is 720 m³. If air in the house is to be exchanged once every 20 minutes, determine (a) the required flow rate of the fan and (b) the average discharge speed of air if the fan diameter is 0.5 m.

5–7E A garden hose attached with a nozzle is used to fill a 20-gal bucket. The inner diameter of the hose is 1 in and it reduces to 0.5 in at the nozzle exit. If the average velocity in the hose is 8 ft/s, determine (a) the volume and mass flow rates of water through the hose, (b) how long it will take to fill the bucket with water, and (c) the average velocity of water at the nozzle exit.

5–8E Air whose density is 0.082 lbm/ft³ enters the duct of an air-conditioning system at a volume flow rate of 450 ft³/min. If the diameter of the duct is 16 in, determine the velocity of the air at the duct inlet and the mass flow rate of air.

5–9 A 0.75-m³ rigid tank initially contains air whose density is 1.18 kg/m³. The tank is connected to a high-pressure supply line through a valve. The valve is opened, and air is allowed to enter the tank until the density in the tank rises to 4.95 kg/m³. Determine the mass of air that has entered the tank. *Answer:* 2.83 kg

5–10 Consider the flow of an incompressible Newtonian fluid between two parallel plates. If the upper plate moves to right with $u_1 = 3$ m/s while the bottom one moves to the left with

$u_2 = 0.75$ m/s, what would be the net flow rate at a cross-section between two plates? Take the plate width to be $b = 5$ cm.

5–11 Consider a fully filled tank of semi-circular cross section tank with radius R and width of b into the page, as shown in Fig. P5-11. If the water is pumped out of the tank at flow rate of $\dot{V} = Kh^2$, where K is a positive constant and h is the water depth at time t. Determine the time needed to drop the water level to a specified h value of h_o in terms of R, K, and h_o.

FIGURE P5–11

5–12 A desktop computer is to be cooled by a fan whose flow rate is 0.40 m³/min. Determine the mass flow rate of air through the fan at an elevation of 3400 m where the air density is 0.7 kg/m³. Also, if the average velocity of air is not to exceed 110 m/min, determine the minimum diameter of the casing of the fan.   *Answers:* 0.00467 kg/s, 0.0569 m

5–13 A smoking lounge is to accommodate 40 heavy smokers. The minimum fresh air requirement for smoking lounges is specified to be 30 L/s per person (ASHRAE, Standard 62, 1989). Determine the minimum required flow rate of fresh air that needs to be supplied to the lounge, and the minimum diameter of the duct if the air velocity is not to exceed 8 m/s.

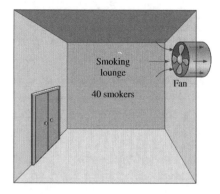

FIGURE P5–13

5–14 The minimum fresh air requirement of a residential building is specified to be 0.35 air changes per hour (ASHRAE, Standard 62, 1989). That is, 35 percent of the

---

* Problems designated by a "C" are concept questions, and students are encouraged to answer them all. Problems designated by an "E" are in English units, and the SI users can ignore them. Problems with the 🖭 icon are solved using EES, and complete solutions together with parametric studies are included on the text website. Problems with the 🖭 icon are comprehensive in nature and are intended to be solved with an equation solver such as EES.

entire air contained in a residence should be replaced by fresh outdoor air every hour. If the ventilation requirement of a 2.7-m-high, 200-m² residence is to be met entirely by a fan, determine the flow capacity in L/min of the fan that needs to be installed. Also determine the minimum diameter of the duct if the average air velocity is not to exceed 5 m/s.

5–15 Air enters a nozzle steadily at 2.21 kg/m³ and 20 m/s and leaves at 0.762 kg/m³ and 150 m/s. If the inlet area of the nozzle is 60 cm², determine (a) the mass flow rate through the nozzle, and (b) the exit area of the nozzle. *Answers:* (a) 0.265 kg/s, (b) 23.2 cm²

5–16 Air at 40°C flow steadily through the pipe shown in Fig. P5–16. If $P_1 = 50$ kPa (gage), $P_2 = 10$ kPa (gage), $D = 3d$, $P_{atm} \cong 100$ kPa, the average velocity at section 2 is $V_2 = 30$ m/s, and air temperature remains nearly constant, determine the average speed at section 1.

FIGURE P5–16

5–17 A hair dryer is basically a duct of constant diameter in which a few layers of electric resistors are placed. A small fan pulls the air in and forces it through the resistors where it is heated. If the density of air is 1.20 kg/m³ at the inlet and 1.05 kg/m³ at the exit, determine the percent increase in the velocity of air as it flows through the hair dryer.

FIGURE P5–17

Mechanical Energy and Efficiency

5–18C Define turbine efficiency, generator efficiency, and combined turbine–generator efficiency.

5–19C What is mechanical efficiency? What does a mechanical efficiency of 100 percent mean for a hydraulic turbine?

5–20C How is the combined pump–motor efficiency of a pump and motor system defined? Can the combined pump–motor efficiency be greater than either the pump or the motor efficiency?

5–21C What is mechanical energy? How does it differ from thermal energy? What are the forms of mechanical energy of a fluid stream?

5–22 At a certain location, wind is blowing steadily at 8 m/s. Determine the mechanical energy of air per unit mass and the power generation potential of a wind turbine with 50-m-diameter blades at that location. Also determine the actual electric power generation assuming an overall efficiency of 30 percent. Take the air density to be 1.25 kg/m³.

5–23 [EES] Reconsider Prob. 5–22. Using EES (or other) software, investigate the effect of wind velocity and the blade span diameter on wind power generation. Let the velocity vary from 5 to 20 m/s in increments of 5 m/s, and the diameter to vary from 20 to 80 m in increments of 20 m. Tabulate the results, and discuss their significance.

5–24E A differential thermocouple with sensors at the inlet and exit of a pump indicates that the temperature of water rises 0.048°F as it flows through the pump at a rate of 1.5 ft³/s. If the shaft power input to the pump is 23 hp and the heat loss to the surrounding air is negligible, determine the mechanical efficiency of the pump. *Answer:* 72.4 percent

FIGURE P5–24E

5–25 Electric power is to be generated by installing a hydraulic turbine–generator at a site 110 m below the free surface of a large water reservoir that can supply water steadily at a rate of 900 kg/s. If the mechanical power output of the turbine is 800 kW and the electric power generation is 750 kW, determine the turbine efficiency and the combined turbine–generator efficiency of this plant. Neglect losses in the pipes.

5–26 Consider a river flowing toward a lake at an average speed of 4 m/s at a rate of 500 m³/s at a location 70 m above the lake surface. Determine the total mechanical energy of the river water per unit mass and the power generation potential of the entire river at that location. *Answer:* 347 MW

FIGURE P5–26

5–27 Water is pumped from a lake to a storage tank 18 m above at a rate of 70 L/s while consuming 20.4 kW of electric power. Disregarding any frictional losses in the pipes and any changes in kinetic energy, determine (a) the overall efficiency of the pump–motor unit and (b) the pressure difference between the inlet and the exit of the pump.

FIGURE P5–27

Bernoulli Equation

5–28C What is stagnation pressure? Explain how it can be measured.

5–29C Express the Bernoulli equation in three different ways using (a) energies, (b) pressures, and (c) heads.

5–30C What are the three major assumptions used in the derivation of the Bernoulli equation?

5–31C Define static, dynamic, and hydrostatic pressure. Under what conditions is their sum constant for a flow stream?

5–32C What is streamwise acceleration? How does it differ from normal acceleration? Can a fluid particle accelerate in steady flow?

5–33C Define pressure head, velocity head, and elevation head for a fluid stream and express them for a fluid stream whose pressure is P, velocity is V, and elevation is z.

5–34C Explain how and why a siphon works. Someone proposes siphoning cold water over a 7-m-high wall. Is this feasible? Explain.

5–35C How is the location of the hydraulic grade line determined for open-channel flow? How is it determined at the outlet of a pipe discharging to the atmosphere?

5–36C In a certain application, a siphon must go over a high wall. Can water or oil with a specific gravity of 0.8 go over a higher wall? Why?

5–37C What is the hydraulic grade line? How does it differ from the energy grade line? Under what conditions do both lines coincide with the free surface of a liquid?

5–38C A glass manometer with oil as the working fluid is connected to an air duct as shown in Fig. P5–38C. Will the oil levels in the manometer be as in Fig. P5–38Ca or b? Explain. What would your response be if the flow direction is reversed?

FIGURE P5–38C

5–39C The velocity of a fluid flowing in a pipe is to be measured by two different Pitot-type mercury manometers shown in Fig. P5–39C. Would you expect both manometers to predict the same velocity for flowing water? If not, which would be more accurate? Explain. What would your response be if air were flowing in the pipe instead of water?

FIGURE P5–39C

5–40C The water level of a tank on a building roof is 20 m above the ground. A hose leads from the tank bottom to the ground. The end of the hose has a nozzle, which is pointed straight up. What is the maximum height to which the water could rise? What factors would reduce this height?

5–41C A student siphons water over a 8.5-m-high wall at sea level. She then climbs to the summit of Mount Shasta (elevation 4390 m, P_{atm} = 58.5 kPa) and attempts the same experiment. Comment on her prospects for success.

5–42 In a hydroelectric power plant, water enters the turbine nozzles at 800 kPa absolute with a low velocity. If the nozzle outlets are exposed to atmospheric pressure of 100 kPa, determine the maximum velocity to which water can be accelerated by the nozzles before striking the turbine blades.

5–43 A Pitot-static probe is used to measure the speed of an aircraft flying at 3000 m. If the differential pressure reading is 3 kPa, determine the speed of the aircraft.

5–44 The air velocity in the duct of a heating system is to be measured by a Pitot-static probe inserted into the duct parallel to the flow. If the differential height between the water columns connected to the two outlets of the probe is 2.4 cm, determine (a) the flow velocity and (b) the pressure rise at the tip of the probe. The air temperature and pressure in the duct are 45°C and 98 kPa, respectively.

5–45E The drinking water needs of an office are met by large water bottles. One end of a 0.25-in-diameter plastic hose is inserted into the bottle placed on a high stand, while the other end with an on/off valve is maintained 2 ft below the bottom of the bottle. If the water level in the bottle is 1.5 ft when it is full, determine how long it will take at the minimum to fill an 8-oz glass ($= 0.00835 \text{ ft}^3$) (*a*) when the bottle is first opened and (*b*) when the bottle is almost empty. Neglect frictional losses.

FIGURE P5–45E

5–46 A piezometer and a Pitot tube are tapped into a 4-cm-diameter horizontal water pipe, and the height of the water columns are measured to be 26 cm in the piezometer and 35 cm in the Pitot tube (both measured from the top surface of the pipe). Determine the velocity at the center of the pipe.

5–47 The diameter of a cylindrical water tank is D_o and its height is H. The tank is filled with water, which is open to the atmosphere. An orifice of diameter D with a smooth entrance (i.e., negligible losses) is open at the bottom. Develop a relation for the time required for the tank (*a*) to empty halfway and (*b*) to empty completely.

5–48E A siphon pumps water from a large reservoir to a lower tank that is initially empty. The tank also has a rounded orifice 20 ft below the reservoir surface where the water leaves the tank. Both the siphon and the orifice diameters are 2 in. Ignoring frictional losses, determine to what height the water will rise in the tank at equilibrium.

5–49 Water enters a tank of diameter D_T steadily at a mass flow rate of $\dot{m}_{\text{in}}$. An orifice at the bottom with diameter D_o allows water to escape. The orifice has a rounded entrance, so the frictional losses are negligible. If the tank is initially empty, (*a*) determine the maximum height that the water will

reach in the tank and (*b*) obtain a relation for water height z as a function of time.

FIGURE P5–49

5–50E Water flows through a horizontal pipe at a rate of 2.4 gal/s. The pipe consists of two sections of diameters 4 in and 2 in with a smooth reducing section. The pressure difference between the two pipe sections is measured by a mercury manometer. Neglecting frictional effects, determine the differential height of mercury between the two pipe sections. *Answer: 3.0 in*

FIGURE P5–50E

5–51 An airplane is flying at an altitude of 12,000 m. Determine the gage pressure at the stagnation point on the nose of the plane if the speed of the plane is 300 km/h. How would you solve this problem if the speed were 1050 km/h? Explain.

5–52 While traveling on a dirt road, the bottom of a car hits a sharp rock and a small hole develops at the bottom of its gas tank. If the height of the gasoline in the tank is 30 cm, determine the initial velocity of the gasoline at the hole. Discuss how the velocity will change with time and how the flow will be affected if the lid of the tank is closed tightly. *Answer: 2.43 m/s*

5–53 The water in an 8-m-diameter, 3-m-high above-ground swimming pool is to be emptied by unplugging a 3-cm-diameter, 25-m-long horizontal pipe attached to the bottom of the pool. Determine the maximum discharge rate of water through the pipe. Also, explain why the actual flow rate will be less.

5–54 Reconsider Prob. 5–53. Determine how long it will take to empty the swimming pool completely. *Answer:* 15.4 h

5–55 [EES] Reconsider Prob. 5–54. Using EES (or other) software, investigate the effect of the discharge pipe diameter on the time required to empty the pool completely. Let the diameter vary from 1 to 10 cm in increments of 1 cm. Tabulate and plot the results.

5–56 Air at 105 kPa and 37°C flows upward through a 6-cm-diameter inclined duct at a rate of 65 L/s. The duct diameter is then reduced to 4 cm through a reducer. The pressure change across the reducer is measured by a water manometer. The elevation difference between the two points on the pipe where the two arms of the manometer are attached is 0.20 m. Determine the differential height between the fluid levels of the two arms of the manometer.

FIGURE P5–56

5–57 A handheld bicycle pump can be used as an atomizer to generate a fine mist of paint or pesticide by forcing air at a high velocity through a small hole and placing a short tube between the liquid reservoir and the high-speed air jet. The pressure across a subsonic jet exposed to the atmosphere is nearly atmospheric, and the surface of the liquid in the reservoir is also open to atmospheric pressure. In light of this, explain how the liquid is sucked up the tube. *Hint:* Read Sec. 5-4 carefully.

FIGURE P5–57

5–58 Water at 20°C is siphoned from a reservoir as shown in Fig. P5–58. For $d = 10$ cm and $D = 16$ cm, determine (a) the minimum flow rate that can be achieved without cavitation occurring in the piping system and (b) the maximum elevation of the highest point of the piping system to avoid cavitation.

FIGURE P5–58

5–59 The water pressure in the mains of a city at a particular location is 270 kPa gage. Determine if this main can serve water to neighborhoods that are 25 m above this location.

5–60 A pressurized tank of water has a 10-cm-diameter orifice at the bottom, where water discharges to the atmosphere. The water level is 2.5 m above the outlet. The tank air pressure above the water level is 250 kPa (absolute) while the atmospheric pressure is 100 kPa. Neglecting frictional effects, determine the initial discharge rate of water from the tank. *Answer:* 0.147 m³/s

FIGURE P5–60

5–61 [EES] Reconsider Prob. 5–60. Using EES (or other) software, investigate the effect of water height in the tank on the discharge velocity. Let the water height vary from 0 to 5 m in increments of 0.5 m. Tabulate and plot the results.

5–62E Air is flowing through a venturi meter whose diameter is 2.6 in at the entrance part (location 1) and 1.8 in at the throat (location 2). The gage pressure is measured to be 12.2 psia at the entrance and 11.8 psia at the throat. Neglecting frictional effects, show that the volume flow rate can be expressed as

$$\dot{V} = A_2 \sqrt{\frac{2(P_1 - P_2)}{\rho(1 - A_2^2/A_1^2)}}$$

and determine the flow rate of air. Take the air density to be 0.075 lbm/ft³.

12.2 psia   11.8 psia

Air   2.6 in   1.8 in

FIGURE P5–62E

5–63 The water level in a tank is 15 m above the ground. A hose is connected to the bottom of the tank, and the nozzle at the end of the hose is pointed straight up. The tank cover is airtight, and the air pressure above the water surface is 3 atm gage. The system is at sea level. Determine the maximum height to which the water stream could rise.   *Answer:* 46.0 m

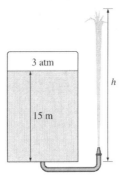

3 atm

h

15 m

FIGURE P5–63

5–64 A Pitot-static probe connected to a water manometer is used to measure the velocity of air. If the deflection (the vertical distance between the fluid levels in the two arms) is 5.5 cm, determine the air velocity. Take the density of air to be 1.16 kg/m³.

Pitot-static
Air   probe

5.5 cm

Manometer

FIGURE P5–64

5–65E The air velocity in a duct is measured by a Pitot-static probe connected to a differential pressure gage. If the air is at 13.4 psia absolute and 70°F and the reading of the differential pressure gage is 0.15 psi, determine the air velocity. *Answer:* 143 ft/s

5–66 In cold climates, water pipes may freeze and burst if proper precautions are not taken. In such an occurrence, the exposed part of a pipe on the ground ruptures, and water shoots up to 42 m. Estimate the gage pressure of water in the pipe. State your assumptions and discuss if the actual pressure is more or less than the value you predicted.

5–67 A well-fitting piston with 4 small holes in a sealed water-filled cylinder, shown in Fig. P5-67, is pushed to the right at a constant speed of 4 mm/s while the pressure in the right compartment remains constant at 50 kPa gage. Disregarding the frictional effects, determine the force F that needs to be applied to the piston to maintain this motion.

Water P = 50 kPa

F Piston d_p = 12 cm

d_r = 3 cm   Cylinder

Hole d_h = 2 mm

FIGURE P5–67

5–68 A fluid of density ρ and viscosity μ flows through a section of horizontal converging–diverging duct. The duct cross-sectional areas A_{inlet}, A_{throat}, and A_{outlet} are known at the inlet, throat (minimum area), and outlet, respectively. Average pressure P_{outlet} is measured at the outlet, and average velocity V_{inlet} is measured at the inlet. (a) Neglecting any irreversibilities such as friction, generate expressions for the average velocity and average pressure at the inlet and the throat in terms of the given variables. (b) In a real flow (with irreversibilities), do you expect the actual pressure at the inlet to be higher or lower than the prediction? Explain.

Energy Equation

5–69C What is useful pump head? How is it related to the power input to the pump?

5–70C Consider the steady adiabatic flow of an incompressible fluid. Can the temperature of the fluid decrease during flow? Explain.

5–71C What is irreversible head loss? How is it related to the mechanical energy loss?

5–72C Consider the steady adiabatic flow of an incompressible fluid. If the temperature of the fluid remains constant during flow, is it accurate to say that the frictional effects are negligible?

5–73C What is the kinetic energy correction factor? Is it significant?

5–74C The water level in a tank is 20 m above the ground. A hose is connected to the bottom of the tank, and the nozzle at the end of the hose is pointed straight up. The water stream from the nozzle is observed to rise 25 m above the ground. Explain what may cause the water from the hose to rise above the tank level.

5–75C A person is filling a knee-high bucket with water using a garden hose and holding it such that water discharges from the hose at the level of his waist. Someone suggests that the bucket will fill faster if the hose is lowered such that water discharges from the hose at the knee level. Do you agree with this suggestion? Explain. Disregard any frictional effects.

5–76C A 3-m-high tank filled with water has a discharge valve near the bottom and another near the top. (*a*) If these two valves are opened, will there be any difference between the discharge velocities of the two water streams? (*b*) If a hose whose discharge end is left open on the ground is first connected to the lower valve and then to the higher valve, will there be any difference between the discharge rates of water for the two cases? Disregard any frictional effects.

5–77E In a hydroelectric power plant, water flows from an elevation of 400 ft to a turbine, where electric power is generated. For an overall turbine–generator efficiency of 85 percent, determine the minimum flow rate required to generate 100 kW of electricity. *Answer:* 217 lbm/s

5–78E Reconsider Prob. 5–77E. Determine the flow rate of water if the irreversible head loss of the piping system between the free surfaces of the inlet and the outlet is 36 ft.

5–79 An oil pump is drawing 25 kW of electric power while pumping oil with $\rho = 860$ kg/m^3 at a rate of 0.1 m^3/s. The inlet and outlet diameters of the pipe are 8 cm and 12 cm, respectively. If the pressure rise of oil in the pump is measured to be 250 kPa and the motor efficiency is 90 percent, determine the mechanical efficiency of the pump. Take the kinetic energy correction factor to be 1.05.

5–80 Water is being pumped from a large lake to a reservoir 25 m above at a rate of 25 L/s by a 10-kW (shaft) pump. If the irreversible head loss of the piping system is 5 m, determine the mechanical efficiency of the pump. *Answer:* 73.6 percent

5–81 Reconsider Prob. 5–80. Using EES (or other) software, investigate the effect of irreversible head loss on the mechanical efficiency of the pump. Let the head loss vary from 0 to 15 m in increments of 1 m. Plot the results, and discuss them.

5–82 A 15-hp (shaft) pump is used to raise water to a 45-m higher elevation. If the mechanical efficiency of the pump is 82 percent, determine the maximum volume flow rate of water.

5–83 Water flows at a rate of 0.035 m^3/s in a horizontal pipe whose diameter is reduced from 15 cm to 8 cm by a reducer. If the pressure at the centerline is measured to be 480 kPa and 445 kPa before and after the reducer, respectively, determine the irreversible head loss in the reducer. Take the kinetic energy correction factors to be 1.05. *Answer:* 1.18 m

5–84 The water level in a tank is 20 m above the ground. A hose is connected to the bottom of the tank, and the nozzle at the end of the hose is pointed straight up. The tank is at sea level, and the water surface is open to the atmosphere. In the line leading from the tank to the nozzle is a pump, which increases the pressure of water. If the water jet rises to a height of 27 m from the ground, determine the minimum pressure rise supplied by the pump to the water line.

FIGURE P5–84

5–85 A hydraulic turbine has 50 m of head available at a flow rate of 1.30 m^3/s, and its overall turbine–generator efficiency is 78 percent. Determine the electric power output of this turbine.

5–86 A fan is to be selected to ventilate a bathroom whose dimensions are 2 m × 3 m × 3 m. The air velocity is not to exceed 8 m/s to minimize vibration and noise. The combined efficiency of the fan–motor unit to be

FIGURE P5–79

used can be taken to be 50 percent. If the fan is to replace the entire volume of air in 10 min, determine (*a*) the wattage of the fan–motor unit to be purchased, (*b*) the diameter of the fan casing, and (*c*) the pressure difference across the fan. Take the air density to be 1.25 kg/m^3 and disregard the effect of the kinetic energy correction factors.

FIGURE P5–86

5–87 Water flows at a rate of 20 L/s through a horizontal pipe whose diameter is constant at 3 cm. The pressure drop across a valve in the pipe is measured to be 2 kPa, as shown in Fig P5–87. Determine the irreversible head loss of the valve, and the useful pumping power needed to overcome the resulting pressure drop. *Answers:* 0.204 m, 40 W

FIGURE P5–87

5–88E The water level in a tank is 34 ft above the ground. A hose is connected to the bottom of the tank at the ground level and the nozzle at the end of the hose is pointed straight up. The tank cover is airtight, but the pressure over the water surface is unknown. Determine the minimum tank air pressure (gage) that will cause a water stream from the nozzle to rise 72 ft from the ground.

5–89 A large tank is initially filled with water 5 m above the center of a sharp-edged 10-cm-diameter orifice. The tank water surface is open to the atmosphere, and the orifice drains to the atmosphere. If the total irreversible head loss in the system is 0.3 m, determine the initial discharge velocity of water from the tank. Take the kinetic energy correction factor at the orifice to be 1.2.

5–90 Water enters a hydraulic turbine through a 30-cm-diameter pipe at a rate of 0.6 m^3/s and exits through a 25-cm-diameter pipe. The pressure drop in the turbine is measured by a mercury manometer to be 1.2 m. For a combined turbine–generator efficiency of 83 percent, determine the net electric power output. Disregard the effect of the kinetic energy correction factors.

FIGURE P5–90

5–91 The velocity profile for turbulent flow in a circular pipe is approximated as $u(r) = u_{max}(1 - r/R)^{1/n}$, where $n = 9$. Determine the kinetic energy correction factor for this flow. *Answer:* 1.04

5–92 Water is pumped from a lower reservoir to a higher reservoir by a pump that provides 20 kW of useful mechanical power to the water. The free surface of the upper reservoir is 45 m higher than the surface of the lower reservoir. If the flow rate of water is measured to be 0.03 m^3/s, determine the irreversible head loss of the system and the lost mechanical power during this process.

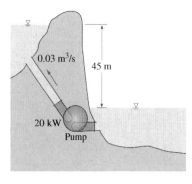

FIGURE P5–92

5–93 Water in a partially filled large tank is to be supplied to the roof top, which is 8 m above the water level in the tank, through a 2.5-cm-internal-diameter pipe by maintaining a constant air pressure of 300 kPa (gage) in the tank. If the head loss in the piping is 2 m of water, determine the discharge rate of the supply of water to the roof top.

5–94 Underground water is to be pumped by a 78 percent efficient 5-kW submerged pump to a pool whose free surface is 30 m above the underground water level. The diameter of the pipe is 7 cm on the intake side and 5 cm on the discharge side. Determine (*a*) the maximum flow rate of water and (*b*) the pressure difference across the pump. Assume the elevation difference between the pump inlet and the outlet and the effect of the kinetic energy correction factors to be negligible.

FIGURE P5–94

5–95 Reconsider Prob. 5–94. Determine the flow rate of water and the pressure difference across the pump if the irreversible head loss of the piping system is 4 m.

5–96E A 73-percent efficient 12-hp pump is pumping water from a lake to a nearby pool at a rate of 1.2 ft³/s through a constant-diameter pipe. The free surface of the pool is 35 ft above that of the lake. Determine the irreversible head loss of the piping system, in ft, and the mechanical power used to overcome it.

5–97 The demand for electric power is usually much higher during the day than it is at night, and utility companies often sell power at night at much lower prices to encourage consumers to use the available power generation capacity and to avoid building new expensive power plants that will be used only a short time during peak periods. Utilities are also willing to purchase power produced during the day from private parties at a high price.

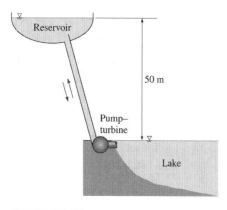

FIGURE P5–97

Suppose a utility company is selling electric power for $0.06/kWh at night and is willing to pay $0.13/kWh for power produced during the day. To take advantage of this opportunity, an entrepreneur is considering building a large reservoir 50 m above the lake level, pumping water from the lake to the reservoir at night using cheap power, and letting the water flow from the reservoir back to the lake during the day, producing power as the pump–motor operates as a turbine–generator during reverse flow. Preliminary analysis shows that a water flow rate of 2 m³/s can be used in either direction, and the irreversible head loss of the piping system is 4 m. The combined pump–motor and turbine–generator efficiencies are expected to be 75 percent each. Assuming the system operates for 10 h each in the pump and turbine modes during a typical day, determine the potential revenue this pump–turbine system can generate per year.

5–98 When a system is subjected to a linear rigid body motion with constant linear acceleration a along a distance L, the modified Bernoulli Equation takes the form

$$\left(\frac{P_1}{\rho} + \frac{V_1^2}{2} + gz_1\right) - \left(\frac{P_2}{\rho} + \frac{V_2^2}{2} + gz_2\right) = aL + \text{Losses}$$

where V_1 and V_2 are velocities relative to a fixed point and 'Losses' which represents frictional losses is zero when the frictional effects are negligible. The tank with two discharge pipes shown in Fig. P5–98 accelerates to the left at a constant linear acceleration of 3 m/s². If volumetric flow rates from both pipes are to be identical, determine the diameter D of the inclined pipe. Disregard any frictional effects.

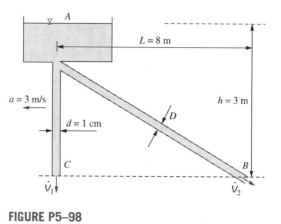

FIGURE P5–98

5–99 A fireboat is to fight fires at coastal areas by drawing seawater with a density of 1030 kg/m³ through a 10-cm-diameter pipe at a rate of 0.04 m³/s and discharging it through a hose nozzle with an exit diameter of 5 cm. The total irreversible head loss of the system is 3 m, and the position of the nozzle is 3 m above sea level. For a pump efficiency of 70 percent, determine the required shaft power input to the pump and the water discharge velocity. *Answers:* 39.2 kW, 20.4 m/s

FIGURE P5–99

Review Problems

5–100 The velocity of a liquid flowing in a circular pipe of radius R varies from zero at the wall to a maximum at the pipe center. The velocity distribution in the pipe can be represented as $V(r)$, where r is the radial distance from the pipe center. Based on the definition of mass flow rate $\dot{m}$, obtain a relation for the average velocity in terms of $V(r)$, R, and r.

5–101 Air at 2.50 kg/m³ enters a nozzle that has an inlet-to-exit area ratio of 2:1 with a velocity of 120 m/s and leaves with a velocity of 330 m/s. Determine the density of air at the exit. *Answer:* 1.82 kg/m³

5–102E The water level in a tank is 55 ft above the ground. A hose is connected to the bottom of the tank, and the nozzle at the end of the hose is pointed straight up. The tank is at sea level, and the water surface is open to the atmosphere. In the line leading from the tank to the nozzle is a pump, which increases the water pressure by 10 psia. Determine the maximum height to which the water stream could rise.

5–103 A pressurized 2-m-diameter tank of water has a 10-cm-diameter orifice at the bottom, where water discharges to the atmosphere. The water level initially is 3 m above the outlet. The tank air pressure above the water level is maintained at 450 kPa absolute and the atmospheric pressure is 100 kPa. Neglecting frictional effects, determine (a) how long it will take for half of the water in the tank to be discharged and (b) the water level in the tank after 10 s.

5–104 Air flows through a pipe at a rate of 120 L/s. The pipe consists of two sections of diameters 22 cm and 10 cm with a smooth reducing section that connects them. The pressure difference between the two pipe sections is measured by a water manometer. Neglecting frictional effects, determine the differential height of water between the two pipe sections. Take the air density to be 1.20 kg/m³. *Answer:* 1.37 cm

FIGURE P5–104

5–105 Air at 100 kPa and 25°C flows in a horizontal duct of variable cross section. The water column in the manometer that measures the difference between two sections has a vertical displacement of 8 cm. If the velocity in the first section is low and the friction is negligible, determine the velocity at the second section. Also, if the manometer reading has a possible error of ±2 mm, conduct an error analysis to estimate the range of validity for the velocity found.

5–106 A very large tank contains air at 102 kPa at a location where the atmospheric air is at 100 kPa and 20°C. Now a 2-cm-diameter tap is opened. Determine the maximum flow rate of air through the hole. What would your response be if air is discharged through a 2-m-long, 4-cm-diameter tube with a 2-cm-diameter nozzle? Would you solve the problem the same way if the pressure in the storage tank were 300 kPa?

FIGURE P5–106

5–107 Water is flowing through a Venturi meter whose diameter is 7 cm at the entrance part and 4 cm at the throat. The pressure is measured to be 380 kPa at the entrance and 150 kPa at the throat. Neglecting frictional effects, determine the flow rate of water. *Answer:* 0.0285 m³/s

5–108 Water flows at a rate of 0.011 m³/s in a horizontal pipe whose diameter increases from 6 to 11 cm by an enlargement section. If the head loss across the enlargement section is 0.65 m and the kinetic energy correction factor at both the inlet and the outlet is 1.05, determine the pressure change.

5–109 The air in a 6-m × 5-m × 4-m hospital room is to be completely replaced by conditioned air every 20 min. If the average air velocity in the circular air duct leading to the room is not to exceed 5 m/s, determine the minimum diameter of the duct.

5–110 Underground water is being pumped into a pool whose cross section is 3 m × 4 m while water is discharged through a 5-cm-diameter orifice at a constant average velocity of 5 m/s. If the water level in the pool rises at a rate of 1.5 cm/min, determine the rate at which water is supplied to the pool, in m³/s.

5–111 A 3-m-high large tank is initially filled with water. The tank water surface is open to the atmosphere, and a sharp-edged 10-cm-diameter orifice at the bottom drains to the atmosphere through a horizontal 80-m-long pipe. If the total irreversible head loss of the system is determined to be 1.5 m, determine the initial velocity of the water from the tank. Disregard the effect of the kinetic energy correction factors. *Answer:* 5.42 m/s

FIGURE P5–111

5–112 Reconsider Prob. 5–111. Using EES (or other) software, investigate the effect of the tank height on the initial discharge velocity of water from the completely filled tank. Let the tank height vary from 2 to 15 m in increments of 1 m, and assume the irreversible head loss to remain constant. Tabulate and plot the results.

5–113 Reconsider Prob. 5–111. In order to drain the tank faster, a pump is installed near the tank exit. Determine the pump head input necessary to establish an average water velocity of 6.5 m/s when the tank is full.

5–114 A $D_0 = 8$-m-diameter tank is initially filled with water 2 m above the center of a $D = 10$-cm-diameter valve near the bottom. The tank surface is open to the atmosphere, and the tank drains through a $L = 80$-m-long pipe connected to the valve. The friction factor of the pipe is given to be $f = 0.015$, and the discharge velocity is expressed as

$$V = \sqrt{\frac{2gz}{1.5 + fL/D}}$$ where z is the water height above the center

of the valve. Determine (*a*) the initial discharge velocity from the tank and (*b*) the time required to empty the tank. The tank can be considered to be empty when the water level drops to the center of the valve.

5–115 In some applications, elbow-type flow meters like the one shown in Fig. P5–115 are used to measure flow rates. The pipe radius is R, the radius of curvature of the elbow is λ, and the pressure difference ΔP across the curvature inside the pipe is measured. From the potential flow theory, it is known that $Vr = C$, where V is the fluid velocity at a distance r from the center of curvature O, and C is a constant. Assuming frictionless steady-state flow and thus the Bernoulli equation across streamlines to be applicable, obtain a relation for the flow rate as a function of ρ, g, ΔP, λ, and R.

Answer: $\dot{V} = \pi\sqrt{\dfrac{2\Delta P}{\rho g \lambda R}}\,(\lambda^2 - R^2)(\lambda - \sqrt{\lambda^2 - R^2})$

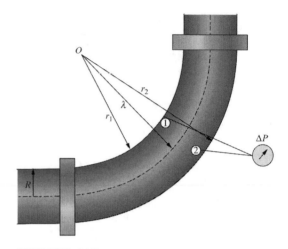

FIGURE P5–115

5–116 The cylindrical water tank with a valve at the bottom shown in Fig. P5–116 contains air at the top part at the local atmospheric pressure of 100 kPa and water as shown. Is it possible to completely empty this tank by fully opening the valve? If not, determine the water height in the tank when water stops flowing out of the fully open valve. Assume the temperature of the air inside the cylinder to remain constant during the discharging process.

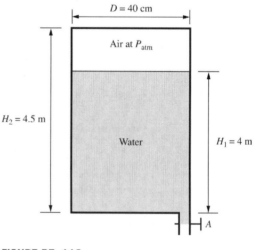

FIGURE P5–116

5–117 A rigid tank of volume 1.5 m³ initially contains atmospheric air at 20°C and 150 kPa. Now a compressor is turned on, and atmospheric air at a constant rate of 0.05 m³/s is supplied to the tank. If the pressure and density in the tank varies as $P/\rho^{1.4} = $ constant during charging, (*a*) obtain a relation for the variation of pressure in the tank with time and (*b*) calculate how long it will take for the absolute pressure in the tank to triple.

5–118 A wind tunnel draws atmospheric air at 20°C and 101.3 kPa by a large fan located near the exit of the tunnel. If the air velocity in the tunnel is 80 m/s, determine the pressure in the tunnel.

20°C
101.3 kPa
Wind tunnel
80 m/s

FIGURE P5–118

Fundamentals of Engineering (FE) Exam Problems

5–119 Water flows in a 5-cm-diameter pipe at a velocity of 0.75 m/s. The mass flow rate of water in the pipe is
(*a*) 353 kg/min   (*b*) 75 kg/min   (*c*) 37.5 kg/min
(*d*) 1.47 kg/min   (*e*) 88.4 kg/min

5–120 Air at 100 kPa and 20°C flows in a 12-cm-diameter pipe at a rate of 9.5 kg/min. The velocity of air in the pipe is
(*a*) 1.4 m/s   (*b*) 6.0 m/s   (*c*) 9.5 m/s   (*d*) 11.8 m/s
(*e*) 14.0 m/s

5–121 A water tank initially contains 140 L of water. Now, equal rates of cold and hot water enter the tank for a period of 30 minutes while warm water is discharged from the tank at a rate of 25 L/min. The amount of water in the tank at the end of this 30-min period is 50 L. The rate of hot water entering the tank is
(*a*) 33 L/min   (*b*) 25 L/min   (*c*) 11 L/min   (*d*) 7 L/min
(*e*) 5 L/min

5–122 Water enters a 4-cm-diameter pipe at a velocity of 1 m/s. The diameter of the pipe is reduced to 3 cm at the exit. The velocity of the water at the exit is
(*a*) 1.78 m/s   (*b*) 1.25 m/s   (*c*) 1 m/s   (*d*) 0.75 m/s
(*e*) 0.50 m/s

5–123 The pressure of water is increased from 100 kPa to 900 kPa by a pump. The mechanical energy increase of water is
(*a*) 0.9 kJ/kg   (*b*) 0.5 kJ/kg   (*c*) 500 kJ/kg   (*d*) 0.8 kJ/kg
(*e*) 800 kJ/kg

5–124 A 75-m-high water body that is open to the atmosphere is available. Water is run through a turbine at a rate of 200 L/s at the bottom of the water body. The pressure difference across the turbine is
(*a*) 736 kPa   (*b*) 0.736 kPa   (*c*) 1.47 kPa   (*d*) 1470 kPa
(*e*) 368 kPa

5–125 A pump is used to increase the pressure of water from 100 kPa to 900 kPa at a rate of 160 L/min. If the shaft power input to the pump is 3 kW, the efficiency of the pump is
(*a*) 0.532   (*b*) 0.660   (*c*) 0.711   (*d*) 0.747   (*e*) 0.855

5–126 A hydraulic turbine is used to generate power by using the water in a dam. The elevation difference between the free surfaces upstream and downstream of the dam is 120 m. The water is supplied to the turbine at a rate of 150 kg/s. If the shaft power output from the turbine is 155 kW, the efficiency of the turbine is
(*a*) 0.77   (*b*) 0.80   (*c*) 0.82   (*d*) 0.85   (*e*) 0.88

5–127 The motor of a pump consumes 1.05 hp of electricity. The pump increases the pressure of water from 120 kPa to 1100 kPa at a rate of 35 L/min. If the motor efficiency is 94 percent, the pump efficiency is
(*a*) 0.75   (*b*) 0.78   (*c*) 0.82   (*d*) 0.85   (*e*) 0.88

5–128 The efficiency of a hydraulic turbine-generator unit is specified to be 85 percent. If the generator efficiency is 96 percent, the turbine efficiency is
(*a*) 0.816   (*b*) 0.850   (*c*) 0.862   (*d*) 0.885   (*e*) 0.960

5–129 Which parameter is *not* related in the Bernoulli equation?
(*a*) Density   (*b*) Velocity   (*c*) Time   (*d*) Pressure
(*e*) Elevation

5–130 Consider incompressible, frictionless flow of a fluid in a horizontal piping. The pressure and velocity of a fluid is measured to be 150 kPa and 1.25 m/s at a specified point. The density of the fluid is 700 kg/m^3. If the pressure is 140 kPa at another point, the velocity of the fluid at that point is
(*a*) 1.26 m/s   (*b*) 1.34 m/s   (*c*) 3.75 m/s   (*d*) 5.49 m/s
(*e*) 7.30 m/s

5–131 Consider incompressible, frictionless flow of water in a vertical piping. The pressure is 240 kPa at 2 m from the ground level. The velocity of water does not change during this flow. The pressure at 15 m from the ground level is
(*a*) 227 kPa   (*b*) 174 kPa   (*c*) 127 kPa   (*d*) 120 kPa
(*e*) 113 kPa

5–132 Consider water flow in a piping network. The pressure, velocity, and elevation at a specified point (point 1) of the flow are 150 kPa, 1.8 m/s, and 14 m. The pressure and velocity at point 2 are 165 kPa and 2.4 m/s. Neglecting frictional effects, the elevation at point 2 is
(*a*) 12.4 m   (*b*) 9.3 m   (*c*) 14.2 m   (*d*) 10.3 m   (*e*) 7.6 m

5–133 The static and stagnation pressures of a fluid in a pipe are measured by a piezometer and a pitot tube to be 200 kPa and 210 kPa, respectively. If the density of the fluid is 550 kg/m^3, the velocity of the fluid is
(*a*) 10 m/s   (*b*) 6.03 m/s   (*c*) 5.55 m/s   (*d*) 3.67 m/s
(*e*) 0.19 m/s

5–134 The static and stagnation pressures of a fluid in a pipe are measured by a piezometer and a pitot tube. The heights of

the fluid in the piozemeter and pitot tube are measured to be 2.2 m and 2.0 m, respectively. If the density of the fluid is 5000 kg/m³, the velocity of the fluid in the pipe is
(a) 0.92 m/s   (b) 1.43 m/s   (c) 1.65 m/s   (d) 1.98 m/s
(e) 2.39 m/s

5–135   The difference between the heights of energy grade line (EGL) and hydraulic grade line (HGL) is equal to
(a) z (b) $P/\rho g$ (c) $V^2/2g$ (d) $z + P/\rho g$ (e) $z + V^2/2g$

5–136   Water at 120 kPa (gage) is flowing in a horizontal pipe at a velocity of 1.15 m/s. The pipe makes a 90° angle at the exit and the water exits the pipe vertically into the air. The maximum height the water jet can rise is
(a) 6.9 m   (b) 7.8 m   (c) 9.4 m   (d) 11.5 m   (e) 12.3 m

5–137   Water is withdrawn at the bottom of a large tank open to the atmosphere. The water velocity is 6.6 m/s. The minimum height of the water in the tank is
(a) 2.22 m   (b) 3.04 m   (c) 4.33 m   (d) 5.75 m   (e) 6.60 m

5–138   Water at 80 kPa (gage) enters a horizontal pipe at a velocity of 1.7 m/s. The pipe makes a 90° angle at the exit and the water exits the pipe vertically into the air. Take the correction factor to be 1. If the irreversible head loss between the inlet and exit of the pipe is 3 m, the height the water jet can rise is
(a) 3.4 m   (b) 5.3 m   (c) 8.2 m   (d) 10.5 m   (e) 12.3 m

5–139   Seawater is to be pumped into a large tank at a rate of 165 kg/min. The tank is open to the atmosphere and the water enters the tank from a 80-m-height. The overall efficiency of the motor-pump unit is 75 percent and the motor consumes electricity at a rate of 3.2 kW. Take the correction factor to be 1. If the irreversible head loss in the piping is 7 m, the velocity of the water at the tank inlet is
(a) 2.34 m/s   (b) 4.05 m/s   (c) 6.21 m/s   (d) 8.33 m/s
(e) 10.7 m/s

5–140   Water enters a pump at 350 kPa at a rate of 1 kg/s. The water leaving the pump enters a turbine in which the pressure is reduced and electricity is produced. The shaft power input to the pump is 1 kW and the shaft power output from the turbine is 1 kW. Both the pump and turbine are 90 percent efficient. If the elevation and velocity of the water remain constant throughout the flow and the irreversible head loss is 1 m, the pressure of water at the turbine exit is
(a) 350 kPa   (b) 100 kPa   (c) 173 kPa   (d) 218 kPa
(e) 129 kPa

5–141   An adiabatic pump is used to increase the pressure of water from 100 kPa to 500 kPa at a rate of 400 L/min. If the efficiency of the pump is 75 percent, the maximum temperature rise of the water across the pump is
(a) 0.096°C   (b) 0.058°C   (c) 0.035°C   (d) 1.52°C
(e) 1.27°C

5–142   The shaft power from a 90 percent-efficient turbine is 500 kW. If the mass flow rate through the turbine

is 575 kg/s, the extracted head removed from the fluid by the turbine is
(a) 48.7 m   (b) 57.5 m   (c) 147 m   (d) 139 m   (e) 98.5 m

Design and Essay Problems

5–143   Using a large bucket whose volume is known and measuring the time it takes to fill the bucket with water from a garden hose, determine the mass flow rate and the average velocity of water through the hose.

5–144   Your company is setting up an experiment that involves the measurement of airflow rate in a duct, and you are to come up with proper instrumentation. Research the available techniques and devices for airflow rate measurement, discuss the advantages and disadvantages of each technique, and make a recommendation.

5–145   Computer-aided designs, the use of better materials, and better manufacturing techniques have resulted in a tremendous increase in the efficiency of pumps, turbines, and electric motors. Contact one or more pump, turbine, and motor manufacturers and obtain information about the efficiency of their products. In general, how does efficiency vary with rated power of these devices?

5–146   Using a handheld bicycle pump to generate an air jet, a soda can as the water reservoir, and a straw as the tube, design and build an atomizer. Study the effects of various parameters such as the tube length, the diameter of the exit hole, and the pumping speed on performance.

5–147   Using a flexible drinking straw and a ruler, explain how you would measure the water flow velocity in a river.

5–148   The power generated by a wind turbine is proportional to the cube of the wind velocity. Inspired by the acceleration of a fluid in a nozzle, someone proposes to install a reducer casing to capture the wind energy from a larger area and accelerate it before the wind strikes the turbine blades, as shown in Fig. P5–148. Evaluate if the proposed modification should be given a consideration in the design of new wind turbines.

FIGURE P5–148

MOMENTUM ANALYSIS OF FLOW SYSTEMS

hen dealing with engineering problems, it is desirable to obtain fast and accurate solutions at minimal cost. Most engineering problems, including those associated with fluid flow, can be analyzed using one of three basic approaches: differential, experimental, and control volume. In *differential approaches*, the problem is formulated accurately using differential quantities, but the solution of the resulting differential equations is difficult, usually requiring the use of numerical methods with extensive computer codes. *Experimental approaches* complemented with dimensional analysis are highly accurate, but they are typically time consuming and expensive. The *finite control volume approach* described in this chapter is remarkably fast and simple and usually gives answers that are sufficiently accurate for most engineering purposes. Therefore, despite the approximations involved, the basic finite control volume analysis performed with paper and pencil has always been an indispensable tool for engineers.

In Chap. 5, the control volume mass and energy analysis of fluid flow systems was presented. In this chapter, we present the finite control volume momentum analysis of fluid flow problems. First we give an overview of Newton's laws and the conservation relations for linear and angular momentum. Then using the Reynolds transport theorem, we develop the linear momentum and angular momentum equations for control volumes and use them to determine the forces and torques associated with fluid flow.

■ ■ ■ ■ ■ ■ ■
OBJECTIVES
When you finish reading this chapter, you should be able to

■ Identify the various kinds of forces and moments acting on a control volume

■ Use control volume analysis to determine the forces associated with fluid flow

■ Use control volume analysis to determine the moments caused by fluid flow and the torque transmitted

Steady swimming of the jellyfish Aurelia aurita. Fluorescent dye placed directly upstream of the animal is drawn underneath the bell as the body relaxes and forms vortex rings below the animal as the body contracts and ejects fluid. The vortex rings simultaneously induce flows for both feeding and propulsion.

Adapted from Dabiri et al., J. Exp. Biol. 208: 1257–1265. Photo credit: Sean P. Colin and John H. Costello.

6–1 · NEWTON'S LAWS

Newton's laws are relations between motions of bodies and the forces acting on them. Newton's first law states that *a body at rest remains at rest, and a body in motion remains in motion at the same velocity in a straight path when the net force acting on it is zero*. Therefore, a body tends to preserve its state of inertia. Newton's second law states that *the acceleration of a body is proportional to the net force acting on it and is inversely proportional to its mass*. Newton's third law states that *when a body exerts a force on a second body, the second body exerts an equal and opposite force on the first*. Therefore, the direction of an exposed reaction force depends on the body taken as the system.

For a rigid body of mass m, Newton's second law is expressed as

Newton's second law:
$$\vec{F} = m\vec{a} = m\,\frac{d\vec{V}}{dt} = \frac{d(m\vec{V})}{dt} \qquad (6\text{--}1)$$

where $\vec{F}$ is the net force acting on the body and $\vec{a}$ is the acceleration of the body under the influence of $\vec{F}$.

The product of the mass and the velocity of a body is called the *linear momentum* or just the *momentum* of the body. The momentum of a rigid body of mass m moving with velocity $\vec{V}$ is $m\vec{V}$ (Fig. 6–1). Then Newton's second law expressed in Eq. 6–1 can also be stated as *the rate of change of the momentum of a body is equal to the net force acting on the body* (Fig. 6–2). This statement is more in line with Newton's original statement of the second law, and it is more appropriate for use in fluid mechanics when studying the forces generated as a result of velocity changes of fluid streams. Therefore, in fluid mechanics, Newton's second law is usually referred to as the *linear momentum equation*.

The momentum of a system remains constant only when the net force acting on it is zero, and thus the momentum of such a system is conserved. This is known as the *conservation of momentum principle*. This principle has proven to be a very useful tool when analyzing collisions such as those between balls; between balls and rackets, bats, or clubs; and between atoms or subatomic particles; and explosions such as those that occur in rockets, missiles, and guns. In fluid mechanics, however, the net force acting on a system is typically *not* zero, and we prefer to work with the linear momentum equation rather than the conservation of momentum principle.

Note that force, acceleration, velocity, and momentum are vector quantities, and as such they have direction as well as magnitude. Also, momentum is a constant multiple of velocity, and thus the direction of momentum is the direction of velocity as shown in Fig 6–1. Any vector equation can be written in scalar form for a specified direction using magnitudes, e.g., $F_x = ma_x = d(mV_x)/dt$ in the x-direction.

The counterpart of Newton's second law for rotating rigid bodies is expressed as $\vec{M} = I\vec{\alpha}$, where $\vec{M}$ is the net moment or torque applied on the body, I is the moment of inertia of the body about the axis of rotation, and $\vec{\alpha}$ is the angular acceleration. It can also be expressed in terms of the rate of change of angular momentum $d\vec{H}/dt$ as

Angular momentum equation:
$$\vec{M} = I\vec{\alpha} = I\,\frac{d\vec{\omega}}{dt} = \frac{d(I\vec{\omega})}{dt} = \frac{d\vec{H}}{dt} \qquad (6\text{--}2)$$

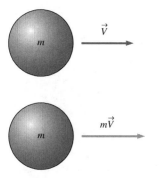

FIGURE 6–1
Linear momentum is the product of mass and velocity, and its direction is the direction of velocity.

FIGURE 6–2
Newton's second law is also expressed as *the rate of change of the momentum of a body is equal to the net force acting on it.*

where $\vec{\omega}$ is the angular velocity. For a rigid body rotating about a fixed x-axis, the angular momentum equation is written in scalar form as

Angular momentum about x-axis: $\qquad M_x = I_x \dfrac{d\omega_x}{dt} = \dfrac{dH_x}{dt}$ $\qquad$ **(6–3)**

The angular momentum equation can be stated as *the rate of change of the angular momentum of a body is equal to the net torque acting on it* (Fig. 6–3).

The total angular momentum of a rotating body remains constant when the net torque acting on it is zero, and thus the angular momentum of such systems is conserved. This is known as the *conservation of angular momentum principle* and is expressed as $I\omega =$ constant. Many interesting phenomena such as ice skaters spinning faster when they bring their arms close to their bodies and divers rotating faster when they curl after the jump can be explained easily with the help of the conservation of angular momentum principle (in both cases, the moment of inertia I is decreased and thus the angular velocity ω is increased as the outer parts of the body are brought closer to the axis of rotation).

6–2 ▪ CHOOSING A CONTROL VOLUME

We now briefly discuss how to *wisely* select a control volume. A control volume can be selected as any arbitrary region in space through which fluid flows, and its bounding control surface can be fixed, moving, and even deforming during flow. The application of a basic conservation law is a systematic procedure for bookkeeping or accounting of the quantity under consideration, and thus it is extremely important that the boundaries of the control volume are well defined during an analysis. Also, the flow rate of any quantity into or out of a control volume depends on the flow velocity *relative to the control surface*, and thus it is essential to know if the control volume remains at rest during flow or if it moves.

Many flow systems involve stationary hardware firmly fixed to a stationary surface, and such systems are best analyzed using *fixed* control volumes. When determining the reaction force acting on a tripod holding the nozzle of a hose, for example, a natural choice for the control volume is one that passes perpendicularly through the nozzle exit flow and through the bottom of the tripod legs (Fig. 6–4a). This is a fixed control volume, and the water velocity relative to a fixed point on the ground is the same as the water velocity relative to the nozzle exit plane.

When analyzing flow systems that are moving or deforming, it is usually more convenient to allow the control volume to *move* or *deform*. When determining the thrust developed by the jet engine of an airplane cruising at constant velocity, for example, a wise choice of control volume is one that encloses the airplane and cuts through the nozzle exit plane (Fig. 6–4b). The control volume in this case moves with velocity $\vec{V}_{CV}$, which is identical to the cruising velocity of the airplane relative to a fixed point on earth. When determining the flow rate of exhaust gases leaving the nozzle, the proper velocity to use is the velocity of the exhaust gases relative to the nozzle exit plane, that is, the *relative velocity* $\vec{V}_r$. Since the entire control volume moves at velocity $\vec{V}_{CV}$, the relative velocity becomes $\vec{V}_r = \vec{V} - \vec{V}_{CV}$, where $\vec{V}$ is the *absolute velocity* of the exhaust gases, i.e., the velocity relative to a fixed

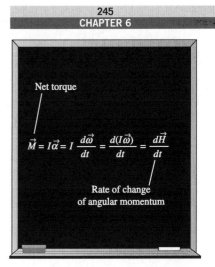

FIGURE 6–3
The rate of change of the angular momentum of a body is equal to the net torque acting on it.

FIGURE 6–4
Examples of (a) fixed, (b) moving, and (c) deforming control volumes.

point on earth. Note that $\vec{V}_r$ is the fluid velocity expressed relative to a coordinate system moving *with* the control volume. Also, this is a vector equation, and velocities in opposite directions have opposite signs. For example, if the airplane is cruising at 500 km/h to the left, and the velocity of the exhaust gases is 800 km/h to the right relative to the ground, the velocity of the exhaust gases relative to the nozzle exit is

$$\vec{V}_r = \vec{V} - \vec{V}_{CV} = 800\vec{i} - (-500\vec{i}) = 1300\vec{i} \text{ km/h}$$

That is, the exhaust gases leave the nozzle at 1300 km/h to the right relative to the nozzle exit (in the direction opposite to that of the airplane); this is the velocity that should be used when evaluating the outflow of exhaust gases through the control surface (Fig. 6–4b). Note that the exhaust gases would appear motionless to an observer on the ground if the relative velocity were equal in magnitude to the airplane velocity.

When analyzing the purging of exhaust gases from a reciprocating internal combustion engine, a wise choice for the control volume is one that comprises the space between the top of the piston and the cylinder head (Fig. 6–4c). This is a *deforming* control volume, since part of the control surface moves relative to other parts. The relative velocity for an inlet or outlet on the deforming part of a control surface (there are no such inlets or outlets in Fig. 6–4c) is then given by $\vec{V}_r = \vec{V} - \vec{V}_{CS}$ where $\vec{V}$ is the absolute fluid velocity and $\vec{V}_{CS}$ is the control surface velocity, both relative to a fixed point outside the control volume. Note that $\vec{V}_{CS} = \vec{V}_{CV}$ for moving but nondeforming control volumes, and $\vec{V}_{CS} = \vec{V}_{CV} = 0$ for fixed ones.

6–3 · FORCES ACTING ON A CONTROL VOLUME

The forces acting on a control volume consist of **body forces** that act throughout the entire body of the control volume (such as gravity, electric, and magnetic forces) and **surface forces** that act on the control surface (such as pressure and viscous forces and reaction forces at points of contact). Only external forces are considered in the analysis. Internal forces (such as the pressure force between a fluid and the inner surfaces of the flow section) are not considered in a control volume analysis unless they are exposed by passing the control surface through that area.

In control volume analysis, the sum of all forces acting on the control volume at a particular instant in time is represented by $\Sigma\vec{F}$ and is expressed as

Total force acting on control volume: $$\sum\vec{F} = \sum\vec{F}_{body} + \sum\vec{F}_{surface} \qquad (6\text{–}4)$$

Body forces act on each volumetric portion of the control volume. The body force acting on a differential element of fluid of volume dV within the control volume is shown in Fig. 6–5, and we must perform a volume integral to account for the net body force on the entire control volume. *Surface forces* act on each portion of the control surface. A differential surface element of area dA and unit outward normal $\vec{n}$ on the control surface is shown in Fig. 6–5, along with the surface force acting on it. We must perform an area integral to obtain the net surface force acting on the entire control surface. As sketched, the surface force may act in a direction independent of that of the outward normal vector.

The most common body force is that of **gravity,** which exerts a downward force on every differential element of the control volume. While other body forces, such as electric and magnetic forces, may be important in some analyses, we consider only gravitational forces here.

The differential body force $d\vec{F}_{\text{body}} = d\vec{F}_{\text{gravity}}$ acting on the small fluid element shown in Fig. 6–6 is simply its weight,

Gravitational force acting on a fluid element: $\qquad d\vec{F}_{\text{gravity}} = \rho\vec{g}\,dV$ **(6–5)**

where ρ is the average density of the element and $\vec{g}$ is the gravitational vector. In Cartesian coordinates we adopt the convention that $\vec{g}$ acts in the negative z-direction, as in Fig. 6–6, so that

Gravitational vector in Cartesian coordinates: $\qquad \vec{g} = -g\vec{k}$ **(6–6)**

Note that the coordinate axes in Fig. 6–6 are oriented so that the gravity vector acts *downward* in the $-z$-direction. On earth at sea level, the gravitational constant g is equal to 9.807 m/s². Since gravity is the only body force being considered, integration of Eq. 6–5 yields

Total body force acting on control volume: $\sum \vec{F}_{\text{body}} = \int_{\text{CV}} \rho\vec{g}\,dV = m_{\text{CV}}\vec{g}$ **(6–7)**

Surface forces are not as simple to analyze since they consist of both *normal* and *tangential* components. Furthermore, while the physical force acting on a surface is independent of orientation of the coordinate axes, the *description* of the force in terms of its coordinate components changes with orientation (Fig. 6–7). In addition, we are rarely fortunate enough to have each of the control surfaces aligned with one of the coordinate axes. While not desiring to delve too deeply into tensor algebra, we are forced to define a **second-order tensor** called the **stress tensor** σ_{ij} in order to adequately describe the surface stresses at a point in the flow,

Stress tensor in Cartesian coordinates: $\sigma_{ij} = \begin{pmatrix} \sigma_{xx} & \sigma_{xy} & \sigma_{xz} \\ \sigma_{yx} & \sigma_{yy} & \sigma_{yz} \\ \sigma_{zx} & \sigma_{zy} & \sigma_{zz} \end{pmatrix}$ **(6–8)**

The diagonal components of the stress tensor, σ_{xx}, σ_{yy}, and σ_{zz}, are called **normal stresses;** they are composed of pressure (which always acts inwardly normal) and viscous stresses. Viscous stresses are discussed in more detail in Chap. 9. The off-diagonal components, σ_{xy}, σ_{zx}, etc., are called **shear stresses;** since pressure can act only normal to a surface, shear stresses are composed entirely of viscous stresses.

When the face is not parallel to one of the coordinate axes, mathematical laws for axes rotation and tensors can be used to calculate the normal and tangential components acting at the face. In addition, an alternate notation called **tensor notation** is convenient when working with tensors but is usually reserved for graduate studies. (For a more in-depth analysis of tensors and tensor notation see, for example, Kundu and Cohen, 2011.)

In Eq. 6–8, σ_{ij} is defined as the stress (force per unit area) in the j-direction acting on a face whose normal is in the i-direction. Note that i and j are merely *indices* of the tensor and are not the same as unit vectors $\vec{i}$ and $\vec{j}$. For example, σ_{xy} is defined as positive for the stress pointing in the y-direction on a face whose outward normal is in the x-direction. This component of the

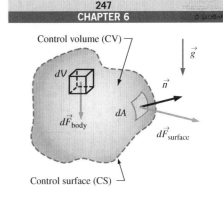

FIGURE 6–5

The total force acting on a control volume is composed of body forces and surface forces; body force is shown on a differential volume element, and surface force is shown on a differential surface element.

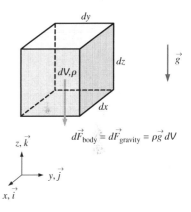

FIGURE 6–6

The gravitational force acting on a differential volume element of fluid is equal to its weight; the axes are oriented so that the gravity vector acts *downward* in the negative z-direction.

(a)

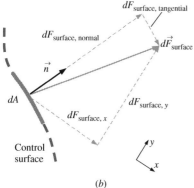

(b)

FIGURE 6–7

When coordinate axes are rotated (*a*) to (*b*), the components of the surface force change, even though the force itself remains the same; only two dimensions are shown here.

stress tensor, along with the other eight components, is shown in Fig. 6–8 for the case of a differential fluid element aligned with the axes in Cartesian coordinates. All the components in Fig. 6–8 are shown on positive faces (right, top, and front) and in their positive orientation by definition. Positive stress components on the *opposing* faces of the fluid element (not shown) point in exactly opposite directions.

The dot product of a second-order tensor and a vector yields a second vector; this operation is often called the **contracted product** or the **inner product** of a tensor and a vector. In our case, it turns out that the inner product of the stress tensor σ_{ij} and the unit outward normal vector $\vec{n}$ of a differential surface element yields a vector whose magnitude is the force per unit area acting on the surface element and whose direction is the direction of the surface force itself. Mathematically we write

Surface force acting on a differential surface element: $d\vec{F}_{\text{surface}} = \sigma_{ij} \cdot \vec{n} \, dA$ **(6–9)**

Finally, we integrate Eq. 6–9 over the entire control surface,

Total surface force acting on control surface: $\sum \vec{F}_{\text{surface}} = \int_{\text{CS}} \sigma_{ij} \cdot \vec{n} \, dA$ **(6–10)**

Substitution of Eqs. 6–7 and 6–10 into Eq. 6–4 yields

$$\sum \vec{F} = \sum \vec{F}_{\text{body}} + \sum \vec{F}_{\text{surface}} = \int_{\text{CV}} \rho \vec{g} \, dV + \int_{\text{CS}} \sigma_{ij} \cdot \vec{n} \, dA \qquad \textbf{(6–11)}$$

This equation turns out to be quite useful in the derivation of the differential form of conservation of linear momentum, as discussed in Chap. 9. For practical control volume analysis, however, it is rare that we need to use Eq. 6–11, especially the cumbersome surface integral that it contains.

A careful selection of the control volume enables us to write the total force acting on the control volume, $\sum \vec{F}$, as the sum of more readily available quantities like weight, pressure, and reaction forces. We recommend the following for control volume analysis:

Total force: $\underbrace{\sum \vec{F}}_{\text{total force}} = \underbrace{\sum \vec{F}_{\text{gravity}}}_{\text{body force}} + \underbrace{\sum \vec{F}_{\text{pressure}} + \sum \vec{F}_{\text{viscous}} + \sum \vec{F}_{\text{other}}}_{\text{surface forces}}$ **(6–12)**

The first term on the right-hand side of Eq. 6–12 is the body force *weight*, since gravity is the only body force we are considering. The other three terms combine to form the net surface force; they are pressure forces, viscous forces, and "other" forces acting on the control surface. $\sum \vec{F}_{\text{other}}$ is composed of reaction forces required to turn the flow; forces at bolts, cables, struts, or walls through which the control surface cuts; etc.

All these surface forces arise as the control volume is isolated from its surroundings for analysis, and the effect of any detached object is accounted for by a force at that location. This is similar to drawing a free-body diagram in your statics and dynamics classes. We should choose the control volume such that forces that are not of interest remain internal, and thus they do not complicate the analysis. A well-chosen control volume exposes only the forces that are to be determined (such as reaction forces) and a minimum number of other forces.

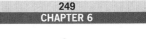

A common simplication in the application of Newton's laws of motion is to subtract the *atmospheric pressure* and work with gage pressures. This is because atmospheric pressure acts in all directions, and its effect cancels out in every direction (Fig. 6–9). This means we can also ignore the pressure forces at outlet sections where the fluid is discharged at subsonic velocities to the atmosphere since the discharge pressures in such cases are very near atmospheric pressure.

As an example of how to wisely choose a control volume, consider control volume analysis of water flowing steadily through a faucet with a partially closed gate valve spigot (Fig. 6–10). It is desired to calculate the net force on the flange to ensure that the flange bolts are strong enough. There are many possible choices for the control volume. Some engineers restrict their control volumes to the fluid itself, as indicated by CV A (the purple control volume) in Fig 6–10. With this control volume, there are pressure forces that vary along the control surface, there are viscous forces along the pipe wall and at locations inside the valve, and there is a body force, namely, the weight of the water in the control volume. Fortunately, to calculate the net force on the flange, we do *not* need to integrate the pressure and viscous stresses all along the control surface. Instead, we can lump the unknown pressure and viscous forces together into one reaction force, representing the net force of the walls on the water. This force, plus the weight of the faucet and the water, is equal to the net force on the flange. (We must be very careful with our signs, of course.)

When choosing a control volume, you are not limited to the fluid alone. Often it is more convenient to slice the control surface *through* solid objects such as walls, struts, or bolts as illustrated by CV B (the red control volume) in Fig. 6–10. A control volume may even surround an entire object, like the one shown here. Control volume B is a wise choice because we are not concerned with any details of the flow or even the geometry inside the control volume. For the case of CV B, we assign a net reaction force acting at the portions of the control surface that slice through the flange bolts. Then, the only other things we need to know are the gage pressure of the water at the flange (the inlet to the control volume) and the weights of the water and the faucet assembly. The pressure everywhere else along the control surface is atmospheric (zero gage pressure) and cancels out. This problem is revisited in Section 6–4, Example 6–7.

6–4 · THE LINEAR MOMENTUM EQUATION

Newton's second law for a system of mass m subjected to net force $\sum \vec{F}$ is expressed as

$$\sum \vec{F} = m\vec{a} = m\frac{d\vec{V}}{dt} = \frac{d}{dt}(m\vec{V}) \qquad \text{(6–13)}$$

where $m\vec{V}$ is the **linear momentum** of the system. Noting that both the density and velocity may change from point to point within the system, Newton's second law can be expressed more generally as

$$\sum \vec{F} = \frac{d}{dt}\int_{\text{sys}} \rho \vec{V}\, dV \qquad \text{(6–14)}$$

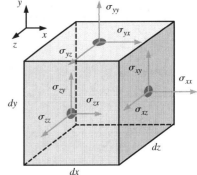

FIGURE 6–8

Components of the stress tensor in Cartesian coordinates on the right, top, and front faces.

FIGURE 6–9

Atmospheric pressure acts in all directions, and thus it can be ignored when performing force balances since its effect cancels out in every direction.

FIGURE 6–10

Cross section through a faucet assembly, illustrating the importance of choosing a control volume wisely; CV B is much easier to work with than CV A.

FIGURE 6–11
The linear momentum equation
is obtained by replacing B in the
Reynolds transport theorem by
the momentum $m\vec{V}$, and b by
the momentum per unit mass $\vec{V}$.

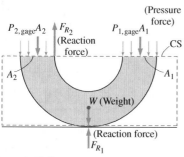

An 180° elbow supported by the ground

FIGURE 6–12
In most flow systems, the sum of
forces $\Sigma \vec{F}$ consists of weights,
pressure forces, and reaction forces.
Gage pressures are used here since
atmospheric pressure cancels out on
all sides of the control surface.

where $\rho \vec{V} \, dV$ is the momentum of a differential element dV, which has mass $\delta m = \rho \, dV$. Therefore, Newton's second law can be stated as *the sum of all external forces acting on a system is equal to the time rate of change of linear momentum of the system*. This statement is valid for a coordinate system that is at rest or moves with a constant velocity, called an *inertial coordinate system* or *inertial reference frame*. Accelerating systems such as aircraft during takeoff are best analyzed using noninertial (or accelerating) coordinate systems fixed to the aircraft. Note that Eq. 6–14 is a vector relation, and thus the quantities $\vec{F}$ and $\vec{V}$ have direction as well as magnitude.

Equation 6–14 is for a given mass of a solid or fluid and is of limited use in fluid mechanics since most flow systems are analyzed using control volumes. The *Reynolds transport theorem* developed in Section 4–6 provides the necessary tools to shift from the system formulation to the control volume formulation. Setting $b = \vec{V}$ and thus $B = m\vec{V}$, the Reynolds transport theorem is expressed for linear momentum as (Fig. 6–11)

$$\frac{d(m\vec{V})_{sys}}{dt} = \frac{d}{dt}\int_{CV} \rho\vec{V} \, dV + \int_{CS} \rho\vec{V} \, (\vec{V}_r \cdot \vec{n}) \, dA \qquad \text{(6–15)}$$

The left-hand side of this equation is, from Eq. 6–13, equal to $\Sigma\vec{F}$. Substituting, the general form of the linear momentum equation that applies to fixed, moving, or deforming control volumes is

General:
$$\sum \vec{F} = \frac{d}{dt}\int_{CV} \rho\vec{V} \, dV + \int_{CS} \rho\vec{V}(\vec{V}_r \cdot \vec{n}) \, dA \qquad \text{(6–16)}$$

which is stated in words as

$$\begin{pmatrix} \text{The sum of all} \\ \text{external forces} \\ \text{acting on a CV} \end{pmatrix} = \begin{pmatrix} \text{The time rate of change} \\ \text{of the linear momentum} \\ \text{of the contents of the CV} \end{pmatrix} + \begin{pmatrix} \text{The net flow rate of} \\ \text{linear momentum out of the} \\ \text{control surface by mass flow} \end{pmatrix}$$

Here $\vec{V}_r = \vec{V} - \vec{V}_{CS}$ is the fluid velocity relative to the control surface (for use in mass flow rate calculations at all locations where the fluid crosses the control surface), and $\vec{V}$ is the fluid velocity as viewed from an inertial reference frame. The product $\rho(\vec{V}_r \cdot \vec{n}) \, dA$ represents the mass flow rate through area element dA into or out of the control volume.

For a fixed control volume (no motion or deformation of the control volume), $\vec{V}_r = \vec{V}$ and the linear momentum equation becomes

Fixed CV:
$$\sum \vec{F} = \frac{d}{dt}\int_{CV} \rho\vec{V} \, dV + \int_{CS} \rho\vec{V}(\vec{V} \cdot \vec{n}) \, dA \qquad \text{(6–17)}$$

Note that the momentum equation is a *vector equation*, and thus each term should be treated as a vector. Also, the components of this equation can be resolved along orthogonal coordinates (such as x, y, and z in the Cartesian coordinate system) for convenience. The sum of forces $\Sigma\vec{F}$ in most cases consists of weights, pressure forces, and reaction forces (Fig. 6–12). The momentum equation is commonly used to calculate the forces (usually on support systems or connectors) induced by the flow.

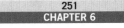
Special Cases

Most momentum problems considered in this text are steady. During *steady flow*, the amount of momentum within the control volume remains constant, and thus the time rate of change of linear momentum of the contents of the control volume (the second term of Eq. 6–16) is zero. Thus,

Steady flow:
$$\sum \vec{F} = \int_{CS} \rho \vec{V} (\vec{V}_r \cdot \vec{n}) \, dA \qquad (6\text{–}18)$$

For a case in which a non-deforming control volume moves at constant velocity (an inertial reference frame), the *first* $\vec{V}$ in Eq. 6-18 may *also* be taken relative to the moving control surface.

While Eq. 6–17 is exact for fixed control volumes, it is not always convenient when solving practical engineering problems because of the integrals. Instead, as we did for conservation of mass, we would like to rewrite Eq. 6–17 in terms of average velocities and mass flow rates through inlets and outlets. In other words, our desire is to rewrite the equation in *algebraic* rather than *integral* form. In many practical applications, fluid crosses the boundaries of the control volume at one or more inlets and one or more outlets, and carries with it some momentum into or out of the control volume. For simplicity, we always draw our control surface such that it slices normal to the inflow or outflow velocity at each such inlet or outlet (Fig. 6–13).

The mass flow rate $\dot{m}$ into or out of the control volume across an inlet or outlet at which ρ is nearly constant is

Mass flow rate across an inlet or outlet:
$$\dot{m} = \int_{A_c} \rho(\vec{V}\cdot\vec{n}) \, dA_c = \rho V_{avg} A_c \qquad (6\text{–}19)$$

Comparing Eq. 6–19 to Eq. 6–17, we notice an extra velocity in the control surface integral of Eq. 6–17. If $\vec{V}$ were uniform ($\vec{V} = \vec{V}_{avg}$) across the inlet or outlet, we could simply take it outside the integral. Then we could write the rate of inflow or outflow of momentum through the inlet or outlet in simple algebraic form,

Momentum flow rate across a uniform inlet or outlet:
$$\int_{A_c} \rho \vec{V}(\vec{V}\cdot\vec{n}) \, dA_c = \rho V_{avg} A_c \vec{V}_{avg} = \dot{m}\vec{V}_{avg} \qquad (6\text{–}20)$$

The uniform flow approximation is reasonable at some inlets and outlets, e.g., the well-rounded entrance to a pipe, the flow at the entrance to a wind tunnel test section, and a slice through a water jet moving at nearly uniform speed through air (Fig. 6–14). At each such inlet or outlet, Eq. 6–20 can be applied directly.

Momentum-Flux Correction Factor, β

Unfortunately, the velocity across most inlets and outlets of practical engineering interest is *not* uniform. Nevertheless, it turns out that we can still convert the control surface integral of Eq. 6–17 into algebraic form, but a dimensionless correction factor β, called the **momentum-flux correction factor,** is required, as first shown by the French scientist Joseph Boussinesq

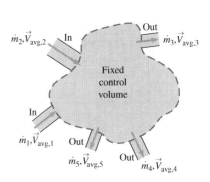

FIGURE 6–13

In a typical engineering problem, the control volume may contain multiple inlets and outlets; at each inlet or outlet we define the mass flow rate $\dot{m}$ and the average velocity $\vec{V}_{avg}$.

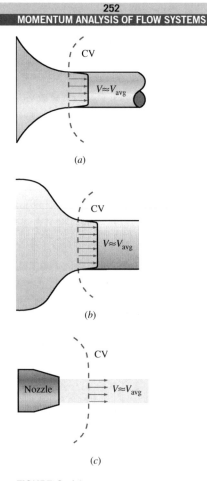

FIGURE 6–14
Examples of inlets or outlets
in which the uniform flow
approximation is reasonable:
(*a*) the well-rounded entrance to
a pipe, (*b*) the entrance to a wind
tunnel test section, and (*c*) a slice
through a free water jet in air.

(1842–1929). The algebraic form of Eq. 6–17 for a fixed control volume is then written as

$$\sum \vec{F} = \frac{d}{dt} \int_{CV} \rho \vec{V} \, dV + \sum_{out} \beta \dot{m} \vec{V}_{avg} - \sum_{in} \beta \dot{m} \vec{V}_{avg} \qquad (6\text{–}21)$$

where a unique value of momentum-flux correction factor is applied to each inlet and outlet in the control surface. Note that $\beta = 1$ *for the case of uniform flow* over an inlet or outlet, as in Fig. 6–14. For the general case, we define β such that the integral form of the momentum flux into or out of the control surface at an inlet or outlet of cross-sectional area A_c can be expressed in terms of mass flow rate $\dot{m}$ through the inlet or outlet and average velocity $\vec{V}_{avg}$ through the inlet or outlet,

Momentum flux across an inlet or outlet: $\qquad \int_{A_c} \rho \vec{V}(\vec{V} \cdot \vec{n}) \, dA_c = \beta \dot{m} \vec{V}_{avg} \qquad (6\text{–}22)$

For the case in which density is uniform over the inlet or outlet and $\vec{V}$ is in the same direction as $\vec{V}_{avg}$ over the inlet or outlet, we solve Eq. 6–22 for β,

$$\beta = \frac{\int_{A_c} \rho V(\vec{V} \cdot \vec{n}) \, dA_c}{\dot{m} V_{avg}} = \frac{\int_{A_c} \rho V(\vec{V} \cdot \vec{n}) \, dA_c}{\rho V_{avg} A_c V_{avg}} \qquad (6\text{–}23)$$

where we have substituted $\rho V_{avg} A_c$ for $\dot{m}$ in the denominator. The densities cancel and since V_{avg} is constant, it can be brought inside the integral. Furthermore, if the control surface slices normal to the inlet or outlet area, $(\vec{V} \cdot \vec{n}) \, dA_c = V \, dA_c$. Then, Eq. 6–23 simplifies to

Momentum-flux correction factor: $\qquad \beta = \frac{1}{A_c} \int_{A_c} \left(\frac{V}{V_{avg}} \right)^2 dA_c \qquad (6\text{–}24)$

It may be shown that β is always greater than or equal to unity.

EXAMPLE 6–1   Momentum-Flux Correction Factor for Laminar Pipe Flow

Consider laminar flow through a very long straight section of round pipe. It is shown in Chap. 8 that the velocity profile through a cross-sectional area of the pipe is parabolic (Fig. 6–15), with the axial velocity component given by

$$V = 2V_{avg}\left(1 - \frac{r^2}{R^2}\right) \qquad (1)$$

where R is the radius of the inner wall of the pipe and V_{avg} is the average velocity. Calculate the momentum-flux correction factor through a cross section of the pipe for the case in which the pipe flow represents an outlet of the control volume, as sketched in Fig. 6–15.

SOLUTION   For a given velocity distribution we are to calculate the momentum-flux correction factor.

Assumptions **1** The flow is incompressible and steady. **2** The control volume slices through the pipe normal to the pipe axis, as sketched in Fig. 6–15.
Analysis We substitute the given velocity profile for V in Eq. 6–24 and integrate, noting that $dA_c = 2\pi r\, dr$,

$$\beta = \frac{1}{A_c}\int_{A_c}\left(\frac{V}{V_{avg}}\right)^2 dA_c = \frac{4}{\pi R^2}\int_0^R\left(1-\frac{r^2}{R^2}\right)^2 2\pi r\, dr \qquad (2)$$

Defining a new integration variable $y = 1 - r^2/R^2$ and thus $dy = -2r\, dr/R^2$ (also, $y = 1$ at $r = 0$, and $y = 0$ at $r = R$) and performing the integration, the momentum-flux correction factor for fully developed laminar flow becomes

Laminar flow: $$\beta = -4\int_1^0 y^2\, dy = -4\left[\frac{y^3}{3}\right]_1^0 = \frac{4}{3} \qquad (3)$$

Discussion We have calculated β for an outlet, but the same result would have been obtained if we had considered the cross section of the pipe as an inlet to the control volume.

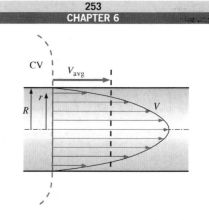

FIGURE 6–15
Velocity profile over a cross section of a pipe in which the flow is fully developed and laminar.

From Example 6–1 we see that β is not very close to unity for fully developed laminar pipe flow, and ignoring β could potentially lead to significant error. If we were to perform the same kind of integration as in Example 6–1 but for fully developed *turbulent* rather than laminar pipe flow, we would find that β ranges from about 1.01 to 1.04. Since these values are so close to unity, many practicing engineers completely disregard the momentum-flux correction factor. While the neglect of β in turbulent flow calculations may have an insignificant effect on the final results, it is wise to keep it in our equations. Doing so not only improves the accuracy of our calculations, but reminds us to include the momentum-flux correction factor when solving laminar flow control volume problems.

For turbulent flow β may have an insignificant effect at inlets and outlets, but for laminar flow β may be important and should not be neglected. It is wise to include β in all momentum control volume problems.

Steady Flow

If the flow is also *steady*, the time derivative term in Eq. 6–21 vanishes and we are left with

Steady linear momentum equation: $$\sum \vec{F} = \sum_{out}\beta\dot{m}\vec{V} - \sum_{in}\beta\dot{m}\vec{V} \qquad (6\text{–}25)$$

where we have dropped the subscript "avg" from average velocity. Equation 6–25 states that *the net force acting on the control volume during steady flow is equal to the difference between the rates of outgoing and incoming momentum flows*. This statement is illustrated in Fig. 6–16. It can also be expressed for any direction, since Eq. 6–25 is a vector equation.

$$\sum \vec{F} = \sum_{out}\beta\dot{m}\vec{V} - \sum_{in}\beta\dot{m}\vec{V}$$

FIGURE 6–16
The net force acting on the control volume during steady flow is equal to the difference between the outgoing and the incoming momentum fluxes.

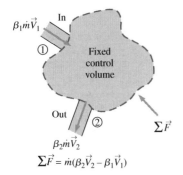

$$\sum \vec{F} = \dot{m}(\beta_2 \vec{V}_2 - \beta_1 \vec{V}_1)$$

FIGURE 6–17
A control volume with only one inlet and one outlet.

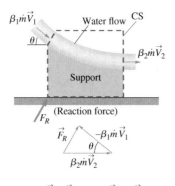

Note: $\vec{V}_2 \neq \vec{V}_1$ even if $|\vec{V}_2| = |\vec{V}_1|$

FIGURE 6–18
The determination by vector addition of the reaction force on the support caused by a change of direction of water.

Steady Flow with One Inlet and One Outlet

Many practical engineering problems involve just one inlet and one outlet (Fig. 6–17). The mass flow rate for such **single-stream systems** remains constant, and Eq. 6–25 reduces to

One inlet and one outlet: $$\sum \vec{F} = \dot{m} \, (\beta_2 \vec{V}_2 - \beta_1 \vec{V}_1) \qquad (6\text{–}26)$$

where we have adopted the usual convention that subscript 1 implies the inlet and subscript 2 the outlet, and $\vec{V}_1$ and $\vec{V}_2$ denote the *average* velocities across the inlet and outlet, respectively.

We emphasize again that all the preceding relations are *vector* equations, and thus all the additions and subtractions are *vector* additions and subtractions. Recall that subtracting a vector is equivalent to adding it after reversing its direction (Fig. 6–18). When writing the momentum equation for a specified coordinate direction (such as the *x*-axis), we use the projections of the vectors on that axis. For example, Eq. 6–26 is written along the *x*-coordinate as

Along x-coordinate: $$\sum F_x = \dot{m}(\beta_2 V_{2,x} - \beta_1 V_{1,x}) \qquad (6\text{–}27)$$

where $\sum F_x$ is the vector sum of the *x*-components of the forces, and $V_{2,x}$ and $V_{1,x}$ are the *x*-components of the outlet and inlet velocities of the fluid stream, respectively. The force or velocity components in the positive *x*-direction are positive quantities, and those in the negative *x*-direction are negative quantities. Also, it is good practice to take the direction of unknown forces in the positive directions (unless the problem is very straightforward). A negative value obtained for an unknown force indicates that the assumed direction is wrong and should be reversed.

Flow with No External Forces

An interesting situation arises when there are no external forces (such as weight, pressure, and reaction forces) acting on the body in the direction of motion—a common situation for space vehicles and satellites. For a control volume with multiple inlets and outlets, Eq. 6–21 reduces in this case to

No external forces: $$0 = \frac{d(m\vec{V})_{CV}}{dt} + \sum_{out} \beta \dot{m} \vec{V} - \sum_{in} \beta \dot{m} \vec{V} \qquad (6\text{–}28)$$

This is an expression of the conservation of momentum principle, which is stated in words as *in the absence of external forces, the rate of change of the momentum of a control volume is equal to the difference between the rates of incoming and outgoing momentum flow rates.*

When the mass *m* of the control volume remains nearly constant, the first term of Eq. 6–28 becomes simply mass times acceleration, since

$$\frac{d(m\vec{V})_{CV}}{dt} = m_{CV} \frac{d\vec{V}_{CV}}{dt} = (m\vec{a})_{CV} = m_{CV}\vec{a}$$

Therefore, the control volume in this case can be treated as a solid body (a fixed-mass system) with a net thrusting force (or just **thrust**) of

Thrust: $$\vec{F}_{thrust} = m_{body}\vec{a} = \sum_{in} \beta \dot{m} \vec{V} - \sum_{out} \beta \dot{m} \vec{V} \qquad (6\text{–}29)$$

acting on the body. In Eq 6–29, fluid velocities are relative to an inertial reference frame—that is, a coordinate system that is fixed in space or is moving uniformly at constant velocity on a straight path. When analyzing the motion of bodies moving at constant velocity on a straight path, it is convenient to choose an inertial reference frame that moves with the body at the same velocity on the same path. In this case the velocities of fluid streams relative to the inertial reference frame are identical to the velocities relative to the moving body, which are much easier to apply. This approach, while not strictly valid for noninertial reference frames, can also be used to calculate the *initial* acceleration of a space vehicle when its rocket is fired (Fig. 6–19).

Recall that thrust is a mechanical force typically generated through the reaction of an accelerating fluid. In the jet engine of an aircraft, for example, hot exhaust gases are accelerated by the action of expansion and outflow of gases through the back of the engine, and a thrusting force is produced by a reaction in the opposite direction. The generation of thrust is based on Newton's third law of motion, which states that *for every action at a point there is an equal and opposite reaction*. In the case of a jet engine, if the engine exerts a force on exhaust gases, then the exhaust gases exert an equal force on the engine in the opposite direction. That is, the pushing force exerted on the departing gases by the engine is equal to the thrusting force the departing gases exert on the remaining mass of the aircraft in the opposite direction $\vec{F}_{\text{thrust}} = -\vec{F}_{\text{push}}$. On the free-body diagram of an aircraft, the effect of outgoing exhaust gases is accounted for by the insertion of a force in the opposite direction of motion of the exhaust gases.

FIGURE 6–19
The thrust needed to lift the space shuttle is generated by the rocket engines as a result of momentum change of the fuel as it is accelerated from about zero to an exit speed of about 2000 m/s after combustion.

NASA

EXAMPLE 6–2   The Force to Hold a Deflector Elbow in Place

A reducing elbow is used to deflect water flow at a rate of 14 kg/s in a horizontal pipe upward 30° while accelerating it (Fig. 6–20). The elbow discharges water into the atmosphere. The cross-sectional area of the elbow is 113 cm² at the inlet and 7 cm² at the outlet. The elevation difference between the centers of the outlet and the inlet is 30 cm. The weight of the elbow and the water in it is considered to be negligible. Determine (*a*) the gage pressure at the center of the inlet of the elbow and (*b*) the anchoring force needed to hold the elbow in place.

FIGURE 6–20
Schematic for Example 6–2.

SOLUTION A reducing elbow deflects water upward and discharges it to the atmosphere. The pressure at the inlet of the elbow and the force needed to hold the elbow in place are to be determined.
Assumptions **1** The flow is steady, and the frictional effects are negligible. **2** The weight of the elbow and the water in it is negligible. **3** The water is discharged to the atmosphere, and thus the gage pressure at the outlet is zero. **4** The flow is turbulent and fully developed at both the inlet and outlet of the control volume, and we take the momentum-flux correction factor to be $\beta = 1.03$ (as a conservative estimate) at both the inlet and the outlet.
Properties We take the density of water to be 1000 kg/m³.
Analysis (*a*) We take the elbow as the control volume and designate the inlet by 1 and the outlet by 2. We also take the *x*- and *z*-coordinates as shown. The continuity equation for this one-inlet, one-outlet, steady-flow system is $\dot{m}_1 = \dot{m}_2 = \dot{m} = 14$ kg/s. Noting that $\dot{m} = \rho AV$, the inlet and outlet velocities of water are

$$V_1 = \frac{\dot{m}}{\rho A_1} = \frac{14 \text{ kg/s}}{(1000 \text{ kg/m}^3)(0.0113 \text{ m}^2)} = 1.24 \text{ m/s}$$

$$V_2 = \frac{\dot{m}}{\rho A_2} = \frac{14 \text{ kg/s}}{(1000 \text{ kg/m}^3)(7 \times 10^{-4} \text{ m}^2)} = 20.0 \text{ m/s}$$

We use the Bernoulli equation (Chap. 5) as a first approximation to calculate the pressure. In Chap. 8 we will learn how to account for frictional losses along the walls. Taking the center of the inlet cross section as the reference level ($z_1 = 0$) and noting that $P_2 = P_{atm}$, the Bernoulli equation for a streamline going through the center of the elbow is expressed as

$$\frac{P_1}{\rho g} + \frac{V_1^2}{2g} + z_1 = \frac{P_2}{\rho g} + \frac{V_2^2}{2g} + z_2$$

$$P_1 - P_2 = \rho g \left(\frac{V_2^2 - V_1^2}{2g} + z_2 - z_1 \right)$$

$$P_1 - P_{atm} = (1000 \text{ kg/m}^3)(9.81 \text{ m/s}^2)$$

$$\times \left(\frac{(20 \text{ m/s})^2 - (1.24 \text{ m/s})^2}{2(9.81 \text{ m/s}^2)} + 0.3 - 0 \right) \left(\frac{1 \text{ kN}}{1000 \text{ kg·m/s}^2} \right)$$

$$P_{1, gage} = 202.2 \text{ kN/m}^2 = \mathbf{202.2 \text{ kPa}} \quad \text{(gage)}$$

(b) The momentum equation for steady flow is

$$\sum \vec{F} = \sum_{out} \beta \dot{m} \vec{V} - \sum_{in} \beta \dot{m} \vec{V}$$

We let the x- and z-components of the anchoring force of the elbow be F_{Rx} and F_{Rz}, and assume them to be in the positive direction. We also use gage pressure since the atmospheric pressure acts on the entire control surface. Then the momentum equations along the x- and z-axes become

$$F_{Rx} + P_{1, gage} A_1 = \beta \dot{m} V_2 \cos \theta - \beta \dot{m} V_1$$

$$F_{Rz} = \beta \dot{m} V_2 \sin \theta$$

where we have set $\beta = \beta_1 = \beta_2$. Solving for F_{Rx} and F_{Rz}, and substituting the given values,

$$F_{Rx} = \beta \dot{m}(V_2 \cos \theta - V_1) - P_{1, gage} A_1$$

$$= 1.03(14 \text{ kg/s})[(20 \cos 30° - 1.24) \text{ m/s}]\left(\frac{1 \text{ N}}{1 \text{ kg·m/s}^2} \right)$$

$$- (202,200 \text{ N/m}^2)(0.0113 \text{ m}^2)$$

$$= 232 - 2285 = \mathbf{-2053 \text{ N}}$$

$$F_{Rz} = \beta \dot{m} V_2 \sin \theta = (1.03)(14 \text{ kg/s})(20 \sin 30° \text{ m/s})\left(\frac{1 \text{ N}}{1 \text{ kg·m/s}^2} \right) = \mathbf{144 \text{ N}}$$

The negative result for F_{Rx} indicates that the assumed direction is wrong, and it should be reversed. Therefore, F_{Rx} acts in the negative x-direction.

Discussion There is a nonzero pressure distribution along the inside walls of the elbow, but since the control volume is outside the elbow, these pressures do not appear in our analysis. The weight of the elbow and the water in it could be added to the vertical force for better accuracy. The actual value of $P_{1, gage}$ will be higher than that calculated here because of frictional and other irreversible losses in the elbow.

FIGURE 6–21
Schematic for Example 6–3.

EXAMPLE 6–3   The Force to Hold a Reversing Elbow in Place

The deflector elbow in Example 6–2 is replaced by a reversing elbow such that the fluid makes a 180° U-turn before it is discharged, as shown in Fig. 6–21. The elevation difference between the centers of the inlet and the exit sections is still 0.3 m. Determine the anchoring force needed to hold the elbow in place.

SOLUTION The inlet and the outlet velocities and the pressure at the inlet of the elbow remain the same, but the vertical component of the anchoring force at the connection of the elbow to the pipe is zero in this case ($F_{Rz} = 0$) since there is no other force or momentum flux in the vertical direction (we are neglecting the weight of the elbow and the water). The horizontal component of the anchoring force is determined from the momentum equation written in the x-direction. Noting that the outlet velocity is negative since it is in the negative x-direction, we have

$$F_{Rx} + P_{1, \text{gage}}A_1 = \beta_2\dot{m}(-V_2) - \beta_1\dot{m}V_1 = -\beta\dot{m}(V_2 + V_1)$$

Solving for F_{Rx} and substituting the known values,

$$F_{Rx} = -\beta\dot{m}(V_2 + V_1) - P_{1, \text{gage}}A_1$$

$$= -(1.03)(14 \text{ kg/s})[(20 + 1.24) \text{ m/s}]\left(\frac{1 \text{ N}}{1 \text{ kg·m/s}^2}\right) - (202,200 \text{ N/m}^2)(0.0113 \text{ m}^2)$$

$$= -306 - 2285 = -\mathbf{2591\ N}$$

Therefore, the horizontal force on the flange is 2591 N acting in the negative x-direction (the elbow is trying to separate from the pipe). This force is equivalent to the weight of about 260 kg mass, and thus the connectors (such as bolts) used must be strong enough to withstand this force.

Discussion The reaction force in the x-direction is larger than that of Example 6–2 since the walls turn the water over a much greater angle. If the reversing elbow is replaced by a straight nozzle (like one used by firefighters) such that water is discharged in the positive x-direction, the momentum equation in the x-direction becomes

$$F_{Rx} + P_{1, \text{gage}}A_1 = \beta\dot{m}V_2 - \beta\dot{m}V_1 \quad \rightarrow \quad F_{Rx} = \beta\dot{m}(V_2 - V_1) - P_{1, \text{gage}}A_1$$

since both V_1 and V_2 are in the positive x-direction. This shows the importance of using the correct sign (positive if in the positive direction and negative if in the opposite direction) for velocities and forces.

EXAMPLE 6–4   Water Jet Striking a Stationary Plate

Water is accelerated by a nozzle to an average speed of 20 m/s, and strikes a stationary vertical plate at a rate of 10 kg/s with a normal velocity of 20 m/s (Fig. 6–22). After the strike, the water stream splatters off in all directions in the plane of the plate. Determine the force needed to prevent the plate from moving horizontally due to the water stream.

SOLUTION A water jet strikes a vertical stationary plate normally. The force needed to hold the plate in place is to be determined.

Assumptions **1** The flow of water at the nozzle outlet is steady. **2** The water splatters in directions normal to the approach direction of the water jet.

FIGURE 6–22
Schematic for Example 6–4.

3 The water jet is exposed to the atmosphere, and thus the pressure of the water jet and the splattered water leaving the control volume is atmospheric pressure, which is disregarded since it acts on the entire system. **4** The vertical forces and momentum fluxes are not considered since they have no effect on the horizontal reaction force. **5** The effect of the momentum-flux correction factor is negligible, and thus $\beta \cong 1$ at the inlet.

Analysis　We draw the control volume for this problem such that it contains the entire plate and cuts through the water jet and the support bar normally. The momentum equation for steady flow is given as

$$\sum \vec{F} = \sum_{\text{out}} \beta \dot{m} \vec{V} - \sum_{\text{in}} \beta \dot{m} \vec{V} \qquad (1)$$

Writing Eq. 1 for this problem along the *x*-direction (without forgetting the negative sign for forces and velocities in the negative *x*-direction) and noting that $V_{1,\,x} = V_1$ and $V_{2,\,x} = 0$ gives

$$-F_R = 0 - \beta \dot{m} V_1$$

Substituting the given values,

$$F_R = \beta \dot{m} V_1 = (1)(10 \text{ kg/s})(20 \text{ m/s})\left(\frac{1 \text{ N}}{1 \text{ kg·m/s}^2}\right) = 200 \text{ N}$$

Therefore, the support must apply a 200-N horizontal force (equivalent to the weight of about a 20-kg mass) in the negative *x*-direction (the opposite direction of the water jet) to hold the plate in place. A similar situation occurs in the downwash of a helicopter (Fig. 6–23).

Discussion　The plate absorbs the full brunt of the momentum of the water jet since the *x*-direction momentum at the outlet of the control volume is zero. If the control volume were drawn instead along the interface between the water and the plate, there would be additional (unknown) pressure forces in the analysis. By cutting the control volume through the support, we avoid having to deal with this additional complexity. This is an example of a "wise" choice of control volume.

FIGURE 6–23
The downwash of a helicopter is similar to the jet discussed in Example 6–4. The jet impinges on the surface of the water in this case, causing circular waves as seen here.
© Purestock/SuperStock RF

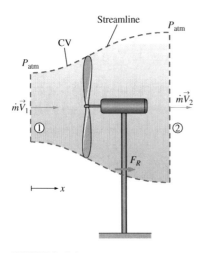

FIGURE 6–24
Schematic for Example 6–5.

EXAMPLE 6–5　**Power Generation and Wind Loading of a Wind Turbine**

A wind generator with a 30-ft-diameter blade span has a cut-in wind speed (minimum speed for power generation) of 7 mph, at which velocity the turbine generates 0.4 kW of electric power (Fig. 6–24). Determine (*a*) the efficiency of the wind turbine–generator unit and (*b*) the horizontal force exerted by the wind on the supporting mast of the wind turbine. What is the effect of doubling the wind velocity to 14 mph on power generation and the force exerted? Assume the efficiency remains the same, and take the density of air to be 0.076 lbm/ft^3.

SOLUTION　The power generation and loading of a wind turbine are to be analyzed. The efficiency and the force exerted on the mast are to be determined, and the effects of doubling the wind velocity are to be investigated.
Assumptions　**1** The wind flow is steady and incompressible. **2** The efficiency of the turbine–generator is independent of wind speed. **3** The frictional effects are negligible, and thus none of the incoming kinetic energy is converted to

thermal energy. **4** The average velocity of air through the wind turbine is the same as the wind velocity (actually, it is considerably less—see Chap. 14). **5** The wind flow is nearly uniform upstream and downstream of the wind turbine and thus the momentum-flux correction factor is $\beta = \beta_1 = \beta_2 \cong 1$.

Properties The density of air is given to be 0.076 lbm/ft^3.

Analysis Kinetic energy is a mechanical form of energy, and thus it can be converted to work entirely. Therefore, the power potential of the wind is proportional to its kinetic energy, which is $V^2/2$ per unit mass, and thus the maximum power is $\dot{m}V^2/2$ for a given mass flow rate:

$$V_1 = (7 \text{ mph})\left(\frac{1.4667 \text{ ft/s}}{1 \text{ mph}}\right) = 10.27 \text{ ft/s}$$

$$\dot{m} = \rho_1 V_1 A_1 = \rho_1 V_1 \frac{\pi D^2}{4} = (0.076 \text{ lbm/ft}^3)(10.27 \text{ ft/s})\frac{\pi(30 \text{ ft})^2}{4} = 551.7 \text{ lbm/s}$$

$$\dot{W}_{max} = \dot{m}ke_1 = \dot{m}\frac{V_1^2}{2}$$

$$= (551.7 \text{ lbm/s})\frac{(10.27 \text{ ft/s})^2}{2}\left(\frac{1 \text{ lbf}}{32.2 \text{ lbm·ft/s}^2}\right)\left(\frac{1 \text{ kW}}{737.56 \text{ lbf·ft/s}}\right)$$

$$= 1.225 \text{ kW}$$

Therefore, the available power to the wind turbine is 1.225 kW at the wind velocity of 7 mph. Then the turbine–generator efficiency becomes

$$\eta_{\text{wind turbine}} = \frac{\dot{W}_{act}}{\dot{W}_{max}} = \frac{0.4 \text{ kW}}{1.225 \text{ kW}} = 0.327 \qquad (\text{or } 32.7\%)$$

(*b*) The frictional effects are assumed to be negligible, and thus the portion of incoming kinetic energy not converted to electric power leaves the wind turbine as outgoing kinetic energy. Noting that the mass flow rate remains constant, the exit velocity is determined to be

$$\dot{m}ke_2 = \dot{m}ke_1(1 - \eta_{\text{wind turbine}}) \rightarrow \dot{m}\frac{V_2^2}{2} = \dot{m}\frac{V_1^2}{2}(1 - \eta_{\text{wind turbine}}) \qquad (1)$$

or

$$V_2 = V_1\sqrt{1 - \eta_{\text{wind turbine}}} = (10.27 \text{ ft/s})\sqrt{1 - 0.327} = 8.43 \text{ ft/s}$$

To determine the force on the mast (Fig. 6–25), we draw a control volume around the wind turbine such that the wind is normal to the control surface at the inlet and the outlet and the entire control surface is at atmospheric pressure (Fig. 6–23). The momentum equation for steady flow is given as

$$\sum \vec{F} = \sum_{out} \beta \dot{m}\vec{V} - \sum_{in} \beta \dot{m}\vec{V} \qquad (2)$$

Writing Eq. 2 along the *x*-direction and noting that $\beta = 1$, $V_{1,x} = V_1$, and $V_{2,x} = V_2$ give

$$F_R = \dot{m}V_2 - \dot{m}V_1 = \dot{m}(V_2 - V_1) \qquad (3)$$

Substituting the known values into Eq. 3 gives

$$F_R = \dot{m}(V_2 - V_1) = (551.7 \text{ lbm/s})(8.43 - 10.27 \text{ ft/s})\left(\frac{1 \text{ lbf}}{32.2 \text{ lbm·ft/s}^2}\right)$$

$$= -31.5 \text{ lbf}$$

FIGURE 6–25
Forces and moments on the supporting mast of a modern wind turbine can be substantial, and increase like V^2; thus the mast is typically quite large and strong.
© *Ingram Publishing/SuperStock RF*

The negative sign indicates that the reaction force acts in the negative x-direction, as expected. Then the force exerted by the wind on the mast becomes $F_{mast} = -F_R =$ **31.5 lbf.**

The power generated is proportional to V^3 since the mass flow rate is proportional to V and the kinetic energy to V^2. Therefore, doubling the wind velocity to 14 mph will increase the power generation by a factor of $2^3 = 8$ to $0.4 \times 8 = 3.2$ kW. The force exerted by the wind on the support mast is proportional to V^2. Therefore, doubling the wind velocity to 14 mph will increase the wind force by a factor of $2^2 = 4$ to $31.5 \times 4 = 126$ lbf.

Discussion Wind turbines are treated in more detail in Chap. 14.

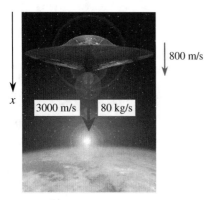

FIGURE 6–26
Schematic for Example 6–6.
© Brand X Pictures/PunchStock

EXAMPLE 6–6   Deceleration of a Spacecraft

A spacecraft with a mass of 12,000 kg is dropping vertically towards a planet at a constant speed of 800 m/s (Fig. 6–26). To slow down the spacecraft, a solid-fuel rocket at the bottom is fired, and combustion gases leave the rocket at a constant rate of 80 kg/s and at a velocity of 3000 m/s relative to the spacecraft in the direction of motion of the spacecraft for a period of 5 s. Disregarding the small change in the mass of the spacecraft, determine (a) the deceleration of the spacecraft during this period, (b) the change of velocity of the spacecraft, and (c) the thrust exerted on the spacecraft.

SOLUTION The rocket of a spacecraft is fired in the direction of motion. The deceleration, the velocity change, and the thrust are to be determined.
Assumptions **1** The flow of combustion gases is steady and one-dimensional during the firing period, but the flight of the spacecraft is unsteady. **2** There are no external forces acting on the spacecraft, and the effect of pressure force at the nozzle outlet is negligible. **3** The mass of discharged fuel is negligible relative to the mass of the spacecraft, and thus, the spacecraft may be treated as a solid body with a constant mass. **4** The nozzle is well designed such that the effect of the momentum-flux correction factor is negligible, and thus, $\beta \cong 1$.
Analysis (a) For convenience, we choose an inertial reference frame that moves with the spacecraft at the same initial velocity. Then the velocities of the fluid stream relative to an inertial reference frame become simply the velocities relative to the spacecraft. We take the direction of motion of the spacecraft as the positive direction along the x-axis. There are no external forces acting on the spacecraft, and its mass is essentially constant. Therefore, the spacecraft can be treated as a solid body with constant mass, and the momentum equation in this case is, from Eq. 6–29,

$$\vec{F}_{thrust} = m_{spacecraft}\,\vec{a}_{spacecraft} = \sum_{in} \beta \dot{m} \vec{V} - \sum_{out} \beta \dot{m} \vec{V}$$

where the fluid stream velocities relative to the inertial reference frame in this case are identical to the velocities relative to the spacecraft. Noting that the motion is on a straight line and the discharged gases move in the positive x-direction, we write the momentum equation using magnitudes as

$$m_{spacecraft} a_{spacecraft} = m_{spacecraft} \frac{dV_{spacecraft}}{dt} = -\dot{m}_{gas} V_{gas}$$

Noting that gases leave in the positive *x*-direction and substituting, the acceleration of the spacecraft during the first 5 seconds is determined to be

$$a_{\text{spacecraft}} = \frac{dV_{\text{spacecraft}}}{dt} = -\frac{\dot{m}_{\text{gas}}}{m_{\text{spacecraft}}}V_{\text{gas}} = -\frac{80 \text{ kg/s}}{12{,}000 \text{ kg}}(+3000 \text{ m/s}) = -20 \text{ m/s}^2$$

The negative value confirms that the spacecraft is decelerating in the positive *x* direction at a rate of 20 m/s^2.

(*b*) Knowing the deceleration, which is constant, the velocity change of the spacecraft during the first 5 seconds is determined from the definition of acceleration to be

$$dV_{\text{spacecraft}} = a_{\text{spacecraft}}dt \rightarrow \Delta V_{\text{spacecraft}} = a_{\text{spacecraft}}\Delta t = (-20 \text{ m/s}^2)(5 \text{ s})$$
$$= -100 \text{ m/s}$$

(*c*) The thrusting force exerted on the space aircraft is, from Eq. 6-29,

$$F_{\text{thrust}} = 0 - \dot{m}_{\text{gas}}V_{\text{gas}} = 0 - (80 \text{ kg/s})(+3000 \text{ m/s})\left(\frac{1 \text{ kN}}{1000 \text{ kg·m/s}^2}\right) = -240 \text{ kN}$$

The negative sign indicates that the trusting force due to firing of the rocket acts on the aircraft in the negative *x*-direction.

Discussion Note that if this fired rocket were attached somewhere on a test stand, it would exert a force of 240 kN (equivalent to the weight of about 24 tons of mass) to its support in the opposite direction of the discharged gases.

■ ***EXAMPLE 6–7***   **Net Force on a Flange**

■ Water flows at a rate of 18.5 gal/min through a flanged faucet with a par-
■ tially closed gate valve spigot (Fig. 6–27). The inner diameter of the pipe
■ at the location of the flange is 0.780 in (= 0.0650 ft), and the pressure
at that location is measured to be 13.0 psig. The total weight of the faucet
assembly plus the water within it is 12.8 lbf. Calculate the net force on the
flange.

SOLUTION   Water flow through a flanged faucet is considered. The net force acting on the flange is to be calculated.

Assumptions   **1** The flow is steady and incompressible. **2** The flow at the inlet and at the outlet is turbulent and fully developed so that the momentum-flux correction factor is about 1.03. **3** The pipe diameter at the outlet of the faucet is the same as that at the flange.

Properties   The density of water at room temperature is 62.3 lbm/ft^3.

Analysis   We choose the faucet and its immediate surroundings as the control volume, as shown in Fig. 6–27 along with all the forces acting on it. These forces include the weight of the water and the weight of the faucet assembly, the gage pressure force at the inlet to the control volume, and the net force of the flange on the control volume, which we call $\vec{F}_R$. We use gage pressure for convenience since the gage pressure on the rest of the control surface is zero (atmospheric pressure). Note that the pressure through the outlet of the control volume is also atmospheric since we are assuming incompressible flow; hence, the gage pressure is also zero through the outlet.

FIGURE 6–27
Control volume for Example 6–7
with all forces shown; gage pressure
is used for convenience.

We now apply the control volume conservation laws. Conservation of mass is trivial here since there is only one inlet and one outlet; namely, the mass flow rate into the control volume is equal to the mass flow rate out of the control volume. Also, the outflow and inflow average velocities are identical since the inner diameter is constant and the water is incompressible, and are determined to be

$$V_2 = V_1 = V = \frac{\dot{V}}{A_c} = \frac{\dot{V}}{\pi D^2/4} = \frac{18.5 \text{ gal/min}}{\pi(0.065 \text{ ft})^2/4}\left(\frac{0.1337 \text{ ft}^3}{1 \text{ gal}}\right)\left(\frac{1 \text{ min}}{60 \text{ s}}\right) = 12.42 \text{ ft/s}$$

Also,

$$\dot{m} = \rho\dot{V} = (62.3 \text{ lbm/ft}^3)(18.5 \text{ gal/min})\left(\frac{0.1337 \text{ ft}^3}{1 \text{ gal}}\right)\left(\frac{1 \text{ min}}{60 \text{ s}}\right) = 2.568 \text{ lbm/s}$$

Next we apply the momentum equation for steady flow,

$$\sum \vec{F} = \sum_{\text{out}} \beta\dot{m}\vec{V} - \sum_{\text{in}} \beta\dot{m}\vec{V} \tag{1}$$

We let the x- and z-components of the force acting on the flange be F_{Rx} and F_{Rz}, and assume them to be in the positive directions. The magnitude of the velocity in the x-direction is $+V_1$ at the inlet, but zero at the outlet. The magnitude of the velocity in the z-direction is zero at the inlet, but $-V_2$ at the outlet. Also, the weight of the faucet assembly and the water within it acts in the $-z$-direction as a body force. No pressure or viscous forces act on the chosen (wise) control volume in the z-direction.

The components of Eq. 1 along the x- and z-directions become

$$F_{Rx} + P_{1,\text{gage}}A_1 = 0 - \dot{m}(+V_1)$$

$$F_{Rz} - W_{\text{faucet}} - W_{\text{water}} = \dot{m}(-V_2) - 0$$

Solving for F_{Rx} and F_{Rz}, and substituting the given values,

$$F_{Rx} = -\dot{m}V_1 - P_{1,\text{gage}}A_1$$

$$= -(2.568 \text{ lbm/s})(12.42 \text{ ft/s})\left(\frac{1 \text{ lbf}}{32.2 \text{ lbm·ft/s}^2}\right) - (13 \text{ lbf/in}^2)\frac{\pi(0.780 \text{ in})^2}{4}$$

$$= -7.20 \text{ lbf}$$

$$F_{Rz} = -\dot{m}V_2 + W_{\text{faucet + water}}$$

$$= -(2.568 \text{ lbm/s})(12.42 \text{ ft/s})\left(\frac{1 \text{ lbf}}{32.2 \text{ lbm·ft/s}^2}\right) + 12.8 \text{ lbf} = 11.8 \text{ lbf}$$

Then the net force of the flange on the control volume is expressed in vector form as

$$\vec{F}_R = F_{Rx}\vec{i} + F_{Rz}\vec{k} = -7.20\vec{i} + 11.8\vec{k} \text{ lbf}$$

From Newton's third law, the force the faucet assembly exerts on the flange is the negative of $\vec{F}_R$,

$$\vec{F}_{\text{faucet on flange}} = -\vec{F}_R = 7.20\vec{i} - 11.8\vec{k} \text{ lbf}$$

Discussion The faucet assembly pulls to the right and down; this agrees with our intuition. Namely, the water exerts a high pressure at the inlet, but the outlet pressure is atmospheric. In addition, the momentum of the water at the inlet in the *x*-direction is lost in the turn, causing an additional force to the right on the pipe walls. The faucet assembly weighs much more than the momentum effect of the water, so we expect the force to be downward. Note that labeling forces such as "faucet on flange" clarifies the direction of the force.

6–5 ■ REVIEW OF ROTATIONAL MOTION AND ANGULAR MOMENTUM

The motion of a rigid body can be considered to be the combination of translational motion of its center of mass and rotational motion about its center of mass. The translational motion is analyzed using the linear momentum equation, Eq. 6–1. Now we discuss the rotational motion—a motion during which all points in the body move in circles about the axis of rotation. Rotational motion is described with angular quantities such as angular distance θ, angular velocity $\vec{\omega}$, and angular acceleration $\vec{\alpha}$.

The amount of rotation of a point in a body is expressed in terms of the angle θ swept by a line of length r that connects the point to the axis of rotation and is perpendicular to the axis. The angle θ is expressed in radians (rad), which is the arc length corresponding to θ on a circle of unit radius. Noting that the circumference of a circle of radius r is $2\pi r$, the angular distance traveled by any point in a rigid body during a complete rotation is 2π rad. The physical distance traveled by a point along its circular path is $l = \theta r$, where r is the normal distance of the point from the axis of rotation and θ is the angular distance in rad. Note that 1 rad corresponds to $360/(2\pi) \cong 57.3°$.

The magnitude of angular velocity ω is the angular distance traveled per unit time, and the magnitude of angular acceleration α is the rate of change of angular velocity. They are expressed as (Fig. 6–28),

$$\omega = \frac{d\theta}{dt} = \frac{d(l/r)}{dt} = \frac{1}{r}\frac{dl}{dt} = \frac{V}{r} \quad \text{and} \quad \alpha = \frac{d\omega}{dt} = \frac{d^2\theta}{dt^2} = \frac{1}{r}\frac{dV}{dt} = \frac{a_t}{r} \quad (6\text{-}30)$$

or

$$V = r\omega \quad \text{and} \quad a_t = r\alpha \quad (6\text{-}31)$$

where V is the linear velocity and a_t is the linear acceleration in the tangential direction for a point located at a distance r from the axis of rotation. Note that ω and α are the same for all points of a rotating rigid body, but V and a_t are not (they are proportional to r).

Newton's second law requires that there must be a force acting in the tangential direction to cause angular acceleration. The strength of the rotating effect, called the *moment* or *torque*, is proportional to the magnitude of the force and its distance from the axis of rotation. The perpendicular distance from the axis of rotation to the line of action of the force is called the *moment arm*, and the magnitude of torque M acting on a point mass m at normal distance r from the axis of rotation is expressed as

$$M = rF_t = rma_t = mr^2\alpha \quad (6\text{-}32)$$

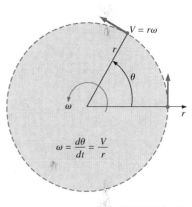

FIGURE 6–28
The relations between angular distance θ, angular velocity ω, and linear velocity V in a plane.

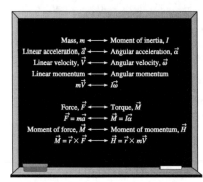

FIGURE 6–29
Analogy between corresponding linear and angular quantities.

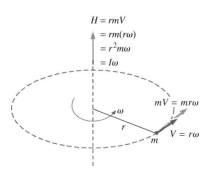

FIGURE 6–30
Angular momentum of point mass m rotating at angular velocity ω at distance r from the axis of rotation.

The total torque acting on a rotating rigid body about an axis is determined by integrating the torque acting on differential mass δm over the entire body to give

Magnitude of torque:
$$M = \int_{\text{mass}} r^2 \alpha \, \delta m = \left[\int_{\text{mass}} r^2 \, \delta m \right] \alpha = I\alpha \qquad \textbf{(6–33)}$$

where I is the *moment of inertia* of the body about the axis of rotation, which is a measure of the inertia of a body against rotation. The relation $M = I\alpha$ is the counterpart of Newton's second law, with torque replacing force, moment of inertia replacing mass, and angular acceleration replacing linear acceleration (Fig. 6–29). Note that unlike mass, the rotational inertia of a body also depends on the distribution of the mass of the body with respect to the axis of rotation. Therefore, a body whose mass is closely packed about its axis of rotation has a small resistance against angular acceleration, while a body whose mass is concentrated at its periphery has a large resistance against angular acceleration. A flywheel is a good example of the latter.

The linear momentum of a body of mass m having velocity $\vec{V}$ is $m\vec{V}$, and the direction of linear momentum is identical to the direction of velocity. Noting that the moment of a force is equal to the product of the force and the normal distance, the magnitude of the moment of momentum, called the **angular momentum,** of a point mass m about an axis is expressed as $H = rmV = r^2 m\omega$, where r is the normal distance from the axis of rotation to the line of action of the momentum vector (Fig. 6–30). Then the total angular momentum of a rotating rigid body is determined by integration to be

Magnitude of angular momentum:
$$H = \int_{\text{mass}} r^2 \omega \, \delta m = \left[\int_{\text{mass}} r^2 \, \delta m \right] \omega = I\omega \qquad \textbf{(6–34)}$$

where again I is the *moment of inertia* of the body about the axis of rotation. It can also be expressed more generally in vector form as

$$\vec{H} = I\vec{\omega} \qquad \textbf{(6–35)}$$

Note that the angular velocity $\vec{\omega}$ is the same at every point of a rigid body.

Newton's second law $\vec{F} = m\vec{a}$ was expressed in terms of the rate of change of linear momentum in Eq. 6–1 as $\vec{F} = d(m\vec{V})/dt$. Likewise, the counterpart of Newton's second law for rotating bodies $\vec{M} = I\vec{\alpha}$ is expressed in Eq. 6–2 in terms of the rate of change of angular momentum as

Angular momentum equation:
$$\vec{M} = I\vec{\alpha} = I\frac{d\vec{\omega}}{dt} = \frac{d(I\vec{\omega})}{dt} = \frac{d\vec{H}}{dt} \qquad \textbf{(6–36)}$$

where $\vec{M}$ is the net torque applied on the body about the axis of rotation.

The angular velocity of rotating machinery is typically expressed in rpm (number of revolutions per minute) and denoted by $\dot{n}$. Noting that velocity is distance traveled per unit time and the angular distance traveled during each revolution is 2π, the angular velocity of rotating machinery is $\omega = 2\pi\dot{n}$ rad/min or

Angular velocity versus rpm:
$$\omega = 2\pi\dot{n} \; (\text{rad/min}) = \frac{2\pi\dot{n}}{60} \quad (\text{rad/s}) \qquad \textbf{(6–37)}$$

Consider a constant force F acting in the tangential direction on the outer surface of a shaft of radius r rotating at an rpm of $\dot{n}$. Noting that work W is force times distance, and power $\dot{W}$ is work done per unit time and thus force times velocity, we have $\dot{W}_{\text{shaft}} = FV = Fr\omega = M\omega$. Therefore, the power transmitted by a shaft rotating at an rpm of $\dot{n}$ under the influence of an applied torque M is (Fig. 6–31)

Shaft power: $$\dot{W}_{\text{shaft}} = \omega M = 2\pi\dot{n}M \qquad (6\text{–}38)$$

The kinetic energy of a body of mass m during translational motion is $\text{KE} = \frac{1}{2}mV^2$. Noting that $V = r\omega$, the rotational kinetic energy of a body of mass m at a distance r from the axis of rotation is $\text{KE} = \frac{1}{2}mr^2\omega^2$. The total rotational kinetic energy of a rotating rigid body about an axis is determined by integrating the rotational kinetic energies of differential masses dm over the entire body to give

Rotational kinetic energy: $$\text{KE}_r = \frac{1}{2}I\omega^2 \qquad (6\text{–}39)$$

where again I is the moment of inertia of the body and ω is the angular velocity.

During rotational motion, the direction of velocity changes even when its magnitude remains constant. Velocity is a vector quantity, and thus a change in direction constitutes a change in velocity with time, and thus acceleration. This is called **centripetal acceleration.** Its magnitude is

$$a_r = \frac{V^2}{r} = r\omega^2$$

Centripetal acceleration is directed toward the axis of rotation (opposite direction of radial acceleration), and thus the radial acceleration is negative. Noting that acceleration is a constant multiple of force, centripetal acceleration is the result of a force acting on the body toward the axis of rotation, known as the **centripetal force,** whose magnitude is $F_r = mV^2/r$. Tangential and radial accelerations are perpendicular to each other (since the radial and tangential directions are perpendicular), and the total linear acceleration is determined by their vector sum, $\vec{a} = \vec{a}_t + \vec{a}_r$. For a body rotating at constant angular velocity, the only acceleration is the centripetal acceleration. The centripetal force does not produce torque since its line of action intersects the axis of rotation.

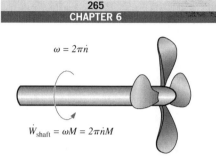

FIGURE 6–31
The relations between angular velocity, rpm, and the power transmitted through a rotating shaft.

6–6 ▪ THE ANGULAR MOMENTUM EQUATION

The linear momentum equation discussed in Section 6–4 is useful for determining the relationship between the linear momentum of flow streams and the resultant forces. Many engineering problems involve the moment of the linear momentum of flow streams, and the rotational effects caused by them. Such problems are best analyzed by the *angular momentum equation,* also called the *moment of momentum equation.* An important class of fluid devices, called *turbomachines,* which include centrifugal pumps, turbines, and fans, is analyzed by the angular momentum equation.

The *moment of a force* $\vec{F}$ about a point O is the vector (or cross) product (Fig. 6–32)

Moment of a force: $$\vec{M} = \vec{r} \times \vec{F} \qquad \textbf{(6–40)}$$

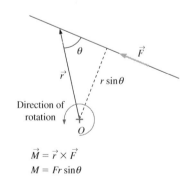

FIGURE 6–32
The moment of a force $\vec{F}$ about a point O is the vector product of the position vector $\vec{r}$ and $\vec{F}$.

FIGURE 6–33
The determination of the direction of the moment by the right-hand rule.

FIGURE 6–34
The angular momentum equation is obtained by replacing B in the Reynolds transport theorem by the angular momentum $\vec{H}$, and b by the angular momentum per unit mass $\vec{r} \times \vec{V}$.

where $\vec{r}$ is the position vector from point O to any point on the line of action of $\vec{F}$. The vector product of two vectors is a vector whose line of action is normal to the plane that contains the crossed vectors ($\vec{r}$ and $\vec{F}$ in this case) and whose magnitude is

Magnitude of the moment of a force: $\qquad M = Fr \sin \theta$ **(6–41)**

where θ is the angle between the lines of action of the vectors $\vec{r}$ and $\vec{F}$. Therefore, the magnitude of the moment about point O is equal to the magnitude of the force multiplied by the normal distance of the line of action of the force from the point O. The sense of the moment vector $\vec{M}$ is determined by the right-hand rule: when the fingers of the right hand are curled in the direction that the force tends to cause rotation, the thumb points the direction of the moment vector (Fig. 6–33). Note that a force whose line of action passes through point O produces zero moment about point O.

The vector product of $\vec{r}$ and the momentum vector $m\vec{V}$ gives the *moment of momentum*, also called the *angular momentum*, about a point O as

Moment of momentum: $\qquad \vec{H} = \vec{r} \times m\vec{V}$ **(6–42)**

Therefore, $\vec{r} \times \vec{V}$ represents the angular momentum per unit mass, and the angular momentum of a differential mass $\delta m = \rho \, dV$ is $d\vec{H} = (\vec{r} \times \vec{V})\rho \, dV$. Then the angular momentum of a system is determined by integration to be

Moment of momentum (system): $\qquad \vec{H}_{sys} = \displaystyle\int_{sys} (\vec{r} \times \vec{V})\rho \, dV$ **(6–43)**

The rate of change of the moment of momentum is

Rate of change of moment of momentum: $\dfrac{d\vec{H}_{sys}}{dt} = \dfrac{d}{dt}\displaystyle\int_{sys} (\vec{r} \times \vec{V})\rho \, dV$ **(6–44)**

The angular momentum equation for a system was expressed in Eq. 6–2 as

$$\sum \vec{M} = \frac{d\vec{H}_{sys}}{dt}$$ **(6–45)**

where $\sum \vec{M} = \sum(\vec{r} \times \vec{F})$ is the net torque or moment applied on the system, which is the vector sum of the moments of all forces acting on the system, and $d\vec{H}_{sys}/dt$ is the rate of change of the angular momentum of the system. Equation 6–45 is stated as *the rate of change of angular momentum of a system is equal to the net torque acting on the system.* This equation is valid for a fixed quantity of mass and an inertial reference frame, i.e., a reference frame that is fixed or moves with a constant velocity in a straight path.

The general control volume formulation of the angular momentum equation is obtained by setting $b = \vec{r} \times \vec{V}$ and thus $B = \vec{H}$ in the general Reynolds transport theorem. It gives (Fig. 6–34)

$$\frac{d\vec{H}_{sys}}{dt} = \frac{d}{dt}\int_{CV} (\vec{r} \times \vec{V})\rho \, dV + \int_{CS} (\vec{r} \times \vec{V})\rho(\vec{V}_r \cdot \vec{n}) \, dA$$ **(6–46)**

The left-hand side of this equation is, from Eq. 6–45, equal to $\sum \vec{M}$. Substituting, the angular momentum equation for a general control volume (stationary or moving, fixed shape or distorting) is

General: $\sum \vec{M} = \dfrac{d}{dt}\displaystyle\int_{CV} (\vec{r} \times \vec{V})\rho \, dV + \int_{CS} (\vec{r} \times \vec{V})\rho(\vec{V}_r \cdot \vec{n}) \, dA$ **(6–47)**

which is stated in words as

$$\begin{pmatrix} \text{The sum of all} \\ \text{external moments} \\ \text{acting on a CV} \end{pmatrix} = \begin{pmatrix} \text{The time rate of change} \\ \text{of the angular momentum} \\ \text{of the contents of the CV} \end{pmatrix} + \begin{pmatrix} \text{The net flow rate of} \\ \text{angular momentum} \\ \text{out of the control} \\ \text{surface by mass flow} \end{pmatrix}$$

Again, $\vec{V}_r = \vec{V} - \vec{V}_{CS}$ is the fluid velocity relative to the control surface (for use in mass flow rate calculations at all locations where the fluid crosses the control surface), and $\vec{V}$ is the fluid velocity as viewed from a fixed reference frame. The product $\rho(\vec{V}_r \cdot \vec{n})\, dA$ represents the mass flow rate through dA into or out of the control volume, depending on the sign.

For a fixed control volume (no motion or deformation of the control volume), $\vec{V}_r = \vec{V}$ and the angular momentum equation becomes

Fixed CV: $$\sum \vec{M} = \frac{d}{dt} \int_{CV} (\vec{r} \times \vec{V})\rho \, dV + \int_{CS} (\vec{r} \times \vec{V})\rho(\vec{V} \cdot \vec{n})\, dA \qquad \text{(6–48)}$$

Also, note that the forces acting on the control volume consist of *body forces* that act throughout the entire body of the control volume such as gravity, and *surface forces* that act on the control surface such as the pressure and reaction forces at points of contact. The net torque consists of the moments of these forces as well as the torques applied on the control volume.

Special Cases

During *steady flow*, the amount of angular momentum within the control volume remains constant, and thus the time rate of change of angular momentum of the contents of the control volume is zero. Then,

Steady flow: $$\sum \vec{M} = \int_{CS} (\vec{r} \times \vec{V})\rho(\vec{V}_r \cdot \vec{n})\, dA \qquad \text{(6–49)}$$

In many practical applications, the fluid crosses the boundaries of the control volume at a certain number of inlets and outlets, and it is convenient to replace the area integral by an algebraic expression written in terms of the average properties over the cross-sectional areas where the fluid enters or leaves the control volume. In such cases, the angular momentum flow rate can be expressed as the difference in the angular momentum of outgoing and incoming streams. Furthermore, in many cases the moment arm $\vec{r}$ is either constant along the inlet or outlet (as in radial flow turbomachines) or is large compared to the diameter of the inlet or outlet pipe (as in rotating lawn sprinklers, Fig. 6–35). In such cases, the *average* value of $\vec{r}$ is used throughout the cross-sectional area of the inlet or outlet. Then, an approximate form of the angular momentum equation in terms of average properties at inlets and outlets becomes

$$\sum \vec{M} \cong \frac{d}{dt} \int_{CV} (\vec{r} \times \vec{V})\rho \, dV + \sum_{out}(\vec{r} \times \dot{m}\vec{V}) - \sum_{in}(\vec{r} \times \dot{m}\vec{V}) \qquad \text{(6–50)}$$

You may be wondering why we don't introduce a correction factor into Eq. 6–50, like we did for conservation of energy (Chap. 5) and for conservation of linear momentum (Section 6–4). The reason is that the cross product of $\vec{r}$ and $\dot{m}\vec{V}$ is dependent on problem geometry, and thus, such a correction

FIGURE 6–35
A rotating lawn sprinkler is a good example of application of the angular momentum equation.
© John A. Rizzo/Getty RF

$$\sum \vec{M} = \sum_{out} \vec{r} \times \dot{m}\vec{V} - \sum_{in} \vec{r} \times \dot{m}\vec{V}$$

FIGURE 6–36
The net torque acting on a control volume during steady flow is equal to the difference between the outgoing and incoming angular momentum flow rates.

factor would vary from problem to problem. Therefore, whereas we can readily calculate a kinetic energy flux correction factor and a momentum flux correction factor for fully developed pipe flow that can be applied to various problems, we cannot do so for angular momentum. Fortunately, in many problems of practical engineering interest, the error associated with using average values of radius and velocity is small, and the approximation of Eq. 6–50 is reasonable.

If the flow is *steady*, Eq. 6–50 further reduces to (Fig. 6–36)

$$\textit{Steady flow:} \qquad \sum \vec{M} = \sum_{out} (\vec{r} \times \dot{m}\vec{V}) - \sum_{in} (\vec{r} \times \dot{m}\vec{V}) \qquad \text{(6–51)}$$

Equation 6–51 states that *the net torque acting on the control volume during steady flow is equal to the difference between the outgoing and incoming angular momentum flow rates.* This statement can also be expressed for any specified direction. Note that velocity $\vec{V}$ in Eq. 6–51 is the velocity relative to an inertial coordinate system.

In many problems, all the significant forces and momentum flows are in the same plane, and thus all give rise to moments in the same plane and about the same axis. For such cases, Eq. 6–51 can be expressed in scalar form as

$$\sum M = \sum_{out} r\dot{m}V - \sum_{in} r\dot{m}V \qquad \text{(6–52)}$$

where r represents the average normal distance between the point about which moments are taken and the line of action of the force or velocity, provided that the sign convention for the moments is observed. That is, all moments in the counterclockwise direction are positive, and all moments in the clockwise direction are negative.

Flow with No External Moments

When there are no external moments applied, the angular momentum equation Eq. 6–50 reduces to

$$\textit{No external moments:} \qquad 0 = \frac{d\vec{H}_{CV}}{dt} + \sum_{out} (\vec{r} \times \dot{m}\vec{V}) - \sum_{in} (\vec{r} \times \dot{m}\vec{V}) \qquad \text{(6–53)}$$

This is an expression of the conservation of angular momentum principle, which can be stated as *in the absence of external moments, the rate of change of the angular momentum of a control volume is equal to the difference between the incoming and outgoing angular momentum fluxes.*

When the moment of inertia I of the control volume remains constant, the first term on the right side of Eq. 6–53 becomes simply moment of inertia times angular acceleration, $I\vec{\alpha}$. Therefore, the control volume in this case can be treated as a solid body, with a net torque of

$$\vec{M}_{body} = I_{body}\vec{\alpha} = \sum_{in} (\vec{r} \times \dot{m}\vec{V}) - \sum_{out} (\vec{r} \times \dot{m}\vec{V}) \qquad \text{(6–54)}$$

(due to a change of angular momentum) acting on it. This approach can be used to determine the angular acceleration of space vehicles and aircraft when a rocket is fired in a direction different than the direction of motion.

FIGURE 6–37
Side and frontal views of a typical
centrifugal pump.

Radial-Flow Devices

Many rotary-flow devices such as centrifugal pumps and fans involve flow in the radial direction normal to the axis of rotation and are called *radial-flow devices* (Chap. 14). In a centrifugal pump, for example, the fluid enters the device in the axial direction through the eye of the impeller, turns outward as it flows through the passages between the blades of the impeller, collects in the scroll, and is discharged in the tangential direction, as shown in Fig. 6–37. Axial-flow devices are easily analyzed using the linear momentum equation. But radial-flow devices involve large changes in angular momentum of the fluid and are best analyzed with the help of the angular momentum equation.

To analyze a centrifugal pump, we choose the annular region that encloses the impeller section as the control volume, as shown in Fig. 6–38. Note that the average flow velocity, in general, has normal and tangential components at both the inlet and the outlet of the impeller section. Also, when the shaft rotates at angular velocity ω, the impeller blades have tangential velocity ωr_1 at the inlet and ωr_2 at the outlet. For steady, incompressible flow, the conservation of mass equation is written as

$$\dot{V}_1 = \dot{V}_2 = \dot{V} \quad \rightarrow \quad (2\pi r_1 b_1)V_{1,n} = (2\pi r_2 b_2)V_{2,n} \qquad \textbf{(6–55)}$$

where b_1 and b_2 are the flow widths at the inlet where $r = r_1$ and at the outlet where $r = r_2$, respectively. (Note that the actual circumferential cross-sectional area is somewhat less than $2\pi rb$ since the blade thickness is not zero.) Then the average normal components $V_{1,n}$ and $V_{2,n}$ of absolute velocity can be expressed in terms of the volumetric flow rate $\dot{V}$ as

$$V_{1,n} = \frac{\dot{V}}{2\pi r_1 b_1} \quad \text{and} \quad V_{2,n} = \frac{\dot{V}}{2\pi r_2 b_2} \qquad \textbf{(6–56)}$$

The normal velocity components $V_{1,n}$ and $V_{2,n}$ as well as pressure acting on the inner and outer circumferential areas pass through the shaft center, and thus they do not contribute to torque about the origin. Then only the tangential velocity components contribute to torque, and the application of the angular momentum equation $\sum M = \sum_{\text{out}} r\dot{m}V - \sum_{\text{in}} r\dot{m}V$ to the control volume gives

Euler's turbine equation: $T_{\text{shaft}} = \dot{m}(r_2 V_{2,t} - r_1 V_{1,t})$ **(6–57)**

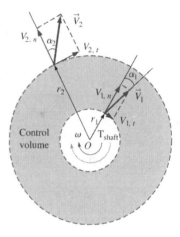

FIGURE 6–38
An annular control volume that encloses the impeller section of a centrifugal pump.

which is known as **Euler's turbine equation.** When the angles α_1 and α_2 between the direction of absolute flow velocities and the radial direction are known, Eq. 6–57 becomes

$$T_{shaft} = \dot{m}(r_2 V_2 \sin \alpha_2 \quad r_1 V_1 \sin \alpha_1) \qquad (6\text{-}58)$$

In the idealized case of the tangential fluid velocity being equal to the blade angular velocity both at the inlet and the exit, we have $V_{1,t} = \omega r_1$ and $V_{2,t} = \omega r_2$, and the torque becomes

$$T_{shaft,\,ideal} = \dot{m}\omega(r_2^2 - r_1^2) \qquad (6\text{-}59)$$

where $\omega = 2\pi\dot{n}$ is the angular velocity of the blades. When the torque is known, the shaft power is determined from $\dot{W}_{shaft} = \omega T_{shaft} = 2\pi\dot{n}T_{shaft}$.

FIGURE 6–39
Schematic for Example 6–8 and the free-body diagram.

EXAMPLE 6–8   Bending Moment Acting at the Base of a Water Pipe

Underground water is pumped through a 10-cm-diameter pipe that consists of a 2-m-long vertical and 1-m-long horizontal section, as shown in Fig. 6–39. Water discharges to atmospheric air at an average velocity of 3 m/s, and the mass of the horizontal pipe section when filled with water is 12 kg per meter length. The pipe is anchored on the ground by a concrete base. Determine the bending moment acting at the base of the pipe (point A) and the required length of the horizontal section that would make the moment at point A zero.

SOLUTION Water is pumped through a piping section. The moment acting at the base and the required length of the horizontal section to make this moment zero is to be determined.

Assumptions **1** The flow is steady. **2** The water is discharged to the atmosphere, and thus the gage pressure at the outlet is zero. **3** The pipe diameter is small compared to the moment arm, and thus we use average values of radius and velocity at the outlet.

Properties We take the density of water to be 1000 kg/m³.

Analysis We take the entire L-shaped pipe as the control volume, and designate the inlet by 1 and the outlet by 2. We also take the x- and z-coordinates as shown. The control volume and the reference frame are fixed.

The conservation of mass equation for this one-inlet, one-outlet, steady-flow system is $\dot{m}_1 = \dot{m}_2 = \dot{m}$, and $V_1 = V_2 = V$ since A_c = constant. The mass flow rate and the weight of the horizontal section of the pipe are

$$\dot{m} = \rho A_c V = (1000 \text{ kg/m}^3)[\pi(0.10 \text{ m})^2/4](3 \text{ m/s}) = 23.56 \text{ kg/s}$$

$$W = mg = (12 \text{ kg/m})(1 \text{ m})(9.81 \text{ m/s}^2)\left(\frac{1 \text{ N}}{1 \text{ kg·m/s}^2}\right) = 117.7 \text{ N}$$

To determine the moment acting on the pipe at point A, we need to take the moment of all forces and momentum flows about that point. This is a steady-flow problem, and all forces and momentum flows are in the same plane. Therefore, the angular momentum equation in this case is expressed as

$$\sum M = \sum_{out} r\dot{m}V - \sum_{in} r\dot{m}V$$

where r is the average moment arm, V is the average speed, all moments in the counterclockwise direction are positive, and all moments in the clockwise direction are negative.

The free-body diagram of the L-shaped pipe is given in Fig. 6–39. Noting that the moments of all forces and momentum flows passing through point A are zero, the only force that yields a moment about point A is the weight W of the horizontal pipe section, and the only momentum flow that yields a moment is the outlet stream (both are negative since both moments are in the clockwise direction). Then the angular momentum equation about point A becomes

$$M_A - r_1 W = -r_2 \dot{m} V_2$$

Solving for M_A and substituting give

$$M_A = r_1 W - r_2 \dot{m} V_2$$

$$= (0.5\ \text{m})(118\ \text{N}) - (2\ \text{m})(23.56\ \text{kg/s})(3\ \text{m/s})\left(\frac{1\ \text{N}}{1\ \text{kg·m/s}^2}\right)$$

$$= -82.5\ \text{N·m}$$

The negative sign indicates that the assumed direction for M_A is wrong and should be reversed. Therefore, a moment of 82.5 N·m acts at the stem of the pipe in the clockwise direction. That is, the concrete base must apply a 82.5 N·m moment on the pipe stem in the clockwise direction to counteract the excess moment caused by the exit stream.

The weight of the horizontal pipe is $w = W/L = 117.7$ N per m length. Therefore, the weight for a length of Lm is Lw with a moment arm of $r_1 = L/2$. Setting $M_A = 0$ and substituting, the length L of the horizontal pipe that would cause the moment at the pipe stem to vanish is determined to be

$$0 = r_1 W - r_2 \dot{m} V_2 \quad \rightarrow \quad 0 = (L/2)Lw - r_2 \dot{m} V_2$$

or

$$L = \sqrt{\frac{2 r_2 \dot{m} V_2}{w}} = \sqrt{\frac{2(2\ \text{m})(23.56\ \text{kg/s})(3\ \text{m/s})}{117.7\ \text{N/m}}\left(\frac{\text{N}}{\text{kg·m/s}^2}\right)} = 1.55\ \text{m}$$

Discussion Note that the pipe weight and the momentum of the exit stream cause opposing moments at point A. This example shows the importance of accounting for the moments of momentums of flow streams when performing a dynamic analysis and evaluating the stresses in pipe materials at critical cross sections.

FIGURE 6–40
Lawn sprinklers often have rotating heads to spread the water over a large area.
© Andy Sotiriou/Getty RF

■ *EXAMPLE 6–9*   **Power Generation from a Sprinkler System**

A large lawn sprinkler (Fig. 6–40) with four identical arms is to be converted into a turbine to generate electric power by attaching a generator to its rotating head, as shown in Fig. 6–41. Water enters the sprinkler from the base along the axis of rotation at a rate of 20 L/s and leaves the nozzles in the tangential direction. The sprinkler rotates at a rate of 300 rpm in a horizontal plane. The diameter of each jet is 1 cm, and the normal distance between the axis of rotation and the center of each nozzle is 0.6 m. Estimate the electric power produced.

SOLUTION A four-armed sprinkler is used to generate electric power. For a specified flow rate and rotational speed, the power produced is to be determined.

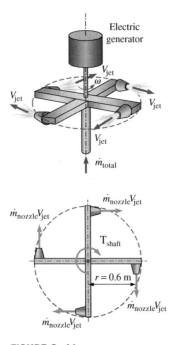

FIGURE 6–41
Schematic for Example 6–9 and the
free-body diagram.

Assumptions **1** The flow is cyclically steady (i.e., steady from a frame of reference rotating with the sprinkler head). **2** The water is discharged to the atmosphere, and thus the gage pressure at the nozzle exit is zero. **3** Generator losses and air drag of rotating components are neglected. **4** The nozzle diameter is small compared to the moment arm, and thus we use average values of radius and velocity at the outlet.

Properties We take the density of water to be 1000 kg/m³ = 1 kg/L.

Analysis We take the disk that encloses the sprinkler arms as the control volume, which is a stationary control volume.

The conservation of mass equation for this steady-flow system is $\dot{m}_1 = \dot{m}_2 = \dot{m}_{total}$. Noting that the four nozzles are identical, we have $\dot{m}_{nozzle} = \dot{m}_{total}/4$ or $\dot{V}_{nozzle} = \dot{V}_{total}/4$ since the density of water is constant. The average jet exit velocity relative to the rotating nozzle is

$$V_{jet,r} = \frac{\dot{V}_{nozzle}}{A_{jet}} = \frac{5\ \text{L/s}}{[\pi(0.01\ \text{m})^2/4]}\left(\frac{1\ \text{m}^3}{1000\ \text{L}}\right) = 63.66\ \text{m/s}$$

The angular and tangential velocities of the nozzles are

$$\omega = 2\pi\dot{n} = 2\pi(300\ \text{rev/min})\left(\frac{1\ \text{min}}{60\ \text{s}}\right) = 31.42\ \text{rad/s}$$

$$V_{nozzle} = r\omega = (0.6\ \text{m})(31.42\ \text{rad/s}) = 18.85\ \text{m/s}$$

Note that water in the nozzle is also moving at an average velocity of 18.85 m/s in the opposite direction when it is discharged. The average absolute velocity of the water jet (velocity relative to a fixed location on earth) is the vector sum of its relative velocity (jet velocity relative to the nozzle) and the absolute nozzle velocity,

$$\vec{V}_{jet} = \vec{V}_{jet,r} + \vec{V}_{nozzle}$$

All of these three velocities are in the tangential direction, and taking the direction of jet flow as positive, the vector equation can be written in scalar form using magnitudes as

$$V_{jet} = V_{jet,r} - V_{nozzle} = 63.66 - 18.85 = 44.81\ \text{m/s}$$

Noting that this is a cyclically steady-flow problem, and all forces and momentum flows are in the same plane, the angular momentum equation is approximated as $\sum M = \sum_{out} r\dot{m}V - \sum_{in} r\dot{m}V$, where r is the moment arm, all moments in the counterclockwise direction are positive, and all moments in the clockwise direction are negative.

The free-body diagram of the disk that contains the sprinkler arms is given in Fig. 6–41. Note that the moments of all forces and momentum flows passing through the axis of rotation are zero. The momentum flows via the water jets leaving the nozzles yield a moment in the clockwise direction and the effect of the generator on the control volume is a moment also in the clockwise direction (thus both are negative). Then the angular momentum equation about the axis of rotation becomes

$$-T_{shaft} = -4r\dot{m}_{nozzle}V_{jet} \quad \text{or} \quad T_{shaft} = r\dot{m}_{total}V_{jet}$$

Substituting, the torque transmitted through the shaft is

$$T_{shaft} = r\dot{m}_{total}V_{jet} = (0.6\ \text{m})(20\ \text{kg/s})(44.81\ \text{m/s})\left(\frac{1\ \text{N}}{1\ \text{kg·m/s}^2}\right) = 537.7\ \text{N·m}$$

since $\dot{m}_{\text{total}} = \rho \dot{V}_{\text{total}} = (1 \text{ kg/L})(20 \text{ L/s}) = 20 \text{ kg/s}$.

Then the power generated becomes

$$\dot{W} = \omega T_{\text{shaft}} = (31.42 \text{ rad/s})(537.7 \text{ N·m})\left(\frac{1 \text{ kW}}{1000 \text{ N·m/s}}\right) = \textbf{16.9 kW}$$

Therefore, this sprinkler-type turbine has the potential to produce 16.9 kW of power.

Discussion To put the result obtained in perspective, we consider two limiting cases. In the first limiting case, the sprinkler is stuck, and thus, the angular velocity is zero. The torque developed is maximum in this case, since $V_{\text{nozzle}} = 0$. Thus $V_{\text{jet}} = V_{\text{jet, }r} = 63.66 \text{ m/s}$, giving $T_{\text{shaft, max}} = 764 \text{ N·m}$. The power generated is zero since the generator shaft does not rotate.

In the second limiting case, the sprinkler shaft is disconnected from the generator (and thus both the useful torque and power generation are zero), and the shaft accelerates until it reaches an equilibrium velocity. Setting $T_{\text{shaft}} = 0$ in the angular momentum equation gives the absolute water-jet velocity (jet velocity relative to an observer on earth) to be zero, $V_{\text{jet}} = 0$. Therefore, the relative velocity $V_{\text{jet, }r}$ and absolute velocity V_{nozzle} are equal but in opposite direction. So, the absolute tangential velocity of the jet (and thus torque) is zero, and the water mass drops straight down like a waterfall under gravity with zero angular momentum (around the axis of rotation). The angular speed of the sprinkler in this case is

$$\dot{n} = \frac{\omega}{2\pi} = \frac{V_{\text{nozzle}}}{2\pi r} = \frac{63.66 \text{ m/s}}{2\pi(0.6 \text{ m})}\left(\frac{60 \text{ s}}{1 \text{ min}}\right) = 1013 \text{ rpm}$$

Of course, the $T_{\text{shaft}} = 0$ case is possible only for an ideal, frictionless nozzle (i.e., 100 percent nozzle efficiency, as a no-load ideal turbine). Otherwise, there would be a resisting torque due to friction of the water, shaft, and surrounding air.

The variation of power produced with angular speed is plotted in Fig. 6–42. Note that the power produced increases with increasing rpm, reaches a maximum (at about 500 rpm in this case), and then decreases. The actual power produced would be less than this due to generator inefficiency (Chap. 5) and other irreversible losses such as fluid friction within the nozzle (Chap. 8), shaft friction, and aerodynamic drag (Chap. 11).

FIGURE 6–42
The variation of power produced with angular speed for the turbine of Example 6–9.

Guest Authors: Alexander Smits, Keith Moored and Peter Dewey, Princeton University

Aquatic animals propel themselves using a wide variety of mechanisms. Most fish flap their tail to produce thrust, and in doing so they shed two single vortices per flapping cycle, creating a wake that resembles a reverse von Kármán vortex street. The non-dimensional number that describes this vortex shedding is the Strouhal number St, where $St = fA/U_\infty$, where f is the frequency of actuation, A is the peak-to-peak amplitude of the trailing edge motion at the half-span, and U_∞ is the steady swimming velocity. Remarkably, a wide variety of fish and mammals swim in the range $0.2 < St < 0.35$.

In manta rays (Fig. 6–43), propulsion is achieved by combining oscillatory and undulatory motions of flexible pectoral fins. That is, as the manta ray

FIGURE 6–43
The manta ray is the largest of the rays, reaching up to 8 m in span. They swim with a motion that is a combination of flapping and undulation of their large pectoral fins.

© Frank & Joyce Burek/Getty RF

274
MOMENTUM ANALYSIS OF FLOW SYSTEMS

FIGURE 6–44

Manta ray fin mechanism, showing the vortex pattern produced in the wake when it is swimming in a range where two single vortices are shed into the wake per flapping cycle. The artificial flexible fin is actuated by four rigid spars; by changing the relative phase differences between adjacent actuators, undulations of varying wavelength can be produced.

flaps its fins, it is also generating a traveling wave motion along the chord, opposite to the direction of its motion. This wave motion is not readily apparent because the wavelength is 6 to 10 times greater than the chord length. A similar undulation is observed in sting rays, but there it is more obvious because the wavelength is less than the chord length. Field observations indicate that many species of manta ray are migratory, and that they are very efficient swimmers. They are difficult to study in the laboratory because they are a protected and somewhat fragile creature. However, it is possible to study many aspects of their swimming behavior by mimicking their propulsive techniques using robots or mechanical devices such as that shown in Fig. 6–44. The flow field generated by such a fin displays the vortex shedding seen in other fish studies, and when time-averaged displays a high momentum jet that contributes to the thrust (Fig 6–45). The thrust and efficiencies can also be measured directly, and it appears that the undulatory motion due to the traveling wave is most important to thrust production at high efficiency in the manta ray.

References

G. S. Triantafyllou, M. S. Triantafyllou, and M. A. Grosenbaugh. Optimal thrust development in oscillating foils with application to fish propulsion. *J. Fluid. Struct.*, 7:205–224, 1993.

Clark, R.P. and Smits, A.J., Thrust production and wake structure of a batoid-inspired oscillating fin. *Journal of Fluid Mechanics*, 562, 415–429, 2006.

Moored, K. W., Dewey, P. A., Leftwich, M. C., Bart-Smith, H. and Smits, A. J., "Bio-inspired propulsion mechanisms based on lamprey and manta ray locomotion." *The Marine Technology Society Journal*, Vol. 45(4), pp. 110–118, 2011.

Dewey, P. A., Carriou, A. and Smits, A. J. "On the relationship between efficiency and wake structure of a batoid-inspired oscillating fin." *Journal of Fluid Mechanics*, Vol. 691, pp. 245–266, 2011.

FIGURE 6–45

Measurements of the wake of the manta ray fin mechanism, with the flow going from bottom to top. On the left, we see the vortices shed in the wake, alternating between positive vorticity (red) and negative vorticity (blue). The induced velocities are shown by the black arrows, and in this case we see that thrust is being produced. On the right, we see the time-averaged velocity field. The unsteady velocity field induced by the vortices produces a high velocity jet in the time-averaged field. The momentum flux associated with this jet contributes to the total thrust on the fin.

Image courtesy of Peter Dewey, Keith Moored and Alexander Smits. Used by permission.

SUMMARY

This chapter deals mainly with the conservation of momentum for finite control volumes. The forces acting on the control volume consist of *body forces* that act throughout the entire body of the control volume (such as gravity, electric, and magnetic forces) and *surface forces* that act on the control surface (such as the pressure forces and reaction forces at points of contact). The sum of all forces acting on the control volume at a particular instant in time is represented by $\Sigma \vec{F}$ and is expressed as

$$\underbrace{\sum \vec{F}}_{\text{total force}} = \underbrace{\sum \vec{F}_{\text{gravity}}}_{\text{body force}} + \underbrace{\sum \vec{F}_{\text{pressure}} + \sum \vec{F}_{\text{viscous}} + \sum \vec{F}_{\text{other}}}_{\text{surface forces}}$$

Newton's second law can be stated as *the sum of all external forces acting on a system is equal to the time rate of change of linear momentum of the system*. Setting $b = \vec{V}$ and thus $B = m\vec{V}$ in the Reynolds transport theorem and utilizing Newton's second law gives the *linear momentum equation* for a control volume as

$$\sum \vec{F} = \frac{d}{dt} \int_{\text{CV}} \rho \vec{V}\, dV + \int_{\text{CS}} \rho \vec{V}(\vec{V}_r \cdot \vec{n})\, dA$$

which reduces to the following special cases:

Steady flow: $\sum \vec{F} = \int_{\text{CS}} \rho \vec{V}(\vec{V}_r \cdot \vec{n})\, dA$

Unsteady flow (algebraic form):

$$\sum \vec{F} = \frac{d}{dt} \int_{\text{CV}} \rho \vec{V}\, dV + \sum_{\text{out}} \beta \dot{m} \vec{V} - \sum_{\text{in}} \beta \dot{m} \vec{V}$$

Steady flow (algebraic form): $\sum \vec{F} = \sum_{\text{out}} \beta \dot{m} \vec{V} - \sum_{\text{in}} \beta \dot{m} \vec{V}$

No external forces: $0 = \dfrac{d(m\vec{V})_{\text{CV}}}{dt} + \sum_{\text{out}} \beta \dot{m} \vec{V} - \sum_{\text{in}} \beta \dot{m} \vec{V}$

where β is the momentum-flux correction factor. A control volume whose mass m remains constant can be treated as a solid body (a fixed-mass system) with a *net thrusting force* (also called simply the *thrust*) of

$$\vec{F}_{\text{thrust}} = m_{\text{CV}} \vec{a} = \sum_{\text{in}} \beta \dot{m} \vec{V} - \sum_{\text{out}} \beta \dot{m} \vec{V}$$

acting on the body.

Newton's second law can also be stated as *the rate of change of angular momentum of a system is equal to the net torque acting on the system*. Setting $b = \vec{r} \times \vec{V}$ and thus $B = \vec{H}$ in the general Reynolds transport theorem gives the *angular momentum equation* as

$$\sum \vec{M} = \frac{d}{dt} \int_{\text{CV}} (\vec{r} \times \vec{V}) \rho\, dV + \int_{\text{CS}} (\vec{r} \times \vec{V}) \rho (\vec{V}_r \cdot \vec{n})\, dA$$

which reduces to the following special cases:

Steady flow: $\sum \vec{M} = \int_{\text{CS}} (\vec{r} \times \vec{V}) \rho (\vec{V}_r \cdot \vec{n})\, dA$

Unsteady flow (algebraic form):

$$\sum \vec{M} = \frac{d}{dt} \int_{\text{CV}} (\vec{r} \times \vec{V}) \rho\, dV + \sum_{\text{out}} \vec{r} \times \dot{m} \vec{V} - \sum_{\text{in}} \vec{r} \times \dot{m} \vec{V}$$

Steady and uniform flow:

$$\sum \vec{M} = \sum_{\text{out}} \vec{r} \times \dot{m} \vec{V} - \sum_{\text{in}} \vec{r} \times \dot{m} \vec{V}$$

Scalar form for one direction:

$$\sum M = \sum_{\text{out}} r \dot{m} V - \sum_{\text{in}} r \dot{m} V$$

No external moments:

$$0 = \frac{d\vec{H}_{\text{CV}}}{dt} + \sum_{\text{out}} \vec{r} \times \dot{m} \vec{V} - \sum_{\text{in}} \vec{r} \times \dot{m} \vec{V}$$

A control volume whose moment of inertia I remains constant can be treated as a solid body (a fixed-mass system), with a net torque of

$$\vec{M}_{\text{CV}} = I_{\text{CV}} \vec{\alpha} = \sum_{\text{in}} \vec{r} \times \dot{m} \vec{V} - \sum_{\text{out}} \vec{r} \times \dot{m} \vec{V}$$

acting on the body. This relation is used to determine the angular acceleration of a spacecraft when a rocket is fired.

The linear and angular momentum equations are of fundamental importance in the analysis of turbomachinery and are used extensively in Chap. 14.

REFERENCES AND SUGGESTED READING

1. P. K. Kundu, I. M. Cohen, and D. R. Dowling. *Fluid Mechanics,* ed. 5. San Diego, CA: Academic Press, 2011.

2. Terry Wright, *Fluid Machinery: Performance, Analysis, and Design,* Boca Raton, FL: CRC Press, 1999.

PROBLEMS*

Newton's Laws and Conservation of Momentum

6–1C Express Newton's first, second, and third laws.

6–2C Express Newton's second law of motion for rotating bodies. What can you say about the angular velocity and angular momentum of a rotating nonrigid body of constant mass if the net torque acting on it is zero?

6–3C Is momentum a vector? If so, in what direction does it point?

6–4C Express the conservation of momentum principle. What can you say about the momentum of a body if the net force acting on it is zero?

Linear Momentum Equation

6–5C Two firefighters are fighting a fire with identical water hoses and nozzles, except that one is holding the hose straight so that the water leaves the nozzle in the same direction it comes, while the other holds it backward so that the water makes a U-turn before being discharged. Which firefighter will experience a greater reaction force?

6–6C How do surface forces arise in the momentum analysis of a control volume? How can we minimize the number of surface forces exposed during analysis?

6–7C Explain the importance of the Reynolds transport theorem in fluid mechanics, and describe how the linear momentum equation is obtained from it.

6–8C What is the importance of the momentum-flux correction factor in the momentum analysis of flow systems? For which type(s) of flow is it significant and must it be considered in analysis: laminar flow, turbulent flow, or jet flow?

6–9C Write the momentum equation for steady one-dimensional flow for the case of no external forces and explain the physical significance of its terms.

6–10C In the application of the momentum equation, explain why we can usually disregard the atmospheric pressure and work with gage pressures only.

6–11C A rocket in space (no friction or resistance to motion) can expel gases relative to itself at some high velocity *V*. Is *V* the upper limit to the rocket's ultimate velocity?

6–12C Describe in terms of momentum and airflow how a helicopter is able to hover.

FIGURE P6–12C
© JupiterImages/Thinkstock/Alamy RF

6–13C Does it take more, equal, or less power for a helicopter to hover at the top of a high mountain than it does at sea level? Explain.

6–14C In a given location, would a helicopter require more energy in summer or winter to achieve a specified performance? Explain.

6–15C A horizontal water jet from a nozzle of constant exit cross section impinges normally on a stationary vertical flat plate. A certain force *F* is required to hold the plate against the water stream. If the water velocity is doubled, will the necessary holding force also be doubled? Explain.

6–16C Describe body forces and surface forces, and explain how the net force acting on a control volume is determined. Is fluid weight a body force or surface force? How about pressure?

6–17C A constant-velocity horizontal water jet from a stationary nozzle impinges normally on a vertical flat plate that rides on a nearly frictionless track. As the water jet hits the plate, it begins to move due to the water force. Will the acceleration of the plate remain constant or change? Explain.

Nozzle

Water jet

FIGURE P6–17C

6–18C A horizontal water jet of constant velocity *V* from a stationary nozzle impinges normally on a vertical flat plate that rides on a nearly frictionless track. As the water jet hits the plate, it begins to move due to the water force. What is the highest velocity the plate can attain? Explain.

6–19 Water enters a 10-cm-diameter pipe steadily with a uniform velocity of 3 m/s and exits with the turbulent flow velocity distribution given by $u = u_{max}(1 - r/R)^{1/7}$. If the pressure drop along the pipe is 10 kPa, determine the drag force exerted on the pipe by water flow.

6–20 A 2.5-cm-diameter horizontal water jet with a speed of $V_j = 40$ m/s relative to the ground is deflected by a 60° stationary cone whose base diameter is 25 cm. Water velocity along the cone varies linearly from zero at the cone surface to the incoming jet speed of 40 m/s at the free surface. Disregarding the effect of gravity and the shear forces, determine the horizontal force F needed to hold the cone stationary.

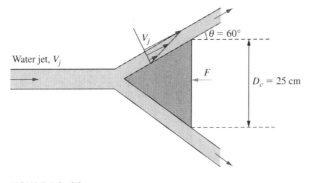

FIGURE P6–20

6–21 A horizontal water jet of constant velocity V impinges normally on a vertical flat plate and splashes off the sides in the vertical plane. The plate is moving toward the oncoming water jet with velocity $\frac{1}{2}V$. If a force F is required to maintain the plate stationary, how much force is required to move the plate toward the water jet?

FIGURE P6–21

6–22 A 90° elbow in a horizontal pipe is used to direct water flow upward at a rate of 40 kg/s. The diameter of the entire elbow is 10 cm. The elbow discharges water into the atmosphere, and thus the pressure at the exit is the local atmospheric pressure. The elevation difference between the centers of the exit and the inlet of the elbow is 50 cm. The weight of the elbow and the water in it is considered to be

negligible. Determine (a) the gage pressure at the center of the inlet of the elbow and (b) the anchoring force needed to hold the elbow in place. Take the momentum-flux correction factor to be 1.03 at both the inlet and the outlet.

FIGURE P6–22

6–23 Repeat Prob. 6–22 for the case of another (identical) elbow attached to the existing elbow so that the fluid makes a U-turn. *Answers:* (a) 9.81 kPa, (b) 497 N

6–24E A horizontal water jet impinges against a vertical flat plate at 25 ft/s and splashes off the sides in the vertical plane. If a horizontal force of 350 lbf is required to hold the plate against the water stream, determine the volume flow rate of the water.

6–25 A reducing elbow in a horizontal pipe is used to deflect water flow by an angle $\theta = 45°$ from the flow direction while accelerating it. The elbow discharges water into the atmosphere. The cross-sectional area of the elbow is 150 cm² at the inlet and 25 cm² at the exit. The elevation difference between the centers of the exit and the inlet is 40 cm. The mass of the elbow and the water in it is 50 kg. Determine the anchoring force needed to hold the elbow in place. Take the momentum-flux correction factor to be 1.03 at both the inlet and outlet.

FIGURE P6–25

6–26 Repeat Prob. 6–25 for the case of $\theta = 110°$.

6–27 Water accelerated by a nozzle to 35 m/s strikes the vertical back surface of a cart moving horizontally at a constant velocity of 10 m/s in the flow direction. The mass flow rate of water through the stationary nozzle is 30 kg/s. After the strike, the water stream splatters off in all directions in the plane of the back surface. (*a*) Determine the force that needs to be applied by the brakes of the cart to prevent it from accelerating. (*b*) If this force were used to generate power instead of wasting it on the brakes, determine the maximum amount of power that could ideally be generated. *Answers:* (*a*) −536 N, (*b*) 5.36 kW

FIGURE P6–27

6–28 Reconsider Prob. 6–27. If the mass of the cart is 400 kg and the brakes fail, determine the acceleration of the cart when the water first strikes it. Assume the mass of water that wets the back surface is negligible.

6–29E A 100-ft³/s water jet is moving in the positive *x*-direction at 18 ft/s. The stream hits a stationary splitter, such that half of the flow is diverted upward at 45° and the other half is directed downward, and both streams have a final average speed of 18 ft/s. Disregarding gravitational effects, determine the *x*- and *z*-components of the force required to hold the splitter in place against the water force.

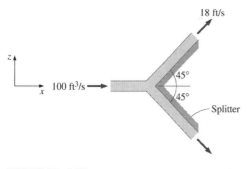

FIGURE P6–29E

6–30E Reconsider Prob. 6–29E. Using EES (or other) software, investigate the effect of the splitter angle on the force exerted on the splitter in the incoming flow direction. Let the half splitter angle vary from 0° to 180° in increments of 10°. Tabulate and plot your results, and draw some conclusions.

6–31 A horizontal 5-cm-diameter water jet with a velocity of 18 m/s impinges normally upon a vertical plate of mass 1000 kg. The plate rides on a nearly frictionless track and is initially stationary. When the jet strikes the plate, the plate begins to move in the direction of the jet. The water always

splatters in the plane of the retreating plate. Determine (*a*) the acceleration of the plate when the jet first strikes it (time = 0), (*b*) the time it takes for the plate to reach a velocity of 9 m/s, and (*c*) the plate velocity 20 s after the jet first strikes the plate. For simplicity, assume the velocity of the jet is increased as the cart moves such that the impulse force exerted by the water jet on the plate remains constant.

6–32E A fan with 24-in-diameter blades moves 2000 cfm (cubic feet per minute) of air at 70°F at sea level. Determine (*a*) the force required to hold the fan and (*b*) the minimum power input required for the fan. Choose a control volume sufficiently large to contain the fan, with the inlet sufficiently far upstream so that the gage pressure at the inlet is nearly zero. Assume air approaches the fan through a large area with negligible velocity and air exits the fan with a uniform velocity at atmospheric pressure through an imaginary cylinder whose diameter is the fan blade diameter. *Answers:* (*a*) 0.820 lbf, (*b*) 5.91 W

6–33E A 3-in-diameter horizontal jet of water, with velocity 140 ft/s, strikes a bent plate, which deflects the water by 135° from its original direction. How much force is required to hold the plate against the water stream and what is its direction? Disregard frictional and gravitational effects.

6–34 Firefighters are holding a nozzle at the end of a hose while trying to extinguish a fire. If the nozzle exit diameter is 8 cm and the water flow rate is 12 m³/min, determine (*a*) the average water exit velocity and (*b*) the horizontal resistance force required of the firefighters to hold the nozzle. *Answers:* (*a*) 39.8 m/s, (*b*) 7958 N

FIGURE P6–34

6–35 A 5-cm-diameter horizontal jet of water with a velocity of 40 m/s relative to the ground strikes a flat plate that is moving in the same direction as the jet at a velocity of 10 m/s. The water splatters in all directions in the plane of the plate. How much force does the water stream exert on the plate?

6–36 Reconsider Prob. 6–35. Using EES (or other) software, investigate the effect of the plate velocity on the force exerted on the plate. Let the plate velocity vary from 0 to 30 m/s, in increments of 3 m/s. Tabulate and plot your results.

6–37E A 3-in-diameter horizontal water jet having a velocity of 90 ft/s strikes a curved plate, which deflects the water 180° at the same speed. Ignoring the frictional effects, determine the force required to hold the plate against the water stream.

Water jet
90 ft/s ←
90 ft/s →   ↓3 in

FIGURE P6–37

6–38 An unloaded helicopter of mass 12,000 kg hovers at sea level while it is being loaded. In the unloaded hover mode, the blades rotate at 550 rpm. The horizontal blades above the helicopter cause a 18-m-diameter air mass to move downward at an average velocity proportional to the overhead blade rotational velocity (rpm). A load of 14,000 kg is loaded onto the helicopter, and the helicopter slowly rises. Determine (a) the volumetric airflow rate downdraft that the helicopter generates during unloaded hover and the required power input and (b) the rpm of the helicopter blades to hover with the 14,000-kg load and the required power input. Take the density of atmospheric air to be 1.18 kg/m³. Assume air approaches the blades from the top through a large area with negligible velocity and air is forced by the blades to move down with a uniform velocity through an imaginary cylinder whose base is the blade span area.

18 m

Load
14,000 kg

FIGURE P6–38

6–39 Reconsider the helicopter in Prob. 6–38, except that it is hovering on top of a 2800-m-high mountain where the air density is 0.928 kg/m³. Noting that the unloaded helicopter blades must rotate at 550 rpm to hover at sea level, determine the blade rotational velocity to hover at the higher altitude. Also determine the percent increase in the required power input to hover at 3000-m altitude relative to that at sea level. *Answers:* 620 rpm, 12.8 percent

6–40 Water is flowing through a 10-cm-diameter water pipe at a rate of 0.1 m³/s. Now a diffuser with an outlet diameter of 20 cm is bolted to the pipe in order to slow down water, as shown in Fig. P6–40. Disregarding frictional effects, determine the force exerted on the bolts due to the water flow.

$d = 10$ cm $D = 20$ cm
Diffuser

FIGURE P6–40

6–41 The weight of a water tank open to the atmosphere is balanced by a counterweight, as shown in Fig. P6–41. There is a 4-cm hole at the bottom of the tank with a discharge coefficient of 0.90, and water level in the tank is maintained constant at 50 cm by water entering the tank horizontally. Determine how much mass must be added to or removed from the counterweight to maintain balance when the hole at the bottom is opened.

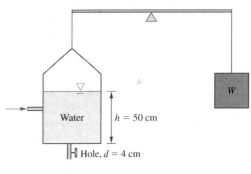

W

Water $h = 50$ cm

Hole, $d = 4$ cm

FIGURE P6–41

6–42 Commercially available large wind turbines have blade span diameters larger than 100 m and generate over 3 MW of electric power at peak design conditions.

30 km/h

60 m

FIGURE P6–42

Consider a wind turbine with a 60-m blade span subjected to 30-km/h steady winds. If the combined turbine–generator efficiency of the wind turbine is 32 percent, determine (a) the power generated by the turbine and (b) the horizontal force exerted by the wind on the supporting mast of the turbine. Take the density of air to be 1.25 kg/m^3, and disregard frictional effects on mast.

6–43 Water enters a centrifugal pump axially at atmospheric pressure at a rate of 0.09 m^3/s and at a velocity of 5 m/s, and leaves in the normal direction along the pump casing, as shown in Fig. P6–43. Determine the force acting on the shaft (which is also the force acting on the bearing of the shaft) in the axial direction.

0.09 m^3/s

Blade

Impeller shroud

$\dot{n}$

Shaft

FIGURE P6–43

6–44 An incompressible fluid of density ρ and viscosity μ flows through a curved duct that turns the flow 180°. The duct cross-sectional area remains constant. The average velocity, momentum flux correction factor, and gage pressure are known at the inlet (1) and outlet (2), as in Fig. P6–44. (a) Write an expression for the horizontal force F_x of the fluid on the walls of the duct in terms of the given variables. (b) Verify your expression by plugging in the following values: $\rho = 998.2$ kg/m^3, $\mu = 1.003 \times 10^{-3}$ kg/m·s, $A_1 = A_2 = 0.025$ m^2, $\beta_1 = 1.01$, $\beta_2 = 1.03$, $V_1 = 10$ m/s, $P_{1,\text{gage}} = 78.47$ kPa, and $P_{2,\text{gage}} = 65.23$ kPa. *Answer:* (b) $F_x = 8680$ N to the right

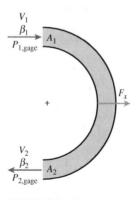

V_1
β_1
$P_{1,\text{gage}}$
A_1

F_x

V_2
β_2
$P_{2,\text{gage}}$
A_2

FIGURE P6–44

6–45 Consider the curved duct of Prob. 6–44, except allow the cross-sectional area to vary along the duct ($A_1 \neq A_2$). (a) Write an expression for the horizontal force F_x of the fluid on the walls of the duct in terms of the given variables. (b) Verify your expression by plugging in the following values: $\rho = 998.2$ kg/m^3, $A_1 = 0.025$ m^2, $A_2 = 0.015$ m^2, $\beta_1 = 1.02$, $\beta_2 = 1.04$, $V_1 = 20$ m/s, $P_{1,\text{gage}} = 88.34$ kPa, and $P_{2,\text{gage}} = 67.48$ kPa. *Answer:* (b) $F_x = 30,700$ N to the right

6–46 As a follow-up to Prob. 6–44, it turns out that for a large enough area ratio A_2/A_1, the inlet pressure is actually *smaller* than the outlet pressure! Explain how this can be true in light of the fact that there is friction and other irreversibilities due to turbulence, and pressure must be lost along the axis of the duct to overcome these irreversibilities.

6–47 An incompressible fluid of density ρ and viscosity μ flows through a curved duct that turns the flow through angle θ. The cross-sectional area also changes. The average velocity, momentum flux correction factor, gage pressure, and area are known at the inlet (1) and outlet (2), as in Fig. P6–47. (a) Write an expression for the horizontal force F_x of the fluid on the walls of the duct in terms of the given variables. (b) Verify your expression by plugging in the following values: $\theta = 135°$, $\rho = 998.2$ kg/m^3, $\mu = 1.003 \times 10^{-3}$ kg/m·s, $A_1 = 0.025$ m^2, $A_2 = 0.050$ m^2, $\beta_1 = 1.01$, $\beta_2 = 1.03$, $V_1 = 6$ m/s, $P_{1,\text{gage}} = 78.47$ kPa, and $P_{2,\text{gage}} = 65.23$ kPa. (*Hint:* You will first need to solve for V_2.) (c) At what turning angle is the force maximized? *Answers:* (b) $F_x = 5500$ N to the right, (c) 180°

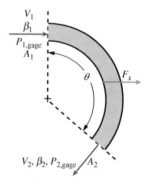

V_1
β_1
$P_{1,\text{gage}}$
A_1

θ

F_x

$V_2, \beta_2, P_{2,\text{gage}}$ A_2

FIGURE P6–47

6–48 Water of density $\rho = 998.2$ kg/m^3 flows through a fireman's nozzle—a converging section of pipe that accelerates the flow. The inlet diameter is $d_1 = 0.100$ m, and the outlet diameter is $d_2 = 0.050$ m. The average velocity, momentum flux correction factor, and gage pressure are known at the inlet (1) and outlet (2), as in Fig. P6–48. (a) Write an expression for the horizontal force F_x of the fluid on the walls of the nozzle in terms of the given variables. (b) Verify your expression by plugging in the following values: $\beta_1 = 1.03$, $\beta_2 = 1.02$, $V_1 = 4$ m/s, $P_{1,\text{gage}} = 123,000$ Pa, and $P_{2,\text{gage}} = 0$ Pa. *Answer:* (b) $F_x = 583$ N to the right

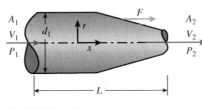

FIGURE P6–48

6–49 Water flowing in a horizontal 25-cm-diameter pipe at 8 m/s and 300 kPa gage enters a 90° bend reducing section, which connects to a 15-cm-diameter vertical pipe. The inlet of the bend is 50 cm above the exit. Neglecting any frictional and gravitational effects, determine the net resultant force exerted on the reducer by the water. Take the momentum-flux correction factor to be 1.04.

6–50 A sluice gate, which controls flow rate in a channel by simply raising or lowering a vertical plate, is commonly used in irrigation systems. A force is exerted on the gate due to the difference between the water heights y_1 and y_2 and the flow velocities V_1 and V_2 upstream and downstream from the gate, respectively. Take the width of the sluice gate (into the page) to be w. Wall shear stresses along the channel walls may be ignored, and for simplicity, we assume steady, uniform flow at locations 1 and 2. Develop a relationship for the force F_R acting on the sluice gate as a function of depths y_1 and y_2, mass flow rate $\dot{m}$, gravitational constant g, gate width w, and water density ρ.

FIGURE P6–50

Angular Momentum Equation

6–51C How is the angular momentum equation obtained from Reynolds transport equations?

6–52C Express the angular momentum equation in scalar form about a specified axis of rotation for a fixed control volume for steady and uniform flow.

6–53C Express the unsteady angular momentum equation in vector form for a control volume that has a constant moment of inertia I, no external moments applied, one outgoing uniform flow stream of velocity $\vec{V}$, and mass flow rate $\dot{m}$.

6–54C Consider two rigid bodies having the same mass and angular speed. Do you think these two bodies must have the same angular momentum? Explain.

6–55 Water is flowing through a 15-cm-diameter pipe that consists of a 3-m-long vertical and 2-m-long horizontal section with a 90° elbow at the exit to force the water to be discharged downward, as shown in Fig. P6–55, in the vertical direction. Water discharges to atmospheric air at a velocity of 7 m/s, and the mass of the pipe section when filled with water is 15 kg per meter length. Determine the moment acting at the intersection of the vertical and horizontal sections of the pipe (point A). What would your answer be if the flow were discharged upward instead of downward?

FIGURE P6–55

6–56E A large lawn sprinkler with two identical arms is used to generate electric power by attaching a generator to its rotating head. Water enters the sprinkler from the base along the axis of rotation at a rate of 5 gal/s and leaves the nozzles in the tangential direction. The sprinkler rotates at a rate of 180 rpm in a horizontal plane. The diameter of each jet is 0.5 in, and the normal distance between the axis of rotation and the center of each nozzle is 2 ft. Determine the maximum possible electrical power produced.

6–57E Reconsider the lawn sprinkler in Prob. 6–56E. If the rotating head is somehow stuck, determine the moment acting on the head.

6–58 The impeller of a centrifugal pump has inner and outer diameters of 13 and 30 cm, respectively, and a flow rate of 0.15 m³/s at a rotational speed of 1200 rpm. The blade width of the impeller is 8 cm at the inlet and 3.5 cm at the outlet. If water enters the impeller in the radial direction and exits at an angle of 60° from the radial direction, determine the minimum power requirement for the pump.

6–59 The impeller of a centrifugal blower has a radius of 18 cm and a blade width of 6.1 cm at the inlet, and a radius of 30 cm and a blade width of 3.4 cm at the outlet. The blower delivers atmospheric air at 20°C and 95 kPa. Disregarding any losses and assuming the tangential components of air velocity at the inlet and the outlet to be equal to the impeller velocity at respective locations, determine the volumetric flow rate of air when the rotational speed of the shaft is 900 rpm and the power consumption of the blower is 120 W. Also determine the normal components of velocity at the inlet and outlet of the impeller.

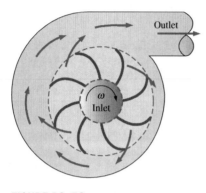

FIGURE P6–59

6–60 Water enters vertically and steadily at a rate of 35 L/s into the sprinkler shown in Fig. P6–60 with unequal arms and unequal discharge areas. The smaller jet has a discharge area of 3 cm² and a normal distance of 50 cm from the axis of rotation. The larger jet has a discharge area of 5 cm² and a normal distance of 35 cm from the axis of rotation. Disregarding any frictional effects, determine (a) the rotational speed of the sprinkler in rpm and (b) the torque required to prevent the sprinkler from rotating.

FIGURE P6–60

6–61 Repeat Prob. 6–60 for a water flow rate of 50 L/s.

6–62 Consider a centrifugal blower that has a radius of 20 cm and a blade width of 8.2 cm at the impeller inlet, and a radius of 45 cm and a blade width of 5.6 cm at the outlet. The blower delivers air at a rate of 0.70 m³/s at a rotational speed of 700 rpm. Assuming the air to enter the impeller in the radial direction and to exit at an angle of 50° from the radial direction, determine the minimum power consumption of the blower. Take the density of air to be 1.25 kg/m³.

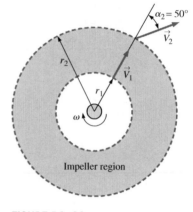

FIGURE P6–62

6–63 Reconsider Prob. 6–62. For the specified flow rate, investigate the effect of discharge angle α_2 on the minimum power input requirements. Assume the air to enter the impeller in the radial direction ($\alpha_1 = 0°$), and vary α_2 from 0° to 85° in increments of 5°. Plot the variation of power input versus α_2, and discuss your results.

6–64E Water enters the impeller of a centrifugal pump radially at a rate of 45 cfm (cubic feet per minute) when the shaft is rotating at 500 rpm. The tangential component of absolute velocity of water at the exit of the 2-ft outer diameter impeller is 110 ft/s. Determine the torque applied to the impeller and the minimum power input to the pump. *Answers:* 160 lbf·ft, 11.3 kW

6–65 A lawn sprinkler with three identical arms is used to water a garden by rotating in a horizontal plane by the impulse caused by water flow. Water enters the sprinkler along the axis of rotation at a rate of 60 L/s and leaves the 1.5-cm-diameter nozzles in the tangential direction. The bearing applies a retarding torque of $T_0 = 50$ N·m due to friction at the anticipated operating speeds. For a normal distance of 40 cm between the axis of rotation and the center of the nozzles, determine the angular velocity of the sprinkler shaft.

6–66 Pelton wheel turbines are commonly used in hydroelectric power plants to generate electric power. In these turbines, a high-speed jet at a velocity of V_j impinges on buckets, forcing the wheel to rotate. The buckets reverse the direction of the jet, and the jet leaves the bucket making an angle β with the direction of the jet, as shown in Fig. P6–66. Show that the power produced by a Pelton wheel of radius r rotating steadily at an angular velocity of ω is $\dot{W}_{shaft} = \rho \omega r \dot{V}(V_j - \omega r)(1 - \cos \beta)$, where ρ is the density and $\dot{V}$ is the volume flow rate of the fluid. Obtain the numerical value for $\rho = 1000$ kg/m³, $r = 2$ m, $\dot{V} = 10$ m³/s, $\dot{n} = 150$ rpm, $\beta = 160°$, and $V_j = 50$ m/s.

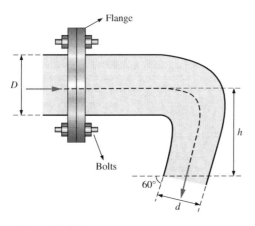

FIGURE P6–66

6–67 Reconsider Prob. 6–66. The maximum efficiency of the turbine occurs when $\beta = 180°$, but this is not practical. Investigate the effect of β on the power generation by allowing it to vary from 0° to 180°. Do you think we are wasting a large fraction of power by using buckets with a β of 160°?

Review Problems

6–68 Water flowing steadily at a rate of 0.16 m³/s is deflected downward by an angled elbow as shown in Fig. P6–68. For $D = 30$ cm, $d = 10$ cm, and $h = 50$ cm, determine the force acting on the flanges of the elbow and the angle its line of action makes with the horizontal. Take the internal volume of the elbow to be 0.03 m³ and disregard the weight of the elbow material and the frictional effects.

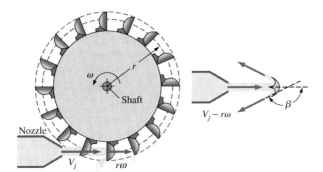

FIGURE P6–68

6–69 Repeat Prob. 6–68 by taking into consideration the weight of the elbow whose mass is 5 kg.

6–70 A 12-cm diameter horizontal water jet with a speed of $V_j = 25$ m/s relative to the ground is deflected by a 40° cone

moving to the left at $V_c = 10$ m/s. Determine the external force, F, needed to maintain the motion of the cone. Disregard the gravity and surface shear effects and assume the cross-sectional area of water jet normal to the direction of motion remains constant throughout the flow. *Answer:* 3240 N

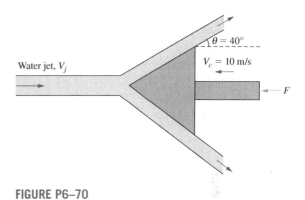

FIGURE P6–70

6–71 Water enters vertically and steadily at a rate of 10 L/s into the sprinkler shown in Fig. P6–71. Both water jets have a diameter of 1.2 cm. Disregarding any frictional effects, determine (*a*) the rotational speed of the sprinkler in rpm and (*b*) the torque required to prevent the sprinkler from rotating.

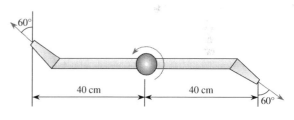

FIGURE P6–71

6–72 Repeat Prob. 6–71 for the case of unequal arms—the left one being 60 cm and the right one 20 cm from the axis of rotation.

6–73 A 6-cm-diameter horizontal water jet having a velocity of 25 m/s strikes a vertical stationary flat plate. The water splatters in all directions in the plane of the plate. How much force is required to hold the plate against the water stream? *Answers:* 1770 N

6–74 Consider steady developing laminar flow of water in a constant-diameter horizontal discharge pipe attached to a tank. The fluid enters the pipe with nearly uniform velocity V and pressure P_1. The velocity profile becomes parabolic after a certain distance with a momentum correction factor of 2 while the pressure drops to P_2. Obtain a relation for the horizontal force acting on the bolts that hold the pipe attached to the tank.

FIGURE P6–74

6–75 A tripod holding a nozzle, which directs a 5-cm-diameter stream of water from a hose, is shown in Fig. P6–75. The nozzle mass is 10 kg when filled with water. The tripod is rated to provide 1800 N of holding force. A firefighter was standing 60 cm behind the nozzle and was hit by the nozzle when the tripod suddenly failed and released the nozzle. You have been hired as an accident reconstructionist and, after testing the tripod, have determined that as water flow rate increased, it did collapse at 1800 N. In your final report you must state the water velocity and the flow rate consistent with the failure and the nozzle velocity when it hit the firefighter. For simplicity, ignore pressure and momentum effects in the upstream portion of the hose. *Answers:* 30.3 m/s, 0.0595 m3/s, 14.7 m/s

FIGURE P6–75

6–76 Consider an airplane with a jet engine attached to the tail section that expels combustion gases at a rate of 18 kg/s with a velocity of $V = 300$ m/s relative to the plane. During landing, a thrust reverser (which serves as a brake for the aircraft and facilitates landing on a short runway) is lowered in the path of the exhaust jet, which deflects the exhaust from rearward to 150°. Determine (*a*) the thrust (forward force) that the engine produces prior to the insertion of the thrust reverser and (*b*) the braking force produced after the thrust reverser is deployed.

FIGURE P6–76

6–77 Reconsider Prob. 6–76. Using EES (or other) software, investigate the effect of thrust reverser angle on the braking force exerted on the airplane. Let the reverser angle vary from 0° (no reversing) to 180° (full reversing) in increments of 10°. Tabulate and plot your results and draw conclusions.

6–78E A spacecraft cruising in space at a constant velocity of 2000 ft/s has a mass of 25,000 lbm. To slow down the spacecraft, a solid fuel rocket is fired, and the combustion gases leave the rocket at a constant rate of 150 lbm/s at a velocity of 5000 ft/s in the same direction as the spacecraft for a period of 5 s. Assuming the mass of the spacecraft remains constant, determine (*a*) the deceleration of the spacecraft during this 5-s period, (*b*) the change of velocity of the spacecraft during this time period, and (*c*) the thrust exerted on the spacecraft.

6–79 A 60-kg ice skater is standing on ice with ice skates (negligible friction). She is holding a flexible hose (essentially weightless) that directs a 2-cm-diameter stream of water horizontally parallel to her skates. The water velocity at the hose outlet is 10 m/s relative to the skater. If she is initially standing still, determine (*a*) the velocity of the skater and the distance she travels in 5 s and (*b*) how long it will take to move 5 m and the velocity at that moment. *Answers:* (*a*) 2.62 m/s, 6.54 m, (*b*) 4.4 s, 2.3 m/s

FIGURE P6–79

6–80 A 5-cm-diameter horizontal jet of water, with velocity 30 m/s, strikes the tip of a horizontal cone, which deflects the water by 45° from its original direction. How much force is required to hold the cone against the water stream?

6–81 Water is flowing into and discharging from a pipe U-section as shown in Fig. P6–81. At flange (1), the total absolute pressure is 200 kPa, and 55 kg/s flows into the pipe. At flange (2), the total pressure is 150 kPa. At location (3), 15 kg/s of water discharges to the atmosphere, which is at 100 kPa. Determine the total *x*- and *z*-forces at the two flanges connecting the pipe. Discuss the significance of gravity force for this problem. Take the momentum-flux correction factor to be 1.03 throughout the pipes.

FIGURE P6–81

6–82 Indiana Jones needs to ascend a 10-m-high building. There is a large hose filled with pressurized water hanging

FIGURE P6–82

down from the building top. He builds a square platform and mounts four 4-cm-diameter nozzles pointing down at each corner. By connecting hose branches, a water jet with 15-m/s velocity can be produced from each nozzle. Jones, the platform, and the nozzles have a combined mass of 150 kg. Determine (*a*) the minimum water jet velocity needed to raise the system, (*b*) how long it takes for the system to rise 10 m when the water jet velocity is 18 m/s and the velocity of the platform at that moment, and (*c*) how much higher will the momentum raise Jones if he shuts off the water at the moment the platform reaches 10 m above the ground. How much time does he have to jump from the platform to the roof? *Answers:* (*a*) 17.1 m/s, (*b*) 4.37 s, 4.57 m/s, (*c*) 1.07 m, 0.933 s

6–83E An engineering student considers using a fan as a levitation demonstration. She plans to face the box-enclosed fan so the air blast is directed face down through a 3-ft-diameter blade span area. The system weighs 5 lbf, and the student will secure the system from rotating. By increasing the power to the fan, she plans to increase the blade rpm and air exit velocity until the exhaust provides sufficient upward force to cause the box fan to hover in the air. Determine (*a*) the air exit velocity to produce 5 lbf, (*b*) the volumetric flow rate needed, and (*c*) the minimum mechanical power that must be supplied to the airstream. Take the air density to be 0.078 lbm/ft³.

FIGURE P6–83E

6–84 Nearly frictionless vertical guide rails maintain a plate of mass m_p in a horizontal position, such that it can slide freely in the vertical direction. A nozzle directs a water stream of area *A* against the plate underside. The water jet splatters in the plate plane, applying an upward force against the plate. The water flow rate $\dot{m}$ (kg/s) can be controlled. Assume that distances are short, so the velocity of the rising jet can be considered constant with height. (*a*) Determine the minimum mass flow rate $\dot{m}_{min}$ necessary to just levitate the plate and obtain a relation for the steady-state velocity of the upward moving plate for $\dot{m} > \dot{m}_{min}$. (*b*) At time $t = 0$, the plate is at rest, and the water jet with $\dot{m} > \dot{m}_{min}$ is suddenly turned on. Apply a force balance to the plate and obtain the integral that relates velocity to time (do not solve).

FIGURE P6–84

6–85 A walnut with a mass of 50 g requires a force of 200 N applied continuously for 0.002 s to be cracked. If walnuts are to be cracked by dropping them from a high place onto a hard surface, determine the minimum height required. Disregard air friction.

6–86 A 7-cm diameter vertical water jet is injected upwards by a nozzle at a speed of 15 m/s. Determine the maximum weight of a flat plate that can be supported by this water jet at a height of 2 m from the nozzle.

6–87 Repeat Prob. 6–86 for a height of 8 m from the nozzle.

6–88 Show that the force exerted by a liquid jet on a stationary nozzle as it leaves with a velocity V is proportional to V^2 or, alternatively, to $\dot{m}^2$. Assume the jet stream is perpendicular to the incoming liquid flow line.

6–89 A soldier jumps from a plane and opens his parachute when his velocity reaches the terminal velocity V_T. The parachute slows him down to his landing velocity of V_F. After the parachute is deployed, the air resistance is proportional to the velocity squared (i.e., $F = kV^2$). The soldier, his parachute, and his gear have a total mass of m. Show that $k = mg/V_F^2$

FIGURE P6–89
© Corbis RF

and develop a relation for the soldier's velocity after he opens the parachute at time $t = 0$.

Answer: $V = V_F \dfrac{V_T + V_F + (V_T - V_F)e^{-2gt/V_f}}{V_T + V_F - (V_T - V_F)e^{-2gt/V_f}}$

6–90 A horizontal water jet with a flow rate of $\dot{V}$ and cross-sectional area of A drives a covered cart of mass m_c along a level and nearly frictionless path. The jet enters a hole at the rear of the cart and all water that enters the cart is retained, increasing the system mass. The relative velocity between the jet of constant velocity V_J and the cart of variable velocity V is $V_J - V$. If the cart is initially empty and stationary when the jet action is initiated, develop a relation (integral form is acceptable) for cart velocity versus time.

FIGURE P6–90

6–91 Water accelerated by a nozzle enters the impeller of a turbine through its outer edge of diameter D with a velocity of V making an angle α with the radial direction at a mass flow rate of $\dot{m}$. Water leaves the impeller in the radial direction. If the angular speed of the turbine shaft is $\dot{n}$, show that the maximum power that can be generated by this radial turbine is $\dot{W}_{\text{shaft}} = \pi \dot{n} \dot{m} DV \sin \alpha$.

6–92 Water enters a two-armed lawn sprinkler along the vertical axis at a rate of 75 L/s, and leaves the sprinkler nozzles as 2-cm diameter jets at an angle of θ from the tangential direction, as shown in Fig. P6–92. The length of each sprinkler

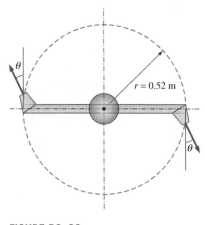

FIGURE P6–92

arm is 0.52 m. Disregarding any frictional effects, determine the rate of rotation $\dot{n}$ of the sprinkler in rev/min for (a) $\theta = 0°$, (b) $\theta = 30°$, and (c) $\theta = 60°$.

6–93 [EES] Reconsider Prob. 6–92. For the specified flow rate, investigate the effect of discharge angle θ on the rate of rotation $\dot{n}$ by varying θ from 0° to 90° in increments of 10°. Plot the rate of rotation versus θ, and discuss your results.

6–94 A stationary water tank of diameter D is mounted on wheels and is placed on a nearly frictionless level surface. A smooth hole of diameter D_o near the bottom of the tank allows water to jet horizontally and rearward and the water jet force propels the system forward. The water in the tank is much heavier than the tank-and-wheel assembly, so only the mass of water remaining in the tank needs to be considered in this problem. Considering the decrease in the mass of water with time, develop relations for (a) the acceleration, (b) the velocity, and (c) the distance traveled by the system as a function of time.

6–95 An orbiting satellite has a mass of 3400 kg and is traveling at a constant velocity of V_0. To alter its orbit, an attached rocket discharges 100 kg of gases from the reaction of solid fuel at a speed of 3000 m/s relative to the satellite in a direction opposite V_0. The fuel discharge rate is constant for 3s. Determine (a) the thrust exerted on the satellite, (b) the acceleration of the satellite during this 3-s period, and (c) the change of velocity of the satellite during this time period.

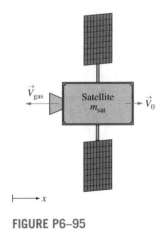

FIGURE P6–95

6–96 Water enters a mixed flow pump axially at a rate of 0.3 m³/s and at a velocity of 7 m/s, and is discharged to the atmosphere at an angle of 75° from the horizontal, as shown in Fig. P6–96. If the discharge flow area is half the inlet area, determine the force acting on the shaft in the axial direction.

FIGURE P6–96

6–97 Water flows steadily through a splitter as shown in Fig. P6–97 with $\dot{V}_1 = 0.08$ m³/s, $\dot{V}_2 = 0.05$ m³/s, $D_1 = D_2 = 12$ cm, $D_3 = 10$ cm. If the pressure readings at the inlet and outlets of the splitter are $P_1 = 100$ kPa, $P_2 = 90$ kPa and $P_3 = 80$ kPa, determine external force needed to hold the device fixed. Disregard the weight effects.

FIGURE P6–97

6–98 Water is discharged from a pipe through a 1.2-m long 5-mm wide rectangular slit underneath of the pipe. Water discharge velocity profile is parabolic, varying from 3 m/s on one end of the slit to 7 m/s on the other, as shown in Fig. P6–98.

FIGURE P6–98

Determine (a) the rate of discharge through the slit and (b) the vertical force acting on the pipe due to this discharge process.

Fundamentals of Engineering (FE) Exam Problems

6–99   When determining the thrust developed by a jet engine, a wise choice of control volume is
(a) Fixed control volume   (b) Moving control volume
(c) Deforming control volume   (d) Moving or deforming control volume   (e) None of these

6–100   Consider an airplane cruising at 850 km/h to the right. If the velocity of exhaust gases is 700 km/h to the left relative to the ground, the velocity of the exhaust gases relative to the nozzle exit is
(a) 1550 km/h   (b) 850 km/h   (c) 700 km/h
(d) 350 km/h   (e) 150 km/h

6–101   Consider water flow through a horizontal, short garden hose at a rate of 30 kg/min. The velocity at the inlet is 1.5 m/s and that at the outlet is 14.5 m/s. Disregard the weight of the hose and water. Taking the momentum-flux correction factor to be 1.04 at both the inlet and the outlet, the anchoring force required to hold the hose in place is
(a) 2.8 N   (b) 8.6 N   (c) 17.5 N   (d) 27.9 N   (e) 43.3 N

6–102   Consider water flow through a horizontal, short garden hose at a rate of 30 kg/min. The velocity at the inlet is 1.5 m/s and that at the outlet is 11.5 m/s. The hose makes a 180° turn before the water is discharged. Disregard the weight of the hose and water. Taking the momentum-flux correction factor to be 1.04 at both the inlet and the outlet, the anchoring force required to hold the hose in place is
(a) 7.6 N   (b) 28.4 N   (c) 16.6 N   (d) 34.1 N
(e) 11.9 N

6–103   A water jet strikes a stationary vertical plate horizontally at a rate of 5 kg/s with a velocity of 35 km/h. Assume the water stream moves in the vertical direction after the strike. The force needed to prevent the plate from moving horizontally is
(a) 15.5 N   (b) 26.3 N   (c) 19.7 N   (d) 34.2 N   (e) 48.6 N

6–104   Consider water flow through a horizontal, short garden hose at a rate of 40 kg/min. The velocity at the inlet is 1.5 m/s and that at the outlet is 16 m/s. The hose makes a 90° turn to a vertical direction before the water is discharged. Disregard the weight of the hose and water. Taking the momentum-flux correction factor to be 1.04 at both the inlet and the outlet, the reaction force in the vertical direction required to hold the hose in place is
(a) 11.1 N   (b) 10.1 N   (c) 9.3 N   (d) 27.2 N   (e) 28.9 N

6–105   Consider water flow through a horizontal, short pipe at a rate of 80 kg/min. The velocity at the inlet is 1.5 m/s and that at the outlet is 16.5 m/s. The pipe makes a 90° turn to a vertical direction before the water is discharged. Disregard the weight of the pipe and water. Taking the momentum-flux correction factor to be 1.04 at both the inlet and the outlet, the reaction force in the horizontal direction required to hold the pipe in place is
(a) 73.7 N   (b) 97.1 N   (c) 99.2 N   (d) 122 N   (e) 153 N

6–106   A water jet strikes a stationary horizontal plate vertically at a rate of 18 kg/s with a velocity of 24 m/s. The mass of the plate is 10 kg. Assume the water stream moves in the horizontal direction after the strike. The force needed to prevent the plate from moving vertically is
(a) 192 N   (b) 240 N   (c) 334 N   (d) 432 N   (e) 530 N

6–107   The velocity of wind at a wind turbine is measured to be 6 m/s. The blade span diameter is 24 m and the efficiency of the wind turbine is 29 percent. The density of air is 1.22 kg/m^3. The horizontal force exerted by the wind on the supporting mast of the wind turbine is
(a) 2524 N   (b) 3127 N   (c) 3475 N   (d) 4138 N
(e) 4313 N

6–108   The velocity of wind at a wind turbine is measured to be 8 m/s. The blade span diameter is 12 m. The density of air is 1.2 kg/m^3. If the horizontal force exerted by the wind on the supporting mast of the wind turbine is 1620 N, the efficiency of the wind turbine is
(a) 27.5%   (b) 31.7%   (c) 29.5%   (d) 35.1%   (e) 33.8%

6–109   The shaft of a turbine rotates at a speed of 800 rpm. If the torque of the shaft is 350 N·m, the shaft power is
(a) 112 kW   (b) 176 kW   (c) 293 kW   (d) 350 kW
(e) 405 kW

6–110   A 3-cm-diameter horizontal pipe attached to a surface makes a 90° turn to a vertical upward direction before the water is discharged at a velocity of 9 m/s. The horizontal section is 5 m long and the vertical section is 4 m long. Neglecting the mass of the water contained in the pipe, the bending moment acting on the base of the pipe on the wall is
(a) 286 N·m   (b) 229 N·m   (c) 207 N·m
(d) 175 N·m   (e) 124 N·m

6–111   A 3-cm-diameter horizontal pipe attached to a surface makes a 90° turn to a vertical upward direction before the water is discharged at a velocity of 6 m/s. The horizontal section is 5 m long and the vertical section is 4 m long. Neglecting the mass of the pipe and considering the weight of the water contained in the pipe, the bending moment acting on the base of the pipe on the wall is
(a) 11.9 N·m   (b) 46.7 N·m   (c) 127 N·m
(d) 104 N·m   (e) 74.8 N·m

6–112   A large lawn sprinkler with four identical arms is to be converted into a turbine to generate electric power by attaching a generator to its rotating head. Water enters the sprinkler from the base along the axis of rotation at a rate of 15 kg/s and leaves the nozzles in the tangential direction at a velocity of 50 m/s relative to the rotating nozzle. The sprinkler rotates at a rate of 400 rpm in a horizontal plane. The normal distance between the axis of rotation and the center of each nozzle is 30 cm. Estimate the electric power produced.

(a) 5430 W   (b) 6288 W   (c) 6634 W   (d) 7056 W
(e) 7875 W

6–113 Consider the impeller of a centrifugal pump with a rotational speed of 900 rpm and a flow rate of 95 kg/min. The impeller radii at the inlet and outlet are 7 cm and 16 cm, respectively. Assuming that the tangential fluid velocity is equal to the blade angular velocity both at the inlet and the exit, the power requirement of the pump is
(a) 83 W   (b) 291 W   (c) 409 W   (d) 756 W   (e) 1125 W

6–114 Water enters the impeller of a centrifugal pump radially at a rate of 450 L/min when the shaft is rotating at 400 rpm. The tangential component of absolute velocity of water at the exit of the 70-cm outer diameter impeller is 55 m/s. The torque applied to the impeller is
(a) 144 N·m   (b) 93.6 N·m   (c) 187 N·m
(d) 112 N·m   (e) 235 N·m

Design and Essay Problem

6–115 Visit a fire station and obtain information about flow rates through hoses and discharge diameters. Using this information, calculate the impulse force to which the firefighters are subjected when holding a fire hose.

DIMENSIONAL ANALYSIS AND MODELING

In this chapter, we first review the concepts of *dimensions* and *units*. We then review the fundamental principle of *dimensional homogeneity*, and show how it is applied to equations in order to *nondimensionalize* them and to identify *dimensionless groups*. We discuss the concept of *similarity* between a *model* and a *prototype*. We also describe a powerful tool for engineers and scientists called *dimensional analysis*, in which the combination of dimensional variables, nondimensional variables, and dimensional constants into *nondimensional parameters* reduces the number of necessary independent parameters in a problem. We present a step-by-step method for obtaining these nondimensional parameters, called the *method of repeating variables*, which is based solely on the dimensions of the variables and constants. Finally, we apply this technique to several practical problems to illustrate both its utility and its limitations.

OBJECTIVES

When you finish reading this chapter, you should be able to

■ Develop a better understanding of dimensions, units, and dimensional homogeneity of equations

■ Understand the numerous benefits of dimensional analysis

■ Know how to use the method of repeating variables to identify nondimensional parameters

■ Understand the concept of dynamic similarity and how to apply it to experimental modeling

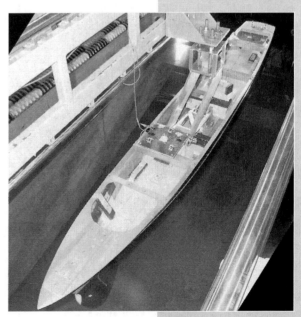

A 1:46.6 scale model of an Arleigh Burke class U.S. Navy fleet destroyer being tested in the 100-m-long towing tank at the University of Iowa. The model is 3.048 m long. In tests like this, the Froude number is the most important nondimensional parameter.

Photograph courtesy of IIHR-Hydroscience & Engineering, University of Iowa. Used by permission.

Length

3.2 cm

cm   1   2   3

FIGURE 7–1

A *dimension* is a measure of a physical quantity without numerical values, while a *unit* is a way to assign a number to the dimension. For example, length is a dimension, but centimeter is a unit.

7–1 ▪ DIMENSIONS AND UNITS

A **dimension** is a measure of a physical quantity (without numerical values), while a **unit** is a way to assign a *number* to that dimension. For example, length is a dimension that is measured in units such as microns (μm), feet (ft), centimeters (cm), meters (m), kilometers (km), etc. (Fig. 7–1). There are seven **primary dimensions** (also called **fundamental** or **basic dimensions**)—mass, length, time, temperature, electric current, amount of light, and amount of matter.

> All nonprimary dimensions can be formed by some combination of the seven primary dimensions.

For example, force has the same dimensions as mass times acceleration (by Newton's second law). Thus, in terms of primary dimensions,

Dimensions of force: $\{\text{Force}\} = \left\{\text{Mass}\,\dfrac{\text{Length}}{\text{Time}^2}\right\} = \{mL/t^2\}$ **(7–1)**

where the brackets indicate "the dimensions of" and the abbreviations are taken from Table 7–1. You should be aware that some authors prefer force instead of mass as a primary dimension—we do not follow that practice.

TABLE 7–1

Primary dimensions and their associated primary SI and English units

Dimension	Symbol*	SI Unit	English Unit
Mass	m	kg (kilogram)	lbm (pound-mass)
Length	L	m (meter)	ft (foot)
Time†	t	s (second)	s (second)
Temperature	T	K (kelvin)	R (rankine)
Electric current	I	A (ampere)	A (ampere)
Amount of light	C	cd (candela)	cd (candela)
Amount of matter	N	mol (mole)	mol (mole)

* We italicize symbols for variables, but not symbols for dimensions.

† Note that some authors use the symbol T for the time dimension and the symbol θ for the temperature dimension. We do not follow this convention to avoid confusion between time and temperature.

FIGURE 7–2

The water strider is an insect that can walk on water due to surface tension.

NPS Photo by Rosalie LaRue.

EXAMPLE 7–1   **Primary Dimensions of Surface Tension**

An engineer is studying how some insects are able to walk on water (Fig. 7–2). A fluid property of importance in this problem is surface tension (σ_s), which has dimensions of force per unit length. Write the dimensions of surface tension in terms of primary dimensions.

SOLUTION The primary dimensions of surface tension are to be determined. *Analysis* From Eq. 7–1, force has dimensions of mass times acceleration, or $\{mL/t^2\}$. Thus,

Dimensions of surface tension: $\{\sigma_s\} = \left\{\dfrac{\text{Force}}{\text{Length}}\right\} = \left\{\dfrac{m \cdot L/t^2}{L}\right\} = \{m/t^2\}$ **(1)**

Discussion The usefulness of expressing the dimensions of a variable or constant in terms of primary dimensions will become clearer in the discussion of the method of repeating variables in Section 7–4.

7–2 · DIMENSIONAL HOMOGENEITY

We've all heard the old saying, You can't add apples and oranges (Fig. 7–3). This is actually a simplified expression of a far more global and fundamental mathematical law for equations, the **law of dimensional homogeneity**, stated as

Every additive term in an equation must have the same dimensions.

Consider, for example, the change in total energy of a simple compressible closed system from one state and/or time (1) to another (2), as illustrated in Fig. 7–4. The change in total energy of the system (ΔE) is given by

Change of total energy of a system: $\Delta E = \Delta U + \Delta KE + \Delta PE$ **(7–2)**

where E has three components: internal energy (U), kinetic energy (KE), and potential energy (PE). These components can be written in terms of the system mass (m); measurable quantities and thermodynamic properties at each of the two states, such as speed (V), elevation (z), and specific internal energy (u); and the gravitational acceleration constant (g),

$$\Delta U = m(u_2 - u_1) \qquad \Delta KE = \frac{1}{2} m(V_2^2 - V_1^2) \qquad \Delta PE = mg(z_2 - z_1) \quad \textbf{(7–3)}$$

It is straightforward to verify that the left side of Eq. 7–2 and all three additive terms on the right side of Eq. 7–2 have the same dimensions—energy. Using the definitions of Eq. 7–3, we write the primary dimensions of each term,

$$\{\Delta E\} = \{\text{Energy}\} = \{\text{Force} \cdot \text{Length}\} \ \rightarrow \ \{\Delta E\} = \{mL^2/t^2\}$$

$$\{\Delta U\} = \left\{ \text{Mass} \frac{\text{Energy}}{\text{Mass}} \right\} = \{\text{Energy}\} \ \rightarrow \ \{\Delta U\} = \{mL^2/t^2\}$$

$$\{\Delta KE\} = \left\{ \text{Mass} \frac{\text{Length}^2}{\text{Time}^2} \right\} \quad \rightarrow \quad \{\Delta KE\} = \{mL^2/t^2\}$$

$$\{\Delta PE\} = \left\{ \text{Mass} \frac{\text{Length}}{\text{Time}^2} \text{Length} \right\} \ \rightarrow \ \{\Delta PE\} = \{mL^2/t^2\}$$

If at some stage of an analysis we find ourselves in a position in which two additive terms in an equation have *different* dimensions, this would be a clear indication that we have made an error at some earlier stage in the analysis (Fig. 7–5). In addition to dimensional homogeneity, calculations are valid only when the *units* are also homogeneous in each additive term. For example, units of energy in the above terms may be J, N·m, or kg·m²/s², all of which are equivalent. Suppose, however, that kJ were used in place of J for one of the terms. This term would be off by a factor of 1000 compared to the other terms. It is wise to write out *all* units when performing mathematical calculations in order to avoid such errors.

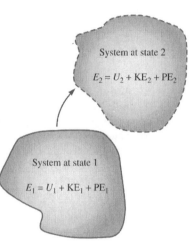

FIGURE 7–3
You can't add apples and oranges!

FIGURE 7–4
Total energy of a system at state 1 and at state 2.

FIGURE 7–5
An equation that is not dimensionally homogeneous is a sure sign of an error.

294
DIMENSIONAL ANALYSIS AND MODELING

FIGURE 7–6
The Bernoulli equation is a good example of a *dimensionally homogeneous* equation. All additive terms, including the constant, have the *same* dimensions, namely that of pressure. In terms of primary dimensions, each term has dimensions $\{m/(t^2L)\}$.

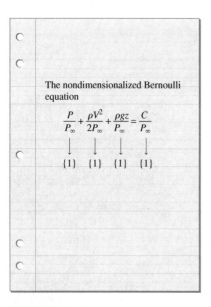

FIGURE 7–7
A *nondimensionalized* form of the Bernoulli equation is formed by dividing each additive term by a pressure (here we use P_∞). Each resulting term is *dimensionless* (dimensions of $\{1\}$).

EXAMPLE 7–2   **Dimensional Homogeneity of the Bernoulli Equation**

Probably the most well-known (and most misused) equation in fluid mechanics is the Bernoulli equation (Fig. 7–6), discussed in Chap. 5. One standard form of the Bernoulli equation for incompressible irrotational fluid flow is

Bernoulli equation:
$$P + \frac{1}{2}\rho V^2 + \rho g z = C \tag{1}$$

(*a*) Verify that each additive term in the Bernoulli equation has the same dimensions. (*b*) What are the dimensions of the constant C?

SOLUTION   We are to verify that the primary dimensions of each additive term in Eq. 1 are the same, and we are to determine the dimensions of constant C.

Analysis   (*a*) Each term is written in terms of primary dimensions,

$$\{P\} = \{\text{Pressure}\} = \left\{\frac{\text{Force}}{\text{Area}}\right\} = \left\{\text{Mass}\frac{\text{Length}}{\text{Time}^2}\frac{1}{\text{Length}^2}\right\} = \left\{\frac{\text{m}}{\text{t}^2\text{L}}\right\}$$

$$\left\{\frac{1}{2}\rho V^2\right\} = \left\{\frac{\text{Mass}}{\text{Volume}}\left(\frac{\text{Length}}{\text{Time}}\right)^2\right\} = \left\{\frac{\text{Mass} \times \text{Length}^2}{\text{Length}^3 \times \text{Time}^2}\right\} = \left\{\frac{\text{m}}{\text{t}^2\text{L}}\right\}$$

$$\{\rho g z\} = \left\{\frac{\text{Mass}}{\text{Volume}}\frac{\text{Length}}{\text{Time}^2}\text{Length}\right\} = \left\{\frac{\text{Mass} \times \text{Length}^2}{\text{Length}^3 \times \text{Time}^2}\right\} = \left\{\frac{\text{m}}{\text{t}^2\text{L}}\right\}$$

Indeed, **all three additive terms have the same dimensions.**

(*b*) From the law of dimensional homogeneity, the constant must have the same dimensions as the other additive terms in the equation. Thus,

Primary dimensions of the Bernoulli constant:
$$\{C\} = \left\{\frac{\text{m}}{\text{t}^2\text{L}}\right\}$$

Discussion   If the dimensions of any of the terms were different from the others, it would indicate that an error was made somewhere in the analysis.

Nondimensionalization of Equations

The law of dimensional homogeneity guarantees that every additive term in an equation has the same dimensions. It follows that if we divide each term in the equation by a collection of variables and constants whose product has those same dimensions, the equation is rendered **nondimensional** (Fig. 7–7). If, in addition, the nondimensional terms in the equation are of order unity, the equation is called **normalized.** Normalization is thus more restrictive than nondimensionalization, even though the two terms are sometimes (incorrectly) used interchangeably.

Each term in a nondimensional equation is dimensionless.

In the process of nondimensionalizing an equation of motion, **nondimensional parameters** often appear—most of which are named after a notable scientist or engineer (e.g., the Reynolds number and the Froude number). This process is referred to by some authors as **inspectional analysis.**

w = component of velocity in the *z*-direction

z = vertical distance

g = gravitational acceleration in the negative *z*-direction

FIGURE 7–8
Object falling in a vacuum. Vertical velocity is drawn positively, so $w < 0$ for a falling object.

As a simple example, consider the equation of motion describing the elevation z of an object falling by gravity through a vacuum (no air drag), as in Fig. 7–8. The initial location of the object is z_0 and its initial velocity is w_0 in the z-direction. From high school physics,

Equation of motion:
$$\frac{d^2z}{dt^2} = -g \qquad (7\text{–}4)$$

Dimensional variables are defined as dimensional quantities that change or vary in the problem. For the simple differential equation given in Eq. 7–4, there are two dimensional variables: z (dimension of length) and t (dimension of time). **Nondimensional** (or **dimensionless**) **variables** are defined as quantities that change or vary in the problem, but have no dimensions; an example is angle of rotation, measured in degrees or radians which are dimensionless units. Gravitational constant g, while dimensional, remains constant and is called a **dimensional constant.** Two additional dimensional constants are relevant to this particular problem, initial location z_0 and initial vertical speed w_0. While dimensional constants may change from problem to problem, they are fixed for a particular problem and are thus distinguished from dimensional variables. We use the term **parameters** for the combined set of dimensional variables, nondimensional variables, and dimensional constants in the problem.

Equation 7–4 is easily solved by integrating twice and applying the initial conditions. The result is an expression for elevation z at any time t:

Dimensional result:
$$z = z_0 + w_0 t - \frac{1}{2} g t^2 \qquad (7\text{–}5)$$

The constant $\frac{1}{2}$ and the exponent 2 in Eq. 7–5 are dimensionless results of the integration. Such constants are called **pure constants.** Other common examples of pure constants are π and e.

To nondimensionalize Eq. 7–4, we need to select **scaling parameters**, based on the primary dimensions contained in the original equation. In fluid flow problems there are typically at least *three* scaling parameters, e.g., L, V, and $P_0 - P_\infty$ (Fig. 7–9), since there are at least three primary dimensions in the general problem (e.g., mass, length, and time). In the case of the falling object being discussed here, there are only two primary dimensions, length and time, and thus we are limited to selecting only *two* scaling parameters. We have some options in the selection of the scaling parameters since we have three available dimensional constants g, z_0, and w_0. We choose z_0 and w_0. You are invited to repeat the analysis with g and z_0 and/or with g and w_0. With these two chosen scaling parameters we nondimensionalize the dimensional variables z and t. The first step is to list the primary dimensions of *all* dimensional variables and dimensional constants in the problem,

Primary dimensions of all parameters:

$$\{z\} = \{L\} \quad \{t\} = \{t\} \quad \{z_0\} = \{L\} \quad \{w_0\} = \{L/t\} \quad \{g\} = \{L/t^2\}$$

The second step is to use our two scaling parameters to nondimensionalize z and t (by inspection) into nondimensional variables z^* and t^*,

Nondimensionalized variables:
$$z^* = \frac{z}{z_0} \quad t^* = \frac{w_0 t}{z_0} \qquad (7\text{–}6)$$

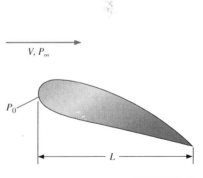

V, P_∞

P_0

L

FIGURE 7–9
In a typical fluid flow problem, the scaling parameters usually include a characteristic length L, a characteristic velocity V, and a reference pressure difference $P_0 - P_\infty$. Other parameters and fluid properties such as density, viscosity, and gravitational acceleration enter the problem as well.

FIGURE 7–10
The *Froude number* is important in free-surface flows such as flow in open channels. Shown here is flow through a sluice gate. The Froude number upstream of the sluice gate is $Fr_1 = V_1/\sqrt{gy_1}$, and it is $Fr_2 = V_2/\sqrt{gy_2}$ downstream of the sluice gate.

Substitution of Eq. 7–6 into Eq. 7–4 gives

$$\frac{d^2z}{dt^2} = \frac{d^2(z_0z^*)}{d(z_0t^*/w_0)^2} = \frac{w_0^2}{z_0}\frac{d^2z^*}{dt^{*2}} = -g \quad \rightarrow \quad \frac{w_0^2}{gz_0}\frac{d^2z^*}{dt^{*2}} = -1 \quad \textbf{(7–7)}$$

which is the desired nondimensional equation. The grouping of dimensional constants in Eq. 7–7 is the square of a well-known **nondimensional parameter** or **dimensionless group** called the **Froude number,**

$$\text{Froude number:} \qquad \qquad Fr = \frac{w_0}{\sqrt{gz_0}} \qquad \qquad \textbf{(7–8)}$$

The Froude (pronounced "Frude") number also appears as a nondimensional parameter in free-surface flows (Chap. 13), and can be thought of as the ratio of inertial force to gravitational force (Fig. 7–10). You should note that in some older textbooks, Fr is defined as the *square* of the parameter shown in Eq. 7–8. Substitution of Eq. 7–8 into Eq. 7–7 yields

$$\text{Nondimensionalized equation of motion:} \qquad \frac{d^2z^*}{dt^{*2}} = -\frac{1}{Fr^2} \qquad \textbf{(7–9)}$$

In dimensionless form, only one parameter remains, namely the Froude number. Equation 7–9 is easily solved by integrating twice and applying the initial conditions. The result is an expression for dimensionless elevation z^* as a function of dimensionless time t^*:

$$\text{Nondimensional result:} \qquad z^* = 1 + t^* - \frac{1}{2Fr^2}t^{*2} \qquad \textbf{(7–10)}$$

Comparison of Eqs. 7–5 and 7–10 reveals that they are equivalent. In fact, for practice, substitute Eqs. 7–6 and 7–8 into Eq. 7–5 to verify Eq. 7–10.

It seems that we went through a lot of extra algebra to generate the same final result. *What then is the advantage of nondimensionalizing the equation?* Before answering this question, we note that the advantages are not so clear in this simple example because we were able to analytically integrate the differential equation of motion. In more complicated problems, the differential equation (or more generally the coupled *set* of differential equations) *cannot* be integrated analytically, and engineers must either integrate the equations numerically, or design and conduct physical experiments to obtain the needed results, both of which can incur considerable time and expense. In such cases, the nondimensional parameters generated by nondimensionalizing the equations are extremely useful and can save much effort and expense in the long run.

There are two key advantages of nondimensionalization (Fig. 7–11). First, *it increases our insight about the relationships between key parameters.* Equation 7–8 reveals, for example, that doubling w_0 has the same effect as decreasing z_0 by a factor of 4. Second, *it reduces the number of parameters in the problem.* For example, the original problem contains one dependent variable, z; one independent variable, t; and *three* additional dimensional constants, g, w_0, and z_0. The nondimensionalized problem contains one dependent parameter, z^*; one independent parameter, t^*; and only *one* additional parameter, namely the dimensionless Froude number, Fr. The number of additional parameters has been reduced from three to one! Example 7–3 further illustrates the advantages of nondimensionalization.

Relationships between key parameters in the problem are identified.

The number of parameters in a nondimensionalized equation is less than the number of parameters in the original equation.

FIGURE 7–11
The two key advantages of nondimensionalization of an equation.

EXAMPLE 7–3   Illustration of the Advantages of Nondimensionalization

Your little brother's high school physics class conducts experiments in a large vertical pipe whose inside is kept under vacuum conditions. The students are able to remotely release a steel ball at initial height z_0 between 0 and 15 m (measured from the bottom of the pipe), and with initial vertical speed w_0 between 0 and 10 m/s. A computer coupled to a network of photosensors along the pipe enables students to plot the trajectory of the steel ball (height z plotted as a function of time t) for each test. The students are unfamiliar with dimensional analysis or nondimensionalization techniques, and therefore conduct several "brute force" experiments to determine how the trajectory is affected by initial conditions z_0 and w_0. First they hold w_0 fixed at 4 m/s and conduct experiments at five different values of z_0: 3, 6, 9, 12, and 15 m. The experimental results are shown in Fig. 7–12a. Next, they hold z_0 fixed at 10 m and conduct experiments at five different values of w_0: 2, 4, 6, 8, and 10 m/s. These results are shown in Fig. 7–12b. Later that evening, your brother shows you the data and the trajectory plots and tells you that they plan to conduct more experiments at different values of z_0 and w_0. You explain to him that by first nondimensionalizing the data, the problem can be reduced to just *one* parameter, and no further experiments are required. Prepare a nondimensional plot to prove your point and discuss.

SOLUTION   A nondimensional plot is to be generated from all the available trajectory data. Specifically, we are to plot z^* as a function of t^*.

Assumptions   The inside of the pipe is subjected to strong enough vacuum pressure that aerodynamic drag on the ball is negligible.

Properties   The gravitational constant is 9.81 m/s².

Analysis   Equation 7–4 is valid for this problem, as is the nondimensionalization that resulted in Eq. 7–9. As previously discussed, this problem combines three of the original dimensional parameters (g, z_0, and w_0) into *one* nondimensional parameter, the Froude number. After converting to the dimensionless variables of Eq. 7–6, the 10 trajectories of Fig. 7–12a and b are replotted in dimensionless format in Fig. 7–13. It is clear that all the trajectories are of the same family, with the Froude number as the only remaining parameter. Fr^2 varies from about 0.041 to about 1.0 in these experiments. If any more experiments are to be conducted, they should include combinations of z_0 and w_0 that produce Froude numbers outside of this range. A large number of additional experiments would be unnecessary, since all the trajectories would be of the same family as those plotted in Fig. 7–13.

Discussion   At low Froude numbers, gravitational forces are much larger than inertial forces, and the ball falls to the floor in a relatively short time. At large values of Fr on the other hand, inertial forces dominate initially, and the ball rises a significant distance before falling; it takes much longer for the ball to hit the ground. The students are obviously not able to adjust the gravitational constant, but if they could, the brute force method would require many more experiments to document the effect of g. If they nondimensionalize first, however, the dimensionless trajectory plots already obtained and shown in Fig. 7–13 would be valid for *any* value of g; no further experiments would be required unless Fr were outside the range of tested values.

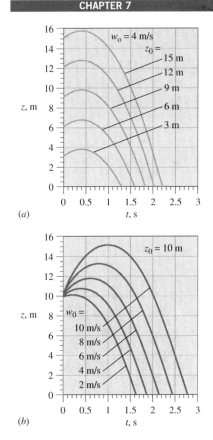

FIGURE 7–12

Trajectories of a steel ball falling in a vacuum: (*a*) w_0 fixed at 4 m/s, and (*b*) z_0 fixed at 10 m (Example 7–3).

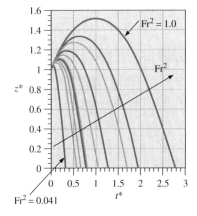

FIGURE 7–13

Trajectories of a steel ball falling in a vacuum. Data of Fig. 7–12a and b are nondimensionalized and combined onto one plot.

If you are still not convinced that nondimensionalizing the equations and the parameters has many advantages, consider this: In order to reasonably document the trajectories of Example 7–3 for a range of all three of the dimensional parameters g, z_0, and w_0, the brute force method would require several (say a minimum of four) additional plots like Fig. 7–12a at various values (levels) of w_0, plus several additional *sets* of such plots for a range of g. A complete data set for three parameters with five levels of each parameter would require $5^3 = 125$ experiments! Nondimensionalization reduces the number of parameters from three to one—a total of only $5^1 = 5$ experiments are required for the same resolution. (For five levels, only five dimensionless trajectories like those of Fig. 7–13 are required, at carefully chosen values of Fr.)

Another advantage of nondimensionalization is that extrapolation to untested values of one or more of the dimensional parameters is possible. For example, the data of Example 7–3 were taken at only one value of gravitational acceleration. Suppose you wanted to extrapolate these data to a different value of g. Example 7–4 shows how this is easily accomplished via the dimensionless data.

FIGURE 7–14
Throwing a baseball on the moon (Example 7–4).

EXAMPLE 7–4   Extrapolation of Nondimensionalized Data

The gravitational constant at the surface of the moon is only about one-sixth of that on earth. An astronaut on the moon throws a baseball at an initial speed of 21.0 m/s at a 5° angle above the horizon and at 2.0 m above the moon's surface (Fig. 7–14). (a) Using the dimensionless data of Example 7–3 shown in Fig. 7–13, predict how long it takes for the baseball to fall to the ground. (b) Do an *exact* calculation and compare the result to that of part (a).

SOLUTION   Experimental data obtained on earth are to be used to predict the time required for a baseball to fall to the ground on the moon.
Assumptions   **1** The horizontal velocity of the baseball is irrelevant. **2** The surface of the moon is perfectly flat near the astronaut. **3** There is no aerodynamic drag on the ball since there is no atmosphere on the moon. **4** Moon gravity is one-sixth that of earth.
Properties   The gravitational constant on the moon is $g_{moon} \cong 9.81/6 = 1.63$ m/s^2.
Analysis   (a) The Froude number is calculated based on the value of g_{moon} and the vertical component of initial speed,

$$w_0 = (21.0 \text{ m/s}) \sin(5°) = 1.830 \text{ m/s}$$

from which

$$\text{Fr}^2 = \frac{w_0^2}{g_{moon} z_0} = \frac{(1.830 \text{ m/s})^2}{(1.63 \text{ m/s}^2)(2.0 \text{ m})} = 1.03$$

This value of Fr^2 is nearly the same as the largest value plotted in Fig. 7–13. Thus, in terms of dimensionless variables, the baseball strikes the ground at $t^* \cong 2.75$, as determined from Fig. 7–13. Converting back to dimensional variables using Eq. 7–6,

Estimated time to strike the ground: $t = \dfrac{t^* z_0}{w_0} = \dfrac{2.75(2.0 \text{ m})}{1.830 \text{ m/s}} = 3.01 \text{ s}$

(b) An exact calculation is obtained by setting z equal to zero in Eq. 7–5 and solving for time t (using the quadratic formula),

Exact time to strike the ground:

$$t = \frac{w_0 + \sqrt{w_0^2 + 2z_0 g}}{g}$$

$$= \frac{1.830 \text{ m/s} + \sqrt{(1.830 \text{ m/s})^2 + 2(2.0 \text{ m})(1.63 \text{ m/s}^2)}}{1.63 \text{ m/s}^2} = 3.05 \text{ s}$$

Discussion If the Froude number had landed between two of the trajectories of Fig. 7–13, interpolation would have been required. Since some of the numbers are precise to only two significant digits, the small difference between the results of part (*a*) and part (*b*) is of no concern. The final result is *t* = 3.0 s to two significant digits.

The differential equations of motion for fluid flow are derived and discussed in Chap. 9. In Chap. 10 you will find an analysis similar to that presented here, but applied to the differential equations for fluid flow. It turns out that the Froude number also appears in that analysis, as do three other important dimensionless parameters—the Reynolds number, Euler number, and Strouhal number (Fig. 7–15).

7–3 · DIMENSIONAL ANALYSIS AND SIMILARITY

Nondimensionalization of an equation by inspection is useful only when we know the equation to begin with. However, in many cases in real-life engineering, the equations are either not known or too difficult to solve; oftentimes *experimentation* is the only method of obtaining reliable information. In most experiments, to save time and money, tests are performed on a geometrically scaled **model**, rather than on the full-scale **prototype.** In such cases, care must be taken to properly scale the results. We introduce here a powerful technique called **dimensional analysis.** While typically taught in fluid mechanics, dimensional analysis is useful in *all* disciplines, especially when it is necessary to design and conduct experiments. You are encouraged to use this powerful tool in other subjects as well, not just in fluid mechanics. The three primary purposes of dimensional analysis are

- To generate nondimensional parameters that help in the design of experiments (physical and/or numerical) and in the reporting of experimental results

- To obtain scaling laws so that prototype performance can be predicted from model performance

- To (sometimes) predict trends in the relationship between parameters

Before discussing the *technique* of dimensional analysis, we first explain the underlying *concept* of dimensional analysis—the principle of **similarity.** There are three necessary conditions for complete similarity between a model and a prototype. The first condition is **geometric similarity**—the model must be the same shape as the prototype, but may be scaled by some constant scale factor. The second condition is **kinematic similarity,** which means that the velocity at any point in the model flow must be proportional

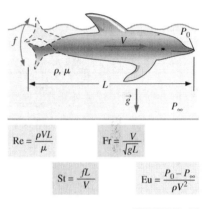

FIGURE 7–15

In a general unsteady fluid flow problem with a free surface, the scaling parameters include a characteristic length L, a characteristic velocity V, a characteristic frequency f, and a reference pressure difference $P_0 - P_\infty$. Nondimensionalization of the differential equations of fluid flow produces four dimensionless parameters: the Reynolds number, Froude number, Strouhal number, and Euler number (see Chap. 10).

(by a constant scale factor) to the velocity at the corresponding point in the prototype flow (Fig. 7–16). Specifically, for kinematic similarity the velocity at corresponding points must scale in magnitude and must point in the same relative direction. You may think of geometric similarity as *length-scale* equivalence and kinematic similarity as *time-scale* equivalence. *Geometric similarity is a prerequisite for kinematic similarity.* Just as the geometric scale factor can be less than, equal to, or greater than one, so can the velocity scale factor. In Fig. 7–16, for example, the geometric scale factor is less than one (model smaller than prototype), but the velocity scale is greater than one (velocities around the model are greater than those around the prototype). You may recall from Chap. 4 that streamlines are kinematic phenomena; hence, the streamline pattern in the model flow is a geometrically scaled copy of that in the prototype flow when kinematic similarity is achieved.

The third and most restrictive similarity condition is that of **dynamic similarity.** Dynamic similarity is achieved when all *forces* in the model flow scale by a constant factor to corresponding forces in the prototype flow (*force-scale* equivalence). As with geometric and kinematic similarity, the scale factor for forces can be less than, equal to, or greater than one. In Fig. 7–16 for example, the force-scale factor is less than one since the force on the model building is less than that on the prototype. *Kinematic similarity is a necessary but insufficient condition for dynamic similarity.* It is thus possible for a model flow and a prototype flow to achieve both geometric and kinematic similarity, yet not dynamic similarity. All three similarity conditions must exist for complete similarity to be ensured.

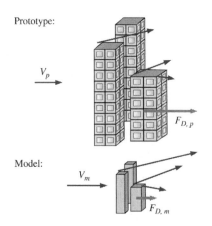

Prototype:

V_p

$F_{D,p}$

Model:

V_m

$F_{D,m}$

FIGURE 7–16

Kinematic similarity is achieved when, at all locations, the speed in the model flow is proportional to that at corresponding locations in the prototype flow, and points in the same direction.

> In a general flow field, complete similarity between a model and prototype is achieved only when there is geometric, kinematic, and dynamic similarity.

We let uppercase Greek letter Pi (Π) denote a nondimensional parameter. In Sec. 7–2, we have already discussed one Π, namely the Froude number, Fr. In a general dimensional analysis problem, there is one Π that we call the **dependent** Π, giving it the notation Π_1. The parameter Π_1 is in general a function of several other Π's, which we call **independent** Π's. The functional relationship is

Functional relationship between Π's: $\quad \Pi_1 = f(\Pi_2, \Pi_3, \ldots, \Pi_k)$ $\qquad$ (7–11)

where k is the total number of Π's.

Consider an experiment in which a scale model is tested to simulate a prototype flow. To ensure complete similarity between the model and the prototype, each independent Π of the model (subscript m) must be identical to the corresponding independent Π of the prototype (subscript p), i.e., $\Pi_{2,m} = \Pi_{2,p}, \Pi_{3,m} = \Pi_{3,p}, \ldots, \Pi_{k,m} = \Pi_{k,p}$.

> To ensure complete similarity, the model and prototype must be geometrically similar, and all independent Π groups must match between model and prototype.

Under these conditions the *dependent* Π of the model ($\Pi_{1,m}$) is guaranteed to also equal the dependent Π of the prototype ($\Pi_{1,p}$). Mathematically, we write a conditional statement for achieving similarity,

If $\qquad \Pi_{2,m} = \Pi_{2,p}$ and $\Pi_{3,m} = \Pi_{3,p} \ldots$ $\qquad$ and $\qquad \Pi_{k,m} = \Pi_{k,p}$,

then $\quad \Pi_{1,m} = \Pi_{1,p}$ $\qquad$ (7–12)

Consider, for example, the design of a new sports car, the aerodynamics of which is to be tested in a wind tunnel. To save money, it is desirable to test a small, geometrically scaled model of the car rather than a full-scale prototype of the car (Fig. 7–17). In the case of aerodynamic drag on an automobile, it turns out that if the flow is approximated as incompressible, there are only two Π's in the problem,

$$\Pi_1 = f(\Pi_2) \quad \text{where} \quad \Pi_1 = \frac{F_D}{\rho V^2 L^2} \quad \text{and} \quad \Pi_2 = \frac{\rho VL}{\mu} \quad (7\text{--}13)$$

The procedure used to generate these Π's is discussed in Section 7–4. In Eq. 7–13, F_D is the magnitude of the aerodynamic drag on the car, ρ is the air density, V is the car's speed (or the speed of the air in the wind tunnel), L is the length of the car, and μ is the viscosity of the air. Π_1 is a nonstandard form of the drag coefficient, and Π_2 is the **Reynolds number,** Re. You will find that many problems in fluid mechanics involve a Reynolds number (Fig. 7–18).

> The Reynolds number is the most well known and useful dimensionless parameter in all of fluid mechanics.

In the problem at hand there is only one independent Π, and Eq. 7–12 ensures that if the independent Π's match (the Reynolds numbers match: $\Pi_{2,m} = \Pi_{2,p}$), then the dependent Π's also match ($\Pi_{1,m} = \Pi_{1,p}$). This enables engineers to measure the aerodynamic drag on the model car and then use this value to predict the aerodynamic drag on the prototype car.

Prototype car

Model car

FIGURE 7–17
Geometric similarity between a prototype car of length L_p and a model car of length L_m.

$$\mathrm{Re} = \frac{\rho VL}{\mu} = \frac{VL}{\nu}$$

FIGURE 7–18
The *Reynolds number* Re is formed by the ratio of density, characteristic speed, and characteristic length to viscosity. Alternatively, it is the ratio of characteristic speed and length to *kinematic viscosity*, defined as $\nu = \mu/\rho$.

■ **EXAMPLE 7–5   Similarity between Model and Prototype Cars**

The aerodynamic drag of a new sports car is to be predicted at a speed of 50.0 mi/h at an air temperature of 25°C. Automotive engineers build a one-fifth scale model of the car to test in a wind tunnel. It is winter and the wind tunnel is located in an unheated building; the temperature of the wind tunnel air is only about 5°C. Determine how fast the engineers should run the wind tunnel in order to achieve similarity between the model and the prototype.

SOLUTION We are to utilize the concept of similarity to determine the speed of the wind tunnel.
Assumptions **1** Compressibility of the air is negligible (the validity of this approximation is discussed later). **2** The wind tunnel walls are far enough away so as to not interfere with the aerodynamic drag on the model car. **3** The model is geometrically similar to the prototype. **4** The wind tunnel has a moving belt to simulate the ground under the car, as in Fig. 7–19. (The moving belt is necessary in order to achieve kinematic similarity everywhere in the flow, in particular underneath the car.)
Properties For air at atmospheric pressure and at $T = 25°C$, $\rho = 1.184$ kg/m³ and $\mu = 1.849 \times 10^{-5}$ kg/m·s. Similarly, at $T = 5°C$, $\rho = 1.269$ kg/m³ and $\mu = 1.754 \times 10^{-5}$ kg/m·s.
Analysis Since there is only one independent Π in this problem, the similarity equation (Eq. 7–12) holds if $\Pi_{2,m} = \Pi_{2,p}$, where Π_2 is given by Eq. 7–13, and we call it the Reynolds number. Thus, we write

$$\Pi_{2,m} = \mathrm{Re}_m = \frac{\rho_m V_m L_m}{\mu_m} = \Pi_{2,p} = \mathrm{Re}_p = \frac{\rho_p V_p L_p}{\mu_p}$$

Wind tunnel test section

FIGURE 7–19

A *drag balance* is a device used in a wind tunnel to measure the aerodynamic drag of a body. When testing automobile models, a *moving belt* is often added to the floor of the wind tunnel to simulate the moving ground (from the car's frame of reference).

which we solve for the unknown wind tunnel speed for the model tests, V_m,

$$V_m = V_p \left(\frac{\mu_m}{\mu_p} \right) \left(\frac{\rho_p}{\rho_m} \right) \left(\frac{L_p}{L_m} \right)$$

$$= (50.0 \text{ mi/h}) \left(\frac{1.754 \times 10^{-5} \text{ kg/m·s}}{1.849 \times 10^{-5} \text{ kg/m·s}} \right) \left(\frac{1.184 \text{ kg/m}^3}{1.269 \text{ kg/m}^3} \right) (5) = 221 \text{ mi/h}$$

Thus, to ensure similarity, the wind tunnel should be run at 221 mi/h (to three significant digits). Note that we were never given the actual length of either car, but the ratio of L_p to L_m is known because the prototype is five times larger than the scale model. When the dimensional parameters are rearranged as nondimensional ratios (as done here), the unit system is irrelevant. Since the units in each numerator cancel those in each denominator, no unit conversions are necessary.

Discussion This speed is quite high (about 100 m/s), and the wind tunnel may not be able to run at that speed. Furthermore, the incompressible approximation may come into question at this high speed (we discuss this in more detail in Example 7–8).

Once we are convinced that complete similarity has been achieved between the model tests and the prototype flow, Eq. 7–12 can be used again to predict the performance of the prototype based on measurements of the performance of the model. This is illustrated in Example 7–6.

EXAMPLE 7–6   **Prediction of Aerodynamic Drag Force on a Prototype Car**

This example is a follow-up to Example 7–5. Suppose the engineers run the wind tunnel at 221 mi/h to achieve similarity between the model and the prototype. The aerodynamic drag force on the model car is measured with a **drag balance** (Fig. 7–19). Several drag readings are recorded, and the average drag force on the model is 21.2 lbf. Predict the aerodynamic drag force on the prototype (at 50 mi/h and 25°C).

SOLUTION   Because of similarity, the model results are to be scaled up to predict the aerodynamic drag force on the prototype.

Analysis   The similarity equation (Eq. 7–12) shows that since $\Pi_{2, m} = \Pi_{2, p}$, $\Pi_{1, m} = \Pi_{1, p}$, where Π_1 is given for this problem by Eq. 7–13. Thus, we write

$$\Pi_{1, m} = \frac{F_{D, m}}{\rho_m V_m^2 L_m^2} = \Pi_{1, p} = \frac{F_{D, p}}{\rho_p V_p^2 L_p^2}$$

which we solve for the unknown aerodynamic drag force on the prototype car, $F_{D, p}$,

$$F_{D, p} = F_{D, m} \left(\frac{\rho_p}{\rho_m} \right) \left(\frac{V_p}{V_m} \right)^2 \left(\frac{L_p}{L_m} \right)^2$$

$$= (21.2 \text{ lbf}) \left(\frac{1.184 \text{ kg/m}^3}{1.269 \text{ kg/m}^3} \right) \left(\frac{50.0 \text{ mi/h}}{221 \text{ mi/h}} \right)^2 (5)^2 = 25.3 \text{ lbf}$$

Discussion By arranging the dimensional parameters as nondimensional ratios, the units cancel nicely even though they are a mixture of SI and English units. Because both velocity and length are squared in the equation for Π_1, the higher speed in the wind tunnel nearly compensates for the model's smaller size, and the drag force on the model is nearly the same as that on the prototype. In fact, if the density and viscosity of the air in the wind tunnel were *identical* to those of the air flowing over the prototype, the two drag forces would be identical as well (Fig. 7–20).

The power of using dimensional analysis and similarity to supplement experimental analysis is further illustrated by the fact that the actual values of the dimensional parameters (density, velocity, etc.) are irrelevant. As long as the corresponding independent Π's are set equal to each other, similarity is achieved—*even if different fluids are used.* This explains why automobile or aircraft performance can be simulated in a water tunnel, and the performance of a submarine can be simulated in a wind tunnel (Fig. 7–21). Suppose, for example, that the engineers in Examples 7–5 and 7–6 use a water tunnel instead of a wind tunnel to test their one-fifth scale model. Using the properties of water at room temperature (20°C is assumed), the water tunnel speed required to achieve similarity is easily calculated as

$$V_m = V_p \left(\frac{\mu_m}{\mu_p}\right)\left(\frac{\rho_p}{\rho_m}\right)\left(\frac{L_p}{L_m}\right)$$
$$= (50.0 \text{ mi/h})\left(\frac{1.002 \times 10^{-3}\,\text{kg/m·s}}{1.849 \times 10^{-5}\,\text{kg/m·s}}\right)\left(\frac{1.184\,\text{kg/m}^3}{998.0\,\text{kg/m}^3}\right)(5) = 16.1 \text{ mi/h}$$

As can be seen, one advantage of a water tunnel is that the required water tunnel speed is much lower than that required for a wind tunnel using the same size model.

7–4 · THE METHOD OF REPEATING VARIABLES AND THE BUCKINGHAM PI THEOREM

We have seen several examples of the usefulness and power of dimensional analysis. Now we are ready to learn how to *generate* the nondimensional parameters, i.e., the Π's. There are several methods that have been developed for this purpose, but the most popular (and simplest) method is the **method of repeating variables,** popularized by Edgar Buckingham (1867–1940). The method was first published by the Russian scientist Dimitri Riabouchinsky (1882–1962) in 1911. We can think of this method as a step-by-step procedure or "recipe" for obtaining nondimensional parameters. There are six steps, listed concisely in Fig. 7–22, and in more detail in Table 7–2. These steps are explained in further detail as we work through a number of example problems.

As with most new procedures, the best way to learn is by example and practice. As a simple first example, consider a ball falling in a vacuum as discussed in Section 7–2. Let us pretend that we do not know that Eq. 7–4 is appropriate for this problem, nor do we know much physics concerning falling objects. In fact, suppose that all we know is that the instantaneous

Prototype

Model

$V_m = V_p \dfrac{L_p}{L_m}$

FIGURE 7–20
For the special case in which the wind tunnel air and the air flowing over the prototype have the same properties ($\rho_m = \rho_p$, $\mu_m = \mu_p$), and under similarity conditions ($V_m = V_p L_p/L_m$), the aerodynamic drag force on the prototype is equal to that on the scale model. If the two fluids do *not* have the same properties, the two drag forces are *not* necessarily the same, even under dynamically similar conditions.

FIGURE 7–21
Similarity can be achieved even when the model fluid is different than the prototype fluid. Here a submarine model is tested in a wind tunnel.
Courtesy NASA Langley Research Center.

304
DIMENSIONAL ANALYSIS AND MODELING

The Method of Repeating Variables

Step 1: List the parameters in the problem and count their total number n.

Step 2: List the primary dimensions of each of the n parameters.

Step 3: Set the *reduction j* as the number of primary dimensions. Calculate k, the expected number of Π's,
$$k = n - j$$

Step 4: Choose j *repeating parameters*.

Step 5: Construct the k Π's, and manipulate as necessary.

Step 6: Write the final functional relationship and check your algebra.

FIGURE 7–22
A concise summary of the six steps that comprise the *method of repeating variables*.

TABLE 7–2

Detailed description of the six steps that comprise the *method of repeating variables**

Step 1 List the parameters (dimensional variables, nondimensional variables, and dimensional constants) and count them. Let n be the total number of parameters in the problem, including the dependent variable. Make sure that any listed independent parameter is indeed independent of the others, i.e., it cannot be expressed in terms of them. (E.g., don't include radius r and area $A = \pi r^2$, since r and A are *not* independent.)

Step 2 List the primary dimensions for each of the n parameters.

Step 3 Guess the **reduction** j. As a first guess, set j equal to the number of primary dimensions represented in the problem. The expected number of Π's (k) is equal to n minus j, according to the **Buckingham Pi theorem,**

> *The Buckingham Pi theorem:* $k = n - j$ (7–14)

If at this step or during any subsequent step, the analysis does not work out, verify that you have included enough parameters in step 1. Otherwise, go back and *reduce j by one* and try again.

Step 4 Choose j **repeating parameters** that will be used to construct each Π. Since the repeating parameters have the potential to appear in each Π, be sure to choose them *wisely* (Table 7–3).

Step 5 Generate the Π's one at a time by grouping the j repeating parameters with one of the remaining parameters, forcing the product to be dimensionless. In this way, construct all k Π's. By convention the first Π, designated as Π_1, is the *dependent* Π (the one on the left side of the list). Manipulate the Π's as necessary to achieve established dimensionless groups (Table 7–5).

Step 6 Check that all the Π's are indeed dimensionless. Write the final functional relationship in the form of Eq. 7–11.

* This is a step-by-step method for finding the dimensionless Π groups when performing a dimensional analysis.

FIGURE 7–23
Setup for dimensional analysis of a ball falling in a vacuum. Elevation z is a function of time t, initial vertical speed w_0, initial elevation z_0, and gravitational constant g.

---

elevation z of the ball must be a function of time t, initial vertical speed w_0, initial elevation z_0, and gravitational constant g (Fig. 7–23). The beauty of dimensional analysis is that the only other thing we need to know is the primary dimensions of each of these quantities. As we go through each step of the method of repeating variables, we explain some of the subtleties of the technique in more detail using the falling ball as an example.

Step 1
There are five parameters (dimensional variables, nondimensional variables, and dimensional constants) in this problem; $n = 5$. They are listed in functional form, with the dependent variable listed as a function of the independent variables and constants:

List of relevant parameters: $z = f(t, w_0, z_0, g)$ $n = 5$

Step 2

The primary dimensions of each parameter are listed here. We recommend writing each dimension with exponents since this helps with later algebra.

z	t	w_0	z_0	g
$\{L^1\}$	$\{t^1\}$	$\{L^1 t^{-1}\}$	$\{L^1\}$	$\{L^1 t^{-2}\}$

Step 3

As a first guess, j is set equal to 2, the number of primary dimensions represented in the problem (L and t).

Reduction: $j = 2$

If this value of j is correct, the number of Π's predicted by the Buckingham Pi theorem is

Number of expected Π's: $k = n - j = 5 - 2 = 3$

Step 4

We need to choose two repeating parameters since $j = 2$. Since this is often the hardest (or at least the most mysterious) part of the method of repeating variables, several guidelines about choosing repeating parameters are listed in Table 7–3.

Following the guidelines of Table 7–3 on the next page, the wisest choice of two repeating parameters is w_0 and z_0.

Repeating parameters: w_0 and z_0

Step 5

Now we combine these repeating parameters into products with each of the remaining parameters, one at a time, to create the Π's. The first Π is always the *dependent* Π and is formed with the dependent variable z.

Dependent Π: $\Pi_1 = z w_0^{a_1} z_0^{b_1}$ (7–15)

where a_1 and b_1 are constant exponents that need to be determined. We apply the primary dimensions of step 2 into Eq. 7–15 and *force* the Π to be dimensionless by setting the exponent of each primary dimension to zero:

Dimensions of Π_1: $\{\Pi_1\} = \{L^0 t^0\} = \{z w_0^{a_1} z_0^{b_1}\} = \{L^1 (L^1 t^{-1})^{a_1} L^{b_1}\}$

Since primary dimensions are by definition independent of each other, we equate the exponents of each primary dimension independently to solve for exponents a_1 and b_1 (Fig. 7–24).

Time: $\{t^0\} = \{t^{-a_1}\}$ $0 = -a_1$ $a_1 = 0$

Length: $\{L^0\} = \{L^1 L^{a_1} L^{b_1}\}$ $0 = 1 + a_1 + b_1$ $b_1 = -1 - a_1$ $b_1 = -1$

Equation 7–15 thus becomes

$$\Pi_1 = \frac{z}{z_0}$$ (7–16)

FIGURE 7–24

The mathematical rules for adding and subtracting exponents during multiplication and division, respectively.

TABLE 7–3

Guidelines for choosing *repeating parameters* in step 4 of the method of repeating variables[*]

Guideline	Comments and Application to Present Problem
1. Never pick the *dependent* variable. Otherwise, it may appear in all the Π's, which is undesirable.	In the present problem we cannot choose z, but we must choose from among the remaining four parameters. Therefore, we must choose two of the following parameters: t, w_0, z_0, and g.
2. The chosen repeating parameters must not *by themselves* be able to form a dimensionless group. Otherwise, it would be impossible to generate the rest of the Π's.	In the present problem, any two of the independent parameters would be valid according to this guideline. For illustrative purposes, however, suppose we have to pick three instead of two repeating parameters. We could not, for example, choose t, w_0, and z_0, because these can form a Π all by themselves (tw_0/z_0).
3. The chosen repeating parameters must represent *all* the primary dimensions in the problem.	Suppose for example that there were *three* primary dimensions (m, L, and t) and *two* repeating parameters were to be chosen. You could not choose, say, a length and a time, since primary dimension mass would not be represented in the dimensions of the repeating parameters. An appropriate choice would be a density and a time, which together represent all three primary dimensions in the problem.
4. Never pick parameters that are already dimensionless. These are Π's already, all by themselves.	Suppose an angle θ were one of the independent parameters. We could not choose θ as a repeating parameter since angles have no dimensions (radian and degree are dimensionless units). In such a case, one of the Π's is already known, namely θ.
5. Never pick two parameters with the *same* dimensions or with dimensions that differ by only an exponent.	In the present problem, two of the parameters, z and z_0, have the same dimensions (length). We cannot choose both of these parameters. (Note that dependent variable z has already been eliminated by guideline 1.) Suppose one parameter has dimensions of length and another parameter has dimensions of volume. In dimensional analysis, volume contains only one primary dimension (length) and *is not dimensionally distinct from length*—we cannot choose both of these parameters.
6. Whenever possible, choose dimensional constants over dimensional variables so that only *one* Π contains the dimensional variable.	If we choose time t as a repeating parameter in the present problem, it would appear in all three Π's. While this would not be *wrong*, it would not be *wise* since we know that ultimately we want some nondimensional height as a function of some nondimensional time and other nondimensional parameter(s). From the original four independent parameters, this restricts us to w_0, z_0, and g.
7. Pick common parameters since they may appear in each of the Π's.	In fluid flow problems we generally pick a length, a velocity, and a mass or density (Fig. 7–25). It is unwise to pick less common parameters like viscosity μ or surface tension σ_s, since we would in general not want μ or σ_s to appear in each of the Π's. In the present problem, w_0 and z_0 are wiser choices than g.
8. Pick simple parameters over complex parameters whenever possible.	It is better to pick parameters with only one or two basic dimensions (e.g., a length, a time, a mass, or a velocity) instead of parameters that are composed of several basic dimensions (e.g., an energy or a pressure).

[*] These guidelines, while not infallible, help you to pick repeating parameters that usually lead to established nondimensional Π groups with minimal effort.

In similar fashion we create the first independent Π (Π_2) by combining the repeating parameters with independent variable t.

First independent Π: $$\Pi_2 = tw_0^{a_2}z_0^{b_2}$$

Dimensions of Π_2: $\{\Pi_2\} = \{L^0 t^0\} = \{tw_0^{a_2}z_0^{b_2}\} = \{t(L^1 t^{-1})^{a_2}L^{b_2}\}$

Equating exponents,

Time: $\{t^0\} = \{t^1 t^{-a_2}\}$ $0 = 1 - a_2$ $a_2 = 1$

Length: $\{L^0\} = \{L^{a_2} L^{b_2}\}$ $0 = a_2 + b_2$ $b_2 = -a_2$ $b_2 = -1$

Π_2 is thus

$$\Pi_2 = \frac{w_0 t}{z_0} \tag{7-17}$$

Finally we create the second independent Π (Π_3) by combining the repeating parameters with g and *forcing* the Π to be dimensionless (Fig. 7–26).

Second independent Π: $\Pi_3 = g w_0^{a_3} z_0^{b_3}$

Dimensions of Π_3: $\{\Pi_3\} = \{L^0 t^0\} = \{g w_0^{a_3} z_0^{b_3}\} = \{L^1 t^{-2}(L^1 t^{-1})^{a_3} L^{b_3}\}$

Equating exponents,

Time: $\{t^0\} = \{t^{-2} t^{-a_3}\}$ $0 = -2 - a_3$ $a_3 = -2$

Length: $\{L^0\} = \{L^1 L^{a_3} L^{b_3}\}$ $0 = 1 + a_3 + b_3$ $b_3 = -1 - a_3$ $b_3 = 1$

Π_3 is thus

$$\Pi_3 = \frac{g z_0}{w_0^2} \tag{7-18}$$

All three Π's have been found, but at this point it is prudent to examine them to see if any manipulation is required. We see immediately that Π_1 and Π_2 are the same as the nondimensionalized variables z^* and t^* defined by Eq. 7–6—no manipulation is necessary for these. However, we recognize that the third Π must be raised to the power of $-\frac{1}{2}$ to be of the same form as an established dimensionless parameter, namely the Froude number of Eq. 7–8:

Modified Π_3: $\Pi_{3,\text{ modified}} = \left(\dfrac{g z_0}{w_0^2}\right)^{-1/2} = \dfrac{w_0}{\sqrt{g z_0}} = \text{Fr} \tag{7-19}$

Such manipulation is often necessary to put the Π's into proper established form. The Π of Eq. 7–18 is not *wrong*, and there is certainly no mathematical advantage of Eq. 7–19 over Eq. 7–18. Instead, we like to say that Eq. 7–19 is more "socially acceptable" than Eq. 7–18, since it is a named, established nondimensional parameter that is commonly used in the literature. In Table 7–4 are listed some guidelines for manipulation of nondimensional Π groups into established nondimensional parameters.

Table 7–5 lists some established nondimensional parameters, most of which are named after a notable scientist or engineer (see Fig. 7–27 and the Historical Spotlight on p. 311). This list is by no means exhaustive. Whenever possible, you should manipulate your Π's as necessary in order to convert them into established nondimensional parameters.

FIGURE 7–25

It is wise to choose *common* parameters as repeating parameters since they may appear in each of your dimensionless Π groups.

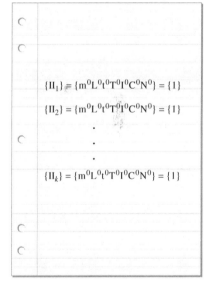

$$\{\Pi_1\} = \{m^0 L^0 t^0 T^0 I^0 C^0 N^0\} = \{1\}$$
$$\{\Pi_2\} = \{m^0 L^0 t^0 T^0 I^0 C^0 N^0\} = \{1\}$$
$$\vdots$$
$$\{\Pi_k\} = \{m^0 L^0 t^0 T^0 I^0 C^0 N^0\} = \{1\}$$

FIGURE 7–26

The Π groups that result from the method of repeating variables are guaranteed to be dimensionless because we *force* the overall exponent of all seven primary dimensions to be zero.

TABLE 7–4

Guidelines for manipulation of the Π's resulting from the method of repeating variables[*]

Guideline	Comments and Application to Present Problem
1. We may impose a constant (dimensionless) exponent on a Π or perform a functional operation on a Π.	We can raise a Π to any exponent n (changing it to Π^n) without changing the dimensionless stature of the Π. For example, in the present problem, we imposed an exponent of $-1/2$ on Π_3. Similarly we can perform the functional operation $\sin(\Pi)$, $\exp(\Pi)$, etc., without influencing the dimensions of the Π.
2. We may multiply a Π by a pure (dimensionless) constant.	Sometimes dimensionless factors of π, 1/2, 2, 4, etc., are included in a Π for convenience. This is perfectly okay since such factors do not influence the dimensions of the Π.
3. We may form a product (or quotient) of any Π with any other Π in the problem to replace one of the Π's.	We could replace Π_3 by $\Pi_3\Pi_1$, Π_3/Π_2, etc. Sometimes such manipulation is necessary to convert our Π into an established Π. In many cases, the established Π would have been produced if we would have chosen different repeating parameters.
4. We may use any of guidelines 1 to 3 in combination.	In general, we can replace any Π with some new Π such as $A\Pi_3{}^B \sin(\Pi_1{}^C)$, where A, B, and C are pure constants.
5. We may substitute a dimensional parameter in the Π with other parameter(s) of the same dimensions.	For example, the Π may contain the square of a length or the cube of a length, for which we may substitute a known area or volume, respectively, in order to make the Π agree with established conventions.

[*] These guidelines are useful in step 5 of the method of repeating variables and are listed to help you convert your nondimensional Π groups into standard, *established* nondimensional parameters, many of which are listed in Table 7–5.

FIGURE 7–27
Established nondimensional parameters are usually named after a notable scientist or engineer.

Step 6

We should double-check that the Π's are indeed dimensionless (Fig. 7–28). You can verify this on your own for the present example. We are finally ready to write the functional relationship between the nondimensional parameters. Combining Eqs. 7–16, 7–17, and 7–19 into the form of Eq. 7–11,

Relationship between Π's: $\Pi_1 = f(\Pi_2, \Pi_3)$ $\rightarrow$ $\dfrac{z}{z_0} = f\left(\dfrac{w_0 t}{z_0}, \dfrac{w_0}{\sqrt{gz_0}} \right)$

Or, in terms of the nondimensional variables z^* and t^* defined previously by Eq. 7–6 and the definition of the Froude number,

Final result of dimensional analysis: $z^* = f(t^*, \text{Fr})$ **(7–20)**

It is useful to compare the result of dimensional analysis, Eq. 7–20, to the exact analytical result, Eq. 7–10. The method of repeating variables properly predicts the functional relationship between dimensionless groups. However,

> The method of repeating variables cannot predict the exact mathematical form of the equation.

This is a fundamental limitation of dimensional analysis and the method of repeating variables. For some simple problems, however, the form of the equation *can* be predicted to within an unknown constant, as is illustrated in Example 7–7.

TABLE 7–5

Some common established nondimensional parameters or Π's encountered in fluid mechanics and heat transfer[*]

Name	Definition	Ratio of Significance
Archimedes number	$Ar = \dfrac{\rho_s g L^3}{\mu^2}(\rho_s - \rho)$	$\dfrac{\text{Gravitational force}}{\text{Viscous force}}$
Aspect ratio	$AR = \dfrac{L}{W} \ \text{or} \ \dfrac{L}{D}$	$\dfrac{\text{Length}}{\text{Width}} \ \text{or} \ \dfrac{\text{Length}}{\text{Diameter}}$
Biot number	$Bi = \dfrac{hL}{k}$	$\dfrac{\text{Surface thermal resistance}}{\text{Internal thermal resistance}}$
Bond number	$Bo = \dfrac{g(\rho_f - \rho_v)L^2}{\sigma_s}$	$\dfrac{\text{Gravitational force}}{\text{Surface tension force}}$
Cavitation number	$Ca \ (\text{sometimes } \sigma_c) = \dfrac{P - P_v}{\rho V^2}$ $\left(\text{sometimes } \dfrac{2(P - P_v)}{\rho V^2}\right)$	$\dfrac{\text{Pressure} - \text{Vapor pressure}}{\text{Inertial pressure}}$
Darcy friction factor	$f = \dfrac{8\tau_w}{\rho V^2}$	$\dfrac{\text{Wall friction force}}{\text{Inertial force}}$
Drag coefficient	$C_D = \dfrac{F_D}{\frac{1}{2}\rho V^2 A}$	$\dfrac{\text{Drag force}}{\text{Dynamic force}}$
Eckert number	$Ec = \dfrac{V^2}{c_p T}$	$\dfrac{\text{Kinetic energy}}{\text{Enthalpy}}$
Euler number	$Eu = \dfrac{\Delta P}{\rho V^2} \left(\text{sometimes } \dfrac{\Delta P}{\frac{1}{2}\rho V^2}\right)$	$\dfrac{\text{Pressure difference}}{\text{Dynamic pressure}}$
Fanning friction factor	$C_f = \dfrac{2\tau_w}{\rho V^2}$	$\dfrac{\text{Wall friction force}}{\text{Inertial force}}$
Fourier number	$Fo \ (\text{sometimes } \tau) = \dfrac{\alpha t}{L^2}$	$\dfrac{\text{Physical time}}{\text{Thermal diffusion time}}$
Froude number	$Fr = \dfrac{V}{\sqrt{gL}} \left(\text{sometimes } \dfrac{V^2}{gL}\right)$	$\dfrac{\text{Inertial force}}{\text{Gravitational force}}$
Grashof number	$Gr = \dfrac{g\beta \lvert \Delta \rvert T L^3 \rho^2}{\mu^2}$	$\dfrac{\text{Buoyancy force}}{\text{Viscous force}}$
Jakob number	$Ja = \dfrac{c_p(T - T_{\text{sat}})}{h_{fg}}$	$\dfrac{\text{Sensible energy}}{\text{Latent energy}}$
Knudsen number	$Kn = \dfrac{\lambda}{L}$	$\dfrac{\text{Mean free path length}}{\text{Characteristic length}}$
Lewis number	$Le = \dfrac{k}{\rho c_p D_{AB}} = \dfrac{\alpha}{D_{AB}}$	$\dfrac{\text{Thermal diffusion}}{\text{Species diffusion}}$
Lift coefficient	$C_L = \dfrac{F_L}{\frac{1}{2}\rho V^2 A}$	$\dfrac{\text{Lift force}}{\text{Dynamic force}}$

(Continued)

ARE YOUR PI'S DIMENSIONLESS?

FIGURE 7–28
A quick check of your algebra
is always wise.

TABLE 7–5 (Continued)

Name	Definition	Ratio of Significance
Mach number	Ma (sometimes M) $= \dfrac{V}{c}$	$\dfrac{\text{Flow speed}}{\text{Speed of sound}}$
Nusselt number	Nu $= \dfrac{Lh}{k}$	$\dfrac{\text{Convection heat transfer}}{\text{Conduction heat transfer}}$
Peclet number	Pe $= \dfrac{\rho L V c_p}{k} = \dfrac{LV}{\alpha}$	$\dfrac{\text{Bulk heat transfer}}{\text{Conduction heat transfer}}$
Power number	$N_P = \dfrac{\dot{W}}{\rho D^5 \omega^3}$	$\dfrac{\text{Power}}{\text{Rotational inertia}}$
Prandtl number	Pr $= \dfrac{\nu}{\alpha} = \dfrac{\mu c_p}{k}$	$\dfrac{\text{Viscous diffusion}}{\text{Thermal diffusion}}$
Pressure coefficient	$C_p = \dfrac{P - P_\infty}{\frac{1}{2}\rho V^2}$	$\dfrac{\text{Static pressure difference}}{\text{Dynamic pressure}}$
Rayleigh number	Ra $= \dfrac{g\beta\lvert\Delta T\rvert L^3 \rho^2 c_p}{k\mu}$	$\dfrac{\text{Buoyancy force}}{\text{Viscous force}}$
Reynolds number	Re $= \dfrac{\rho V L}{\mu} = \dfrac{VL}{\nu}$	$\dfrac{\text{Inertial force}}{\text{Viscous force}}$
Richardson number	Ri $= \dfrac{L^5 g \Delta\rho}{\rho \dot{V}^2}$	$\dfrac{\text{Buoyancy force}}{\text{Inertial force}}$
Schmidt number	Sc $= \dfrac{\mu}{\rho D_{AB}} = \dfrac{\nu}{D_{AB}}$	$\dfrac{\text{Viscous diffusion}}{\text{Species diffusion}}$
Sherwood number	Sh $= \dfrac{VL}{D_{AB}}$	$\dfrac{\text{Overall mass diffusion}}{\text{Species diffusion}}$
Specific heat ratio	k (sometimes γ) $= \dfrac{c_p}{c_V}$	$\dfrac{\text{Enthalpy}}{\text{Internal energy}}$
Stanton number	St $= \dfrac{h}{\rho c_p V}$	$\dfrac{\text{Heat transfer}}{\text{Thermal capacity}}$
Stokes number	Stk (sometimes St) $= \dfrac{\rho_p D_p^2 V}{18\mu L}$	$\dfrac{\text{Particle relaxation time}}{\text{Characteristic flow time}}$
Strouhal number	St (sometimes S or Sr) $= \dfrac{fL}{V}$	$\dfrac{\text{Characteristic flow time}}{\text{Period of oscillation}}$
Weber number	We $= \dfrac{\rho V^2 L}{\sigma_s}$	$\dfrac{\text{Inertial force}}{\text{Surface tension force}}$

* A is a characteristic area, D is a characteristic diameter, f is a characteristic frequency (Hz), L is a characteristic length, t is a characteristic time, T is a characteristic (absolute) temperature, V is a characteristic velocity, W is a characteristic width, $\dot{W}$ is a characteristic power, ω is a characteristic angular velocity (rad/s). Other parameters and fluid properties in these Π's include: c = speed of sound, c_p, c_v = specific heats, D_p = particle diameter, D_{AB} = species diffusion coefficient, h = convective heat transfer coefficient, h_{fg} = latent heat of evaporation, k = thermal conductivity, P = pressure, T_{sat} = saturation temperature, $\dot{V}$ = volume flow rate, α = thermal diffusivity, β = coefficient of thermal expansion, λ = mean free path length, μ = viscosity, ν = kinematic viscosity, ρ = fluid density, ρ_f = liquid density, ρ_p = particle density, ρ_s = solid density, ρ_v = vapor density, σ_s = surface tension, and τ_w = shear stress along a wall.

HISTORICAL SPOTLIGHT ■ Persons Honored by Nondimensional Parameters

Guest Author: Glenn Brown, Oklahoma State University

Commonly used, established dimensionless numbers have been given names for convenience, and to honor persons who have contributed in the development of science and engineering. In many cases, the namesake was not the first to define the number, but usually he/she used it or a similar parameter in his/her work. The following is a list of some, but not all, such persons. Also keep in mind that some numbers may have more than one name.

Archimedes (287–212 BC) Greek mathematician who defined buoyant forces.

Biot, Jean-Baptiste (1774–1862) French mathematician who did pioneering work in heat, electricity, and elasticity. He also helped measure the arc of the meridian as part of the metric system development.

Darcy, Henry P. G. (1803–1858) French engineer who performed extensive experiments on pipe flow and the first quantifiable filtration tests.

Eckert, Ernst R. G. (1904–2004) German–American engineer and student of Schmidt who did early work in boundary layer heat transfer.

Euler, Leonhard (1707–1783) Swiss mathematician and associate of Daniel Bernoulli who formulated equations of fluid motion and introduced the concept of centrifugal machinery.

Fanning, John T. (1837–1911) American engineer and textbook author who published in 1877 a modified form of Weisbach's equation with a table of resistance values computed from Darcy's data.

Fourier, Jean B. J. (1768–1830) French mathematician who did pioneering work in heat transfer and several other topics.

Froude, William (1810–1879) English engineer who developed naval modeling methods and the transfer of wave and boundary resistance from model to prototype.

Grashof, Franz (1826–1893) German engineer and educator known as a prolific author, editor, corrector, and dispatcher of publications.

Jakob, Max (1879–1955) German–American physicist, engineer, and textbook author who did pioneering work in heat transfer.

Knudsen, Martin (1871–1949) Danish physicist who helped develop the kinetic theory of gases.

Lewis, Warren K. (1882–1975) American engineer who researched distillation, extraction, and fluidized bed reactions.

Mach, Ernst (1838–1916) Austrian physicist who was first to realize that bodies traveling faster than the speed of sound would drastically alter the properties of the fluid. His ideas had great influence on twentieth-century thought, both in physics and in philosophy, and influenced Einstein's development of the theory of relativity.

Nusselt, Wilhelm (1882–1957) German engineer who was the first to apply similarity theory to heat transfer.

Peclet, Jean C. E. (1793–1857) French educator, physicist, and industrial researcher.

Prandtl, Ludwig (1875–1953) German engineer and developer of boundary layer theory who is considered the founder of modern fluid mechanics.

Lord Raleigh, John W. Strutt (1842–1919) English scientist who investigated dynamic similarity, cavitation, and bubble collapse.

Reynolds, Osborne (1842–1912) English engineer who investigated flow in pipes and developed viscous flow equations based on mean velocities.

Richardson, Lewis F. (1881–1953) English mathematician, physicist, and psychologist who was a pioneer in the application of fluid mechanics to the modeling of atmospheric turbulence.

Schmidt, Ernst (1892–1975) German scientist and pioneer in the field of heat and mass transfer. He was the first to measure the velocity and temperature field in a free convection boundary layer.

Sherwood, Thomas K. (1903–1976) American engineer and educator. He researched mass transfer and its interaction with flow, chemical reactions, and industrial process operations.

Stanton, Thomas E. (1865–1931) English engineer and student of Reynolds who contributed to a number of areas of fluid flow.

Stokes, George G. (1819–1903) Irish scientist who developed equations of viscous motion and diffusion.

Strouhal, Vincenz (1850–1922) Czech physicist who showed that the period of oscillations shed by a wire are related to the velocity of the air passing over it.

Weber, Moritz (1871–1951) German professor who applied similarity analysis to capillary flows.

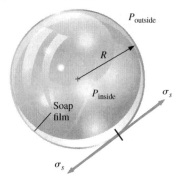

FIGURE 7–29

The pressure inside a soap bubble is greater than that surrounding the soap bubble due to surface tension in the soap film.

What happens if
$$k = n - j = 0?$$

Do the following:
• Check your list of parameters.
• Check your algebra.
• If all else fails, reduce *j* by one.

FIGURE 7–30

If the method of repeating variables indicates zero Π's, we have either made an error, or we need to reduce *j* by one and start over.

EXAMPLE 7–7   **Pressure in a Soap Bubble**

Some children are playing with soap bubbles, and you become curious as to the relationship between soap bubble radius and the pressure inside the soap bubble (Fig. 7–29). You reason that the pressure inside the soap bubble must be greater than atmospheric pressure, and that the shell of the soap bubble is under tension, much like the skin of a balloon. You also know that the property surface tension must be important in this problem. Not knowing any other physics, you decide to approach the problem using dimensional analysis. Establish a relationship between pressure difference $\Delta P = P_{inside} - P_{outside}$, soap bubble radius R, and the surface tension σ_s of the soap film.

SOLUTION   The pressure difference between the inside of a soap bubble and the outside air is to be analyzed by the method of repeating variables.
Assumptions   **1** The soap bubble is neutrally buoyant in the air, and gravity is not relevant. **2** No other variables or constants are important in this problem.
Analysis   The step-by-step method of repeating variables is employed.

Step 1   There are three variables and constants in this problem; $n = 3$. They are listed in functional form, with the dependent variable listed as a function of the independent variables and constants:

List of relevant parameters: $\Delta P = f(R, \sigma_s)$ $n = 3$

Step 2   The primary dimensions of each parameter are listed. The dimensions of surface tension are obtained from Example 7–1, and those of pressure from Example 7–2.

$$\begin{array}{ccc} \Delta P & R & \sigma_s \\ \{m^1 L^{-1} t^{-2}\} & \{L^1\} & \{m^1 t^{-2}\} \end{array}$$

Step 3   As a first guess, j is set equal to 3, the number of primary dimensions represented in the problem (m, L, and t).

Reduction (first guess): $j = 3$

If this value of j is correct, the expected number of Π's is $k = n - j = 3 - 3 = 0$. But how can we have zero Π's? Something is obviously not right (Fig. 7–30). At times like this, we need to first go back and make sure that we are not neglecting some important variable or constant in the problem. Since we are confident that the pressure difference should depend only on soap bubble radius and surface tension, we reduce the value of j by one,

Reduction (second guess): $j = 2$

If this value of j is correct, $k = n - j = 3 - 2 = 1$. Thus we expect *one* Π, which is more physically realistic than zero Π's.

Step 4   We need to choose two repeating parameters since $j = 2$. Following the guidelines of Table 7–3, our only choices are R and σ_s, since ΔP is the dependent variable.

Step 5   We combine these repeating parameters into a product with the dependent variable ΔP to create the dependent Π,

Dependent Π: $\Pi_1 = \Delta P R^{a_1} \sigma_s^{b_1}$ (1)

We apply the primary dimensions of step 2 into Eq. 1 and force the Π to be dimensionless.

Dimensions of Π_1:

$$\{\Pi_1\} = \{m^0 L^0 t^0\} = \{\Delta P R^{a_1} \sigma_s^{b_1}\} = \{(m^1 L^{-1} t^{-2}) L^{a_1} (m^1 t^{-2})^{b_1}\}$$

We equate the exponents of each primary dimension to solve for a_1 and b_1:

Time: $\qquad \{t^0\} = \{t^{-2} t^{-2b_1}\} \qquad 0 = -2 - 2b_1 \qquad b_1 = -1$

Mass: $\qquad \{m^0\} = \{m^1 m^{b_1}\} \qquad 0 = 1 + b_1 \qquad b_1 = -1$

Length: $\qquad \{L^0\} = \{L^{-1} L^{a_1}\} \qquad 0 = -1 + a_1 \qquad a_1 = 1$

Fortunately, the first two results agree with each other, and Eq. 1 thus becomes

$$\Pi_1 = \frac{\Delta P R}{\sigma_s} \qquad (2)$$

From Table 7–5, the established nondimensional parameter most similar to Eq. 2 is the **Weber number,** defined as a pressure (ρV^2) times a length divided by surface tension. There is no need to further manipulate this Π.

Step 6 We write the final functional relationship. In the case at hand, there is only one Π, which is a function of *nothing*. This is possible only if the Π is constant. Putting Eq. 2 into the functional form of Eq. 7–11,

Relationship between Π's:

$$\Pi_1 = \frac{\Delta P R}{\sigma_s} = f(\text{nothing}) = \text{constant} \quad \to \quad \Delta P = \text{constant} \frac{\sigma_s}{R} \qquad (3)$$

Discussion This is an example of how we can sometimes predict *trends* with dimensional analysis, even without knowing much of the physics of the problem. For example, we know from our result that if the radius of the soap bubble doubles, the pressure difference decreases by a factor of 2. Similarly, if the value of surface tension doubles, ΔP increases by a factor of 2. Dimensional analysis cannot predict the value of the constant in Eq. 3; further analysis (or *one* experiment) reveals that the constant is equal to 4 (Chap. 2).

EXAMPLE 7–8 Lift on a Wing

Some aeronautical engineers are designing an airplane and wish to predict the lift produced by their new wing design (Fig. 7–31). The chord length L_c of the wing is 1.12 m, and its **planform area** A (area viewed from the top when the wing is at zero angle of attack) is 10.7 m². The prototype is to fly at $V = 52.0$ m/s close to the ground where $T = 25°C$. They build a one-tenth scale model of the wing to test in a pressurized wind tunnel. The wind tunnel can be pressurized to a maximum of 5 atm. At what speed and pressure should they run the wind tunnel in order to achieve dynamic similarity?

SOLUTION We are to determine the speed and pressure at which to run the wind tunnel in order to achieve dynamic similarity.

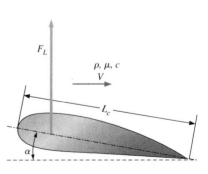

FIGURE 7–31

Lift F_L on a wing of chord length L_c at angle of attack α in a flow of free-stream speed V with density ρ, viscosity μ, and speed of sound c. The angle of attack α is measured relative to the free-stream flow direction.

Assumptions   **1** The prototype wing flies through the air at standard atmospheric pressure. **2** The model is geometrically similar to the prototype.

Analysis   First, the step-by-step method of repeating variables is employed to obtain the nondimensional parameters. Then, the dependent Π's are matched between prototype and model.

Step 1   There are seven parameters (variables and constants) in this problem; $n = 7$. They are listed in functional form, with the dependent variable listed as a function of the independent parameters:

List of relevant parameters: $F_L = f(V, L_c, \rho, \mu, c, \alpha)$ $n = 7$

where F_L is the lift force on the wing, V is the fluid speed, L_c is the chord length, ρ is the fluid density, μ is the fluid viscosity, c is the speed of sound in the fluid, and α is the angle of attack of the wing.

Step 2   The primary dimensions of each parameter are listed; angle α is dimensionless:

$$
\begin{array}{ccccccc}
F_L & V & L_c & \rho & \mu & c & \alpha \\
\{m^1 L^1 t^{-2}\} & \{L^1 t^{-1}\} & \{L^1\} & \{m^1 L^{-3}\} & \{m^1 L^{-1} t^{-1}\} & \{L^1 t^{-1}\} & \{1\}
\end{array}
$$

Step 3   As a first guess, j is set equal to 3, the number of primary dimensions represented in the problem (m, L, and t).

Reduction: $j = 3$

If this value of j is correct, the expected number of Π's is $k = n - j = 7 - 3 = 4$.

Step 4   We need to choose three repeating parameters since $j = 3$. Following the guidelines listed in Table 7–3, we cannot pick the dependent variable F_L. Nor can we pick α since it is already dimensionless. We cannot choose both V and c since their dimensions are identical. It would not be desirable to have μ appear in all the Π's. The best choice of repeating parameters is thus either V, L_c, and ρ or c, L_c, and ρ. Of these, the former is the better choice since the speed of sound appears in only one of the established nondimensional parameters of Table 7–5, whereas the velocity scale is more "common" and appears in several of the parameters (Fig. 7–32).

Repeating parameters: V, L_c, and ρ

Step 5   The dependent Π is generated:

$$\Pi_1 = F_L V^{a_1} L_c^{b_1} \rho^{c_1} \quad \rightarrow \quad \{\Pi_1\} = \{(m^1 L^1 t^{-2})(L^1 t^{-1})^{a_1}(L^1)^{b_1}(m^1 L^{-3})^{c_1}\}$$

The exponents are calculated by forcing the Π to be dimensionless (algebra not shown). We get $a_1 = -2$, $b_1 = -2$, and $c_1 = -1$. The dependent Π is thus

$$\Pi_1 = \frac{F_L}{\rho V^2 L_c^2}$$

From Table 7–5, the established nondimensional parameter most similar to our Π_1 is the **lift coefficient**, defined in terms of planform area A rather than the square of chord length, and with a factor of 1/2 in the denominator. Thus, we may manipulate this Π according to the guidelines listed in Table 7–4 as follows:

Modified Π_1: $\Pi_{1,\text{modified}} = \dfrac{F_L}{\frac{1}{2}\rho V^2 A} = \text{Lift coefficient} = C_L$

FIGURE 7–32

Oftentimes when performing the method of repeating variables, the most difficult part of the procedure is choosing the repeating parameters. With practice, however, you will learn to choose these parameters wisely.

Similarly, the first independent Π is generated:

$$\Pi_2 = \mu V^{a_2} L_c^{b_2} \rho^{c_2} \quad \rightarrow \quad \{\Pi_2\} = \{(m^1 L^{-1} t^{-1})(L^1 t^{-1})^{a_2}(L^1)^{b_2}(m^1 L^{-3})^{c_2}\}$$

from which $a_2 = -1$, $b_2 = -1$, and $c_2 = -1$, and thus

$$\Pi_2 = \frac{\mu}{\rho V L_c}$$

We recognize this Π as the inverse of the Reynolds number. So, after inverting,

Modified Π_2: $\qquad \Pi_{2,\,\text{modified}} = \dfrac{\rho V L_c}{\mu} = \text{Reynolds number} = \text{Re}$

The third Π is formed with the speed of sound, the details of which are left for you to generate on your own. The result is

$$\Pi_3 = \frac{V}{c} = \text{Mach number} = \text{Ma}$$

Finally, since the angle of attack α is already dimensionless, it is a dimensionless Π group all by itself (Fig. 7–33). You are invited to go through the algebra; you will find that all the exponents turn out to be zero, and thus

$$\Pi_4 = \alpha = \text{Angle of attack}$$

Step 6 We write the final functional relationship as

$$C_L = \frac{F_L}{\frac{1}{2}\rho V^2 A} = f(\text{Re}, \text{Ma}, \alpha) \tag{1}$$

To achieve dynamic similarity, Eq. 7–12 requires that all three of the dependent nondimensional parameters in Eq. 1 match between the model and the prototype. While it is trivial to match the angle of attack, it is not so simple to simultaneously match the Reynolds number and the Mach number. For example, if the wind tunnel were run at the same temperature and pressure as those of the prototype, such that ρ, μ, and c of the air flowing over the model were the same as ρ, μ, and c of the air flowing over the prototype, Reynolds number similarity would be achieved by setting the wind tunnel air speed to 10 times that of the prototype (since the model is one-tenth scale). But then the Mach numbers would differ by a factor of 10. At 25°C, c is approximately 346 m/s, and the Mach number of the prototype airplane wing is $\text{Ma}_p = 52.0/346 = 0.150$—subsonic. At the required wind tunnel speed, Ma_m would be 1.50—supersonic! This is clearly unacceptable since the physics of the flow changes dramatically from subsonic to supersonic conditions. At the other extreme, if we were to match Mach numbers, the Reynolds number of the model would be 10 times too small.

What should we do? A common rule of thumb is that for Mach numbers less than about 0.3, as is the fortunate case here, compressibility effects are practically negligible. Thus, it is not necessary to exactly match the Mach number; rather, as long as Ma_m is kept below about 0.3, approximate dynamic similarity can be achieved by matching the Reynolds number. Now the problem shifts to one of how to match Re while maintaining a low Mach number. This is where the pressurization feature of the wind tunnel comes in. At constant temperature, density is proportional to pressure, while viscosity and speed of sound are very weak functions of pressure. If the wind tunnel pressure could be pumped to 10 atm, we could run the model test at the

FIGURE 7–33
A parameter that is dimensionless (like an angle) is already a nondimensional Π all by itself— we know this Π without doing any further algebra.

A parameter that is already dimensionless becomes a Π parameter all by itself.

same speed as the prototype and achieve a nearly perfect match in both Re and Ma. However, at the maximum wind tunnel pressure of 5 atm, the required wind tunnel speed would be twice that of the prototype, or 104 m/s. The Mach number of the wind tunnel model would thus be $Ma_m = 104/346 = 0.301$—approximately at the limit of incompressibility according to our rule of thumb. In summary, the wind tunnel should be run at approximately **100 m/s, 5 atm, and 25°C**.

Discussion This example illustrates one of the (frustrating) limitations of dimensional analysis; namely, *You may not always be able to match all the dependent Π's simultaneously in a model test.* Compromises must be made in which only the most important Π's are matched. In many practical situations in fluid mechanics, the Reynolds number is not critical for dynamic similarity, provided that Re is high enough. If the Mach number of the prototype were significantly larger than about 0.3, we would be wise to precisely match the Mach number rather than the Reynolds number in order to ensure reasonable results. Furthermore, if a different gas were used to test the model, we would also need to match the specific heat ratio (k), since compressible flow behavior is strongly dependent on k (Chap. 12). We discuss such model testing problems in more detail in Section 7–5.

Recall that in Examples 7–5 and 7–6 the air speed of the prototype car is 50.0 mi/h, and that of the wind tunnel is 221 mi/h. At 25°C, this corresponds to a prototype Mach number of $Ma_p = 0.065$, and at 5°C, the Mach number of the wind tunnel is 0.29—on the borderline of the incompressible limit. In hindsight, we should have included the speed of sound in our dimensional analysis, which would have generated the Mach number as an additional Π. Another way to match the Reynolds number while keeping the Mach number low is to use a *liquid* such as water, since liquids are nearly incompressible, even at fairly high speeds.

FIGURE 7–34

Friction on the inside wall of a pipe. The shear stress τ_w on the pipe walls is a function of average fluid speed V, average wall roughness height ε, fluid density ρ, fluid viscosity μ, and inside pipe diameter D.

EXAMPLE 7–9 **Friction in a Pipe**

Consider flow of an incompressible fluid of density ρ and viscosity μ through a long, horizontal section of round pipe of diameter D. The velocity profile is sketched in Fig. 7–34; V is the average speed across the pipe cross section, which by conservation of mass remains constant down the pipe. For a very long pipe, the flow eventually becomes hydrodynamically **fully developed**, which means that the velocity profile also remains uniform down the pipe. Because of frictional forces between the fluid and the pipe wall, there exists a shear stress τ_w on the inside pipe wall as sketched. The shear stress is also constant down the pipe in the fully developed region. We assume some constant average roughness height ε along the inside wall of the pipe. In fact, the only parameter that is *not* constant down the length of pipe is the pressure, which must decrease (linearly) down the pipe in order to "push" the fluid through the pipe to overcome friction. Develop a nondimensional relationship between shear stress τ_w and the other parameters in the problem.

SOLUTION We are to generate a nondimensional relationship between shear stress and other parameters.

Assumptions   **1** The flow is hydrodynamically fully developed. **2** The fluid is incompressible. **3** No other parameters are significant in the problem.
Analysis   The step-by-step method of repeating variables is employed to obtain the nondimensional parameters.

Step 1   There are six variables and constants in this problem; $n = 6$. They are listed in functional form, with the dependent variable listed as a function of the independent variables and constants:

List of relevant parameters: $\tau_w = f(V, \varepsilon, \rho, \mu, D)$ $n = 6$

Step 2   The primary dimensions of each parameter are listed. Note that shear stress is a force per unit area, and thus has the same dimensions as pressure.

$$\begin{array}{cccccc} \tau_w & V & \varepsilon & \rho & \mu & D \\ \{m^1 L^{-1} t^{-2}\} & \{L^1 t^{-1}\} & \{L^1\} & \{m^1 L^{-3}\} & \{m^1 L^{-1} t^{-1}\} & \{L^1\} \end{array}$$

Step 3   As a first guess, j is set equal to 3, the number of primary dimensions represented in the problem (m, L, and t).

Reduction: $j = 3$

If this value of j is correct, the expected number of Π's is $k = n - j = 6 - 3 = 3$.

Step 4   We choose three repeating parameters since $j = 3$. Following the guidelines of Table 7–3, we cannot pick the dependent variable τ_w. We cannot choose both ε and D since their dimensions are identical, and it would not be desirable to have μ or ε appear in all the Π's. The best choice of repeating parameters is thus V, D, and ρ.

Repeating parameters: V, D, and ρ

Step 5   The dependent Π is generated:

$$\Pi_1 = \tau_w V^{a_1} D^{b_1} \rho^{c_1} \quad \rightarrow \quad \{\Pi_1\} = \{(m^1 L^{-1} t^{-2})(L^1 t^{-1})^{a_1}(L^1)^{b_1}(m^1 L^{-3})^{c_1}\}$$

from which $a_1 = -2$, $b_1 = 0$, and $c_1 = -1$, and thus the dependent Π is

$$\Pi_1 = \frac{\tau_w}{\rho V^2}$$

From Table 7–5, the established nondimensional parameter most similar to this Π_1 is the **Darcy friction factor**, defined with a factor of 8 in the numerator (Fig. 7–35). Thus, we manipulate this Π according to the guidelines listed in Table 7–4 as follows:

Modified Π_1: $\Pi_{1, \text{modified}} = \dfrac{8\tau_w}{\rho V^2}$ = Darcy friction factor = f

Similarly, the two independent Π's are generated, the details of which are left for you to do on your own:

$$\Pi_2 = \mu V^{a_2} D^{b_2} \rho^{c_2} \quad \rightarrow \quad \Pi_2 = \frac{\rho V D}{\mu} = \text{Reynolds number} = \text{Re}$$

$$\Pi_3 = \varepsilon V^{a_3} D^{b_3} \rho^{c_3} \quad \rightarrow \quad \Pi_3 = \frac{\varepsilon}{D} = \text{Roughness ratio}$$

Darcy friction factor: $f = \dfrac{8\tau_w}{\rho V^2}$

Fanning friction factor: $C_f = \dfrac{2\tau_w}{\rho V^2}$

FIGURE 7–35
Although the *Darcy friction factor* for pipe flows is most common, you should be aware of an alternative, less common friction factor called the *Fanning friction factor*. The relationship between the two is $f = 4C_f$.

Step 6 We write the final functional relationship as

$$f = \frac{8\tau_w}{\rho V^2} = f\left(Re, \frac{\varepsilon}{D}\right) \tag{1}$$

Discussion The result applies to both laminar and turbulent fully developed pipe flow; it turns out, however, that the second independent Π (roughness ratio ε/D) is not nearly as important in laminar pipe flow as in turbulent pipe flow. This problem presents an interesting connection between geometric similarity and dimensional analysis. Namely, it is necessary to match ε/D since it is an independent Π in the problem. From a different perspective, thinking of roughness as a geometric property, it is necessary to match ε/D to ensure *geometric similarity* between two pipes.

To verify the validity of Eq. 1 of Example 7–9, we use **computational fluid dynamics (CFD)** to predict the velocity profiles and the values of wall shear stress for two physically different but dynamically similar pipe flows:

- *Air* at 300 K flowing at an average speed of 14.5 ft/s through a pipe of inner diameter 1.00 ft and average roughness height 0.0010 ft.

- *Water* at 300 K flowing at an average speed of 3.09 m/s through a pipe of inner diameter 0.0300 m and average roughness height 0.030 mm.

The two pipes are clearly geometrically similar since they are both round pipes. They have the same average roughness ratio ($\varepsilon/D = 0.0010$ in both cases). We have carefully chosen the values of average speed and diameter such that the two flows are also *dynamically* similar. Specifically, the other independent Π (the Reynolds number) also matches between the two flows.

$$Re_{air} = \frac{\rho_{air} V_{air} D_{air}}{\mu_{air}} = \frac{(1.225 \text{ kg/m}^3)(14.5 \text{ ft/s})(1.00 \text{ ft})}{1.789 \times 10^{-5} \text{ kg/m·s}}\left(\frac{0.3048 \text{ m}}{\text{ft}}\right)^2 = 9.22 \times 10^4$$

where the fluid properties are those built into the CFD code, and

$$Re_{water} = \frac{\rho_{water} V_{water} D_{water}}{\mu_{water}} = \frac{(998.2 \text{ kg/m}^3)(3.09 \text{ m/s})(0.0300 \text{ m})}{0.001003 \text{ kg/m·s}} = 9.22 \times 10^4$$

Hence by Eq. 7–12, we expect that the *dependent* Π's should match between the two flows as well. We generate a computational mesh for each of the two flows, and use a commercial CFD code to generate the velocity profile, from which the shear stress is calculated. Fully developed, time-averaged, turbulent velocity profiles near the far end of both pipes are compared. Although the pipes are of different diameters and the fluids are vastly different, the velocity profile shapes look quite similar. In fact, when we plot *normalized* axial velocity (u/V) as a function of *normalized* radius (r/R), we find that the two profiles fall on top of each other (Fig. 7–36).

Wall shear stress is also calculated from the CFD results for each flow, a comparison of which is shown in Table 7–6. There are several reasons why the wall shear stress in the water pipe is orders of magnitude larger than that in the air pipe. Namely, water is over 800 times as dense as air and over 50 times as viscous. Furthermore, shear stress is proportional to the *gradient* of velocity, and the water pipe diameter is less than one-tenth that of the air

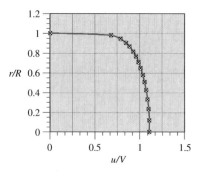

FIGURE 7–36
Normalized axial velocity profiles for fully developed flow through a pipe as predicted by CFD; profiles of air (circles) and water (crosses) are shown on the same plot.

TABLE 7–6

Comparison of wall shear stress and nondimensionalized wall shear stress for fully developed flow through an air pipe and a water pipe as predicted by CFD*

Parameter	Air Flow	Water Flow
Wall shear stress	$\tau_{w,\,air} = 0.0557 \text{ N/m}^2$	$\tau_{w,\,water} = 22.2 \text{ N/m}^2$
Dimensionless wall shear stress (Darcy friction factor)	$f_{air} = \dfrac{8\tau_{w,\,air}}{\rho_{air} V_{air}^2} = 0.0186$	$f_{water} = \dfrac{8\tau_{w,\,water}}{\rho_{water} V_{water}^2} = 0.0186$

* Data obtained with ANSYS-FLUENT using the standard k-ε turbulence model with wall functions.

pipe, leading to steeper velocity gradients. In terms of the *nondimensionalized* wall shear stress, f, however, Table 7–6 shows that the results are identical due to dynamic similarity between the two flows. Note that although the values are reported to three significant digits, the reliability of turbulence models in CFD is accurate to at most two significant digits (Chap. 15).

7–5 ▪ EXPERIMENTAL TESTING, MODELING, AND INCOMPLETE SIMILARITY

One of the most useful applications of dimensional analysis is in designing physical and/or numerical experiments, and in reporting the results of such experiments. In this section we discuss both of these applications, and point out situations in which complete dynamic similarity is not achievable.

Setup of an Experiment and Correlation of Experimental Data

As a generic example, consider a problem in which there are five original parameters (one of which is the *dependent* parameter). A complete set of experiments (called a **full factorial** test matrix) is conducted by testing every possible combination of several levels of each of the four independent parameters. A full factorial test with five levels of each of the four independent parameters would require $5^4 = 625$ experiments. While experimental design techniques (**fractional factorial** test matrices; see Montgomery, 2013) can significantly reduce the size of the test matrix, the number of required experiments would still be large. However, assuming that three primary dimensions are represented in the problem, we can reduce the number of parameters from five to two ($k = 5 - 3 = 2$ nondimensional Π groups), and the number of *independent* parameters from four to one. Thus, for the same resolution (five tested levels of each independent parameter) we would then need to conduct a total of only $5^1 = 5$ experiments. You don't have to be a genius to realize that replacing 625 experiments by 5 experiments is cost effective. You can see why it is wise to perform a dimensional analysis *before* conducting an experiment.

Continuing our discussion of this generic example (a two-Π problem), once the experiments are complete, we plot the dependent dimensionless parameter (Π_1) as a function of the independent dimensionless parameter (Π_2), as in Fig. 7–37. We then determine the functional form of the relationship by

(a)

(b)

FIGURE 7–37

For a two-Π problem, we plot dependent dimensionless parameter (Π_1) as a function of independent dimensionless parameter (Π_2). The resulting plot can be (a) linear or (b) nonlinear. In either case, regression and curve-fitting techniques are available to determine the relationship between the Π's.

performing a **regression analysis** on the data. If we are lucky, the data may correlate linearly. If not, we can try linear regression on log–linear or log–log coordinates, polynomial curve fitting, etc., to establish an approximate relationship between the two Π's. See Holman (2001) for details about these curve-fitting techniques.

If there are more than two Π's in the problem (e.g., a three-Π problem or a four-Π problem), we need to set up a test matrix to determine the relationship between the dependent Π and the independent Π's. In many cases we discover that one or more of the dependent Π's has negligible effect and can be removed from the list of necessary dimensionless parameters.

As we have seen (Example 7–7), dimensional analysis sometimes yields only *one* Π. In a one-Π problem, we know the form of the relationship between the original parameters to within some unknown constant. In such a case, only *one* experiment is needed to determine that constant.

Incomplete Similarity

We have shown several examples in which the nondimensional Π groups are easily obtained with paper and pencil through straightforward use of the method of repeating variables. In fact, after sufficient practice, you should be able to obtain the Π's with ease—sometimes in your head or on the "back of an envelope." Unfortunately, it is often a much different story when we go to apply the results of our dimensional analysis to experimental data. The problem is that it is not always possible to match *all* the Π's of a model to the corresponding Π's of the prototype, even if we are careful to achieve geometric similarity. This situation is called **incomplete similarity.** Fortunately, in some cases of incomplete similarity, we are still able to extrapolate model test data to obtain reasonable full-scale predictions.

Wind Tunnel Testing

We illustrate incomplete similarity with the problem of measuring the aerodynamic drag force on a model truck in a wind tunnel (Fig. 7–38). Suppose we purchase a one-sixteenth scale die-cast model of a tractor-trailer rig (18-wheeler). The model is geometrically similar to the prototype—even in the details such as side mirrors, mud flaps, etc. The model truck is 0.991 m long, corresponding to a full-scale prototype length of 15.9 m. The model truck is to be tested in a wind tunnel that has a maximum speed of 70 m/s. The wind tunnel test section is 1.0 m tall and 1.2 m wide—big enough to accommodate the model without needing to worry about wall interference or blockage effects. The air in the wind tunnel is at the same temperature and pressure as the air flowing around the prototype. We want to simulate flow at $V_p = 60$ mi/h (26.8 m/s) over the full-scale prototype truck.

The first thing we do is match the Reynolds numbers,

$$\text{Re}_m = \frac{\rho_m V_m L_m}{\mu_m} = \text{Re}_p = \frac{\rho_p V_p L_p}{\mu_p}$$

which can be solved for the required wind tunnel speed for the model tests V_m,

$$V_m = V_p \left(\frac{\mu_m}{\mu_p}\right)\left(\frac{\rho_p}{\rho_m}\right)\left(\frac{L_p}{L_m}\right) = (26.8 \text{ m/s})(1)(1)\left(\frac{16}{1}\right) = 429 \text{ m/s}$$

Wind tunnel test section

V

Model

F_D

Moving belt   Drag balance

FIGURE 7–38
Measurement of aerodynamic drag on a model truck in a wind tunnel equipped with a *drag balance* and a *moving belt* ground plane.

Thus, to match the Reynolds number between model and prototype, the wind tunnel should be run at 429 m/s (to three significant digits). We obviously have a problem here, since this speed is more than six times greater than the maximum achievable wind tunnel speed. Moreover, even if we *could* run the wind tunnel that fast, the flow would be *supersonic*, since the speed of sound in air at room temperature is about 346 m/s. While the Mach number of the prototype truck moving through the air is 26.8/335 = 0.080, that of the wind tunnel air moving over the model would be 429/335 = 1.28 (if the wind tunnel could go that fast).

It is clearly not possible to match the model Reynolds number to that of the prototype with this model and wind tunnel facility. What do we do? There are several options:

- If we had a bigger wind tunnel, we could test with a larger model. Automobile manufacturers typically test with three-eighths scale model cars and with one-eighth scale model trucks and buses in very large wind tunnels. Some wind tunnels are even large enough for full-scale automobile tests (Fig. 7–39a). As you can imagine, however, the bigger the wind tunnel and the model the more expensive the tests. We must also be careful that the model is not too big for the wind tunnel. A useful rule of thumb is that the **blockage** (ratio of the model frontal area to the cross-sectional area of the test section) should be less than 7.5 percent. Otherwise, the wind tunnel walls adversely affect both geometric and kinematic similarity.

- We could use a different fluid for the model tests. For example, water tunnels can achieve higher Reynolds numbers than can wind tunnels of the same size, but they are much more expensive to build and operate (Fig. 7–39b).

- We could pressurize the wind tunnel and/or adjust the air temperature to increase the maximum Reynolds number capability. While these techniques can help, the increase in the Reynolds number is limited.

- If all else fails, we could run the wind tunnel at several speeds near the maximum speed, and then extrapolate our results to the full-scale Reynolds number.

Fortunately, it turns out that for many wind tunnel tests the last option is quite viable. While drag coefficient C_D is a strong function of the Reynolds number at low values of Re, C_D often levels off for Re above some value. In other words, for flow over many objects, especially "bluff" objects like trucks, buildings, etc., the flow is **Reynolds number independent** above some threshold value of Re (Fig. 7–40), typically when the boundary layer and the wake are both fully turbulent.

■ *EXAMPLE 7–10*   **Model Truck Wind Tunnel Measurements**
■
■ A one-sixteenth scale model tractor-trailer truck (18-wheeler) is tested in a
■ wind tunnel as sketched in Fig. 7–38. The model truck is 0.991 m long,
■ 0.257 m tall, and 0.159 m wide. During the tests, the moving ground belt
speed is adjusted so as to always match the speed of the air moving through
the test section. Aerodynamic drag force F_D is measured as a function of

---

(a)

(b)

FIGURE 7–39

(a) The Langley full-scale wind tunnel (LFST) is large enough that full-scale vehicles can be tested. (b) For the same scale model and speed, water tunnels achieve higher Reynolds numbers than wind tunnels.

(b) NASA/Eric James

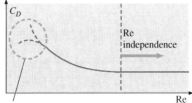

FIGURE 7–40

For many objects, the drag coefficient levels off at Reynolds numbers above some threshold value. This fortunate situation is called *Reynolds number independence*. It enables us to extrapolate to prototype Reynolds numbers that are outside of the range of our experimental facility.

TABLE 7–7

Wind tunnel data: aerodynamic drag force on a model truck as a function of wind tunnel speed

V, m/s	F_D, N
20	12.4
25	19.0
30	22.1
35	29.0
40	34.3
45	39.9
50	47.2
55	55.5
60	66.0
65	77.6
70	89.9

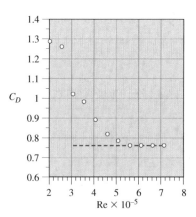

FIGURE 7–41

Aerodynamic drag coefficient as a function of the Reynolds number. The values are calculated from wind tunnel test data on a model truck (Table 7–7).

wind tunnel speed; the experimental results are listed in Table 7–7. Plot the drag coefficient C_D as a function of the Reynolds number Re, where the area used for the calculation of C_D is the frontal area of the model truck (the area you see when you look at the model from upstream), and the length scale used for calculation of Re is truck width W. Have we achieved dynamic similarity? Have we achieved Reynolds number independence in our wind tunnel test? Estimate the aerodynamic drag force on the prototype truck traveling on the highway at 26.8 m/s. Assume that both the wind tunnel air and the air flowing over the prototype car are at 25°C and standard atmospheric pressure.

SOLUTION We are to calculate and plot C_D as a function of Re for a given set of wind tunnel measurements and determine if dynamic similarity and/or Reynolds number independence have been achieved. Finally, we are to estimate the aerodynamic drag force acting on the prototype truck.

Assumptions 1 The model truck is geometrically similar to the prototype truck. 2 The aerodynamic drag on the strut(s) holding the model truck is negligible.

Properties For air at atmospheric pressure and at $T = 25°C$, $\rho = 1.184$ kg/m^3 and $\mu = 1.849 \times 10^{-5}$ kg/m·s.

Analysis We calculate C_D and Re for the last data point listed in Table 7–7 (at the fastest wind tunnel speed),

$$C_{D,m} = \frac{F_{D,m}}{\frac{1}{2}\rho_m V_m^2 A_m} = \frac{89.9\,\text{N}}{\frac{1}{2}(1.184\,\text{kg/m}^3)(70\,\text{m/s})^2(0.159\,\text{m})(0.257\,\text{m})}\left(\frac{1\,\text{kg·m/s}^2}{1\,\text{N}}\right)$$
$$= 0.758$$

and

$$\text{Re}_m = \frac{\rho_m V_m W_m}{\mu_m} = \frac{(1.184\,\text{kg/m}^3)(70\,\text{m/s})(0.159\,\text{m})}{1.849 \times 10^{-5}\,\text{kg/m·s}} = 7.13 \times 10^5 \quad (1)$$

We repeat these calculations for all the data points in Table 7–7, and we plot C_D versus Re in Fig. 7–41.

Have we achieved dynamic similarity? Well, we have *geometric* similarity between model and prototype, but the Reynolds number of the prototype truck is

$$\text{Re}_p = \frac{\rho_p V_p W_p}{\mu_p} = \frac{(1.184\,\text{kg/m}^3)(26.8\,\text{m/s})[16(0.159\,\text{m})]}{1.849 \times 10^{-5}\,\text{kg/m·s}} = 4.37 \times 10^6 \quad (2)$$

where the width of the prototype is specified as 16 times that of the model. Comparison of Eqs. 1 and 2 reveals that the prototype Reynolds number is more than six times larger than that of the model. Since we cannot match the independent Π's in the problem, **dynamic similarity has not been achieved.**

Have we achieved Reynolds number independence? From Fig. 7–41 we see that **Reynolds number independence has indeed been achieved**—at Re greater than about 5×10^5, C_D has leveled off to a value of about 0.76 (to two significant digits).

Since we have achieved Reynolds number independence, we can extrapolate to the full-scale prototype, assuming that C_D remains constant as Re is increased to that of the full-scale prototype.

Predicted aerodynamic drag on the prototype:

$$F_{D,p} = \tfrac{1}{2}\rho_p V_p^2 A_p C_{D,p}$$

$$= \tfrac{1}{2}(1.184 \text{ kg/m}^3)(26.8 \text{ m/s})^2[16^2(0.159 \text{ m})(0.257 \text{ m})](0.76)\left(\frac{1 \text{ N}}{1 \text{ kg·m/s}^2}\right)$$

$$= \textbf{3400 N}$$

Discussion We give our final result to two significant digits. More than that cannot be justified. As always, we must exercise caution when performing an extrapolation, since we have no guarantee that the extrapolated results are correct.

Flows with Free Surfaces

For the case of model testing of flows with free surfaces (boats and ships, floods, river flows, aqueducts, hydroelectric dam spillways, interaction of waves with piers, soil erosion, etc.), complications arise that preclude complete similarity between model and prototype. For example, if a model river is built to study flooding, the model is often several hundred times smaller than the prototype due to limited lab space. If the vertical dimensions of the model were scaled proportionately, the depth of the model river would be so small that surface tension effects (and the Weber number) would become important, and would perhaps even dominate the model flow, even though surface tension effects are negligible in the prototype flow. In addition, although the flow in the actual river may be turbulent, the flow in the model river may be laminar, especially if the slope of the riverbed is geometrically similar to that of the prototype. To avoid these problems, researchers often use a **distorted model** in which the vertical scale of the model (e.g., river depth) is exaggerated in comparison to the horizontal scale of the model (e.g., river width). In addition, the model riverbed slope is often made proportionally steeper than that of the prototype. These modifications result in incomplete similarity due to lack of geometric similarity. Model tests are still useful under these circumstances, but other tricks (like deliberately roughening the model surfaces) and empirical corrections and correlations are required to properly scale up the model data.

In many practical problems involving free surfaces, both the Reynolds number and Froude number appear as relevant independent Π groups in the dimensional analysis (Fig. 7–42). It is difficult (often impossible) to match both of these dimensionless parameters simultaneously. For a free-surface flow with length scale L, velocity scale V, and kinematic viscosity ν, the Reynolds number is matched between model and prototype when

$$\text{Re}_p = \frac{V_p L_p}{\nu_p} = \text{Re}_m = \frac{V_m L_m}{\nu_m} \tag{7-21}$$

The Froude number is matched between model and prototype when

$$\text{Fr}_p = \frac{V_p}{\sqrt{gL_p}} = \text{Fr}_m = \frac{V_m}{\sqrt{gL_m}} \tag{7-22}$$

$$\text{Re} = \frac{\rho V L}{\mu} = \frac{V L}{\nu} \qquad \text{Fr} = \frac{V}{\sqrt{gL}}$$

FIGURE 7–42

In many flows involving a liquid with a free surface, both the Reynolds number and Froude number are relevant nondimensional parameters. Since it is not always possible to match both Re and Fr between model and prototype, we are sometimes forced to settle for incomplete similarity.

To match both Re and Fr, we solve Eqs. 7–21 and 7–22 simultaneously for the required length scale factor L_m/L_p,

$$\frac{L_m}{L_p} = \frac{\nu_m}{\nu_p}\frac{V_p}{V_m} = \left(\frac{V_m}{V_p}\right)^2 \qquad (7\text{–}23)$$

Eliminating the ratio V_m/V_p from Eq. 7–23, we see that

Required ratio of kinematic viscosities to match both Re and Fr:

$$\frac{\nu_m}{\nu_p} = \left(\frac{L_m}{L_p}\right)^{3/2} \qquad (7\text{–}24)$$

Thus, to ensure complete similarity (assuming geometric similarity is achievable without unwanted surface tension effects as discussed previously), we would need to use a liquid whose kinematic viscosity satisfies Eq. 7–24. Although it is sometimes possible to find an appropriate liquid for use with the model, in most cases it is either impractical or impossible, as Example 7–11 illustrates. In such cases, it is more important to match Froude number than Reynolds number (Fig. 7–43).

(a)

(b)

(c)

FIGURE 7–43
A NACA 0024 airfoil being tested in a towing tank at Fr = (*a*) 0.19, (*b*) 0.37, and (*c*) 0.55. In tests like this, the Froude number is the most important nondimensional parameter.

Photograph courtesy of IIHR-Hydroscience & Engineering, University of Iowa. Used by permission.

EXAMPLE 7–11   Model Lock and River

In the late 1990s the U.S. Army Corps of Engineers designed an experiment to model the flow of the Tennessee River downstream of the Kentucky Lock and Dam (Fig. 7–44). Because of laboratory space restrictions, they built a scale model with a length scale factor of $L_m/L_p = 1/100$. Suggest a liquid that would be appropriate for the experiment.

SOLUTION We are to suggest a liquid to use in an experiment involving a one-hundredth scale model of a lock, dam, and river.
Assumptions **1** The model is geometrically similar to the prototype. **2** The model river is deep enough that surface tension effects are not significant.
Properties For water at atmospheric pressure and at $T = 20°C$, the prototype kinematic viscosity is $\nu_p = 1.002 \times 10^{-6}$ m²/s.
Analysis From Eq. 7–24,

Required kinematic viscosity of model liquid:

$$\nu_m = \nu_p\left(\frac{L_m}{L_p}\right)^{3/2} = (1.002 \times 10^{-6}\,\text{m}^2/\text{s})\left(\frac{1}{100}\right)^{3/2} = 1.00 \times 10^{-9}\,\text{m}^2/\text{s} \qquad (1)$$

Thus, we need to find a liquid that has a viscosity of 1.00×10^{-9} m²/s. A quick glance through the appendices yields no such liquid. Hot water has a lower kinematic viscosity than cold water, but only by a factor of about 3. Liquid mercury has a very small kinematic viscosity, but it is of order 10^{-7} m²/s—still two orders of magnitude too large to satisfy Eq. 1. Even if liquid mercury would work, it would be too expensive and too hazardous to use in such a test. What do we do? The bottom line is that *we cannot match both the Froude number and the Reynolds number in this model test.*

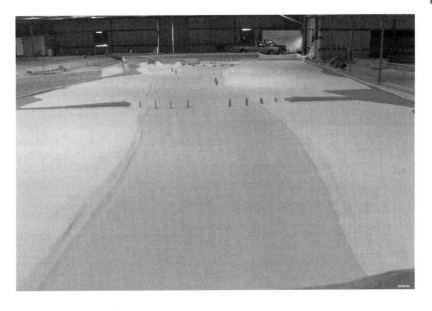

FIGURE 7–44
A 1:100 scale model constructed to investigate navigation conditions in the lower lock approach for a distance of 2 mi downstream of the dam. The model includes a scaled version of the spillway, powerhouse, and existing lock. In addition to navigation, the model was used to evaluate environmental issues associated with the new lock and required railroad and highway bridge relocations. The view here is looking upstream toward the lock and dam. At this scale, 52.8 ft on the model represents 1 mi on the prototype. A (real, full-scale) pickup truck in the background gives you a feel for the model scale.

Photo courtesy of the U.S. Army Corps of Engineers, Nashville.

In other words, it is impossible to achieve complete similarity between model and prototype in this case. Instead, we do the best job we can under conditions of incomplete similarity. Water is typically used in such tests for convenience.

Discussion It turns out that for this kind of experiment, Froude number matching is more critical than Reynolds number matching. As discussed previously for wind tunnel testing, Reynolds number independence is achieved at high enough values of Re. Even if we are unable to achieve Reynolds number independence, we can often extrapolate our low Reynolds number model data to predict full-scale Reynolds number behavior (Fig. 7–45). A high level of confidence in using this kind of extrapolation comes only after much laboratory experience with similar problems.

In closing this section on experiments and incomplete similarity, we mention the importance of similarity in the production of Hollywood movies in which model boats, trains, airplanes, buildings, monsters, etc., are blown up or burned. Movie producers must pay attention to dynamic similarity in order to make the small-scale fires and explosions appear as realistic as possible. You may recall some low-budget movies where the special effects are unconvincing. In most cases this is due to lack of dynamic similarity between the small model and the full-scale prototype. If the model's Froude number and/or Reynolds number differ too much from those of the prototype, the special effects don't look right, even to the untrained eye. The next time you watch a movie, be on the alert for incomplete similarity!

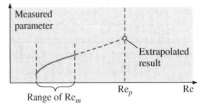

FIGURE 7–45
In many experiments involving free surfaces, we cannot match both the Froude number and the Reynolds number. However, we can often *extrapolate* low Re model test data to predict high Re prototype behavior.

APPLICATION SPOTLIGHT ■ How a Fly Flies

(a)

(b)

FIGURE 7–46

(a) The fruit fly, *Drosophila melanogaster*, flaps its tiny wings back and forth 200 times a second, creating a blurred image of the stroke plane. *(b)* The dynamically scaled model, *Robofly*, flaps its wings once every 5 s in 2 tons of mineral oil. Sensors at the base of the wings record aerodynamic forces, while fine bubbles are used to visualize the flow. The size and speed of the robot, as well as the properties of the oil, were carefully chosen to match the Reynolds number of a real fly.

Photos © Courtesy of Michael Dickinson, CALTECH.

Guest Author: Michael Dickinson,
California Institute of Technology

An interesting application of dimensional analysis is in the study of how insects fly. The small size and fast wing speed of an insect, such as a tiny fruit fly, make it difficult to directly measure the forces or visualize the air motion created by the fly's wings. However, using principles of dimensional analysis, it is possible to study insect aerodynamics on a larger-scale, slowly moving model—a mechanical robot. The forces created by a hovering fly and flapping robot are dynamically similar if the Reynolds number is the same for each case. For a flapping wing, Re is calculated as $2\Phi R L_c \omega / \nu$, where Φ is the angular amplitude of the wing stroke, R is the wing length, L_c is the average wing width (chord length), ω is the angular frequency of the stroke, and ν is the kinematic viscosity of the surrounding fluid. A fruit fly flaps its 2.5-mm-long, 0.7-mm-wide wings 200 times per second over a 2.8-rad stroke in air with a kinematic viscosity of 1.5×10^{-5} m²/s. The resulting Reynolds number is approximately 130. By choosing mineral oil with a kinematic viscosity of 1.15×10^{-4} m²/s, it is possible to match this Reynolds number on a robotic fly that is 100 times larger, flapping its wings over 1000 times more slowly! If the fly is not stationary, but rather moving through the air, it is necessary to match another dimensionless parameter to ensure dynamic similarity, the reduced frequency, $\sigma = 2\Phi R \omega / V$, which measures the ratio of the flapping velocity of the wing tip ($2\Phi R \omega$) to the forward velocity of the body (V). To simulate forward flight, a set of motors tows *Robofly* through its oil tank at an appropriately scaled speed.

Dynamically scaled robots have helped show that insects use a variety of different mechanisms to produce forces as they fly. During each back-and-forth stroke, insect wings travel at high angles of attack, generating a prominent leading-edge vortex. The low pressure of this large vortex pulls the wings upward. Insects can further augment the strength of the leading-edge vortex by rotating their wings at the end of each stroke. After the wing changes direction, it can also generate forces by quickly running through the wake of the previous stroke.

Figure 7–46a shows a real fly flapping its wings, and Fig. 7–46b shows *Robofly* flapping its wings. Because of the larger length scale and shorter time scale of the model, measurements and flow visualizations are possible. Experiments with dynamically scaled model insects continue to teach researchers how insects manipulate wing motion to steer and maneuver.

References
Dickinson, M. H., Lehmann, F.-O., and Sane, S., "Wing rotation and the aerodynamic basis of insect flight," *Science*, 284, p. 1954, 1999.
Dickinson, M. H., "Solving the mystery of insect flight," *Scientific American*, 284, No. 6, pp. 35–41, June 2001.
Fry, S. N., Sayaman, R., and Dickinson, M. H., "The aerodynamics of free-flight maneuvers in *Drosophila*," *Science*, 300, pp. 495–498, 2003.

SUMMARY

There is a difference between *dimensions* and *units*; a *dimension* is a measure of a physical quantity (without numerical values), while a *unit* is a way to assign a number to that dimension. There are seven *primary dimensions*—not just in fluid mechanics, but in all fields of science and engineering. They are mass, length, time, temperature, electric current, amount of light, and amount of matter. *All other dimensions can be formed by combination of these seven primary dimensions.*

All mathematical equations must be *dimensionally homogeneous*; this fundamental principle can be applied to equations in order to nondimensionalize them and to identify *dimensionless groups*, also called *nondimensional parameters*. A powerful tool to reduce the number of necessary independent parameters in a problem is called *dimensional analysis*. The *method of repeating variables* is a step-by-step procedure for finding the nondimensional parameters, or Π's, based simply on the dimensions of the variables and constants in the problem. The six steps in the method of repeating variables are summarized here.

Step 1   List the n parameters (variables and constants) in the problem.

Step 2   List the primary dimensions of each parameter.

Step 3   Guess the *reduction j*, usually equal to the number of primary dimensions in the problem. If the analysis does not work out, reduce j by one and try again. The expected number of Π's (k) is equal to n minus j.

Step 4   Wisely choose j *repeating parameters* for construction of the Π's.

Step 5   Generate the k Π's one at a time by grouping the j repeating parameters with each of the remaining variables or constants, forcing the product to be dimensionless, and manipulating the Π's as necessary to achieve established nondimensional parameters.

Step 6   Check your work and write the final functional relationship.

When all the dimensionless groups match between a model and a prototype, *dynamic similarity* is achieved, and we are able to directly predict prototype performance based on model experiments. However, it is not always possible to match *all* the Π groups when trying to achieve similarity between a model and a prototype. In such cases, we run the model tests under conditions of *incomplete similarity*, matching the most important Π groups as best we can, and then extrapolating the model test results to prototype conditions.

We use the concepts presented in this chapter throughout the remainder of the book. For example, dimensional analysis is applied to fully developed pipe flows in Chap. 8 (friction factors, loss coefficients, etc.). In Chap. 10, we normalize the differential equations of fluid flow derived in Chap. 9, producing several dimensionless parameters. Drag and lift coefficients are used extensively in Chap. 11, and dimensionless parameters also appear in the chapters on compressible flow and open-channel flow (Chaps. 12 and 13). We learn in Chap. 14 that dynamic similarity is often the basis for design and testing of pumps and turbines. Finally, dimensionless parameters are also used in computations of fluid flows (Chap. 15).

REFERENCES AND SUGGESTED READING

1. D. C. Montgomery. *Design and Analysis of Experiments*, 8th ed. New York: Wiley, 2013.

2. J. P. Holman. *Experimental Methods for Engineers*, 7th ed. New York: McGraw-Hill, 2001.

PROBLEMS*

Dimensions and Units, Primary Dimensions

7–1C   List the seven *primary dimensions*. What is significant about these seven?

---

* Problems designated by a "C" are concept questions, and students are encouraged to answer them all. Problems designated by an "E" are in English units, and the SI users can ignore them. Problems with the ⚙ icon are solved using EES, and complete solutions together with parametric studies are included on the text website. Problems with the 📷 icon are comprehensive in nature and are intended to be solved with an equation solver such as EES.

7–2C   What is the difference between a *dimension* and a *unit*? Give three examples of each.

7–3   Write the primary dimensions of the *universal ideal gas constant* R_u. (Hint: Use the *ideal gas law*, $PV = nR_uT$ where P is pressure, V is volume, T is absolute temperature, and n is the number of moles of the gas.)   *Answer:* $\{m^1L^2t^{-2}T^{-1}N^{-1}\}$

7–4   Write the primary dimensions of each of the following variables from the field of thermodynamics, showing all your work: (*a*) energy E; (*b*) specific energy $e = E/m$; (*c*) power $\dot{W}$.
Answers: (*a*) $\{m^1L^2t^{-2}\}$, (*b*) $\{L^2t^{-2}\}$, (*c*) $\{m^1L^2t^{-3}\}$

7–5 When performing a dimensional analysis, one of the first steps is to list the primary dimensions of each relevant parameter. It is handy to have a table of parameters and their primary dimensions. We have started such a table for you (Table P7–5), in which we have included some of the basic parameters commonly encountered in fluid mechanics. As you work through homework problems in this chapter, add to this table. You should be able to build up a table with dozens of parameters.

TABLE P7–5

Parameter Name	Parameter Symbol	Primary Dimensions
Acceleration	a	L^1t^{-2}
Angle	θ, ϕ, etc.	1 (none)
Density	ρ	m^1L^{-3}
Force	F	$m^1L^1t^{-2}$
Frequency	f	t^{-1}
Pressure	P	$m^1L^{-1}t^{-2}$
Surface tension	σ_s	m^1t^{-2}
Velocity	V	L^1t^{-1}
Viscosity	μ	$m^1L^{-1}t^{-1}$
Volume flow rate	$\dot{V}$	L^3t^{-1}

7–6 Consider the table of Prob. 7–5 where the primary dimensions of several variables are listed in the mass–length–time system. Some engineers prefer the force–length–time system (force replaces mass as one of the primary dimensions). Write the primary dimensions of three of these (density, surface tension, and viscosity) in the force–length–time system.

7–7 On a periodic chart of the elements, molar mass (M), also called *atomic weight*, is often listed as though it were a dimensionless quantity (Fig. P7–7). In reality, atomic weight is the mass of 1 mol of the element. For example, the atomic weight of nitrogen $M_{\text{nitrogen}} = 14.0067$. We interpret this as 14.0067 g/mol of elemental nitrogen, or in the English system, 14.0067 lbm/lbmol of elemental nitrogen. What are the primary dimensions of atomic weight?

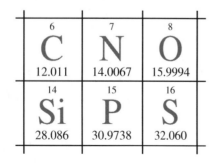

FIGURE P7–7

7–8 Some authors prefer to use *force* as a primary dimension in place of mass. In a typical fluid mechanics problem, then, the four represented primary dimensions m, L, t, and T are replaced by F, L, t, and T. The primary dimension of force in this system is {force} = {F}. Using the results of Prob. 7–3, rewrite the primary dimensions of the universal gas constant in this alternate system of primary dimensions.

7–9 We define the *specific ideal gas constant* R_{gas} for a particular gas as the ratio of the universal gas constant and the molar mass (also called *molecular weight*) of the gas, $R_{\text{gas}} = R_u/M$. For a particular gas, then, the ideal gas law is written as follows:

$$P\mathcal{V} = mR_{\text{gas}}T \quad \text{or} \quad P = \rho R_{\text{gas}}T$$

where P is pressure, $\mathcal{V}$ is volume, m is mass, T is absolute temperature, and ρ is the density of the particular gas. What are the primary dimensions of R_{gas}? For air, $R_{\text{air}} = 287.0$ J/kg·K in standard SI units. Verify that these units agree with your result.

7–10 The *moment of force* ($\vec{M}$) is formed by the cross product of a moment arm ($\vec{r}$) and an applied force ($\vec{F}$), as sketched in Fig. P7–10. What are the primary dimensions of moment of force? List its units in primary SI units and in primary English units.

FIGURE P7–10

7–11 What are the primary dimensions of electric voltage (E)? (*Hint*: Make use of the fact that electric power is equal to voltage times current.)

7–12 You are probably familiar with *Ohm's law* for electric circuits (Fig. P7–12), where ΔE is the voltage difference or *potential* across the resistor, I is the electric current passing through the resistor, and R is the electrical resistance. What are the primary dimensions of electrical resistance? *Answer:* $\{m^1L^2t^{-3}I^{-2}\}$

FIGURE P7–12

7–13 Write the primary dimensions of each of the following variables, showing all your work: (*a*) acceleration *a*; (*b*) angular velocity *ω*; (*c*) angular acceleration *α*.

7–14 *Angular momentum*, also called *moment of momentum* ($\vec{H}$), is formed by the cross product of a moment arm ($\vec{r}$) and the linear momentum ($m\vec{V}$) of a fluid particle, as sketched in Fig. P7–14. What are the primary dimensions of angular momentum? List the units of angular momentum in primary SI units and in primary English units. *Answers:* $\{m^1 L^2 t^{-1}\}$, kg·m²/s, lbm·ft²/s

FIGURE P7–14

7–15 Write the primary dimensions of each of the following variables, showing all your work: (*a*) specific heat at constant pressure c_p; (*b*) specific weight ρg; (*c*) specific enthalpy *h*.

7–16 **Thermal conductivity** *k* is a measure of the ability of a material to conduct heat (Fig. P7–16). For conduction heat transfer in the *x*-direction through a surface normal to the *x*-direction, **Fourier's law of heat conduction** is expressed as

$$\dot{Q}_{\text{conduction}} = -kA\frac{dT}{dx}$$

where $\dot{Q}_{\text{conduction}}$ is the rate of heat transfer and *A* is the area normal to the direction of heat transfer. Determine the primary dimensions of thermal conductivity (*k*). Look up a value of *k* in the appendices and verify that its SI units are consistent with your result. In particular, write the primary SI units of *k*.

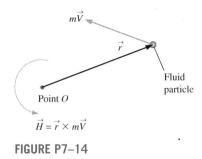

FIGURE P7–16

7–17 Write the primary dimensions of each of the following variables from the study of convection heat transfer (Fig. P7–17), showing all your work: (*a*) heat generation rate $\dot{g}$ (*Hint*: rate of conversion of thermal energy per unit volume); (*b*) heat flux $\dot{q}$ (*Hint*: rate of heat transfer per unit area); (*c*) heat transfer coefficient *h* (*Hint*: heat flux per unit temperature difference).

FIGURE P7–17

7–18 Thumb through the appendices of your thermodynamics book, and find three properties or constants not mentioned in Probs. 7–1 to 7–17. List the name of each property or constant and its SI units. Then write out the primary dimensions of each property or constant.

7–19E Thumb through the appendices of this book and/or your thermodynamics book, and find three properties or constants not mentioned in Probs. 7–1 to 7–17. List the name of each property or constant and its English units. Then write out the primary dimensions of each property or constant.

Dimensional Homogeneity

7–20C Explain the *law of dimensional homogeneity* in simple terms.

7–21 In Chap. 4, we defined the *material acceleration*, which is the acceleration following a fluid particle,

$$\vec{a}(x, y, z, t) = \frac{\partial \vec{V}}{\partial t} + (\vec{V}\cdot\vec{\nabla})\vec{V}$$

(*a*) What are the primary dimensions of the gradient operator $\vec{\nabla}$? (*b*) Verify that each additive term in the equation has the same dimensions. *Answers:* (*a*) $\{L^{-1}\}$; (*b*) $\{L^1 t^{-2}\}$

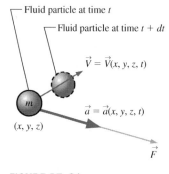

FIGURE P7–21

7–22 Newton's second law is the foundation for the differential equation of conservation of linear momentum (to be discussed in Chap. 9). In terms of the material acceleration following a fluid particle (Fig. P7–21), we write Newton's second law as follows:

$$\vec{F} = m\vec{a} = m\left(\frac{\partial \vec{V}}{\partial t} + (\vec{V} \cdot \vec{\nabla})\vec{V}\right)$$

Or, dividing both sides by the mass m of the fluid particle,

$$\frac{\vec{F}}{m} = \frac{\partial \vec{V}}{\partial t} + (\vec{V} \cdot \vec{\nabla})\vec{V}$$

Write the primary dimensions of each additive term in the (second) equation, and verify that the equation is dimensionally homogeneous. Show all your work.

7–23 In Chap. 9, we discuss the differential equation for conservation of mass, the *continuity equation*. In cylindrical coordinates, and for steady flow,

$$\frac{1}{r}\frac{\partial(ru_r)}{\partial r} + \frac{1}{r}\frac{\partial u_\theta}{\partial \theta} + \frac{\partial u_z}{\partial z} = 0$$

Write the primary dimensions of each additive term in the equation, and verify that the equation is dimensionally homogeneous. Show all your work.

7–24 The *Reynolds transport theorem* (RTT) is discussed in Chap. 4. For the general case of a moving and/or deforming control volume, we write the RTT as follows:

$$\frac{dB_{sys}}{dt} = \frac{d}{dt}\int_{CV} \rho b \, d\mathcal{V} + \int_{CS} \rho b \vec{V}_r \cdot \vec{n} \, dA$$

where $\vec{V}_r$ is the *relative velocity*, i.e., the velocity of the fluid relative to the control surface. Write the primary dimensions of each additive term in the equation, and verify that the equation is dimensionally homogeneous. Show all your work. (*Hint*: Since B can be any property of the flow—scalar, vector, or even tensor—it can have a variety of dimensions. So, just let the dimensions of B be those of B itself, $\{B\}$. Also, b is defined as B per unit mass.)

7–25 An important application of fluid mechanics is the study of room ventilation. In particular, suppose there is a **source** S (mass per unit time) of air pollution in a room of volume $\mathcal{V}$ (Fig. P7–25). Examples include carbon monoxide from cigarette smoke or an unvented kerosene heater, gases like ammonia from household cleaning products, and vapors given off by evaporation of **volatile organic compounds** (VOCs) from an open container. We let c represent the **mass concentration** (mass of contaminant per unit volume of air). $\dot{\mathcal{V}}$ is the volume flow rate of fresh air entering the room. If the room air is well mixed so that the mass concentration c is uniform throughout the room, but varies with time, the differential equation for mass concentration in the room as a function of time is

$$\mathcal{V}\frac{dc}{dt} = S - \dot{\mathcal{V}}c - cA_s k_w$$

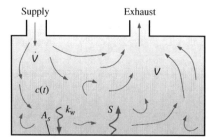

FIGURE P7–25

where k_w is an **adsorption coefficient** and A_s is the surface area of walls, floors, furniture, etc., that adsorb some of the contaminant. Write the primary dimensions of the first three terms in the equation (including the term on the left side), and verify that those terms are dimensionally homogeneous. Then determine the dimensions of k_w. Show all your work.

7–26 In Chap. 4 we defined *volumetric strain rate* as the rate of increase of volume of a fluid element per unit volume (Fig. P7–26). In Cartesian coordinates we write the volumetric strain rate as

$$\frac{1}{\mathcal{V}}\frac{D\mathcal{V}}{Dt} = \frac{\partial u}{\partial x} + \frac{\partial v}{\partial y} + \frac{\partial w}{\partial z}$$

Write the primary dimensions of each additive term, and verify that the equation is dimensionally homogeneous. Show all your work.

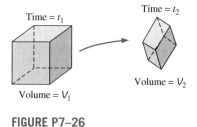

FIGURE P7–26

7–27 Cold water enters a pipe, where it is heated by an external heat source (Fig. P7–27). The inlet and outlet water temperatures are T_{in} and T_{out}, respectively. The total rate of heat transfer $\dot{Q}$ from the surroundings into the water in the pipe is

$$\dot{Q} = \dot{m}c_p(T_{out} - T_{in})$$

where $\dot{m}$ is the mass flow rate of water through the pipe, and c_p is the specific heat of the water. Write the primary dimensions of each additive term in the equation, and verify that the equation is dimensionally homogeneous. Show all your work.

$$\dot{Q} = \dot{m}c_p(T_{out} - T_{in})$$

FIGURE P7–27

Nondimensionalization of Equations

7–28C What is the primary reason for *nondimensionalizing* an equation?

7–29 Recall from Chap. 4 that the volumetric strain rate is zero for a steady incompressible flow. In Cartesian coordinates we express this as

$$\frac{\partial u}{\partial x} + \frac{\partial v}{\partial y} + \frac{\partial w}{\partial z} = 0$$

Suppose the characteristic speed and characteristic length for a given flow field are V and L, respectively (Fig. P7–29). Define the following dimensionless variables,

$$x^* = \frac{x}{L}, \quad y^* = \frac{y}{L}, \quad z^* = \frac{z}{L},$$

$$u^* = \frac{u}{V}, \quad v^* = \frac{v}{V}, \quad \text{and} \quad w^* = \frac{w}{V}$$

Nondimensionalize the equation, and identify any established (named) dimensionless parameters that may appear. Discuss.

FIGURE P7–29

7–30 In an oscillating compressible flow field the volumetric strain rate is *not* zero, but varies with time following a fluid particle. In Cartesian coordinates we express this as

$$\frac{1}{V}\frac{DV}{Dt} = \frac{\partial u}{\partial x} + \frac{\partial v}{\partial y} + \frac{\partial w}{\partial z}$$

Suppose the characteristic speed and characteristic length for a given flow field are V and L, respectively. Also suppose that f is a characteristic frequency of the oscillation (Fig. P7–30). Define the following dimensionless variables,

$$t^* = ft, \quad V^* = \frac{V}{L^3}, \quad x^* = \frac{x}{L}, \quad y^* = \frac{y}{L},$$

$$z^* = \frac{z}{L}, \quad u^* = \frac{u}{V}, \quad v^* = \frac{v}{V}, \quad \text{and} \quad w^* = \frac{w}{V}$$

Nondimensionalize the equation and identify any established (named) dimensionless parameters that may appear.

Time t_1 Time t_2 Time t_3

f = frequency of oscillation

FIGURE P7–30

7–31 In Chap. 9, we define the **stream function** ψ for two-dimensional incompressible flow in the *xy*-plane,

$$u = \frac{\partial \psi}{\partial y} \qquad v = -\frac{\partial \psi}{\partial x}$$

where u and v are the velocity components in the *x*- and *y*-directions, respectively. (*a*) What are the primary dimensions of ψ? (*b*) Suppose a certain two-dimensional flow has a characteristic length scale L and a characteristic time scale t. Define dimensionless forms of variables x, y, u, v, and ψ. (*c*) Rewrite the equations in nondimensional form, and identify any established dimensionless parameters that may appear.

7–32 In an oscillating incompressible flow field the force per unit mass acting on a fluid particle is obtained from Newton's second law in intensive form (see Prob. 7–22),

$$\frac{\vec{F}}{m} = \frac{\partial \vec{V}}{\partial t} + (\vec{V}\cdot\vec{\nabla})\vec{V}$$

Suppose the characteristic speed and characteristic length for a given flow field are V_∞ and L, respectively. Also suppose that ω is a characteristic angular frequency (rad/s) of the oscillation (Fig. P7–32). Define the following nondimensionalized variables,

$$t^* = \omega t, \quad \vec{x}^* = \frac{\vec{x}}{L}, \quad \vec{\nabla}^* = L\vec{\nabla}, \quad \text{and} \quad \vec{V}^* = \frac{\vec{V}}{V_\infty}$$

Since there is no given characteristic scale for the force per unit mass acting on a fluid particle, we assign one, noting that $\{\vec{F}/m\} = \{L/t^2\}$. Namely, we let

$$(\vec{F}/m)^* = \frac{1}{\omega^2 L}\,\vec{F}/m$$

Nondimensionalize the equation of motion and identify any established (named) dimensionless parameters that may appear.

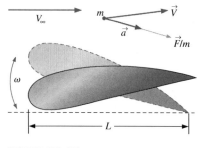

FIGURE P7–32

7–33 A wind tunnel is used to measure the pressure distribution in the airflow over an airplane model (Fig. P7–33). The air speed in the wind tunnel is low enough that compressible effects are negligible. As discussed in Chap. 5, the Bernoulli equation approximation is valid in such a flow situation everywhere except very close to the body surface or wind tunnel wall surfaces and in the wake region behind the model. Far away from the model, the air flows at speed V_∞ and pressure P_∞, and the air density ρ is approximately constant. Gravitational effects are generally negligible in airflows, so we write the Bernoulli equation as

$$P + \frac{1}{2}\rho V^2 = P_\infty + \frac{1}{2}\rho V_\infty^2$$

Wind tunnel test section

FIGURE P7–33

Nondimensionalize the equation, and generate an expression for the **pressure coefficient** C_p at any point in the flow where the Bernoulli equation is valid. C_p is defined as

$$C_p = \frac{P - P_\infty}{\frac{1}{2}\rho V_\infty^2}$$

Answer: $C_p = 1 - V^2/V_\infty^2$

7–34 Consider ventilation of a well-mixed room as in Fig. P7–25. The differential equation for mass concentration in the room as a function of time is given in Prob. 7–25 and is repeated here for convenience,

$$V\frac{dc}{dt} = S - \dot{V}c - cA_s k_w$$

There are three characteristic parameters in such a situation: L, a characteristic length scale of the room (assume $L = V^{1/3}$); $\dot{V}$, the volume flow rate of fresh air into the room, and c_{limit}, the maximum mass concentration that is not harmful. (a) Using these three characteristic parameters, define dimensionless forms of all the variables in the equation. (*Hint*: For example, define $c^* = c/c_{limit}$.) (b) Rewrite the equation in dimensionless form, and identify any established dimensionless groups that may appear.

Dimensional Analysis and Similarity

7–35C List the three primary purposes of dimensional analysis.

7–36C List and describe the three necessary conditions for complete similarity between a model and a prototype.

7–37 A student team is to design a human-powered submarine for a design competition. The overall length of the prototype submarine is 4.85 m, and its student designers hope that it can travel fully submerged through water at 0.440 m/s. The water is freshwater (a lake) at $T = 15°C$. The design team builds a one-fifth scale model to test in their university's wind tunnel (Fig. P7–37). A shield surrounds the drag balance strut so that the aerodynamic drag of the strut itself does not influence the measured drag. The air in the wind tunnel is at

Wind tunnel test section

Drag balance

FIGURE P7–37

25°C and at one standard atmosphere pressure. At what air speed do they need to run the wind tunnel in order to achieve similarity? *Answer: 30.2 m/s*

7–38 Repeat Prob. 7–37 with all the same conditions except that the only facility available to the students is a much smaller wind tunnel. Their model submarine is a one-twenty-fourth scale model instead of a one-fifth scale model. At what air speed do they need to run the wind tunnel in order to achieve similarity? Do you notice anything disturbing or suspicious about your result? Discuss your results.

7–39 This is a follow-up to Prob. 7–37. The students measure the aerodynamic drag on their model submarine in the wind tunnel (Fig. P7–37). They are careful to run the wind tunnel at conditions that ensure similarity with the prototype submarine. Their measured drag force is 5.70 N. Estimate the drag force on the prototype submarine at the conditions given in Prob. 7–37.   *Answer: 25.5 N*

7–40E A lightweight parachute is being designed for military use (Fig. P7–40E). Its diameter D is 24 ft and the total weight W of the falling payload, parachute, and equipment is 230 lbf. The design *terminal settling speed* V_t of the parachute at this weight is 18 ft/s. A one-twelfth scale model of the parachute is tested in a wind tunnel. The wind tunnel temperature and pressure are the same as those of the prototype, namely 60°F and standard atmospheric pressure. (*a*) Calculate the drag coefficient of the prototype. (*Hint*: At terminal settling speed, weight is balanced by aerodynamic drag.) (*b*) At what wind tunnel speed should the wind tunnel be run in order to achieve dynamic similarity? (*c*) Estimate the aerodynamic drag of the model parachute in the wind tunnel (in lbf).

FIGURE P7–40E

7–41 Some wind tunnels are *pressurized*. Discuss why a research facility would go through all the extra trouble and expense to pressurize a wind tunnel. If the air pressure in the tunnel increases by a factor of 1.8, all else being equal (same wind speed, same model, etc.), by what factor will the Reynolds number increase?

7–42E The aerodynamic drag of a new sports car is to be predicted at a speed of 60.0 mi/h at an air temperature of 25°C. Automotive engineers build a one-third scale model of the car (Fig. P7–42E) to test in a wind tunnel. The temperature of the wind tunnel air is also 25°C. The drag force is measured with a drag balance, and the moving belt is used to simulate the moving ground (from the car's frame of reference). Determine how fast the engineers should run the wind tunnel to achieve similarity between the model and the prototype.

FIGURE P7–42E

7–43E This is a follow-up to Prob. 7–42E. The aerodynamic drag on the model in the wind tunnel (Fig. P7–42E) is measured to be 33.5 lbf when the wind tunnel is operated at the speed that ensures similarity with the prototype car. Estimate the drag force (in lbf) on the prototype car at the conditions given in Prob. 7–42E.

7–44 Consider the common situation in which a researcher is trying to match the Reynolds number of a large prototype vehicle with that of a small-scale model in a wind tunnel. Is it better for the air in the wind tunnel to be cold or hot? Why? Support your argument by comparing wind tunnel air at 10°C and at 40°C, all else being equal.

7–45E Some students want to visualize flow over a spinning baseball. Their fluids laboratory has a nice water tunnel into which they can inject multicolored dye streaklines, so they decide to test a spinning baseball in the water tunnel (Fig. P7–45E). Similarity requires that they match both the Reynolds number and the Strouhal number between their model test and the actual baseball that moves through the air at 85 mi/h and spins at 320 rpm. Both the air and the water are at 68°F. At what speed should they run the water in the water tunnel, and at what rpm should they spin their baseball? *Answers: 5.63 mi/h, 21.2 rpm*

Water tunnel test section

FIGURE P7–45E

Dimensionless Parameters and the Method of Repeating Variables

7–46 Using primary dimensions, verify that the Archimedes number (Table 7–5) is indeed dimensionless.

7–47 Using primary dimensions, verify that the Grashof number (Table 7–5) is indeed dimensionless.

7–48 Using primary dimensions, verify that the Rayleigh number (Table 7–5) is indeed dimensionless. What other established nondimensional parameter is formed by the ratio of Ra and Gr? *Answer:* the Prandtl number

7–49 A periodic *Kármán vortex street* is formed when a uniform stream flows over a circular cylinder (Fig. P7–49). Use the method of repeating variables to generate a dimensionless relationship for Kármán vortex shedding frequency f_k as a function of free-stream speed V, fluid density ρ, fluid viscosity μ, and cylinder diameter D. Show all your work. *Answer:* St $= f$ (Re)

FIGURE P7–49

7–50 Repeat Prob. 7–49, but with an additional independent parameter included, namely, the speed of sound c in the fluid. Use the method of repeating variables to generate a dimensionless relationship for Kármán vortex shedding frequency f_k as a function of free-stream speed V, fluid density ρ, fluid viscosity μ, cylinder diameter D, and speed of sound c. Show all your work.

7–51 A stirrer is used to mix chemicals in a large tank (Fig. P7–51). The shaft power $\dot{W}$ supplied to the stirrer blades is a function of stirrer diameter D, liquid density ρ, liquid viscosity μ, and the angular velocity ω of the spinning

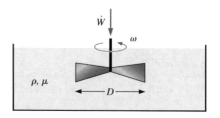

FIGURE P7–51

blades. Use the method of repeating variables to generate a dimensionless relationship between these parameters. Show all your work and be sure to identify your Π groups, modifying them as necessary. *Answer:* $N_p = f$ (Re)

7–52 Repeat Prob. 7–51 except do not assume that the tank is large. Instead, let tank diameter D_{tank} and average liquid depth h_{tank} be additional relevant parameters.

7–53 Albert Einstein is pondering how to write his (soon-to-be-famous) equation. He knows that energy E is a function of mass m and the speed of light c, but he doesn't know the functional relationship ($E = m^2c$? $E = mc^4$?). Pretend that Albert knows nothing about dimensional analysis, but since you are taking a fluid mechanics class, you help Albert come up with his equation. Use the step-by-step method of repeating variables to generate a dimensionless relationship between these parameters, showing all of your work. Compare this to Einstein's famous equation—does dimensional analysis give you the correct form of the equation?

FIGURE P7–53

7–54 The *Richardson number* is defined as

$$\text{Ri} = \frac{L^5 g \, \Delta\rho}{\rho \dot{V}^2}$$

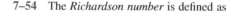

Miguel is working on a problem that has a characteristic length scale L, a characteristic velocity V, a characteristic density difference $\Delta\rho$, a characteristic (average) density ρ, and of course the gravitational constant g, which is always available. He wants to define a Richardson number, but does not have a characteristic volume flow rate. Help Miguel define a characteristic volume flow rate based on the parameters available to him, and then define an appropriate Richardson number in terms of the given parameters.

7–55 Consider fully developed **Couette flow**—flow between two infinite parallel plates separated by distance h, with the top plate moving and the bottom plate stationary as illustrated in Fig. P7–55. The flow is steady, incompressible, and two-dimensional in the xy-plane. Use the method of repeating variables to generate a dimensionless relationship for the x-component of fluid velocity u as a function of fluid viscosity μ, top plate speed V, distance h, fluid density ρ, and distance y. Show all your work.   *Answer: $u/V = f(\mathrm{Re}, y/h)$*

FIGURE P7–55

7–56 Consider *developing* Couette flow—the same flow as Prob. 7–55 except that the flow is not yet steady-state, but is developing with time. In other words, time t is an additional parameter in the problem. Generate a dimensionless relationship between all the variables.

7–57 The speed of sound c in an ideal gas is known to be a function of the ratio of specific heats k, absolute temperature T, and specific ideal gas constant R_{gas} (Fig. P7–57). Showing all your work, use dimensional analysis to find the functional relationship between these parameters.

FIGURE P7–57

7–58 Repeat Prob. 7–57, except let the speed of sound c in an ideal gas be a function of absolute temperature T, universal ideal gas constant R_u, molar mass (molecular weight) M of the gas, and ratio of specific heats k. Showing all your work, use dimensional analysis to find the functional relationship between these parameters.

7–59 Repeat Prob. 7–57, except let the speed of sound c in an ideal gas be a function only of absolute temperature T and specific ideal gas constant R_{gas}. Showing all your work, use dimensional analysis to find the functional relationship between these parameters.   *Answer: $c/\sqrt{R_{gas}T}$ = constant*

7–60 Repeat Prob. 7–57, except let speed of sound c in an ideal gas be a function only of pressure P and gas density ρ. Showing all your work, use dimensional analysis to find the functional relationship between these parameters. Verify that your results are consistent with the equation for speed of sound in an ideal gas, $c = \sqrt{kR_{gas}T}$.

7–61 When small aerosol particles or microorganisms move through air or water, the Reynolds number is very small ($\mathrm{Re} \ll 1$). Such flows are called **creeping flows.** The aerodynamic drag on an object in creeping flow is a function only of its speed V, some characteristic length scale L of the object, and fluid viscosity μ (Fig. P7–61). Use dimensional analysis to generate a relationship for F_D as a function of the independent variables.

FIGURE P7–61

7–62 A tiny aerosol particle of density ρ_p and characteristic diameter D_p falls in air of density ρ and viscosity μ (Fig. P7–62). If the particle is small enough, the creeping flow approximation is valid, and the terminal settling speed of the particle V depends only on D_p, μ, gravitational constant g, and the density difference $(\rho_p - \rho)$. Use dimensional analysis to generate a relationship for V as a function of the independent variables. Name any established dimensionless parameters that appear in your analysis.

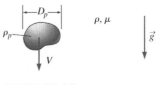

FIGURE P7–62

7–63 Combine the results of Probs. 7–61 and 7–62 to generate an equation for the settling speed V of an aerosol particle falling in air (Fig. P7–62). Verify that your result is consistent with the functional relationship obtained in Prob. 7–62. For consistency, use the notation of Prob. 7–62. (*Hint*: For a particle falling at constant settling speed, the particle's net weight must equal its aerodynamic drag. Your final result should be an equation for V that is valid to within some unknown constant.)

7–64   You will need the results of Prob. 7–63 to do this problem. A tiny aerosol particle falls at steady settling speed V. The Reynolds number is small enough that the creeping flow approximation is valid. If the particle size is doubled, all else being equal, by what factor will the settling speed go up? If the density difference $(\rho_p - \rho)$ is doubled, all else being equal, by what factor will the settling speed go up?

7–65   An incompressible fluid of density ρ and viscosity μ flows at average speed V through a long, horizontal section of round pipe of length L, inner diameter D, and inner wall roughness height ε (Fig. P7–65). The pipe is long enough that the flow is fully developed, meaning that the velocity profile does not change down the pipe. Pressure decreases (linearly) down the pipe in order to "push" the fluid through the pipe to overcome friction. Using the method of repeating variables, develop a nondimensional relationship between pressure drop $\Delta P = P_1 - P_2$ and the other parameters in the problem. Be sure to modify your Π groups as necessary to achieve established nondimensional parameters, and name them. (*Hint*: For consistency, choose D rather than L or ε as one of your repeating parameters.)   *Answer*: Eu = f (Re, ε/D, L/D)

FIGURE P7–65

7–66   Consider *laminar* flow through a long section of pipe, as in Fig. P7–65. For laminar flow it turns out that wall roughness is not a relevant parameter unless ε is very large. The volume flow rate $\dot{V}$ through the pipe is a function of pipe diameter D, fluid viscosity μ, and axial pressure gradient dP/dx. If pipe diameter is doubled, all else being equal, by what factor will volume flow rate increase? Use dimensional analysis.

7–67   One of the first things you learn in physics class is the law of universal gravitation, $F = G\dfrac{m_1 m_2}{r^2}$, where F is the attractive force between two bodies, m_1 and m_2 are the masses of the two bodies, r is the distance between the two bodies, and G is the universal gravitational constant equal to $(6.67428 \pm 0.00067) \times 10^{-11}$ [the units of G are not given here]. (*a*) Calculate the SI units of G. For consistency, give your answer in terms of kg, m, and s. (*b*) Suppose you don't remember the law of universal gravitation, but you are clever enough to know that F is a function of G, m_1, m_2, and r. Use dimensional analysis and the method of repeating variables (show all your work) to generate a nondimensional expression for $F = F(G, m_1, m_2, r)$. Give your answer as $\Pi_1 =$ function of $(\Pi_2, \Pi_3, \ldots)$. (*c*) Dimensional analysis cannot yield the exact form of the function. However, compare your

result to the law of universal gravitation to find the form of the function (e.g., $\Pi_1 = \Pi_2{}^2$ or some other functional form).

7–68   Jen is working on a spring–mass–damper system, as shown in Fig. P7–68. She remembers from her dynamic systems class that the damping ratio ζ is a nondimensional property of such systems and that ζ is a function of spring constant k, mass m, and damping coefficient c. Unfortunately, she does not recall the exact form of the equation for ζ. However, she is taking a fluid mechanics class and decides to use her newly acquired knowledge about dimensional analysis to recall the form of the equation. Help Jen develop the equation for ζ using the method of repeating variables, showing all of your work. (*Hint*: Typical units for k are N/m and those for c are N·s/m.)

FIGURE P7–68

7–69   Bill is working on an electrical circuit problem. He remembers from his electrical engineering class that voltage drop ΔE is a function of electrical current I and electrical resistance R. Unfortunately, he does not recall the exact form of the equation for ΔE. However, he is taking a fluid mechanics class and decides to use his newly acquired knowledge about dimensional analysis to recall the form of the equation. Help Bill develop the equation for ΔE using the method of repeating variables, showing all of your work. Compare this to Ohm's law—does dimensional analysis give you the correct form of the equation?

7–70   A boundary layer is a thin region (usually along a wall) in which viscous forces are significant and within which the flow is rotational. Consider a boundary layer growing along a thin flat plate (Fig. P7–70). The flow is steady. The boundary layer thickness δ at any downstream distance x is a function of x, free-stream velocity V_∞, and fluid properties ρ (density)

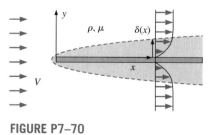

FIGURE P7–70

and μ (viscosity). Use the method of repeating variables to generate a dimensionless relationship for δ as a function of the other parameters. Show all your work.

7–71 A liquid of density ρ and viscosity μ is pumped at volume flow rate $\dot{V}$ through a pump of diameter D. The blades of the pump rotate at angular velocity ω. The pump supplies a pressure rise ΔP to the liquid. Using dimensional analysis, generate a dimensionless relationship for ΔP as a function of the other parameters in the problem. Identify any established nondimensional parameters that appear in your result. *Hint*: For consistency (and whenever possible), it is wise to choose a length, a density, and a velocity (or angular velocity) as repeating variables.

7–72 A propeller of diameter D rotates at angular velocity ω in a liquid of density ρ and viscosity μ. The required torque T is determined to be a function of D, ω, ρ, and μ. Using dimensional analysis, generate a dimensionless relationship. Identify any established nondimensional parameters that appear in your result. *Hint*: For consistency (and whenever possible), it is wise to choose a length, a density, and a velocity (or angular velocity) as repeating variables.

7–73 Repeat Prob. 7–72 for the case in which the propeller operates in a compressible gas instead of a liquid.

7–74 In the study of turbulent flow, turbulent viscous dissipation rate ε (rate of energy loss per unit mass) is known to be a function of length scale l and velocity scale u' of the large-scale turbulent eddies. Using dimensional analysis (Buckingham pi and the method of repeating variables) and showing all of your work, generate an expression for ε as a function of l and u'.

7–75 The rate of heat transfer to water flowing in a pipe was analyzed in Prob. 7–27. Let us approach that same problem, but now with dimensional analysis. Cold water enters a pipe, where it is heated by an external heat source (Fig. P7–75). The inlet and outlet water temperatures are T_{in} and T_{out}, respectively. The total rate of heat transfer $\dot{Q}$ from the surroundings into the water in the pipe is known to be a function of mass flow rate $\dot{m}$, the specific heat c_p of the water, and the temperature difference between the incoming and outgoing water. Showing all your work, use dimensional analysis to find the functional relationship between these parameters, and compare to the analytical equation given in Prob. 7–27. (*Note*: We are pretending that we do not know the analytical equation.)

7–76 Consider a liquid in a cylindrical container in which both the container and the liquid are rotating as a rigid body (solid-body rotation). The elevation difference h between the center of the liquid surface and the rim of the liquid surface is a function of angular velocity ω, fluid density ρ, gravitational acceleration g, and radius R (Fig. P7–76). Use the method of repeating variables to find a dimensionless relationship between the parameters. Show all your work.   *Answer: $h/R = f$ (Fr)*

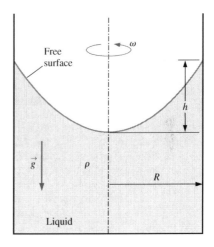

FIGURE P7–76

7–77 Consider the case in which the container and liquid of Prob. 7–76 are initially at rest. At $t = 0$ the container begins to rotate. It takes some time for the liquid to rotate as a rigid body, and we expect that the liquid's viscosity is an additional relevant parameter in the unsteady problem. Repeat Prob. 7–76, but with two additional independent parameters included, namely, fluid viscosity μ and time t. (We are interested in the development of height h as a function of time and the other parameters.)

Experimental Testing and Incomplete Similarity

7–78C Although we usually think of a model as being smaller than the prototype, describe at least three situations in which it is better for the model to be *larger* than the prototype.

7–79C Discuss the purpose of a moving ground belt in wind tunnel tests of flow over model automobiles. Think of an alternative if a moving ground belt is unavailable.

7–80C Consider again the model truck example discussed in Section 7–5, except that the maximum speed of the wind tunnel is only 50 m/s. Aerodynamic force data are taken for wind tunnel speeds between $V = 20$ and 50 m/s—assume the same data for these speeds as those listed in Table 7–7. Based on these data *alone,* can the researchers be confident that they have reached Reynolds number independence?

c_p = specific heat of the water

FIGURE P7–75

7–81C   Define *wind tunnel blockage*. What is the rule of thumb about the maximum acceptable blockage for a wind tunnel test? Explain why there would be measurement errors if the blockage were significantly higher than this value.

7–82C   What is the rule of thumb about the Mach number limit in order that the incompressible flow approximation is reasonable? Explain why wind tunnel results would be incorrect if this rule of thumb were violated.

7–83   A one-sixteenth scale model of a new sports car is tested in a wind tunnel. The prototype car is 4.37 m long, 1.30 m tall, and 1.69 m wide. During the tests, the moving ground belt speed is adjusted so as to always match the speed of the air moving through the test section. Aerodynamic drag force F_D is measured as a function of wind tunnel speed; the experimental results are listed in Table P7–83. Plot drag coefficient C_D as a function of the Reynolds number Re, where the area used for calculation of C_D is the frontal area of the model car (assume A = width × height), and the length scale used for calculation of Re is car width W. Have we achieved dynamic similarity? Have we achieved Reynolds number independence in our wind tunnel test? Estimate the aerodynamic drag force on the prototype car traveling on the highway at 31.3 m/s (70 mi/h). Assume that both the wind tunnel air and the air flowing over the prototype car are at 25°C and atmospheric pressure.   *Answers:* no, yes, 408 N

TABLE P7–83

V, m/s	F_D, N
10	0.29
15	0.64
20	0.96
25	1.41
30	1.55
35	2.10
40	2.65
45	3.28
50	4.07
55	4.91

7–84   Water at 20°C flows through a long, straight pipe. The pressure drop is measured along a section of the pipe of length $L = 1.3$ m as a function of average velocity V through the pipe (Table P7–84). The inner diameter of the pipe is $D = 10.4$ cm. (*a*) Nondimensionalize the data and plot the Euler number as a function of the Reynolds number. Has the experiment been run at high enough speeds to achieve Reynolds number independence? (*b*) Extrapolate the experimental data to predict the pressure drop at an average speed of 80 m/s.   *Answer:* 1,940,000 N/m²

TABLE P7–84

V, m/s	ΔP, N/m²
0.5	77.0
1	306
2	1218
4	4865
6	10,920
8	19,440
10	30,340
15	68,330
20	121,400
25	189,800
30	273,200
35	372,100
40	485,300
45	614,900
50	758,700

7–85   In the model truck example discussed in Section 7–5, the wind tunnel test section is 3.5 m long, 0.85 m tall, and 0.90 m wide. The one-sixteenth scale model truck is 0.991 m long, 0.257 m tall, and 0.159 m wide. What is the wind tunnel blockage of this model truck? Is it within acceptable limits according to the standard rule of thumb?

7–86E   A small wind tunnel in a university's undergraduate fluid flow laboratory has a test section that is 20 by 20 in in cross section and is 4.0 ft long. Its maximum speed is 145 ft/s. Some students wish to build a model 18-wheeler to study how aerodynamic drag is affected by rounding off the back of the trailer. A full-size (prototype) tractor-trailer rig is 52 ft long, 8.33 ft wide, and 12 ft high. Both the air in the wind tunnel and the air flowing over the prototype are at 80°F and atmospheric pressure. (*a*) What is the largest scale model they can build to stay within the rule-of-thumb guidelines for blockage? What are the dimensions of the model truck in inches? (*b*) What is the maximum model truck Reynolds number achievable by the students? (*c*) Are the students able to achieve Reynolds number independence? Discuss.

7–87   Use dimensional analysis to show that in a problem involving shallow water waves (Fig. P7–87), both the Froude number and the Reynolds number are relevant dimensionless parameters. The wave speed c of waves on the surface of a liquid is a function of depth h, gravitational acceleration g, fluid density ρ, and fluid viscosity μ. Manipulate your Π's to get the parameters into the following form:

$$\text{Fr} = \frac{c}{\sqrt{gh}} = f(\text{Re}) \quad \text{where Re} = \frac{\rho ch}{\mu}$$

FIGURE P7–87

Review Problems

7–88C There are many established nondimensional parameters besides those listed in Table 7–5. Do a literature search or an Internet search and find at least three established, named nondimensional parameters that are *not* listed in Table 7–5. For each one, provide its definition and its ratio of significance, following the format of Table 7–5. If your equation contains any variables not identified in Table 7–5, be sure to identify those variables.

7–89C Think about and describe a prototype flow and a corresponding model flow that have geometric similarity, but not kinematic similarity, even though the Reynolds numbers match. Explain.

7–90C For each statement, choose whether the statement is true or false and discuss your answer briefly.

(*a*) Kinematic similarity is a necessary and sufficient condition for dynamic similarity.

(*b*) Geometric similarity is a necessary condition for dynamic similarity.

(*c*) Geometric similarity is a necessary condition for kinematic similarity.

(*d*) Dynamic similarity is a necessary condition for kinematic similarity.

7–91 Write the primary dimensions of each of the following variables from the field of solid mechanics, showing all your work: (*a*) moment of inertia I; (*b*) modulus of elasticity E, also called Young's modulus; (*c*) strain ε; (*d*) stress σ. (*e*) Finally, show that the relationship between stress and strain (Hooke's law) is a dimensionally homogeneous equation.

7–92 Force F is applied at the tip of a cantilever beam of length L and moment of inertia I (Fig. P7–92). The modulus of elasticity of the beam material is E. When the force is applied, the tip deflection of the beam is z_d. Use dimensional analysis to generate a relationship for z_d as a function of the independent variables. Name any established dimensionless parameters that appear in your analysis.

FIGURE P7–92

7–93 An explosion occurs in the atmosphere when an anti-aircraft missile meets its target (Fig. P7–93). A **shock wave** (also called a **blast wave**) spreads out radially from the explosion. The pressure difference across the blast wave ΔP and its radial distance r from the center are functions of time t, speed of sound c, and the total amount of energy E released by the explosion. (*a*) Generate dimensionless relationships between ΔP and the other parameters and between r and the other parameters. (*b*) For a given explosion, if the time t since the explosion doubles, all else being equal, by what factor will ΔP decrease?

FIGURE P7–93

7–94 The Archimedes number listed in Table 7–5 is appropriate for buoyant *particles* in a fluid. Do a literature search or an Internet search and find an alternative definition of the Archimedes number that is appropriate for buoyant *fluids* (e.g., buoyant jets and buoyant plumes, heating and air-conditioning applications). Provide its definition and its ratio of significance, following the format of Table 7–5. If your equation contains any variables not identified in Table 7–5, be sure to identify those variables. Finally, look through the established dimensionless parameters listed in Table 7–5 and find one that is similar to this alternate form of the Archimedes number.

7–95 Consider steady, laminar, fully developed, two-dimensional **Poiseuille flow**—flow between two infinite parallel plates separated by distance h, with both the top plate and bottom plate stationary, and a forced pressure gradient dP/dx driving the flow as illustrated in Fig. P7–95. (dP/dx is constant and negative.) The flow is steady, incompressible, and two-dimensional in the xy-plane. The flow is also *fully developed*, meaning that the velocity profile does not change with downstream distance x. Because of the fully developed nature of the flow, there are no inertial effects and density does not enter the problem. It turns out that u, the velocity component in the x-direction, is a function of distance h, pressure gradient dP/dx, fluid viscosity μ, and vertical coordinate y. Perform a dimensional analysis (showing all your work), and generate a dimensionless relationship between the given variables.

340
DIMENSIONAL ANALYSIS AND MODELING

FIGURE P7–95

7–96 Consider the steady, laminar, fully developed, two-dimensional Poiseuille flow of Prob. 7–95. The maximum velocity u_{max} occurs at the center of the channel. (*a*) Generate a dimensionless relationship for u_{max} as a function of distance between plates h, pressure gradient dP/dx, and fluid viscosity μ. (*b*) If the plate separation distance h is doubled, all else being equal, by what factor will u_{max} change? (*c*) If the pressure gradient dP/dx is doubled, all else being equal, by what factor will u_{max} change? (*d*) How many experiments are required to describe the complete relationship between u_{max} and the other parameters in the problem?

7–97 The pressure drop $\Delta P = P_1 - P_2$ through a long section of round pipe can be written in terms of the shear stress τ_w along the wall. Shown in Fig. P7–97 is the shear stress acting by the wall on the fluid. The shaded region is a control volume composed of the fluid in the pipe between axial locations 1 and 2. There are two dimensionless parameters related to the pressure drop: the Euler number Eu and the Darcy friction factor f. (*a*) Using the control volume sketched in Fig. P7–97, generate a relationship for f in terms of Eu (and any other properties or parameters in the problem as needed). (*b*) Using the experimental data and conditions of Prob. 7–84 (Table P7–84), plot the Darcy friction factor as a function of Re. Does f show Reynolds number independence at large values of Re? If so, what is the value of f at very high Re?

Answers: (a) $f = 2\dfrac{D}{L}$Eu*; (b) yes, 0.0487*

FIGURE P7–97

7–98 Oftentimes it is desirable to work with an established dimensionless parameter, but the characteristic scales available do not match those used to define the parameter. In such cases, we *create* the needed characteristic scales based on dimensional reasoning (usually by inspection). Suppose for example that we have a characteristic velocity scale V, characteristic area A, fluid density ρ, and fluid viscosity μ,

and we wish to define a Reynolds number. We create a length scale $L = \sqrt{A}$, and define

$$\text{Re} = \frac{\rho V \sqrt{A}}{\mu}$$

In similar fashion, define the desired established dimensionless parameter for each case: (*a*) Define a Froude number, given $\dot{V}'$ = volume flow rate per unit depth, length scale L, and gravitational constant g. (*b*) Define a Reynolds number, given $\dot{V}'$ = volume flow rate per unit depth and kinematic viscosity ν. (*c*) Define a Richardson number (see Table 7–5), given $\dot{V}'$ = volume flow rate per unit depth, length scale L, characteristic density difference $\Delta\rho$, characteristic density ρ, and gravitational constant g.

7–99 A liquid of density ρ and viscosity μ flows by gravity through a hole of diameter d in the bottom of a tank of diameter D (Fig. P7–99). At the start of the experiment, the liquid surface is at height h above the bottom of the tank, as sketched. The liquid exits the tank as a jet with average velocity V straight down as also sketched. Using dimensional analysis, generate a dimensionless relationship for V as a function of the other parameters in the problem. Identify any established nondimensional parameters that appear in your result. (*Hint:* There are three length scales in this problem. For consistency, choose h as your length scale.)

FIGURE P7–99

7–100 Repeat Prob. 7–99 except for a different dependent parameter, namely, the time required to empty the tank t_{empty}. Generate a dimensionless relationship for t_{empty} as a function of the following independent parameters: hole diameter d, tank diameter D, density ρ, viscosity μ, initial liquid surface height h, and gravitational acceleration g.

7–101 A liquid delivery system is being designed such that ethylene glycol flows out of a hole in the bottom of a large tank, as in Fig. P7–99. The designers need to predict how long it will take for the ethylene glycol to completely drain. Since it would be very expensive to run tests with a full-scale prototype using ethylene glycol, they decide to build a one-quarter scale model for experimental testing, and they plan to use *water* as their test liquid. The model is geometrically similar to the prototype (Fig. P7–101). (*a*) The temperature

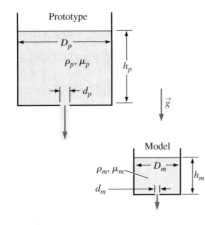

FIGURE P7–101

of the ethylene glycol in the prototype tank is 60°C, at which $\nu = 4.75 \times 10^{-6}$ m²/s. At what temperature should the water in the model experiment be set in order to ensure complete similarity between model and prototype? (b) The experiment is run with water at the proper temperature as calculated in part (a). It takes 3.27 min to drain the model tank. Predict how long it will take to drain the ethylene glycol from the prototype tank. *Answers:* (a) 45.8°C, (b) 6.54 min

7–102 Liquid flows out of a hole in the bottom of a tank as in Fig. P7–99. Consider the case in which the hole is very small compared to the tank ($d \ll D$). Experiments reveal that average jet velocity V is nearly independent of d, D, ρ, or μ. In fact, for a wide range of these parameters, it turns out that V depends only on liquid surface height h and gravitational acceleration g. If the liquid surface height is doubled, all else being equal, by what factor will the average jet velocity increase? *Answer:* $\sqrt{2}$

7–103 An aerosol particle of characteristic size D_p moves in an airflow of characteristic length L and characteristic velocity V. The characteristic time required for the particle to adjust to a sudden change in air speed is called the **particle relaxation time** τ_p,

$$\tau_p = \frac{\rho_p D_p^2}{18\mu}$$

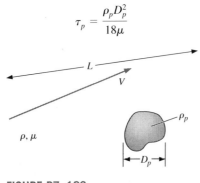

FIGURE P7–103

Verify that the primary dimensions of τ_p are time. Then create a dimensionless form of τ_p, based on some characteristic velocity V and some characteristic length L of the airflow (Fig. P7–103). What established dimensionless parameter do you create?

7–104 Compare the primary dimensions of each of the following properties in the *mass-based* primary dimension system (m, L, t, T, I, C, N) to those in the *force-based* primary dimension system (F, L, t, T, I, C, N): (a) pressure or stress; (b) moment or torque; (c) work or energy. Based on your results, explain when and why some authors prefer to use force as a primary dimension in place of mass.

7–105 The Stanton number is listed as a named, established nondimensional parameter in Table 7–5. However, careful analysis reveals that it can actually be formed by a combination of the Reynolds number, Nusselt number, and Prandtl number. Find the relationship between these four dimensionless groups, showing all your work. Can you also form the Stanton number by some combination of only *two* other established dimensionless parameters?

7–106 Consider a variation of the fully developed Couette flow problem of Prob. 7–55—flow between two infinite parallel plates separated by distance h, with the top plate moving at speed V_{top} and the bottom plate moving at speed V_{bottom} as illustrated in Fig. P7–106. The flow is steady, incompressible, and two-dimensional in the xy-plane. Generate a dimensionless relationship for the x-component of fluid velocity u as a function of fluid viscosity μ, plate speeds V_{top} and V_{bottom}, distance h, fluid density ρ, and distance y. (*Hint:* Think carefully about the list of parameters before rushing into the algebra.)

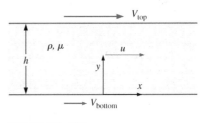

FIGURE P7–106

7–107 What are the primary dimensions of electric charge q, the units of which are **coulombs** (C)? (*Hint:* Look up the fundamental definition of electric current.)

7–108 What are the primary dimensions of electrical capacitance C, the units of which are **farads**? (*Hint:* Look up the fundamental definition of electrical capacitance.)

7–109 In many electronic circuits in which some kind of time scale is involved, such as filters and time-delay circuits (Fig. P7–109—a *low-pass filter*), you often see a resistor (R) and a capacitor (C) in series. In fact, the product of R and C is called the *electrical time constant*, RC. Showing all your work,

FIGURE P7–109

what are the primary dimensions of *RC*? Using dimensional reasoning alone, explain why a resistor and capacitor are often found together in timing circuits.

7–110 From fundamental electronics, the current flowing through a capacitor at any instant of time is equal to the capacitance times the rate of change of voltage (electromotive force) across the capacitor,

$$I = C \frac{dE}{dt}$$

Write the primary dimensions of both sides of this equation, and verify that the equation is dimensionally homogeneous. Show all your work.

7–111 An **electrostatic precipitator** (ESP) is a device used in various applications to clean particle-laden air. First, the dusty air passes through the *charging stage* of the ESP, where dust particles are given a positive charge q_p (coulombs) by charged ionizer wires (Fig. P7–111). The dusty air then enters the *collector stage* of the device, where it flows between two oppositely charged plates. The applied *electric field strength* between the plates is E_f (voltage difference per unit distance). Shown in Fig. P7–111 is a charged dust particle of diameter D_p. It is attracted to the negatively charged plate and moves toward that plate at a speed called the **drift velocity** w. If the plates are long enough, the dust particle impacts the negatively charged plate and adheres to it. Clean air exits the device. It turns out that for very small particles the drift velocity depends only on q_p, E_f, D_p, and air viscosity μ. (*a*) Generate a dimensionless relationship between the drift velocity through the collector stage of the ESP and the

given parameters. Show all your work. (*b*) If the electric field strength is doubled, all else being equal, by what factor will the drift velocity change? (*c*) For a given ESP, if the particle diameter is doubled, all else being equal, by what factor will the drift velocity change?

7–112 Experiments are being designed to measure the horizontal force F on a fireman's nozzle, as shown in Fig. P7–112. Force F is a function of velocity V_1, pressure drop $\Delta P = P_1 - P_2$, density ρ, viscosity μ, inlet area A_1, outlet area A_2, and length L. Perform a dimensional analysis for $F = f(V_1, \Delta P, \rho, \mu, A_1, A_2, L)$. For consistency, use V_1, A_1, and ρ as the repeating parameters and generate a dimensionless relationship. Identify any established nondimensional parameters that appear in your result.

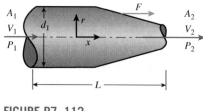

FIGURE P7–112

7–113 When a capillary tube of small diameter D is inserted into a container of liquid, the liquid rises to height h inside the tube (Fig. P7–113). h is a function of liquid density ρ, tube diameter D, gravitational constant g, contact angle ϕ, and the surface tension σ_s of the liquid. (*a*) Generate a dimensionless relationship for h as a function of the given parameters. (*b*) Compare your result to the exact analytical equation for h given in Chap. 2. Are your dimensional analysis results consistent with the exact equation? Discuss.

FIGURE P7–111

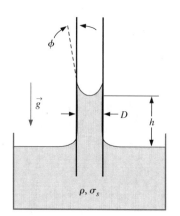

FIGURE P7–113

7–114 Repeat part (*a*) of Prob. 7–113, except instead of height *h*, find a functional relationship for the time scale t_{rise} needed for the liquid to climb up to its final height in the capillary tube. (*Hint*: Check the list of independent parameters in Prob. 7–113. Are there any additional relevant parameters?)

7–115 Sound intensity *I* is defined as the acoustic power per unit area emanating from a sound source. We know that *I* is a function of sound pressure level *P* (dimensions of pressure) and fluid properties ρ (density) and speed of sound *c*. (*a*) Use the method of repeating variables in mass-based primary dimensions to generate a dimensionless relationship for *I* as a function of the other parameters. Show all your work. What happens if you choose three repeating variables? Discuss. (*b*) Repeat part (*a*), but use the force-based primary dimension system. Discuss.

7–116 Repeat Prob. 7–115, but with the distance *r* from the sound source as an additional independent parameter.

7–117 Engineers at MIT have developed a mechanical model of a tuna fish to study its locomotion. The "Robotuna" shown in Fig. P7–117 is 1.0 m long and swims at speeds up to 2.0 m/s. Real bluefin tuna can exceed 3.0 m in length and have been clocked at speeds greater than 13 m/s. How fast would the 1.0-m Robotuna need to swim in order to match the Reynolds number of a real tuna that is 2.0 m long and swims at 10 m/s?

FIGURE P7–117

Photo by David Barrett of MIT, used by permission.

7–118 In Example 7–7, the mass-based system of primary dimensions was used to establish a relationship for the pressure difference $\Delta P = P_{\text{inside}} - P_{\text{outside}}$ between the inside and outside of a soap bubble as a function of soap bubble radius *R* and surface tension σ_s of the soap film (Fig. P7–118). Repeat the dimensional analysis using the method of repeating variables, but use the *force-based* system of primary dimensions instead. Show all your work. Do you get the same result?

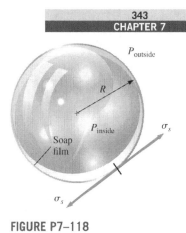

FIGURE P7–118

7–119 Many of the established nondimensional parameters listed in Table 7–5 can be formed by the product or ratio of two other established nondimensional parameters. For each pair of nondimensional parameters listed, find a *third* established nondimensional parameter that is formed by some manipulation of the two given parameters: (*a*) Reynolds number and Prandtl number; (*b*) Schmidt number and Prandtl number; (*c*) Reynolds number and Schmidt number.

7–120 A common device used in various applications to clean particle-laden air is the **reverse-flow cyclone** (Fig. P7–120). Dusty air (volume flow rate $\dot{V}$ and density ρ) enters tangentially through an opening in the side of the cyclone and swirls around in the tank. Dust particles are flung outward and fall out the bottom, while clean air is drawn out the top. The reverse-flow cyclones being studied are all geometrically similar; hence, diameter *D* represents the only length scale required to fully specify the entire cyclone geometry. Engineers are concerned about the pressure drop δP through the cyclone. (*a*) Generate a dimensionless relationship between the pressure drop through the cyclone and the given parameters. Show all your work. (*b*) If the cyclone size is doubled, all else being equal, by what factor will the pressure drop change? (*c*) If the volume flow rate is doubled, all else being equal, by what factor will the pressure drop change?
Answers: (*a*) $D^4 \delta P / \rho \dot{V}^2$ = constant, (*b*) 1/16, (*c*) 4

FIGURE P7–120

344
DIMENSIONAL ANALYSIS AND MODELING

Fundamentals of Engineering (FE) Exam Problems

7–121 Which one is *not* a primary dimension?
(a) Velocity   (b) Time   (c) Electric current
(d) Temperature   (e) Mass

7–122 The primary dimensions of kinematic viscosity are
(a) $m \cdot L/t^2$ (b) $m/L \cdot t$ (c) L^2/t (d) $L^2/m \cdot t$ (e) $L/m \cdot t^2$

7–123 The thermal conductivity of a substance may be defined as the rate of heat transfer per unit length per unit temperature difference. The primary dimensions of thermal conductivity are
(a) $m^2 \cdot L/t^2 \cdot T$ (b) $m^2 \cdot L^2/t \cdot T$ (c) $L^2/m \cdot t^2 \cdot T$
(d) $m \cdot L/t^3 \cdot T$ (e) $m \cdot L^2/t^3 \cdot T$

7–124 The primary dimensions of the gas constant over the universal gas constant R/R_u are
(a) $L^2/t^2 \cdot T$ (b) $m \cdot L/N$ (c) $m/t \cdot N \cdot T$
(d) m/L^3 (e) N/m

7–125 The primary dimensions of the universal gas constant R_u are
(a) $m \cdot L/t^2 \cdot T$ (b) $m^2 \cdot L/N$ (c) $m \cdot L^2/t^2 \cdot N \cdot T$
(d) $L^2/t^2 \cdot T$ (e) $N/m \cdot t$

7–126 There are four additive terms in an equation, and their units are given below. Which one is not consistent with this equation?
(a) J   (b) W/m   (c) $kg \cdot m^2/s^2$ (d) $Pa \cdot m^3$ (e) $N \cdot m$

7–127 The heat transfer coefficient is a nondimensional parameter which is a function of viscosity μ, specific heat c_p (kJ/kg·K), and thermal conductivity k (W/m·K). This nondimensional parameter is expressed as
(a) $c_p/\mu k$ (b) $k/\mu c_p$ (c) $\mu/c_p k$ (d) $\mu c_p/k$ (e) $c_p k/\mu$

7–128 The nondimensional heat transfer coefficient is a function of convection coefficient h (W/m²·K), thermal conductivity k (W/m·K), and characteristic length L. This nondimensional parameter is expressed as
(a) hL/k (b) h/kL (c) L/hk (d) hk/L (e) kL/h

7–129 The drag coefficient C_D is a nondimensional parameter and is a function of drag force F_D, density ρ, velocity V, and area A. The drag coefficient is expressed as
(a) $\dfrac{F_D V^2}{2\rho A}$ (b) $\dfrac{2F_D}{\rho VA}$ (c) $\dfrac{\rho VA^2}{F_D}$ (d) $\dfrac{F_D A}{\rho V}$ (e) $\dfrac{2F_D}{\rho V^2 A}$

7–130 Which similarity condition is related to force-scale equivalence?
(a) Geometric   (b) Kinematic   (c) Dynamic
(d) Kinematic and dynamic   (e) Geometric and kinematic

7–131 A one-third scale model of a car is to be tested in a wind tunnel. The conditions of the actual car are $V = 75$ km/h and $T = 0°C$ and the air temperature in the wind tunnel is 20°C.

The properties of air at 1 atm and 0°C: $\rho = 1.292$ kg/m³, $v = 1.338 \times 10^{-5}$ m²/s.

The properties of air at 1 atm and 20°C: $\rho = 1.204$ kg/m³, $v = 1.516 \times 10^{-5}$ m²/s.

In order to achieve similarity between the model and the prototype, the wind tunnel velocity should be
(a) 255 km/h   (b) 225 km/h   (c) 147 km/h   (d) 75 km/h
(e) 25 km/h

7–132 A one-fourth scale model of a car is to be tested in a wind tunnel. The conditions of the actual car are $V = 45$ km/h and $T = 0°C$ and the air temperature in the wind tunnel is 20°C. In order to achieve similarity between the model and the prototype, the wind tunnel is run at 204 km/h.

The properties of air at 1 atm and 0°C: $\rho = 1.292$ kg/m³, $v = 1.338 \times 10^{-5}$ m²/s.

The properties of air at 1 atm and 20°C: $\rho = 1.204$ kg/m³, $v = 1.516 \times 10^{-5}$ m²/s.

If the average drag force on the model is measured to be 70 N, the drag force on the prototype is
(a) 17.5 N   (b) 58.5 N   (c) 70 N   (d) 93.2 N   (e) 280 N

7–133 A one-third scale model of an airplane is to be tested in water. The airplane has a velocity of 900 km/h in air at −50°C. The water temperature in the test section is 10°C.

The properties of air at 1 atm and −50°C: $\rho = 1.582$ kg/m³, $\mu = 1.474 \times 10^{-5}$ kg/m·s.

The properties of water at 1 atm and 10°C: $\rho = 999.7$ kg/m³, $\mu = 1.307 \times 10^{-3}$ kg/m·s.

In order to achieve similarity between the model and the prototype, the water velocity on the model should be
(a) 97 km/h   (b) 186 km/h   (c) 263 km/h   (d) 379 km/h
(e) 450 km/h

7–134 A one-fourth scale model of an airplane is to be tested in water. The airplane has a velocity of 700 km/h in air at −50°C. The water temperature in the test section is 10°C. In order to achieve similarity between the model and the prototype, the test is done at a water velocity of 393 km/h.

The properties of air at 1 atm and −50°C: $\rho = 1.582$ kg/m³, $\mu = 1.474 \times 10^{-5}$ kg/m·s.

The properties of water at 1 atm and 10°C: $\rho = 999.7$ kg/m³, $\mu = 1.307 \times 10^{-3}$ kg/m·s.

If the average drag force on the model is measured to be 13,800 N, the drag force on the prototype is
(a) 590 N   (b) 862 N   (c) 1109 N   (d) 4655 N
(e) 3450 N

7–135 Consider a boundary layer growing along a thin flat plate. This problem involves the following parameters: boundary layer thickness δ, downstream distance x, free-stream velocity V, fluid density ρ, and fluid viscosity μ. The number of expected nondimensional parameters Πs for this problem is
(a) 5   (b) 4   (c) 3   (d) 2   (e) 1

7–136 Consider unsteady fully developed Coutte flow-flow between two infinite parallel plates. This problem involves the following parameters: velocity component u, distance between the plates h, vertical distance y, top plate speed V, fluid density ρ, fluid viscosity μ, and time t. The number of expected nondimensional parameters Πs for this problem is
(*a*) 6   (*b*) 5   (*c*) 4   (*d*) 3   (*e*) 2

7–137 Consider a boundary layer growing along a thin flat plate. This problem involves the following parameters: boundary layer thickness δ, downstream distance x, free-stream velocity V, fluid density ρ, and fluid viscosity μ. The number of primary dimensions represented in this problem is
(*a*) 1   (*b*) 2   (*c*) 3   (*d*) 4   (*e*) 5

7–138 Consider a boundary layer growing along a thin flat plate. This problem involves the following parameters: boundary layer thickness δ, downstream distance x, free-stream velocity V, fluid density ρ, and fluid viscosity μ. The dependent parameter is δ. If we choose three repeating parameters as x, ρ, and V, the dependent Π is
(*a*) $\delta x^2/V$ (*b*) $\delta V^2/x\rho$ (*c*) $\delta\rho/xV$ (*d*) $x/\delta V$ (*e*) δ/x

INTERNAL FLOW

Fluid flow is classified as *external* or *internal*, depending on whether the fluid is forced to flow over a surface or in a conduit. Internal and external flows exhibit very different characteristics. In this chapter we consider *internal flow* where the conduit is completely filled with the fluid, and the flow is driven primarily by a pressure difference. This should not be confused with *open-channel flow* (Chap. 13) where the conduit is partially filled by the fluid and thus the flow is partially bounded by solid surfaces, as in an irrigation ditch, and the flow is driven by gravity alone.

We start this chapter with a general physical description of internal flow through pipes and ducts including the *entrance region* and the *fully developed* region. We continue with a discussion of the dimensionless *Reynolds number* and its physical significance. We then introduce the *pressure drop* correlations associated with pipe flow for both laminar and turbulent flows. Then, we discuss minor losses and determine the pressure drop and pumping power requirements for real-world piping systems. Finally, we present a brief overview of flow measurement devices.

OBJECTIVES

When you finish reading this chapter, you should be able to

- Have a deeper understanding of laminar and turbulent flow in pipes and the analysis of fully developed flow

- Calculate the major and minor losses associated with pipe flow in piping networks and determine the pumping power requirements

- Understand various velocity and flow rate measurement techniques and learn their advantages and disadvantages

Internal flows through pipes, elbows, tees, valves, etc., as in this oil refinery, are found in nearly every industry.
Royalty Free/CORBIS

FIGURE 8–1
Circular pipes can withstand large pressure differences between the inside and the outside without undergoing any significant distortion, but noncircular pipes cannot.

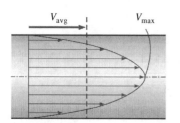

FIGURE 8–2
Average velocity V_{avg} is defined as the average speed through a cross section. For fully developed laminar pipe flow, V_{avg} is half of the maximum velocity.

8–1 · INTRODUCTION

Liquid or gas flow through *pipes* or *ducts* is commonly used in heating and cooling applications and fluid distribution networks. The fluid in such applications is usually forced to flow by a fan or pump through a flow section. We pay particular attention to *friction,* which is directly related to the *pressure drop* and *head loss* during flow through pipes and ducts. The pressure drop is then used to determine the pumping power requirement. A typical piping system involves pipes of different diameters connected to each other by various fittings or elbows to route the fluid, valves to control the flow rate, and pumps to pressurize the fluid.

The terms *pipe, duct,* and *conduit* are usually used interchangeably for flow sections. In general, flow sections of circular cross section are referred to as *pipes* (especially when the fluid is a liquid), and flow sections of noncircular cross section as *ducts* (especially when the fluid is a gas). Small-diameter pipes are usually referred to as *tubes.* Given this uncertainty, we will use more descriptive phrases (such as *a circular pipe* or *a rectangular duct*) whenever necessary to avoid any misunderstandings.

You have probably noticed that most fluids, especially liquids, are transported in *circular pipes.* This is because pipes with a circular cross section can withstand large pressure differences between the inside and the outside without undergoing significant distortion. *Noncircular pipes* are usually used in applications such as the heating and cooling systems of buildings where the pressure difference is relatively small, the manufacturing and installation costs are lower, and the available space is limited for ductwork (Fig. 8–1).

Although the theory of fluid flow is reasonably well understood, theoretical solutions are obtained only for a few simple cases such as fully developed laminar flow in a circular pipe. Therefore, we must rely on experimental results and empirical relations for most fluid flow problems rather than closed-form analytical solutions. Noting that the experimental results are obtained under carefully controlled laboratory conditions and that no two systems are exactly alike, we must not be so naive as to view the results obtained as "exact." An error of 10 percent (or more) in friction factors calculated using the relations in this chapter is the "norm" rather than the "exception."

The fluid velocity in a pipe changes from *zero* at the wall because of the no-slip condition to a maximum at the pipe center. In fluid flow, it is convenient to work with an *average* velocity V_{avg}, which remains constant in incompressible flow when the cross-sectional area of the pipe is constant (Fig. 8–2). The average velocity in heating and cooling applications may change somewhat because of changes in density with temperature. But, in practice, we evaluate the fluid properties at some average temperature and treat them as constants. The convenience of working with constant properties usually more than justifies the slight loss in accuracy.

Also, the friction between the fluid particles in a pipe does cause a slight rise in fluid temperature as a result of the mechanical energy being converted to sensible thermal energy. But this temperature rise due to *frictional heating* is usually too small to warrant any consideration in calculations and thus is disregarded. For example, in the absence of any heat transfer, no noticeable difference can be detected between the inlet and outlet temperatures of water flowing in a pipe. The primary consequence of friction in fluid flow is pressure drop, and thus any significant temperature change in the fluid is due to heat transfer.

The value of the average velocity V_{avg} at some streamwise cross-section is determined from the requirement that the *conservation of mass* principle be satisfied (Fig. 8–2). That is,

$$\dot{m} = \rho V_{avg} A_c = \int_{A_c} \rho u(r) \, dA_c \tag{8-1}$$

where $\dot{m}$ is the mass flow rate, ρ is the density, A_c is the cross-sectional area, and $u(r)$ is the velocity profile. Then the average velocity for incompressible flow in a circular pipe of radius R is expressed as

$$V_{avg} = \frac{\int_{A_c} \rho u(r) \, dA_c}{\rho A_c} = \frac{\int_0^R \rho u(r) 2\pi r \, dr}{\rho \pi R^2} = \frac{2}{R^2} \int_0^R u(r) r \, dr \tag{8-2}$$

Therefore, when we know the flow rate or the velocity profile, the average velocity can be determined easily.

8–2 · LAMINAR AND TURBULENT FLOWS

If you have been around smokers, you probably noticed that the cigarette smoke rises in a smooth plume for the first few centimeters and then starts fluctuating randomly in all directions as it continues its rise. Other plumes behave similarly (Fig. 8–3). Likewise, a careful inspection of flow in a pipe reveals that the fluid flow is streamlined at low velocities but turns chaotic as the velocity is increased above a critical value, as shown in Fig. 8–4. The flow regime in the first case is said to be **laminar,** characterized by *smooth streamlines* and *highly ordered motion,* and **turbulent** in the second case, where it is characterized by *velocity fluctuations* and *highly disordered motion.* The **transition** from laminar to turbulent flow does not occur suddenly; rather, it occurs over some region in which the flow fluctuates between laminar and turbulent flows before it becomes fully turbulent. Most flows encountered in practice are turbulent. Laminar flow is encountered when highly viscous fluids such as oils flow in small pipes or narrow passages.

We can verify the existence of these laminar, transitional, and turbulent flow regimes by injecting some dye streaks into the flow in a glass pipe, as the British engineer Osborne Reynolds (1842–1912) did over a century ago. We observe that the dye streak forms a *straight and smooth line* at low velocities when the flow is laminar (we may see some blurring because of molecular diffusion), has *bursts of fluctuations* in the transitional regime, and *zigzags rapidly and disorderly* when the flow becomes fully turbulent. These zigzags and the dispersion of the dye are indicative of the fluctuations in the main flow and the rapid mixing of fluid particles from adjacent layers.

The *intense mixing* of the fluid in turbulent flow as a result of rapid fluctuations enhances momentum transfer between fluid particles, which increases the friction force on the pipe wall and thus the required pumping power. The friction factor reaches a maximum when the flow becomes fully turbulent.

FIGURE 8–3
Laminar and turbulent flow regimes of a candle smoke plume.

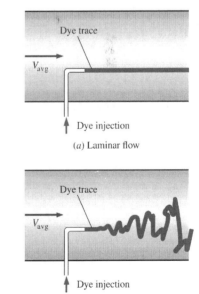

FIGURE 8–4
The behavior of colored fluid injected into the flow in (*a*) laminar and (*b*) turbulent flow in a pipe.

FIGURE 8–5
The Reynolds number can be viewed as the ratio of inertial forces to viscous forces acting on a fluid element.

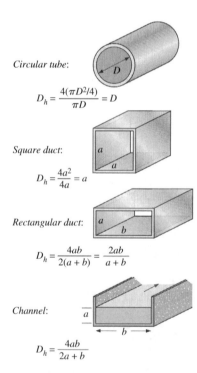

FIGURE 8–6
The hydraulic diameter $D_h = 4A_c/p$ is defined such that it reduces to ordinary diameter for circular tubes. When there is a free surface, such as in open-channel flow, the wetted perimeter includes only the walls in contact with the fluid.

Reynolds Number

The transition from laminar to turbulent flow depends on the *geometry, surface roughness, flow velocity, surface temperature,* and *type of fluid,* among other things. After exhaustive experiments in the 1880s, Osborne Reynolds discovered that the flow regime depends mainly on the ratio of *inertial forces* to *viscous forces* in the fluid (Fig. 8–5). This ratio is called the **Reynolds number** and is expressed for internal flow in a circular pipe as

$$\text{Re} = \frac{\text{Inertial forces}}{\text{Viscous forces}} = \frac{V_{avg}D}{\nu} = \frac{\rho V_{avg}D}{\mu} \qquad (8\text{–}3)$$

where V_{avg} = average flow velocity (m/s), D = characteristic length of the geometry (diameter in this case, in m), and $\nu = \mu/\rho$ = kinematic viscosity of the fluid (m^2/s). Note that the Reynolds number is a *dimensionless* quantity (Chap. 7). Also, kinematic viscosity has units m^2/s, and can be viewed as *viscous diffusivity* or *diffusivity for momentum.*

At large Reynolds numbers, the inertial forces, which are proportional to the fluid density and the square of the fluid velocity, are large relative to the viscous forces, and thus the viscous forces cannot prevent the random and rapid fluctuations of the fluid. At *small* or *moderate* Reynolds numbers, however, the viscous forces are large enough to suppress these fluctuations and to keep the fluid "in line." Thus the flow is *turbulent* in the first case and *laminar* in the second.

The Reynolds number at which the flow becomes turbulent is called the **critical Reynolds number,** Re_{cr}. The value of the critical Reynolds number is different for different geometries and flow conditions. For internal flow in a circular pipe, the generally accepted value of the critical Reynolds number is $\text{Re}_{cr} = 2300$.

For flow through noncircular pipes, the Reynolds number is based on the **hydraulic diameter** D_h defined as (Fig. 8–6)

Hydraulic diameter: $$D_h = \frac{4A_c}{p} \qquad (8\text{–}4)$$

where A_c is the cross-sectional area of the pipe and p is its wetted perimeter. The hydraulic diameter is defined such that it reduces to ordinary diameter D for circular pipes,

Circular pipes: $$D_h = \frac{4A_c}{p} = \frac{4(\pi D^2/4)}{\pi D} = D$$

It certainly is desirable to have precise values of Reynolds numbers for laminar, transitional, and turbulent flows, but this is not the case in practice. It turns out that the transition from laminar to turbulent flow also depends on the degree of disturbance of the flow by *surface roughness, pipe vibrations,* and *fluctuations in the upstream flow.* Under most practical conditions, the flow in a circular pipe is laminar for Re $\lesssim$ 2300, turbulent for Re $\gtrsim$ 4000, and transitional in between. That is,

$$\text{Re} \lesssim 2300 \qquad \text{laminar flow}$$
$$2300 \lesssim \text{Re} \lesssim 4000 \qquad \text{transitional flow}$$
$$\text{Re} \gtrsim 4000 \qquad \text{turbulent flow}$$

In transitional flow, the flow switches between laminar and turbulent in a disorderly fashion (Fig. 8–7). It should be kept in mind that laminar flow can be maintained at much higher Reynolds numbers in very smooth pipes by avoiding flow disturbances and pipe vibrations. In such carefully controlled laboratory experiments, laminar flow has been maintained at Reynolds numbers of up to 100,000.

8–3 · THE ENTRANCE REGION

Consider a fluid entering a circular pipe at a uniform velocity. Because of the no-slip condition, the fluid particles in the layer in contact with the wall of the pipe come to a complete stop. This layer also causes the fluid particles in the adjacent layers to slow down gradually as a result of friction. To make up for this velocity reduction, the velocity of the fluid at the midsection of the pipe has to increase to keep the mass flow rate through the pipe constant. As a result, a velocity gradient develops along the pipe.

The region of the flow in which the effects of the viscous shearing forces caused by fluid viscosity are felt is called the **velocity boundary layer** or just the **boundary layer.** The hypothetical boundary surface divides the flow in a pipe into two regions: the **boundary layer region,** in which the viscous effects and the velocity changes are significant, and the **irrotational (core) flow region,** in which the frictional effects are negligible and the velocity remains essentially constant in the radial direction.

The thickness of this boundary layer increases in the flow direction until the boundary layer reaches the pipe center and thus fills the entire pipe, as shown in Fig. 8–8, and the velocity becomes fully developed a little farther downstream. The region from the pipe inlet to the point at which the velocity profile is fully developed is called the **hydrodynamic entrance region,** and the length of this region is called the **hydrodynamic entry length** L_h. Flow in the entrance region is called *hydrodynamically developing flow* since this is the region where the velocity profile develops. The region beyond the entrance region in which the velocity profile is fully developed and remains unchanged is called the **hydrodynamically fully developed region.** The flow is said to be **fully developed** when the normalized temperature profile remains unchanged as well. Hydrodynamically fully developed flow is equivalent to fully developed flow when the fluid in the pipe is not heated or cooled since the fluid temperature in this case remains

FIGURE 8–7
In the transitional flow region of $2300 \leq \mathrm{Re} \leq 4000$, the flow switches between laminar and turbulent somewhat randomly.

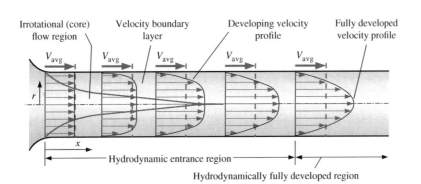

FIGURE 8–8
The development of the velocity boundary layer in a pipe. (The developed average velocity profile is parabolic in laminar flow, as shown, but much flatter or fuller in turbulent flow.)

FIGURE 8–9
In the fully developed flow region of a pipe, the velocity profile does not change downstream, and thus the wall shear stress remains constant as well.

essentially constant throughout. The velocity profile in the fully developed region is *parabolic* in laminar flow and much *flatter* (or *fuller*) in turbulent flow due to eddy motion and more vigorous mixing in the radial direction. The time-averaged velocity profile remains unchanged when the flow is fully developed, and thus

Hydrodynamically fully developed: $\dfrac{\partial u(r, x)}{\partial x} = 0 \rightarrow u = u(r)$ **(8–5)**

The shear stress at the pipe wall τ_w is related to the slope of the velocity profile at the surface. Noting that the velocity profile remains unchanged in the hydrodynamically fully developed region, the wall shear stress also remains constant in that region (Fig. 8–9).

Consider fluid flow in the hydrodynamic entrance region of a pipe. The wall shear stress is the *highest* at the pipe inlet where the thickness of the boundary layer is smallest, and decreases gradually to the fully developed value, as shown in Fig. 8–10. Therefore, the pressure drop is *higher* in the entrance regions of a pipe, and the effect of the entrance region is always to *increase* the average friction factor for the entire pipe. This increase may be significant for short pipes but is negligible for long ones.

Entry Lengths

The hydrodynamic entry length is usually taken to be the distance from the pipe entrance to where the wall shear stress (and thus the friction factor) reaches within about 2 percent of the fully developed value. In *laminar flow*, the nondimensional hydrodynamic entry length is given approximately as [see Kays and Crawford (2004) and Shah and Bhatti (1987)]

$$\frac{L_{h,\,\text{laminar}}}{D} \cong 0.05\text{Re}$$ **(8–6)**

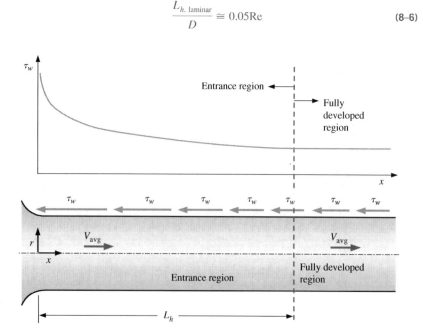

FIGURE 8–10
The variation of wall shear stress in the flow direction for flow in a pipe from the entrance region into the fully developed region.

For Re = 20, the hydrodynamic entry length is about the size of the diameter, but increases linearly with velocity. In the limiting laminar case of Re = 2300, the hydrodynamic entry length is 115D.

In *turbulent flow,* the intense mixing during random fluctuations usually overshadows the effects of molecular diffusion. The nondimensional hydrodynamic entry length for turbulent flow is approximated as [see Bhatti and Shah (1987) and Zhi-qing (1982)]

$$\frac{L_{h, \text{ turbulent}}}{D} = 1.359 \text{Re}^{1/4} \tag{8–7}$$

The entry length is much shorter in turbulent flow, as expected, and its dependence on the Reynolds number is weaker. In many pipe flows of practical engineering interest, the entrance effects become insignificant beyond a pipe length of about 10 diameters, and the nondimensional hydrodynamic entry length is approximated as

$$\frac{L_{h, \text{ turbulent}}}{D} \approx 10 \tag{8–8}$$

Precise correlations for calculating the frictional head losses in entrance regions are available in the literature. However, the pipes used in practice are usually several times the length of the entrance region, and thus the flow through the pipes is often assumed to be fully developed for the entire length of the pipe. This simplistic approach gives *reasonable* results for long pipes but sometimes poor results for short ones since it underpredicts the wall shear stress and thus the friction factor.

8–4 ▪ LAMINAR FLOW IN PIPES

We mentioned in Section 8–2 that flow in pipes is laminar for Re ≲ 2300, and that the flow is fully developed if the pipe is sufficiently long (relative to the entry length) so that the entrance effects are negligible. In this section, we consider the steady, laminar, incompressible flow of fluid with constant properties in the fully developed region of a straight circular pipe. We obtain the momentum equation by applying a momentum balance to a differential volume element, and we obtain the velocity profile by solving it. Then we use it to obtain a relation for the friction factor. An important aspect of the analysis here is that it is one of the few available for viscous flow.

In fully developed laminar flow, each fluid particle moves at a constant axial velocity along a streamline and the velocity profile $u(r)$ remains unchanged in the flow direction. There is no motion in the radial direction, and thus the velocity component in the direction normal to the pipe axis is everywhere zero. There is no acceleration since the flow is steady and fully developed.

Now consider a ring-shaped differential volume element of radius r, thickness dr, and length dx oriented coaxially with the pipe, as shown in Fig. 8–11. The volume element involves only pressure and viscous effects and thus the pressure and shear forces must balance each other. The pressure force acting on a submerged plane surface is the product of the pressure at the centroid of the surface and the surface area. A force balance on the volume element in the flow direction gives

$$(2\pi r \, dr \, P)_x - (2\pi r \, dr \, P)_{x+dx} + (2\pi r \, dx \, \tau)_r - (2\pi r \, dx \, \tau)_{r+dr} = 0 \tag{8–9}$$

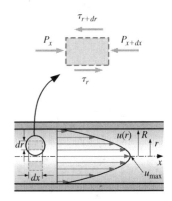

FIGURE 8–11

Free-body diagram of a ring-shaped differential fluid element of radius r, thickness dr, and length dx oriented coaxially with a horizontal pipe in fully developed laminar flow. (The size of the fluid element is greatly exaggerated for clarity.)

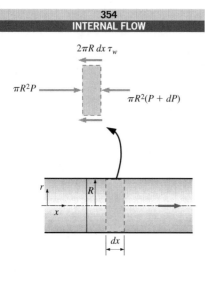

Force balance:

$$\pi R^2 P - \pi R^2 (P + dP) - 2\pi R\, dx\, \tau_w = 0$$

Simplifying:

$$\frac{dP}{dx} = -\frac{2\tau_w}{R}$$

FIGURE 8–12

Free-body diagram of a fluid disk element of radius R and length dx in fully developed laminar flow in a horizontal pipe.

which indicates that in fully developed flow in a horizontal pipe, the viscous and pressure forces balance each other. Dividing by $2\pi dr dx$ and rearranging,

$$r\frac{P_{x+dx} - P_x}{dx} + \frac{(r\tau)_{r+dr} - (r\tau)_r}{dr} = 0 \qquad (8\text{–}10)$$

Taking the limit as $dr, dx \rightarrow 0$ gives

$$r\frac{dP}{dx} + \frac{d(r\tau)}{dr} = 0 \qquad (8\text{–}11)$$

Substituting $\tau = -\mu(du/dr)$, dividing by r, and taking $\mu = $ constant gives the desired equation,

$$\frac{\mu}{r}\frac{d}{dr}\left(r\frac{du}{dr}\right) = \frac{dP}{dx} \qquad (8\text{–}12)$$

The quantity du/dr is negative in pipe flow, and the negative sign is included to obtain positive values for τ. (Or, $du/dr = -du/dy$ if we define $y = R - r$.) The left side of Eq. 8–12 is a function of r, and the right side is a function of x. The equality must hold for any value of r and x, and an equality of the form $f(r) = g(x)$ can be satisfied only if both $f(r)$ and $g(x)$ are equal to the same constant. Thus we conclude that $dP/dx = $ constant. This is verified by writing a force balance on a volume element of radius R and thickness dx (a slice of the pipe as in Fig. 8–12), which gives

$$\frac{dP}{dx} = -\frac{2\tau_w}{R} \qquad (8\text{–}13)$$

Here τ_w is constant since the viscosity and the velocity profile are constants in the fully developed region. Therefore, $dP/dx = $ constant.

Equation 8–12 is solved by rearranging and integrating it twice to give

$$u(r) = \frac{r^2}{4\mu}\left(\frac{dP}{dx}\right) + C_1 \ln r + C_2 \qquad (8\text{–}14)$$

The velocity profile $u(r)$ is obtained by applying the boundary conditions $\partial u/\partial r = 0$ at $r = 0$ (because of symmetry about the centerline) and $u = 0$ at $r = R$ (the no-slip condition at the pipe wall),

$$u(r) = -\frac{R^2}{4\mu}\left(\frac{dP}{dx}\right)\left(1 - \frac{r^2}{R^2}\right) \qquad (8\text{–}15)$$

Therefore, the velocity profile in fully developed laminar flow in a pipe is *parabolic* with a maximum at the centerline and a minimum (zero) at the pipe wall. Also, the axial velocity u is positive for any r, and thus the axial pressure gradient dP/dx must be negative (i.e., pressure must decrease in the flow direction because of viscous effects—it takes pressure to push the fluid through the pipe).

The average velocity is determined from its definition by substituting Eq. 8–15 into Eq. 8–2, and performing the integration, yielding

$$V_{avg} = \frac{2}{R^2}\int_0^R u(r)r\,dr = \frac{-2}{R^2}\int_0^R \frac{R^2}{4\mu}\left(\frac{dP}{dx}\right)\left(1 - \frac{r^2}{R^2}\right)r\,dr = -\frac{R^2}{8\mu}\left(\frac{dP}{dx}\right) \qquad (8\text{–}16)$$

Combining the last two equations, the velocity profile is rewritten as

$$u(r) = 2V_{avg}\left(1 - \frac{r^2}{R^2}\right) \qquad (8\text{–}17)$$

This is a convenient form for the velocity profile since V_{avg} can be determined easily from the flow rate information.

The maximum velocity occurs at the centerline and is determined from Eq. 8–17 by substituting $r = 0$,

$$u_{max} = 2V_{avg} \tag{8–18}$$

Therefore, *the average velocity in fully developed laminar pipe flow is one-half of the maximum velocity.*

Pressure Drop and Head Loss

A quantity of interest in the analysis of pipe flow is the *pressure drop* ΔP since it is directly related to the power requirements of the fan or pump to maintain flow. We note that dP/dx = constant, and integrating from $x = x_1$ where the pressure is P_1 to $x = x_1 + L$ where the pressure is P_2 gives

$$\frac{dP}{dx} = \frac{P_2 - P_1}{L} \tag{8–19}$$

Substituting Eq. 8–19 into the V_{avg} expression in Eq. 8–16, the pressure drop is expressed as

Laminar flow: $\quad \Delta P = P_1 - P_2 = \dfrac{8\mu L V_{avg}}{R^2} = \dfrac{32\mu L V_{avg}}{D^2} \tag{8–20}$

The symbol Δ is typically used to indicate the difference between the final and initial values, like $\Delta y = y_2 - y_1$. But in fluid flow, ΔP is used to designate pressure drop, and thus it is $P_1 - P_2$. A pressure drop due to viscous effects represents an irreversible pressure loss, and it is sometimes called **pressure loss** ΔP_L to emphasize that it is a *loss* (just like the head loss h_L, which as we shall see is proportional to ΔP_L.)

Note from Eq. 8–20 that the pressure drop is proportional to the viscosity μ of the fluid, and ΔP would be zero if there were no friction. Therefore, the drop of pressure from P_1 to P_2 in this case is due entirely to viscous effects, and Eq. 8–20 represents the pressure loss ΔP_L when a fluid of viscosity μ flows through a pipe of constant diameter D and length L at average velocity V_{avg}.

In practice, it is convenient to express the pressure loss for all types of fully developed internal flows (laminar or turbulent flows, circular or noncircular pipes, smooth or rough surfaces, horizontal or inclined pipes) as (Fig. 8–13)

Pressure loss: $\quad \Delta P_L = f \dfrac{L}{D} \dfrac{\rho V_{avg}^2}{2} \tag{8–21}$

where $\rho V_{avg}^2/2$ is the *dynamic pressure* and f is the **Darcy friction factor,**

$$f = \frac{8\tau_w}{\rho V_{avg}^2} \tag{8–22}$$

It is also called the **Darcy–Weisbach friction factor,** named after the Frenchman Henry Darcy (1803–1858) and the German Julius Weisbach (1806–1871), the two engineers who provided the greatest contribution to its development. It should not be confused with the *friction coefficient* C_f

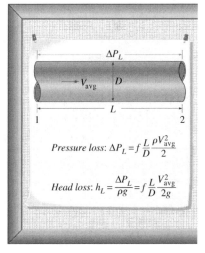

FIGURE 8–13

The relation for pressure loss (and head loss) is one of the most general relations in fluid mechanics, and it is valid for laminar or turbulent flows, circular or noncircular pipes, and pipes with smooth or rough surfaces.

[also called the *Fanning friction factor*, named after the American engineer John Fanning (1837–1911)], which is defined as $C_f = 2\tau_w/(\rho V_{avg}^2) = f/4$.

Setting Eqs. 8–20 and 8–21 equal to each other and solving for f gives the friction factor for fully developed laminar flow in a circular pipe,

Circular pipe, laminar:
$$f = \frac{64\mu}{\rho D V_{avg}} = \frac{64}{\text{Re}} \qquad (8\text{–}23)$$

This equation shows that *in laminar flow, the friction factor is a function of the Reynolds number only and is independent of the roughness of the pipe surface (assuming, of course, that the roughness is not extreme).*

In the analysis of piping systems, pressure losses are commonly expressed in terms of the *equivalent fluid column height*, called the **head loss** h_L. Noting from fluid statics that $\Delta P = \rho gh$ and thus a pressure difference of ΔP corresponds to a fluid height of $h = \Delta P/\rho g$, the *pipe head loss* is obtained by dividing ΔP_L by ρg to give

Head loss:
$$h_L = \frac{\Delta P_L}{\rho g} = f \frac{L}{D} \frac{V_{avg}^2}{2g} \qquad (8\text{–}24)$$

The head loss h_L represents *the additional height that the fluid needs to be raised by a pump in order to overcome the frictional losses in the pipe.* The head loss is caused by viscosity, and it is directly related to the wall shear stress. Equations 8–21 and 8–24 are valid for both laminar and turbulent flows in both circular and noncircular pipes, but Eq. 8–23 is valid only for fully developed laminar flow in circular pipes.

Once the pressure loss (or head loss) is known, the required pumping power *to overcome the pressure loss* is determined from

$$\dot{W}_{pump,L} = \dot{V}\,\Delta P_L = \dot{V}\rho g h_L = \dot{m} g h_L \qquad (8\text{–}25)$$

where $\dot{V}$ is the volume flow rate and $\dot{m}$ is the mass flow rate.

The average velocity for laminar flow in a horizontal pipe is, from Eq. 8–20,

Horizontal pipe:
$$V_{avg} = \frac{(P_1 - P_2)R^2}{8\mu L} = \frac{(P_1 - P_2)D^2}{32\mu L} = \frac{\Delta P\, D^2}{32\mu L} \qquad (8\text{–}26)$$

Then the volume flow rate for laminar flow through a horizontal pipe of diameter D and length L becomes

$$\dot{V} = V_{avg}A_c = \frac{(P_1 - P_2)R^2}{8\mu L}\pi R^2 = \frac{(P_1 - P_2)\pi D^4}{128\mu L} = \frac{\Delta P\,\pi D^4}{128\mu L} \qquad (8\text{–}27)$$

This equation is known as **Poiseuille's law,** and this flow is called *Hagen–Poiseuille flow* in honor of the works of G. Hagen (1797–1884) and J. Poiseuille (1799–1869) on the subject. Note from Eq. 8–27 that *for a specified flow rate, the pressure drops and thus the required pumping power is proportional to the length of the pipe and the viscosity of the fluid, but it is inversely proportional to the fourth power of the radius (or diameter) of the pipe.* Therefore, the pumping power requirement for a laminar-flow piping system can be reduced by a factor of 16 by doubling the pipe diameter (Fig. 8–14). Of course the benefits of the reduction in the energy costs must be weighed against the increased cost of construction due to using a larger-diameter pipe.

The pressure drop ΔP equals the pressure loss ΔP_L in the case of a horizontal pipe, but this is not the case for inclined pipes or pipes with variable cross-sectional area. This can be demonstrated by writing the energy

FIGURE 8–14
The pumping power requirement for a laminar-flow piping system can be reduced by a factor of 16 by doubling the pipe diameter.

equation for steady, incompressible one-dimensional flow in terms of heads as (see Chap. 5)

$$\frac{P_1}{\rho g} + \alpha_1 \frac{V_1^2}{2g} + z_1 + h_{\text{pump}, u} = \frac{P_2}{\rho g} + \alpha_2 \frac{V_2^2}{2g} + z_2 + h_{\text{turbine}, e} + h_L \qquad \text{(8–28)}$$

where $h_{\text{pump}, u}$ is the useful pump head delivered to the fluid, $h_{\text{turbine}, e}$ is the turbine head extracted from the fluid, h_L is the irreversible head loss between sections 1 and 2, V_1 and V_2 are the average velocities at sections 1 and 2, respectively, and α_1 and α_2 are the *kinetic energy correction factors* at sections 1 and 2 (it can be shown that $\alpha = 2$ for fully developed laminar flow and about 1.05 for fully developed turbulent flow). Equation 8–28 can be rearranged as

$$P_1 - P_2 = \rho(\alpha_2 V_2^2 - \alpha_1 V_1^2)/2 + \rho g[(z_2 - z_1) + h_{\text{turbine}, e} - h_{\text{pump}, u} + h_L] \qquad \text{(8–29)}$$

Therefore, the pressure drop $\Delta P = P_1 - P_2$ and pressure loss $\Delta P_L = \rho g h_L$ for a given flow section are equivalent if (1) the flow section is horizontal so that there are no hydrostatic or gravity effects ($z_1 = z_2$), (2) the flow section does not involve any work devices such as a pump or a turbine since they change the fluid pressure ($h_{\text{pump}, u} = h_{\text{turbine}, e} = 0$), (3) the cross-sectional area of the flow section is constant and thus the average flow velocity is constant ($V_1 = V_2$), and (4) the velocity profiles at sections 1 and 2 are the same shape ($\alpha_1 = \alpha_2$).

Effect of Gravity on Velocity and Flow Rate in Laminar Flow

Gravity has no affect on flow in horizontal pipes, but it has a significant effect on both the velocity and the flow rate in uphill or downhill pipes. Relations for inclined pipes can be obtained in a similar manner from a force balance in the direction of flow. The only additional force in this case is the component of the fluid weight in the flow direction, whose magnitude is

$$W_x = W \sin\theta = \rho g V_{\text{element}} \sin\theta = \rho g(2\pi r\, dr\, dx) \sin\theta \qquad \text{(8–30)}$$

where θ is the angle between the horizontal and the flow direction (Fig. 8–15). The force balance in Eq. 8–9 now becomes

$$(2\pi r\, dr\, P)_x - (2\pi r\, dr\, P)_{x+dx} + (2\pi r\, dx\, \tau)_r$$
$$- (2\pi r\, dx\, \tau)_{r+dr} - \rho g(2\pi r\, dr\, dx)\sin\theta = 0 \qquad \text{(8–31)}$$

which results in the differential equation

$$\frac{\mu}{r}\frac{d}{dr}\left(r\frac{du}{dr}\right) = \frac{dP}{dx} + \rho g \sin\theta \qquad \text{(8–32)}$$

Following the same solution procedure as previously, the velocity profile is

$$u(r) = -\frac{R^2}{4\mu}\left(\frac{dP}{dx} + \rho g \sin\theta\right)\left(1 - \frac{r^2}{R^2}\right) \qquad \text{(8–33)}$$

From Eq. 8–33, the *average velocity* and the *volume flow rate* relations for laminar flow through inclined pipes are, respectively,

$$V_{\text{avg}} = \frac{(\Delta P - \rho g L \sin\theta)D^2}{32\mu L} \quad \text{and} \quad \dot{V} = \frac{(\Delta P - \rho g L \sin\theta)\pi D^4}{128\mu L} \qquad \text{(8–34)}$$

which are identical to the corresponding relations for horizontal pipes, except that ΔP is replaced by $\Delta P - \rho g L \sin\theta$. Therefore, the results already obtained for horizontal pipes can also be used for inclined pipes provided

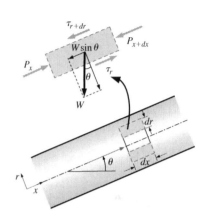

FIGURE 8–15
Free-body diagram of a ring-shaped differential fluid element of radius r, thickness dr, and length dx oriented coaxially with an inclined pipe in fully developed laminar flow.

Laminar Flow in Circular Pipes

(Fully developed flow with no pump or turbine in the flow section, and

$$\Delta P = P_1 - P_2)$$

Horizontal pipe: $\dot{V} = \dfrac{\Delta P\, \pi D^4}{128\mu L}$

Inclined pipe: $\dot{V} = \dfrac{(\Delta P - \rho g L \sin\theta)\pi D^4}{128\mu L}$

Uphill flow: $\theta > 0$ and $\sin\theta > 0$

Downhill flow: $\theta < 0$ and $\sin\theta < 0$

FIGURE 8–16
The relations developed for fully developed laminar flow through horizontal pipes can also be used for inclined pipes by replacing ΔP with $\Delta P - \rho g L \sin\theta$.

that ΔP is replaced by $\Delta P - \rho g L \sin \theta$ (Fig. 8–16). Note that $\theta > 0$ and thus $\sin \theta > 0$ for uphill flow, and $\theta < 0$ and thus $\sin \theta < 0$ for downhill flow.

In inclined pipes, the combined effect of pressure difference and gravity drives the flow. Gravity helps downhill flow but opposes uphill flow. Therefore, much greater pressure differences need to be applied to maintain a specified flow rate in uphill flow although this becomes important only for liquids, because the density of gases is generally low. In the special case of *no flow* ($\dot{V} = 0$), Eq. 8–34 yields $\Delta P = \rho g L \sin \theta$, which is what we would obtain from fluid statics (Chap. 3).

Laminar Flow in Noncircular Pipes

The friction factor f relations are given in Table 8–1 for *fully developed laminar flow* in pipes of various cross sections. The Reynolds number for flow in these pipes is based on the hydraulic diameter $D_h = 4A_c/p$, where A_c is the cross-sectional area of the pipe and p is its wetted perimeter.

TABLE 8–1

Friction factor for fully developed *laminar flow* in pipes of various cross sections ($D_h = 4A_c/p$ and $\text{Re} = V_{\text{avg}} D_h / \nu$)

Tube Geometry	a/b or $\theta°$	Friction Factor f
Circle	—	64.00/Re
Rectangle	a/b	
	1	56.92/Re
	2	62.20/Re
	3	68.36/Re
	4	72.92/Re
	6	78.80/Re
	8	82.32/Re
	∞	96.00/Re
Ellipse	a/b	
	1	64.00/Re
	2	67.28/Re
	4	72.96/Re
	8	76.60/Re
	16	78.16/Re
Isosceles triangle	θ	
	10°	50.80/Re
	30°	52.28/Re
	60°	53.32/Re
	90°	52.60/Re
	120°	50.96/Re

EXAMPLE 8–1   Laminar Flow in Horizontal and Inclined Pipes

Consider the fully developed flow of glycerin at 40°C through a 70-m-long, 4-cm-diameter, horizontal, circular pipe. If the flow velocity at the centerline is measured to be 6 m/s, determine the velocity profile and the pressure difference across this 70-m-long section of the pipe, and the useful pumping power required to maintain this flow. For the same useful pumping power input, determine the percent increase of the flow rate if the pipe is inclined 15° downward and the percent decrease if it is inclined 15° upward. The pump is located outside this pipe section.

FIGURE 8–17
Schematic for Example 8–1.

SOLUTION   The centerline velocity in a horizontal pipe in fully developed flow is measured. The velocity profile, the pressure difference across the pipe, and the pumping power required are to be determined. The effects of downward and upward tilting of the pipe on the flow rate is to be investigated.
Assumptions   **1** The flow is steady, laminar, incompressible, and fully developed. **2** There are no pumps or turbines in the flow section. **3** There are no valves, elbows, or other devices that may cause local losses.
Properties   The density and dynamic viscosity of glycerin at 40°C are $\rho = 1252 \text{ kg/m}^3$ and $\mu = 0.3073 \text{ kg/m·s}$, respectively.
Analysis   The velocity profile in fully developed laminar flow in a circular pipe is expressed as

$$u(r) = u_{\text{max}}\left(1 - \frac{r^2}{R^2}\right)$$

Substituting, the velocity profile is determined to be

$$u(r) = (6 \text{ m/s})\left(1 - \frac{r^2}{(0.02 \text{ m})^2}\right) = 6(1 - 2500r^2)$$

where u is in m/s and r is in m. The average velocity, the flow rate, and the Reynolds number are

$$V = V_{\text{avg}} = \frac{u_{\text{max}}}{2} = \frac{6 \text{ m/s}}{2} = 3 \text{ m/s}$$

$$\dot{V} = V_{\text{avg}}A_c = V(\pi D^2/4) = (3 \text{ m/s})[\pi(0.04 \text{ m})^2/4] = 3.77 \times 10^{-3} \text{ m}^3/\text{s}$$

$$\text{Re} = \frac{\rho V D}{\mu} = \frac{(1252 \text{ kg/m}^3)(3 \text{ m/s})(0.04 \text{ m})}{0.3073 \text{ kg/m·s}} = 488.9$$

which is less than 2300. Therefore, the flow is indeed laminar. Then the friction factor and the head loss become

$$f = \frac{64}{\text{Re}} = \frac{64}{488.9} = 0.1309$$

$$h_L = f\frac{L V^2}{D\, 2g} = 0.1309\frac{(70 \text{ m})}{(0.04 \text{ m})}\frac{(3 \text{ m/s})^2}{2(9.81 \text{ m/s}^2)} = 105.1\text{m}$$

The energy balance for steady, incompressible one-dimensional flow is given by Eq. 8–28 as

$$\frac{P_1}{\rho g} + \alpha_1\frac{V_1^2}{2g} + z_1 + h_{\text{pump, u}} = \frac{P_2}{\rho g} + \alpha_2\frac{V_2^2}{2g} + z_2 + h_{\text{turbine, e}} + h_L$$

For fully developed flow in a constant diameter pipe with no pumps or turbines, it reduces to

$$\Delta P = P_1 - P_2 = \rho g(z_2 - z_1 + h_L)$$

Then the pressure difference and the required useful pumping power for the horizontal case become

$$\Delta P = \rho g(z_2 - z_1 + h_L)$$

$$= (1252 \text{ kg/m}^3)(9.81 \text{ m/s}^2)(0 + 105.1 \text{ m})\left(\frac{1 \text{ kPa}}{1000 \text{ kg/m·s}^2}\right)$$

$$= 1291 \text{ kPa}$$

$$\dot{W}_{\text{pump, u}} = \dot{V}\Delta P = (3.77 \times 10^3 \text{ m}^3/\text{s})(1291 \text{ kPa})\left(\frac{1 \text{ kW}}{\text{kPa·m}^3/\text{s}}\right) = \mathbf{4.87 \text{ kW}}$$

The elevation difference and the pressure difference for a pipe inclined upwards 15° is

$$\Delta z = z_2 - z_1 = L\sin 15° = (70 \text{ m})\sin 15° = 18.1 \text{ m}$$

$$\Delta P_{\text{upward}} = (1252 \text{ kg/m}^3)(9.81 \text{ m/s}^2)(18.1 \text{ m} + 105.1 \text{ m})\left(\frac{1 \text{ kPa}}{1000 \text{ kg/m·s}^2}\right)$$

$$= 1366 \text{ kPa}$$

Then the flow rate through the upward inclined pipe becomes

$$\dot{V}_{\text{upward}} = \frac{\dot{W}_{\text{pump, u}}}{\Delta P_{\text{upward}}} = \frac{4.87 \text{ kW}}{1366 \text{ kPa}}\left(\frac{1 \text{ kPa·m}^3/\text{s}}{1 \text{ kW}}\right) = 3.57 \times 10^{-3} \text{ m}^3/\text{s}$$

which is a decrease of **5.6 percent** in flow rate. It can be shown similarly that when the pipe is inclined 15° downward from the horizontal, the flow rate will increase by **5.6 percent.**

Discussion Note that the flow is driven by the combined effect of pumping power and gravity. As expected, gravity opposes uphill flow, enhances downhill flow, and has no effect on horizontal flow. Downhill flow can occur even in the absence of a pressure difference applied by a pump. For the case of $P_1 = P_2$ (i.e., no applied pressure difference), the pressure throughout the entire pipe would remain constant, and the fluid would flow through the pipe under the influence of gravity at a rate that depends on the angle of inclination, reaching its maximum value when the pipe is vertical. When solving pipe flow problems, it is always a good idea to calculate the Reynolds number to verify the flow regime—laminar or turbulent.

FIGURE 8–18
Schematic for Example 8–2.

EXAMPLE 8–2   Pressure Drop and Head Loss in a Pipe

Water at 40°F ($\rho = 62.42 \text{ lbm/ft}^3$ and $\mu = 1.038 \times 10^{-3} \text{ lbm/ft·s}$) is flowing steadily through a 0.12-in- (= 0.010 ft) diameter 30-ft-long horizontal pipe at an average velocity of 3.0 ft/s (Fig. 8–18). Determine (*a*) the head loss, (*b*) the pressure drop, and (*c*) the pumping power requirement to overcome this pressure drop.

SOLUTION The average flow velocity in a pipe is given. The head loss, the pressure drop, and the pumping power are to be determined.

Assumptions **1** The flow is steady and incompressible. **2** The entrance effects are negligible, and thus the flow is fully developed. **3** The pipe involves no components such as bends, valves, and connectors.

Properties The density and dynamic viscosity of water are given to be $\rho = 62.42$ lbm/ft^3 and $\mu = 1.038 \times 10^{-3}$ lbm/ft·s, respectively.

Analysis (a) First we need to determine the flow regime. The Reynolds number is

$$\text{Re} = \frac{\rho V_{\text{avg}} D}{\mu} = \frac{(62.42 \text{ lbm/ft}^3)(3 \text{ ft/s})(0.01 \text{ ft})}{1.038 \times 10^{-3} \text{ lbm/ft·s}} = 1803$$

which is less than 2300. Therefore, the flow is laminar. Then the friction factor and the head loss become

$$f = \frac{64}{\text{Re}} = \frac{64}{1803} = 0.0355$$

$$h_L = f \frac{L}{D} \frac{V_{\text{avg}}^2}{2g} = 0.0355 \frac{30 \text{ ft}}{0.01 \text{ ft}} \frac{(3 \text{ ft/s})^2}{2(32.2 \text{ ft/s}^2)} = \mathbf{14.9 \text{ ft}}$$

(b) Noting that the pipe is horizontal and its diameter is constant, the pressure drop in the pipe is due entirely to the frictional losses and is equivalent to the pressure loss,

$$\Delta P = \Delta P_L = f \frac{L}{D} \frac{\rho V_{\text{avg}}^2}{2} = 0.0355 \frac{30 \text{ ft}}{0.01 \text{ ft}} \frac{(62.42 \text{ lbm/ft}^3)(3 \text{ ft/s})^2}{2} \left(\frac{1 \text{ lbf}}{32.2 \text{ lbm·ft/s}^2} \right)$$

$$= \mathbf{929 \text{ lbf/ft}^2 = 6.45 \text{ psi}}$$

(c) The volume flow rate and the pumping power requirements are

$$\dot{V} = V_{\text{avg}} A_c = V_{\text{avg}}(\pi D^2/4) = (3 \text{ ft/s})[\pi(0.01 \text{ ft})^2/4] = 0.000236 \text{ ft}^3/\text{s}$$

$$\dot{W}_{\text{pump}} = \dot{V} \Delta P = (0.000236 \text{ ft}^3/\text{s})(929 \text{ lbf/ft}^2) \left(\frac{1 \text{ W}}{0.737 \text{ lbf·ft/s}} \right) = \mathbf{0.30 \text{ W}}$$

Therefore, power input in the amount of 0.30 W is needed to overcome the frictional losses in the flow due to viscosity.

Discussion The pressure rise provided by a pump is often listed by a pump manufacturer in units of head (Chap. 14). Thus, the pump in this flow needs to provide 14.9 ft of water head in order to overcome the irreversible head loss.

(a)

(b)

(c)

FIGURE 8–19
Water exiting a tube: (a) laminar flow at low flow rate, (b) turbulent flow at high flow rate, and (c) same as (b) but with a short shutter exposure to capture individual eddies.

Photos by Alex Wouden.

8–5 ■ TURBULENT FLOW IN PIPES

Most flows encountered in engineering practice are turbulent, and thus it is important to understand how turbulence affects wall shear stress. However, turbulent flow is a complex mechanism dominated by fluctuations, and despite tremendous amounts of work done in this area by researchers, turbulent flow still is not fully understood. Therefore, we must rely on experiments and the empirical or semi-empirical correlations developed for various situations.

Turbulent flow is characterized by disorderly and rapid fluctuations of swirling regions of fluid, called **eddies,** throughout the flow (Fig. 8–19). These fluctuations provide an additional mechanism for momentum and energy

(a) Before
turbulence

(b) After
turbulence

FIGURE 8–20
The intense mixing in turbulent flow brings fluid particles at different momentums into close contact and thus enhances momentum transfer.

FIGURE 8–21
Fluctuations of the velocity component u with time at a specified location in turbulent flow.

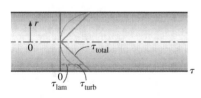

FIGURE 8–22
The velocity profile and the variation of shear stress with radial distance for turbulent flow in a pipe.

transfer. In laminar flow, fluid particles flow in an orderly manner along pathlines, and momentum and energy are transferred across streamlines by molecular diffusion. In turbulent flow, the swirling eddies transport mass, momentum, and energy to other regions of flow much more rapidly than molecular diffusion, greatly enhancing mass, momentum, and heat transfer. As a result, turbulent flow is associated with much higher values of friction, heat transfer, and mass transfer coefficients (Fig. 8–20).

Even when the average flow is steady, the eddy motion in turbulent flow causes significant fluctuations in the values of velocity, temperature, pressure, and even density (in compressible flow). Figure 8–21 shows the variation of the instantaneous velocity component u with time at a specified location, as can be measured with a hot-wire anemometer probe or other sensitive device. We observe that the instantaneous values of the velocity fluctuate about an average value, which suggests that the velocity can be expressed as the sum of an *average value* $\bar{u}$ and a *fluctuating component* u',

$$u = \bar{u} + u' \tag{8–35}$$

This is also the case for other properties such as the velocity component v in the y-direction, and thus $v = \bar{v} + v'$, $P = \bar{P} + P'$, and $T = \bar{T} + T'$. The average value of a property at some location is determined by averaging it over a time interval that is sufficiently large so that the time average levels off to a constant. Therefore, the time average of fluctuating components is zero, e.g., $\bar{u}' = 0$. The magnitude of u' is usually just a few percent of $\bar{u}$, but the high frequencies of eddies (on the order of a thousand per second) make them very effective for the transport of momentum, thermal energy, and mass. In time-averaged *stationary* turbulent flow, the average values of properties (indicated by an overbar) are independent of time. The chaotic fluctuations of fluid particles play a dominant role in pressure drop, and these random motions must be considered in analyses together with the average velocity.

Perhaps the first thought that comes to mind is to determine the shear stress in an analogous manner to laminar flow from $\tau = -\mu \, d\bar{u}/dr$, where $\bar{u}(r)$ is the average velocity profile for turbulent flow. But the experimental studies show that this is not the case, and the effective shear stress is much larger due to the turbulent fluctuations. Therefore, it is convenient to think of the turbulent shear stress as consisting of two parts: the *laminar component*, which accounts for the friction between layers in the flow direction (expressed as $\tau_{\text{lam}} = -\mu \, d\bar{u}/dr$), and the *turbulent component*, which accounts for the friction between the fluctuating fluid particles and the fluid body (denoted as τ_{turb} and is related to the fluctuation components of velocity). Then the *total shear stress* in turbulent flow can be expressed as

$$\tau_{\text{total}} = \tau_{\text{lam}} + \tau_{\text{turb}} \tag{8–36}$$

The typical average velocity profile and relative magnitudes of laminar and turbulent components of shear stress for turbulent flow in a pipe are given in Fig. 8–22. Note that although the velocity profile is approximately parabolic in laminar flow, it becomes flatter or "fuller" in turbulent flow, with a sharp drop near the pipe wall. The fullness increases with the Reynolds number, and the velocity profile becomes more nearly uniform, lending support to the commonly utilized uniform velocity profile approximation for fully developed turbulent pipe flow. Keep in mind, however, that the flow speed at the wall of a stationary pipe is always zero (no-slip condition).

Turbulent Shear Stress

Consider turbulent flow in a horizontal pipe, and the upward eddy motion of a fluid particle from a layer of lower velocity to an adjacent layer of higher velocity through a differential area dA as a result of the velocity fluctuation v', as shown in Fig. 8–23. The mass flow rate of the fluid particle rising through dA is $\rho v' dA$, and its net effect on the layer above dA is a reduction in its average flow velocity because of momentum transfer to the fluid particle with lower average flow velocity. This momentum transfer causes the horizontal velocity of the fluid particle to increase by u', and thus its momentum in the horizontal direction to increase at a rate of $(\rho v' dA)u'$, which must be equal to the decrease in the momentum of the upper fluid layer. Noting that force in a given direction is equal to the rate of change of momentum in that direction, the horizontal force acting on a fluid element above dA due to the passing of fluid particles through dA is $\delta F = (\rho v' dA)(-u') = -\rho u' v' dA$. Therefore, the shear force per unit area due to the eddy motion of fluid particles $\delta F/dA = -\rho u'v'$ can be viewed as the instantaneous turbulent shear stress. Then the **turbulent shear stress** can be expressed as

$$\tau_{\text{turb}} = -\rho \overline{u'v'} \qquad (8\text{–}37)$$

where $\overline{u'v'}$ is the time average of the product of the fluctuating velocity components u' and v'. Note that $\overline{u'v'} \neq 0$ even though $\overline{u'} = 0$ and $\overline{v'} = 0$ (and thus $\overline{u'}\,\overline{v'} = 0$), and experimental results show that $\overline{u'v'}$ is usually a negative quantity. Terms such as $-\rho\overline{u'v'}$ or $-\rho\overline{u'^2}$ are called **Reynolds stresses** or **turbulent stresses.**

Many semi-empirical formulations have been developed that model the Reynolds stress in terms of average velocity gradients in order to provide mathematical *closure* to the equations of motion. Such models are called **turbulence models** and are discussed in more detail in Chap. 15.

The random eddy motion of groups of particles resembles the random motion of molecules in a gas—colliding with each other after traveling a certain distance and exchanging momentum in the process. Therefore, momentum transport by eddies in turbulent flows is analogous to the molecular momentum diffusion. In many of the simpler turbulence models, turbulent shear stress is expressed in an analogous manner as suggested by the French mathematician Joseph Boussinesq (1842–1929) in 1877 as

$$\tau_{\text{turb}} = -\rho\overline{u'v'} = \mu_t \frac{\partial \overline{u}}{\partial y} \qquad (8\text{–}38)$$

where μ_t is the **eddy viscosity** or **turbulent viscosity,** which accounts for momentum transport by turbulent eddies. Then the total shear stress can be expressed conveniently as

$$\tau_{\text{total}} = (\mu + \mu_t)\frac{\partial u}{\partial y} = \rho(\nu + \nu_t)\frac{\partial u}{\partial y} \qquad (8\text{–}39)$$

where $\nu_t = \mu_t/\rho$ is the **kinematic eddy viscosity** or **kinematic turbulent viscosity** (also called the *eddy diffusivity of momentum*). The concept of eddy viscosity is very appealing, but it is of no practical use unless its value can be determined. In other words, eddy viscosity must be modeled as a function of the average flow variables; we call this *eddy viscosity closure*. For example, in the early 1900s, the German engineer L. Prandtl introduced the concept of

FIGURE 8–23
Fluid particle moving upward through a differential area dA as a result of the velocity fluctuation v'.

Laminar flow

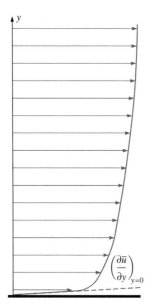

Turbulent flow

FIGURE 8–24
The velocity gradients at the wall, and thus the wall shear stress, are much larger for turbulent flow than they are for laminar flow, even though the turbulent boundary layer is thicker than the laminar one for the same value of free-stream velocity.

mixing length l_m, which is related to the average size of the eddies that are primarily responsible for mixing, and expressed the turbulent shear stress as

$$\tau_{\text{turb}} = \mu_t \frac{\partial \overline{u}}{\partial y} = \rho l_m^2 \left(\frac{\partial \overline{u}}{\partial y} \right)^2 \qquad \text{(8–40)}$$

But this concept is also of limited use since l_m is not a constant for a given flow (in the vicinity of the wall, for example, l_m is nearly proportional to the distance from the wall) and its determination is not easy. Final mathematical closure is obtained only when l_m is written as a function of average flow variables, distance from the wall, etc.

Eddy motion and thus eddy diffusivities are much larger than their molecular counterparts in the core region of a turbulent boundary layer. The eddy motion loses its intensity close to the wall and diminishes at the wall because of the no-slip condition (u' and v' are identically zero at a stationary wall). Therefore, the velocity profile is very slowly changing in the core region of a turbulent boundary layer, but very steep in the thin layer adjacent to the wall, resulting in large velocity gradients at the wall surface. So it is no surprise that the wall shear stress is much larger in turbulent flow than it is in laminar flow (Fig. 8–24).

Note that the molecular diffusivity of momentum ν (as well as μ) is a fluid property, and its value is listed in fluid handbooks. Eddy diffusivity ν_t (as well as μ_t), however, is *not* a fluid property, and its value depends on flow conditions. Eddy diffusivity ν_t decreases toward the wall, becoming zero at the wall. Its value ranges from zero at the wall to several thousand times the value of the molecular diffusivity in the core region.

Turbulent Velocity Profile

Unlike laminar flow, the expressions for the velocity profile in a turbulent flow are based on both analysis and measurements, and thus they are semi-empirical in nature with constants determined from experimental data. Consider fully developed turbulent flow in a pipe, and let u denote the time-averaged velocity in the axial direction (and thus drop the overbar from $\overline{u}$ for simplicity).

Typical velocity profiles for fully developed laminar and turbulent flows are given in Fig. 8–25. Note that the velocity profile is parabolic in laminar flow but is much fuller in turbulent flow, with a sharp drop near the pipe wall. Turbulent flow along a wall can be considered to consist of four regions, characterized by the distance from the wall (Fig. 8–25). The very thin layer next to the wall where viscous effects are dominant is the **viscous** (or **laminar** or **linear** or **wall**) sublayer. The velocity profile in this layer is very nearly *linear*, and the flow is streamlined. Next to the viscous sublayer is the **buffer layer,** in which turbulent effects are becoming significant, but the flow is still dominated by viscous effects. Above the buffer layer is the **overlap** (or **transition**) **layer,** also called the **inertial sublayer,** in which the turbulent effects are much more significant, but still not dominant. Above that is the **outer** (or **turbulent**) **layer** in the remaining part of the flow in which turbulent effects dominate over molecular diffusion (viscous) effects.

Flow characteristics are quite different in different regions, and thus it is difficult to come up with an analytic relation for the velocity profile for the entire flow as we did for laminar flow. The best approach in the turbulent case turns out to be to identify the key variables and functional forms using dimensional analysis, and then to use experimental data to determine the numerical values of any constants.

The thickness of the viscous sublayer is very small (typically, much less than 1 percent of the pipe diameter), but this thin layer next to the wall plays a dominant role on flow characteristics because of the large velocity gradients it involves. The wall dampens any eddy motion, and thus the flow in this layer is essentially laminar and the shear stress consists of laminar shear stress which is proportional to the fluid viscosity. Considering that velocity changes from zero to nearly the core region value across a layer that is sometimes no thicker than a hair (almost like a step function), we would expect the velocity profile in this layer to be very nearly linear, and experiments confirm that. Then the velocity gradient in the viscous sublayer remains nearly constant at $du/dy = u/y$, and the wall shear stress can be expressed as

$$\tau_w = \mu \frac{u}{y} = \rho \nu \frac{u}{y} \quad \text{or} \quad \frac{\tau_w}{\rho} = \frac{\nu u}{y} \tag{8-41}$$

where y is the distance from the wall (note that $y = R - r$ for a circular pipe). The quantity τ_w/ρ is frequently encountered in the analysis of turbulent velocity profiles. The square root of τ_w/ρ has the dimensions of velocity, and thus it is convenient to view it as a fictitious velocity called the **friction velocity** expressed as $u_* = \sqrt{\tau_w/\rho}$. Substituting this into Eq. 8–41, the velocity profile in the viscous sublayer is expressed in dimensionless form as

Viscous sublayer:
$$\frac{u}{u_*} = \frac{yu_*}{\nu} \tag{8-42}$$

This equation is known as the **law of the wall,** and it is found to satisfactorily correlate with experimental data for smooth surfaces for $0 \le yu_*/\nu \le 5$. Therefore, the thickness of the viscous sublayer is roughly

Thickness of viscous sublayer:
$$y = \delta_{sublayer} = \frac{5\nu}{u_*} = \frac{25\nu}{u_\delta} \tag{8-43}$$

where u_δ is the flow velocity at the edge of the viscous sublayer (where $u_\delta \approx 5u_*$), which is closely related to the average velocity in a pipe. Thus we conclude that *the thickness of the viscous sublayer is proportional to the kinematic viscosity and inversely proportional to the average flow velocity.* In other words, the viscous sublayer is suppressed and it gets thinner as the velocity (and thus the Reynolds number) increases. Consequently, the velocity profile becomes nearly flat and thus the velocity distribution becomes more uniform at very high Reynolds numbers.

The quantity ν/u_* has dimensions of length and is called the **viscous length;** it is used to nondimensionalize the distance y from the surface. In boundary layer analysis, it is convenient to work with nondimensionalized distance and nondimensionalized velocity defined as

Nondimensionalized variables:
$$y^+ = \frac{yu_*}{\nu} \quad \text{and} \quad u^+ = \frac{u}{u_*} \tag{8-44}$$

Then the law of the wall (Eq. 8–42) becomes simply

Normalized law of the wall:
$$u^+ = y^+ \tag{8-45}$$

Note that the friction velocity u_* is used to nondimensionalize both y and u, and y^+ resembles the Reynolds number expression.

Laminar flow

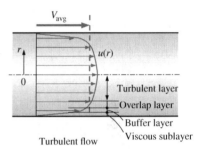

Turbulent layer
Overlap layer
Buffer layer
Viscous sublayer

Turbulent flow

FIGURE 8–25

The velocity profile in fully developed pipe flow is parabolic in laminar flow, but much fuller in turbulent flow. Note that $u(r)$ in the turbulent case is the *time-averaged* velocity component in the axial direction (the overbar on u has been dropped for simplicity).

FIGURE 8–26
Comparison of the law of the wall and the logarithmic-law velocity profiles with experimental data for fully developed turbulent flow in a pipe.

FIGURE 8–27
Power-law velocity profiles for fully developed turbulent flow in a pipe for different exponents, and its comparison with the laminar velocity profile.

In the overlap layer, the experimental data for velocity are observed to line up on a straight line when plotted against the logarithm of distance from the wall. Dimensional analysis indicates and the experiments confirm that the velocity in the overlap layer is proportional to the logarithm of distance, and the velocity profile can be expressed as

The logarithmic law:
$$\frac{u}{u_*} = \frac{1}{\kappa} \ln \frac{y u_*}{\nu} + B \qquad \text{(8–46)}$$

where κ and B are constants whose values are determined experimentally to be about 0.40 and 5.0, respectively. Equation 8–46 is known as the **logarithmic law.** Substituting the values of the constants, the velocity profile is determined to be

Overlap layer:
$$\frac{u}{u_*} = 2.5 \ln \frac{y u_*}{\nu} + 5.0 \quad \text{or} \quad u^+ = 2.5 \ln y^+ + 5.0 \qquad \text{(8–47)}$$

It turns out that the logarithmic law in Eq. 8–47 satisfactorily represents experimental data for the entire flow region except for the regions very close to the wall and near the pipe center, as shown in Fig. 8–26, and thus it is viewed as a *universal velocity profile* for turbulent flow in pipes or over surfaces. Note from the figure that the logarithmic-law velocity profile is quite accurate for $y^+ > 30$, but neither velocity profile is accurate in the buffer layer, i.e., the region $5 < y^+ < 30$. Also, the viscous sublayer appears much larger in the figure than it is since we used a logarithmic scale for distance from the wall.

A good approximation for the outer turbulent layer of pipe flow can be obtained by evaluating the constant B in Eq. 8–46 from the requirement that maximum velocity in a pipe occurs at the centerline where $r = 0$. Solving for B from Eq. 8–46 by setting $y = R - r = R$ and $u = u_{max}$, and substituting it back into Eq. 8–46 together with $\kappa = 0.4$ gives

Outer turbulent layer:
$$\frac{u_{max} - u}{u_*} = 2.5 \ln \frac{R}{R - r} \qquad \text{(8–48)}$$

The deviation of velocity from the centerline value $u_{max} - u$ is called the **velocity defect,** and Eq. 8–48 is called the **velocity defect law.** This relation shows that the normalized velocity profile in the core region of turbulent flow in a pipe depends on the distance from the centerline and is independent of the viscosity of the fluid. This is not surprising since the eddy motion is dominant in this region, and the effect of fluid viscosity is negligible.

Numerous other empirical velocity profiles exist for turbulent pipe flow. Among those, the simplest and the best known is the **power-law velocity profile** expressed as

Power-law velocity profile:
$$\frac{u}{u_{max}} = \left(\frac{y}{R}\right)^{1/n} \quad \text{or} \quad \frac{u}{u_{max}} = \left(1 - \frac{r}{R}\right)^{1/n} \qquad \text{(8–49)}$$

where the exponent n is a constant whose value depends on the Reynolds number. The value of n increases with increasing Reynolds number. The value $n = 7$ generally approximates many flows in practice, giving rise to the term *one-seventh power-law velocity profile.*

Various power-law velocity profiles are shown in Fig. 8–27 for $n = 6$, 8, and 10 together with the velocity profile for fully developed laminar flow for comparison. Note that the turbulent velocity profile is fuller than the laminar one, and it becomes more flat as n (and thus the Reynolds number) increases. Also note that the power-law profile cannot be used to calculate wall shear stress since it gives a velocity gradient of infinity there, and it fails to give zero slope at the centerline. But these regions of discrepancy constitute a small portion of the overall flow, and the power-law profile gives highly accurate results for turbulent flow through a pipe.

Despite the small thickness of the viscous sublayer (usually much less than 1 percent of the pipe diameter), the characteristics of the flow in this layer are very important since they set the stage for flow in the rest of the pipe. Any irregularity or roughness on the surface disturbs this layer and affects the flow. Therefore, unlike laminar flow, the friction factor in turbulent flow is a strong function of surface roughness.

It should be kept in mind that roughness is a relative concept, and it has significance when its height ε is comparable to the thickness of the viscous sublayer (which is a function of the Reynolds number). All materials appear "rough" under a microscope with sufficient magnification. In fluid mechanics, a surface is characterized as being rough when the hills of roughness protrude out of the viscous sublayer. A surface is said to be *hydrodynamically smooth* when the sublayer submerges the roughness elements. Glass and plastic surfaces are generally considered to be hydrodynamically smooth.

The Moody Chart and the Colebrook Equation

The friction factor in fully developed turbulent pipe flow depends on the Reynolds number and the **relative roughness** ε/D, which is the ratio of the mean height of roughness of the pipe to the pipe diameter. The functional form of this dependence cannot be obtained from a theoretical analysis, and all available results are obtained from painstaking experiments using artificially roughened surfaces (usually by gluing sand grains of a known size on the inner surfaces of the pipes). Most such experiments were conducted by Prandtl's student J. Nikuradse in 1933, followed by the works of others. The friction factor was calculated from measurements of the flow rate and the pressure drop.

The experimental results are presented in tabular, graphical, and functional forms obtained by curve-fitting experimental data. In 1939, Cyril F. Colebrook (1910–1997) combined the available data for transition and turbulent flow in smooth as well as rough pipes into the following implicit relation (Fig. 8–28) known as the **Colebrook equation:**

$$\frac{1}{\sqrt{f}} = -2.0 \log\left(\frac{\varepsilon/D}{3.7} + \frac{2.51}{\mathrm{Re}\sqrt{f}}\right) \quad \text{(turbulent flow)} \tag{8-50}$$

We note that the logarithm in Eq. 8–50 is a base 10 rather than a natural logarithm. In 1942, the American engineer Hunter Rouse (1906–1996) verified Colebrook's equation and produced a graphical plot of f as a function of Re and the product $\mathrm{Re}\sqrt{f}$. He also presented the laminar flow relation and a table of commercial pipe roughness. Two years later, Lewis F. Moody (1880–1953) redrew Rouse's diagram into the form commonly used today. The now famous **Moody chart** is given in the appendix as Fig. A–12. It presents the Darcy friction factor for pipe flow as a function of Reynolds number and ε/D over a wide range. It is probably one of the most widely accepted and used charts in engineering. Although it is developed for circular pipes, it can also be used for noncircular pipes by replacing the diameter with the hydraulic diameter.

$$\frac{1}{\sqrt{f}} = -2.0 \log\left(\frac{\varepsilon/D}{3.7} + \frac{2.51}{Re\sqrt{f}}\right)$$

The Colebrook equation is *implicit* in f since f appears on both sides of the equation. It must be solved iteratively.

FIGURE 8–28
The Colebrook equation.

TABLE 8–2

Equivalent roughness values for new commercial pipes*

Material	Roughness, ε	
	ft	mm
Glass, plastic	0 (smooth)	
Concrete	0.003–0.03	0.9–9
Wood stave	0.0016	0.5
Rubber, smoothed	0.000033	0.01
Copper or brass tubing	0.000005	0.0015
Cast iron	0.00085	0.26
Galvanized iron	0.0005	0.15
Wrought iron	0.00015	0.046
Stainless steel	0.000007	0.002
Commercial steel	0.00015	0.045

*The uncertainty in these values can be as much as ±60 percent.

Relative Roughness, ε/D	Friction Factor, f
0.0*	0.0119
0.00001	0.0119
0.0001	0.0134
0.0005	0.0172
0.001	0.0199
0.005	0.0305
0.01	0.0380
0.05	0.0716

*Smooth surface. All values are for Re = 10^6 and are calculated from the Colebrook equation.

FIGURE 8–29

The friction factor is minimum for a smooth pipe and increases with roughness.

Commercially available pipes differ from those used in the experiments in that the roughness of pipes in the market is not uniform and it is difficult to give a precise description of it. Equivalent roughness values for some commercial pipes are given in Table 8–2 as well as on the Moody chart. But it should be kept in mind that these values are for new pipes, and the relative roughness of pipes may increase with use as a result of corrosion, scale buildup, and precipitation. As a result, the friction factor may increase by a factor of 5 to 10. Actual operating conditions must be considered in the design of piping systems. Also, the Moody chart and its equivalent Colebrook equation involve several uncertainties (the roughness size, experimental error, curve fitting of data, etc.), and thus the results obtained should not be treated as "exact." They are is usually considered to be accurate to ±15 percent over the entire range in the figure.

The Colebrook equation is implicit in f, and thus the determination of the friction factor requires iteration. An approximate *explicit* relation for f was given by S. E. Haaland in 1983 as

$$\frac{1}{\sqrt{f}} \cong -1.8 \log\left[\frac{6.9}{Re} + \left(\frac{\varepsilon/D}{3.7}\right)^{1.11}\right] \tag{8–51}$$

The results obtained from this relation are within 2 percent of those obtained from the Colebrook equation. If more accurate results are desired, Eq. 8–51 can be used as a good *first guess* in a Newton iteration when using a programmable calculator or a spreadsheet to solve for f with Eq. 8–50.

We make the following observations from the Moody chart:

• For laminar flow, the friction factor decreases with increasing Reynolds number, and it is independent of surface roughness.

• The friction factor is a minimum for a smooth pipe (but still not zero because of the no-slip condition) and increases with roughness (Fig. 8–29). The Colebrook equation in this case ($\varepsilon = 0$) reduces to the **Prandtl equation** expressed as $1/\sqrt{f} = 2.0 \log(Re\sqrt{f}) - 0.8$.

• The transition region from the laminar to turbulent regime (2300 < Re < 4000) is indicated by the shaded area in the Moody chart (Figs. 8–30 and A–12). The flow in this region may be laminar or turbulent, depending on flow disturbances, or it may alternate between laminar and turbulent, and thus the friction factor may also alternate between the values for laminar and turbulent flow. The data in this range are the least reliable. At small relative roughnesses, the friction factor increases in the transition region and approaches the value for smooth pipes.

• At very large Reynolds numbers (to the right of the dashed line on the Moody chart) the friction factor curves corresponding to specified relative roughness curves are nearly horizontal, and thus the friction factors are independent of the Reynolds number (Fig. 8–30). The flow in that region is called *fully rough turbulent flow* or just *fully rough flow* because the thickness of the viscous sublayer decreases with increasing Reynolds number, and it becomes so thin that it is negligibly small compared to the surface roughness height. The viscous effects in this case are produced in the main flow primarily by the protruding roughness elements, and the contribution of the viscous sublayer is negligible. The Colebrook equation in the *fully rough* zone (Re → ∞) reduces to the **von Kármán equation**

FIGURE 8–30
At very large Reynolds numbers, the friction factor curves on the Moody chart are nearly horizontal, and thus the friction factors are independent of the Reynolds number. See Fig. A–12 for a full-page Moody chart.

expressed as $1/\sqrt{f} = -2.0 \log[(\varepsilon/D)/3.7]$, which is explicit in f. Some authors call this zone *completely* (or *fully*) *turbulent flow,* but this is misleading since the flow to the left of the dashed blue line in Fig. 8–30 is also fully turbulent.

In calculations, we should make sure that we use the actual internal diameter of the pipe, which may be different than the nominal diameter. For example, the internal diameter of a steel pipe whose nominal diameter is 1 in is 1.049 in (Table 8–3).

Types of Fluid Flow Problems

In the design and analysis of piping systems that involve the use of the Moody chart (or the Colebrook equation), we usually encounter three types of problems (the fluid and the roughness of the pipe are assumed to be specified in all cases) (Fig. 8–31):

1. Determining the **pressure drop** (or head loss) when the pipe length and diameter are given for a specified flow rate (or velocity)
2. Determining the **flow rate** when the pipe length and diameter are given for a specified pressure drop (or head loss)
3. Determining the **pipe diameter** when the pipe length and flow rate are given for a specified pressure drop (or head loss)

Problems of the *first type* are straightforward and can be solved directly by using the Moody chart. Problems of the *second type* and *third type* are commonly encountered in engineering design (in the selection of pipe diameter, for example, that minimizes the sum of the construction and pumping costs), but the use of the Moody chart with such problems requires an iterative approach—an equation solver (such as EES) is recommended.

In problems of the *second type*, the diameter is given but the flow rate is unknown. A good guess for the friction factor in that case is obtained from the completely turbulent flow region for the given roughness. This is true for large Reynolds numbers, which is often the case in practice. Once the flow rate is obtained, the friction factor is corrected using the Moody chart or the Colebrook equation, and the process is repeated until the solution

TABLE 8–3

Standard sizes for Schedule 40 steel pipes

Nominal Size, in	Actual Inside Diameter, in
$\frac{1}{8}$	0.269
$\frac{1}{4}$	0.364
$\frac{3}{8}$	0.493
$\frac{1}{2}$	0.622
$\frac{3}{4}$	0.824
1	1.049
$1\frac{1}{2}$	1.610
2	2.067
$2\frac{1}{2}$	2.469
3	3.068
5	5.047
10	10.02

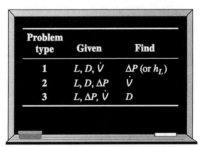

Problem type	Given	Find
1	$L, D, \dot{V}$	ΔP (or h_L)
2	$L, D, \Delta P$	$\dot{V}$
3	$L, \Delta P, \dot{V}$	D

FIGURE 8–31
The three types of problems encountered in pipe flow.

converges. (Typically only a few iterations are required for convergence to three or four digits of precision.)

In problems of the *third type*, the diameter is not known and thus the Reynolds number and the relative roughness cannot be calculated. Therefore, we start calculations by assuming a pipe diameter. The pressure drop calculated for the assumed diameter is then compared to the specified pressure drop, and calculations are repeated with another pipe diameter in an iterative fashion until convergence.

To avoid tedious iterations in head loss, flow rate, and diameter calculations, Swamee and Jain (1976) proposed the following explicit relations that are accurate to within 2 percent of the Moody chart:

$$h_L = 1.07 \frac{\dot{V}^2 L}{g D^5} \left\{ \ln\left[\frac{\varepsilon}{3.7D} + 4.62\left(\frac{\nu D}{\dot{V}}\right)^{0.9} \right] \right\}^{-2} \qquad \begin{array}{l} 10^{-6} < \varepsilon/D < 10^{-2} \\ 3000 < \text{Re} < 3 \times 10^8 \end{array} \qquad (8\text{–}52)$$

$$\dot{V} = -0.965\left(\frac{g D^5 h_L}{L}\right)^{0.5} \ln\left[\frac{\varepsilon}{3.7D} + \left(\frac{3.17\nu^2 L}{g D^3 h_L}\right)^{0.5} \right] \qquad \text{Re} > 2000 \qquad (8\text{–}53)$$

$$D = 0.66\left[\varepsilon^{1.25}\left(\frac{L\dot{V}^2}{g h_L}\right)^{4.75} + \nu \dot{V}^{9.4}\left(\frac{L}{g h_L}\right)^{5.2} \right]^{0.04} \qquad \begin{array}{l} 10^{-6} < \varepsilon/D < 10^{-2} \\ 5000 < \text{Re} < 3 \times 10^8 \end{array} \qquad (8\text{–}54)$$

Note that all quantities are dimensional and the units simplify to the desired unit (for example, to m or ft in the last relation) when consistent units are used. Noting that the Moody chart is accurate to within 15 percent of experimental data, we should have no reservation in using these approximate relations in the design of piping systems.

FIGURE 8–32
Schematic for Example 8–3.

EXAMPLE 8–3 **Determining the Head Loss in a Water Pipe**

Water at 60°F ($\rho = 62.36$ lbm/ft³ and $\mu = 7.536 \times 10^{-4}$ lbm/ft·s) is flowing steadily in a 2-in-diameter horizontal pipe made of stainless steel at a rate of 0.2 ft³/s (Fig. 8–32). Determine the pressure drop, the head loss, and the required pumping power input for flow over a 200-ft-long section of the pipe.

SOLUTION The flow rate through a specified water pipe is given. The pressure drop, the head loss, and the pumping power requirements are to be determined.
Assumptions **1** The flow is steady and incompressible. **2** The entrance effects are negligible, and thus the flow is fully developed. **3** The pipe involves no components such as bends, valves, and connectors. **4** The piping section involves no work devices such as a pump or a turbine.
Properties The density and dynamic viscosity of water are given to be $\rho = 62.36$ lbm/ft³ and $\mu = 7.536 \times 10^{-4}$ lbm/ft·s, respectively.
Analysis We recognize this as a problem of the first type, since flow rate, pipe length, and pipe diameter are known. First we calculate the average velocity and the Reynolds number to determine the flow regime:

$$V = \frac{\dot{V}}{A_c} = \frac{\dot{V}}{\pi D^2/4} = \frac{0.2 \text{ ft}^3/\text{s}}{\pi(2/12 \text{ ft})^2/4} = 9.17 \text{ ft/s}$$

$$\text{Re} = \frac{\rho V D}{\mu} = \frac{(62.36 \text{ lbm/ft}^3)(9.17 \text{ ft/s})(2/12 \text{ ft})}{7.536 \times 10^{-4} \text{ lbm/ft·s}} = 126{,}400$$

Since Re is greater than 4000, the flow is turbulent. The relative roughness of the pipe is estimated using Table 8–2

$$\varepsilon/D = \frac{0.000007 \text{ ft}}{2/12 \text{ ft}} = 0.000042$$

The friction factor corresponding to this relative roughness and Reynolds number is determined from the Moody chart. To avoid any reading error, we determine f from the Colebrook equation on which the Moody chart is based:

$$\frac{1}{\sqrt{f}} = -2.0 \log\left(\frac{\varepsilon/D}{3.7} + \frac{2.51}{\text{Re}\sqrt{f}}\right) \rightarrow \frac{1}{\sqrt{f}} = -2.0 \log\left(\frac{0.000042}{3.7} + \frac{2.51}{126{,}400\sqrt{f}}\right)$$

Using an equation solver or an iterative scheme, the friction factor is determined to be $f = 0.0174$. Then the pressure drop (which is equivalent to pressure loss in this case), head loss, and the required power input become

$$\Delta P = \Delta P_L = f\frac{L}{D}\frac{\rho V^2}{2} = 0.0174\frac{200 \text{ ft}}{2/12 \text{ ft}}\frac{(62.36 \text{ lbm/ft}^3)(9.17 \text{ ft/s})^2}{2}\left(\frac{1 \text{ lbf}}{32.2 \text{ lbm·ft/s}^2}\right)$$

$$= 1700 \text{ lbf/ft}^2 = 11.8 \text{ psi}$$

$$h_L = \frac{\Delta P_L}{\rho g} = f\frac{L}{D}\frac{V^2}{2g} = 0.0174\frac{200 \text{ ft}}{2/12 \text{ ft}}\frac{(9.17 \text{ ft/s})^2}{2(32.2 \text{ ft/s}^2)} = 27.3 \text{ ft}$$

$$\dot{W}_{\text{pump}} = \dot{V}\,\Delta P = (0.2 \text{ ft}^3/\text{s})(1700 \text{ lbf/ft}^2)\left(\frac{1 \text{ W}}{0.737 \text{ lbf·ft/s}}\right) = 461 \text{ W}$$

Therefore, power input in the amount of 461 W is needed to overcome the frictional losses in the pipe.

Discussion It is common practice to write our final answers to three significant digits, even though we know that the results are accurate to at most two significant digits because of inherent inaccuracies in the Colebrook equation, as discussed previously. The friction factor could also be determined easily from the explicit Haaland relation (Eq. 8–51). It would give $f = 0.0172$, which is sufficiently close to 0.0174. Also, the friction factor corresponding to $\varepsilon = 0$ in this case is 0.0171, which indicates that this stainless-steel pipe can be approximated as smooth with minimal error.

■ *EXAMPLE 8–4*   Determining the Diameter of an Air Duct

Heated air at 1 atm and 35°C is to be transported in a 150-m-long circular plastic duct at a rate of 0.35 m³/s (Fig. 8–33). If the head loss in the pipe is not to exceed 20 m, determine the minimum diameter of the duct.

SOLUTION The flow rate and the head loss in an air duct are given. The diameter of the duct is to be determined.

Assumptions **1** The flow is steady and incompressible. **2** The entrance effects are negligible, and thus the flow is fully developed. **3** The duct involves no components such as bends, valves, and connectors. **4** Air is an ideal gas. **5** The duct is smooth since it is made of plastic. **6** The flow is turbulent (to be verified).

Properties The density, dynamic viscosity, and kinematic viscosity of air at 35°C are $\rho = 1.145 \text{ kg/m}^3$, $\mu = 1.895 \times 10^{-5} \text{ kg/m·s}$, and $\nu = 1.655 \times 10^{-5} \text{ m}^2/\text{s}$.

FIGURE 8–33
Schematic for Example 8–4.

Analysis This is a problem of the third type since it involves the determination of diameter for specified flow rate and head loss. We can solve this problem using three different approaches: (1) an iterative approach by assuming a pipe diameter, calculating the head loss, comparing the result to the specified head loss, and repeating calculations until the calculated head loss matches the specified value; (2) writing all the relevant equations (leaving the diameter as an unknown) and solving them simultaneously using an equation solver; and (3) using the third Swamee–Jain formula. We will demonstrate the use of the last two approaches.

The average velocity, the Reynolds number, the friction factor, and the head loss relations are expressed as (D is in m, V is in m/s, and Re and f are dimensionless)

$$V = \frac{\dot{V}}{A_c} = \frac{\dot{V}}{\pi D^2/4} = \frac{0.35 \text{ m}^3/\text{s}}{\pi D^2/4}$$

$$\text{Re} = \frac{VD}{\nu} = \frac{VD}{1.655 \times 10^{-5} \text{ m}^2/\text{s}}$$

$$\frac{1}{\sqrt{f}} = -2.0 \log\left(\frac{\varepsilon/D}{3.7} + \frac{2.51}{\text{Re}\sqrt{f}}\right) = -2.0 \log\left(\frac{2.51}{\text{Re}\sqrt{f}}\right)$$

$$h_L = f\frac{L}{D}\frac{V^2}{2g} \quad \rightarrow \quad 20 \text{ m} = f\frac{150 \text{ m}}{D}\frac{V^2}{2(9.81 \text{ m/s}^2)}$$

The roughness is approximately zero for a plastic pipe (Table 8–2). Therefore, this is a set of four equations and four unknowns, and solving them with an equation solver such as EES gives

$$D = 0.267 \text{ m}, \quad f = 0.0180, \quad V = 6.24 \text{ m/s}, \quad \text{and} \quad \text{Re} = 100,800$$

Therefore, the diameter of the duct should be more than 26.7 cm if the head loss is not to exceed 20 m. Note that Re > 4000, and thus the turbulent flow assumption is verified.

The diameter can also be determined directly from the third Swamee–Jain formula to be

$$D = 0.66\left[\varepsilon^{1.25}\left(\frac{L\dot{V}^2}{gh_L}\right)^{4.75} + \nu\dot{V}^{9.4}\left(\frac{L}{gh_L}\right)^{5.2}\right]^{0.04}$$

$$= 0.66\left[0 + (1.655 \times 10^{-5} \text{ m}^2/\text{s})(0.35 \text{ m}^3/\text{s})^{9.4}\left(\frac{150 \text{ m}}{(9.81 \text{ m/s}^2)(20 \text{ m})}\right)^{5.2}\right]^{0.04}$$

$$= 0.271 \text{ m}$$

Discussion Note that the difference between the two results is less than 2 percent. Therefore, the simple Swamee–Jain relation can be used with confidence. Finally, the first (iterative) approach requires an initial guess for D. If we use the Swamee–Jain result as our initial guess, the diameter converges to $D = 0.267$ m in short order.

EXAMPLE 8–5   Determining the Flow Rate of Air in a Duct

Reconsider Example 8–4. Now the duct length is doubled while its diameter is maintained constant. If the total head loss is to remain constant, determine the drop in the flow rate through the duct.

SOLUTION The diameter and the head loss in an air duct are given. The drop in the flow rate is to be determined.

Analysis This is a problem of the second type since it involves the determination of the flow rate for a specified pipe diameter and head loss. The solution involves an iterative approach since the flow rate (and thus the flow velocity) is not known.

The average velocity, Reynolds number, friction factor, and the head loss relations are expressed as (D is in m, V is in m/s, and Re and f are dimensionless)

$$V = \frac{\dot{V}}{A_c} = \frac{\dot{V}}{\pi D^2/4} \quad \rightarrow \quad V = \frac{\dot{V}}{\pi (0.267 \text{ m})^2/4}$$

$$\text{Re} = \frac{VD}{\nu} \quad \rightarrow \quad \text{Re} = \frac{V(0.267 \text{ m})}{1.655 \times 10^{-5} \text{ m}^2/\text{s}}$$

$$\frac{1}{\sqrt{f}} = -2.0 \log\left(\frac{\varepsilon/D}{3.7} + \frac{2.51}{\text{Re}\sqrt{f}}\right) \quad \rightarrow \quad \frac{1}{\sqrt{f}} = -2.0 \log\left(\frac{2.51}{\text{Re}\sqrt{f}}\right)$$

$$h_L = f \frac{L}{D} \frac{V^2}{2g} \quad \rightarrow \quad 20 \text{ m} = f \frac{300 \text{ m}}{0.267 \text{ m}} \frac{V^2}{2(9.81 \text{ m/s}^2)}$$

This is a set of four equations in four unknowns and solving them with an equation solver such as EES (Fig. 8–34) gives

$$\dot{V} = 0.24 \text{ m}^3/\text{s}, \quad f = 0.0195, \quad V = 4.23 \text{ m/s}, \quad \text{and} \quad \text{Re} = 68{,}300$$

Then the drop in the flow rate becomes

$$\dot{V}_{drop} = \dot{V}_{old} - \dot{V}_{new} = 0.35 - 0.24 = \textbf{0.11 m}^3/\textbf{s} \quad \text{(a drop of 31 percent)}$$

Therefore, for a specified head loss (or available head or fan pumping power), the flow rate drops by about 31 percent from 0.35 to 0.24 m³/s when the duct length doubles.

Alternative Solution If a computer is not available (as in an exam situation), another option is to set up a *manual iteration loop*. We have found that the best convergence is usually realized by first guessing the friction factor *f*, and then solving for the velocity *V*. The equation for *V* as a function of *f* is

Average velocity through the pipe: $\quad V = \sqrt{\dfrac{2gh_L}{fL/D}}$

Once *V* is calculated, the Reynolds number can be calculated, from which a *corrected* friction factor is obtained from the Moody chart or the Colebrook equation. We repeat the calculations with the corrected value of *f* until convergence. We guess *f* = 0.04 for illustration:

Iteration	f (guess)	V, m/s	Re	Corrected f
1	0.04	2.955	4.724×10^4	0.0212
2	0.0212	4.059	6.489×10^4	0.01973
3	0.01973	4.207	6.727×10^4	0.01957
4	0.01957	4.224	6.754×10^4	0.01956
5	0.01956	4.225	6.756×10^4	0.01956

Notice that the iteration has converged to three digits in only three iterations and to four digits in only four iterations. The final results are identical to those obtained with EES, yet do not require a computer.

FIGURE 8–34
EES solution for Example 8–5.

Discussion The new flow rate can also be determined directly from the second Swamee–Jain formula to be

$$\dot{V} = -0.965 \left(\frac{gD^5 h_L}{L} \right)^{0.5} \ln \left[\frac{\varepsilon}{3.7D} + \left(\frac{3.17 \nu^2 L}{gD^3 h_L} \right)^{0.5} \right]$$

$$= -0.965 \left(\frac{(9.81 \text{ m/s}^2)(0.267 \text{ m})^5(20 \text{ m})}{300 \text{ m}} \right)^{0.5}$$

$$\times \ln \left[0 + \left(\frac{3.17(1.655 \times 10^{-5} \text{ m}^2/\text{s})^2(300 \text{ m})}{(9.81 \text{ m/s}^2)(0.267 \text{ m})^3(20 \text{ m})} \right)^{0.5} \right]$$

$$= 0.24 \text{ m}^3/\text{s}$$

Note that the result from the Swamee–Jain relation is the same (to two significant digits) as that obtained with the Colebrook equation using EES or using our manual iteration technique. Therefore, the simple Swamee–Jain relation can be used with confidence.

8–6 · MINOR LOSSES

The fluid in a typical piping system passes through various fittings, valves, bends, elbows, tees, inlets, exits, expansions, and contractions in addition to the straight sections of piping. These components interrupt the smooth flow of the fluid and cause additional losses because of the flow separation and mixing they induce. In a typical system with long pipes, these losses are minor compared to the head loss in the straight sections (the **major losses**) and are called **minor losses.** Although this is generally true, in some cases the minor losses may be greater than the major losses. This is the case, for example, in systems with several turns and valves in a short distance. The head loss introduced by a completely open valve, for example, may be negligible. But a partially closed valve may cause the largest head loss in the system, as evidenced by the drop in the flow rate. Flow through valves and fittings is very complex, and a theoretical analysis is generally not plausible. Therefore, minor losses are determined experimentally, usually by the manufacturers of the components.

Minor losses are usually expressed in terms of the **loss coefficient** K_L (also called the **resistance coefficient**), defined as (Fig. 8–35)

Pipe section with valve:

$(P_1 - P_2)_{\text{valve}}$

Pipe section without valve:

$(P_1 - P_2)_{\text{pipe}}$

$$\Delta P_L = (P_1 - P_2)_{\text{valve}} - (P_1 - P_2)_{\text{pipe}}$$

FIGURE 8–35

For a constant-diameter section of a pipe with a minor loss component, the loss coefficient of the component (such as the gate valve shown) is determined by measuring the additional pressure loss it causes and dividing it by the dynamic pressure in the pipe.

Loss coefficient:
$$K_L = \frac{h_L}{V^2/(2g)} \qquad (8\text{–}55)$$

where h_L is the *additional* irreversible head loss in the piping system caused by insertion of the component, and is defined as $h_L = \Delta P_L/\rho g$. For example, imagine replacing the valve in Fig. 8–35 with a section of constant diameter pipe from location 1 to location 2. ΔP_L is defined as the pressure drop from 1 to 2 for the case *with* the valve, $(P_1 - P_2)_{\text{valve}}$, *minus* the pressure drop that would occur in the imaginary straight pipe section from 1 to 2 *without* the valve, $(P_1 - P_2)_{\text{pipe}}$ at the same flow rate. While the majority of the irreversible head loss occurs locally near the valve, some of it occurs downstream of the valve due to induced swirling turbulent eddies that are produced in the valve and continue downstream. These eddies "waste" mechanical energy because they are ultimately dissipated into heat while the flow in the downstream section of pipe eventually returns to fully developed conditions. When measuring minor losses in some minor loss components, such as *elbows,* for example, location 2 must be considerably far downstream (tens of pipe diameters) in order to fully account for the additional irreversible losses due to these decaying eddies.

When the pipe diameter downstream of the component *changes,* determination of the minor loss is even more complicated. In all cases, however, it is based on the *additional* irreversible loss of mechanical energy that would otherwise not exist if the minor loss component were not there. For simplicity, you may think of the minor loss as occurring *locally* across the minor loss component, but keep in mind that the component influences the flow for several pipe diameters downstream. By the way, this is the reason why most flow meter manufacturers recommend installing their flow meter at least 10 to 20 pipe diameters downstream of any elbows or valves—this allows the swirling turbulent eddies generated by the elbow or valve to largely disappear and the velocity profile to become fully developed before entering the flow meter. (Most flow meters are calibrated with a fully developed velocity profile at the flow meter inlet, and yield the best accuracy when such conditions also exist in the actual application.)

When the inlet diameter equals the outlet diameter, the loss coefficient of a component can also be determined by measuring the pressure loss across the component and dividing it by the dynamic pressure, $K_L = \Delta P_L/(\frac{1}{2}\rho V^2)$. When the loss coefficient for a component is available, the head loss for that component is determined from

Minor loss:
$$h_L = K_L \frac{V^2}{2g} \tag{8–56}$$

The loss coefficient, in general, depends on the geometry of the component and the Reynolds number, just like the friction factor. However, it is usually assumed to be independent of the Reynolds number. This is a reasonable approximation since most flows in practice have large Reynolds numbers and the loss coefficients (including the friction factor) tend to be independent of the Reynolds number at large Reynolds numbers.

Minor losses are also expressed in terms of the **equivalent length** L_{equiv}, defined as (Fig. 8–36)

Equivalent length:
$$h_L = K_L \frac{V^2}{2g} = f \frac{L_{equiv}}{D} \frac{V^2}{2g} \quad \rightarrow \quad L_{equiv} = \frac{D}{f} K_L \tag{8–57}$$

where f is the friction factor and D is the diameter of the pipe that contains the component. The head loss caused by the component is equivalent to the head loss caused by a section of the pipe whose length is L_{equiv}. Therefore, the contribution of a component to the head loss is accounted for by simply adding L_{equiv} to the total pipe length.

Both approaches are used in practice, but the use of loss coefficients is more common. Therefore, we also use that approach in this book. Once all the loss coefficients are available, the total head loss in a piping system is determined from

Total head loss (general): $\quad h_{L, total} = h_{L, major} + h_{L, minor}$

$$= \sum_i f_i \frac{L_i}{D_i} \frac{V_i^2}{2g} + \sum_j K_{L,j} \frac{V_j^2}{2g} \tag{8–58}$$

$\Delta P = P_1 - P_2 = P_3 - P_4$

FIGURE 8–36

The head loss caused by a component (such as the angle valve shown) is equivalent to the head loss caused by a section of the pipe whose length is the equivalent length.

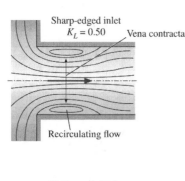

Sharp-edged inlet
$K_L = 0.50$ Vena contracta

Recirculating flow

Well-rounded inlet
$K_L = 0.03$

D

r

FIGURE 8–37

The head loss at the inlet of a pipe is almost negligible for well-rounded inlets ($K_L = 0.03$ for $r/D > 0.2$) but increases to about 0.50 for sharp-edged inlets.

where i represents each pipe section with constant diameter and j represents each component that causes a minor loss. If the entire piping system being analyzed has a constant diameter, Eq. 8–58 reduces to

Total head loss $(D = constant)$: $h_{L,\,total} = \left(f\dfrac{L}{D} + \sum K_L\right)\dfrac{V^2}{2g}$ (8–59)

where V is the average flow velocity through the entire system (note that V = constant since D = constant).

Representative loss coefficients K_L are given in Table 8–4 for inlets, exits, bends, sudden and gradual area changes, and valves. There is considerable uncertainty in these values since the loss coefficients, in general, vary with the pipe diameter, the surface roughness, the Reynolds number, and the details of the design. The loss coefficients of two seemingly identical valves by two different manufacturers, for example, can differ by a factor of 2 or more. Therefore, the particular manufacturer's data should be consulted in the final design of piping systems rather than relying on the representative values in handbooks.

The head loss at the inlet of a pipe is a strong function of geometry. It is almost negligible for well-rounded inlets ($K_L = 0.03$ for $r/D > 0.2$), but increases to about 0.50 for sharp-edged inlets (Fig. 8–37). That is, a sharp-edged inlet causes half of the velocity head to be lost as the fluid enters the pipe. This is because the fluid cannot make sharp 90° turns easily, especially at high velocities. As a result, the flow separates at the corners, and the flow is constricted into the **vena contracta** region formed in the midsection of the pipe (Fig. 8–38). Therefore, a sharp-edged inlet acts like a flow constriction. The velocity increases in the vena contracta region (and the pressure decreases) because of the reduced effective flow area and then decreases as the flow fills the entire cross section of the pipe. There would be negligible loss if the pressure were increased in accordance with Bernoulli's equation (the velocity head would simply be converted into pressure head). However, this deceleration process is far from ideal and the viscous dissipation caused by intense mixing and the turbulent eddies converts part of the kinetic energy into frictional heating, as evidenced by a slight rise in fluid temperature. The end result is a drop in velocity without much pressure recovery, and the inlet loss is a measure of this irreversible pressure drop.

Even slight rounding of the edges can result in significant reduction of K_L, as shown in Fig. 8–39. The loss coefficient rises sharply (to about $K_L = 0.8$) when the pipe protrudes into the reservoir since some fluid near the edge in this case is forced to make a 180° turn.

The loss coefficient for a submerged pipe exit is often listed in handbooks as $K_L = 1$. More precisely, however, K_L is equal to the kinetic energy correction factor α at the exit of the pipe. Although α is indeed close to 1 for fully developed *turbulent* pipe flow, it is equal to 2 for fully developed *laminar* pipe flow. To avoid possible errors when analyzing laminar pipe flow, then, it is best to always set $K_L = \alpha$ at a submerged pipe exit. At any such exit, whether laminar or turbulent, the fluid leaving the pipe loses *all* of its kinetic energy as it mixes with the reservoir fluid and eventually comes to rest through the irreversible action of viscosity. This is true regardless of the shape of the exit (Table 8–4 and Fig. 8–40). Therefore, there is no advantage to rounding off the sharp edges of pipe exits.

TABLE 8–4

Loss coefficients K_L of various pipe components for turbulent flow (for use in the relation $h_L = K_L V^2/(2g)$, where V is the average velocity in the pipe that contains the component)*

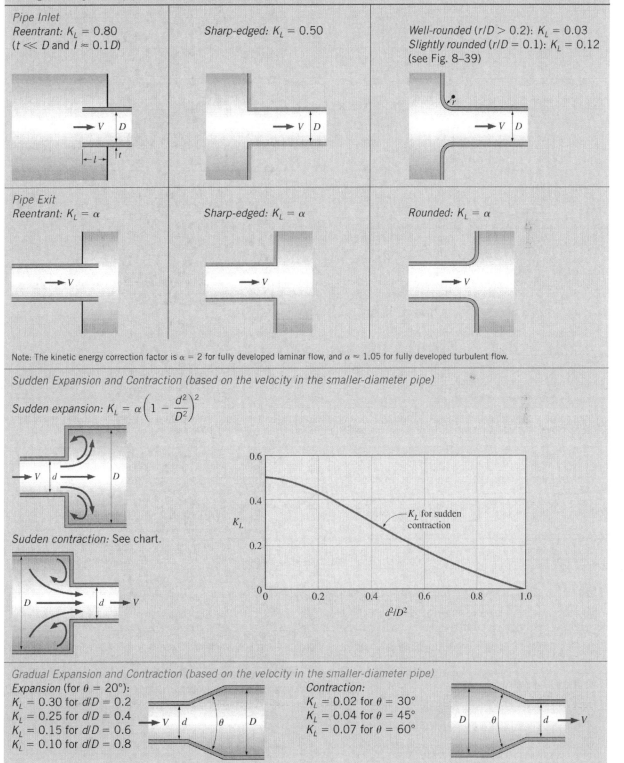

Pipe Inlet
Reentrant: $K_L = 0.80$
($t \ll D$ and $l \approx 0.1D$)

Sharp-edged: $K_L = 0.50$

Well-rounded ($r/D > 0.2$): $K_L = 0.03$
Slightly rounded ($r/D = 0.1$): $K_L = 0.12$
(see Fig. 8–39)

Pipe Exit
Reentrant: $K_L = \alpha$

Sharp-edged: $K_L = \alpha$

Rounded: $K_L = \alpha$

Note: The kinetic energy correction factor is $\alpha = 2$ for fully developed laminar flow, and $\alpha \approx 1.05$ for fully developed turbulent flow.

Sudden Expansion and Contraction (based on the velocity in the smaller-diameter pipe)

Sudden expansion: $K_L = \alpha\left(1 - \dfrac{d^2}{D^2}\right)^2$

Sudden contraction: See chart.

K_L for sudden contraction

Gradual Expansion and Contraction (based on the velocity in the smaller-diameter pipe)
Expansion (for $\theta = 20°$):
$K_L = 0.30$ for $d/D = 0.2$
$K_L = 0.25$ for $d/D = 0.4$
$K_L = 0.15$ for $d/D = 0.6$
$K_L = 0.10$ for $d/D = 0.8$

Contraction:
$K_L = 0.02$ for $\theta = 30°$
$K_L = 0.04$ for $\theta = 45°$
$K_L = 0.07$ for $\theta = 60°$

TABLE 8–4 (CONCLUDED)

Bends and Branches

90° smooth bend:
Flanged: $K_L = 0.3$
Threaded: $K_L = 0.9$

$V \rightarrow$

90° miter bend
(without vanes): $K_L = 1.1$

$V \rightarrow$

90° miter bend
(with vanes): $K_L = 0.2$

$V \rightarrow$

45° threaded elbow:
$K_L = 0.4$

$V \rightarrow$ 45°

180° return bend:
Flanged: $K_L = 0.2$
Threaded: $K_L = 1.5$

$V \rightarrow$

Tee (branch flow):
Flanged: $K_L = 1.0$
Threaded: $K_L = 2.0$

$V \rightarrow$

Tee (line flow):
Flanged: $K_L = 0.2$
Threaded: $K_L = 0.9$

$V \rightarrow$

Threaded union:
$K_L = 0.08$

$V \rightarrow$

Valves

Globe valve, fully open: $K_L = 10$	Gate valve, fully open: $K_L = 0.2$
Angle valve, fully open: $K_L = 5$	$\frac{1}{4}$ closed: $K_L = 0.3$
Ball valve, fully open: $K_L = 0.05$	$\frac{1}{2}$ closed: $K_L = 2.1$
Swing check valve: $K_L = 2$	$\frac{3}{4}$ closed: $K_L = 17$

* These are representative values for loss coefficients. Actual values strongly depend on the design and manufacture of the components and may differ from the given values considerably (especially for valves). Actual manufacturer's data should be used in the final design.

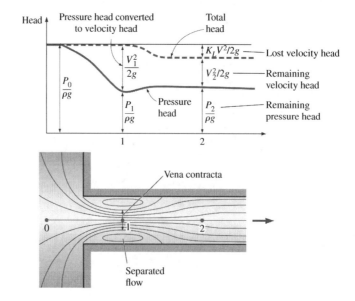

FIGURE 8–38
Graphical representation of flow contraction and the associated head loss at a sharp-edged pipe inlet.

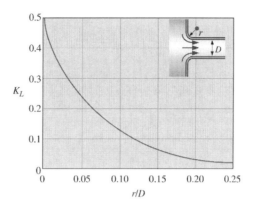

FIGURE 8–39
The effect of rounding of a pipe inlet on the loss coefficient.
Data from ASHRAE Handbook of Fundamentals.

Piping systems often involve *sudden* or *gradual* expansion or contraction sections to accommodate changes in flow rates or properties such as density and velocity. The losses are usually much greater in the case of *sudden* expansion and contraction (or wide-angle expansion) because of flow separation. By combining the equations of mass, momentum, and energy balance, the loss coefficient for the case of a **sudden expansion** is approximated as

$$K_L = \alpha \left(1 - \frac{A_{small}}{A_{large}} \right)^2 \quad \text{(sudden expansion)} \qquad (8\text{–}60)$$

where A_{small} and A_{large} are the cross-sectional areas of the small and large pipes, respectively. Note that $K_L = 0$ when there is no area change ($A_{small} = A_{large}$) and $K_L = \alpha$ when a pipe discharges into a reservoir ($A_{large} \gg A_{small}$). No such relation exists for a sudden contraction, and the K_L values in that case must be read from a chart or table (e.g., Table 8–4). The losses due to expansions and contractions can be reduced significantly by installing conical gradual area changers (nozzles and diffusers) between the small and large pipes. The K_L values for representative cases of gradual expansion and contraction are given in Table 8–4. Note that in head loss calculations, the velocity in the *small pipe* is to be used as the reference velocity in Eq. 8–56. Losses during expansion are usually much higher than the losses during contraction because of flow separation.

Piping systems also involve changes in direction without a change in diameter, and such flow sections are called *bends* or *elbows*. The losses in these devices are due to flow separation (just like a car being thrown off the road when it enters a turn too fast) on the inner side and the swirling secondary flows that result. The losses during changes of direction can be minimized by making the turn "easy" on the fluid by using circular arcs (like 90° elbows) instead of sharp turns (like miter bends) (Fig. 8–41). But the use of sharp turns (and thus suffering a penalty in loss coefficient) may be necessary when the turning space is limited. In such cases, the losses can be minimized by utilizing properly placed guide vanes to help the flow turn in an orderly manner without being thrown off the course. The loss coefficients for some elbows and miter bends as well as tees are given in Table 8–4. These coefficients do not include the frictional losses along the pipe bend. Such losses should be calculated as in straight pipes (using the length of the centerline as the pipe length) and added to other losses.

Mixing
Submerged outlet
Entrained ambient fluid

FIGURE 8–40
All the kinetic energy of the flow is "lost" (turned into thermal energy) through friction as the jet decelerates and mixes with ambient fluid downstream of a submerged outlet.

Flanged elbow
$K_L = 0.3$

Sharp turn
$K_L = 1.1$

FIGURE 8–41
The losses during changes of direction can be minimized by making the turn "easy" on the fluid by using circular arcs instead of sharp turns.

380
INTERNAL FLOW

(a)

A globe
valve

V_1 V_2

Constriction

$V_2 = V_1$

$V_{constriction} > V_1$

(b)

FIGURE 8–42

(a) The large head loss in a partially
closed globe valve is due to
irreversible deceleration, flow
separation, and mixing of high-
velocity fluid coming from the narrow
valve passage. (b) The head loss
through a fully-open ball valve, on the
other hand, is quite small.

Photo by John M. Cimbala.

①   6 cm   9 cm   ②

Water
7 m/s
150 kPa

FIGURE 8–43
Schematic for Example 8–6.

Valves are commonly used in piping systems to control flow rates by
simply altering the head loss until the desired flow rate is achieved. For
valves it is desirable to have a very low loss coefficient when they are fully
open, such as with a *ball valve,* so that they cause minimal head loss during
full-load operation (Fig. 8–42b). Several different valve designs, each with
its own advantages and disadvantages, are in common use today. The *gate
valve* slides up and down like a gate, the *globe valve* (Fig. 8–42a) closes a
hole placed in the valve, the *angle valve* is a globe valve with a 90° turn, and
the *check valve* allows the fluid to flow only in one direction like a diode
in an electric circuit. Table 8–4 lists the representative loss coefficients of
the popular designs. Note that the loss coefficient increases drastically as
a valve is closed. Also, the deviation in the loss coefficients for different
manufacturers is greatest for valves because of their complex geometries.

EXAMPLE 8–6   **Head Loss and Pressure Rise
during Gradual Expansion**

A 6-cm-diameter horizontal water pipe expands gradually to a 9-cm-diameter
pipe (Fig. 8–43). The walls of the expansion section are angled 10° from the
axis. The average velocity and pressure of water before the expansion section
are 7 m/s and 150 kPa, respectively. Determine the head loss in the expansion section and the pressure in the larger-diameter pipe.

SOLUTION A horizontal water pipe expands gradually into a larger-diameter
pipe. The head loss and pressure after the expansion are to be determined.
Assumptions 1 The flow is steady and incompressible. 2 The flow at sections 1
and 2 is fully developed and turbulent with $\alpha_1 = \alpha_2 \cong 1.06$.
Properties We take the density of water to be $\rho = 1000$ kg/m^3. The loss coefficient for a gradual expansion of total included angle $\theta = 20°$ and diameter
ratio $d/D = 6/9$ is $K_L = 0.133$ (by interpolation using Table 8–4).
Analysis Noting that the density of water remains constant, the downstream
velocity of water is determined from conservation of mass to be

$$\dot{m}_1 = \dot{m}_2 \;\rightarrow\; \rho V_1 A_1 = \rho V_2 A_2 \;\rightarrow\; V_2 = \frac{A_1}{A_2} V_1 = \frac{D_1^2}{D_2^2} V_1$$

$$V_2 = \frac{(0.06\ \text{m})^2}{(0.09\ \text{m})^2} (7\ \text{m/s}) = 3.11\ \text{m/s}$$

Then the irreversible head loss in the expansion section becomes

$$h_L = K_L \frac{V_1^2}{2g} = (0.133) \frac{(7\ \text{m/s})^2}{2(9.81\ \text{m/s}^2)} = 0.333\ \text{m}$$

Noting that $z_1 = z_2$ and there are no pumps or turbines involved, the energy
equation for the expansion section is expressed in terms of heads as

$$\frac{P_1}{\rho g} + \alpha_1 \frac{V_1^2}{2g} + \cancel{z_1} + \cancelto{0}{h_{pump,\,u}} = \frac{P_2}{\rho g} + \alpha_2 \frac{V_2^2}{2g} + \cancel{z_2} + \cancelto{0}{h_{turbine,\,e}} + h_L$$

or

$$\frac{P_1}{\rho g} + \alpha_1 \frac{V_1^2}{2g} = \frac{P_2}{\rho g} + \alpha_2 \frac{V_2^2}{2g} + h_L$$

Solving for P_2 and substituting,

$$P_2 = P_1 + \rho \left\{ \frac{\alpha_1 V_1^2 - \alpha_2 V_2^2}{2} - g h_L \right\} = (150 \text{ kPa}) + (1000 \text{ kg/m}^3)$$

$$\times \left\{ \frac{1.06(7 \text{ m/s})^2 - 1.06(3.11 \text{ m/s})^2}{2} - (9.81 \text{ m/s}^2)(0.333 \text{ m}) \right\}$$

$$\times \left(\frac{1 \text{ kN}}{1000 \text{ kg·m/s}^2} \right) \left(\frac{1 \text{ kPa}}{1 \text{ kN/m}^2} \right)$$

$$= \mathbf{168 \text{ kPa}}$$

Therefore, despite the head (and pressure) loss, the pressure *increases* from 150 to 168 kPa after the expansion. This is due to the conversion of dynamic pressure to static pressure when the average flow velocity is decreased in the larger pipe.

Discussion It is common knowledge that higher pressure upstream is necessary to cause flow, and it may come as a surprise to you that the downstream pressure has *increased* after the expansion, despite the loss. This is because the flow is driven by the sum of the three heads that comprise the total head (namely, pressure head, velocity head, and elevation head). During flow expansion, the higher velocity head upstream is converted to pressure head downstream, and this increase outweighs the nonrecoverable head loss. Also, you may be tempted to solve this problem using the Bernoulli equation. Such a solution would ignore the head loss (and the associated pressure loss) and result in an incorrect higher pressure for the fluid downstream.

8–7 · PIPING NETWORKS AND PUMP SELECTION

Series and Parallel Pipes

Most piping systems encountered in practice such as the water distribution systems in cities or commercial or residential establishments involve numerous parallel and series connections as well as several sources (supply of fluid into the system) and loads (discharges of fluid from the system) (Fig. 8–44). A piping project may involve the design of a new system or the expansion of an existing system. The engineering objective in such projects is to design a piping system that will reliably deliver the specified flow rates at specified pressures at minimum total (initial plus operating and maintenance) cost. Once the layout of the system is prepared, the determination of the pipe diameters and the pressures throughout the system, while remaining within the budget constraints, typically requires solving the system repeatedly until the optimal solution is reached. Computer modeling and analysis of such systems make this tedious task a simple chore.

Piping systems typically involve several pipes connected to each other in series and/or in parallel, as shown in Figs. 8–45 and 8–46. When the pipes are connected **in series,** the flow rate through the entire system remains constant regardless of the diameters of the individual pipes in the system. This is a natural consequence of the conservation of mass principle for steady incompressible flow. The total head loss in this case is equal to the sum of the head losses in individual pipes in the system, including the minor losses. The

FIGURE 8–44

A piping network in an industrial facility.

Courtesy UMDE Engineering, Contracting, and Trading. Used by permission.

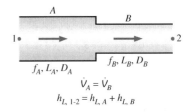

$$\dot{V}_A = \dot{V}_B$$
$$h_{L,\,1\text{-}2} = h_{L,\,A} + h_{L,\,B}$$

FIGURE 8–45

For pipes *in series*, the flow rate is the same in each pipe, and the total head loss is the sum of the head losses in the individual pipes.

FIGURE 8–46

For pipes *in parallel*, the head loss is the same in each pipe, and the total flow rate is the sum of the flow rates in individual pipes.

expansion or contraction losses at connections are considered to belong to the smaller-diameter pipe since the expansion and contraction loss coefficients are defined on the basis of the average velocity in the smaller-diameter pipe.

For a pipe that branches out into two (or more) **parallel pipes** and then rejoins at a junction downstream, the total flow rate is the sum of the flow rates in the individual pipes. The pressure drop (or head loss) in each individual pipe connected in parallel must be the same since $\Delta P = P_A - P_B$ and the junction pressures P_A and P_B are the same for all the individual pipes. For a system of two parallel pipes 1 and 2 between junctions A and B with negligible minor losses, this is expressed as

$$h_{L,1} = h_{L,2} \quad \rightarrow \quad f_1 \frac{L_1}{D_1} \frac{V_1^2}{2g} = f_2 \frac{L_2}{D_2} \frac{V_2^2}{2g}$$

Then the ratio of the average velocities and the flow rates in the two parallel pipes become

$$\frac{V_1}{V_2} = \left(\frac{f_2 L_2 D_1}{f_1 L_1 D_2} \right)^{1/2} \quad \text{and} \quad \frac{\dot{V}_1}{\dot{V}_2} = \frac{A_{c,1}V_1}{A_{c,2}V_2} = \frac{D_1^2}{D_2^2} \left(\frac{f_2 L_2 D_1}{f_1 L_1 D_2} \right)^{1/2}$$

Therefore, the relative flow rates in parallel pipes are established from the requirement that the head loss in each pipe be the same. This result can be extended to any number of pipes connected in parallel. The result is also valid for pipes for which the minor losses are significant if the equivalent lengths for components that contribute to minor losses are added to the pipe length. Note that the flow rate in one of the parallel branches is proportional to its diameter to the power 5/2 and is inversely proportional to the square root of its length and friction factor.

The analysis of piping networks, no matter how complex they are, is based on two simple principles:

1. *Conservation of mass throughout the system must be satisfied.* This is done by requiring the total flow into a junction to be equal to the total flow out of the junction for all junctions in the system. Also, the flow rate must remain constant in pipes connected in series regardless of the changes in diameters.

2. *Pressure drop (and thus head loss) between two junctions must be the same for all paths between the two junctions.* This is because pressure is a point function and it cannot have two values at a specified point. In practice this rule is used by requiring that the algebraic sum of head losses in a loop (for all loops) be equal to zero. (A head loss is taken to be positive for flow in the clockwise direction and negative for flow in the counterclockwise direction.)

Therefore, the analysis of piping networks is very similar to the analysis of electric circuits (Kirchhoff's laws), with flow rate corresponding to electric current and pressure corresponding to electric potential. However, the situation is much more complex here since, unlike the electrical resistance, the "flow resistance" is a highly nonlinear function. Therefore, the analysis of piping networks requires the simultaneous solution of a system of nonlinear equations, which requires software such as EES, Mathcad, Matlab, etc., or commercially available software designed specifically for such applications.

Piping Systems with Pumps and Turbines

When a piping system involves a pump and/or turbine, the steady-flow energy equation on a unit-mass basis is expressed as (see Section 5–6)

$$\frac{P_1}{\rho} + \alpha_1 \frac{V_1^2}{2} + gz_1 + w_{\text{pump},u} = \frac{P_2}{\rho} + \alpha_2 \frac{V_2^2}{2} + gz_2 + w_{\text{turbine},e} + gh_L \qquad (8\text{–}61)$$

or in terms of heads as

$$\frac{P_1}{\rho g} + \alpha_1 \frac{V_1^2}{2g} + z_1 + h_{\text{pump},u} = \frac{P_2}{\rho g} + \alpha_2 \frac{V_2^2}{2g} + z_2 + h_{\text{turbine},e} + h_L \qquad (8\text{–}62)$$

where $h_{\text{pump},u} = w_{\text{pump},u}/g$ is the useful pump head delivered to the fluid, $h_{\text{turbine},e} = w_{\text{turbine},e}/g$ is the turbine head extracted from the fluid, α is the kinetic energy correction factor whose value is about 1.05 for most (turbulent) flows encountered in practice, and h_L is the total head loss in the piping (including the minor losses if they are significant) between points 1 and 2. The pump head is zero if the piping system does not involve a pump or a fan, the turbine head is zero if the system does not involve a turbine, and both are zero if the system does not involve any mechanical work-producing or work-consuming devices.

Many practical piping systems involve a pump to move a fluid from one reservoir to another. Taking points 1 and 2 to be at the *free surfaces* of the reservoirs (Fig. 8–47), the energy equation is solved for the required useful pump head, yielding

$$h_{\text{pump},u} = (z_2 - z_1) + h_L \qquad (8\text{–}63)$$

since the velocities at free surfaces are negligible for large reservoirs and the pressures are at atmospheric pressure. Therefore, the useful pump head is equal to the elevation difference between the two reservoirs plus the head loss. If the head loss is negligible compared to $z_2 - z_1$, the useful pump head is equal to the elevation difference between the two reservoirs. In the case of $z_1 > z_2$ (the first reservoir being at a higher elevation than the second one) with no pump, the flow is driven by gravity at a flow rate that causes a head loss equal to the elevation difference. A similar argument can be given for the turbine head for a hydroelectric power plant by replacing $h_{\text{pump},u}$ in Eq. 8–63 by $-h_{\text{turbine},e}$.

Once the useful pump head is known, the *mechanical power that needs to be delivered by the pump to the fluid* and the *electric power consumed by the motor of the pump* for a specified flow rate are determined from

$$\dot{W}_{\text{pump,shaft}} = \frac{\rho \dot{V} g h_{\text{pump},u}}{\eta_{\text{pump}}} \quad \text{and} \quad \dot{W}_{\text{elect}} = \frac{\rho \dot{V} g h_{\text{pump},u}}{\eta_{\text{pump–motor}}} \qquad (8\text{–}64)$$

where $\eta_{\text{pump–motor}}$ is the *efficiency of the pump–motor combination,* which is the product of the pump and the motor efficiencies (Fig. 8–48). The pump–motor efficiency is defined as the ratio of the net mechanical energy delivered to the fluid by the pump to the electric energy consumed by the motor of the pump, and it typically ranges between 50 and 85 percent.

The head loss of a piping system increases (usually quadratically) with the flow rate. A plot of required useful pump head $h_{\text{pump},u}$ as a function of flow rate is called the **system** (or **demand**) **curve.** The head produced by a pump is not a constant either. Both the pump head and the pump efficiency vary

FIGURE 8–47

When a pump moves a fluid from one reservoir to another, the useful pump head requirement is equal to the elevation difference between the two reservoirs plus the head loss.

$$h_{\text{pump},u} = (z_2 - z_1) + h_L$$
$$\dot{W}_{\text{pump},u} = \rho \dot{V} g h_{\text{pump},u}$$

$$\eta_{\text{pump–motor}} = \eta_{\text{pump}} \eta_{\text{motor}}$$
$$= 0.70 \times 0.90 = 0.63$$

FIGURE 8–48

The efficiency of the pump–motor combination is the product of the pump and the motor efficiencies.

Photo by Yunus Çengel.

FIGURE 8–49

Characteristic pump curves for centrifugal pumps, the system curve for a piping system, and the operating point.

with the flow rate, and pump manufacturers supply this variation in tabular or graphical form, as shown in Fig. 8–49. These experimentally determined $h_{pump,\,u}$ and $\eta_{pump,\,u}$ versus $\dot{V}$ curves are called **characteristic (or supply or performance) curves.** Note that the flow rate of a pump increases as the required head decreases. The intersection point of the pump head curve with the vertical axis typically represents the *maximum head* (called the **shutoff head**) the pump can provide, while the intersection point with the horizontal axis indicates the *maximum flow rate* (called the **free delivery**) that the pump can supply.

The *efficiency* of a pump is highest at a certain combination of head and flow rate. Therefore, a pump that can supply the required head and flow rate is not necessarily a good choice for a piping system unless the efficiency of the pump at those conditions is sufficiently high. The pump installed in a piping system will operate at the point where the *system curve* and the *characteristic curve* intersect. This point of intersection is called the **operating point,** as shown in Fig. 8–46. The useful head produced by the pump at this point matches the head requirements of the system at that flow rate. Also, the efficiency of the pump during operation is the value corresponding to that flow rate.

EXAMPLE 8–7   **Pumping Water through Two Parallel Pipes**

Water at 20°C is to be pumped from a reservoir ($z_A = 5$ m) to another reservoir at a higher elevation ($z_B = 13$ m) through two 36-m-long pipes connected in parallel, as shown in Fig. 8–50. The pipes are made of commercial steel, and the diameters of the two pipes are 4 and 8 cm. Water is to be pumped by a 70 percent efficient motor–pump combination that draws 8 kW of electric power during operation. The minor losses and the head loss in pipes that connect the parallel pipes to the two reservoirs are considered to be negligible. Determine the total flow rate between the reservoirs and the flow rate through each of the parallel pipes.

FIGURE 8–50
The piping system discussed in
Example 8–7.

SOLUTION   The pumping power input to a piping system with two parallel pipes is given. The flow rates are to be determined.

Assumptions   **1** The flow is steady (since the reservoirs are large) and incompressible. **2** The entrance effects are negligible, and thus the flow is fully developed. **3** The elevations of the reservoirs remain constant. **4** The minor losses and the head loss in pipes other than the parallel pipes are negligible. **5** Flows through both pipes are turbulent (to be verified).

Properties   The density and dynamic viscosity of water at 20°C are $\rho = 998 \text{ kg/m}^3$ and $\mu = 1.002 \times 10^{-3}$ kg/m·s. The roughness of commercial steel pipe is $\varepsilon = 0.000045$ m (Table 8–2).

Analysis   This problem cannot be solved directly since the velocities (or flow rates) in the pipes are not known. Therefore, we would normally use a trial-and-error approach here. However, equation solvers such as EES are widely available, and thus, we simply set up the equations to be solved by an equation solver. The useful head supplied by the pump to the fluid is determined from

$$\dot{W}_{elect} = \frac{\rho \dot{V} g h_{pump,u}}{\eta_{pump-motor}} \quad \rightarrow \quad 8000 \text{ W} = \frac{(998 \text{ kg/m}^3)\dot{V}(9.81 \text{ m/s}^2)h_{pump,u}}{0.70} \quad (1)$$

We choose points A and B at the free surfaces of the two reservoirs. Noting that the fluid at both points is open to the atmosphere (and thus $P_A = P_B = P_{atm}$) and that the fluid velocities at both points are nearly zero ($V_A \approx V_B \approx 0$) since the reservoirs are large, the energy equation for a control volume between these two points simplifies to

$$\frac{P_A}{\rho g} + \alpha_A \frac{V_A^2}{2g}\!\!\nearrow^{0} + z_A + h_{pump,u} = \frac{P_B}{\rho g} + \alpha_B \frac{V_B^2}{2g}\!\!\nearrow^{0} + z_B + h_L$$

or

$$h_{pump,u} = (z_B - z_A) + h_L$$

or

$$h_{pump,u} = (13 \text{ m} - 5 \text{ m}) + h_L \quad (2)$$

where

$$h_L = h_{L,1} = h_{L,2} \quad (3)(4)$$

We designate the 4-cm-diameter pipe by 1 and the 8-cm-diameter pipe by 2. Equations for the average velocity, the Reynolds number, the friction factor, and the head loss in each pipe are

$$V_1 = \frac{\dot{V}_1}{A_{c,1}} = \frac{\dot{V}_1}{\pi D_1^2/4} \quad \rightarrow \quad V_1 = \frac{\dot{V}_1}{\pi(0.04\text{ m})^2/4} \tag{5}$$

$$V_2 = \frac{\dot{V}_2}{A_{c,2}} = \frac{\dot{V}_2}{\pi D_2^2/4} \quad \rightarrow \quad V_2 = \frac{\dot{V}_2}{\pi(0.08\text{ m})^2/4} \tag{6}$$

$$\text{Re}_1 = \frac{\rho V_1 D_1}{\mu} \quad \rightarrow \quad \text{Re}_1 = \frac{(998\text{ kg/m}^3)V_1(0.04\text{ m})}{1.002 \times 10^{-3}\text{ kg/m·s}} \tag{7}$$

$$\text{Re}_2 = \frac{\rho V_2 D_2}{\mu} \quad \rightarrow \quad \text{Re}_2 = \frac{(998\text{ kg/m}^3)V_2(0.08\text{ m})}{1.002 \times 10^{-3}\text{ kg/m·s}} \tag{8}$$

$$\frac{1}{\sqrt{f_1}} = -2.0\log\left(\frac{\varepsilon/D_1}{3.7} + \frac{2.51}{\text{Re}_1\sqrt{f_1}}\right)$$

$$\rightarrow \quad \frac{1}{\sqrt{f_1}} = -2.0\log\left(\frac{0.000045}{3.7 \times 0.04} + \frac{2.51}{\text{Re}_1\sqrt{f_1}}\right) \tag{9}$$

$$\frac{1}{\sqrt{f_2}} = -2.0\log\left(\frac{\varepsilon/D_2}{3.7} + \frac{2.51}{\text{Re}_2\sqrt{f_2}}\right)$$

$$\rightarrow \quad \frac{1}{\sqrt{f_2}} = -2.0\log\left(\frac{0.000045}{3.7 \times 0.08} + \frac{2.51}{\text{Re}_2\sqrt{f_2}}\right) \tag{10}$$

$$h_{L,1} = f_1\frac{L_1}{D_1}\frac{V_1^2}{2g} \quad \rightarrow \quad h_{L,1} = f_1\frac{36\text{ m}}{0.04\text{ m}}\frac{V_1^2}{2(9.81\text{ m/s}^2)} \tag{11}$$

$$h_{L,2} = f_2\frac{L_2}{D_2}\frac{V_2^2}{2g} \quad \rightarrow \quad h_{L,2} = f_2\frac{36\text{ m}}{0.08\text{ m}}\frac{V_2^2}{2(9.81\text{ m/s}^2)} \tag{12}$$

$$\dot{V} = \dot{V}_1 + \dot{V}_2 \tag{13}$$

This is a system of 13 equations in 13 unknowns, and their simultaneous solution by an equation solver gives

$$\dot{V} = 0.0300\text{ m}^3\text{/s}, \quad \dot{V}_1 = 0.00415\text{ m}^3\text{/s}, \quad \dot{V}_2 = 0.0259\text{ m}^3\text{/s}$$

$$V_1 = 3.30\text{ m/s}, \quad V_2 = 5.15\text{ m/s}, \quad h_L = h_{L,1} = h_{L,2} = 11.1\text{ m}, \quad h_{\text{pump}} = 19.1\text{ m}$$

$$\text{Re}_1 = 131,600, \quad \text{Re}_2 = 410,000, \quad f_1 = 0.0221, \quad f_2 = 0.0182$$

Note that Re > 4000 for both pipes, and thus the assumption of turbulent flow is verified.

Discussion The two parallel pipes have the same length and roughness, but the diameter of the first pipe is half the diameter of the second one. Yet only 14 percent of the water flows through the first pipe. This shows the strong dependence of the flow rate on diameter. Also, it can be shown that if the free surfaces of the two reservoirs were at the same elevation (and thus $z_A = z_B$), the flow rate would increase by 20 percent from 0.0300 to 0.0361 m³/s. Alternately, if the reservoirs were as given but the irreversible head losses were negligible, the flow rate would become 0.0715 m³/s (an increase of 138 percent).

EXAMPLE 8–8   Gravity-Driven Water Flow in a Pipe

Water at 10°C flows from a large reservoir to a smaller one through a 5-cm-diameter cast iron piping system, as shown in Fig. 8–51. Determine the elevation z_1 for a flow rate of 6 L/s.

SOLUTION   The flow rate through a piping system connecting two reservoirs is given. The elevation of the source is to be determined.
Assumptions   **1** The flow is steady and incompressible. **2** The elevations of the reservoirs remain constant. **3** There are no pumps or turbines in the line.
Properties   The density and dynamic viscosity of water at 10°C are $\rho = 999.7$ kg/m³ and $\mu = 1.307 \times 10^{-3}$ kg/m·s. The roughness of cast iron pipe is $\varepsilon = 0.00026$ m (Table 8–2).
Analysis   The piping system involves 89 m of piping, a sharp-edged entrance ($K_L = 0.5$), two standard flanged elbows ($K_L = 0.3$ each), a fully open gate valve ($K_L = 0.2$), and a submerged exit ($K_L = 1.06$). We choose points 1 and 2 at the free surfaces of the two reservoirs. Noting that the fluid at both points is open to the atmosphere (and thus $P_1 = P_2 = P_{atm}$) and that the fluid velocities at both points are nearly zero ($V_1 \approx V_2 \approx 0$), the energy equation for a control volume between these two points simplifies to

$$\cancel{\frac{P_1}{\rho g}} + \alpha_1 \overset{0}{\cancel{\frac{V_1^2}{2g}}} + z_1 = \cancel{\frac{P_2}{\rho g}} + \alpha_2 \overset{0}{\cancel{\frac{V_2^2}{2g}}} + z_2 + h_L \quad \rightarrow \quad z_1 = z_2 + h_L$$

where

$$h_L = h_{L,\,total} = h_{L,\,major} + h_{L,\,minor} = \left(f\frac{L}{D} + \sum K_L\right)\frac{V^2}{2g}$$

since the diameter of the piping system is constant. The average velocity in the pipe and the Reynolds number are

$$V = \frac{\dot{V}}{A_c} = \frac{\dot{V}}{\pi D^2/4} = \frac{0.006 \text{ m}^3/\text{s}}{\pi (0.05 \text{ m})^2/4} = 3.06 \text{ m/s}$$

$$\text{Re} = \frac{\rho V D}{\mu} = \frac{(999.7 \text{ kg/m}^3)(3.06 \text{ m/s})(0.05 \text{ m})}{1.307 \times 10^{-3} \text{ kg/m·s}} = 117,000$$

The flow is turbulent since Re > 4000. Noting that $\varepsilon/D = 0.00026/0.05 = 0.0052$, the friction factor is determined from the Colebrook equation (or the Moody chart),

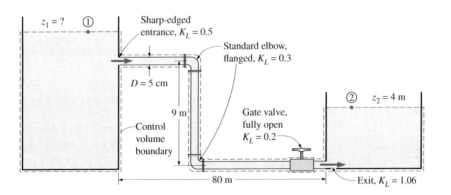

$z_1 = ?$ ①   Sharp-edged entrance, $K_L = 0.5$

Standard elbow, flanged, $K_L = 0.3$

$D = 5$ cm

Control volume boundary

9 m

Gate valve, fully open $K_L = 0.2$

② $z_2 = 4$ m

80 m

Exit, $K_L = 1.06$

FIGURE 8–51
The piping system discussed in Example 8–8.

$$\frac{1}{\sqrt{f}} = -2.0 \log\left(\frac{\varepsilon/D}{3.7} + \frac{2.51}{\text{Re}\sqrt{f}}\right) \rightarrow \frac{1}{\sqrt{f}} = -2.0 \log\left(\frac{0.0052}{3.7} + \frac{2.51}{117{,}000\sqrt{f}}\right)$$

It gives $f = 0.0315$. The sum of the loss coefficients is

$$\sum K_L = K_{L,\,\text{entrance}} + 2K_{L,\,\text{elbow}} + K_{L,\,\text{valve}} + K_{L,\,\text{exit}}$$

$$= 0.5 + 2 \times 0.3 + 0.2 + 1.06 = 2.36$$

Then the total head loss and the elevation of the source become

$$h_L = \left(f\frac{L}{D} + \sum K_L\right)\frac{V^2}{2g} = \left(0.0315\,\frac{89\text{ m}}{0.05\text{ m}} + 2.36\right)\frac{(3.06\text{ m/s})^2}{2(9.81\text{ m/s}^2)} = 27.9\text{ m}$$

$$z_1 = z_2 + h_L = 4 + 27.9 = \mathbf{31.9\text{ m}}$$

Therefore, the free surface of the first reservoir must be 31.9 m above the ground level to ensure water flow between the two reservoirs at the specified rate.

Discussion Note that $fL/D = 56.1$ in this case, which is about 24 times the total minor loss coefficient. Therefore, ignoring the sources of minor losses in this case would result in about 4 percent error. It can be shown that at the same flow rate, the total head loss would be 35.9 m (instead of 27.9 m) if the valve were three-fourths closed, and it would drop to 24.8 m if the pipe between the two reservoirs were straight at the ground level (thus eliminating the elbows and the vertical section of the pipe). The head loss could be reduced further (from 24.8 to 24.6 m) by rounding the entrance. The head loss can be reduced significantly (from 27.9 to 16.0 m) by replacing the cast iron pipes by smooth pipes such as those made of plastic.

EXAMPLE 8–9   Effect of Flushing on Flow Rate from a Shower

The bathroom plumbing of a building consists of 1.5-cm-diameter copper pipes with threaded connectors, as shown in Fig. 8–52. (*a*) If the gage pressure at the inlet of the system is 200 kPa during a shower and the toilet reservoir is full (no flow in that branch), determine the flow rate of water through the shower head. (*b*) Determine the effect of flushing of the toilet on the flow rate through the shower head. Take the loss coefficients of the shower head and the reservoir to be 12 and 14, respectively.

FIGURE 8–52
Schematic for Example 8–9.

SOLUTION The cold-water plumbing system of a bathroom is given. The flow rate through the shower and the effect of flushing the toilet on the flow rate are to be determined.

Assumptions **1** The flow is steady and incompressible. **2** The flow is turbulent and fully developed. **3** The reservoir is open to the atmosphere. **4** The velocity heads are negligible.

Properties The properties of water at 20°C are $\rho = 998$ kg/m^3, $\mu = 1.002 \times 10^{-3}$ kg/m·s, and $\nu = \mu/\rho = 1.004 \times 10^{-6}$ m^2/s. The roughness of copper pipes is $\varepsilon = 1.5 \times 10^{-6}$ m.

Analysis This is a problem of the second type since it involves the determination of the flow rate for a specified pipe diameter and pressure drop. The solution involves an iterative approach since the flow rate (and thus the flow velocity) is not known.

(*a*) The piping system of the shower alone involves 11 m of piping, a tee with line flow ($K_L = 0.9$), two standard elbows ($K_L = 0.9$ each), a fully open globe valve ($K_L = 10$), and a shower head ($K_L = 12$). Therefore, $\sum K_L = 0.9 + 2 \times 0.9 + 10 + 12 = 24.7$. Noting that the shower head is open to the atmosphere, and the velocity heads are negligible, the energy equation for a control volume between points 1 and 2 simplifies to

$$\frac{P_1}{\rho g} + \alpha_1 \frac{V_1^2}{2g} + z_1 + h_{\text{pump}, u} = \frac{P_2}{\rho g} + \alpha_2 \frac{V_2^2}{2g} + z_2 + h_{\text{turbine}, e} + h_L$$

$$\rightarrow \quad \frac{P_{1, \text{gage}}}{\rho g} = (z_2 - z_1) + h_L$$

Therefore, the head loss is

$$h_L = \frac{200{,}000 \text{ N/m}^2}{(998 \text{ kg/m}^3)(9.81 \text{ m/s}^2)} - 2 \text{ m} = 18.4 \text{ m}$$

Also,

$$h_L = \left(f \frac{L}{D} + \sum K_L \right) \frac{V^2}{2g} \quad \rightarrow \quad 18.4 = \left(f \frac{11 \text{ m}}{0.015 \text{ m}} + 24.7 \right) \frac{V^2}{2(9.81 \text{ m/s}^2)}$$

since the diameter of the piping system is constant. Equations for the average velocity in the pipe, the Reynolds number, and the friction factor are

$$V = \frac{\dot{V}}{A_c} = \frac{\dot{V}}{\pi D^2/4} \quad \rightarrow \quad V = \frac{\dot{V}}{\pi(0.015 \text{ m})^2/4}$$

$$\text{Re} = \frac{VD}{\nu} \quad \rightarrow \quad \text{Re} = \frac{V(0.015 \text{ m})}{1.004 \times 10^{-6} \text{ m}^2/\text{s}}$$

$$\frac{1}{\sqrt{f}} = -2.0 \log\left(\frac{\varepsilon/D}{3.7} + \frac{2.51}{\text{Re}\sqrt{f}} \right)$$

$$\rightarrow \quad \frac{1}{\sqrt{f}} = -2.0 \log\left(\frac{1.5 \times 10^{-6} \text{ m}}{3.7(0.015 \text{ m})} + \frac{2.51}{\text{Re}\sqrt{f}} \right)$$

This is a set of four equations with four unknowns, and solving them with an equation solver such as EES gives

$$\dot{V} = 0.00053 \text{ m}^3/\text{s}, \quad f = 0.0218, \quad V = 2.98 \text{ m/s}, \quad \text{and} \quad Re = 44{,}550$$

Therefore, the flow rate of water through the shower head is **0.53 L/s.**

(*b*) When the toilet is flushed, the float moves and opens the valve. The discharged water starts to refill the reservoir, resulting in parallel flow after the tee connection. The head loss and minor loss coefficients for the shower branch were determined in (*a*) to be $h_{L,2} = 18.4$ m and $\sum K_{L,2} = 24.7$, respectively. The corresponding quantities for the reservoir branch can be determined similarly to be

$$h_{L,3} = \frac{200{,}000 \text{ N/m}^2}{(998 \text{ kg/m}^3)(9.81 \text{ m/s}^2)} - 1 \text{ m} = 19.4 \text{ m}$$

$$\sum K_{L,3} = 2 + 10 + 0.9 + 14 = 26.9$$

The relevant equations in this case are

$$\dot{V}_1 = \dot{V}_2 + \dot{V}_3$$

$$h_{L,2} = f_1 \frac{5 \text{ m}}{0.015 \text{ m}} \frac{V_1^2}{2(9.81 \text{ m/s}^2)} + \left(f_2 \frac{6 \text{ m}}{0.015 \text{ m}} + 24.7 \right) \frac{V_2^2}{2(9.81 \text{ m/s}^2)} = 18.4$$

$$h_{L,3} = f_1 \frac{5 \text{ m}}{0.015 \text{ m}} \frac{V_1^2}{2(9.81 \text{ m/s}^2)} + \left(f_3 \frac{1 \text{ m}}{0.015 \text{ m}} + 26.9 \right) \frac{V_3^2}{2(9.81 \text{ m/s}^2)} = 19.4$$

$$V_1 = \frac{\dot{V}_1}{\pi(0.015 \text{ m})^2/4}, \quad V_2 = \frac{\dot{V}_2}{\pi(0.015 \text{ m})^2/4}, \quad V_3 = \frac{\dot{V}_3}{\pi(0.015 \text{ m})^2/4}$$

$$Re_1 = \frac{V_1(0.015 \text{ m})}{1.004 \times 10^{-6} \text{m}^2/\text{s}}, \quad Re_2 = \frac{V_2(0.015 \text{ m})}{1.004 \times 10^{-6} \text{m}^2/\text{s}}, \quad Re_3 = \frac{V_3(0.015 \text{ m})}{1.004 \times 10^{-6} \text{m}^2/\text{s}}$$

$$\frac{1}{\sqrt{f_1}} = -2.0 \log\left(\frac{1.5 \times 10^{-6} \text{ m}}{3.7(0.015 \text{ m})} + \frac{2.51}{Re_1 \sqrt{f_1}} \right)$$

$$\frac{1}{\sqrt{f_2}} = -2.0 \log\left(\frac{1.5 \times 10^{-6} \text{ m}}{3.7(0.015 \text{ m})} + \frac{2.51}{Re_2 \sqrt{f_2}} \right)$$

$$\frac{1}{\sqrt{f_3}} = -2.0 \log\left(\frac{1.5 \times 10^{-6} \text{ m}}{3.7(0.015 \text{ m})} + \frac{2.51}{Re_3 \sqrt{f_3}} \right)$$

Solving these 12 equations in 12 unknowns simultaneously using an equation solver, the flow rates are determined to be

$$\dot{V}_1 = 0.00090 \text{ m}^3/\text{s}, \quad \dot{V}_2 = 0.00042 \text{ m}^3/\text{s}, \quad \text{and} \quad \dot{V}_3 = 0.00048 \text{ m}^3/\text{s}$$

Therefore, the flushing of the toilet **reduces the flow rate of cold water through the shower by 21 percent** from 0.53 to 0.42 L/s, causing the shower water to suddenly get very hot (Fig. 8–53).

Discussion If the velocity heads were considered, the flow rate through the shower would be 0.43 instead of 0.42 L/s. Therefore, the assumption of negligible velocity heads is reasonable in this case. Note that a leak in a piping system would cause the same effect, and thus an unexplained drop in flow rate at an end point may signal a leak in the system.

Ouch!

FIGURE 8–53
Flow rate of cold water through a shower may be affected significantly by the flushing of a nearby toilet.

8–8 ▪ FLOW RATE AND VELOCITY MEASUREMENT

A major application area of fluid mechanics is the determination of the flow rate of fluids, and numerous devices have been developed over the years for the purpose of flow metering. Flowmeters range widely in their level of sophistication, size, cost, accuracy, versatility, capacity, pressure drop, and the operating principle. We give an overview of the meters commonly used to measure the flow rate of liquids and gases flowing through pipes or ducts. We limit our consideration to incompressible flow.

Some flowmeters measure the flow rate directly by discharging and recharging a measuring chamber of known volume continuously and keeping track of the number of discharges per unit time. But most flowmeters measure the flow rate indirectly—they measure the average velocity V or a quantity that is related to average velocity such as pressure and drag, and determine the volume flow rate $\dot{V}$ from

$$\dot{V} = VA_c \qquad (8\text{--}65)$$

where A_c is the cross-sectional area of flow. Therefore, measuring the flow rate is usually done by measuring flow velocity, and many flowmeters are simply velocimeters used for the purpose of metering flow.

The velocity in a pipe varies from zero at the wall to a maximum at the center, and it is important to keep this in mind when taking velocity measurements. For laminar flow, for example, the average velocity is half the centerline velocity. But this is not the case in turbulent flow, and it may be necessary to take the weighted average or an integral of several local velocity measurements to determine the average velocity.

The flow rate measurement techniques range from very crude to very elegant. The flow rate of water through a garden hose, for example, can be measured simply by collecting the water in a bucket of known volume and dividing the amount collected by the collection time (Fig. 8–54). A crude way of estimating the flow velocity of a river is to drop a float on the river and measure the drift time between two specified locations. At the other extreme, some flowmeters use the propagation of sound in flowing fluids while others use the electromotive force generated when a fluid passes through a magnetic field. In this section we discuss devices that are commonly used to measure velocity and flow rate, starting with the Pitot-static probe introduced in Chap. 5.

FIGURE 8–54
A primitive (but fairly accurate) way of measuring the flow rate of water through a garden hose involves collecting water in a bucket and recording the collection time.

Pitot and Pitot-Static Probes

Pitot probes (also called *Pitot tubes*) and **Pitot-static probes,** named after the French engineer Henri de Pitot (1695–1771), are widely used for flow speed measurement. A Pitot probe is just a tube with a pressure tap at the stagnation point that measures stagnation pressure, while a Pitot-static probe has both a stagnation pressure tap and several circumferential static pressure taps and it measures both stagnation and static pressures (Figs. 8–55 and 8–56). Pitot was the first person to measure velocity with the upstream pointed tube, while French engineer Henry Darcy (1803–1858) developed most of the features of the instruments we use today, including the use of small openings and the placement of the static tube on the same assembly. Therefore, it is more appropriate to call the Pitot-static probes **Pitot–Darcy probes.**

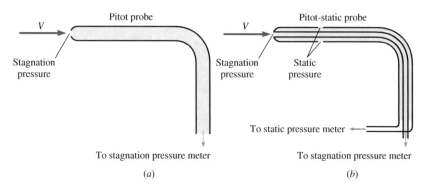

(a)   (b)

FIGURE 8–55

(a) A Pitot probe measures stagnation pressure at the nose of the probe, while (b) a Pitot-static probe measures both stagnation pressure and static pressure, from which the flow speed is calculated.

FIGURE 8–55

(a) A Pitot probe measures stagnation pressure at the nose of the probe, while (b) a Pitot-static probe measures both stagnation pressure and static pressure, from which the flow speed is calculated.

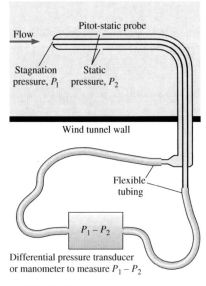

FIGURE 8–56

Measuring flow velocity with a Pitot-static probe. (A manometer may be used in place of the differential pressure transducer.)

FIGURE 8–57

Close-up of a Pitot-static probe, showing the stagnation pressure hole and two of the five static circumferential pressure holes.

Photo by Po-Ya Abel Chuang.

The Pitot-static probe measures local velocity by measuring the pressure difference in conjunction with the Bernoulli equation. It consists of a slender double-tube aligned with the flow and connected to a differential pressure meter. The inner tube is fully open to flow at the nose, and thus it measures the stagnation pressure at that location (point 1). The outer tube is sealed at the nose, but it has holes on the side of the outer wall (point 2) and thus it measures the static pressure. For incompressible flow with sufficiently high velocities (so that the frictional effects between points 1 and 2 are negligible), the Bernoulli equation is applicable and is expressed as

$$\frac{P_1}{\rho g} + \frac{V_1^2}{2g} + z_1 = \frac{P_2}{\rho g} + \frac{V_2^2}{2g} + z_2 \tag{8-66}$$

Noting that $z_1 \cong z_2$ since the static pressure holes of the Pitot-static probe are arranged circumferentially around the tube and $V_1 = 0$ because of the stagnation conditions, the flow velocity $V = V_2$ becomes

Pitot formula: $$V = \sqrt{\frac{2(P_1 - P_2)}{\rho}} \tag{8-67}$$

which is known as the **Pitot formula.** If the velocity is measured at a location where the local velocity is equal to the average flow velocity, the volume flow rate can be determined from $\dot{V} = VA_c$.

The Pitot-static probe is a simple, inexpensive, and highly reliable device since it has no moving parts (Fig. 8–57). It also causes very small pressure drop and usually does not disturb the flow appreciably. However, it is important that it be properly aligned with the flow to avoid significant errors that may be caused by misalignment. Also, the difference between the static and stagnation pressures (which is the dynamic pressure) is proportional to the density of the fluid and the square of the flow velocity. It is used to measure velocity in both liquids and gases. Noting that gases have low densities, the flow velocity should be sufficiently high when the Pitot-static probe is used for gas flow such that a measurable dynamic pressure develops.

Obstruction Flowmeters: Orifice, Venturi, and Nozzle Meters

Consider incompressible steady flow of a fluid in a horizontal pipe of diameter D that is constricted to a flow area of diameter d, as shown in Fig. 8–58. The mass balance and the Bernoulli equations between a location before the constriction (point 1) and the location where constriction occurs (point 2) are written as

Mass balance: $\dot{V} = A_1 V_1 = A_2 V_2 \rightarrow V_1 = (A_2/A_1)V_2 = (d/D)^2 V_2$ **(8–68)**

Bernoulli equation $(z_1 = z_2)$: $\dfrac{P_1}{\rho g} + \dfrac{V_1^2}{2g} = \dfrac{P_2}{\rho g} + \dfrac{V_2^2}{2g}$ **(8–69)**

Combining Eqs. 8–68 and 8–69 and solving for velocity V_2 gives

Obstruction (with no loss): $V_2 = \sqrt{\dfrac{2(P_1 - P_2)}{\rho(1 - \beta^4)}}$ **(8–70)**

FIGURE 8–58
Flow through a constriction in a pipe.

where $\beta = d/D$ is the diameter ratio. Once V_2 is known, the flow rate can be determined from $\dot{V} = A_2 V_2 = (\pi d^2/4)V_2$.

This simple analysis shows that the flow rate through a pipe can be determined by constricting the flow and measuring the decrease in pressure due to the increase in velocity at the constriction site. Noting that the pressure drop between two points along the flow is measured easily by a differential pressure transducer or manometer, it appears that a simple flow rate measurement device can be built by obstructing the flow. Flowmeters based on this principle are called **obstruction flowmeters** and are widely used to measure flow rates of gases and liquids.

The velocity in Eq. 8–70 is obtained by assuming no loss, and thus it is the maximum velocity that can occur at the constriction site. In reality, some pressure losses due to frictional effects are inevitable, and thus the actual velocity is less. Also, the fluid stream continues to contract past the obstruction, and the vena contracta area is less than the flow area of the obstruction. Both losses can be accounted for by incorporating a correction factor called the **discharge coefficient** C_d whose value (which is less than 1) is determined experimentally. Then the flow rate for obstruction flowmeters is expressed as

Obstruction flowmeters: $\dot{V} = A_0 C_d \sqrt{\dfrac{2(P_1 - P_2)}{\rho(1 - \beta^4)}}$ **(8–71)**

where $A_0 = A_2 = \pi d^2/4$ is the cross-sectional area of the throat or orifice and $\beta = d/D$ is the ratio of throat diameter to pipe diameter. The value of C_d depends on both β and the Reynolds number $\mathrm{Re} = V_1 D/\nu$, and charts and curve-fit correlations for C_d are available for various types of obstruction meters.

Of the numerous types of obstruction meters available, those most widely used are orifice meters, flow nozzles, and Venturi meters (Fig. 8–59). For standardized geometries, the experimentally determined data for discharge coefficients are expressed as (Miller, 1997)

Orifice meters: $C_d = 0.5959 + 0.0312\beta^{2.1} - 0.184\beta^8 + \dfrac{91.71\beta^{2.5}}{\mathrm{Re}^{0.75}}$ **(8–72)**

Nozzle meters: $C_d = 0.9975 - \dfrac{6.53\beta^{0.5}}{\mathrm{Re}^{0.5}}$ **(8–73)**

These relations are valid for $0.25 < \beta < 0.75$ and $10^4 < \mathrm{Re} < 10^7$. Precise values of C_d depend on the particular design of the obstruction, and thus the

(a) Orifice meter

(b) Flow nozzle

(c) Venturi meter

FIGURE 8–59
Common types of obstruction meters.

$$V_2 > V_1 \rightarrow P_1 > P_2 \quad \text{Orifice}$$

FIGURE 8–60

An orifice meter and schematic showing its built-in pressure transducer and digital readout.

Courtesy KOBOLD Instruments, Pittsburgh, PA. www.koboldusa.com. Used by permission.

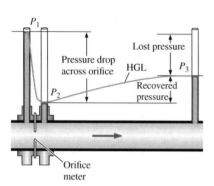

FIGURE 8–61

The variation of pressure along a flow section with an orifice meter as measured with piezometer tubes; the lost pressure and the pressure recovery are shown.

manufacturer's data should be consulted when available. Also, the Reynolds number depends on the flow velocity, which is not known a priori. Therefore, the solution is iterative in nature when curve-fit correlations are used for C_d. For flows with high Reynolds numbers (Re > 30,000), the value of C_d can be taken to be 0.96 for flow nozzles and 0.61 for orifices.

Owing to its streamlined design, the discharge coefficients of Venturi meters are very high, ranging between 0.95 and 0.99 (the higher values are for the higher Reynolds numbers) for most flows. In the absence of specific data, we can take $C_d = 0.98$ for Venturi meters.

The orifice meter has the simplest design and it occupies minimal space as it consists of a plate with a hole in the middle, but there are considerable variations in design (Fig. 8–60). Some orifice meters are sharp-edged, while others are beveled or rounded. The sudden change in the flow area in orifice meters causes considerable swirl and thus significant head loss or permanent pressure loss, as shown in Fig. 8–61. In nozzle meters, the plate is replaced by a nozzle, and thus the flow in the nozzle is streamlined. As a result, the vena contracta is practically eliminated and the head loss is smaller. However, flow nozzle meters are more expensive than orifice meters.

The Venturi meter, invented by the American engineer Clemens Herschel (1842–1930) and named by him after the Italian Giovanni Venturi (1746–1822) for his pioneering work on conical flow sections, is the most accurate flowmeter in this group, but it is also the most expensive. Its gradual contraction and expansion prevent flow separation and swirling, and it suffers only frictional losses on the inner wall surfaces. Venturi meters cause very low head losses, and thus, they should be preferred for applications that cannot allow large pressure drops.

When an obstruction flowmeter is placed in a piping system, its net effect on the flow system is like that of a minor loss. The minor loss coefficient of the flowmeter is available from the manufacturer, and should be included when summing minor losses in the system. In general, orifice meters have the highest minor loss coefficients, while Venturi meters have the lowest. Note that the pressure drop $P_1 - P_2$ measured to calculate the flow rate is *not* the same as the total pressure drop caused by the obstruction flowmeter because of the locations of the pressure taps.

Finally, obstruction flowmeters are also used to measure compressible-gas flow rates, but an additional correction factor must be inserted into Eq. 8–71 to account for compressibility effects. In such cases, the equation is written for *mass flow* rate instead of volume flow rate, and the compressible correction factor is typically an empirically curve-fitted equation (like the one for C_d) and is available from the flowmeter manufacturer.

■ **EXAMPLE 8–10**   **Measuring Flow Rate with an Orifice Meter**

The flow rate of methanol at 20°C ($\rho = 788.4$ kg/m³ and $\mu = 5.857 \times 10^{-4}$ kg/m·s) through a 4-cm-diameter pipe is to be measured with a 3-cm-diameter orifice meter equipped with a mercury manometer across the orifice plate, as shown in Fig. 8–62. If the differential height of the manometer is 11 cm, determine the flow rate of methanol through the pipe and the average flow velocity.

SOLUTION The flow rate of methanol is to be measured with an orifice meter. For a given pressure drop across the orifice plate, the flow rate and the average flow velocity are to be determined.
Assumptions **1** The flow is steady and incompressible. **2** Our first guess for the discharge coefficient of the orifice meter is $C_d = 0.61$.
Properties The density and dynamic viscosity of methanol are given to be $\rho = 788.4$ kg/m³ and $\mu = 5.857 \times 10^{-4}$ kg/m·s, respectively. We take the density of mercury to be 13,600 kg/m³.
Analysis The diameter ratio and the throat area of the orifice are

$$\beta = \frac{d}{D} = \frac{3}{4} = 0.75$$

$$A_0 = \frac{\pi d^2}{4} = \frac{\pi (0.03 \text{ m})^2}{4} = 7.069 \times 10^{-4} \text{ m}^2$$

The pressure drop across the orifice plate is

$$\Delta P = P_1 - P_2 = (\rho_{Hg} - \rho_{met})gh$$

Then the flow rate relation for obstruction meters becomes

$$\dot{V} = A_0 C_d \sqrt{\frac{2(P_1 - P_2)}{\rho(1 - \beta^4)}} = A_0 C_d \sqrt{\frac{2(\rho_{Hg} - \rho_{met})gh}{\rho_{met}(1 - \beta^4)}} = A_0 C_d \sqrt{\frac{2(\rho_{Hg}/\rho_{met} - 1)gh}{1 - \beta^4}}$$

Substituting, the flow rate is determined to be

$$\dot{V} = (7.069 \times 10^{-4} \text{ m}^2)(0.61)\sqrt{\frac{2(13,600/788.4 - 1)(9.81 \text{ m/s}^2)(0.11 \text{ m})}{1 - 0.75^4}}$$
$$= 3.09 \times 10^{-3} \text{ m}^3/\text{s}$$

which is equivalent to 3.09 L/s. The average flow velocity in the pipe is determined by dividing the flow rate by the cross-sectional area of the pipe,

$$V = \frac{\dot{V}}{A_c} = \frac{\dot{V}}{\pi D^2/4} = \frac{3.09 \times 10^{-3} \text{ m}^3/\text{s}}{\pi (0.04 \text{ m})^2/4} = 2.46 \text{ m/s}$$

The Reynolds number of flow through the pipe is

$$Re = \frac{\rho V D}{\mu} = \frac{(788.4 \text{ kg/m}^3)(2.46 \text{ m/s})(0.04 \text{ m})}{5.857 \times 10^{-4} \text{ kg/m·s}} = 1.32 \times 10^5$$

FIGURE 8–62
Schematic for the orifice meter considered in Example 8–10.

Substituting $\beta = 0.75$ and Re $= 1.32 \times 10^5$ into the orifice discharge coefficient relation

$$C_d = 0.5959 + 0.0312\beta^{2.1} - 0.184\beta^8 + \frac{91.71\beta^{2.5}}{\text{Re}^{0.75}}$$

gives $C_d = 0.601$, which differs from the original guessed value of 0.61. Using this refined value of C_d, the flow rate becomes 3.04 L/s, which differs from our original result by 1.6 percent. After a couple iterations, the final converged flow rate is **3.04 L/s,** and the average velocity is **2.42 m/s** (to three significant digits).

Discussion If the problem is solved using an equation solver such as EES, then it can be formulated using the curve-fit formula for C_d (which depends on the Reynolds number), and all equations can be solved simultaneously by letting the equation solver perform the iterations as necessary.

FIGURE 8–63
A positive displacement flowmeter with double helical three-lobe impeller design.

Courtesy Flow Technology, Inc.
Source: www.ftimeters.com.

FIGURE 8–64
A nutating disk flowmeter.
(Top) Courtesy Badger Meter, Inc.
Used by Permission.

Positive Displacement Flowmeters

When we buy gasoline for our car, we are interested in the total amount of gasoline that flows through the nozzle during the period we fill the tank rather than the flow rate of gasoline. Likewise, we care about the total amount of water or natural gas we use in our homes during a billing period. In these and many other applications, the quantity of interest is the total amount of mass or volume of a fluid that passes through a cross section of a pipe over a certain period of time rather than the instantaneous value of flow rate, and **positive displacement flowmeters** are well suited for such applications. There are numerous types of displacement meters, and they are based on continuous filling and discharging of the measuring chamber. They operate by trapping a certain amount of incoming fluid, displacing it to the discharge side of the meter, and counting the number of such discharge–recharge cycles to determine the total amount of fluid displaced.

Figure 8–63 shows a positive displacement flowmeter with two rotating impellers driven by the flowing liquid. Each impeller has three gear lobes, and a pulsed output signal is generated each time a lobe passes by a nonintrusive sensor. Each pulse represents a known volume of liquid that is captured in between the lobes of the impellers, and an electronic controller converts the pulses to volume units. The clearance between the impeller and its casing must be controlled carefully to prevent leakage and thus to avoid error. This particular meter has a quoted accuracy of 0.1 percent, has a low pressure drop, and can be used with high- or low-viscosity liquids at temperatures up to 230°C and pressures up to 7 MPa for flow rates of up to 700 gal/min (or 50 L/s).

The most widely used flowmeters to measure liquid volumes are **nutating disk flowmeters,** shown in Fig. 8–64. They are commonly used as water and gasoline meters. The liquid enters the nutating disk meter through the chamber (A). This causes the disk (B) to nutate or wobble and results in the rotation of a spindle (C) and the excitation of a magnet (D). This signal is transmitted through the casing of the meter to a second magnet (E). The total volume is obtained by counting the number of these signals during a discharge process.

Quantities of gas flows, such as the amount of natural gas used in buildings, are commonly metered by using **bellows flowmeters** that displace a certain amount of gas volume (or mass) during each revolution.

(a)   (b)   (c)

Turbine Flowmeters

We all know from experience that a propeller held against the wind rotates, and the rate of rotation increases as the wind velocity increases. You may also have seen that the turbine blades of wind turbines rotate rather slowly at low winds, but quite fast at high winds. These observations suggest that the flow velocity in a pipe can be measured by placing a freely rotating propeller inside a pipe section and doing the necessary calibration. Flow measurement devices that work on this principle are called **turbine flowmeters** or sometimes **propeller flowmeters,** although the latter is a misnomer since, by definition, propellers add energy to a fluid, while turbines extract energy from a fluid.

A turbine flowmeter consists of a cylindrical flow section that houses a turbine (a vaned rotor) that is free to rotate, additional stationary vanes at the inlet to straighten the flow, and a sensor that generates a pulse each time a marked point on the turbine passes by to determine the rate of rotation. The rotational speed of the turbine is nearly proportional to the flow rate of the fluid. Turbine flowmeters give highly accurate results (as accurate as 0.25 percent) over a wide range of flow rates when calibrated properly for the anticipated flow conditions. Turbine flowmeters have very few blades (sometimes just two blades) when used to measure liquid flow, but several blades when used to measure gas flow to ensure adequate torque generation. The head loss caused by the turbine is very small.

Turbine flowmeters have been used extensively for flow measurement since the 1940s because of their simplicity, low cost, and accuracy over a wide range of flow conditions. They are commercially available for both liquids and gases and for pipes of practically all sizes. Turbine flowmeters are also commonly used to measure flow velocities in unconfined flows such as winds, rivers, and ocean currents. The handheld device shown in Fig. 8–65c is used to measure wind velocity.

Paddlewheel Flowmeters

Paddlewheel flowmeters are low-cost alternatives to turbine flowmeters for flows where very high accuracy is not required. In paddlewheel flowmeters,

FIGURE 8–65
(a) An in-line turbine flowmeter to measure liquid flow, with flow from left to right, (b) a cutaway view of the turbine blades inside the flowmeter, and (c) a handheld turbine flowmeter to measure wind speed, measuring no flow at the time the photo was taken so that the turbine blades are visible. The flowmeter in (c) also measures the air temperature for convenience.

Photos (a) and (c) by John M. Cimbala.
Photo (b) Courtesy Hoffer Flow Controls.

Retainer cap

Paddlewheel
sensor

Sensor
housing

Truseal
locknut

Flow ⟶

FIGURE 8–66
Paddlewheel flowmeter to measure
liquid flow, with flow from left to
right, and a schematic diagram of
its operation.
Photo by John M. Cimbala.

the paddlewheel (the rotor and the blades) is perpendicular to the flow, as shown in Fig. 8–66, rather than parallel as was the case with turbine flowmeters. The paddles cover only a portion of the flow cross section (typically less than half), and thus the head loss is smaller compared to that of turbine flowmeters, but the depth of insertion of the paddlewheel into the flow is of critical importance for accuracy. Also, no strainers are required since the paddlewheels are less susceptible to fouling. A sensor detects the passage of each of the paddlewheel blades and transmits a signal. A microprocessor then converts this rotational speed information to flow rate or integrated flow quantity.

Variable-Area Flowmeters (Rotameters)

A simple, reliable, inexpensive, and easy-to-install flowmeter with reasonably low pressure drop and no electrical connections that gives a direct reading of flow rate for a wide range of liquids and gases is the **variable-area flowmeter,** also called a **rotameter** or **floatmeter.** A variable-area flowmeter consists of a vertical tapered conical transparent tube made of glass or plastic with a float inside that is free to move, as shown in Fig. 8–67. As fluid flows through the tapered tube, the float rises within the tube to a location where the float weight, drag force, and buoyancy force balance each other and the net force acting on the float is zero. The flow rate is determined by simply matching the position of the float against the graduated flow scale outside the tapered transparent tube. The float itself is typically either a sphere or a loose-fitting piston-like cylinder (as in Fig. 8–67a).

We know from experience that high winds knock down trees, break power lines, and blow away hats or umbrellas. This is because the drag force increases with flow velocity. The weight and the buoyancy force acting on the float are constant, but the drag force changes with flow velocity. Also, the velocity along the tapered tube decreases in the flow direction because of the increase in the cross-sectional area. There is a certain velocity that generates enough drag to balance the float weight and the buoyancy force, and the location at which this velocity occurs around the float is the location where the float settles. The degree of tapering of the tube can be made such that the vertical rise changes linearly with flow rate, and thus the tube can be calibrated linearly for flow rates. The transparent tube also allows the fluid to be seen during flow.

There are several kinds of variable-area flowmeters. The gravity-based flowmeter, as shown in Fig. 8–67a must be positioned vertically, with fluid entering from the bottom and leaving from the top. In spring-opposed flowmeters (Fig. 8–67b), the drag force is balanced by the spring force, and such flowmeters can be installed horizontally.

The accuracy of variable-area flowmeters is typically ±5 percent. Therefore, these flowmeters are not appropriate for applications that require precision measurements. However, some manufacturers quote accuracies of the order of 1 percent. Also, these meters depend on visual checking of the location of the float, and thus they cannot be used to measure the flow rate of fluids that are opaque or dirty, or fluids that coat the float since such fluids block visual access. Finally, glass tubes are prone to breakage and thus they pose a safety hazard if toxic fluids are handled. In such applications, variable-area flowmeters should be installed at locations with minimum traffic.

Ultrasonic Flowmeters

It is a common observation that when a stone is dropped into calm water, the waves that are generated spread out as concentric circles uniformly in all directions. But when a stone is thrown into flowing water such as a river, the waves move much faster in the flow direction (the wave and flow velocities are added since they are in the same direction) compared to the waves moving in the upstream direction (the wave and flow velocities are subtracted since they are in opposite directions). As a result, the waves appear spread out downstream while they appear tightly packed upstream. The difference between the number of waves in the upstream and downstream parts of the flow per unit length is proportional to the flow velocity, and this suggests that flow velocity can be measured by comparing the propagation of waves in the forward and backward directions with respect to the flow. **Ultrasonic flowmeters** operate on this principle, using sound waves in the ultrasonic range (beyond human hearing ability, typically at a frequency of 1 MHz).

Ultrasonic (or acoustic) flowmeters operate by generating sound waves with a transducer and measuring the propagation of those waves through a flowing fluid. There are two basic kinds of ultrasonic flowmeters: *transit time* and *Doppler-effect* (or *frequency shift*) flowmeters. The transit time flowmeter transmits sound waves in the upstream and downstream directions and measures the difference in travel time. A typical transit time ultrasonic meter is shown schematically in Fig. 8–68. It involves two transducers that alternately transmit and receive ultrasonic waves, one in the direction of flow and the other in the opposite direction. The travel time for each direction can be measured accurately, and the difference in the travel time is calculated. The average flow velocity V in the pipe is proportional to this travel time difference Δt, and is determined from

$$V = KL\,\Delta t \tag{8–74}$$

where L is the distance between the transducers and K is a constant.

Doppler-Effect Ultrasonic Flowmeters

You have probably noticed that when a fast-moving car approaches with its horn blowing, the tone of the high-pitched sound of the horn drops to a lower pitch as the car passes by. This is due to the sonic waves being compressed in front of the car and being spread out behind it. This shift in frequency is called the **Doppler effect,** and it forms the basis for the operation of most ultrasonic flowmeters.

Doppler-effect ultrasonic flowmeters measure the average flow velocity along the sonic path. This is done by clamping a piezoelectric transducer on the outside surface of a pipe (or pressing the transducer against the pipe for handheld units). The transducer transmits a sound wave at a fixed frequency through the pipe wall and into the flowing liquid. The waves reflected by impurities, such as suspended solid particles or entrained gas bubbles, are relayed to a receiving transducer. The change in the frequency of the reflected waves is proportional to the flow velocity, and a microprocessor determines the flow velocity by comparing the frequency shift between the transmitted and reflected signals (Figs. 8–69 and 8–70). The flow rate and the total amount of flow can also be determined using the measured velocity by properly configuring the flowmeter for the given pipe and flow conditions.

(b)

(a)

FIGURE 8–67

Two types of variable-area flowmeters: (a) an ordinary gravity-based meter and (b) a spring-opposed meter.

(a) Photo by Luke A. Cimbala and (b) Courtesy Insite, Universal Flow Monitors, Inc. Used by permission.

FIGURE 8–68

The operation of a transit time ultrasonic flowmeter equipped with two transducers.

400
INTERNAL FLOW

FIGURE 8–69
The operation of a Doppler-effect ultrasonic flowmeter equipped with a transducer pressed on the outer surface of a pipe.

The operation of ultrasonic flowmeters depends on the ultrasound waves being reflected off discontinuities in density. Ordinary ultrasonic flowmeters require the liquid to contain impurities in concentrations greater than 25 parts per million (ppm) in sizes greater than at least 30 μm. But advanced ultrasonic units can also measure the velocity of clean liquids by sensing the waves reflected off turbulent swirls and eddies in the flow stream, provided that they are installed at locations where such disturbances are nonsymmetrical and at a high level, such as a flow section just downstream of a 90° elbow.

Ultrasonic flowmeters have the following advantages:

- They are easy and quick to install by clamping them on the outside of pipes of 0.6 cm to over 3 m in diameter (Fig. 8–70), and even on open channels.

- They are nonintrusive. Since the meters clamp on, there is no need to stop operation and drill holes into piping, and no production downtime.

- There is no pressure drop since the meters do not interfere with the flow.

- Since there is no direct contact with the fluid, there is no danger of corrosion or clogging.

- They are suitable for a wide range of fluids from toxic chemicals to slurries to clean liquids, for permanent or temporary flow measurement.

- There are no moving parts, and thus the meters provide reliable and maintenance-free operation.

- They can also measure flow quantities in reverse flow.

- The quoted accuracies are 1 to 2 percent.

Ultrasonic flowmeters are noninvasive devices, and the ultrasonic transducers can effectively transmit signals through polyvinyl chloride (PVC), steel, iron,

FIGURE 8–70
Ultrasonic clamp-on flowmeters enable one to measure flow velocity without even contacting (or disturbing) the fluid by simply pressing a transducer on the outer surface of the pipe.

Photo by J. Matthew Deepe.

and glass pipe walls. However, coated pipes and concrete pipes are not suitable for this measurement technique since they absorb ultrasonic waves.

Electromagnetic Flowmeters

It has been known since Faraday's experiments in the 1830s that when a conductor is moved in a magnetic field, an electromotive force develops across that conductor as a result of magnetic induction. Faraday's law states that the voltage induced across any conductor as it moves at right angles through a magnetic field is proportional to the velocity of that conductor. This suggests that we may be able to determine flow velocity by replacing the solid conductor by a conducting fluid, and **electromagnetic flowmeters** do just that. Electromagnetic flowmeters have been in use since the mid-1950s, and they come in various designs such as full-flow and insertion types.

A *full-flow electromagnetic flowmeter* is a nonintrusive device that consists of a magnetic coil that encircles the pipe, and two electrodes drilled into the pipe along a diameter flush with the inner surface of the pipe so that the electrodes are in contact with the fluid but do not interfere with the flow and thus do not cause any head loss (Fig. 8–71a). The electrodes are connected to a voltmeter. The coils generate a magnetic field when subjected to electric current, and the voltmeter measures the electric potential difference between the electrodes. This potential difference is proportional to the flow velocity of the conducting fluid, and thus the flow velocity can be calculated by relating it to the voltage generated.

Insertion electromagnetic flowmeters operate similarly, but the magnetic field is confined within a flow channel at the tip of a rod inserted into the flow, as shown in Fig. 8–71b.

Electromagnetic flowmeters are well-suited for measuring flow velocities of liquid metals such as mercury, sodium, and potassium that are used in some nuclear reactors. They can also be used for liquids that are poor conductors, such as water, provided that they contain an adequate amount of charged particles. Blood and seawater, for example, contain sufficient amounts of ions, and thus electromagnetic flowmeters can be used to measure their flow rates. Electromagnetic flowmeters can also be used to measure the flow rates of chemicals, pharmaceuticals, cosmetics, corrosive

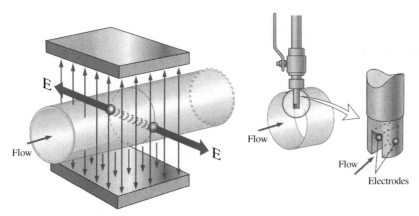

(a) Full-flow electromagnetic flowmeter   (b) Insertion electromagnetic flowmeter

FIGURE 8–71
(a) Full-flow and (b) insertion electromagnetic flowmeters,

www.flocat.com.

liquids, beverages, fertilizers, and numerous slurries and sludges, provided that the substances have high enough electrical conductivities. Electromagnetic flowmeters are not suitable for use with distilled or deionized water.

Electromagnetic flowmeters measure flow velocity indirectly, and thus careful calibration is important during installation. Their use is limited by their relatively high cost, power consumption, and the restrictions on the types of suitable fluids with which they can be used.

Vortex Flowmeters

You have probably noticed that when a flow stream such as a river encounters an obstruction such as a rock, the fluid separates and moves around the rock. But the presence of the rock is felt for some distance downstream via the swirls generated by it.

Most flows encountered in practice are turbulent, and a disk or a short cylinder placed in the flow coaxially sheds vortices (see also Chap. 4). It is observed that these vortices are shed periodically, and the shedding frequency is proportional to the average flow velocity. This suggests that the flow rate can be determined by generating vortices in the flow by placing an obstruction in the flow and measuring the shedding frequency. The flow measurement devices that work on this principle are called **vortex flowmeters.** The *Strouhal number*, defined as St $= fd/V$, where f is the vortex shedding frequency, d is the characteristic diameter or width of the obstruction, and V is the velocity of the flow impinging on the obstruction, also remains constant in this case, provided that the flow velocity is high enough.

A vortex flowmeter consists of a sharp-edged bluff body (strut) placed in the flow that serves as the vortex generator, and a detector (such as a pressure transducer that records the oscillation in pressure) placed a short distance downstream on the inner surface of the casing to measure the shedding frequency. The detector can be an ultrasonic, electronic, or fiber-optic sensor that monitors the changes in the vortex pattern and transmits a pulsating output signal (Fig. 8–72). A microprocessor then uses the frequency information to calculate and display the flow velocity or flow rate. The frequency of vortex shedding is proportional to the average velocity over a wide range of Reynolds numbers, and vortex flowmeters operate reliably and accurately at Reynolds numbers from 10^4 to 10^7.

The vortex flowmeter has the advantage that it has no moving parts and thus is inherently reliable, versatile, and very accurate (usually ± 1 percent over a wide range of flow rates), but it obstructs the flow and thus causes considerable head loss.

Thermal (Hot-Wire and Hot-Film) Anemometers

Thermal anemometers were introduced in the late 1950s and have been in common use since then in fluid research facilities and labs. As the name implies, thermal anemometers involve an electrically heated sensor, as shown in Fig. 8–73, and utilize a thermal effect to measure flow velocity. Thermal anemometers have extremely small sensors, and thus they can be used to measure the instantaneous velocity at any point in the flow without appreciably disturbing the flow. They can take thousands of velocity measurements per second with excellent spatial and temporal resolution, and

Bluff
body (strut)   Receiving
                                transducer

Flow

Vortex   Transmitting
                    transducer

FIGURE 8–72
The operation of a vortex flowmeter.

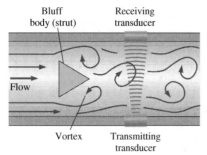

Electric
current I

Flow
velocity V

                            Wire
                            support

Sensor (a thin wire
approximately 1 mm long
with a diameter of 5 μm)

FIGURE 8–73
The electrically heated sensor and its support, which are the components of a hot-wire probe.

thus they can be used to study the details of fluctuations in turbulent flow. They can measure velocities in liquids and gases accurately over a wide range—from a few centimeters to over a hundred meters per second.

A thermal anemometer is called a **hot-wire anemometer** if the sensing element is a wire, and a **hot-film anemometer** if the sensor is a thin metallic film (less than 0.1 μm thick) mounted usually on a relatively thick ceramic support having a diameter of about 50 μm. The hot-wire anemometer is characterized by its very small sensor wire—usually a few microns in diameter and a couple of millimeters in length. The sensor is usually made of platinum, tungsten, or platinum–iridium alloys, and it is attached to the probe through needle-like holders. The fine wire sensor of a hot-wire anemometer is very fragile because of its small size and can easily break if the liquid or gas contains excessive amounts of contaminants or particulate matter. This is especially of consequence at high velocities. In such cases, the more rugged hot-film probes should be used. But the sensor of the hot-film probe is larger, has significantly lower frequency response, and interferes more with the flow; thus it is not always suitable for studying the fine details of turbulent flow.

The operating principle of a constant-temperature anemometer (CTA), which is the most common type and is shown schematically in Fig. 8–74, is as follows: the sensor is electrically heated to a specified temperature (typically about 200°C). The sensor tends to cool as it loses heat to the surrounding flowing fluid, but electronic controls maintain the sensor at a constant temperature by varying the electric current (which is done by varying the voltage) as needed. The higher the flow velocity, the higher the rate of heat transfer from the sensor, and thus the larger the voltage that needs to be applied across the sensor to maintain it at constant temperature. There is a close correlation between the flow velocity and voltage, and the flow velocity is determined by measuring the voltage applied by an amplifier or the electric current passing through the sensor.

The sensor is maintained at a constant temperature during operation, and thus its thermal energy content remains constant. The conservation of energy principle requires that the electrical Joule heating $\dot{W}_{elect} = I^2 R_w = E^2/R_w$ of the sensor must be equal to the total rate of heat loss from the sensor $\dot{Q}_{total}$, which consists of convection heat transfer since conduction to the wire supports and radiation to the surrounding surfaces are small and can be disregarded. Using proper relations for forced convection, the energy balance is expressed by **King's law** as

$$E^2 = a + bV^n \qquad (8\text{–}75)$$

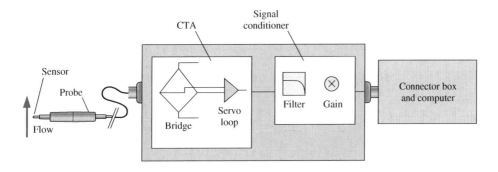

FIGURE 8–74
Schematic of a thermal anemometer system.

FIGURE 8–75
Thermal anemometer probes
with single, double, and triple
sensors to measure (*a*) one-, (*b*) two-,
and (*c*) three-dimensional velocity
components simultaneously.

(*a*)   (*b*)   (*c*)

where E is the voltage, and the values of the constants a, b, and n are calibrated for a given probe. Once the voltage is measured, this relation gives the flow velocity V directly.

Most hot-wire sensors have a diameter of 5 μm and a length of approximately 1 mm and are made of tungsten. The wire is spot-welded to needle-shaped prongs embedded in a probe body, which is connected to the anemometer electronics. Thermal anemometers can be used to measure two- or three-dimensional velocity components simultaneously by using probes with two or three sensors, respectively (Fig. 8–75). When selecting probes, consideration should be given to the type and the contamination level of the fluid, the number of velocity components to be measured, the required spatial and temporal resolution, and the location of measurement.

Laser Doppler Velocimetry

Laser Doppler velocimetry (LDV), also called **laser velocimetry (LV)** or **laser Doppler anemometry (LDA),** is an optical technique to measure flow velocity at any desired point without disturbing the flow. Unlike thermal anemometry, LDV involves no probes or wires inserted into the flow, and thus it is a nonintrusive method. Like thermal anemometry, it can accurately measure velocity at a very small volume, and thus it can also be used to study the details of flow at a locality, including turbulent fluctuations, and it can be traversed through the entire flow field without intrusion.

The LDV technique was developed in the mid-1960s and has found widespread acceptance because of the high accuracy it provides for both gas and liquid flows; the high spatial resolution it offers; and, in recent years, its ability to measure all three velocity components. Its drawbacks are the relatively high cost; the requirement for sufficient transparency between the laser source, the target location in the flow, and the photodetector; and the requirement for careful alignment of emitted and reflected beams for accuracy. The latter drawback is eliminated for the case of a fiber-optic LDV system, since it is aligned at the factory.

The operating principle of LDV is based on sending a highly coherent monochromatic (all waves are in phase and at the same wavelength) light beam toward the target, collecting the light reflected by small particles in the target area, determining the change in frequency of the reflected radiation due to the Doppler effect, and relating this frequency shift to the flow velocity of the fluid at the target area.

LDV systems are available in many different configurations. A basic dual-beam LDV system to measure a single velocity component is shown in Fig. 8–76. In the heart of all LDV systems is a laser power source, which is

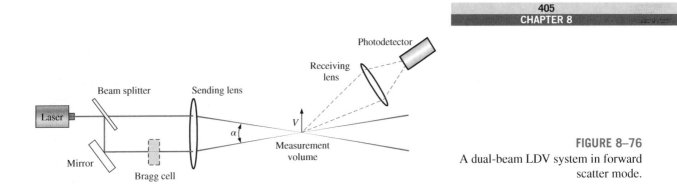

FIGURE 8–76
A dual-beam LDV system in forward scatter mode.

usually a helium–neon or argon-ion laser with a power output of 10 mW to 20 W. Lasers are preferred over other light sources since laser beams are highly coherent and highly focused. The helium–neon laser, for example, emits radiation at a wavelength of 0.6328 μm, which is in the reddish-orange color range. The laser beam is first split into two parallel beams of equal intensity by a half-silvered mirror called a *beam splitter*. Both beams then pass through a converging lens that focuses the beams at a point in the flow (the *target*). The small fluid volume where the two beams intersect is the region where the velocity is measured and is called the *measurement volume* or the *focal volume*. The measurement volume resembles an ellipsoid, typically of 0.1 mm diameter and 0.5 mm in length. The laser light is scattered by particles passing through this measurement volume, and the light scattered in a certain direction is collected by a receiving lens and is passed through a photodetector that converts the fluctuations in light intensity into fluctuations in a voltage signal. Finally, a signal processor determines the frequency of the voltage signal and thus the velocity of the flow.

The waves of the two laser beams that cross in the measurement volume are shown schematically in Fig. 8–77. The waves of the two beams interfere in the measurement volume, creating a bright fringe where they are in phase and thus support each other, and creating a dark fringe where they are out of phase and thus cancel each other. The bright and dark fringes form lines parallel to the midplane between the two incident laser beams. Using trigonometry, the spacing s between the fringe lines, which can be viewed as the wavelength of fringes, can be shown to be $s = \lambda/[2 \sin(\alpha/2)]$, where λ is the wavelength of the laser beam and α is the angle between the two laser beams. When a particle traverses these fringe lines at velocity V, the frequency of the scattered fringe lines is

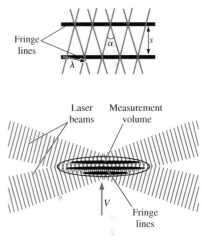

FIGURE 8–77
Fringes that form as a result of the interference at the intersection of two laser beams of an LDV system (lines represent peaks of waves). The top diagram is a close-up view of two fringes.

$$f = \frac{V}{s} = \frac{2V \sin(\alpha/2)}{\lambda} \qquad (8\text{–}76)$$

This fundamental relation shows the flow velocity to be proportional to the frequency and is known as the **LDV equation.** As a particle passes through the measurement volume, the reflected light is bright, then dark, then bright, etc., because of the fringe pattern, and the flow velocity is determined by measuring the frequency of the reflected light. The velocity profile at a cross section of a pipe, for example, can be obtained by mapping the flow across the pipe (Fig. 8–78).

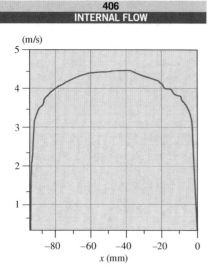

(m/s)

x (mm)

FIGURE 8–78

A time-averaged velocity profile in turbulent pipe flow obtained by an LDV system.

Courtesy Dantec Dynamics,
www.dantecdynamics.com. Used by permission.

The LDV method obviously depends on the presence of scattered fringe lines, and thus the flow must contain a sufficient amount of small particles called *seeds* or *seeding particles*. These particles must be small enough to follow the flow closely so that the particle velocity is equal to the flow velocity, but large enough (relative to the wavelength of the laser light) to scatter an adequate amount of light. Particles with a diameter of 1 μm usually serve the purpose well. Some fluids such as tap water naturally contain an adequate amount of such particles, and no seeding is necessary. Gases such as air are commonly seeded with smoke or with particles made of latex, oil, or other materials. By using three laser beam pairs at different wavelengths, the LDV system is also used to obtain all three velocity components at any point in the flow.

Particle Image Velocimetry

Particle image velocimetry (PIV) is a double-pulsed laser technique used to measure the instantaneous velocity distribution in a plane of flow by photographically determining the displacement of particles in the plane during a very short time interval. Unlike methods like hot-wire anemometry and LDV that measure velocity at a point, PIV provides velocity values simultaneously throughout an entire cross section, and thus it is a whole-field technique. PIV combines the accuracy of LDV with the capability of flow visualization and provides instantaneous flow field mapping. The entire instantaneous velocity profile at a cross section of pipe, for example, can be obtained with a single PIV measurement. A PIV system can be viewed as a camera that can take a snapshot of velocity distribution at any desired plane in a flow. Ordinary flow visualization gives a qualitative picture of the details of flow. PIV also provides an accurate *quantitative* description of various flow quantities such as the velocity field, and thus the capability to analyze the flow numerically using the velocity data provided. Because of its whole-field capability, PIV is also used to validate computational fluid dynamics (CFD) codes (Chap. 15).

The PIV technique has been used since the mid-1980s, and its use and capabilities have grown in recent years with improvements in frame grabber and charge-coupled device (CCD) camera technologies. The accuracy, flexibility, and versatility of PIV systems with their ability to capture whole-field images with submicrosecond exposure time have made them extremely valuable tools in the study of supersonic flows, explosions, flame propagation, bubble growth and collapse, turbulence, and unsteady flow.

The PIV technique for velocity measurement consists of two main steps: visualization and image processing. The first step is to seed the flow with suitable particles in order to trace the fluid motion. Then a pulse of laser light sheet illuminates a thin slice of the flow field at the desired plane, and the positions of particles in that plane are determined by detecting the light scattered by particles on a digital video or photographic camera positioned at right angles to the light sheet (Fig. 8–79). After a very short time period Δt (typically in μs), the particles are illuminated again by a second pulse of laser light sheet, and their new positions are recorded. Using the information on these two superimposed camera images, the particle displacements Δs are determined for all particles, and the magnitude of velocity of each particle in the plane of the laser light sheet is determined from $\Delta s / \Delta t$. The direction of motion of the particles is also determined from the two

FIGURE 8–79
A PIV system to study flame stabilization.

positions, so that two components of velocity in the plane are calculated. The built-in algorithms of PIV systems determine the velocities at hundreds or thousands of area elements called *interrogation regions* throughout the entire plane and display the velocity field on the computer monitor in any desired form (Fig. 8–80).

The PIV technique relies on the laser light scattered by particles, and thus the flow must be seeded if necessary with particles, also called *markers,* in order to obtain an adequate reflected signal. Seed particles must be able to follow the pathlines in the flow for their motion to be representative of

FIGURE 8–80
Instantaneous PIV velocity vectors superimposed on a hummingbird in hover. Color scale is from low velocity (blue) to high velocity (red).
Photo by Douglas Warrick. Used by permission.

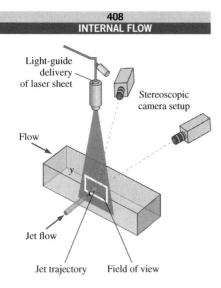

408
INTERNAL FLOW

Light-guide
delivery
of laser sheet

Stereoscopic
camera setup

Flow

y

Jet flow

Jet trajectory   Field of view

FIGURE 8–81
A three-dimensional PIV system
set up to study the mixing of an
air jet with cross duct flow.

the flow, and this requires the particle density to be equal to the fluid density (so that they are neutrally buoyant) or the particles to be so small (typically μm-sized) that their movement relative to the fluid is insignificant. A variety of such particles is available to seed gas or liquid flow. Very small particles must be used in high-speed flows. Silicon carbide particles (mean diameter of 1.5 μm) are suitable for both liquid and gas flow, titanium dioxide particles (mean diameter of 0.2 μm) are usually used for gas flow and are suitable for high-temperature applications, and polystyrene latex particles (nominal diameter of 1.0 μm) are suitable for low-temperature applications. Metallic-coated particles (mean diameter of 9.0 μm) are also used to seed water flows for LDV measurements because of their high reflectivity. Gas bubbles as well as droplets of some liquids such as olive oil or silicon oil are also used as seeding particles after they are atomized to μm-sized spheres.

A variety of laser light sources such as argon, copper vapor, and Nd:YAG can be used with PIV systems, depending on the requirements for pulse duration, power, and time between pulses. Nd:YAG lasers are commonly used in PIV systems over a wide range of applications. A beam delivery system such as a light arm or a fiber-optic system is used to generate and deliver a high-energy pulsed laser sheet at a specified thickness.

With PIV, other flow properties such as vorticity and strain rates can also be obtained, and the details of turbulence can be studied. Recent advances in PIV technology have made it possible to obtain three-dimensional velocity profiles at a cross section of a flow using two cameras (Fig. 8–81). This is done by recording the images of the target plane simultaneously by both cameras at different angles, processing the information to produce two separate two-dimensional velocity maps, and combining these two maps to generate the instantaneous three-dimensional velocity field.

Introduction to Biofluid Mechanics[1]

Biofluid mechanics can cover a number of physiological systems in the human body but the term also applies to all animal species as there are a number of basic fluid systems that are essentially a series of piping networks to transport a fluid (be it liquid or gas or perhaps both). If we focus on humans, these fluid systems are the cardiovascular, respiratory, lymphatic, ocular, and gastrointestinal to name several. We should keep in mind that all these systems are similar to other mechanical piping networks in that the fundamental constituents for the network include a pump, pipes, valves, and a fluid. For our purposes, we will focus more on the cardiovascular system to demonstrate the basic concepts of a piping network within a human.

Figure 8–82 illustrates the cardiovascular system, more specifically, the systemic circulation or the vessels (pipes) that carry the blood (fluid) from the heart, specifically the left ventricle (pump), to the rest of the body. Keep in mind there is a separate network of vessels from the right ventricle to the lungs to oxygenate the blood again. What is unique about the series of pipes in the systemic circulation is that the geometry or cross section is not circular but rather elliptical and in fact, unlike the typical mechanical systems for piping networks that have fittings to transition from one size pipe to

[1] This section was contributed by Professor Keefe Manning of Penn State University.

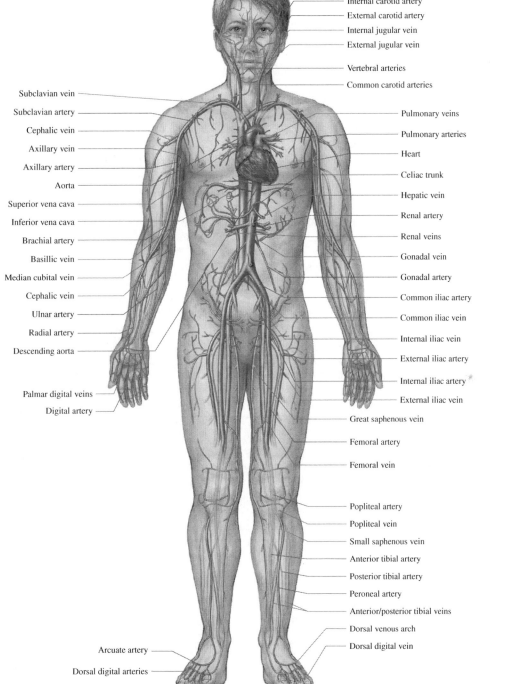

Basilar artery
Internal carotid artery
External carotid artery
Internal jugular vein
External jugular vein
Vertebral arteries
Common carotid arteries

Subclavian vein
Subclavian artery
Cephalic vein
Axillary vein
Axillary artery
Aorta
Superior vena cava
Inferior vena cava
Brachial artery
Basillic vein
Median cubital vein
Cephalic vein
Ulnar artery
Radial artery
Descending aorta

Pulmonary veins
Pulmonary arteries
Heart
Celiac trunk
Hepatic vein
Renal artery
Renal veins
Gonadal vein
Gonadal artery
Common iliac artery
Common iliac vein
Internal iliac vein
External iliac artery
Internal iliac artery
External iliac vein

Palmar digital veins
Digital artery

Great saphenous vein
Femoral artery
Femoral vein

Popliteal artery
Popliteal vein
Small saphenous vein
Anterior tibial artery
Posterior tibial artery
Peroneal artery
Anterior/posterior tibial veins
Dorsal venous arch
Dorsal digital vein

Arcuate artery
Dorsal digital arteries

FIGURE 8–82
The cardiovascular system.
McGraw-Hill Companies, Inc.

another size pipe, the cardiovascular system starting with the aorta (the first vessel from the left ventricle) continually tapers from approximately 25 mm in diameter to 5 microns in diameter at the capillary level and then gradually increases in diameter to approximately 25 mm at the vena cava, which is the vessel connected to the right ventricle. Another important element of the circulation and specifically the vessels is that they are compliant and can expand to accommodate blood volume as needed to regulate pressure changes to maintain homeostasis.

The cardiovascular system is a complex network of pipes that themselves are living and respond to stresses as do blood elements that react when the norm has changed. Even with this network, the system is even more intricate given that the flow is continually moving based on pulses initiated from the heart to drive blood through the network. This pulsatility propagates through the blood and the vessel wall creating an interaction of waves and reflections within the system. Because of the discontinuities associated with the branching, bifurcations, and curvature as seen in Fig. 8-82 initial and boundary conditions are not straightforward. Understanding blood flow is a challenging endeavor given the complexities of the vessel network and the components themselves.

Flow measurement techniques like PIV and LDV are extremely useful in characterizing the flow within and about medical devices, particularly those implanted in the cardiovascular system. Much can be ascertained, and design changes can be made, using these techniques with respect to how blood might flow through or about these cardiovascular devices. Furthermore, we can even use these measurements to then estimate levels of blood damage and the potential for clotting to occur. To ensure we have an accurate representation of the cardiovascular system on the bench, engineers have designed mock circulatory loops or flow loops that allow the experimentalist to simulate cardiac flow and pressure waveforms for bench top studies. For example, Dr. Gus Rosenberg developed the Penn State mock circulatory loop in the early 1970s (Rosenberg et al., 1981). We also need to simulate blood for these particular flow measurement techniques to ensure that the fluid is transparent but also mimics the behavior of blood as a non-Newtonian fluid. We have developed a blood analog that does that and also matches the refractive index of the acrylic models that represent the cardiovascular devices, thus allowing the laser light to pass through the acrylic into the flow field without any refraction. The simulated loop and fluid are critical to ensure that the measurements are acquired under controllable physiological conditions and with sufficient accuracy.

The Pennsylvania State University has been developing mechanical circulatory support devices (blood pumps) since the 1970s, which are devices that help patients stay alive as they await a heart transplant (former Vice President Dick Cheney used such technology while awaiting a heart transplant). Through the years, PIV and LDV have been used quite successfully to measure the flow and make design changes that reduce clotting. Our recent focus has been the development of a pulsatile pediatric ventricular assist device (PVAD) that helps children stay alive until they can receive a donor heart. The device operates pneumatically with air pulsing into a chamber which then causes a diaphragm to inflate against a polyurethane urea sac (the blood contacting surface within the PVAD). The flow is directed into the device from a tube attached to the left ventricle, passes through a mechanical

FIGURE 8–83

(*a*) An artist rendering of the 12-cc pulsatile Penn State pediatric ventricular assist device with the inlet attached to the left atrium and the outlet attached to the ascending aorta (*b*) The direction of blood through the PVAD.

Photo (b) Permission granted from ASME, Cooper et al. JBME, 2008.

heart valve into the PVAD, and then flows through the outlet of the device through another mechanical heart valve and into a tube which is attached to the ascending aorta as shown in Fig. 8–83*a*. Fig. 8–83*b* shows the flow path through the PVAD, and it should be noted that it can be placed within the palm of an adult's hand. One of the first PIV PVAD studies was to determine which type of mechanical heart valve (tilting disc or bileaflet) would be used with the device. Fig. 8–84 illustrates part of the PIV study results

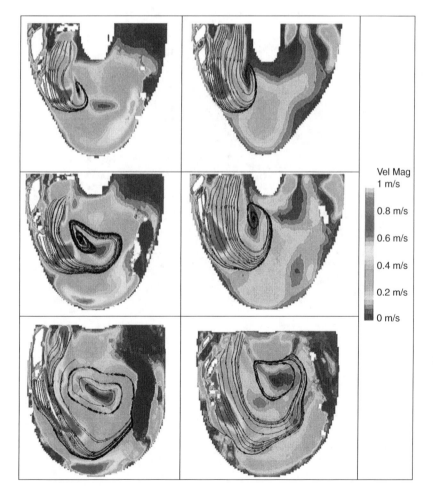

FIGURE 8–84

Particle traces for the BSM valve configuration at 250 ms (left column) and for the CM valve configuration at 350 ms (right column) for the 7 mm (top row), 8.2 mm (middle row), and 11 mm (bottom row) planes. These images highlight the first time step that the rotational flow pattern is fully developed.

Permission granted from ASME, Cooper et al. JBME, 2008.

(Cooper *et al.*, 2008). Here, we used particle traces as a way to examine how the vortical structure would develop inside the device, which for this technology is a way to ensure adequate wall washing (sufficient wall shear) to prevent clotting on the blood contacting surfaces within the device. The tighter rotation would lead to more momentum over the entire cardiac cycle and create a larger vortical structure.

Our research group has also looked at characterizing the flow through mechanical heart valves. In one study (Manning *et al.*, 2008), we focused on the flow within the housing of a Bjork-Shiley Monostrut mechanical heart valve (tilting disc valve) as shown in Fig. 8–85*b*. We removed part of the housing and inserted an optical window to allow access for the LDV system. Instead of using a flow-through loop for this study, we used a single-shot chamber (Fig. 8–85*a*) that mimicked the mitral valve position since we are more interested in the closure fluid dynamics. The mitral valve sits between the left atrium and left ventricle. The native heart valves, like the mitral valve, are passive, similar to a check valve, and

FIGURE 8–85
(*a*) The single shot chamber mimics the closure dynamics of the Bjork-Shiley Monostrut valve.
(*b*) On the lefthand side is a view of the intact Bjork-Shiley Monostrut mechanical heart valve. To the right, the modification to the valve housing is displayed. The window was later filled in with acrylic to maintain similar fluid dynamic patterns and rigidity.

Permission granted from ASME, Manning et al. JBME, 2008.

respond to the pressure changes within the heart's different structures. In this study, we measured how fast the fluid flowed through the small gap between the tilting disc and valve housing, and also how large was the vortex that is created as the tilting disc closes. Figure 8–86 is a schematic illustration of the flow, and Fig. 8–87 is a time sequence of the flow that was measured using LDV within a couple of milliseconds around impact of the valve housing during closure. The intense vortex can be measured right at impact. These data were collected over hundreds of simulated heart beats. We then used these velocity measurements to estimate the amount of potential blood damage by relating time duration and shear magnitude.

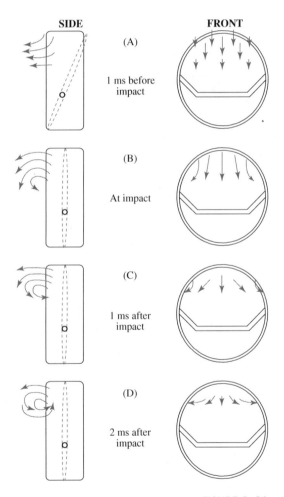

FIGURE 8–86

These schematics depict side and front views of the overall flow structure generated by the closing occluder for four successive times.

Permission granted from ASME, Manning et al. JBME, 2008.

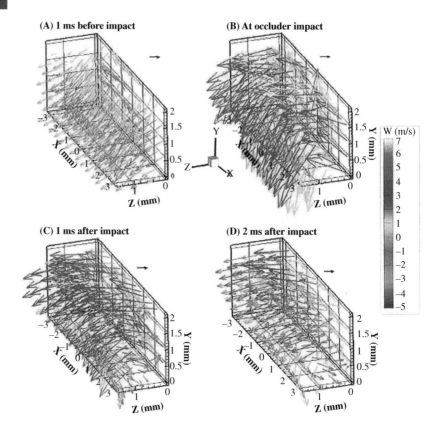

(A) 1 ms before impact

(B) At occluder impact

(C) 1 ms after impact

(D) 2 ms after impact

W (m/s)

FIGURE 8–87

Three-dimensional flow structures are constructed with the vectors indicating direction and the color signifying axial velocity strength. The valve closes right to left, with $x = 0$ representing the centerline of the leaflet. The four plots show the flow (a) 1 ms before impact, (b) at impact, (c) 1 ms following closure, and (d) 2 ms after closure.

Permission granted from ASME, Manning et al. JBME, 2008.

EXAMPLE 8–11   Blood flow through the Aortic Bifurcation

Blood flows from the heart (specifically, the left ventricle) into the aorta to feed the rest of the body oxygen. As blood flow moves from the ascending aorta and downward to the abdominal aorta, some of the volume is directed through a branching network. As the blood reaches the pelvic region, there is a bifurcation (see Fig. 8–88) into the left and right common iliac arteries. This bifurcation is symmetrical but the common iliac vessels are not the same diameter. Given that the kinematic viscosity of blood is 4 cSt (centistokes), the abdominal aorta's diameter is 15 mm, the right common iliac artery's diameter is 10 mm, and the left common iliac artery's diameter is 8 mm, determine the mean flow rate through the right common iliac artery if the abdominal aorta's mean velocity is 30 cm/s and the left common iliac artery's mean velocity is 40 cm/s.

SOLUTION The mean velocities for two of the three vessels is provided along with the diameters of all three vessels. Approximate the vessels as rigid pipes.

Assumptions **1** The flow is steady even though the heart contracts and relaxes approximately 75 beats per minute creating a pulsatile flow. **2** The entrance effects are negligible and the flow is considered fully developed. **3** Blood acts as a Newtonian fluid.

Properties The kinematic viscosity at 37°C is 4 cSt.

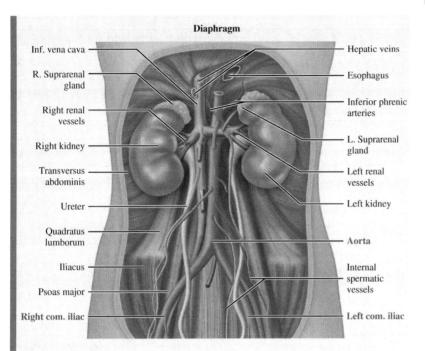

FIGURE 8-88
Anatomy of the human body. Note the aorta and left and right common iliac arteries.

Analysis Using conservation of mass, we can say the flow rate of the abdominal aorta ($\dot{V}_1$) equals the sum of both common iliac arteries ($\dot{V}_2$ for left and $\dot{V}_3$ for right). Thus,

$$\dot{V}_1 = \dot{V}_2 + \dot{V}_3$$

Since we are using the mean velocities, we know the diameters, and the density of blood is the same throughout this section of the circulatory system, we can rewrite the equation to be

$V_1A_1 = V_2A_2 + V_3A_3$ where V are the average velocities and A are the areas.

By rearranging and solving for V_3, the equation becomes,

$$V_3 = (V_1A_1 - V_2A_2)/A_3$$

Inserting the values we know,

$$V_3 = (30 \text{ cm/s} \times (1.5 \text{ cm})^2 - 40 \text{ cm/s} \times (0.8 \text{ cm})^2)/(1.0 \text{ cm})^2$$
$$V_3 = 41.9 \text{ cm/s}$$

Discussion Since we assume a steady flow, the mean velocities are appropriate, but in reality there will be a maximum positive velocity and also some retrograde (or reverse) flow towards the heart as the left ventricle fills during diastole. The velocity profiles through these vessels and many large arteries will vary over a cardiac cycle. It is also assumed that blood will behave as a Newtonian fluid even though it is viscoelastic. Many researchers use this assumption since at this particular location, the shear rate is sufficient to reach the asymptotic value for blood viscosity.

APPLICATION SPOTLIGHT ■ PIV Applied to Cardiac Flow

Guest Authors: Jean Hertzberg[1], Brett Fenster[2],
Jamey Browning[1] and Joyce Schroeder[2]
[1]Department of Mechanical Engineering, University of Colorado, Boulder, CO.
[2]National Jewish Health Center, Denver, CO.

MRI (magnetic resonance imaging) can measure the velocity field of blood moving through the human heart, including all three velocity components (u,v,w) with reasonable resolution in 3-D space and time (Bock *et al.*, 2010). Figure 8–89 shows blood moving from the right atrium into the right ventricle at the peak of diastole (the heart-filling phase) of a normal volunteer subject. The black arrow shows the long axis of the ventricle. The smaller arrows show the velocity vector field and are colored by velocity magnitude, with blue at the slow end of the scale, up to red at 0.5 m/s.

The flow patterns change rapidly with time during the approximately one-second long cardiac cycle, and show complex geometry. The flow moves in a subtle helical path from the atrium into the ventricle, as shown by the white stream tube. The tricuspid valve between the atrium and the ventricle is a set of three thin tissue flaps, which are not visible in this data set. The effect of the valve on flow patterns can be seen as flow curls around one of the flaps, shown by the yellow stream tube. The details of the flow (including *vorticity*, Chapter 4) are expected to reveal information about the underlying physics of the interaction between the heart and lungs, and lead to improved diagnostics for pathologic conditions like pulmonary hypertension (Fenster *et al.*, 2012).

After the right ventricle is filled, the tricuspid valve closes, the ventricle contracts, and blood is ejected into the pulmonary arteries which lead to the lungs, where the blood is oxygenated. After that, the blood goes to the left side of the heart, where the pressure is raised by the contraction of the left ventricle. The oxygenated blood is then ejected into the aorta and is distributed to the body. In this way, the heart functions as two separate positive displacement pumps.

Since calibration of these data is difficult, it's important to check the data for consistency. One useful test, conservation of mass in the ventricle throughout one cardiac cycle, is applied by computing the volume flow of blood entering the ventricle during diastole, and comparing it to the volume that leaves during systole. Similarly, the net flow through the right side of the heart must match the net flow through the left side of the heart in each cycle.

References
Bock J, Frydrychowicz A, Stalder AF, Bley TA, Burkhardt H, Hennig J, and Markl M. 2010. 40 phase contrast MRI at 3 T: Effect of standard and blood-pool contrast agents on SNR, PC-MRA, and blood flow visualization. *Magnetic Resonance in Medicine* 63(2):330-338.

Fenster BE, Schroeder JD, Hertzberg JR, and Chung JH. 2012. 4-Dimensional Cardiac Magnetic Resonance in a Patient With Bicuspid Pulmonic Valve: Characterization of Post-Stenotic Flow. *J Am Coll Cardiol* 59(25):e49.

FIGURE 8–89
MRI-PIV measurements of flow through a human heart.

Photo courtesy of Jean Hertzberg.

SUMMARY

In *internal flow*, a pipe is completely filled with a fluid. *Laminar flow* is characterized by smooth streamlines and highly ordered motion, and *turbulent flow* is characterized by unsteady disorderly velocity fluctuations and highly disordered motion. The *Reynolds number* is defined as

$$\text{Re} = \frac{\text{Inertial forces}}{\text{Viscous forces}} = \frac{V_{avg}D}{\nu} = \frac{\rho V_{avg}D}{\mu}$$

Under most practical conditions, the flow in a pipe is laminar at $\text{Re} < 2300$, turbulent at $\text{Re} > 4000$, and transitional in between.

The region of the flow in which the effects of the viscous shearing forces are felt is called the *velocity boundary layer*. The region from the pipe inlet to the point at which the flow becomes fully developed is called the *hydrodynamic entrance region*, and the length of this region is called the *hydrodynamic entry length* L_h. It is given by

$$\frac{L_{h,\,laminar}}{D} \cong 0.05\,\text{Re} \quad \text{and} \quad \frac{L_{h,\,turbulent}}{D} \cong 10$$

The friction coefficient in the fully developed flow region is constant. The *maximum* and *average* velocities in fully developed laminar flow in a circular pipe are

$$u_{max} = 2V_{avg} \quad \text{and} \quad V_{avg} = \frac{\Delta P D^2}{32\mu L}$$

The *volume flow rate* and the *pressure drop* for laminar flow in a horizontal pipe are

$$\dot{V} = V_{avg}A_c = \frac{\Delta P \pi D^4}{128\mu L} \quad \text{and} \quad \Delta P = \frac{32\mu L V_{avg}}{D^2}$$

The *pressure loss* and *head loss* for all types of internal flows (laminar or turbulent, in circular or noncircular pipes, smooth or rough surfaces) are expressed as

$$\Delta P_L = f\frac{L}{D}\frac{\rho V^2}{2} \quad \text{and} \quad h_L = \frac{\Delta P_L}{\rho g} = f\frac{L}{D}\frac{V^2}{2g}$$

where $\rho V^2/2$ is the *dynamic pressure* and the dimensionless quantity f is the *friction factor*. For fully developed laminar flow in a round pipe, the friction factor is $f = 64/\text{Re}$.

For noncircular pipes, the diameter in the previous relations is replaced by the *hydraulic diameter* defined as $D_h = 4A_c/p$, where A_c is the cross-sectional area of the pipe and p is its wetted perimeter.

In fully developed turbulent flow, the friction factor depends on the Reynolds number and the *relative roughness*

ε/D. The friction factor in turbulent flow is given by the *Colebrook equation*, expressed as

$$\frac{1}{\sqrt{f}} = -2.0\log\!\left(\frac{\varepsilon/D}{3.7} + \frac{2.51}{\text{Re}\sqrt{f}}\right)$$

The plot of this formula is known as the *Moody chart*. The design and analysis of piping systems involve the determination of the head loss, flow rate, or the pipe diameter. Tedious iterations in these calculations can be avoided by the approximate Swamee–Jain formulas expressed as

$$h_L = 1.07\frac{\dot{V}^2 L}{gD^5}\left\{\ln\!\left[\frac{\varepsilon}{3.7D} + 4.62\!\left(\frac{\nu D}{\dot{V}}\right)^{0.9}\right]\right\}^{-2}$$

$$10^{-6} < \varepsilon/D < 10^{-2}$$
$$3000 < \text{Re} < 3 \times 10^8$$

$$\dot{V} = -0.965\left(\frac{gD^5 h_L}{L}\right)^{0.5}\ln\!\left[\frac{\varepsilon}{3.7D} + \left(\frac{3.17\nu^2 L}{gD^3 h_L}\right)^{0.5}\right]$$

$$\text{Re} > 2000$$

$$D = 0.66\left[\varepsilon^{1.25}\!\left(\frac{L\dot{V}^2}{gh_L}\right)^{4.75} + \nu\dot{V}^{9.4}\!\left(\frac{L}{gh_L}\right)^{5.2}\right]^{0.04}$$

$$10^{-6} < \varepsilon/D < 10^{-2}$$
$$5000 < \text{Re} < 3 \times 10^8$$

The losses that occur in piping components such as fittings, valves, bends, elbows, tees, inlets, exits, expansions, and contractions are called *minor losses*. The minor losses are usually expressed in terms of the *loss coefficient* K_L. The head loss for a component is determined from

$$h_L = K_L\frac{V^2}{2g}$$

When all the loss coefficients are available, the total head loss in a piping system is

$$h_{L,\,total} = h_{L,\,major} + h_{L,\,minor} = \sum_i f_i\frac{L_i}{D_i}\frac{V_i^2}{2g} + \sum_j K_{L,j}\frac{V_j^2}{2g}$$

If the entire piping system is of constant diameter, the total head loss reduces to

$$h_{L,\,total} = \left(f\frac{L}{D} + \sum K_L\right)\frac{V^2}{2g}$$

The analysis of a piping system is based on two simple principles: (1) The conservation of mass throughout the system must be satisfied and (2) the pressure drop between two points must be the same for all paths between the two points. When the pipes are connected *in series*, the flow rate through the entire system remains constant regardless of the diameters of the individual pipes. For a pipe that branches out into two (or more) *parallel pipes* and then rejoins at a junction downstream, the total flow rate is the sum of the flow rates in the individual pipes but the head loss in each branch is the same.

When a piping system involves a pump and/or turbine, the steady-flow energy equation is expressed as

$$\frac{P_1}{\rho g} + \alpha_1 \frac{V_1^2}{2g} + z_1 + h_{\text{pump}, u}$$

$$= \frac{P_2}{\rho g} + \alpha_2 \frac{V_2^2}{2g} + z_2 + h_{\text{turbine}, e} + h_L$$

When the useful pump head $h_{\text{pump}, u}$ is known, the mechanical power that needs to be supplied by the pump to the fluid and the electric power consumed by the motor of the pump for a specified flow rate are

$$\dot{W}_{\text{pump, shaft}} = \frac{\rho \dot{V} g h_{\text{pump}, u}}{\eta_{\text{pump}}} \quad \text{and} \quad \dot{W}_{\text{elect}} = \frac{\rho \dot{V} g h_{\text{pump}, u}}{\eta_{\text{pump-motor}}}$$

where $\eta_{\text{pump-motor}}$ is the *efficiency of the pump–motor combination*, which is the product of the pump and the motor efficiencies.

The plot of the head loss versus the flow rate $\dot{V}$ is called the *system curve*. The head produced by a pump is not a constant, and the curves of $h_{\text{pump}, u}$ and η_{pump} versus $\dot{V}$ are called the *characteristic curves*. A pump installed in a piping system operates at the *operating point*, which is the point of intersection of the system curve and the characteristic curve.

Flow measurement techniques and devices can be considered in three major categories: (1) volume (or mass) flow rate measurement techniques and devices such as obstruction flowmeters, turbine meters, positive displacement flowmeters, rotameters, and ultrasonic meters; (2) point velocity measurement techniques such as the Pitot-static probes, hot-wires, and LDV; and (3) whole-field velocity measurement techniques such as PIV.

The emphasis in this chapter has been on flow through pipes, including blood vessels. A detailed treatment of numerous types of pumps and turbines, including their operation principles and performance parameters, is given in Chap. 14.

REFERENCES AND SUGGESTED READING

1. H. S. Bean (ed.). *Fluid Meters: Their Theory and Applications*, 6th ed. New York: American Society of Mechanical Engineers, 1971.

2. M. S. Bhatti and R. K. Shah. "Turbulent and Transition Flow Convective Heat Transfer in Ducts." In *Handbook of Single-Phase Convective Heat Transfer*, ed. S. Kakaç, R. K. Shah, and W. Aung. New York: Wiley Interscience, 1987.

3 B. T. Cooper, B. N. Roszelle, T. C. Long, S. Deutsch, and K. B. Manning. "The 12 cc Penn State pulsatile pediatric ventricular assist device: fluid dynamics associated with valve selection." *J. of Biomechonicol Engineering*. 130 (2008) pp. 041019.

4. C. F. Colebrook. "Turbulent Flow in Pipes, with Particular Reference to the Transition between the Smooth and Rough Pipe Laws," *Journal of the Institute of Civil Engineers London*. 11 (1939), pp. 133–156.

5. F. Durst, A. Melling, and J. H. Whitelaw. *Principles and Practice of Laser-Doppler Anemometry*, 2nd ed. New York: Academic, 1981.

6. *Fundamentals of Orifice Meter Measurement*. Houston, TX: Daniel Measurement and Control, 1997.

7. S. E. Haaland. "Simple and Explicit Formulas for the Friction Factor in Turbulent Pipe Flow," *Journal of Fluids Engineering*, March 1983, pp. 89–90.

8. I. E. Idelchik. *Handbook of Hydraulic Resistance*, 3rd ed. Boca Raton, FL: CRC Press, 1993.

9. W. M. Kays, M. E. Crawford, B. Weigand. *Convective Heat and Mass Transfer*, 4th ed. New York: McGraw-Hill, 2004.

10. K. B. Manning, L. H. Herbertson, A. A. Fontaine, and S. S. Deutsch. "A detailed fluid mechanics study of tilting disk mechanical heart valve closure and the implications to blood damage." *J. Biomech. Eng.* 130(4) (2008), pp. 041001-1-4.

11. R. W. Miller. *Flow Measurement Engineering Handbook*, 3rd ed. New York: McGraw-Hill, 1997.

12. L. F. Moody. "Friction Factors for Pipe Flows," *Transactions of the ASME* 66 (1944), pp. 671–684.

13. G. Rosenberg, W. M. Phillips, D. L. Landis, and W. S. Pierce. "Design and evaluation of the Pennsylvania State University mock circulatory system." *ASAIO J.* 4 (1981) pp. 41–49.

14. O. Reynolds. "On the Experimental Investigation of the Circumstances Which Determine Whether the Motion of Water Shall Be Direct or Sinuous, and the Law of Resistance in Parallel Channels." *Philosophical Transactions of the Royal Society of London*, 174 (1883), pp. 935–982.

15. H. Schlichting. *Boundary Layer Theory*, 7th ed. New York: Springer, 2000.

16. R. K. Shah and M. S. Bhatti. "Laminar Convective Heat Transfer in Ducts." In *Handbook of Single-Phase Convective Heat Transfer*, ed. S. Kakaç, R. K. Shah, and W. Aung. New York: Wiley Interscience, 1987.

17. P. L. Skousen. *Valve Handbook*. New York: McGraw-Hill, 1998.

18. P. K. Swamee and A. K. Jain. "Explicit Equations for Pipe-Flow Problems," *Journal of the Hydraulics Division.* ASCE 102, no. HY5 (May 1976), pp. 657–664.

19. G. Vass. "Ultrasonic Flowmeter Basics," *Sensors*, 14, no. 10 (1997).

20. A. J. Wheeler and A. R. Ganji. *Introduction to Engineering Experimentation*. Englewood Cliffs, NJ: Prentice-Hall, 1996.

21. W. Zhi-qing. "Study on Correction Coefficients of Laminar and Turbulent Entrance Region Effects in Round Pipes," *Applied Mathematical Mechanics*, 3 (1982), p. 433.

PROBLEMS*

Laminar and Turbulent Flow

8–1C Consider the flow of air and water in pipes of the same diameter, at the same temperature, and at the same mean velocity. Which flow is more likely to be turbulent? Why?

8–2C Consider laminar flow in a circular pipe. Is the wall shear stress τ_w higher near the inlet of the pipe or near the exit? Why? What would your response be if the flow were turbulent?

8–3C What is hydraulic diameter? How is it defined? What is it equal to for a circular pipe of diameter D?

8–4C How is the hydrodynamic entry length defined for flow in a pipe? Is the entry length longer in laminar or turbulent flow?

8–5C Why are liquids usually transported in circular pipes?

8–6C What is the physical significance of the Reynolds number? How is it defined for (*a*) flow in a circular pipe of inner diameter D and (*b*) flow in a rectangular duct of cross section $a \times b$?

FIGURE P8–6C

8–7C Consider a person walking first in air and then in water at the same speed. For which motion will the Reynolds number be higher?

8–8C Show that the Reynolds number for flow in a circular pipe of diameter D can be expressed as $Re = 4\dot{m}/(\pi D\mu)$.

8–9C Which fluid at room temperature requires a larger pump to flow at a specified velocity in a given pipe: water or engine oil? Why?

8–10C What is the generally accepted value of the Reynolds number above which the flow in smooth pipes is turbulent?

8–11C How does surface roughness affect the pressure drop in a pipe if the flow is turbulent? What would your response be if the flow were laminar?

8–12E Shown here is a cool picture of water being released at 300,000 gallons per second in the spring of 2008. This was part of a revitalization effort for the ecosystem of the Grand Canyon and the Colorado River. Estimate the Reynolds number of the pipe flow. Is it laminar or turbulent? (*Hint:* For a length scale, approximate the height of the man in the blue shirt directly above the pipe to be 6 ft.)

---

* Problems designated by a "C" are concept questions, and students are encouraged to answer them all. Problems designated by an "E" are in English units, and the SI users can ignore them. Problems with the ⬡ icon are solved using EES, and complete solutions together with parametric studies are included on the text website. Problems with the ⬡ icon are comprehensive in nature and are intended to be solved with an equation solver such as EES.

FIGURE P8–12E

Photo courtesy of Don Becker, U.S. Geological Survey.

Fully Developed Flow in Pipes

8–13C Someone claims that the volume flow rate in a circular pipe with laminar flow can be determined by measuring the velocity at the centerline in the fully developed region, multiplying it by the cross-sectional area, and dividing the result by 2. Do you agree? Explain.

8–14C Someone claims that the average velocity in a circular pipe in fully developed laminar flow can be determined by simply measuring the velocity at $R/2$ (midway between the wall surface and the centerline). Do you agree? Explain.

8–15C Someone claims that the shear stress at the center of a circular pipe during fully developed laminar flow is zero. Do you agree with this claim? Explain.

8–16C Someone claims that in fully developed turbulent flow in a pipe, the shear stress is a maximum at the pipe wall. Do you agree with this claim? Explain.

8–17C How does the wall shear stress τ_w vary along the flow direction in the fully developed region in (a) laminar flow and (b) turbulent flow?

8–18C What fluid property is responsible for the development of the velocity boundary layer? For what kinds of fluids will there be no velocity boundary layer in a pipe?

8–19C In the fully developed region of flow in a circular pipe, does the velocity profile change in the flow direction?

8–20C How is the friction factor for flow in a pipe related to the pressure loss? How is the pressure loss related to the pumping power requirement for a given mass flow rate?

8–21C Discuss whether fully developed pipe flow is one-, two-, or three-dimensional.

8–22C Consider fully developed flow in a circular pipe with negligible entrance effects. If the length of the pipe is doubled, the head loss will (a) double, (b) more than double, (c) less than double, (d) reduce by half, or (e) remain constant.

8–23C Consider fully developed laminar flow in a circular pipe. If the diameter of the pipe is reduced by half while the flow rate and the pipe length are held constant, the head loss will (a) double, (b) triple, (c) quadruple, (d) increase by a factor of 8, or (e) increase by a factor of 16.

8–24C Explain why the friction factor is independent of the Reynolds number at very large Reynolds numbers.

8–25C What is turbulent viscosity? What causes it?

8–26C The head loss for a certain circular pipe is given by $h_L = 0.0826 f L(\dot{V}^2/D^5)$, where f is the friction factor (dimensionless), L is the pipe length, $\dot{V}$ is the volumetric flow rate, and D is the pipe diameter. Determine if the 0.0826 is a dimensional or dimensionless constant. Is this equation dimensionally homogeneous as it stands?

8–27C Consider fully developed laminar flow in a circular pipe. If the viscosity of the fluid is reduced by half by heating while the flow rate is held constant, how does the head loss change?

8–28C How is head loss related to pressure loss? For a given fluid, explain how you would convert head loss to pressure loss.

8–29C Consider laminar flow of air in a circular pipe with perfectly smooth surfaces. Do you think the friction factor for this flow is zero? Explain.

8–30C What is the physical mechanism that causes the friction factor to be higher in turbulent flow?

8–31 The velocity profile for the fully developed laminar flow of a Newtonian fluid between two large parallel plates is given by

$$u(y) = \frac{3u_0}{2}\left[1 - \left(\frac{y}{h}\right)^2\right]$$

where $2h$ is the distance between the two plates, u_0 is the velocity at the center plane, and y is the vertical coordinate from the center plane. For a plate width of b, obtain a relation for the flow rate through the plates.

8–32 Water flows steadily through a reducing pipe section. The flow upstream with a radius of R_1 is laminar with a velocity profile of $u_1(r) = u_{01}(1 - r^2/R_1^2)$ while the flow downstream is turbulent with a velocity profile of $u_2(r) = u_{02}(1 - r/R_2)^{1/7}$. For incompressible flow with $R_2/R_1 = 4/7$, determine the ratio of centerline velocities u_{01}/u_{02}.

8–33 Water at 10°C (ρ = 999.7 kg/m³ and μ = 1.307 × 10^{-3} kg/m·s) is flowing steadily in a 0.12-cm-diameter, 15-m-long pipe at an average velocity of 0.9 m/s. Determine (*a*) the pressure drop, (*b*) the head loss, and (*c*) the pumping power requirement to overcome this pressure drop. *Answers:* (*a*) 392 kPa, (*b*) 40.0 m, (*c*) 0.399 W

8–34 Consider an air solar collector that is 1 m wide and 5 m long and has a constant spacing of 3 cm between the glass cover and the collector plate. Air flows at an average temperature of 45°C at a rate of 0.15 m³/s through the 1-m-wide edge of the collector along the 5-m-long passageway. Disregarding the entrance and roughness effects and the 90° bend, determine the pressure drop in the collector. *Answer:* 32.3 Pa

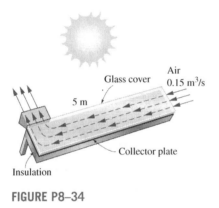

Air
0.15 m³/s
Glass cover
5 m
Collector plate
Insulation

FIGURE P8–34

8–35E Heated air at 1 atm and 100°F is to be transported in a 400-ft-long circular plastic duct at a rate of 12 ft³/s. If the head loss in the pipe is not to exceed 50 ft, determine the minimum diameter of the duct.

8–36 In fully developed laminar flow in a circular pipe, the velocity at *R*/2 (midway between the wall surface and the centerline) is measured to be 11 m/s. Determine the velocity at the center of the pipe. *Answer:* 14.7 m/s

8–37 The velocity profile in fully developed laminar flow in a circular pipe of inner radius R = 2 cm, in m/s, is given by $u(r) = 4(1 - r^2/R^2)$. Determine the average and maximum velocities in the pipe and the volume flow rate.

$$u(r) = 4\left(1 - \frac{r^2}{R^2}\right)$$

$R = 2$ cm

FIGURE P8–37

8–38 Repeat Prob. 8–37 for a pipe of inner radius 7 cm.

8–39 Water at 15°C (ρ = 999.1 kg/m³ and μ = 1.138 × 10^{-3} kg/m·s) is flowing steadily in a 30-m-long and 5-cm-diameter horizontal pipe made of stainless steel at a rate of 9 L/s. Determine (*a*) the pressure drop, (*b*) the head loss, and (*c*) the pumping power requirement to overcome this pressure drop.

9 L/s
5 cm
30 m

FIGURE P8–39

8–40 Consider the flow of oil with ρ = 894 kg/m³ and μ = 2.33 kg/m·s in a 28-cm-diameter pipeline at an average velocity of 0.5 m/s. A 330-m-long section of the pipeline passes through the icy waters of a lake. Disregarding the entrance effects, determine the pumping power required to overcome the pressure losses and to maintain the flow of oil in the pipe.

8–41 Consider laminar flow of a fluid through a square channel with smooth surfaces. Now the average velocity of the fluid is doubled. Determine the change in the head loss of the fluid. Assume the flow regime remains unchanged.

8–42 Repeat Prob. 8–41 for turbulent flow in smooth pipes for which the friction factor is given as f = 0.184Re$^{-0.2}$. What would your answer be for fully turbulent flow in a rough pipe?

8–43 Air enters a 10-m-long section of a rectangular duct of cross section 15 cm × 20 cm made of commercial steel at 1 atm and 35°C at an average velocity of 7 m/s. Disregarding the entrance effects, determine the fan power needed to overcome the pressure losses in this section of the duct. *Answer:* 7.00 W

10 m
15 cm
Air
7 m/s
20 cm

FIGURE P8–43

8–44E Water at 70°F passes through 0.75-in-internal-diameter copper tubes at a rate of 0.5 lbm/s. Determine the pumping power per ft of pipe length required to maintain this flow at the specified rate.

8–45 Oil with $\rho = 876$ kg/m³ and $\mu = 0.24$ kg/m·s is flowing through a 1.5-cm-diameter pipe that discharges into the atmosphere at 88 kPa. The absolute pressure 15 m before the exit is measured to be 135 kPa. Determine the flow rate of oil through the pipe if the pipe is (*a*) horizontal, (*b*) inclined 8° upward from the horizontal, and (*c*) inclined 8° downward from the horizontal.

FIGURE P8–45

8–46 Glycerin at 40°C with $\rho = 1252$ kg/m³ and $\mu = 0.27$ kg/m·s is flowing through a 2-cm-diameter, 25-m-long pipe that discharges into the atmosphere at 100 kPa. The flow rate through the pipe is 0.048 L/s. (*a*) Determine the absolute pressure 25 m before the pipe exit. (*b*) At what angle θ must the pipe be inclined downward from the horizontal for the pressure in the entire pipe to be atmospheric pressure and the flow rate to be maintained the same?

8–47E Air at 1 atm and 60°F is flowing through a 1 ft × 1 ft square duct made of commercial steel at a rate of 1600 cfm. Determine the pressure drop and head loss per ft of the duct.

FIGURE P8–47E

8–48 Water enters into a cone of height H and base radius R through a small hole of cross-sectional area A_h and the discharge coefficient is C_d at the base with a constant uniform velocity of V. Obtain a relation for the variation of water height h from the cone base with time. Air escapes the cone through the tip at the top as water enters the cone from the bottom.

8–49 The velocity profile for incompressible turbulent flow in a pipe of radius R is given by $u(r) = u_{max}(1 - r/R_2)^{1/7}$. Obtain an expression for the average velocity in the pipe.

8–50 Oil with a density of 850 kg/m³ and kinematic viscosity of 0.00062 m²/s is being discharged by a 8-mm-diameter, 40-m-long horizontal pipe from a storage tank open to

the atmosphere. The height of the liquid level above the center of the pipe is 4 m. Disregarding the minor losses, determine the flow rate of oil through the pipe.

FIGURE P8–50

8–51 In an air heating system, heated air at 40°C and 105 kPa absolute is distributed through a 0.2 m × 0.3 m rectangular duct made of commercial steel at a rate of 0.5 m³/s. Determine the pressure drop and head loss through a 40-m-long section of the duct. *Answers:* 124 Pa, 10.8 m

8–52 Glycerin at 40°C with $\rho = 1252$ kg/m³ and $\mu = 0.27$ kg/m·s is flowing through a 4-cm-diameter horizontal smooth pipe with an average velocity of 3.5 m/s. Determine the pressure drop per 10 m of the pipe.

8–53 [EES] Reconsider Prob. 8–52. Using EES (or other) software, investigate the effect of the pipe diameter on the pressure drop for the same constant flow rate. Let the pipe diameter vary from 1 to 10 cm in increments of 1 cm. Tabulate and plot the results, and draw conclusions.

8–54E Oil at 80°F ($\rho = 56.8$ lbm/ft³ and $\mu = 0.0278$ lbm/ft·s) is flowing steadily in a 0.5-in-diameter, 175-ft-long pipe. During the flow, the pressure at the pipe inlet and exit is measured to be 80 psi and 14 psi, respectively. Determine the flow rate of oil through the pipe assuming the pipe is (*a*) horizontal, (*b*) inclined 20° upward, and (*c*) inclined 20° downward.

8–55 Liquid ammonia at −20°C is flowing through a 20-m-long section of a 5-mm-diameter copper tube at a rate of 0.09 kg/s. Determine the pressure drop, the head loss, and the pumping power required to overcome the frictional losses in the tube. *Answers:* 1240 kPa, 189 m, 0.167 kW

Minor Losses

8–56C During a retrofitting project of a fluid flow system to reduce the pumping power, it is proposed to install vanes into the miter elbows or to replace the sharp turns in 90° miter elbows by smooth curved bends. Which approach will result in a greater reduction in pumping power requirements?

8–57C Define equivalent length for minor loss in pipe flow. How is it related to the minor loss coefficient?

FIGURE P8–66

8–58C The effect of rounding of a pipe inlet on the loss coefficient is (*a*) negligible, (*b*) somewhat significant, or (*c*) very significant.

8–59C The effect of rounding of a pipe exit on the loss coefficient is (*a*) negligible, (*b*) somewhat significant, or (*c*) very significant.

8–60C Which has a greater minor loss coefficient during pipe flow: gradual expansion or gradual contraction? Why?

8–61C A piping system involves sharp turns, and thus large minor head losses. One way of reducing the head loss is to replace the sharp turns by circular elbows. What is another way?

8–62C What is minor loss in pipe flow? How is the minor loss coefficient K_L defined?

8–63 Water is to be withdrawn from an 8-m-high water reservoir by drilling a 2.2-cm-diameter hole at the bottom surface. Disregarding the effect of the kinetic energy correction factor, determine the flow rate of water through the hole if (*a*) the entrance of the hole is well-rounded and (*b*) the entrance is sharp-edged.

8–64 Consider flow from a water reservoir through a circular hole of diameter D at the side wall at a vertical distance H from the free surface. The flow rate through an actual hole with a sharp-edged entrance ($K_L = 0.5$) is considerably less than the flow rate calculated assuming "frictionless" flow and thus zero loss for the hole. Disregarding the effect of the kinetic energy correction factor, obtain a relation for the "equivalent diameter" of the sharp-edged hole for use in frictionless flow relations.

Frictionless flow   Actual flow

FIGURE P8–64

8–65 Repeat Prob. 8–64 for a slightly rounded entrance ($K_L = 0.12$).

8–66 A horizontal pipe has an abrupt expansion from $D_1 = 8$ cm to $D_2 = 16$ cm. The water velocity in the smaller section is 10 m/s and the flow is turbulent. The pressure in the smaller section is $P_1 = 410$ kPa. Taking the kinetic energy correction factor to be 1.06 at both the inlet and the outlet, determine the downstream pressure P_2, and estimate the error that would have occurred if Bernoulli's equation had been used. *Answers:* 432 kPa, 25.4 kPa

Piping Systems and Pump Selection

8–67C Water is pumped from a large lower reservoir to a higher reservoir. Someone claims that if the head loss is negligible, the required pump head is equal to the elevation difference between the free surfaces of the two reservoirs. Do you agree?

8–68C A piping system equipped with a pump is operating steadily. Explain how the operating point (the flow rate and the head loss) is established.

8–69C A person filling a bucket with water using a garden hose suddenly remembers that attaching a nozzle to the hose increases the discharge velocity of water and wonders if this increased velocity would decrease the filling time of the bucket. What would happen to the filling time if a nozzle were attached to the hose: increase it, decrease it, or have no effect? Why?

8–70C Consider two identical 2-m-high open tanks filled with water on top of a 1-m-high table. The discharge valve of one of the tanks is connected to a hose whose other end is left open on the ground while the other tank does not have a hose connected to its discharge valve. Now the discharge valves of both tanks are opened. Disregarding any frictional loses in the hose, which tank do you think empties completely first? Why?

8–71C A piping system involves two pipes of different diameters (but of identical length, material, and roughness) connected in series. How would you compare the (*a*) flow rates and (*b*) pressure drops in these two pipes?

8–72C A piping system involves two pipes of different diameters (but of identical length, material, and roughness) connected in parallel. How would you compare the (*a*) flow rates and (*b*) pressure drops in these two pipes?

8–73C A piping system involves two pipes of identical diameters but of different lengths connected in parallel. How would you compare the pressure drops in these two pipes?

8–74C For a piping system, define the system curve, the characteristic curve, and the operating point on a head versus flow rate chart.

8–75 A 4-m-high cylindrical tank having a cross-sectional area of $A_T = 1.5$ m^2 is filled with equal volumes of water and oil whose specific gravity is SG = 0.75. Now a 1-cm-diameter hole at the bottom of the tank is opened, and water starts to flow out. If the discharge coefficient of the hole is $C_d = 0.85$, determine how long it will take for the water in the tank, which is open to the atmosphere to empty completely.

8–76   A semi-spherical tank of radius R is completely filled with water. Now a hole of cross sectional area A_h and discharge coefficient C_d at the bottom of the tank is fully opened and water starts to flow out. Develop an expression for the time needed to empty the tank completely.

FIGURE P8–76

8–77   The water needs of a small farm are to be met by pumping water from a well that can supply water continuously at a rate of 4 L/s. The water level in the well is 20 m below the ground level, and water is to be pumped to a large tank on a hill, which is 58 m above the ground level of the well, using 5-cm internal diameter plastic pipes. The required length of piping is measured to be 420 m, and the total minor loss coefficient due to the use of elbows, vanes, etc. is estimated to be 12. Taking the efficiency of the pump to be 75 percent, determine the rated power of the pump that needs to be purchased, in kW. The density and viscosity of water at anticipated operation conditions are taken to be 1000 kg/m^3 and 0.00131 kg/m·s, respectively. Is it wise to purchase a suitable pump that meets the total power requirements, or is it necessary to also pay particular attention to the large elevation head in this case? Explain. *Answer:* 6.0 kW

8–78E   Water at 70°F flows by gravity from a large reservoir at a high elevation to a smaller one through a 60-ft-long, 2-in-diameter cast iron piping system that includes four standard flanged elbows, a well-rounded entrance, a sharp-edged exit, and a fully open gate valve. Taking the free surface of the lower reservoir as the reference level, determine the elevation z_1 of the higher reservoir for a flow rate of 10 ft^3/min. *Answer:* 12.6 ft

8–79   A 2.4-m-diameter tank is initially filled with water 4 m above the center of a sharp-edged 10-cm-diameter orifice. The tank water surface is open to the atmosphere, and the orifice drains to the atmosphere. Neglecting the effect of the kinetic energy correction factor, calculate (*a*) the initial velocity from the tank and (*b*) the time required to empty the tank. Does the loss coefficient of the orifice cause a significant increase in the draining time of the tank?

FIGURE P8–79

8–80   A 3-m-diameter tank is initially filled with water 2 m above the center of a sharp-edged 10-cm-diameter orifice. The tank water surface is open to the atmosphere, and the orifice drains to the atmosphere through a 100-m-long pipe. The friction coefficient of the pipe is taken to be 0.015 and the effect of the kinetic energy correction factor can be neglected. Determine (*a*) the initial velocity from the tank and (*b*) the time required to empty the tank.

8–81   Reconsider Prob. 8–80. In order to drain the tank faster, a pump is installed near the tank exit as in Fig. P8–81. Determine how much pump power input is necessary to establish an average water velocity of 4 m/s when the tank is full at $z = 2$ m. Also, assuming the discharge velocity to remain constant, estimate the time required to drain the tank.

Someone suggests that it makes no difference whether the pump is located at the beginning or at the end of the pipe, and that the performance will be the same in either case, but another person argues that placing the pump near the end of the pipe may cause cavitation. The water temperature is 30°C, so the water vapor pressure is $P_v = 4.246$ kPa $= 0.43$ m-H$_2$O, and the system is located at sea level. Investigate if there is the possibility of cavitation and if we should be concerned about the location of the pump.

FIGURE P8–81

8–82   Water to a residential area is transported at a rate of 1.5 m^3/s via 70-cm-internal-diameter concrete pipes with a surface roughness of 3 mm and a total length of 1500 m.

In order to reduce pumping power requirements, it is proposed to line the interior surfaces of the concrete pipe with 2-cm-thick petroleum-based lining that has a surface roughness thickness of 0.04 mm. There is a concern that the reduction of pipe diameter to 66 cm and the increase in average velocity may offset any gains. Taking $\rho = 1000$ kg/m³ and $\nu = 1 \times 10^{-6}$ m²/s for water, determine the percent increase or decrease in the pumping power requirements due to pipe frictional losses as a result of lining the concrete pipes.

8–83E A clothes dryer discharges air at 1 atm and 120°F at a rate of 1.2 ft³/s when its 5-in-diameter, well-rounded vent with negligible loss is not connected to any duct. Determine the flow rate when the vent is connected to a 15-ft-long, 5-in-diameter duct made of galvanized iron, with three 90° flanged smooth bends. Take the friction factor of the duct to be 0.019, and assume the fan power input to remain constant.

FIGURE P8–83E

8–84 Oil at 20°C is flowing through a vertical glass funnel that consists of a 20-cm-high cylindrical reservoir and a 1-cm-diameter, 40-cm-high pipe. The funnel is always maintained full by the addition of oil from a tank. Assuming the entrance effects to be negligible, determine the flow rate of oil through the funnel and calculate the "funnel effectiveness,"

FIGURE P8–84

which is defined as the ratio of the actual flow rate through the funnel to the maximum flow rate for the "frictionless" case. *Answers:* 3.83 × 10⁻⁶ m³/s, 1.4 percent

8–85 Repeat Prob. 8–84 assuming (a) the diameter of the pipe is tripled and (b) the length of the pipe is tripled while the diameter is maintained the same.

8–86 Water at 15°C is drained from a large reservoir using two horizontal plastic pipes connected in series. The first pipe is 20 m long and has a 10-cm diameter, while the second pipe is 35 m long and has a 4-cm diameter. The water level in the reservoir is 18 m above the centerline of the pipe. The pipe entrance is sharp-edged, and the contraction between the two pipes is sudden. Neglecting the effect of the kinetic energy correction factor, determine the discharge rate of water from the reservoir.

FIGURE P8–86

8–87E A farmer is to pump water at 70°F from a river to a water storage tank nearby using a 125-ft-long, 5-in-diameter plastic pipe with three flanged 90° smooth bends. The water velocity near the river surface is 6 ft/s, and the pipe inlet is placed in the river normal to the flow direction of water to take advantage of the dynamic pressure. The elevation difference between the river and the free surface of the tank is 12 ft. For a flow rate of 1.5 ft³/s and an overall pump efficiency of 70 percent, determine the required electric power input to the pump.

8–88E Reconsider Prob. 8–89E. Using EES (or other) software, investigate the effect of the pipe diameter on the required electric power input to the pump. Let the pipe diameter vary from 1 to 10 in, in increments of 1 in. Tabulate and plot the results, and draw conclusions.

8–89 A water tank filled with solar-heated water at 40°C is to be used for showers in a field using gravity-driven flow. The system includes 35 m of 1.5-cm-diameter galvanized iron piping with four miter bends (90°) without vanes and a wide-open globe valve. If water is to flow at a rate of 1.2 L/s through the shower head, determine how high the water level in the tank must be from the exit level of the shower. Disregard the losses at the entrance and at the shower head, and neglect the effect of the kinetic energy correction factor.

8–90 Two water reservoirs A and B are connected to each other through a 40-m-long, 2-cm-diameter cast iron pipe with a sharp-edged entrance. The pipe also involves a swing check valve and a fully open gate valve. The water level in both reservoirs is the same, but reservoir A is pressurized by compressed air while reservoir B is open to the atmosphere at 88 kPa. If the initial flow rate through the pipe is 1.2 L/s, determine the absolute air pressure on top of reservoir A. Take the water temperature to be 10°C. *Answer: 733 kPa*

FIGURE P8–90

8–91 A vented tanker is to be filled with fuel oil with $\rho = 920$ kg/m³ and $\mu = 0.045$ kg/m·s from an underground reservoir using a 25-m-long, 4-cm-diameter plastic hose with a slightly rounded entrance and two 90° smooth bends. The elevation difference between the oil level in the reservoir and the top of the tanker where the hose is discharged is 5 m. The capacity of the tanker is 18 m³ and the filling time is 30 min. Taking the kinetic energy correction factor at the hose discharge to be 1.05 and assuming an overall pump efficiency of 82 percent, determine the required power input to the pump.

FIGURE P8–91

8–92 Two pipes of identical length and material are connected in parallel. The diameter of pipe A is twice the diameter of pipe B. Assuming the friction factor to be the same in both cases and disregarding minor losses, determine the ratio of the flow rates in the two pipes.

8–93 A certain part of cast iron piping of a water distribution system involves a parallel section. Both parallel pipes have a diameter of 30 cm, and the flow is fully turbulent. One of the branches (pipe A) is 1500 m long while the other branch (pipe B) is 2500 m long. If the flow rate through pipe A is 0.4 m³/s, determine the flow rate through pipe B. Disregard minor losses and assume the water

temperature to be 15°C. Show that the flow is fully rough, and thus the friction factor is independent of Reynolds number. *Answer: 0.310 m³/s*

FIGURE P8–93

8–94 Repeat Prob. 8–93 assuming pipe A has a halfway-closed gate valve ($K_L = 2.1$) while pipe B has a fully open globe valve ($K_L = 10$), and the other minor losses are negligible.

8–95 A geothermal district heating system involves the transport of geothermal water at 110°C from a geothermal well to a city at about the same elevation for a distance of 12 km at a rate of 1.5 m³/s in 60-cm-diameter stainless-steel pipes. The fluid pressures at the wellhead and the arrival point in the city are to be the same. The minor losses are negligible because of the large length-to-diameter ratio and the relatively small number of components that cause minor losses. (*a*) Assuming the pump–motor efficiency to be 80 percent, determine the electric power consumption of the system for pumping. Would you recommend the use of a single large pump or several smaller pumps of the same total pumping power scattered along the pipeline? Explain. (*b*) Determine the daily cost of power consumption of the system if the unit cost of electricity is $0.06/kWh. (*c*) The temperature of geothermal water is estimated to drop 0.5°C during this long flow. Determine if the frictional heating during flow can make up for this drop in temperature.

8–96 Repeat Prob. 8–95 for cast iron pipes of the same diameter.

8–97 Water is transported by gravity through a 12-cm-diameter 800-m-long plastic pipe with an elevation gradient of 0.01 (i.e., an elevation drop of 1 m per 100 m of pipe length). Taking $\rho = 1000$ kg/m³ and $\nu = 1 \times 10^{-6}$ m²/s for water, determine the flow rate of water through the pipe. If the pipe were horizontal, what would the power requirements be to maintain the same flow rate?

8–98 Gasoline ($\rho = 680$ kg/m³ and $\nu = 4.29 \times 10^{-7}$ m²/s) is transported at a rate of 240 L/s for a distance of 2 km. The surface roughness of the piping is 0.03 mm. If the head loss due to pipe friction is not to exceed 10 m, determine the minimum diameter of the pipe.

8–99 In large buildings, hot water in a water tank is circulated through a loop so that the user doesn't have to wait for all the water in long piping to drain before hot water starts coming out. A certain recirculating loop involves 40-m-long, 1.2-cm-diameter cast iron pipes with six 90° threaded smooth bends and two fully open gate valves. If the average flow velocity through the loop is 2 m/s, determine the required power input for the recirculating pump. Take the average water temperature to be 60°C and the efficiency of the pump to be 70 percent. *Answer: 0.111 kW*

8–100 Reconsider Prob. 8–99. Using EES (or other) software, investigate the effect of the average flow velocity on the power input to the recirculating pump. Let the velocity vary from 0 to 3 m/s in increments of 0.3 m/s. Tabulate and plot the results.

8–101 Repeat Prob. 8–99 for plastic (smooth) pipes.

8–102 Water at 20°C is to be pumped from a reservoir ($z_A = 2$ m) to another reservoir at a higher elevation ($z_B = 9$ m) through two 25-m-long plastic pipes connected in parallel. The diameters of the two pipes are 3 cm and 5 cm. Water is to be pumped by a 68 percent efficient motor–pump unit that draws 7 kW of electric power during operation. The minor losses and the head loss in the pipes that connect the parallel pipes to the two reservoirs are considered to be negligible. Determine the total flow rate between the reservoirs and the flow rates through each of the parallel pipes.

FIGURE P8–102

8–103 A 6-m-tall chimney shown in Fig. P8–103 is to be designed to discharge hot gases from a fireplace at 180°C at a constant rate of 0.15 m³/s when the atmospheric air temperature is 20°C. Assuming no heat transfer from the chimney and taking the chimney entrance loss coefficient to be 1.5 and the friction coefficient of the chimney to be 0.020, determine the chimney diameter that would discharge the hot gases at the desired rate. Note that $P_3 = P_4 = P_{atm}$ and $P_2 = P_1 = P_{atm} + \rho_{atm\ air}\, gh$, and assume the hot gases in the entire chimney are at 180°C.

FIGURE P8–103

8–104 An inverted 3-m-high conical container shown in Fig. P8–104 is initially filled with 2-m-high water. At time $t = 0$, a faucet is opened to supply water into the container at a rate of 3 L/s. At the same time, a 4-cm-diameter hole with a discharge coefficient of 0.90 at the bottom of the container is opened. Determine how long it will take for the water level in the tank to drop to 1-m.

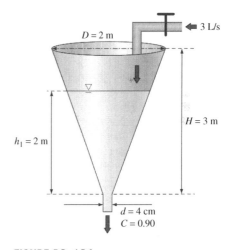

FIGURE P8–104

Flow Rate and Velocity Measurements

8–105C What is the difference between the operating principles of thermal and laser Doppler anemometers?

8–106C What is the difference between laser Doppler velocimetry (LDV) and particle image velocimetry (PIV)?

8–107C What are the primary considerations when selecting a flowmeter to measure the flow rate of a fluid?

8–108C Explain how flow rate is measured with a Pitot-static tube, and discuss its advantages and disadvantages with respect to cost, pressure drop, reliability, and accuracy.

8–109C Explain how flow rate is measured with obstruction-type flowmeters. Compare orifice meters, flow nozzles, and Venturi meters with respect to cost, size, head loss, and accuracy.

8–110C How do positive displacement flowmeters operate? Why are they commonly used to meter gasoline, water, and natural gas?

8–111C Explain how flow rate is measured with a turbine flowmeter, and discuss how they compare to other types of flowmeters with respect to cost, head loss, and accuracy.

8–112C What is the operating principle of variable-area flowmeters (rotameters)? How do they compare to other types of flowmeters with respect to cost, head loss, and reliability?

8–113 The flow rate of water at 20°C ($\rho = 998$ kg/m^3 and $\mu = 1.002 \times 10^{-3}$ kg/m·s) through a 60-cm-diameter pipe is measured with an orifice meter with a 30-cm-diameter opening to be 400 L/s. Determine the pressure difference indicated by the orifice meter and the head loss.

8–114 A Pitot-static probe is mounted in a 2.5-cm-inner diameter pipe at a location where the local velocity is approximately equal to the average velocity. The oil in the pipe has density $\rho = 860$ kg/m^3 and viscosity $\mu = 0.0103$ kg/m·s. The pressure difference is measured to be 95.8 Pa. Calculate the volume flow rate through the pipe in cubic meters per second.

8–115 Calculate the Reynolds number of the flow of Prob. 8–114. Is it laminar or turbulent?

8–116 A flow nozzle equipped with a differential pressure gage is used to measure the flow rate of water at 10°C ($\rho = 999.7$ kg/m^3 and $\mu = 1.307 \times 10^{-3}$ kg/m·s) through a 3-cm-diameter horizontal pipe. The nozzle exit diameter is 1.5 cm, and the measured pressure drop is 3 kPa. Determine the volume flow rate of water, the average velocity through the pipe, and the head loss.

3 cm

1.5 cm

$\Delta P = 3$ kPa

Differential
pressure gage

FIGURE P8–116

8–117 The flow rate of water through a 10-cm-diameter pipe is to be determined by measuring the water velocity at several locations along a cross section. For the set of measurements given in the table, determine the flow rate.

r, cm	V, m/s
0	6.4
1	6.1
2	5.2
3	4.4
4	2.0
5	0.0

8–118E An orifice with a 1.8-in-diameter opening is used to measure the mass flow rate of water at 60°F ($\rho = 62.36$ lbm/ft^3 and $\mu = 7.536 \times 10^{-4}$ lbm/ft·s) through a horizontal 4-in-diameter pipe. A mercury manometer is used to measure the pressure difference across the orifice. If the differential height of the manometer is 7 in, determine the volume flow rate of water through the pipe, the average velocity, and the head loss caused by the orifice meter.

4 in

1.8 in

7 in

FIGURE P8–118E

8–119E Repeat Prob. 8–118E for a differential height of 10 in.

8–120 Air ($\rho = 1.225$ kg/m^3 and $\mu = 1.789 \times 10^{-5}$ kg/m·s) flows in a wind tunnel, and the wind tunnel speed is measured with a Pitot-static probe. For a certain run, the stagnation pressure is measured to be 472.6 Pa gage and the static pressure is 15.43 Pa gage. Calculate the wind-tunnel speed.

8–121 A Venturi meter equipped with a differential pressure gage is used to measure the flow rate of water at 15°C ($\rho = 999.1$ kg/m^3) through a 5-cm-diameter horizontal pipe. The diameter of the Venturi neck is 3 cm, and the measured pressure drop is 5 kPa. Taking the discharge coefficient to be 0.98, determine the volume flow rate of water and the average velocity through the pipe. *Answers: 2.35 L/s and 1.20 m/s*

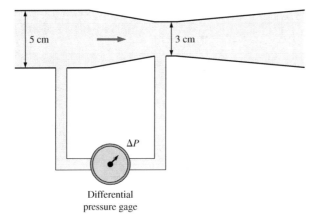

FIGURE P8–121

8–122 Reconsider Prob. 8–121. Letting the pressure drop vary from 1 kPa to 10 kPa, evaluate the flow rate at intervals of 1 kPa, and plot it against the pressure drop.

8–123 The mass flow rate of air at 20°C ($\rho = 1.204$ kg/m^3) through a 18-cm-diameter duct is measured with a Venturi meter equipped with a water manometer. The Venturi neck has a diameter of 5 cm, and the manometer has a maximum differential height of 40 cm. Taking the discharge coefficient to be 0.98, determine the maximum mass flow rate of air this Venturi meter/manometer can measure. *Answer:* 0.188 kg/s

FIGURE P8–123

8–124 Repeat Prob. 8–123 for a Venturi neck diameter of 6 cm.

8–125 A vertical Venturi meter equipped with a differential pressure gage shown in Fig. P8–125 is used to measure the flow rate of liquid propane at 10°C ($\rho = 514.7$ kg/m^3) through an 10-cm-diameter vertical pipe. For a discharge coefficient of 0.98, determine the volume flow rate of propane through the pipe.

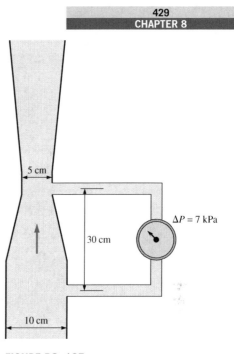

FIGURE P8–125

8–126E The volume flow rate of liquid refrigerant-134a at 10°F ($\rho = 83.31$ lbm/ft^3) is to be measured with a horizontal Venturi meter with a diameter of 5 in at the inlet and 2 in at the throat. If a differential pressure meter indicates a pressure drop of 6.4 psi, determine the flow rate of the refrigerant. Take the discharge coefficient of the Venturi meter to be 0.98.

8–127 A 22-L kerosene tank ($\rho = 820$ kg/m^3) is filled with a 2-cm-diameter hose equipped with a 1.5-cm-diameter nozzle meter. If it takes 20 s to fill the tank, determine the pressure difference indicated by the nozzle meter.

8–128 The flow rate of water at 20°C ($\rho = 998$ kg/m^3 and $\mu = 1.002 \times 10^{-3}$ kg/m·s) through a 4-cm-diameter pipe is measured with a 2-cm-diameter nozzle meter equipped with an inverted air–water manometer. If the manometer indicates a

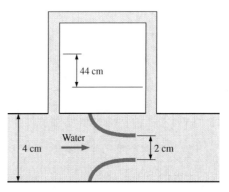

FIGURE P8–128

differential water height of 44 cm, determine the volume flow rate of water and the head loss caused by the nozzle meter.

8–129 The flow rate of ammonia at 10°C ($\rho = 624.6$ kg/m³ and $\mu = 1.697 \times 10^{-4}$ kg/m·s) through a 2-cm-diameter pipe is to be measured with a 1.5-cm-diameter flow nozzle equipped with a differential pressure gage. If the gage reads a pressure differential of 4 kPa, determine the flow rate of ammonia through the pipe, and the average flow velocity.

Review Problems

8–130 In a laminar flow through a circular tube of radius of R, the velocity and temperature profiles at a cross-section are given by $u = u_0(1 - r^2/R^2)$ and $T(r) = A + Br^2 - Cr^4$ where A, B and C are positive constants. Obtain a relation for the bulk fluid temperature at that cross section.

8–131 The conical container with a thin horizontal tube attached at the bottom, shown in Fig. P8–131, is to be used to measure the viscosity of an oil. The flow through the tube is laminar. The discharge time needed for the oil level to drop from h_1 to h_2 is to be measured by a stopwatch. Develop an expression for the viscosity of oil in the container as a function of the discharge time t.

FIGURE P8–131

8–132 Shell-and-tube heat exchangers with hundreds of tubes housed in a shell are commonly used in practice for heat transfer between two fluids. Such a heat exchanger used in an active solar hot-water system transfers heat from a water-antifreeze solution flowing through the shell and the solar collector to fresh water flowing through the tubes at an average temperature of 60°C at a rate of 15 L/s. The heat exchanger contains 80 brass tubes 1 cm in inner diameter and 1.5 m in length. Disregarding inlet, exit, and header losses, determine the pressure drop across a single tube and the pumping power required by the tube-side fluid of the heat exchanger.

After operating a long time, 1-mm-thick scale builds up on the inner surfaces with an equivalent roughness of 0.4 mm.

For the same pumping power input, determine the percent reduction in the flow rate of water through the tubes.

FIGURE P8–132

8–133 The compressed air requirements of a manufacturing facility are met by a 120-hp compressor that draws in air from the outside through an 9-m-long, 22-cm-diameter duct made of thin galvanized iron sheets. The compressor takes in air at a rate of 0.27 m³/s at the outdoor conditions of 15°C and 95 kPa. Disregarding any minor losses, determine the useful power used by the compressor to overcome the frictional losses in this duct.   *Answer:* 6.74 W

FIGURE P8–133

8–134 A house built on a riverside is to be cooled in summer by utilizing the cool water of the river. A 15-m-long section of a circular stainless-steel duct of 20-cm diameter passes through the water. Air flows through the underwater section of the duct at 3 m/s at an average temperature of 15°C. For an overall fan efficiency of 62 percent, determine the fan power needed to overcome the flow resistance in this section of the duct.

FIGURE P8–134

8–135 The velocity profile in fully developed laminar flow in a circular pipe, in m/s, is given by $u(r) = 6(1 - 100r^2)$, where r is the radial distance from the centerline of the pipe in m. Determine (a) the radius of the pipe, (b) the average velocity through the pipe, and (c) the maximum velocity in the pipe.

8–136E The velocity profile in a fully developed laminar flow of water at 40°F in a 250-ft-long horizontal circular pipe, in ft/s, is given by $u(r) = 0.8(1 - 625r^2)$, where r is the radial distance from the centerline of the pipe in ft. Determine (a) the volume flow rate of water through the pipe, (b) the pressure drop across the pipe, and (c) the useful pumping power required to overcome this pressure drop.

8–137E Repeat Prob. 8–136E assuming the pipe is inclined 12° from the horizontal and the flow is uphill.

8–138 Oil at 20°C is flowing steadily through a 5-cm-diameter 40-m-long pipe. The pressures at the pipe inlet and outlet are measured to be 745 and 97.0 kPa, respectively, and the flow is expected to be laminar. Determine the flow rate of oil through the pipe, assuming fully developed flow and that the pipe is (a) horizontal, (b) inclined 15° upward, and (c) inclined 15° downward. Also, verify that the flow through the pipe is laminar.

8–139 Consider flow from a reservoir through a horizontal pipe of length L and diameter D that penetrates into the side wall at a vertical distance H from the free surface. The flow rate through an actual pipe with a reentrant section ($K_L = 0.8$) is considerably less than the flow rate through the hole calculated assuming "frictionless" flow and thus zero loss. Obtain a relation for the "equivalent diameter" of the reentrant pipe for use in relations for frictionless flow through a hole and determine its value for a pipe friction factor, length, and diameter of 0.018, 10 m, and 0.04 m, respectively. Assume the friction factor of the pipe to remain constant and the effect of the kinetic energy correction factor to be negligible.

8–140 A highly viscous liquid discharges from a large container through a small-diameter tube in laminar flow. Disregarding entrance effects and velocity heads, obtain a relation for the variation of fluid depth in the tank with time.

FIGURE P8–140

8–141 A student is to determine the kinematic viscosity of an oil using the system shown in Prob. 8–140. The initial fluid height in the tank is $H = 40$ cm, the tube diameter is $d = 6$ mm, the tube length is $L = 0.65$ m, and the tank diameter is $D = 0.63$ m. The student observes that it takes 1400 s for the fluid level in the tank to drop to 34 cm. Find the fluid viscosity.

8–142 A circular water pipe has an abrupt expansion from diameter $D_1 = 8$ cm to $D_2 = 24$ cm. The pressure and the average water velocity in the smaller pipe are $P_1 = 135$ kPa and 10 m/s, respectively, and the flow is turbulent. By applying the continuity, momentum, and energy equations and disregarding the effects of the kinetic energy and momentum-flux correction factors, show that the loss coefficient for sudden expansion is $K_L = (1 - D_1^2/D_2^2)^2$, and calculate K_L and P_2 for the given case.

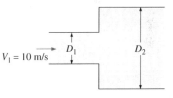

FIGURE P8–142

8–143 In a geothermal district heating system, 10,000 kg/s of hot water must be delivered a distance of 10 km in a horizontal pipe. The minor losses are negligible, and the only significant energy loss arises from pipe friction. The friction factor is taken to be 0.015. Specifying a larger-diameter pipe would reduce water velocity, velocity head, pipe friction, and thus power consumption. But a larger pipe would also cost more money initially to purchase and install. Otherwise stated, there is an optimum pipe diameter that will minimize the sum of pipe cost and future electric power cost.

Assume the system will run 24 h/day, every day, for 30 years. During this time the cost of electricity remains constant at $0.06/kWh. Assume system performance stays constant over the decades (this may not be true, especially if highly mineralized water is passed through the pipeline—scale may form). The pump has an overall efficiency of 80 percent. The cost to purchase, install, and insulate a 10-km pipe depends on the diameter D and is given by $Cost = \$10^6 D^2$, where D is in m. Assuming zero inflation and interest rate for simplicity and zero salvage value and zero maintenance cost, determine the optimum pipe diameter.

8–144 Water at 15°C is to be discharged from a reservoir at a rate of 18 L/s using two horizontal cast iron pipes connected in series and a pump between them. The first pipe is 20 m long and has a 6-cm diameter, while the second pipe is 35 m long and has a 4-cm diameter. The water level in the reservoir is 30 m above the centerline of the pipe. The pipe entrance is sharp-edged, and losses associated with the connection of the pump are negligible. Neglecting the effect of the kinetic energy correction factor, determine the required pumping head and the minimum pumping power to maintain the indicated flow rate.

FIGURE P8–144

8–145 Reconsider Prob. 8–144. Using EES (or other) software, investigate the effect of the second pipe diameter on the required pumping head to maintain the indicated flow rate. Let the diameter vary from 1 to 10 cm in increments of 1 cm. Tabulate and plot the results.

8–146 Two pipes of identical diameter and material are connected in parallel. The length of pipe A is five times the length of pipe B. Assuming the flow is fully turbulent in both pipes and thus the friction factor is independent of the Reynolds number and disregarding minor losses, determine the ratio of the flow rates in the two pipes.
Answer: 0.447

8–147 A pipeline that transports oil at 40°C at a rate of 3 m³/s branches out into two parallel pipes made of commercial steel that reconnect downstream. Pipe A is 500 m long and has a diameter of 30 cm while pipe B is 800 m long and has a diameter of 45 cm. The minor losses are considered to be negligible. Determine the flow rate through each of the parallel pipes.

FIGURE P8–147

8–148 Repeat Prob. 8–147 for hot-water flow of a district heating system at 100°C.

8–149 A system that consists of two interconnected cylindrical tanks with $D_1 = 30$ cm and $D_2 = 12$ cm is to be used to determine the discharge coefficient of a short $D_0 = 5$ mm diameter orifice. At the beginning ($t = 0$ s), the fluid heights in the tanks are $h_1 = 50$ cm and $h_2 = 15$ cm, as shown in Fig. P8–149. If it takes 170 s for the fluid levels in the two tanks to equalize and the flow to stop, determine the discharge coefficient of the orifice. Disregard any other losses associated with this flow.

FIGURE P8–149

8–150 The compressed air requirements of a textile factory are met by a large compressor that draws in 0.6 m³/s air at atmospheric conditions of 20°C and 1 bar (100 kPa) and consumes 300 kW electric power when operating. Air is compressed to a gage pressure of 8 bar (absolute pressure of 900 kPa), and compressed air is transported to the production area through a 15-cm-internal-diameter, 83-m-long, galvanized steel pipe with a surface roughness of 0.15 mm. The average temperature of compressed air in the pipe is 60°C.

The compressed air line has 8 elbows with a loss coefficient of 0.6 each. If the compressor efficiency is 85 percent, determine the pressure drop and the power wasted in the transportation line. *Answers:* 1.40 kPa, 0.125 kW

8–151 Reconsider Prob. 8–150. In order to reduce the head losses in the piping and thus the power wasted, someone suggests doubling the diameter of the 83-m-long compressed air pipes. Calculating the reduction in wasted power, and determine if this is a worthwhile idea. Considering the cost of replacement, does this proposal make sense to you?

8–152E A water fountain is to be installed at a remote location by attaching a cast iron pipe directly to a water main through which water is flowing at 70°F and 60 psig. The entrance to the pipe is sharp-edged, and the 70-ft-long piping system involves three 90° miter bends without vanes, a fully open gate valve, and an angle valve with a loss coefficient of 5 when fully open. If the system is to provide water at a rate of 15 gal/min and the elevation difference between the pipe and the fountain is negligible, determine the minimum diameter of the piping system. *Answer:* 0.713 in

FIGURE P8–152E

8–153E Repeat Prob. 8–152E for plastic (smooth) pipes.

8–154 In a hydroelectric power plant, water at 20°C is supplied to the turbine at a rate of 0.6 m³/s through a 200-m-long, 0.35-m-diameter cast iron pipe. The elevation difference between the free surface of the reservoir and the turbine discharge is 140 m, and the combined turbine–generator efficiency is 80 percent. Disregarding the minor losses because of the large length-to-diameter ratio, determine the electric power output of this plant.

8–155 In Prob. 8–154, the pipe diameter is tripled in order to reduce the pipe losses. Determine the percent increase in the net power output as a result of this modification.

8–156E The drinking water needs of an office are met by large water bottles. One end of a 0.35-in-diameter, 6-ft-long plastic hose is inserted into the bottle placed on a high stand, while the other end with an on/off valve is maintained 3 ft below the bottom of the bottle. If the water level in the bottle is 1 ft when it is full, determine how long it would take to fill an

8-oz glass (= 0.00835 ft³) (*a*) when the bottle is first opened and (*b*) when the bottle is almost empty. Take the total minor loss coefficient, including the on/off valve, to be 2.8 when it is fully open. Assume the water temperature to be the same as the room temperature of 70°F. *Answers:* (*a*) 2.4 s, (*b*) 2.8 s

FIGURE P8–156E

8–157E Reconsider Prob. 8–156E. Using EES (or other) software, investigate the effect of the hose diameter on the time required to fill a glass when the bottle is full. Let the diameter vary from 0.2 to 2 in, in increments of 0.2 in. Tabulate and plot the results.

8–158E Reconsider Prob. 8–156E. The office worker who set up the siphoning system purchased a 12-ft-long reel of the plastic tube and wanted to use the whole thing to avoid cutting it in pieces, thinking that it is the elevation difference that makes siphoning work, and the length of the tube is not important. So he used the entire 12-ft-long tube. Assuming the turns or constrictions in the tube are not significant (being very optimistic) and the same elevation is maintained, determine the time it takes to fill a glass of water for both cases (bottle nearly full and bottle nearly empty).

8–159 Water is to be withdrawn from a 7-m-high water reservoir by drilling a well-rounded 4-cm-diameter hole with negligible loss near the bottom and attaching a horizontal

FIGURE P8–159

90° bend of negligible length. Taking the kinetic energy correction factor to be 1.05, determine the flow rate of water through the bend if (a) the bend is a flanged smooth bend and (b) the bend is a miter bend without vanes. *Answers:* (a) 12.7 L/s, (b) 10.0 L/s

8–160   The water at 20°C in a 10-m-diameter, 2-m-high aboveground swimming pool is to be emptied by unplugging a 5-cm-diameter, 25-m-long horizontal plastic pipe attached to the bottom of the pool. Determine the initial rate of discharge of water through the pipe and the time (hours) it would take to empty the swimming pool completely assuming the entrance to the pipe is well-rounded with negligible loss. Take the friction factor of the pipe to be 0.022. Using the initial discharge velocity, check if this is a reasonable value for the friction factor.   *Answers:* 3.55 L/s, 24.6 h

FIGURE P8–160

8–161   Reconsider Prob. 8–160. Using EES (or other) software, investigate the effect of the discharge pipe diameter on the time required to empty the pool completely. Let the diameter vary from 1 to 10 cm, in increments of 1 cm. Tabulate and plot the results.

8–162   Repeat Prob. 8–160 for a sharp-edged entrance to the pipe with $K_L = 0.5$. Is this "minor loss" truly "minor" or not?

8–163   An elderly woman is rushed to the hospital because she is having a heart attack. The emergency room doctor informs her that she needs immediate coronary artery (a vessel that wraps around the heart) bypass surgery because one coronary artery has 75 percent blockage (caused by atherosclerotic plaque). This surgery involves using an artificial graft (typically made of Dacron) to divert blood from the coronary artery around the blockage and reattach to the coronary artery beyond the blockage site as illustrated in Figure P8-163. The coronary artery diameter is 5.0 mm and its length is 15.0 mm. The bypass graft diameter is 4.0 mm and its length is 20.0 mm.

Bypass graft

Coronary Artery

FIGURE P8-163

The flow rate within the bypass graft is 0.45 liters per minute (recall 1 ml equals 1 cm³). Blood has a density of 1060 kg/m³ and a dynamic viscosity of 3.5 centipoise. Assume that the Dacron and coronary artery have the same material properties and ignore any minor losses. Assume the friction factor is the same in both tubes. Ignoring the plaque in determining the head loss for the coronary artery, calculate the velocity through the small gap between the plaque and the coronary artery.

Fundamentals of Engineering (FE) Exam Problems

8–164   The average velocity for fully developed laminar pipe flow is
(a) $V_{max}/2$ (b) $V_{max}/3$ (c) V_{max} (d) $2V_{max}/3$ (e) $3V_{max}/4$

8–165   The Reynolds number is not a function of
(a) Fluid velocity   (b) Fluid density
(c) Characteristic length   (d) Surface roughness
(e) Fluid viscosity

8–166   Air flows in a 5 cm by 8 cm cross section rectangular duct at a velocity of 4 m/s at 1 atm and 15°C. The Reynolds number for this flow is
(a) 13,605   (b) 16,745   (c) 17,690   (d) 21,770
(e) 23,235

8–167   Air at 1 atm and 20°C flows in a 4-cm-diameter tube. The maximum velocity of air to keep the flow laminar is
(a) 0.872 m/s   (b) 1.52 m/s   (c) 2.14 m/s
(d) 3.11 m/s   (e) 3.79 m/s

8–168   Consider laminar flow of water in a 0.8-cm-diameter pipe at a rate of 1.15 L/min. The velocity of water halfway between the surface and the center of the pipe is
(a) 0.381 m/s   (b) 0.762 m/s   (c) 1.15 m/s
(d) 0.874 m/s   (e) 0.572 m/s

8–169   Consider laminar flow of water at 15°C in a 0.7-cm-diameter pipe at a velocity of 0.4 m/s. The pressure drop of water for a pipe length of 50 m is
(a) 6.8 kPa   (b) 8.7 kPa   (c) 11.5 kPa   (d) 14.9 kPa
(e) 17.3 kPa

8–170   Engine oil at 40°C ($\rho = 876$ kg/m³, $\mu = 0.2177$ kg/m·s) flows in a 20-cm-diameter pipe at a velocity of 1.2 m/s. The pressure drop of oil for a pipe length of 20 m is
(a) 4180 Pa   (b) 5044 Pa   (c) 6236 Pa   (d) 7419 Pa
(e) 8615 Pa

8–171   A fluid flows in a 25-cm-diameter pipe at a velocity of 4.5 m/s. If the pressure drop along the pipe is estimated to be 6400 Pa, the required pumping power to overcome this pressure drop is
(a) 452 W   (b) 640 W   (c) 923 W   (d) 1235 W
(e) 1508 W

8–172   Water flows in a 15-cm-diameter pipe at a velocity of 1.8 m/s. If the head loss along the pipe is estimated to be

16 m, the required pumping power to overcome this head loss is

(a) 3.22 kW   (b) 3.77 kW   (c) 4.45 kW   (d) 4.99 kW
(e) 5.54 kW

8–173   The pressure drop for a given flow is determined to be 100 Pa. For the same flow rate, if we reduce the diameter of the pipe by half, the pressure drop will be

(a) 25 Pa   (b) 50 Pa   (c) 200 Pa   (d) 400 Pa   (e) 1600 Pa

8–174   Air at 1 atm and 25°C ($v = 1.562 \times 10^{-5}$ m^2/s) flows in a 9-cm-diameter cast iron pipe at a velocity of 5 m/s. The roughness of the pipe is 0.26 mm. The head loss for a pipe length of 24 m is

(a) 8.1 m   (b) 10.2 m   (c) 12.9 m   (d) 15.5 m   (e) 23.7 m

8–175   Consider air flow in a 10-cm-diameter pipe at a high velocity so that the Reynolds number is very large. The roughness of the pipe is 0.002 mm. The friction factor for this flow is

(a) 0.0311   (b) 0.0290   (c) 0.0247   (d) 0.0206   (e) 0.0163

8–176   Air at 1 atm and 40°C flows in a 8-cm-diameter pipe at a rate of 2500 L/min. The friction factor is determined from the Moody chart to be 0.027. The required power input to overcome the pressure drop for a pipe length of 150 m is

(a) 310 W   (b) 188 W   (c) 132 W   (d) 81.7 W
(e) 35.9 W

8–177   Water at 10°C ($\rho = 999.7$ kg/m^3, $\mu = 1.307 \times 10^{-3}$ kg/m·s) is to be transported in a 5-cm-diamater, 30-m-long circular pipe. The roughness of the pipe is 0.22 mm. If the pressure drop in the pipe is not to exceed 19 kPa, the maximum flow rate of water is

(a) 324 L/min   (b) 281 L/min   (c) 243 L/min
(d) 195 L/min   (e) 168 L/min

8–178   The valve in a piping system causes a 3.1 m head loss. If the velocity of the flow is 6 m/s, the loss coefficient of this valve is

(a) 0.87   (b) 1.69   (c) 1.25   (d) 0.54   (e) 2.03

8–179   Consider a sharp-edged pipe exit for fully developed laminar flow of a fluid. The velocity of the flow is 4 m/s. This minor loss is equivalent to a head loss of

(a) 0.72 m   (b) 1.16 m   (c) 1.63 m   (d) 2.0 m   (e) 4.0 m

8–180   A water flow system involves a 180° return bend (threaded) and a 90° miter bend (without vanes). The velocity of water is 1.2 m/s. The minor losses due to these bends are equivalent to a pressure loss of

(a) 648 Pa   (b) 933 Pa   (c) 1255 Pa   (d) 1872 Pa
(e) 2600 Pa

8–181   A constant-diameter piping system involves multiple flow restrictions with a total loss coefficient of 4.4. The friction factor of piping is 0.025 and the diameter of the pipe is 7 cm. These minor losses are equivalent to the losses in a pipe of length

(a) 12.3 m   (b) 9.1 m   (c) 7.0 m   (d) 4.4 m   (e) 2.5 m

8–182   Air flows in an 8-cm-diameter, 33-m-long pipe at a velocity of 5.5 m/s. The piping system involves multiple flow restrictions with a total minor loss coefficient of 2.6. The friction factor of pipe is obtained from the Moody chart to be 0.025. The total head loss of this piping system is

(a) 13.5 m   (b) 7.6 m   (c) 19.9 m   (d) 24.5 m
(e) 4.2 m

8–183   Consider a pipe that branches out into two parallel pipes and then rejoins at a junction downstream. The two parallel pipes have the same lengths and friction factors. The diameters of the pipes are 2 cm and 4 cm. If the flow rate in one pipe is 10 L/min, the flow rate in the other pipe is

(a) 10 L/min   (b) 3.3 L/min   (c) 100 L/min   (d) 40 L/min
(e) 56.6 L/min

8–184   Consider a pipe that branches out into two parallel pipes and then rejoins at a junction downstream. The two parallel pipes have the same lengths and friction factors. The diameters of the pipes are 2 cm and 4 cm. If the head loss in one pipe is 0.5 m, the head loss in the other pipe is

(a) 0.5 m   (b) 1 m   (c) 0.25 m   (d) 2 m   (e) 0.125 m

8–185   A pump moves water from a reservoir to another reservoir through a piping system at a rate of 0.15 m^3/min. Both reservoirs are open to the atmosphere. The elevation difference between the two reservoirs is 35 m and the total head loss is estimated to be 4 m. If the efficiency of the motor-pump unit is 65 percent, the electrical power input to the motor of the pump is

(a) 1664 W   (b) 1472 W   (c) 1238 W   (d) 983 W
(e) 805 W

8–186   Consider a pipe that branches out into three parallel pipes and then rejoins at a junction downstream. All three pipes have the same diameters ($D = 3$ cm) and friction factors ($f = 0.018$). The lengths of pipe 1 and pipe 2 are 5 m and 8 m, respectively while the velocities of the fluid in pipe 2 and pipe 3 are 2 m/s and 4 m/s, respectively. The length of pipe 3 is

(a) 8 m   (b) 5 m   (c) 4 m   (d) 2 m   (e) 1 m

Design and Essay Problems

8–187   Electronic boxes such as computers are commonly cooled by a fan. Write an essay on forced air cooling of electronic boxes and on the selection of the fan for electronic devices.

8–188   Design an experiment to measure the viscosity of liquids using a vertical funnel with a cylindrical reservoir of height h and a narrow flow section of diameter D and length L. Making appropriate assumptions, obtain a relation for viscosity in terms of easily measurable quantities such as density and volume flow rate. Is there a need for the use of a correction factor?

8–189 A pump is to be selected for a waterfall in a garden. The water collects in a pond at the bottom, and the elevation difference between the free surface of the pond and the location where the water is discharged is 3 m. The flow rate of water is to be at least 8 L/s. Select an appropriate motor– pump unit for this job and identify three manufacturers with product model numbers and prices. Make a selection and explain why you selected that particular product.

Also estimate the cost of annual power consumption of this unit assuming continuous operation.

8–190 During a camping trip you notice that water is discharged from a high reservoir to a stream in the valley through a 30-cm-diameter plastic pipe. The elevation difference between the free surface of the reservoir and the stream is 70 m. You conceive the idea of generating power from this water. Design a power plant that will produce the most power from this resource. Also, investigate the effect of power generation on the discharge rate of water. What discharge rate maximizes the power production?

PROPERTY TABLES AND CHARTS (SI UNITS)*

TABLE A–1   Molar Mass, Gas Constant, and Ideal-Gas Specfic Heats of Some Substances   940

TABLE A–2   Boiling and Freezing Point Properties   941

TABLE A–3   Properties of Saturated Water   942

TABLE A–4   Properties of Saturated Refrigerant-134a   943

TABLE A–5   Properties of Saturated Ammonia   944

TABLE A–6   Properties of Saturated Propane   945

TABLE A–7   Properties of Liquids   946

TABLE A–8   Properties of Liquid Metals   947

TABLE A–9   Properties of Air at 1 atm Pressure   948

TABLE A–10   Properties of Gases at 1 atm Pressure   949

TABLE A–11   Properties of the Atmosphere at High Altitude   951

FIGURE A–12   The Moody Chart for the Friction Factor for Fully Developed Flow in Circular Pipes   952

TABLE A–13   One-Dimensional Isentropic Compressible Flow Functions for an Ideal Gas with $k = 1.4$ 953

TABLE A–14   One-Dimensional Normal Shock Functions for an Ideal Gas with $k = 1.4$ 954

TABLE A–15   Rayleigh Flow Functions for an Ideal Gas with $k = 1.4$ 955

TABLE A–16   Fanno Flow Functions for an Ideal Gas with $k = 1.4$ 956

* Most properties in the tables are obtained from the property database of EES, and the original sources are listed under the tables. Properties are often listed to more significant digits than the claimed accuracy for the purpose of minimizing accumulated round-off error in hand calculations and ensuring a close match with the results obtained with EES.

TABLE A–1

Molar mass, gas constant, and ideal-gas specfic heats of some substances

Substance	Molar Mass M, kg/kmol	Gas Constant R, kJ/kg·K*	Specific Heat Data at 25°C		
			c_p, kJ/kg·K	c_v, kJ/kg·K	$k = c_p/c_v$
Air	28.97	0.2870	1.005	0.7180	1.400
Ammonia, NH_3	17.03	0.4882	2.093	1.605	1.304
Argon, Ar	39.95	0.2081	0.5203	0.3122	1.667
Bromine, Br_2	159.81	0.05202	0.2253	0.1732	1.300
Isobutane, C_4H_{10}	58.12	0.1430	1.663	1.520	1.094
n-Butane, C_4H_{10}	58.12	0.1430	1.694	1.551	1.092
Carbon dioxide, CO_2	44.01	0.1889	0.8439	0.6550	1.288
Carbon monoxide, CO	28.01	0.2968	1.039	0.7417	1.400
Chlorine, Cl_2	70.905	0.1173	0.4781	0.3608	1.325
Chlorodifluoromethane (R-22), $CHClF_2$	86.47	0.09615	0.6496	0.5535	1.174
Ethane, C_2H_6	30.070	0.2765	1.744	1.468	1.188
Ethylene, C_2H_4	28.054	0.2964	1.527	1.231	1.241
Fluorine, F_2	38.00	0.2187	0.8237	0.6050	1.362
Helium, He	4.003	2.077	5.193	3.116	1.667
n-Heptane, C_7H_{16}	100.20	0.08297	1.649	1.566	1.053
n-Hexane, C_6H_{14}	86.18	0.09647	1.654	1.558	1.062
Hydrogen, H_2	2.016	4.124	14.30	10.18	1.405
Krypton, Kr	83.80	0.09921	0.2480	0.1488	1.667
Methane, CH_4	16.04	0.5182	2.226	1.708	1.303
Neon, Ne	20.183	0.4119	1.030	0.6180	1.667
Nitrogen, N_2	28.01	0.2968	1.040	0.7429	1.400
Nitric oxide, NO	30.006	0.2771	0.9992	0.7221	1.384
Nitrogen dioxide, NO_2	46.006	0.1889	0.8060	0.6171	1.306
Oxygen, O_2	32.00	0.2598	0.9180	0.6582	1.395
n-Pentane, C_5H_{12}	72.15	0.1152	1.664	1.549	1.074
Propane, C_3H_8	44.097	0.1885	1.669	1.480	1.127
Propylene, C_3H_6	42.08	0.1976	1.531	1.333	1.148
Steam, H_2O	18.015	0.4615	1.865	1.403	1.329
Sulfur dioxide, SO_2	64.06	0.1298	0.6228	0.4930	1.263
Tetrachloromethane, CCl_4	153.82	0.05405	0.5415	0.4875	1.111
Tetrafluoroethane (R-134a), $C_2H_2F_4$	102.03	0.08149	0.8334	0.7519	1.108
Trifluoroethane (R-143a), $C_2H_3F_3$	84.04	0.09893	0.9291	0.8302	1.119
Xenon, Xe	131.30	0.06332	0.1583	0.09499	1.667

* The unit kJ/kg·K is equivalent to kPa·m³/kg·K. The gas constant is calculated from $R = R_u/M$, where $R_u = 8.31447$ kJ/kmol·K is the universal gas constant and M is the molar mass.

Source: Specific heat values are obtained primarily from the property routines prepared by The National Institute of Standards and Technology (NIST), Gaithersburg, MD.

TABLE A–2

Boiling and freezing point properties

Substance	Boiling Data at 1 atm		Freezing Data		Liquid Properties		
	Normal Boiling Point, °C	Latent Heat of Vaporization h_{fg}, kJ/kg	Freezing Point, °C	Latent Heat of Fusion h_{if}, kJ/kg	Temperature, °C	Density ρ, kg/m³	Specific Heat c_p, kJ/kg·K
Ammonia	−33.3	1357	−77.7	322.4	−33.3	682	4.43
					−20	665	4.52
					0	639	4.60
					25	602	4.80
Argon	−185.9	161.6	−189.3	28	−185.6	1394	1.14
Benzene	80.2	394	5.5	126	20	879	1.72
Brine (20% sodium chloride by mass)	103.9	—	−17.4	—	20	1150	3.11
n-Butane	−0.5	385.2	−138.5	80.3	−0.5	601	2.31
Carbon dioxide	−78.4*	230.5 (at 0°C)	−56.6		0	298	0.59
Ethanol	78.2	838.3	−114.2	109	25	783	2.46
Ethyl alcohol	78.6	855	−156	108	20	789	2.84
Ethylene glycol	198.1	800.1	−10.8	181.1	20	1109	2.84
Glycerine	179.9	974	18.9	200.6	20	1261	2.32
Helium	−268.9	22.8	—	—	−268.9	146.2	22.8
Hydrogen	−252.8	445.7	−259.2	59.5	−252.8	70.7	10.0
Isobutane	−11.7	367.1	−160	105.7	−11.7	593.8	2.28
Kerosene	204–293	251	−24.9	—	20	820	2.00
Mercury	356.7	294.7	−38.9	11.4	25	13,560	0.139
Methane	−161.5	510.4	−182.2	58.4	−161.5	423	3.49
					−100	301	5.79
Methanol	64.5	1100	−97.7	99.2	25	787	2.55
Nitrogen	−195.8	198.6	−210	25.3	−195.8	809	2.06
					−160	596	2.97
Octane	124.8	306.3	−57.5	180.7	20	703	2.10
Oil (light)					25	910	1.80
Oxygen	−183	212.7	−218.8	13.7	−183	1141	1.71
Petroleum	—	230–384			20	640	2.0
Propane	−42.1	427.8	−187.7	80.0	−42.1	581	2.25
					0	529	2.53
					50	449	3.13
Refrigerant-134a	−26.1	216.8	−96.6	—	−50	1443	1.23
					−26.1	1374	1.27
					0	1295	1.34
					25	1207	1.43
Water	100	2257	0.0	333.7	0	1000	4.22
					25	997	4.18
					50	988	4.18
					75	975	4.19
					100	958	4.22

* Sublimation temperature. (At pressures below the triple-point pressure of 518 kPa, carbon dioxide exists as a solid or gas. Also, the freezing-point temperature of carbon dioxide is the triple-point temperature of −56.5°C.)

TABLE A–3

Properties of saturated water

Temp. T, °C	Saturation Pressure P_{sat}, kPa	Density ρ, kg/m³ Liquid	Density ρ, kg/m³ Vapor	Enthalpy of Vaporization h_{fg}, kJ/kg	Specific Heat c_p, J/kg·K Liquid	Specific Heat c_p, J/kg·K Vapor	Thermal Conductivity k, W/m·K Liquid	Thermal Conductivity k, W/m·K Vapor	Dynamic Viscosity μ, kg/m·s Liquid	Dynamic Viscosity μ, kg/m·s Vapor	Prandtl Number Pr Liquid	Prandtl Number Pr Vapor	Volume Expansion Coefficient β, 1/K Liquid	Surface Tension, N/m Liquid
0.01	0.6113	999.8	0.0048	2501	4217	1854	0.561	0.0171	1.792×10^{-3}	0.922×10^{-5}	13.5	1.00	-0.068×10^{-3}	0.0756
5	0.8721	999.9	0.0068	2490	4205	1857	0.571	0.0173	1.519×10^{-3}	0.934×10^{-5}	11.2	1.00	0.015×10^{-3}	0.0749
10	1.2276	999.7	0.0094	2478	4194	1862	0.580	0.0176	1.307×10^{-3}	0.946×10^{-5}	9.45	1.00	0.733×10^{-3}	0.0742
15	1.7051	999.1	0.0128	2466	4186	1863	0.589	0.0179	1.138×10^{-3}	0.959×10^{-5}	8.09	1.00	0.138×10^{-3}	0.0735
20	2.339	998.0	0.0173	2454	4182	1867	0.598	0.0182	1.002×10^{-3}	0.973×10^{-5}	7.01	1.00	0.195×10^{-3}	0.0727
25	3.169	997.0	0.0231	2442	4180	1870	0.607	0.0186	0.891×10^{-3}	0.987×10^{-5}	6.14	1.00	0.247×10^{-3}	0.0720
30	4.246	996.0	0.0304	2431	4178	1875	0.615	0.0189	0.798×10^{-3}	1.001×10^{-5}	5.42	1.00	0.294×10^{-3}	0.0712
35	5.628	994.0	0.0397	2419	4178	1880	0.623	0.0192	0.720×10^{-3}	1.016×10^{-5}	4.83	1.00	0.337×10^{-3}	0.0704
40	7.384	992.1	0.0512	2407	4179	1885	0.631	0.0196	0.653×10^{-3}	1.031×10^{-5}	4.32	1.00	0.377×10^{-3}	0.0696
45	9.593	990.1	0.0655	2395	4180	1892	0.637	0.0200	0.596×10^{-3}	1.046×10^{-5}	3.91	1.00	0.415×10^{-3}	0.0688
50	12.35	988.1	0.0831	2383	4181	1900	0.644	0.0204	0.547×10^{-3}	1.062×10^{-5}	3.55	1.00	0.451×10^{-3}	0.0679
55	15.76	985.2	0.1045	2371	4183	1908	0.649	0.0208	0.504×10^{-3}	1.077×10^{-5}	3.25	1.00	0.484×10^{-3}	0.0671
60	19.94	983.3	0.1304	2359	4185	1916	0.654	0.0212	0.467×10^{-3}	1.093×10^{-5}	2.99	1.00	0.517×10^{-3}	0.0662
65	25.03	980.4	0.1614	2346	4187	1926	0.659	0.0216	0.433×10^{-3}	1.110×10^{-5}	2.75	1.00	0.548×10^{-3}	0.0654
70	31.19	977.5	0.1983	2334	4190	1936	0.663	0.0221	0.404×10^{-3}	1.126×10^{-5}	2.55	1.00	0.578×10^{-3}	0.0645
75	38.58	974.7	0.2421	2321	4193	1948	0.667	0.0225	0.378×10^{-3}	1.142×10^{-5}	2.38	1.00	0.607×10^{-3}	0.0636
80	47.39	971.8	0.2935	2309	4197	1962	0.670	0.0230	0.355×10^{-3}	1.159×10^{-5}	2.22	1.00	0.653×10^{-3}	0.0627
85	57.83	968.1	0.3536	2296	4201	1977	0.673	0.0235	0.333×10^{-3}	1.176×10^{-5}	2.08	1.00	0.670×10^{-3}	0.0617
90	70.14	965.3	0.4235	2283	4206	1993	0.675	0.0240	0.315×10^{-3}	1.193×10^{-5}	1.96	1.00	0.702×10^{-3}	0.0608
95	84.55	961.5	0.5045	2270	4212	2010	0.677	0.0246	0.297×10^{-3}	1.210×10^{-5}	1.85	1.00	0.716×10^{-3}	0.0599
100	101.33	957.9	0.5978	2257	4217	2029	0.679	0.0251	0.282×10^{-3}	1.227×10^{-5}	1.75	1.00	0.750×10^{-3}	0.0589
110	143.27	950.6	0.8263	2230	4229	2071	0.682	0.0262	0.255×10^{-3}	1.261×10^{-5}	1.58	1.00	0.798×10^{-3}	0.0570
120	198.53	943.4	1.121	2203	4244	2120	0.683	0.0275	0.232×10^{-3}	1.296×10^{-5}	1.44	1.00	0.858×10^{-3}	0.0550
130	270.1	934.6	1.496	2174	4263	2177	0.684	0.0288	0.213×10^{-3}	1.330×10^{-5}	1.33	1.01	0.913×10^{-3}	0.0529
140	361.3	921.7	1.965	2145	4286	2244	0.683	0.0301	0.197×10^{-3}	1.365×10^{-5}	1.24	1.02	0.970×10^{-3}	0.0509
150	475.8	916.6	2.546	2114	4311	2314	0.682	0.0316	0.183×10^{-3}	1.399×10^{-5}	1.16	1.02	1.025×10^{-3}	0.0487
160	617.8	907.4	3.256	2083	4340	2420	0.680	0.0331	0.170×10^{-3}	1.434×10^{-5}	1.09	1.05	1.145×10^{-3}	0.0466
170	791.7	897.7	4.119	2050	4370	2490	0.677	0.0347	0.160×10^{-3}	1.468×10^{-5}	1.03	1.05	1.178×10^{-3}	0.0444
180	1,002.1	887.3	5.153	2015	4410	2590	0.673	0.0364	0.150×10^{-3}	1.502×10^{-5}	0.983	1.07	1.210×10^{-3}	0.0422
190	1,254.4	876.4	6.388	1979	4460	2710	0.669	0.0382	0.142×10^{-3}	1.537×10^{-5}	0.947	1.09	1.280×10^{-3}	0.0399
200	1,553.8	864.3	7.852	1941	4500	2840	0.663	0.0401	0.134×10^{-3}	1.571×10^{-5}	0.910	1.11	1.350×10^{-3}	0.0377
220	2,318	840.3	11.60	1859	4610	3110	0.650	0.0442	0.122×10^{-3}	1.641×10^{-5}	0.865	1.15	1.520×10^{-3}	0.0331
240	3,344	813.7	16.73	1767	4760	3520	0.632	0.0487	0.111×10^{-3}	1.712×10^{-5}	0.836	1.24	1.720×10^{-3}	0.0284
260	4,688	783.7	23.69	1663	4970	4070	0.609	0.0540	0.102×10^{-3}	1.788×10^{-5}	0.832	1.35	2.000×10^{-3}	0.0237
280	6,412	750.8	33.15	1544	5280	4835	0.581	0.0605	0.094×10^{-3}	1.870×10^{-5}	0.854	1.49	2.380×10^{-3}	0.0190
300	8,581	713.8	46.15	1405	5750	5980	0.548	0.0695	0.086×10^{-3}	1.965×10^{-5}	0.902	1.69	2.950×10^{-3}	0.0144
320	11,274	667.1	64.57	1239	6540	7900	0.509	0.0836	0.078×10^{-3}	2.084×10^{-5}	1.00	1.97		0.0099
340	14,586	610.5	92.62	1028	8240	11,870	0.469	0.110	0.070×10^{-3}	2.255×10^{-5}	1.23	2.43		0.0056
360	18,651	528.3	144.0	720	14,690	25,800	0.427	0.178	0.060×10^{-3}	2.571×10^{-5}	2.06	3.73		0.0019
374.14	22,090	317.0	317.0	0	—	—	—	—	0.043×10^{-3}	4.313×10^{-5}				0

Note 1: Kinematic viscosity ν and thermal diffusivity α can be calculated from their definitions, $\nu = \mu/\rho$ and $\alpha = k/\rho c_p = \nu/\text{Pr}$. The temperatures 0.01°C, 100°C, and 374.14°C are the triple-, boiling-, and critical-point temperatures of water, respectively. The properties listed above (except the vapor density) can be used at any pressure with negligible error except at temperatures near the critical-point value.

Note 2: The unit kJ/kg·°C for specific heat is equivalent to kJ/kg·K, and the unit W/m·°C for thermal conductivity is equivalent to W/m·K.

Source: Viscosity and thermal conductivity data are from J. V. Sengers and J. T. R. Watson, *Journal of Physical and Chemical Reference Data* 15 (1986), pp. 1291–1322. Other data are obtained from various sources or calculated.

TABLE A–4

Properties of saturated refrigerant-134a

Temp. T, °C	Saturation Pressure P, kPa	Density ρ, kg/m³ Liquid	Density Vapor	Enthalpy of Vaporization h_{fg}, kJ/kg	Specific Heat c_p, J/kg·K Liquid	Specific Heat Vapor	Thermal Conductivity k, W/m·K Liquid	Thermal Conductivity Vapor	Dynamic Viscosity μ, kg/m·s Liquid	Dynamic Viscosity Vapor	Prandtl Number Pr Liquid	Prandtl Number Vapor	Volume Expansion Coefficient β, 1/K Liquid	Surface Tension, N/m Liquid
−40	51.2	1418	2.773	225.9	1254	748.6	0.1101	0.00811	4.878×10^{-4}	2.550×10^{-6}	5.558	0.235	0.00205	0.01760
−35	66.2	1403	3.524	222.7	1264	764.1	0.1084	0.00862	4.509×10^{-4}	3.003×10^{-6}	5.257	0.266	0.00209	0.01682
−30	84.4	1389	4.429	219.5	1273	780.2	0.1066	0.00913	4.178×10^{-4}	3.504×10^{-6}	4.992	0.299	0.00215	0.01604
−25	106.5	1374	5.509	216.3	1283	797.2	0.1047	0.00963	3.882×10^{-4}	4.054×10^{-6}	4.757	0.335	0.00220	0.01527
−20	132.8	1359	6.787	213.0	1294	814.9	0.1028	0.01013	3.614×10^{-4}	4.651×10^{-6}	4.548	0.374	0.00227	0.01451
−15	164.0	1343	8.288	209.5	1306	833.5	0.1009	0.01063	3.371×10^{-4}	5.295×10^{-6}	4.363	0.415	0.00233	0.01376
−10	200.7	1327	10.04	206.0	1318	853.1	0.0989	0.01112	3.150×10^{-4}	5.982×10^{-6}	4.198	0.459	0.00241	0.01302
−5	243.5	1311	12.07	202.4	1330	873.8	0.0968	0.01161	2.947×10^{-4}	6.709×10^{-6}	4.051	0.505	0.00249	0.01229
0	293.0	1295	14.42	198.7	1344	895.6	0.0947	0.01210	2.761×10^{-4}	7.471×10^{-6}	3.919	0.553	0.00258	0.01156
5	349.9	1278	17.12	194.8	1358	918.7	0.0925	0.01259	2.589×10^{-4}	8.264×10^{-6}	3.802	0.603	0.00269	0.01084
10	414.9	1261	20.22	190.8	1374	943.2	0.0903	0.01308	2.430×10^{-4}	9.081×10^{-6}	3.697	0.655	0.00280	0.01014
15	488.7	1244	23.75	186.6	1390	969.4	0.0880	0.01357	2.281×10^{-4}	9.915×10^{-6}	3.604	0.708	0.00293	0.00944
20	572.1	1226	27.77	182.3	1408	997.6	0.0856	0.01406	2.142×10^{-4}	1.075×10^{-5}	3.521	0.763	0.00307	0.00876
25	665.8	1207	32.34	177.8	1427	1028	0.0833	0.01456	2.012×10^{-4}	1.160×10^{-5}	3.448	0.819	0.00324	0.00808
30	770.6	1188	37.53	173.1	1448	1061	0.0808	0.01507	1.888×10^{-4}	1.244×10^{-5}	3.383	0.877	0.00342	0.00742
35	887.5	1168	43.41	168.2	1471	1098	0.0783	0.01558	1.772×10^{-4}	1.327×10^{-5}	3.328	0.935	0.00364	0.00677
40	1017.1	1147	50.08	163.0	1498	1138	0.0757	0.01610	1.660×10^{-4}	1.408×10^{-5}	3.285	0.995	0.00390	0.00613
45	1160.5	1125	57.66	157.6	1529	1184	0.0731	0.01664	1.554×10^{-4}	1.486×10^{-5}	3.253	1.058	0.00420	0.00550
50	1318.6	1102	66.27	151.8	1566	1237	0.0704	0.01720	1.453×10^{-4}	1.562×10^{-5}	3.231	1.123	0.00456	0.00489
55	1492.3	1078	76.11	145.7	1608	1298	0.0676	0.01777	1.355×10^{-4}	1.634×10^{-5}	3.223	1.193	0.00500	0.00429
60	1682.8	1053	87.38	139.1	1659	1372	0.0647	0.01838	1.260×10^{-4}	1.704×10^{-5}	3.229	1.272	0.00554	0.00372
65	1891.0	1026	100.4	132.1	1722	1462	0.0618	0.01902	1.167×10^{-4}	1.771×10^{-5}	3.255	1.362	0.00624	0.00315
70	2118.2	996.2	115.6	124.4	1801	1577	0.0587	0.01972	1.077×10^{-4}	1.839×10^{-5}	3.307	1.471	0.00716	0.00261
75	2365.8	964	133.6	115.9	1907	1731	0.0555	0.02048	9.891×10^{-5}	1.908×10^{-5}	3.400	1.612	0.00843	0.00209
80	2635.2	928.2	155.3	106.4	2056	1948	0.0521	0.02133	9.011×10^{-5}	1.982×10^{-5}	3.558	1.810	0.01031	0.00160
85	2928.2	887.1	182.3	95.4	2287	2281	0.0484	0.02233	8.124×10^{-5}	2.071×10^{-5}	3.837	2.116	0.01336	0.00114
90	3246.9	837.7	217.8	82.2	2701	2865	0.0444	0.02357	7.203×10^{-5}	2.187×10^{-5}	4.385	2.658	0.01911	0.00071
95	3594.1	772.5	269.3	64.9	3675	4144	0.0396	0.02544	6.190×10^{-5}	2.370×10^{-5}	5.746	3.862	0.03343	0.00033
100	3975.1	651.7	376.3	33.9	7959	8785	0.0322	0.02989	4.765×10^{-5}	2.833×10^{-5}	11.77	8.326	0.10047	0.00004

Note 1: Kinematic viscosity ν and thermal diffusivity α can be calculated from their definitions, $\nu = \mu/\rho$ and $\alpha = k/\rho c_p = \nu/\text{Pr}$. The properties listed here (except the vapor density) can be used at any pressures with negligible error except at temperatures near the critical-point value.

Note 2: The unit kJ/kg·°C for specific heat is equivalent to kJ/kg·K, and the unit W/m·°C for thermal conductivity is equivalent to W/m·K.

Source: Data generated from the EES software developed by S. A. Klein and F. L. Alvarado. Original sources: R. Tillner-Roth and H. D. Baehr, "An International Standard Formulation for the Thermodynamic Properties of 1,1,1,2-Tetrafluoroethane (HFC-134a) for Temperatures from 170 K to 455 K and Pressures up to 70 MPa," *J. Phys. Chem, Ref. Data*, Vol. 23, No. 5, 1994; M. J. Assael, N. K. Dalaouti, A. A. Griva, and J. H. Dymond, "Viscosity and Thermal Conductivity of Halogenated Methane and Ethane Refrigerants," *IJR*, Vol. 22, pp. 525–535, 1999; NIST REFPROP 6 program (M. O. McLinden, S. A. Klein, E. W. Lemmon, and A. P. Peskin, Physical and Chemical Properties Division, National Institute of Standards and Technology, Boulder, CO 80303, 1995).

TABLE A-5

Properties of saturated ammonia

Temp. T, °C	Saturation Pressure P, kPa	Density ρ, kg/m³ Liquid	Density ρ, kg/m³ Vapor	Enthalpy of Vaporization h_{fg}, kJ/kg	Specific Heat c_p, J/kg·K Liquid	Specific Heat c_p, J/kg·K Vapor	Thermal Conductivity k, W/m·K Liquid	Thermal Conductivity k, W/m·K Vapor	Dynamic Viscosity μ, kg/m·s Liquid	Dynamic Viscosity μ, kg/m·s Vapor	Prandtl Number Pr Liquid	Prandtl Number Pr Vapor	Volume Expansion Coefficient β, 1/K Liquid	Surface Tension, N/m Liquid
−40	71.66	690.2	0.6435	1389	4414	2242	—	0.01792	2.926×10^{-4}	7.957×10^{-6}	—	0.9955	0.00176	0.03565
−30	119.4	677.8	1.037	1360	4465	2322	—	0.01898	2.630×10^{-4}	8.311×10^{-6}	—	1.017	0.00185	0.03341
−25	151.5	671.5	1.296	1345	4489	2369	0.5968	0.01957	2.492×10^{-4}	8.490×10^{-6}	1.875	1.028	0.00190	0.03229
−20	190.1	665.1	1.603	1329	4514	2420	0.5853	0.02015	2.361×10^{-4}	8.669×10^{-6}	1.821	1.041	0.00194	0.03118
−15	236.2	658.6	1.966	1313	4538	2476	0.5737	0.02075	2.236×10^{-4}	8.851×10^{-6}	1.769	1.056	0.00199	0.03007
−10	290.8	652.1	2.391	1297	4564	2536	0.5621	0.02138	2.117×10^{-4}	9.034×10^{-6}	1.718	1.072	0.00205	0.02896
−5	354.9	645.4	2.886	1280	4589	2601	0.5505	0.02203	2.003×10^{-4}	9.218×10^{-6}	1.670	1.089	0.00210	0.02786
0	429.6	638.6	3.458	1262	4617	2672	0.5390	0.02270	1.896×10^{-4}	9.405×10^{-6}	1.624	1.107	0.00216	0.02676
5	516	631.7	4.116	1244	4645	2749	0.5274	0.02341	1.794×10^{-4}	9.593×10^{-6}	1.580	1.126	0.00223	0.02566
10	615.3	624.6	4.870	1226	4676	2831	0.5158	0.02415	1.697×10^{-4}	9.784×10^{-6}	1.539	1.147	0.00230	0.02457
15	728.8	617.5	5.729	1206	4709	2920	0.5042	0.02492	1.606×10^{-4}	9.978×10^{-6}	1.500	1.169	0.00237	0.02348
20	857.8	610.2	6.705	1186	4745	3016	0.4927	0.02573	1.519×10^{-4}	1.017×10^{-5}	1.463	1.193	0.00245	0.02240
25	1003	602.8	7.809	1166	4784	3120	0.4811	0.02658	1.438×10^{-4}	1.037×10^{-5}	1.430	1.218	0.00254	0.02132
30	1167	595.2	9.055	1144	4828	3232	0.4695	0.02748	1.361×10^{-4}	1.057×10^{-5}	1.399	1.244	0.00264	0.02024
35	1351	587.4	10.46	1122	4877	3354	0.4579	0.02843	1.288×10^{-4}	1.078×10^{-5}	1.372	1.272	0.00275	0.01917
40	1555	579.4	12.03	1099	4932	3486	0.4464	0.02943	1.219×10^{-4}	1.099×10^{-5}	1.347	1.303	0.00287	0.01810
45	1782	571.3	13.8	1075	4993	3631	0.4348	0.03049	1.155×10^{-4}	1.121×10^{-5}	1.327	1.335	0.00301	0.01704
50	2033	562.9	15.78	1051	5063	3790	0.4232	0.03162	1.094×10^{-4}	1.143×10^{-5}	1.310	1.371	0.00316	0.01598
55	2310	554.2	18.00	1025	5143	3967	0.4116	0.03283	1.037×10^{-4}	1.166×10^{-5}	1.297	1.409	0.00334	0.01493
60	2614	545.2	20.48	997.4	5234	4163	0.4001	0.03412	9.846×10^{-5}	1.189×10^{-5}	1.288	1.452	0.00354	0.01389
65	2948	536.0	23.26	968.9	5340	4384	0.3885	0.03550	9.347×10^{-5}	1.213×10^{-5}	1.285	1.499	0.00377	0.01285
70	3312	526.3	26.39	939.0	5463	4634	0.3769	0.03700	8.879×10^{-5}	1.238×10^{-5}	1.287	1.551	0.00404	0.01181
75	3709	516.2	29.90	907.5	5608	4923	0.3653	0.03862	8.440×10^{-5}	1.264×10^{-5}	1.296	1.612	0.00436	0.01079
80	4141	505.7	33.87	874.1	5780	5260	0.3538	0.04038	8.030×10^{-5}	1.292×10^{-5}	1.312	1.683	0.00474	0.00977
85	4609	494.5	38.36	838.6	5988	5659	0.3422	0.04232	7.645×10^{-5}	1.322×10^{-5}	1.338	1.768	0.00521	0.00876
90	5116	482.8	43.48	800.6	6242	6142	0.3306	0.04447	7.284×10^{-5}	1.354×10^{-5}	1.375	1.871	0.00579	0.00776
95	5665	470.2	49.35	759.8	6561	6740	0.3190	0.04687	6.946×10^{-5}	1.389×10^{-5}	1.429	1.999	0.00652	0.00677
100	6257	456.6	56.15	715.5	6972	7503	0.3075	0.04958	6.628×10^{-5}	1.429×10^{-5}	1.503	2.163	0.00749	0.00579

Note 1: Kinematic viscosity ν and thermal diffusivity α can be calculated from their definitions, $\nu = \mu/\rho$ and $\alpha = k/\rho c_p = \nu/\text{Pr}$. The properties listed here (except the vapor density) can be used at any pressures with negligible error except at temperatures near the critical-point value.

Note 2: The unit kJ/kg·°C for specific heat is equivalent to kJ/kg·K, and the unit W/m·°C for thermal conductivity is equivalent to W/m·K.

Source: Data generated from the EES software developed by S. A. Klein and F. L. Alvarado. Original sources: Tillner-Roth, Harms-Watzenberg, and Baehr, "Eine neue Fundamentalgleichung fur Ammoniak," DKV-Tagungsbericht 20:167–181, 1993; Liley and Desai, "Thermophysical Properties of Refrigerants," *ASHRAE*, 1993, ISBN 1-1883413-10-9.

TABLE A–6

Properties of saturated propane

Temp. T, °C	Saturation Pressure P, kPa	Density ρ, kg/m³		Enthalpy of Vaporization h_{fg}, kJ/kg	Specific Heat c_p, J/kg·K		Thermal Conductivity k, W/m·K		Dynamic Viscosity μ, kg/m·s		Prandtl Number Pr		Volume Expansion Coefficient β, 1/K	Surface Tension, N/m
		Liquid	Vapor		Liquid	Vapor	Liquid	Vapor	Liquid	Vapor	Liquid	Vapor	Liquid	Liquid
−120	0.4053	664.7	0.01408	498.3	2003	1115	0.1802	0.00589	6.136×10^{-4}	4.372×10^{-6}	6.820	0.827	0.00153	0.02630
−110	1.157	654.5	0.03776	489.3	2021	1148	0.1738	0.00645	5.054×10^{-4}	4.625×10^{-6}	5.878	0.822	0.00157	0.02486
−100	2.881	644.2	0.08872	480.4	2044	1183	0.1672	0.00705	4.252×10^{-4}	4.881×10^{-6}	5.195	0.819	0.00161	0.02344
−90	6.406	633.8	0.1870	471.5	2070	1221	0.1606	0.00769	3.635×10^{-4}	5.143×10^{-6}	4.686	0.817	0.00166	0.02202
−80	12.97	623.2	0.3602	462.4	2100	1263	0.1539	0.00836	3.149×10^{-4}	5.409×10^{-6}	4.297	0.817	0.00171	0.02062
−70	24.26	612.5	0.6439	453.1	2134	1308	0.1472	0.00908	2.755×10^{-4}	5.680×10^{-6}	3.994	0.818	0.00177	0.01923
−60	42.46	601.5	1.081	443.5	2173	1358	0.1407	0.00985	2.430×10^{-4}	5.956×10^{-6}	3.755	0.821	0.00184	0.01785
−50	70.24	590.3	1.724	433.6	2217	1412	0.1343	0.01067	2.158×10^{-4}	6.239×10^{-6}	3.563	0.825	0.00192	0.01649
−40	110.7	578.8	2.629	423.1	2258	1471	0.1281	0.01155	1.926×10^{-4}	6.529×10^{-6}	3.395	0.831	0.00201	0.01515
−30	167.3	567.0	3.864	412.1	2310	1535	0.1221	0.01250	1.726×10^{-4}	6.827×10^{-6}	3.266	0.839	0.00213	0.01382
−20	243.8	554.7	5.503	400.3	2368	1605	0.1163	0.01351	1.551×10^{-4}	7.136×10^{-6}	3.158	0.848	0.00226	0.01251
−10	344.4	542.0	7.635	387.8	2433	1682	0.1107	0.01459	1.397×10^{-4}	7.457×10^{-6}	3.069	0.860	0.00242	0.01122
0	473.3	528.7	10.36	374.2	2507	1768	0.1054	0.01576	1.259×10^{-4}	7.794×10^{-6}	2.996	0.875	0.00262	0.00996
5	549.8	521.8	11.99	367.0	2547	1814	0.1028	0.01637	1.195×10^{-4}	7.970×10^{-6}	2.964	0.883	0.00273	0.00934
10	635.1	514.7	13.81	359.5	2590	1864	0.1002	0.01701	1.135×10^{-4}	8.151×10^{-6}	2.935	0.893	0.00286	0.00872
15	729.8	507.5	15.85	351.7	2637	1917	0.0977	0.01767	1.077×10^{-4}	8.339×10^{-6}	2.909	0.905	0.00301	0.00811
20	834.4	500.0	18.13	343.4	2688	1974	0.0952	0.01836	1.022×10^{-4}	8.534×10^{-6}	2.886	0.918	0.00318	0.00751
25	949.7	492.2	20.68	334.8	2742	2036	0.0928	0.01908	9.702×10^{-5}	8.738×10^{-6}	2.866	0.933	0.00337	0.00691
30	1076	484.2	23.53	325.8	2802	2104	0.0904	0.01982	9.197×10^{-5}	8.952×10^{-6}	2.850	0.950	0.00358	0.00633
35	1215	475.8	26.72	316.2	2869	2179	0.0881	0.02061	8.710×10^{-5}	9.178×10^{-6}	2.837	0.971	0.00384	0.00575
40	1366	467.1	30.29	306.1	2943	2264	0.0857	0.02142	8.240×10^{-5}	9.417×10^{-6}	2.828	0.995	0.00413	0.00518
45	1530	458.0	34.29	295.3	3026	2361	0.0834	0.02228	7.785×10^{-5}	9.674×10^{-6}	2.824	1.025	0.00448	0.00463
50	1708	448.5	38.79	283.9	3122	2473	0.0811	0.02319	7.343×10^{-5}	9.950×10^{-6}	2.826	1.061	0.00491	0.00408
60	2110	427.5	49.66	258.4	3283	2769	0.0765	0.02517	6.487×10^{-5}	1.058×10^{-5}	2.784	1.164	0.00609	0.00303
70	2580	403.2	64.02	228.0	3595	3241	0.0717	0.02746	5.649×10^{-5}	1.138×10^{-5}	2.834	1.343	0.00811	0.00204
80	3127	373.0	84.28	189.7	4501	4173	0.0663	0.03029	4.790×10^{-5}	1.249×10^{-5}	3.251	1.722	0.01248	0.00114
90	3769	329.1	118.6	133.2	6977	7239	0.0595	0.03441	3.807×10^{-5}	1.448×10^{-5}	4.465	3.047	0.02847	0.00037

Note 1: Kinematic viscosity ν and thermal diffusivity α can be calculated from their definitions, $\nu = \mu/\rho$ and $\alpha = k/\rho c_p = \nu/\mathrm{Pr}$. The properties listed here (except the vapor density) can be used at any pressures with negligible error except at temperatures near the critical-point value.

Note 2: The unit kJ/kg·°C for specific heat is equivalent to kJ/kg·K, and the unit W/m·°C for thermal conductivity is equivalent to W/m·K.

Source: Data generated from the EES software developed by S. A. Klein and F. L. Alvarado. Original sources: Reiner Tillner-Roth, "Fundamental Equations of State," Shaker, Verlag, Aachan, 1998; B. A. Younglove and J. F. Ely, "Thermophysical Properties of Fluids. II Methane, Ethane, Propane, Isobutane, and Normal Butane," *J. Phys. Chem. Ref. Data*, Vol. 16, No. 4, 1987; G.R. Somayajulu, "A Generalized Equation for Surface Tension from the Triple-Point to the Critical-Point," *International Journal of Thermophysics*, Vol. 9, No. 4, 1988.

TABLE A-7

Properties of liquids

Temp. T, °C	Density ρ, kg/m^3	Specific Heat c_p, J/kg·K	Thermal Conductivity k, W/m·K	Thermal Diffusivity α, m^2/s	Dynamic Viscosity μ, kg/m·s	Kinematic Viscosity ν, m^2/s	Prandtl Number Pr	Volume Expansion Coeff. β, 1/K
\multicolumn{9}{c}{*Methane (CH$_4$)*}
−160	420.2	3492	0.1863	1.270×10^{-7}	1.133×10^{-4}	2.699×10^{-7}	2.126	0.00352
−150	405.0	3580	0.1703	1.174×10^{-7}	9.169×10^{-5}	2.264×10^{-7}	1.927	0.00391
−140	388.8	3700	0.1550	1.077×10^{-7}	7.551×10^{-5}	1.942×10^{-7}	1.803	0.00444
−130	371.1	3875	0.1402	9.749×10^{-8}	6.288×10^{-5}	1.694×10^{-7}	1.738	0.00520
−120	351.4	4146	0.1258	8.634×10^{-8}	5.257×10^{-5}	1.496×10^{-7}	1.732	0.00637
−110	328.8	4611	0.1115	7.356×10^{-8}	4.377×10^{-5}	1.331×10^{-7}	1.810	0.00841
−100	301.0	5578	0.0967	5.761×10^{-8}	3.577×10^{-5}	1.188×10^{-7}	2.063	0.01282
−90	261.7	8902	0.0797	3.423×10^{-8}	2.761×10^{-5}	1.055×10^{-7}	3.082	0.02922
\multicolumn{9}{c}{*Methanol [CH$_3$(OH)]*}
20	788.4	2515	0.1987	1.002×10^{-7}	5.857×10^{-4}	7.429×10^{-7}	7.414	0.00118
30	779.1	2577	0.1980	9.862×10^{-8}	5.088×10^{-4}	6.531×10^{-7}	6.622	0.00120
40	769.6	2644	0.1972	9.690×10^{-8}	4.460×10^{-4}	5.795×10^{-7}	5.980	0.00123
50	760.1	2718	0.1965	9.509×10^{-8}	3.942×10^{-4}	5.185×10^{-7}	5.453	0.00127
60	750.4	2798	0.1957	9.320×10^{-8}	3.510×10^{-4}	4.677×10^{-7}	5.018	0.00132
70	740.4	2885	0.1950	9.128×10^{-8}	3.146×10^{-4}	4.250×10^{-7}	4.655	0.00137
\multicolumn{9}{c}{*Isobutane (R600a)*}
−100	683.8	1881	0.1383	1.075×10^{-7}	9.305×10^{-4}	1.360×10^{-6}	12.65	0.00142
−75	659.3	1970	0.1357	1.044×10^{-7}	5.624×10^{-4}	8.531×10^{-7}	8.167	0.00150
−50	634.3	2069	0.1283	9.773×10^{-8}	3.769×10^{-4}	5.942×10^{-7}	6.079	0.00161
−25	608.2	2180	0.1181	8.906×10^{-8}	2.688×10^{-4}	4.420×10^{-7}	4.963	0.00177
0	580.6	2306	0.1068	7.974×10^{-8}	1.993×10^{-4}	3.432×10^{-7}	4.304	0.00199
25	550.7	2455	0.0956	7.069×10^{-8}	1.510×10^{-4}	2.743×10^{-7}	3.880	0.00232
50	517.3	2640	0.0851	6.233×10^{-8}	1.155×10^{-4}	2.233×10^{-7}	3.582	0.00286
75	478.5	2896	0.0757	5.460×10^{-8}	8.785×10^{-5}	1.836×10^{-7}	3.363	0.00385
100	429.6	3361	0.0669	4.634×10^{-8}	6.483×10^{-5}	1.509×10^{-7}	3.256	0.00628
\multicolumn{9}{c}{*Glycerin*}
0	1276	2262	0.2820	9.773×10^{-8}	10.49	8.219×10^{-3}	84,101	
5	1273	2288	0.2835	9.732×10^{-8}	6.730	5.287×10^{-3}	54,327	
10	1270	2320	0.2846	9.662×10^{-8}	4.241	3.339×10^{-3}	34,561	
15	1267	2354	0.2856	9.576×10^{-8}	2.496	1.970×10^{-3}	20,570	
20	1264	2386	0.2860	9.484×10^{-8}	1.519	1.201×10^{-3}	12,671	
25	1261	2416	0.2860	9.388×10^{-8}	0.9934	7.878×10^{-4}	8,392	
30	1258	2447	0.2860	9.291×10^{-8}	0.6582	5.232×10^{-4}	5,631	
35	1255	2478	0.2860	9.195×10^{-8}	0.4347	3.464×10^{-4}	3,767	
40	1252	2513	0.2863	9.101×10^{-8}	0.3073	2.455×10^{-4}	2,697	
\multicolumn{9}{c}{*Engine Oil (unused)*}
0	899.0	1797	0.1469	9.097×10^{-8}	3.814	4.242×10^{-3}	46,636	0.00070
20	888.1	1881	0.1450	8.680×10^{-8}	0.8374	9.429×10^{-4}	10,863	0.00070
40	876.0	1964	0.1444	8.391×10^{-8}	0.2177	2.485×10^{-4}	2,962	0.00070
60	863.9	2048	0.1404	7.934×10^{-8}	0.07399	8.565×10^{-5}	1,080	0.00070
80	852.0	2132	0.1380	7.599×10^{-8}	0.03232	3.794×10^{-5}	499.3	0.00070
100	840.0	2220	0.1367	7.330×10^{-8}	0.01718	2.046×10^{-5}	279.1	0.00070
120	828.9	2308	0.1347	7.042×10^{-8}	0.01029	1.241×10^{-5}	176.3	0.00070
140	816.8	2395	0.1330	6.798×10^{-8}	0.006558	8.029×10^{-6}	118.1	0.00070
150	810.3	2441	0.1327	6.708×10^{-8}	0.005344	6.595×10^{-6}	98.31	0.00070

Source: Data generated from the EES software developed by S. A. Klein and F. L. Alvarado. Originally based on various sources.

TABLE A–8

Properties of liquid metals

Temp. T, °C	Density ρ, kg/m³	Specific Heat c_p, J/kg·K	Thermal Conductivity k, W/m·K	Thermal Diffusivity α, m²/s	Dynamic Viscosity μ, kg/m·s	Kinematic Viscosity ν, m²/s	Prandtl Number Pr	Volume Expansion Coeff. β, 1/K
\multicolumn{9}{c}{Mercury (Hg) Melting Point: −39°C}
0	13595	140.4	8.18200	4.287×10^{-6}	1.687×10^{-3}	1.241×10^{-7}	0.0289	1.810×10^{-4}
25	13534	139.4	8.51533	4.514×10^{-6}	1.534×10^{-3}	1.133×10^{-7}	0.0251	1.810×10^{-4}
50	13473	138.6	8.83632	4.734×10^{-6}	1.423×10^{-3}	1.056×10^{-7}	0.0223	1.810×10^{-4}
75	13412	137.8	9.15632	4.956×10^{-6}	1.316×10^{-3}	9.819×10^{-8}	0.0198	1.810×10^{-4}
100	13351	137.1	9.46706	5.170×10^{-6}	1.245×10^{-3}	9.326×10^{-8}	0.0180	1.810×10^{-4}
150	13231	136.1	10.07780	5.595×10^{-6}	1.126×10^{-3}	8.514×10^{-8}	0.0152	1.810×10^{-4}
200	13112	135.5	10.65465	5.996×10^{-6}	1.043×10^{-3}	7.959×10^{-8}	0.0133	1.815×10^{-4}
250	12993	135.3	11.18150	6.363×10^{-6}	9.820×10^{-4}	7.558×10^{-8}	0.0119	1.829×10^{-4}
300	12873	135.3	11.68150	6.705×10^{-6}	9.336×10^{-4}	7.252×10^{-8}	0.0108	1.854×10^{-4}
\multicolumn{9}{c}{Bismuth (Bi) Melting Point: 271°C}
350	9969	146.0	16.28	1.118×10^{-5}	1.540×10^{-3}	1.545×10^{-7}	0.01381	
400	9908	148.2	16.10	1.096×10^{-5}	1.422×10^{-3}	1.436×10^{-7}	0.01310	
500	9785	152.8	15.74	1.052×10^{-5}	1.188×10^{-3}	1.215×10^{-7}	0.01154	
600	9663	157.3	15.60	1.026×10^{-5}	1.013×10^{-3}	1.048×10^{-7}	0.01022	
700	9540	161.8	15.60	1.010×10^{-5}	8.736×10^{-4}	9.157×10^{-8}	0.00906	
\multicolumn{9}{c}{Lead (Pb) Melting Point: 327°C}
400	10506	158	15.97	9.623×10^{-6}	2.277×10^{-3}	2.167×10^{-7}	0.02252	
450	10449	156	15.74	9.649×10^{-6}	2.065×10^{-3}	1.976×10^{-7}	0.02048	
500	10390	155	15.54	9.651×10^{-6}	1.884×10^{-3}	1.814×10^{-7}	0.01879	
550	10329	155	15.39	9.610×10^{-6}	1.758×10^{-3}	1.702×10^{-7}	0.01771	
600	10267	155	15.23	9.568×10^{-6}	1.632×10^{-3}	1.589×10^{-7}	0.01661	
650	10206	155	15.07	9.526×10^{-6}	1.505×10^{-3}	1.475×10^{-7}	0.01549	
700	10145	155	14.91	9.483×10^{-6}	1.379×10^{-3}	1.360×10^{-7}	0.01434	
\multicolumn{9}{c}{Sodium (Na) Melting Point: 98°C}
100	927.3	1378	85.84	6.718×10^{-5}	6.892×10^{-4}	7.432×10^{-7}	0.01106	
200	902.5	1349	80.84	6.639×10^{-5}	5.385×10^{-4}	5.967×10^{-7}	0.008987	
300	877.8	1320	75.84	6.544×10^{-5}	3.878×10^{-4}	4.418×10^{-7}	0.006751	
400	853.0	1296	71.20	6.437×10^{-5}	2.720×10^{-4}	3.188×10^{-7}	0.004953	
500	828.5	1284	67.41	6.335×10^{-5}	2.411×10^{-4}	2.909×10^{-7}	0.004593	
600	804.0	1272	63.63	6.220×10^{-5}	2.101×10^{-4}	2.614×10^{-7}	0.004202	
\multicolumn{9}{c}{Potassium (K) Melting Point: 64°C}
200	795.2	790.8	43.99	6.995×10^{-5}	3.350×10^{-4}	4.213×10^{-7}	0.006023	
300	771.6	772.8	42.01	7.045×10^{-5}	2.667×10^{-4}	3.456×10^{-7}	0.004906	
400	748.0	754.8	40.03	7.090×10^{-5}	1.984×10^{-4}	2.652×10^{-7}	0.00374	
500	723.9	750.0	37.81	6.964×10^{-5}	1.668×10^{-4}	2.304×10^{-7}	0.003309	
600	699.6	750.0	35.50	6.765×10^{-5}	1.487×10^{-4}	2.126×10^{-7}	0.003143	
\multicolumn{9}{c}{Sodium–Potassium (%22Na-%78K) Melting Point: −11°C}
100	847.3	944.4	25.64	3.205×10^{-5}	5.707×10^{-4}	6.736×10^{-7}	0.02102	
200	823.2	922.5	26.27	3.459×10^{-5}	4.587×10^{-4}	5.572×10^{-7}	0.01611	
300	799.1	900.6	26.89	3.736×10^{-5}	3.467×10^{-4}	4.339×10^{-7}	0.01161	
400	775.0	879.0	27.50	4.037×10^{-5}	2.357×10^{-4}	3.041×10^{-7}	0.00753	
500	751.5	880.1	27.89	4.217×10^{-5}	2.108×10^{-4}	2.805×10^{-7}	0.00665	
600	728.0	881.2	28.28	4.408×10^{-5}	1.859×10^{-4}	2.553×10^{-7}	0.00579	

Source: Data generated from the EES software developed by S. A. Klein and F. L. Alvarado. Originally based on various sources.

TABLE A–9

Properties of air at 1 atm pressure

Temp. T, °C	Density ρ, kg/m³	Specific Heat c_p J/kg·K	Thermal Conductivity k, W/m·K	Thermal Diffusivity α, m²/s	Dynamic Viscosity μ, kg/m·s	Kinematic Viscosity ν, m²/s	Prandtl Number Pr
−150	2.866	983	0.01171	4.158×10^{-6}	8.636×10^{-6}	3.013×10^{-6}	0.7246
−100	2.038	966	0.01582	8.036×10^{-6}	1.189×10^{-6}	5.837×10^{-6}	0.7263
−50	1.582	999	0.01979	1.252×10^{-5}	1.474×10^{-5}	9.319×10^{-6}	0.7440
−40	1.514	1002	0.02057	1.356×10^{-5}	1.527×10^{-5}	1.008×10^{-5}	0.7436
−30	1.451	1004	0.02134	1.465×10^{-5}	1.579×10^{-5}	1.087×10^{-5}	0.7425
−20	1.394	1005	0.02211	1.578×10^{-5}	1.630×10^{-5}	1.169×10^{-5}	0.7408
−10	1.341	1006	0.02288	1.696×10^{-5}	1.680×10^{-5}	1.252×10^{-5}	0.7387
0	1.292	1006	0.02364	1.818×10^{-5}	1.729×10^{-5}	1.338×10^{-5}	0.7362
5	1.269	1006	0.02401	1.880×10^{-5}	1.754×10^{-5}	1.382×10^{-5}	0.7350
10	1.246	1006	0.02439	1.944×10^{-5}	1.778×10^{-5}	1.426×10^{-5}	0.7336
15	1.225	1007	0.02476	2.009×10^{-5}	1.802×10^{-5}	1.470×10^{-5}	0.7323
20	1.204	1007	0.02514	2.074×10^{-5}	1.825×10^{-5}	1.516×10^{-5}	0.7309
25	1.184	1007	0.02551	2.141×10^{-5}	1.849×10^{-5}	1.562×10^{-5}	0.7296
30	1.164	1007	0.02588	2.208×10^{-5}	1.872×10^{-5}	1.608×10^{-5}	0.7282
35	1.145	1007	0.02625	2.277×10^{-5}	1.895×10^{-5}	1.655×10^{-5}	0.7268
40	1.127	1007	0.02662	2.346×10^{-5}	1.918×10^{-5}	1.702×10^{-5}	0.7255
45	1.109	1007	0.02699	2.416×10^{-5}	1.941×10^{-5}	1.750×10^{-5}	0.7241
50	1.092	1007	0.02735	2.487×10^{-5}	1.963×10^{-5}	1.798×10^{-5}	0.7228
60	1.059	1007	0.02808	2.632×10^{-5}	2.008×10^{-5}	1.896×10^{-5}	0.7202
70	1.028	1007	0.02881	2.780×10^{-5}	2.052×10^{-5}	1.995×10^{-5}	0.7177
80	0.9994	1008	0.02953	2.931×10^{-5}	2.096×10^{-5}	2.097×10^{-5}	0.7154
90	0.9718	1008	0.03024	3.086×10^{-5}	2.139×10^{-5}	2.201×10^{-5}	0.7132
100	0.9458	1009	0.03095	3.243×10^{-5}	2.181×10^{-5}	2.306×10^{-5}	0.7111
120	0.8977	1011	0.03235	3.565×10^{-5}	2.264×10^{-5}	2.522×10^{-5}	0.7073
140	0.8542	1013	0.03374	3.898×10^{-5}	2.345×10^{-5}	2.745×10^{-5}	0.7041
160	0.8148	1016	0.03511	4.241×10^{-5}	2.420×10^{-5}	2.975×10^{-5}	0.7014
180	0.7788	1019	0.03646	4.593×10^{-5}	2.504×10^{-5}	3.212×10^{-5}	0.6992
200	0.7459	1023	0.03779	4.954×10^{-5}	2.577×10^{-5}	3.455×10^{-5}	0.6974
250	0.6746	1033	0.04104	5.890×10^{-5}	2.760×10^{-5}	4.091×10^{-5}	0.6946
300	0.6158	1044	0.04418	6.871×10^{-5}	2.934×10^{-5}	4.765×10^{-5}	0.6935
350	0.5664	1056	0.04721	7.892×10^{-5}	3.101×10^{-5}	5.475×10^{-5}	0.6937
400	0.5243	1069	0.05015	8.951×10^{-5}	3.261×10^{-5}	6.219×10^{-5}	0.6948
450	0.4880	1081	0.05298	1.004×10^{-4}	3.415×10^{-5}	6.997×10^{-5}	0.6965
500	0.4565	1093	0.05572	1.117×10^{-4}	3.563×10^{-5}	7.806×10^{-5}	0.6986
600	0.4042	1115	0.06093	1.352×10^{-4}	3.846×10^{-5}	9.515×10^{-5}	0.7037
700	0.3627	1135	0.06581	1.598×10^{-4}	4.111×10^{-5}	1.133×10^{-4}	0.7092
800	0.3289	1153	0.07037	1.855×10^{-4}	4.362×10^{-5}	1.326×10^{-4}	0.7149
900	0.3008	1169	0.07465	2.122×10^{-4}	4.600×10^{-5}	1.529×10^{-4}	0.7206
1000	0.2772	1184	0.07868	2.398×10^{-4}	4.826×10^{-5}	1.741×10^{-4}	0.7260
1500	0.1990	1234	0.09599	3.908×10^{-4}	5.817×10^{-5}	2.922×10^{-4}	0.7478
2000	0.1553	1264	0.11113	5.664×10^{-4}	6.630×10^{-5}	4.270×10^{-4}	0.7539

Note: For ideal gases, the properties c_p, k, μ, and Pr are independent of pressure. The properties ρ, ν, and α at a pressure P (in atm) other than 1 atm are determined by multiplying the values of ρ at the given temperature by P and by dividing ν and α by P.

Source: Data generated from the EES software developed by S. A. Klein and F. L. Alvarado. Original sources: Keenan, Chao, Keyes, Gas Tables, Wiley, 198; and Thermophysical Properties of Matter, Vol. 3: Thermal Conductivity, Y. S. Touloukian, P. E. Liley, S. C. Saxena, Vol. 11: Viscosity, Y. S. Touloukian, S. C. Saxena, and P. Hestermans, IFI/Plenun, NY, 1970, ISBN 0-306067020-8.

TABLE A–10

Properties of gases at 1 atm pressure

Temp. T, °C	Density ρ, kg/m³	Specific Heat c_p J/kg·K	Thermal Conductivity k, W/m·K	Thermal Diffusivity α, m²/s	Dynamic Viscosity μ, kg/m·s	Kinematic Viscosity ν, m²/s	Prandtl Number Pr
			Carbon Dioxide, CO_2				
−50	2.4035	746	0.01051	5.860×10^{-6}	1.129×10^{-5}	4.699×10^{-6}	0.8019
0	1.9635	811	0.01456	9.141×10^{-6}	1.375×10^{-5}	7.003×10^{-6}	0.7661
50	1.6597	866.6	0.01858	1.291×10^{-5}	1.612×10^{-5}	9.714×10^{-6}	0.7520
100	1.4373	914.8	0.02257	1.716×10^{-5}	1.841×10^{-5}	1.281×10^{-5}	0.7464
150	1.2675	957.4	0.02652	2.186×10^{-5}	2.063×10^{-5}	1.627×10^{-5}	0.7445
200	1.1336	995.2	0.03044	2.698×10^{-5}	2.276×10^{-5}	2.008×10^{-5}	0.7442
300	0.9358	1060	0.03814	3.847×10^{-5}	2.682×10^{-5}	2.866×10^{-5}	0.7450
400	0.7968	1112	0.04565	5.151×10^{-5}	3.061×10^{-5}	3.842×10^{-5}	0.7458
500	0.6937	1156	0.05293	6.600×10^{-5}	3.416×10^{-5}	4.924×10^{-5}	0.7460
1000	0.4213	1292	0.08491	1.560×10^{-4}	4.898×10^{-5}	1.162×10^{-4}	0.7455
1500	0.3025	1356	0.10688	2.606×10^{-4}	6.106×10^{-5}	2.019×10^{-4}	0.7745
2000	0.2359	1387	0.11522	3.521×10^{-4}	7.322×10^{-5}	3.103×10^{-4}	0.8815
			Carbon Monoxide, CO				
−50	1.5297	1081	0.01901	1.149×10^{-5}	1.378×10^{-5}	9.012×10^{-6}	0.7840
0	1.2497	1048	0.02278	1.739×10^{-5}	1.629×10^{-5}	1.303×10^{-5}	0.7499
50	1.0563	1039	0.02641	2.407×10^{-5}	1.863×10^{-5}	1.764×10^{-5}	0.7328
100	0.9148	1041	0.02992	3.142×10^{-5}	2.080×10^{-5}	2.274×10^{-5}	0.7239
150	0.8067	1049	0.03330	3.936×10^{-5}	2.283×10^{-5}	2.830×10^{-5}	0.7191
200	0.7214	1060	0.03656	4.782×10^{-5}	2.472×10^{-5}	3.426×10^{-5}	0.7164
300	0.5956	1085	0.04277	6.619×10^{-5}	2.812×10^{-5}	4.722×10^{-5}	0.7134
400	0.5071	1111	0.04860	8.628×10^{-5}	3.111×10^{-5}	6.136×10^{-5}	0.7111
500	0.4415	1135	0.05412	1.079×10^{-4}	3.379×10^{-5}	7.653×10^{-5}	0.7087
1000	0.2681	1226	0.07894	2.401×10^{-4}	4.557×10^{-5}	1.700×10^{-4}	0.7080
1500	0.1925	1279	0.10458	4.246×10^{-4}	6.321×10^{-5}	3.284×10^{-4}	0.7733
2000	0.1502	1309	0.13833	7.034×10^{-4}	9.826×10^{-5}	6.543×10^{-4}	0.9302
			Methane, CH_4				
−50	0.8761	2243	0.02367	1.204×10^{-5}	8.564×10^{-6}	9.774×10^{-6}	0.8116
0	0.7158	2217	0.03042	1.917×10^{-5}	1.028×10^{-5}	1.436×10^{-5}	0.7494
50	0.6050	2302	0.03766	2.704×10^{-5}	1.191×10^{-5}	1.969×10^{-5}	0.7282
100	0.5240	2443	0.04534	3.543×10^{-5}	1.345×10^{-5}	2.567×10^{-5}	0.7247
150	0.4620	2611	0.05344	4.431×10^{-5}	1.491×10^{-5}	3.227×10^{-5}	0.7284
200	0.4132	2791	0.06194	5.370×10^{-5}	1.630×10^{-5}	3.944×10^{-5}	0.7344
300	0.3411	3158	0.07996	7.422×10^{-5}	1.886×10^{-5}	5.529×10^{-5}	0.7450
400	0.2904	3510	0.09918	9.727×10^{-5}	2.119×10^{-5}	7.297×10^{-5}	0.7501
500	0.2529	3836	0.11933	1.230×10^{-4}	2.334×10^{-5}	9.228×10^{-5}	0.7502
1000	0.1536	5042	0.22562	2.914×10^{-4}	3.281×10^{-5}	2.136×10^{-4}	0.7331
1500	0.1103	5701	0.31857	5.068×10^{-4}	4.434×10^{-5}	4.022×10^{-4}	0.7936
2000	0.0860	6001	0.36750	7.120×10^{-4}	6.360×10^{-5}	7.395×10^{-4}	1.0386
			Hydrogen, H_2				
−50	0.11010	12635	0.1404	1.009×10^{-4}	7.293×10^{-6}	6.624×10^{-5}	0.6562
0	0.08995	13920	0.1652	1.319×10^{-4}	8.391×10^{-6}	9.329×10^{-5}	0.7071
50	0.07603	14349	0.1881	1.724×10^{-4}	9.427×10^{-6}	1.240×10^{-4}	0.7191
100	0.06584	14473	0.2095	2.199×10^{-4}	1.041×10^{-5}	1.582×10^{-4}	0.7196
150	0.05806	14492	0.2296	2.729×10^{-4}	1.136×10^{-5}	1.957×10^{-4}	0.7174
200	0.05193	14482	0.2486	3.306×10^{-4}	1.228×10^{-5}	2.365×10^{-4}	0.7155
300	0.04287	14481	0.2843	4.580×10^{-4}	1.403×10^{-5}	3.274×10^{-4}	0.7149
400	0.03650	14540	0.3180	5.992×10^{-4}	1.570×10^{-5}	4.302×10^{-4}	0.7179
500	0.03178	14653	0.3509	7.535×10^{-4}	1.730×10^{-5}	5.443×10^{-4}	0.7224
1000	0.01930	15577	0.5206	1.732×10^{-3}	2.455×10^{-5}	1.272×10^{-3}	0.7345
1500	0.01386	16553	0.6581	2.869×10^{-3}	3.099×10^{-5}	2.237×10^{-3}	0.7795
2000	0.01081	17400	0.5480	2.914×10^{-3}	3.690×10^{-5}	3.414×10^{-3}	1.1717

(Continued)

TABLE A–10

Properties of gases at 1 atm pressure (*Continued*)

Temp. T, °C	Density ρ, kg/m^3	Specific Heat c_p J/kg·K	Thermal Conductivity k, W/m·K	Thermal Diffusivity α, m^2/s	Dynamic Viscosity μ, kg/m·s	Kinematic Viscosity ν, m^2/s	Prandtl Number Pr
\multicolumn{8}{c}{*Nitrogen, N$_2$*}
−50	1.5299	957.3	0.02001	1.366×10^{-5}	1.390×10^{-5}	9.091×10^{-6}	0.6655
0	1.2498	1035	0.02384	1.843×10^{-5}	1.640×10^{-5}	1.312×10^{-5}	0.7121
50	1.0564	1042	0.02746	2.494×10^{-5}	1.874×10^{-5}	1.774×10^{-5}	0.7114
100	0.9149	1041	0.03090	3.244×10^{-5}	2.094×10^{-5}	2.289×10^{-5}	0.7056
150	0.8068	1043	0.03416	4.058×10^{-5}	2.300×10^{-5}	2.851×10^{-5}	0.7025
200	0.7215	1050	0.03727	4.921×10^{-5}	2.494×10^{-5}	3.457×10^{-5}	0.7025
300	0.5956	1070	0.04309	6.758×10^{-5}	2.849×10^{-5}	4.783×10^{-5}	0.7078
400	0.5072	1095	0.04848	8.727×10^{-5}	3.166×10^{-5}	6.242×10^{-5}	0.7153
500	0.4416	1120	0.05358	1.083×10^{-4}	3.451×10^{-5}	7.816×10^{-5}	0.7215
1000	0.2681	1213	0.07938	2.440×10^{-4}	4.594×10^{-5}	1.713×10^{-4}	0.7022
1500	0.1925	1266	0.11793	4.839×10^{-4}	5.562×10^{-5}	2.889×10^{-4}	0.5969
2000	0.1502	1297	0.18590	9.543×10^{-4}	6.426×10^{-5}	4.278×10^{-4}	0.4483
\multicolumn{8}{c}{*Oxygen, O$_2$*}
−50	1.7475	984.4	0.02067	1.201×10^{-5}	1.616×10^{-5}	9.246×10^{-6}	0.7694
0	1.4277	928.7	0.02472	1.865×10^{-5}	1.916×10^{-5}	1.342×10^{-5}	0.7198
50	1.2068	921.7	0.02867	2.577×10^{-5}	2.194×10^{-5}	1.818×10^{-5}	0.7053
100	1.0451	931.8	0.03254	3.342×10^{-5}	2.451×10^{-5}	2.346×10^{-5}	0.7019
150	0.9216	947.6	0.03637	4.164×10^{-5}	2.694×10^{-5}	2.923×10^{-5}	0.7019
200	0.8242	964.7	0.04014	5.048×10^{-5}	2.923×10^{-5}	3.546×10^{-5}	0.7025
300	0.6804	997.1	0.04751	7.003×10^{-5}	3.350×10^{-5}	4.923×10^{-5}	0.7030
400	0.5793	1025	0.05463	9.204×10^{-5}	3.744×10^{-5}	6.463×10^{-5}	0.7023
500	0.5044	1048	0.06148	1.163×10^{-4}	4.114×10^{-5}	8.156×10^{-5}	0.7010
1000	0.3063	1121	0.09198	2.678×10^{-4}	5.732×10^{-5}	1.871×10^{-4}	0.6986
1500	0.2199	1165	0.11901	4.643×10^{-4}	7.133×10^{-5}	3.243×10^{-4}	0.6985
2000	0.1716	1201	0.14705	7.139×10^{-4}	8.417×10^{-5}	4.907×10^{-4}	0.6873
\multicolumn{8}{c}{*Water Vapor, H$_2$O*}
−50	0.9839	1892	0.01353	7.271×10^{-6}	7.187×10^{-6}	7.305×10^{-6}	1.0047
0	0.8038	1874	0.01673	1.110×10^{-5}	8.956×10^{-6}	1.114×10^{-5}	1.0033
50	0.6794	1874	0.02032	1.596×10^{-5}	1.078×10^{-5}	1.587×10^{-5}	0.9944
100	0.5884	1887	0.02429	2.187×10^{-5}	1.265×10^{-5}	2.150×10^{-5}	0.9830
150	0.5189	1908	0.02861	2.890×10^{-5}	1.456×10^{-5}	2.806×10^{-5}	0.9712
200	0.4640	1935	0.03326	3.705×10^{-5}	1.650×10^{-5}	3.556×10^{-5}	0.9599
300	0.3831	1997	0.04345	5.680×10^{-5}	2.045×10^{-5}	5.340×10^{-5}	0.9401
400	0.3262	2066	0.05467	8.114×10^{-5}	2.446×10^{-5}	7.498×10^{-5}	0.9240
500	0.2840	2137	0.06677	1.100×10^{-4}	2.847×10^{-5}	1.002×10^{-4}	0.9108
1000	0.1725	2471	0.13623	3.196×10^{-4}	4.762×10^{-5}	2.761×10^{-4}	0.8639
1500	0.1238	2736	0.21301	6.288×10^{-4}	6.411×10^{-5}	5.177×10^{-4}	0.8233
2000	0.0966	2928	0.29183	1.032×10^{-3}	7.808×10^{-5}	8.084×10^{-4}	0.7833

Note: For ideal gases, the properties c_p, k, μ, and Pr are independent of pressure. The properties ρ, ν, and α at a pressure P (in atm) other than 1 atm are determined by multiplying the values of ρ at the given temperature by P and by dividing ν and α by P.

Source: Data generated from the EES software developed by S. A. Klein and F. L. Alvarado. Originally based on various sources.

TABLE A–11

Properties of the atmosphere at high altitude

Altitude, m	Temperature, °C	Pressure, kPa	Gravity g, m/s^2	Speed of Sound, m/s	Density, kg/m^3	Viscosity μ, kg/m·s	Thermal Conductivity, W/m·K
0	15.00	101.33	9.807	340.3	1.225	1.789×10^{-5}	0.0253
200	13.70	98.95	9.806	339.5	1.202	1.783×10^{-5}	0.0252
400	12.40	96.61	9.805	338.8	1.179	1.777×10^{-5}	0.0252
600	11.10	94.32	9.805	338.0	1.156	1.771×10^{-5}	0.0251
800	9.80	92.08	9.804	337.2	1.134	1.764×10^{-5}	0.0250
1000	8.50	89.88	9.804	336.4	1.112	1.758×10^{-5}	0.0249
1200	7.20	87.72	9.803	335.7	1.090	1.752×10^{-5}	0.0248
1400	5.90	85.60	9.802	334.9	1.069	1.745×10^{-5}	0.0247
1600	4.60	83.53	9.802	334.1	1.048	1.739×10^{-5}	0.0245
1800	3.30	81.49	9.801	333.3	1.027	1.732×10^{-5}	0.0244
2000	2.00	79.50	9.800	332.5	1.007	1.726×10^{-5}	0.0243
2200	0.70	77.55	9.800	331.7	0.987	1.720×10^{-5}	0.0242
2400	−0.59	75.63	9.799	331.0	0.967	1.713×10^{-5}	0.0241
2600	−1.89	73.76	9.799	330.2	0.947	1.707×10^{-5}	0.0240
2800	−3.19	71.92	9.798	329.4	0.928	1.700×10^{-5}	0.0239
3000	−4.49	70.12	9.797	328.6	0.909	1.694×10^{-5}	0.0238
3200	−5.79	68.36	9.797	327.8	0.891	1.687×10^{-5}	0.0237
3400	−7.09	66.63	9.796	327.0	0.872	1.681×10^{-5}	0.0236
3600	−8.39	64.94	9.796	326.2	0.854	1.674×10^{-5}	0.0235
3800	−9.69	63.28	9.795	325.4	0.837	1.668×10^{-5}	0.0234
4000	−10.98	61.66	9.794	324.6	0.819	1.661×10^{-5}	0.0233
4200	−12.3	60.07	9.794	323.8	0.802	1.655×10^{-5}	0.0232
4400	−13.6	58.52	9.793	323.0	0.785	1.648×10^{-5}	0.0231
4600	−14.9	57.00	9.793	322.2	0.769	1.642×10^{-5}	0.0230
4800	−16.2	55.51	9.792	321.4	0.752	1.635×10^{-5}	0.0229
5000	−17.5	54.05	9.791	320.5	0.736	1.628×10^{-5}	0.0228
5200	−18.8	52.62	9.791	319.7	0.721	1.622×10^{-5}	0.0227
5400	−20.1	51.23	9.790	318.9	0.705	1.615×10^{-5}	0.0226
5600	−21.4	49.86	9.789	318.1	0.690	1.608×10^{-5}	0.0224
5800	−22.7	48.52	9.785	317.3	0.675	1.602×10^{-5}	0.0223
6000	−24.0	47.22	9.788	316.5	0.660	1.595×10^{-5}	0.0222
6200	−25.3	45.94	9.788	315.6	0.646	1.588×10^{-5}	0.0221
6400	−26.6	44.69	9.787	314.8	0.631	1.582×10^{-5}	0.0220
6600	−27.9	43.47	9.786	314.0	0.617	1.575×10^{-5}	0.0219
6800	−29.2	42.27	9.785	313.1	0.604	1.568×10^{-5}	0.0218
7000	−30.5	41.11	9.785	312.3	0.590	1.561×10^{-5}	0.0217
8000	−36.9	35.65	9.782	308.1	0.526	1.527×10^{-5}	0.0212
9000	−43.4	30.80	9.779	303.8	0.467	1.493×10^{-5}	0.0206
10,000	−49.9	26.50	9.776	299.5	0.414	1.458×10^{-5}	0.0201
12,000	−56.5	19.40	9.770	295.1	0.312	1.422×10^{-5}	0.0195
14,000	−56.5	14.17	9.764	295.1	0.228	1.422×10^{-5}	0.0195
16,000	−56.5	10.53	9.758	295.1	0.166	1.422×10^{-5}	0.0195
18,000	−56.5	7.57	9.751	295.1	0.122	1.422×10^{-5}	0.0195

Source: U.S. Standard Atmosphere Supplements, U.S. Government Printing Office, 1966. Based on year-round mean conditions at 45° latitude and varies with the time of the year and the weather patterns. The conditions at sea level ($z = 0$) are taken to be P = 101.325 kPa, T = 15°C, ρ = 1.2250 kg/m^3, g = 9.80665 m^2/s.

FIGURE A–12

The Moody chart for the friction factor for fully developed flow in circular pipes for use in the head loss relation $h_L = f \dfrac{L}{D} \dfrac{V^2}{2g}$. Friction factors in the turbulent flow are evaluated from the Colebrook equation $\dfrac{1}{\sqrt{f}} = -2\log_{10}\left(\dfrac{\varepsilon/D}{3.7} + \dfrac{2.51}{\mathrm{Re}\,\sqrt{f}}\right)$.

$$\text{Ma}^* = \text{Ma}\sqrt{\frac{k+1}{2+(k-1)\text{Ma}^2}}$$

$$\frac{A}{A^*} = \frac{1}{\text{Ma}}\left[\left(\frac{2}{k+1}\right)\left(1+\frac{k-1}{2}\text{Ma}^2\right)\right]^{0.5(k+1)/(k-1)}$$

$$\frac{P}{P_0} = \left(1+\frac{k-1}{2}\text{Ma}^2\right)^{-k/(k-1)}$$

$$\frac{\rho}{\rho_0} = \left(1+\frac{k-1}{2}\text{Ma}^2\right)^{-1/(k-1)}$$

$$\frac{T}{T_0} = \left(1+\frac{k-1}{2}\text{Ma}^2\right)^{-1}$$

TABLE A–13

One-dimensional isentropic compressible flow functions for an ideal gas with $k = 1.4$

Ma	Ma*	A/A*	P/P$_0$	ρ/ρ_0	T/T$_0$
0	0	∞	1.0000	1.0000	1.0000
0.1	0.1094	5.8218	0.9930	0.9950	0.9980
0.2	0.2182	2.9635	0.9725	0.9803	0.9921
0.3	0.3257	2.0351	0.9395	0.9564	0.9823
0.4	0.4313	1.5901	0.8956	0.9243	0.9690
0.5	0.5345	1.3398	0.8430	0.8852	0.9524
0.6	0.6348	1.1882	0.7840	0.8405	0.9328
0.7	0.7318	1.0944	0.7209	0.7916	0.9107
0.8	0.8251	1.0382	0.6560	0.7400	0.8865
0.9	0.9146	1.0089	0.5913	0.6870	0.8606
1.0	1.0000	1.0000	0.5283	0.6339	0.8333
1.2	1.1583	1.0304	0.4124	0.5311	0.7764
1.4	1.2999	1.1149	0.3142	0.4374	0.7184
1.6	1.4254	1.2502	0.2353	0.3557	0.6614
1.8	1.5360	1.4390	0.1740	0.2868	0.6068
2.0	1.6330	1.6875	0.1278	0.2300	0.5556
2.2	1.7179	2.0050	0.0935	0.1841	0.5081
2.4	1.7922	2.4031	0.0684	0.1472	0.4647
2.6	1.8571	2.8960	0.0501	0.1179	0.4252
2.8	1.9140	3.5001	0.0368	0.0946	0.3894
3.0	1.9640	4.2346	0.0272	0.0760	0.3571
5.0	2.2361	25.000	0.0019	0.0113	0.1667
∞	2.2495	∞	0	0	0

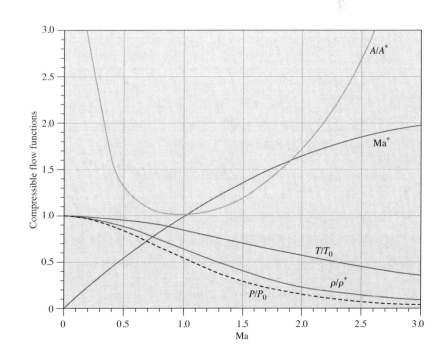

$$T_{01} = T_{02}$$

$$Ma_2 = \sqrt{\frac{(k-1)Ma_1^2 + 2}{2kMa_1^2 - k + 1}}$$

$$\frac{P_2}{P_1} = \frac{1 + kMa_1^2}{1 + kMa_2^2} = \frac{2kMa_1^2 - k + 1}{k + 1}$$

$$\frac{\rho_2}{\rho_1} = \frac{P_2/P_1}{T_2/T_1} = \frac{(k+1)Ma_1^2}{2 + (k-1)Ma_1^2} = \frac{V_1}{V_2}$$

$$\frac{T_2}{T_1} = \frac{2 + Ma_1^2(k-1)}{2 + Ma_2^2(k-1)}$$

$$\frac{P_{02}}{P_{01}} = \frac{Ma_1}{Ma_2}\left[\frac{1 + Ma_2^2(k-1)/2}{1 + Ma_1^2(k-1)/2}\right]^{(k+1)/[2(k-1)]}$$

$$\frac{P_{02}}{P_1} = \frac{(1 + kMa_1^2)[1 + Ma_2^2(k-1)/2]^{k/(k-1)}}{1 + kMa_2^2}$$

TABLE A–14

One-dimensional normal shock functions for an ideal gas with $k = 1.4$

Ma_1	Ma_2	P_2/P_1	ρ_2/ρ_1	T_2/T_1	P_{02}/P_{01}	P_{02}/P_1
1.0	1.0000	1.0000	1.0000	1.0000	1.0000	1.8929
1.1	0.9118	1.2450	1.1691	1.0649	0.9989	2.1328
1.2	0.8422	1.5133	1.3416	1.1280	0.9928	2.4075
1.3	0.7860	1.8050	1.5157	1.1909	0.9794	2.7136
1.4	0.7397	2.1200	1.6897	1.2547	0.9582	3.0492
1.5	0.7011	2.4583	1.8621	1.3202	0.9298	3.4133
1.6	0.6684	2.8200	2.0317	1.3880	0.8952	3.8050
1.7	0.6405	3.2050	2.1977	1.4583	0.8557	4.2238
1.8	0.6165	3.6133	2.3592	1.5316	0.8127	4.6695
1.9	0.5956	4.0450	2.5157	1.6079	0.7674	5.1418
2.0	0.5774	4.5000	2.6667	1.6875	0.7209	5.6404
2.1	0.5613	4.9783	2.8119	1.7705	0.6742	6.1654
2.2	0.5471	5.4800	2.9512	1.8569	0.6281	6.7165
2.3	0.5344	6.0050	3.0845	1.9468	0.5833	7.2937
2.4	0.5231	6.5533	3.2119	2.0403	0.5401	7.8969
2.5	0.5130	7.1250	3.3333	2.1375	0.4990	8.5261
2.6	0.5039	7.7200	3.4490	2.2383	0.4601	9.1813
2.7	0.4956	8.3383	3.5590	2.3429	0.4236	9.8624
2.8	0.4882	8.9800	3.6636	2.4512	0.3895	10.5694
2.9	0.4814	9.6450	3.7629	2.5632	0.3577	11.3022
3.0	0.4752	10.3333	3.8571	2.6790	0.3283	12.0610
4.0	0.4350	18.5000	4.5714	4.0469	0.1388	21.0681
5.0	0.4152	29.000	5.0000	5.8000	0.0617	32.6335
∞	0.3780	∞	6.0000	∞	0	∞

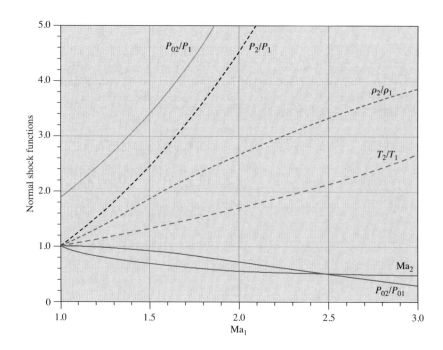

$$\frac{T_0}{T_0^*} = \frac{(k + 1)\text{Ma}^2[2 + (k - 1)\text{Ma}^2]}{(1 + k\text{Ma}^2)^2}$$

$$\frac{P_0}{P_0^*} = \frac{k + 1}{1 + k\text{Ma}^2}\left(\frac{2 + (k - 1)\text{Ma}^2}{k + 1}\right)^{k/(k-1)}$$

$$\frac{T}{T^*} = \left(\frac{\text{Ma}(1 + k)}{1 + k\text{Ma}^2}\right)^2$$

$$\frac{P}{P^*} = \frac{1 + k}{1 + k\text{Ma}^2}$$

$$\frac{V}{V^*} = \frac{\rho^*}{\rho} = \frac{(1 + k)\text{Ma}^2}{1 + k\text{Ma}^2}$$

TABLE A–15

Rayleigh flow functions for an ideal gas with $k = 1.4$

Ma	T_0/T_0^*	P_0/P_0^*	T/T^*	P/P^*	V/V^*
0.0	0.0000	1.2679	0.0000	2.4000	0.0000
0.1	0.0468	1.2591	0.0560	2.3669	0.0237
0.2	0.1736	1.2346	0.2066	2.2727	0.0909
0.3	0.3469	1.1985	0.4089	2.1314	0.1918
0.4	0.5290	1.1566	0.6151	1.9608	0.3137
0.5	0.6914	1.1141	0.7901	1.7778	0.4444
0.6	0.8189	1.0753	0.9167	1.5957	0.5745
0.7	0.9085	1.0431	0.9929	1.4235	0.6975
0.8	0.9639	1.0193	1.0255	1.2658	0.8101
0.9	0.9921	1.0049	1.0245	1.1246	0.9110
1.0	1.0000	1.0000	1.0000	1.0000	1.0000
1.2	0.9787	1.0194	0.9118	0.7958	1.1459
1.4	0.9343	1.0777	0.8054	0.6410	1.2564
1.6	0.8842	1.1756	0.7017	0.5236	1.3403
1.8	0.8363	1.3159	0.6089	0.4335	1.4046
2.0	0.7934	1.5031	0.5289	0.3636	1.4545
2.2	0.7561	1.7434	0.4611	0.3086	1.4938
2.4	0.7242	2.0451	0.4038	0.2648	1.5252
2.6	0.6970	2.4177	0.3556	0.2294	1.5505
2.8	0.6738	2.8731	0.3149	0.2004	1.5711
3.0	0.6540	3.4245	0.2803	0.1765	1.5882

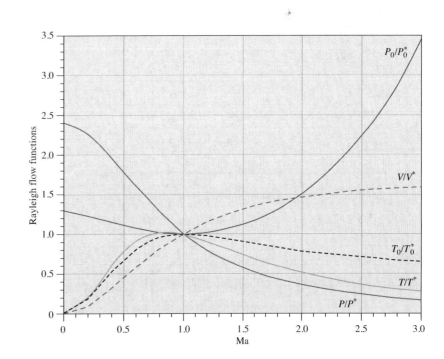

$$T_0 = T_0^*$$

$$\frac{P_0}{P_0^*} = \frac{\rho_0}{\rho_0^*} = \frac{1}{\text{Ma}}\left(\frac{2 + (k-1)\text{Ma}^2}{k+1}\right)^{(k+1)/2(k-1)}$$

$$\frac{T}{T^*} = \frac{k+1}{2 + (k-1)\text{Ma}^2}$$

$$\frac{P}{P^*} = \frac{1}{\text{Ma}}\left(\frac{k+1}{2 + (k-1)\text{Ma}^2}\right)^{1/2}$$

$$\frac{V}{V^*} = \frac{\rho^*}{\rho} = \text{Ma}\left(\frac{k+1}{2 + (k-1)\text{Ma}^2}\right)^{1/2}$$

$$\frac{fL^*}{D} = \frac{1 - \text{Ma}^2}{k\text{Ma}^2} + \frac{k+1}{2k}\ln\frac{(k+1)\text{Ma}^2}{2 + (k-1)\text{Ma}^2}$$

TABLE A–16

Fanno flow functions for an ideal gas with $k = 1.4$

Ma	P_0/P_0^*	T/T^*	P/P^*	V/V^*	fL^*/D
0.0	∞	1.2000	∞	0.0000	∞
0.1	5.8218	1.1976	10.9435	0.1094	66.9216
0.2	2.9635	1.1905	5.4554	0.2182	14.5333
0.3	2.0351	1.1788	3.6191	0.3257	5.2993
0.4	1.5901	1.1628	2.6958	0.4313	2.3085
0.5	1.3398	1.1429	2.1381	0.5345	1.0691
0.6	1.1882	1.1194	1.7634	0.6348	0.4908
0.7	1.0944	1.0929	1.4935	0.7318	0.2081
0.8	1.0382	1.0638	1.2893	0.8251	0.0723
0.9	1.0089	1.0327	1.1291	0.9146	0.0145
1.0	1.0000	1.0000	1.0000	1.0000	0.0000
1.2	1.0304	0.9317	0.8044	1.1583	0.0336
1.4	1.1149	0.8621	0.6632	1.2999	0.0997
1.6	1.2502	0.7937	0.5568	1.4254	0.1724
1.8	1.4390	0.7282	0.4741	1.5360	0.2419
2.0	1.6875	0.6667	0.4082	1.6330	0.3050
2.2	2.0050	0.6098	0.3549	1.7179	0.3609
2.4	2.4031	0.5576	0.3111	1.7922	0.4099
2.6	2.8960	0.5102	0.2747	1.8571	0.4526
2.8	3.5001	0.4673	0.2441	1.9140	0.4898
3.0	4.2346	0.4286	0.2182	1.9640	0.5222

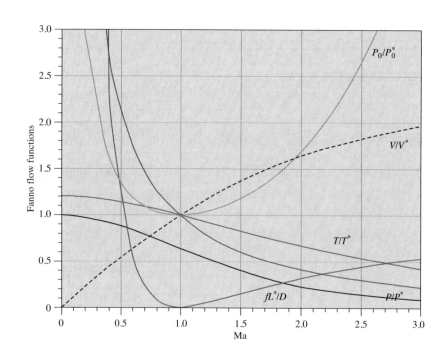

PROPERTY TABLES AND CHARTS (ENGLISH UNITS)*

TABLE A–1E   Molar Mass, Gas Constant, and
Ideal-Gas Specific Heats of Some Substances   958

TABLE A–2E   Boiling and Freezing Point Properties   959

TABLE A–3E   Properties of Saturated Water   960

TABLE A–4E   Properties of Saturated Refrigerant-134a   961

TABLE A–5E   Properties of Saturated Ammonia   962

TABLE A–6E   Properties of Saturated Propane   963

TABLE A–7E   Properties of Liquids   964

TABLE A–8E   Properties of Liquid Metals   965

TABLE A–9E   Properties of Air at 1 atm Pressure   966

TABLE A–10E Properties of Gases at 1 atm Pressure   967

TABLE A–11E Properties of the Atmosphere at High Altitude   969

*Most properties in the tables are obtained from the property database of EES, and the original sources are listed under the tables. Properties are often listed to more significant digits than the claimed accuracy for the purpose of minimizing accumulated round-off error in hand calculations and ensuring a close match with the results obtained with EES.

TABLE A–1E

Molar mass, gas constant, and ideal-gas specific heats of some substances

Substance	Molar Mass, M, lbm/lbmol	Gas Constant R^*		Specific Heat Data at 77°F		
		Btu/ lbm·R	psia·ft³/ lbm·R	c_p, Btu/lbm·R	c_v, Btu/lbm·R	$k = c_p/c_v$
Air	28.97	0.06855	0.3704	0.2400	0.1715	1.400
Ammonia, NH_3	17.03	0.1166	0.6301	0.4999	0.3834	1.304
Argon, Ar	39.95	0.04970	0.2686	0.1243	0.07457	1.667
Bromine, Br_2	159.81	0.01242	0.06714	0.0538	0.04137	1.300
Isobutane, C_4H_{10}	58.12	0.03415	0.1846	0.3972	0.3631	1.094
n-Butane, C_4H_{10}	58.12	0.03415	0.1846	0.4046	0.3705	1.092
Carbon dioxide, CO_2	44.01	0.04512	0.2438	0.2016	0.1564	1.288
Carbon monoxide, CO	28.01	0.07089	0.3831	0.2482	0.1772	1.400
Chlorine, Cl_2	70.905	0.02802	0.1514	0.1142	0.08618	1.325
Chlorodifluoromethane (R-22), $CHClF_2$	86.47	0.02297	0.1241	0.1552	0.1322	1.174
Ethane, C_2H_6	30.070	0.06604	0.3569	0.4166	0.3506	1.188
Ethylene, C_2H_4	28.054	0.07079	0.3826	0.3647	0.2940	1.241
Fluorine, F_2	38.00	0.05224	0.2823	0.1967	0.1445	1.362
Helium, He	4.003	0.4961	2.681	1.2403	0.7442	1.667
n-Heptane, C_7H_{16}	100.20	0.01982	0.1071	0.3939	0.3740	1.053
n-Hexane, C_6H_{14}	86.18	0.02304	0.1245	0.3951	0.3721	1.062
Hydrogen, H_2	2.016	0.9850	5.323	3.416	2.431	1.405
Krypton, Kr	83.80	0.02370	0.1281	0.05923	0.03554	1.667
Methane, CH_4	16.04	0.1238	0.6688	0.5317	0.4080	1.303
Neon, Ne	20.183	0.09838	0.5316	0.2460	0.1476	1.667
Nitrogen, N_2	28.01	0.07089	0.3831	0.2484	0.1774	1.400
Nitric oxide, NO	30.006	0.06618	0.3577	0.2387	0.1725	1.384
Nitrogen dioxide, NO_2	46.006	0.04512	0.2438	0.1925	0.1474	1.306
Oxygen, O_2	32.00	0.06205	0.3353	0.2193	0.1572	1.395
n-Pentane, C_5H_{12}	72.15	0.02752	0.1487	0.3974	0.3700	1.074
Propane, C_3H_8	44.097	0.04502	0.2433	0.3986	0.3535	1.127
Propylene, C_3H_6	42.08	0.04720	0.2550	0.3657	0.3184	1.148
Steam, H_2O	18.015	0.1102	0.5957	0.4455	0.3351	1.329
Sulfur dioxide, SO_2	64.06	0.03100	0.1675	0.1488	0.1178	1.263
Tetrachloromethane, CCl_4	153.82	0.01291	0.06976	0.1293	0.1164	1.111
Tetrafluoroethane (R-134a), $C_2H_2F_4$	102.03	0.01946	0.1052	0.1991	0.1796	1.108
Trifluoroethane (R-143a), $C_2H_3F_3$	84.04	0.02363	0.1277	0.2219	0.1983	1.119
Xenon, Xe	131.30	0.01512	0.08173	0.03781	0.02269	1.667

*The gas constant is calculated from $R = R_u/M$, where $R_u = 1.9859$ Btu/lbmol·R = 10.732 psia·ft³/lbmol·R is the universal gas constant and M is the molar mass.

Source: Specific heat values are mostly obtained from the property routines prepared by The National Institute of Standards and Technology (NIST), Gaithersburg, MD.

TABLE A–2E

Boiling and freezing point properties

Substance	Boiling Data at 1 atm		Freezing Data		Liquid Properties		
	Normal Boiling Point, °F	Latent Heat of Vaporization h_{fg}, Btu/lbm	Freezing Point, °F	Latent Heat of Fusion h_{if}, Btu/lbm	Temperature, °F	Density ρ, lbm/ft³	Specific Heat c_p, Btu/lbm·R
Ammonia	−27.9	24.54	−107.9	138.6	−27.9	42.6	1.06
					0	41.3	1.083
					40	39.5	1.103
					80	37.5	1.135
Argon	−302.6	69.5	−308.7	12.0	−302.6	87.0	0.272
Benzene	176.4	169.4	41.9	54.2	68	54.9	0.411
Brine (20% sodium chloride by mass)	219.0	—	0.7	—	68	71.8	0.743
n-Butane	31.1	165.6	−217.3	34.5	31.1	37.5	0.552
Carbon dioxide	−109.2*	99.6 (at 32°F)	−69.8	—	32	57.8	0.583
Ethanol	172.8	360.5	−173.6	46.9	77	48.9	0.588
Ethyl alcohol	173.5	368	−248.8	46.4	68	49.3	0.678
Ethylene glycol	388.6	344.0	12.6	77.9	68	69.2	0.678
Glycerine	355.8	419	66.0	86.3	68	78.7	0.554
Helium	−452.1	9.80	—	—	−452.1	9.13	5.45
Hydrogen	−423.0	191.7	−434.5	25.6	−423.0	4.41	2.39
Isobutane	10.9	157.8	−255.5	45.5	10.9	37.1	0.545
Kerosene	399–559	108	−12.8	—	68	51.2	0.478
Mercury	674.1	126.7	−38.0	4.90	77	847	0.033
Methane	−258.7	219.6	296.0	25.1	−258.7	26.4	0.834
					−160	20.0	1.074
Methanol	148.1	473	−143.9	42.7	77	49.1	0.609
Nitrogen	−320.4	85.4	−346.0	10.9	−320.4	50.5	0.492
					−260	38.2	0.643
Octane	256.6	131.7	−71.5	77.9	68	43.9	0.502
Oil (light)	—	—			77	56.8	0.430
Oxygen	−297.3	91.5	−361.8	5.9	−297.3	71.2	0.408
Petroleum	—	99–165			68	40.0	0.478
Propane	−43.7	184.0	−305.8	34.4	−43.7	36.3	0.538
					32	33.0	0.604
					100	29.4	0.673
Refrigerant-134a	−15.0	93.2	−141.9	—	−40	88.5	0.283
					−15	86.0	0.294
					32	80.9	0.318
					90	73.6	0.348
Water	212	970.5	32	143.5	32	62.4	1.01
					90	62.1	1.00
					150	61.2	1.00
					212	59.8	1.01

*Sublimation temperature. (At pressures below the triple-point pressure of 75.1 psia, carbon dioxide exists as a solid or gas. Also, the freezing-point temperature of carbon dioxide is the triple-point temperature of −69.8°F.)

TABLE A–3E

Properties of saturated water

Temp. T, °F	Saturation Pressure P_{sat}, psia	Density ρ, lbm/ft³		Enthalpy of Vaporization h_{fg}, Btu/lbm	Specific Heat c_p, Btu/lbm·R		Thermal Conductivity k, Btu/h·ft·R		Dynamic Viscosity μ, lbm/ft·s		Prandtl Number Pr		Volume Expansion Coefficient β, 1/R	Surface Tension, lbf/ft
		Liquid	Vapor		Liquid	Vapor	Liquid	Vapor	Liquid	Vapor	Liquid	Vapor	Liquid	Liquid
32.02	0.0887	62.41	0.00030	1075	1.010	0.446	0.324	0.0099	1.204×10^{-3}	6.194×10^{-6}	13.5	1.00	-0.038×10^{-3}	0.00518
40	0.1217	62.42	0.00034	1071	1.004	0.447	0.329	0.0100	1.038×10^{-3}	6.278×10^{-6}	11.4	1.01	0.003×10^{-3}	0.00514
50	0.1780	62.41	0.00059	1065	1.000	0.448	0.335	0.0102	8.781×10^{-4}	6.361×10^{-6}	9.44	1.01	0.047×10^{-3}	0.00509
60	0.2563	62.36	0.00083	1060	0.999	0.449	0.341	0.0104	7.536×10^{-4}	6.444×10^{-6}	7.95	1.00	0.080×10^{-3}	0.00503
70	0.3632	62.30	0.00115	1054	0.999	0.450	0.347	0.0106	6.556×10^{-4}	6.556×10^{-6}	6.79	1.00	0.115×10^{-3}	0.00497
80	0.5073	62.22	0.00158	1048	0.999	0.451	0.352	0.0108	5.764×10^{-4}	6.667×10^{-6}	5.89	1.00	0.145×10^{-3}	0.00491
90	0.6988	62.12	0.00214	1043	0.999	0.453	0.358	0.0110	5.117×10^{-4}	6.778×10^{-6}	5.14	1.00	0.174×10^{-3}	0.00485
100	0.9503	62.00	0.00286	1037	0.999	0.454	0.363	0.0112	4.578×10^{-4}	6.889×10^{-6}	4.54	1.01	0.200×10^{-3}	0.00479
110	1.2763	61.86	0.00377	1031	0.999	0.456	0.367	0.0115	4.128×10^{-4}	7.000×10^{-6}	4.05	1.00	0.224×10^{-3}	0.00473
120	1.6945	61.71	0.00493	1026	0.999	0.458	0.371	0.0117	3.744×10^{-4}	7.111×10^{-6}	3.63	1.00	0.246×10^{-3}	0.00467
130	2.225	61.55	0.00636	1020	0.999	0.460	0.375	0.0120	3.417×10^{-4}	7.222×10^{-6}	3.28	1.00	0.267×10^{-3}	0.00460
140	2.892	61.38	0.00814	1014	0.999	0.463	0.378	0.0122	3.136×10^{-4}	7.333×10^{-6}	2.98	1.00	0.287×10^{-3}	0.00454
150	3.722	61.19	0.0103	1008	1.000	0.465	0.381	0.0125	2.889×10^{-4}	7.472×10^{-6}	2.73	1.00	0.306×10^{-3}	0.00447
160	4.745	60.99	0.0129	1002	1.000	0.468	0.384	0.0128	2.675×10^{-4}	7.583×10^{-6}	2.51	1.00	0.325×10^{-3}	0.00440
170	5.996	60.79	0.0161	996	1.001	0.472	0.386	0.0131	2.483×10^{-4}	7.722×10^{-6}	2.90	1.00	0.346×10^{-3}	0.00434
180	7.515	60.57	0.0199	990	1.002	0.475	0.388	0.0134	2.317×10^{-4}	7.833×10^{-6}	2.15	1.00	0.367×10^{-3}	0.00427
190	9.343	60.35	0.0244	984	1.004	0.479	0.390	0.0137	2.169×10^{-4}	7.972×10^{-6}	2.01	1.00	0.382×10^{-3}	0.00420
200	11.53	60.12	0.0297	978	1.005	0.483	0.391	0.0141	2.036×10^{-4}	8.083×10^{-6}	1.88	1.00	0.395×10^{-3}	0.00412
210	14.125	59.87	0.0359	972	1.007	0.487	0.392	0.0144	1.917×10^{-4}	8.222×10^{-6}	1.77	1.00	0.412×10^{-3}	0.00405
212	14.698	59.82	0.0373	970	1.007	0.488	0.392	0.0145	1.894×10^{-4}	8.250×10^{-6}	1.75	1.00	0.417×10^{-3}	0.00404
220	17.19	59.62	0.0432	965	1.009	0.492	0.393	0.0148	1.808×10^{-4}	8.333×10^{-6}	1.67	1.00	0.429×10^{-3}	0.00398
230	20.78	59.36	0.0516	959	1.011	0.497	0.394	0.0152	1.711×10^{-4}	8.472×10^{-6}	1.58	1.00	0.443×10^{-3}	0.00390
240	24.97	59.09	0.0612	952	1.013	0.503	0.394	0.0156	1.625×10^{-4}	8.611×10^{-6}	1.50	1.00	0.462×10^{-3}	0.00383
250	29.82	58.82	0.0723	946	1.015	0.509	0.395	0.0160	1.544×10^{-4}	8.611×10^{-6}	1.43	1.00	0.480×10^{-3}	0.00375
260	35.42	58.53	0.0850	939	1.018	0.516	0.395	0.0164	1.472×10^{-4}	8.861×10^{-6}	1.37	1.00	0.497×10^{-3}	0.00367
270	41.85	58.24	0.0993	932	1.020	0.523	0.395	0.0168	1.406×10^{-4}	9.000×10^{-6}	1.31	1.01	0.514×10^{-3}	0.00360
280	49.18	57.94	0.1156	925	1.023	0.530	0.395	0.0172	1.344×10^{-4}	9.111×10^{-6}	1.25	1.01	0.532×10^{-3}	0.00352
290	57.53	57.63	0.3390	918	1.026	0.538	0.395	0.0177	1.289×10^{-4}	9.250×10^{-6}	1.21	1.01	0.549×10^{-3}	0.00344
300	66.98	57.31	0.1545	910	1.029	0.547	0.394	0.0182	1.236×10^{-4}	9.389×10^{-6}	1.16	1.02	0.566×10^{-3}	0.00336
320	89.60	56.65	0.2033	895	1.036	0.567	0.393	0.0191	1.144×10^{-4}	9.639×10^{-6}	1.09	1.03	0.636×10^{-3}	0.00319
340	117.93	55.95	0.2637	880	1.044	0.590	0.391	0.0202	1.063×10^{-4}	9.889×10^{-6}	1.02	1.04	0.656×10^{-3}	0.00303
360	152.92	55.22	0.3377	863	1.054	0.617	0.389	0.0213	9.972×10^{-5}	1.013×10^{-5}	0.973	1.06	0.681×10^{-3}	0.00286
380	195.60	54.46	0.4275	845	1.065	0.647	0.385	0.0224	9.361×10^{-5}	1.041×10^{-5}	0.932	1.08	0.720×10^{-3}	0.00269
400	241.1	53.65	0.5359	827	1.078	0.683	0.382	0.0237	8.833×10^{-5}	1.066×10^{-5}	0.893	1.11	0.771×10^{-3}	0.00251
450	422.1	51.46	0.9082	775	1.121	0.799	0.370	0.0271	7.722×10^{-5}	1.130×10^{-5}	0.842	1.20	0.912×10^{-3}	0.00207
500	680.0	48.95	1.479	715	1.188	0.972	0.352	0.0312	6.833×10^{-5}	1.200×10^{-5}	0.830	1.35	1.111×10^{-3}	0.00162
550	1046.7	45.96	4.268	641	1.298	1.247	0.329	0.0368	6.083×10^{-5}	1.280×10^{-5}	0.864	1.56	1.445×10^{-3}	0.00118
600	1541	42.32	3.736	550	1.509	1.759	0.299	0.0461	5.389×10^{-5}	1.380×10^{-5}	0.979	1.90	1.885×10^{-3}	0.00074
650	2210	37.31	6.152	422	2.086	3.103	0.267	0.0677	4.639×10^{-5}	1.542×10^{-5}	1.30	2.54		0.00034
700	3090	27.28	13.44	168	13.80	25.90	0.254	0.1964	3.417×10^{-5}	2.044×10^{-5}	6.68	9.71		0.00002
705.44	3204	19.79	19.79	0	∞	∞	∞	∞	2.897×10^{-5}	2.897×10^{-5}				0

Note 1: Kinematic viscosity ν and thermal diffusivity α can be calculated from their definitions, $\nu = \mu/\rho$ and $\alpha = k/\rho c_p = \nu/\mathrm{Pr}$. The temperatures 32.02°F, 212°F, and 705.44°F are the triple-, boiling-, and critical-point temperatures of water, respectively. All properties listed above (except the vapor density) can be used at any pressure with negligible error except at temperatures near the critical-point value.

Note 2: The unit Btu/lbm·°F for specific heat is equivalent to Btu/lbm·R, and the unit Btu/h·ft·°F for thermal conductivity is equivalent to Btu/h·ft·R.

Source: Viscosity and thermal conductivity data are from J. V. Sengers and J. T. R. Watson, *Journal of Physical and Chemical Reference Data* 15 (1986), pp. 1291–1322. Other data are obtained from various sources or calculated.

TABLE A–4E

Properties of saturated refrigerant-134a

Temp. T, °F	Saturation Pressure P, psia	Density ρ, lbm/ft³ Liquid	Density ρ, lbm/ft³ Vapor	Enthalpy of Vaporization h_{fg}, Btu/lbm	Specific Heat c_p, Btu/lbm·R Liquid	Specific Heat c_p, Btu/lbm·R Vapor	Thermal Conductivity k, Btu/h·ft·R Liquid	Thermal Conductivity k, Btu/h·ft·R Vapor	Dynamic Viscosity μ, lbm/ft·s Liquid	Dynamic Viscosity μ, lbm/ft·s Vapor	Prandtl Number Pr Liquid	Prandtl Number Pr Vapor	Volume Expansion Coefficient β, 1/R Liquid	Surface Tension, lbf/ft Liquid
−40	7.4	88.51	0.1731	97.1	0.2996	0.1788	0.0636	0.00466	3.278×10^{-4}	1.714×10^{-6}	5.558	0.237	0.00114	0.001206
−30	9.9	87.5	0.2258	95.6	0.3021	0.1829	0.0626	0.00497	3.004×10^{-4}	2.053×10^{-6}	5.226	0.272	0.00117	0.001146
−20	12.9	86.48	0.2905	94.1	0.3046	0.1872	0.0613	0.00529	2.762×10^{-4}	2.433×10^{-6}	4.937	0.310	0.00120	0.001087
−10	16.6	85.44	0.3691	92.5	0.3074	0.1918	0.0602	0.00559	2.546×10^{-4}	2.856×10^{-6}	4.684	0.352	0.00124	0.001029
0	21.2	84.38	0.4635	90.9	0.3103	0.1966	0.0589	0.00589	2.354×10^{-4}	3.314×10^{-6}	4.463	0.398	0.00128	0.000972
10	26.6	83.31	0.5761	89.3	0.3134	0.2017	0.0576	0.00619	2.181×10^{-4}	3.811×10^{-6}	4.269	0.447	0.00132	0.000915
20	33.1	82.2	0.7094	87.5	0.3167	0.2070	0.0563	0.00648	2.024×10^{-4}	4.342×10^{-6}	4.098	0.500	0.00137	0.000859
30	40.8	81.08	0.866	85.8	0.3203	0.2127	0.0550	0.00676	1.883×10^{-4}	4.906×10^{-6}	3.947	0.555	0.00142	0.000803
40	49.8	79.92	1.049	83.9	0.3240	0.2188	0.0536	0.00704	1.752×10^{-4}	5.494×10^{-6}	3.814	0.614	0.00149	0.000749
50	60.2	78.73	1.262	82.0	0.3281	0.2253	0.0522	0.00732	1.633×10^{-4}	6.103×10^{-6}	3.697	0.677	0.00156	0.000695
60	72.2	77.51	1.509	80.0	0.3325	0.2323	0.0507	0.00758	1.522×10^{-4}	6.725×10^{-6}	3.594	0.742	0.00163	0.000642
70	85.9	76.25	1.794	78.0	0.3372	0.2398	0.0492	0.00785	1.420×10^{-4}	7.356×10^{-6}	3.504	0.810	0.00173	0.000590
80	101.4	74.94	2.122	75.8	0.3424	0.2481	0.0476	0.00810	1.324×10^{-4}	7.986×10^{-6}	3.425	0.880	0.00183	0.000538
90	119.1	73.59	2.5	73.5	0.3481	0.2572	0.0460	0.00835	1.234×10^{-4}	8.611×10^{-6}	3.357	0.955	0.00195	0.000488
100	138.9	72.17	2.935	71.1	0.3548	0.2674	0.0444	0.00860	1.149×10^{-4}	9.222×10^{-6}	3.303	1.032	0.00210	0.000439
110	161.2	70.69	3.435	68.5	0.3627	0.2790	0.0427	0.00884	1.068×10^{-4}	9.814×10^{-6}	3.262	1.115	0.00227	0.000391
120	186.0	69.13	4.012	65.8	0.3719	0.2925	0.0410	0.00908	9.911×10^{-5}	1.038×10^{-5}	3.235	1.204	0.00248	0.000344
130	213.5	67.48	4.679	62.9	0.3829	0.3083	0.0392	0.00931	9.175×10^{-5}	1.092×10^{-5}	3.223	1.303	0.00275	0.000299
140	244.1	65.72	5.455	59.8	0.3963	0.3276	0.0374	0.00954	8.464×10^{-5}	1.144×10^{-5}	3.229	1.416	0.00308	0.000255
150	277.8	63.83	6.367	56.4	0.4131	0.3520	0.0355	0.00976	7.778×10^{-5}	1.195×10^{-5}	3.259	1.551	0.00351	0.000212
160	314.9	61.76	7.45	52.7	0.4352	0.3839	0.0335	0.00998	7.108×10^{-5}	1.245×10^{-5}	3.324	1.725	0.00411	0.000171
170	355.8	59.47	8.762	48.5	0.4659	0.4286	0.0314	0.01020	6.450×10^{-5}	1.298×10^{-5}	3.443	1.963	0.00498	0.000132
180	400.7	56.85	10.4	43.7	0.5123	0.4960	0.0292	0.01041	5.792×10^{-5}	1.356×10^{-5}	3.661	2.327	0.00637	0.000095
190	449.9	53.75	12.53	38.0	0.5929	0.6112	0.0267	0.01063	5.119×10^{-5}	1.431×10^{-5}	4.090	2.964	0.00891	0.000061
200	504.0	49.75	15.57	30.7	0.7717	0.8544	0.0239	0.01085	4.397×10^{-5}	1.544×10^{-5}	5.119	4.376	0.01490	0.000031
210	563.8	43.19	21.18	18.9	1.4786	1.6683	0.0199	0.01110	3.483×10^{-5}	1.787×10^{-5}	9.311	9.669	0.04021	0.000006

Note 1: Kinematic viscosity ν and thermal diffusivity α can be calculated from their definitions, $\nu = \mu/\rho$ and $\alpha = k/\rho c_p = \nu/\text{Pr}$. The properties listed here (except the vapor density) can be used at any pressures with negligible error except at temperatures near the critical-point value.

Note 2: The unit Btu/lbm·°F for specific heat is equivalent to Btu/lbm·R, and the unit Btu/h·ft·°F for thermal conductivity is equivalent to Btu/h·ft·R.

Source: Data generated from the EES software developed by S. A. Klein and F. L. Alvarado. Original sources: R. Tillner-Roth and H. D. Baehr, "An International Standard Formulation for the Thermodynamic Properties of 1,1,1,2-Tetrafluoroethane (HFC-134a) for Temperatures from 170 K to 455 K and Pressures up to 70 MPa," *J. Phys. Chem. Ref. Data,* Vol. 23, No. 5, 1994; M. J. Assael, N. K. Dalaouti, A. A. Griva, and J. H. Dymond, "Viscosity and Thermal Conductivity of Halogenated Methane and Ethane Refrigerants," *IJR,* Vol. 22, pp. 525–535, 1999; NIST REFPROP 6 program (M. O. McLinden, S. A. Klein, E. W. Lemmon, and A. P. Peskin, Physicial and Chemical Properties Division, National Institute of Standards and Technology, Boulder, CO 80303, 1995).

TABLE A–5E

Properties of saturated ammonia

Temp. T, °F	Saturation Pressure P, psia	Density ρ, lbm/ft³ Liquid	Density Vapor	Enthalpy of Vaporization h_{fg}, Btu/lbm	Specific Heat c_p, Btu/lbm·R Liquid	Specific Heat Vapor	Thermal Conductivity k, Btu/h·ft·R Liquid	Thermal Conductivity Vapor	Dynamic Viscosity μ, lbm/ft·s Liquid	Dynamic Viscosity Vapor	Prandtl Number Pr Liquid	Prandtl Number Pr Vapor	Volume Expansion Coefficient β, 1/R Liquid	Surface Tension, lbf/ft Liquid
−40	10.4	43.08	0.0402	597.0	1.0542	0.5354	—	0.01026	1.966×10^{-4}	5.342×10^{-6}	—	1.003	0.00098	0.002443
−30	13.9	42.66	0.0527	590.2	1.0610	0.5457	—	0.01057	1.853×10^{-4}	5.472×10^{-6}	—	1.017	0.00101	0.002357
−20	18.3	42.33	0.0681	583.2	1.0677	0.5571	0.3501	0.01089	1.746×10^{-4}	5.600×10^{-6}	1.917	1.031	0.00103	0.002272
−10	23.7	41.79	0.0869	575.9	1.0742	0.5698	0.3426	0.01121	1.645×10^{-4}	5.731×10^{-6}	1.856	1.048	0.00106	0.002187
0	30.4	41.34	0.1097	568.4	1.0807	0.5838	0.3352	0.01154	1.549×10^{-4}	5.861×10^{-6}	1.797	1.068	0.00109	0.002103
10	38.5	40.89	0.1370	560.7	1.0873	0.5992	0.3278	0.01187	1.458×10^{-4}	5.994×10^{-6}	1.740	1.089	0.00112	0.002018
20	48.2	40.43	0.1694	552.6	1.0941	0.6160	0.3203	0.01220	1.371×10^{-4}	6.125×10^{-6}	1.686	1.113	0.00116	0.001934
30	59.8	39.96	0.2075	544.4	1.1012	0.6344	0.3129	0.01254	1.290×10^{-4}	6.256×10^{-6}	1.634	1.140	0.00119	0.001850
40	73.4	39.48	0.2521	535.8	1.1087	0.6544	0.3055	0.01288	1.213×10^{-4}	6.389×10^{-6}	1.585	1.168	0.00123	0.001767
50	89.2	38.99	0.3040	526.9	1.1168	0.6762	0.2980	0.01323	1.140×10^{-4}	6.522×10^{-6}	1.539	1.200	0.00128	0.001684
60	107.7	38.50	0.3641	517.7	1.1256	0.6999	0.2906	0.01358	1.072×10^{-4}	6.656×10^{-6}	1.495	1.234	0.00132	0.001601
70	128.9	37.99	0.4332	508.1	1.1353	0.7257	0.2832	0.01394	1.008×10^{-4}	6.786×10^{-6}	1.456	1.272	0.00137	0.001518
80	153.2	37.47	0.5124	498.2	1.1461	0.7539	0.2757	0.01431	9.486×10^{-5}	6.922×10^{-6}	1.419	1.313	0.00143	0.001436
90	180.8	36.94	0.6029	487.8	1.1582	0.7846	0.2683	0.01468	8.922×10^{-5}	7.056×10^{-6}	1.387	1.358	0.00149	0.001354
100	212.0	36.40	0.7060	477.0	1.1719	0.8183	0.2609	0.01505	8.397×10^{-5}	7.189×10^{-6}	1.358	1.407	0.00156	0.001273
110	247.2	35.83	0.8233	465.8	1.1875	0.8554	0.2535	0.01543	7.903×10^{-5}	7.325×10^{-6}	1.333	1.461	0.00164	0.001192
120	286.5	35.26	0.9564	454.1	1.2054	0.8965	0.2460	0.01582	7.444×10^{-5}	7.458×10^{-6}	1.313	1.522	0.00174	0.001111
130	330.4	34.66	1.1074	441.7	1.2261	0.9425	0.2386	0.01621	7.017×10^{-5}	7.594×10^{-6}	1.298	1.589	0.00184	0.001031
140	379.2	34.04	1.2786	428.8	1.2502	0.9943	0.2312	0.01661	6.617×10^{-5}	7.731×10^{-6}	1.288	1.666	0.00196	0.000951
150	433.2	33.39	1.4730	415.2	1.2785	1.0533	0.2237	0.01702	6.244×10^{-5}	7.867×10^{-6}	1.285	1.753	0.00211	0.000872
160	492.7	32.72	1.6940	400.8	1.3120	1.1214	0.2163	0.01744	5.900×10^{-5}	8.006×10^{-6}	1.288	1.853	0.00228	0.000794
170	558.2	32.01	1.9460	385.4	1.3523	1.2012	0.2089	0.01786	5.578×10^{-5}	8.142×10^{-6}	1.300	1.971	0.00249	0.000716
180	630.1	31.26	2.2346	369.1	1.4015	1.2965	0.2014	0.01829	5.278×10^{-5}	8.281×10^{-6}	1.322	2.113	0.00274	0.000638
190	708.6	30.47	2.5670	351.6	1.4624	1.4128	0.1940	0.01874	5.000×10^{-5}	8.419×10^{-6}	1.357	2.286	0.00306	0.000562
200	794.4	29.62	2.9527	332.7	1.5397	1.5586	0.1866	0.01919	4.742×10^{-5}	8.561×10^{-6}	1.409	2.503	0.00348	0.000486
210	887.9	28.70	3.4053	312.0	1.6411	1.7473	0.1791	0.01966	4.500×10^{-5}	8.703×10^{-6}	1.484	2.784	0.00403	0.000411
220	989.5	27.69	3.9440	289.2	1.7798	2.0022	0.1717	0.02015	4.275×10^{-5}	8.844×10^{-6}	1.595	3.164	0.00480	0.000338
230	1099.8	25.57	4.5987	263.5	1.9824	2.3659	0.1643	0.02065	4.064×10^{-5}	8.989×10^{-6}	1.765	3.707	0.00594	0.000265
240	1219.4	25.28	5.4197	234.0	2.3100	2.9264	0.1568	0.02119	3.864×10^{-5}	9.136×10^{-6}	2.049	4.542	0.00784	0.000194

Note 1: Kinematic viscosity ν and thermal diffusivity α can be calculated from their definitions, $\nu = \mu/\rho$ and $\alpha = k/\rho c_p = \nu/\text{Pr}$. The properties listed here (except the vapor density) can be used at any pressures with negligible error except at temperatures near the critical-point value.

Note 2: The unit Btu/lbm·°F for specific heat is equivalent to Btu/lbm·R, and the unit Btu/h·ft·°F for thermal conductivity is equivalent to Btu/h·ft·R.

Source: Data generated from the EES software developed by S. A. Klein and F. L. Alvarado. Original sources: Tillner-Roth, Harms-Watzenterg, and Baehr, "Eine neue Fundamentalgleichung fur Ammoniak," *DKV-Tagungsbericht* 20: 167–181, 1993; Liley and Desai, "Thermophysical Properties of Refrigerants," *ASHRAE*, 1993, ISBN 1-1883413-10-9.

TABLE A–6E

Properties of saturated propane

Temp. T, °F	Saturation Pressure P, psia	Density ρ, lbm/ft³ Liquid	Density ρ, lbm/ft³ Vapor	Enthalpy of Vaporization h_{fg}, Btu/lbm	Specific Heat c_p, Btu/lbm·R Liquid	Specific Heat c_p, Btu/lbm·R Vapor	Thermal Conductivity k, Btu/h·ft·R Liquid	Thermal Conductivity k, Btu/h·ft·R Vapor	Dynamic Viscosity μ, lbm/ft·s Liquid	Dynamic Viscosity μ, lbm/ft·s Vapor	Prandtl Number Pr Liquid	Prandtl Number Pr Vapor	Volume Expansion Coefficient β, 1/R Liquid	Surface Tension, lbf/ft Liquid
−200	0.0201	42.06	0.0003	217.7	0.4750	0.2595	0.1073	0.00313	5.012×10^{-4}	2.789×10^{-6}	7.991	0.833	0.00083	0.001890
−180	0.0752	41.36	0.0011	213.4	0.4793	0.2680	0.1033	0.00347	3.941×10^{-4}	2.975×10^{-6}	6.582	0.826	0.00086	0.001780
−160	0.2307	40.65	0.0032	209.1	0.4845	0.2769	0.0992	0.00384	3.199×10^{-4}	3.164×10^{-6}	5.626	0.821	0.00088	0.001671
−140	0.6037	39.93	0.0078	204.8	0.4907	0.2866	0.0949	0.00423	2.660×10^{-4}	3.358×10^{-6}	4.951	0.818	0.00091	0.001563
−120	1.389	39.20	0.0170	200.5	0.4982	0.2971	0.0906	0.00465	2.252×10^{-4}	3.556×10^{-6}	4.457	0.817	0.00094	0.001455
−100	2.878	38.46	0.0334	196.1	0.5069	0.3087	0.0863	0.00511	1.934×10^{-4}	3.756×10^{-6}	4.087	0.817	0.00097	0.001349
−90	4.006	38.08	0.0453	193.9	0.5117	0.3150	0.0842	0.00534	1.799×10^{-4}	3.858×10^{-6}	3.936	0.819	0.00099	0.001297
−80	5.467	37.70	0.0605	191.6	0.5169	0.3215	0.0821	0.00559	1.678×10^{-4}	3.961×10^{-6}	3.803	0.820	0.00101	0.001244
−70	7.327	37.32	0.0793	189.3	0.5224	0.3284	0.0800	0.00585	1.569×10^{-4}	4.067×10^{-6}	3.686	0.822	0.00104	0.001192
−60	9.657	36.93	0.1024	186.9	0.5283	0.3357	0.0780	0.00611	1.469×10^{-4}	4.172×10^{-6}	3.582	0.825	0.00106	0.001140
−50	12.54	36.54	0.1305	184.4	0.5345	0.3433	0.0760	0.00639	1.378×10^{-4}	4.278×10^{-6}	3.490	0.828	0.00109	0.001089
−40	16.05	36.13	0.1641	181.9	0.5392	0.3513	0.0740	0.00668	1.294×10^{-4}	4.386×10^{-6}	3.395	0.831	0.00112	0.001038
−30	20.29	35.73	0.2041	179.3	0.5460	0.3596	0.0721	0.00697	1.217×10^{-4}	4.497×10^{-6}	3.320	0.835	0.00115	0.000987
−20	25.34	35.31	0.2512	176.6	0.5531	0.3684	0.0702	0.00728	1.146×10^{-4}	4.611×10^{-6}	3.253	0.840	0.00119	0.000937
−10	31.3	34.89	0.3063	173.8	0.5607	0.3776	0.0683	0.00761	1.079×10^{-4}	4.725×10^{-6}	3.192	0.845	0.00123	0.000887
0	38.28	34.46	0.3703	170.9	0.5689	0.3874	0.0665	0.00794	1.018×10^{-4}	4.842×10^{-6}	3.137	0.850	0.00127	0.000838
10	46.38	34.02	0.4441	167.9	0.5775	0.3976	0.0647	0.00829	9.606×10^{-5}	4.961×10^{-6}	3.088	0.857	0.00132	0.000789
20	55.7	33.56	0.5289	164.8	0.5867	0.4084	0.0629	0.00865	9.067×10^{-5}	5.086×10^{-6}	3.043	0.864	0.00138	0.000740
30	66.35	33.10	0.6259	161.6	0.5966	0.4199	0.0612	0.00903	8.561×10^{-5}	5.211×10^{-6}	3.003	0.873	0.00144	0.000692
40	78.45	32.62	0.7365	158.1	0.6072	0.4321	0.0595	0.00942	8.081×10^{-5}	5.342×10^{-6}	2.967	0.882	0.00151	0.000644
50	92.12	32.13	0.8621	154.6	0.6187	0.4452	0.0579	0.00983	7.631×10^{-5}	5.478×10^{-6}	2.935	0.893	0.00159	0.000597
60	107.5	31.63	1.0046	150.8	0.6311	0.4593	0.0563	0.01025	7.200×10^{-5}	5.617×10^{-6}	2.906	0.906	0.00168	0.000551
70	124.6	31.11	1.1659	146.8	0.6447	0.4746	0.0547	0.01070	6.794×10^{-5}	5.764×10^{-6}	2.881	0.921	0.00179	0.000505
80	143.7	30.56	1.3484	142.7	0.6596	0.4915	0.0532	0.01116	6.406×10^{-5}	5.919×10^{-6}	2.860	0.938	0.00191	0.000460
90	164.8	30.00	1.5549	138.2	0.6762	0.5103	0.0517	0.01165	6.033×10^{-5}	6.081×10^{-6}	2.843	0.959	0.00205	0.000416
100	188.1	29.41	1.7887	133.6	0.6947	0.5315	0.0501	0.01217	5.675×10^{-5}	6.256×10^{-6}	2.831	0.984	0.00222	0.000372
120	241.8	28.13	2.3562	123.2	0.7403	0.5844	0.0472	0.01328	5.000×10^{-5}	6.644×10^{-6}	2.825	1.052	0.00267	0.000288
140	306.1	26.69	3.1003	111.1	0.7841	0.6613	0.0442	0.01454	4.358×10^{-5}	7.111×10^{-6}	2.784	1.164	0.00338	0.000208
160	382.4	24.98	4.1145	96.4	0.8696	0.7911	0.0411	0.01603	3.733×10^{-5}	7.719×10^{-6}	2.845	1.371	0.00459	0.000133
180	472.9	22.79	5.6265	77.1	1.1436	1.0813	0.0376	0.01793	3.083×10^{-5}	8.617×10^{-6}	3.380	1.870	0.00791	0.000065

Note 1: Kinematic viscosity ν and thermal diffusivity α can be calculated from their definitions, $\nu = \mu/\rho$ and $\alpha = k/\rho c_p = \nu/\text{Pr}$. The properties listed here (except the vapor density) can be used at any pressures with negligible error except at temperatures near the critical-point value.

Note 2: The unit Btu/lbm·°F for specific heat is equivalent to Btu/lbm·R, and the unit Btu/h·ft·°F for thermal conductivity is equivalent to Btu/h·ft·R.

Source: Data generated from the EES software developed by S. A. Klein and F. L. Alvarado. Original sources: Reiner Tillner-Roth, "Fundamental Equations of State," Shaker, Verlag, Aachan, 1998; B. A. Younglove and J. F. Ely, "Thermophysical Properties of Fluids. II Methane, Ethane, Propane, Isobutane, and Normal Butane," *J. Phys. Chem. Ref. Data*, Vol. 16, No. 4, 1987; G. R. Somayajulu, "A Generalized Equation for Surface Tension from the Triple-Point to the Critical-Point," *International Journal of Thermophysics*, Vol. 9, No. 4, 1988.

TABLE A–7E

Properties of liquids

Temp. T, °F	Density ρ, lbm/ft³	Specific Heat c_p, Btu/lbm·R	Thermal Conductivity k, Btu/h·ft·R	Thermal Diffusivity α, ft²/s	Dynamic Viscosity μ, lbm/ft·s	Kinematic Viscosity ν, ft²/s	Prandtl Number Pr	Volume Expansion Coeff. β, 1/R
\multicolumn{9}{c}{Methane (CH_4)}
−280	27.41	0.8152	0.1205	1.497×10^{-6}	1.057×10^{-4}	3.857×10^{-6}	2.575	0.00175
−260	26.43	0.8301	0.1097	1.389×10^{-6}	8.014×10^{-5}	3.032×10^{-6}	2.183	0.00192
−240	25.39	0.8523	0.0994	1.276×10^{-6}	6.303×10^{-5}	2.482×10^{-6}	1.945	0.00215
−220	24.27	0.8838	0.0896	1.159×10^{-6}	5.075×10^{-5}	2.091×10^{-6}	1.803	0.00247
−200	23.04	0.9314	0.0801	1.036×10^{-6}	4.142×10^{-5}	1.798×10^{-6}	1.734	0.00295
−180	21.64	1.010	0.0709	9.008×10^{-7}	3.394×10^{-5}	1.568×10^{-6}	1.741	0.00374
−160	19.99	1.158	0.0616	7.397×10^{-7}	2.758×10^{-5}	1.379×10^{-6}	1.865	0.00526
−140	17.84	1.542	0.0518	5.234×10^{-7}	2.168×10^{-5}	1.215×10^{-6}	2.322	0.00943
\multicolumn{9}{c}{Methanol [$CH_3(OH)$]}
70	49.15	0.6024	0.1148	1.076×10^{-6}	3.872×10^{-4}	7.879×10^{-6}	7.317	0.000656
90	48.50	0.6189	0.1143	1.057×10^{-6}	3.317×10^{-4}	6.840×10^{-6}	6.468	0.000671
110	47.85	0.6373	0.1138	1.036×10^{-6}	2.872×10^{-4}	6.005×10^{-6}	5.793	0.000691
130	47.18	0.6576	0.1133	1.014×10^{-6}	2.513×10^{-4}	5.326×10^{-6}	5.250	0.000716
150	46.50	0.6796	0.1128	9.918×10^{-7}	2.218×10^{-4}	4.769×10^{-6}	4.808	0.000749
170	45.80	0.7035	0.1124	9.687×10^{-7}	1.973×10^{-4}	4.308×10^{-6}	4.447	0.000789
\multicolumn{9}{c}{Isobutane (R600a)}
−150	42.75	0.4483	0.0799	1.157×10^{-6}	6.417×10^{-4}	1.500×10^{-5}	12.96	0.000785
−100	41.06	0.4721	0.0782	1.120×10^{-6}	3.669×10^{-4}	8.939×10^{-6}	7.977	0.000836
−50	39.31	0.4986	0.0731	1.036×10^{-6}	2.376×10^{-4}	6.043×10^{-6}	5.830	0.000908
0	37.48	0.5289	0.0664	9.299×10^{-7}	1.651×10^{-4}	4.406×10^{-6}	4.738	0.001012
50	35.52	0.5643	0.0591	8.187×10^{-7}	1.196×10^{-4}	3.368×10^{-6}	4.114	0.001169
100	33.35	0.6075	0.0521	7.139×10^{-7}	8.847×10^{-5}	2.653×10^{-6}	3.716	0.001421
150	30.84	0.6656	0.0457	6.188×10^{-7}	6.558×10^{-5}	2.127×10^{-6}	3.437	0.001883
200	27.73	0.7635	0.0400	5.249×10^{-7}	4.750×10^{-5}	1.713×10^{-6}	3.264	0.002970
\multicolumn{9}{c}{Glycerin}
32	79.65	0.5402	0.163	1.052×10^{-6}	7.047	0.08847	84101	
40	79.49	0.5458	0.1637	1.048×10^{-6}	4.803	0.06042	57655	
50	79.28	0.5541	0.1645	1.040×10^{-6}	2.850	0.03594	34561	
60	79.07	0.5632	0.1651	1.029×10^{-6}	1.547	0.01956	18995	
70	78.86	0.5715	0.1652	1.018×10^{-6}	0.9422	0.01195	11730	
80	78.66	0.5794	0.1652	1.007×10^{-6}	0.5497	0.00699	6941	
90	78.45	0.5878	0.1652	9.955×10^{-7}	0.3756	0.004787	4809	
100	78.24	0.5964	0.1653	9.841×10^{-7}	0.2277	0.00291	2957	
\multicolumn{9}{c}{Engine Oil (unused)}
32	56.12	0.4291	0.0849	9.792×10^{-7}	2.563	4.566×10^{-2}	46636	0.000389
50	55.79	0.4395	0.08338	9.448×10^{-7}	1.210	2.169×10^{-2}	22963	0.000389
75	55.3	0.4531	0.08378	9.288×10^{-7}	0.4286	7.751×10^{-3}	8345	0.000389
100	54.77	0.4669	0.08367	9.089×10^{-7}	0.1630	2.977×10^{-3}	3275	0.000389
125	54.24	0.4809	0.08207	8.740×10^{-7}	7.617×10^{-2}	1.404×10^{-3}	1607	0.000389
150	53.73	0.4946	0.08046	8.411×10^{-7}	3.833×10^{-2}	7.135×10^{-4}	848.3	0.000389
200	52.68	0.5231	0.07936	7.999×10^{-7}	1.405×10^{-2}	2.668×10^{-4}	333.6	0.000389
250	51.71	0.5523	0.07776	7.563×10^{-7}	6.744×10^{-3}	1.304×10^{-4}	172.5	0.000389
300	50.63	0.5818	0.07673	7.236×10^{-7}	3.661×10^{-3}	7.232×10^{-5}	99.94	0.000389

Source: Data generated from the EES software developed by S. A. Klein and F. L. Alvarado. Originally based on various sources.

TABLE A–8E

Properties of liquid metals

Temp. T, °F	Density ρ, lbm/ft^3	Specific Heat c_p, Btu/lbm·R	Thermal Conductivity k, Btu/h·ft·R	Thermal Diffusivity α, ft^2/s	Dynamic Viscosity μ, lbm/ft·s	Kinematic Viscosity ν, ft^2/s	Prandtl Number Pr	Volume Expansion Coeff. β, 1/R
Mercury (Hg) Melting Point: −38°F								
32	848.7	0.03353	4.727	4.614×10^{-5}	1.133×10^{-3}	1.335×10^{-6}	0.02895	1.005×10^{-4}
50	847.2	0.03344	4.805	4.712×10^{-5}	1.092×10^{-3}	1.289×10^{-6}	0.02737	1.005×10^{-4}
100	842.9	0.03319	5.015	4.980×10^{-5}	9.919×10^{-4}	1.176×10^{-6}	0.02363	1.005×10^{-4}
150	838.7	0.03298	5.221	5.244×10^{-5}	9.122×10^{-4}	1.087×10^{-6}	0.02074	1.005×10^{-4}
200	834.5	0.03279	5.422	5.504×10^{-5}	8.492×10^{-4}	1.017×10^{-6}	0.01849	1.005×10^{-4}
300	826.2	0.03252	5.815	6.013×10^{-5}	7.583×10^{-4}	9.180×10^{-7}	0.01527	1.005×10^{-4}
400	817.9	0.03236	6.184	6.491×10^{-5}	6.972×10^{-4}	8.524×10^{-7}	0.01313	1.008×10^{-4}
500	809.6	0.03230	6.518	6.924×10^{-5}	6.525×10^{-4}	8.061×10^{-7}	0.01164	1.018×10^{-4}
600	801.3	0.03235	6.839	7.329×10^{-5}	6.186×10^{-4}	7.719×10^{-7}	0.01053	1.035×10^{-4}
Bismuth (Bi) Melting Point: 520°F								
700	620.7	0.03509	9.361	1.193×10^{-4}	1.001×10^{-3}	1.614×10^{-6}	0.01352	
800	616.5	0.03569	9.245	1.167×10^{-4}	9.142×10^{-4}	1.482×10^{-6}	0.01271	
900	612.2	0.0363	9.129	1.141×10^{-4}	8.267×10^{-4}	1.350×10^{-6}	0.01183	
1000	608.0	0.0369	9.014	1.116×10^{-4}	7.392×10^{-4}	1.215×10^{-6}	0.0109	
1100	603.7	0.0375	9.014	1.105×10^{-4}	6.872×10^{-4}	1.138×10^{-6}	0.01029	
Lead (Pb) Melting Point: 621°F								
700	658	0.03797	9.302	1.034×10^{-4}	1.612×10^{-3}	2.450×10^{-6}	0.02369	
800	654	0.03750	9.157	1.037×10^{-4}	1.453×10^{-3}	2.223×10^{-6}	0.02143	
900	650	0.03702	9.013	1.040×10^{-4}	1.296×10^{-3}	1.994×10^{-6}	0.01917	
1000	645.7	0.03702	8.912	1.035×10^{-4}	1.202×10^{-3}	1.862×10^{-6}	0.01798	
1100	641.5	0.03702	8.810	1.030×10^{-4}	1.108×10^{-3}	1.727×10^{-6}	0.01676	
1200	637.2	0.03702	8.709	1.025×10^{-4}	1.013×10^{-3}	1.590×10^{-6}	0.01551	
Sodium (Na) Melting Point: 208°F								
300	57.13	0.3258	48.19	7.192×10^{-4}	4.136×10^{-4}	7.239×10^{-6}	0.01007	
400	56.28	0.3219	46.58	7.142×10^{-4}	3.572×10^{-4}	6.350×10^{-6}	0.008891	
500	55.42	0.3181	44.98	7.087×10^{-4}	3.011×10^{-4}	5.433×10^{-6}	0.007667	
600	54.56	0.3143	43.37	7.026×10^{-4}	2.448×10^{-4}	4.488×10^{-6}	0.006387	
800	52.85	0.3089	40.55	6.901×10^{-4}	1.772×10^{-4}	3.354×10^{-6}	0.004860	
1000	51.14	0.3057	38.12	6.773×10^{-4}	1.541×10^{-4}	3.014×10^{-6}	0.004449	
Potassium (K) Melting Point: 147°F								
300	50.40	0.1911	26.00	7.500×10^{-4}	2.486×10^{-4}	4.933×10^{-6}	0.006577	
400	49.58	0.1887	25.37	7.532×10^{-4}	2.231×10^{-4}	4.500×10^{-6}	0.005975	
500	48.76	0.1863	24.73	7.562×10^{-4}	1.976×10^{-4}	4.052×10^{-6}	0.005359	
600	47.94	0.1839	24.09	7.591×10^{-4}	1.721×10^{-4}	3.589×10^{-6}	0.004728	
800	46.31	0.1791	22.82	7.643×10^{-4}	1.210×10^{-4}	2.614×10^{-6}	0.003420	
1000	44.62	0.1791	21.34	7.417×10^{-4}	1.075×10^{-4}	2.409×10^{-6}	0.003248	
Sodium–Potassium (%22Na–%78K) Melting Point: 12°F								
200	52.99	0.2259	14.79	3.432×10^{-4}	3.886×10^{-4}	7.331×10^{-6}	0.02136	
300	52.16	0.2230	14.99	3.580×10^{-4}	3.467×10^{-4}	6.647×10^{-6}	0.01857	
400	51.32	0.2201	15.19	3.735×10^{-4}	3.050×10^{-4}	5.940×10^{-6}	0.0159	
600	49.65	0.2143	15.59	4.070×10^{-4}	2.213×10^{-4}	4.456×10^{-6}	0.01095	
800	47.99	0.2100	15.95	4.396×10^{-4}	1.539×10^{-4}	3.207×10^{-6}	0.007296	
1000	46.36	0.2103	16.20	4.615×10^{-4}	1.353×10^{-4}	2.919×10^{-6}	0.006324	

Source: Data generated from the EES software developed by S. A. Klein and F. L. Alvarado. Originally based on various sources.

TABLE A-9E

Properties of air at 1 atm pressure

Temp. T, °F	Density ρ, lbm/ft³	Specific Heat c_p, Btu/lbm·R	Thermal Conductivity k, Btu/h·ft·R	Thermal Diffusivity α, ft²/s	Dynamic Viscosity μ, lbm/ft·s	Kinematic Viscosity ν, ft²/s	Prandtl Number Pr
-300	0.24844	0.5072	0.00508	1.119×10^{-5}	4.039×10^{-6}	1.625×10^{-5}	1.4501
-200	0.15276	0.2247	0.00778	6.294×10^{-5}	6.772×10^{-6}	4.433×10^{-5}	0.7042
-100	0.11029	0.2360	0.01037	1.106×10^{-4}	9.042×10^{-6}	8.197×10^{-5}	0.7404
-50	0.09683	0.2389	0.01164	1.397×10^{-4}	1.006×10^{-5}	1.039×10^{-4}	0.7439
0	0.08630	0.2401	0.01288	1.726×10^{-4}	1.102×10^{-5}	1.278×10^{-4}	0.7403
10	0.08446	0.2402	0.01312	1.797×10^{-4}	1.121×10^{-5}	1.328×10^{-4}	0.7391
20	0.08270	0.2403	0.01336	1.868×10^{-4}	1.140×10^{-5}	1.379×10^{-4}	0.7378
30	0.08101	0.2403	0.01361	1.942×10^{-4}	1.158×10^{-5}	1.430×10^{-4}	0.7365
40	0.07939	0.2404	0.01385	2.016×10^{-4}	1.176×10^{-5}	1.482×10^{-4}	0.7350
50	0.07783	0.2404	0.01409	2.092×10^{-4}	1.194×10^{-5}	1.535×10^{-4}	0.7336
60	0.07633	0.2404	0.01433	2.169×10^{-4}	1.212×10^{-5}	1.588×10^{-4}	0.7321
70	0.07489	0.2404	0.01457	2.248×10^{-4}	1.230×10^{-5}	1.643×10^{-4}	0.7306
80	0.07350	0.2404	0.01481	2.328×10^{-4}	1.247×10^{-5}	1.697×10^{-4}	0.7290
90	0.07217	0.2404	0.01505	2.409×10^{-4}	1.265×10^{-5}	1.753×10^{-4}	0.7275
100	0.07088	0.2405	0.01529	2.491×10^{-4}	1.281×10^{-5}	1.809×10^{-4}	0.7260
110	0.06963	0.2405	0.01552	2.575×10^{-4}	1.299×10^{-5}	1.866×10^{-4}	0.7245
120	0.06843	0.2405	0.01576	2.660×10^{-4}	1.316×10^{-5}	1.923×10^{-4}	0.7230
130	0.06727	0.2405	0.01599	2.746×10^{-4}	1.332×10^{-5}	1.981×10^{-4}	0.7216
140	0.06615	0.2406	0.01623	2.833×10^{-4}	1.349×10^{-5}	2.040×10^{-4}	0.7202
150	0.06507	0.2406	0.01646	2.921×10^{-4}	1.365×10^{-5}	2.099×10^{-4}	0.7188
160	0.06402	0.2406	0.01669	3.010×10^{-4}	1.382×10^{-5}	2.159×10^{-4}	0.7174
170	0.06300	0.2407	0.01692	3.100×10^{-4}	1.398×10^{-5}	2.220×10^{-4}	0.7161
180	0.06201	0.2408	0.01715	3.191×10^{-4}	1.414×10^{-5}	2.281×10^{-4}	0.7148
190	0.06106	0.2408	0.01738	3.284×10^{-4}	1.430×10^{-5}	2.343×10^{-4}	0.7136
200	0.06013	0.2409	0.01761	3.377×10^{-4}	1.446×10^{-5}	2.406×10^{-4}	0.7124
250	0.05590	0.2415	0.01874	3.857×10^{-4}	1.524×10^{-5}	2.727×10^{-4}	0.7071
300	0.05222	0.2423	0.01985	4.358×10^{-4}	1.599×10^{-5}	3.063×10^{-4}	0.7028
350	0.04899	0.2433	0.02094	4.879×10^{-4}	1.672×10^{-5}	3.413×10^{-4}	0.6995
400	0.04614	0.2445	0.02200	5.419×10^{-4}	1.743×10^{-5}	3.777×10^{-4}	0.6971
450	0.04361	0.2458	0.02305	5.974×10^{-4}	1.812×10^{-5}	4.154×10^{-4}	0.6953
500	0.04134	0.2472	0.02408	6.546×10^{-4}	1.878×10^{-5}	4.544×10^{-4}	0.6942
600	0.03743	0.2503	0.02608	7.732×10^{-4}	2.007×10^{-5}	5.361×10^{-4}	0.6934
700	0.03421	0.2535	0.02800	8.970×10^{-4}	2.129×10^{-5}	6.225×10^{-4}	0.6940
800	0.03149	0.2568	0.02986	1.025×10^{-3}	2.247×10^{-5}	7.134×10^{-4}	0.6956
900	0.02917	0.2599	0.03164	1.158×10^{-3}	2.359×10^{-5}	8.087×10^{-4}	0.6978
1000	0.02718	0.2630	0.03336	1.296×10^{-3}	2.467×10^{-5}	9.080×10^{-4}	0.7004
1500	0.02024	0.2761	0.04106	2.041×10^{-3}	2.957×10^{-5}	1.460×10^{-3}	0.7158
2000	0.01613	0.2855	0.04752	2.867×10^{-3}	3.379×10^{-5}	2.095×10^{-3}	0.7308
2500	0.01340	0.2922	0.05309	3.765×10^{-3}	3.750×10^{-5}	2.798×10^{-3}	0.7432
3000	0.01147	0.2972	0.05811	4.737×10^{-3}	4.082×10^{-5}	3.560×10^{-3}	0.7516
3500	0.01002	0.3010	0.06293	5.797×10^{-3}	4.381×10^{-5}	4.373×10^{-3}	0.7543
4000	0.00889	0.3040	0.06789	6.975×10^{-3}	4.651×10^{-5}	5.229×10^{-3}	0.7497

Note: For ideal gases, the properties c_p, k, μ, and Pr are independent of pressure. The properties ρ, ν, and α at a pressure P (in atm) other than 1 atm are determined by multiplying the values of ρ at the given temperature by P and by dividing ν and α by P.

Source: Data generated from the EES software developed by S. A. Klein and F. L. Alvarado. Original sources: Keenan, Chao, Keyes, Gas Tables, Wiley, 198; and *Thermophysical Properties of Matter*, Vol. 3: *Thermal Conductivity*, Y. S. Touloukian, P. E. Liley, S. C. Saxena, Vol. 11: *Viscosity*, Y. S. Touloukian, S. C. Saxena, and P. Hestermans, IFI/Plenum, NY, 1970, ISBN 0-306067020-8.

TABLE A–10E

Properties of gases at 1 atm pressure

Temp. T, °F	Density ρ, lbm/ft³	Specific Heat c_p, Btu/lbm·R	Thermal Conductivity k, Btu/h·ft·R	Thermal Diffusivity α, ft²/s	Dynamic Viscosity μ, lbm/ft·s	Kinematic Viscosity ν, ft²/s	Prandtl Number Pr
Carbon Dioxide, CO_2							
−50	0.14712	0.1797	0.00628	6.600×10^{-5}	7.739×10^{-6}	5.261×10^{-5}	0.7970
0	0.13111	0.1885	0.00758	8.522×10^{-5}	8.661×10^{-6}	6.606×10^{-5}	0.7751
50	0.11825	0.1965	0.00888	1.061×10^{-4}	9.564×10^{-6}	8.086×10^{-5}	0.7621
100	0.10769	0.2039	0.01017	1.286×10^{-4}	1.045×10^{-5}	9.703×10^{-5}	0.7543
200	0.09136	0.2171	0.01273	1.784×10^{-4}	1.217×10^{-5}	1.332×10^{-4}	0.7469
300	0.07934	0.2284	0.01528	2.341×10^{-4}	1.382×10^{-5}	1.743×10^{-4}	0.7445
500	0.06280	0.2473	0.02027	3.626×10^{-4}	1.696×10^{-5}	2.700×10^{-4}	0.7446
1000	0.04129	0.2796	0.03213	7.733×10^{-4}	2.381×10^{-5}	5.767×10^{-4}	0.7458
1500	0.03075	0.2995	0.04281	1.290×10^{-3}	2.956×10^{-5}	9.610×10^{-4}	0.7445
2000	0.02450	0.3124	0.05193	1.885×10^{-3}	3.451×10^{-5}	1.408×10^{-3}	0.7474
Carbon Monoxide, CO							
−50	0.09363	0.2571	0.01118	1.290×10^{-4}	9.419×10^{-6}	1.005×10^{-4}	0.7798
0	0.08345	0.2523	0.01240	1.636×10^{-4}	1.036×10^{-5}	1.242×10^{-4}	0.7593
50	0.07526	0.2496	0.01359	2.009×10^{-4}	1.127×10^{-5}	1.498×10^{-4}	0.7454
100	0.06854	0.2484	0.01476	2.408×10^{-4}	1.214×10^{-5}	1.772×10^{-4}	0.7359
200	0.05815	0.2485	0.01702	3.273×10^{-4}	1.379×10^{-5}	2.372×10^{-4}	0.7247
300	0.05049	0.2505	0.01920	4.217×10^{-4}	1.531×10^{-5}	3.032×10^{-4}	0.7191
500	0.03997	0.2567	0.02331	6.311×10^{-4}	1.802×10^{-5}	4.508×10^{-4}	0.7143
1000	0.02628	0.2732	0.03243	1.254×10^{-3}	2.334×10^{-5}	8.881×10^{-4}	0.7078
1500	0.01957	0.2862	0.04049	2.008×10^{-3}	2.766×10^{-5}	1.413×10^{-3}	0.7038
2000	0.01559	0.2958	0.04822	2.903×10^{-3}	3.231×10^{-5}	2.072×10^{-3}	0.7136
Methane, CH_4							
−50	0.05363	0.5335	0.01401	1.360×10^{-4}	5.861×10^{-6}	1.092×10^{-4}	0.8033
0	0.04779	0.5277	0.01616	1.780×10^{-4}	6.506×10^{-6}	1.361×10^{-4}	0.7649
50	0.04311	0.5320	0.01839	2.228×10^{-4}	7.133×10^{-6}	1.655×10^{-4}	0.7428
100	0.03925	0.5433	0.02071	2.698×10^{-4}	7.742×10^{-6}	1.972×10^{-4}	0.7311
200	0.03330	0.5784	0.02559	3.690×10^{-4}	8.906×10^{-6}	2.674×10^{-4}	0.7245
300	0.02892	0.6226	0.03077	4.748×10^{-4}	1.000×10^{-5}	3.457×10^{-4}	0.7283
500	0.02289	0.7194	0.04195	7.075×10^{-4}	1.200×10^{-5}	5.244×10^{-4}	0.7412
1000	0.01505	0.9438	0.07346	1.436×10^{-3}	1.620×10^{-5}	1.076×10^{-3}	0.7491
1500	0.01121	1.1162	0.10766	2.390×10^{-3}	1.974×10^{-5}	1.760×10^{-3}	0.7366
2000	0.00893	1.2419	0.14151	3.544×10^{-3}	2.327×10^{-5}	2.605×10^{-3}	0.7353
Hydrogen, H_2							
−50	0.00674	3.0603	0.08246	1.110×10^{-3}	4.969×10^{-6}	7.373×10^{-4}	0.6638
0	0.00601	3.2508	0.09049	1.287×10^{-3}	5.381×10^{-6}	8.960×10^{-4}	0.6960
50	0.00542	3.3553	0.09818	1.500×10^{-3}	5.781×10^{-6}	1.067×10^{-3}	0.7112
100	0.00493	3.4118	0.10555	1.742×10^{-3}	6.167×10^{-6}	1.250×10^{-3}	0.7177
200	0.00419	3.4549	0.11946	2.295×10^{-3}	6.911×10^{-6}	1.652×10^{-3}	0.7197
300	0.00363	3.4613	0.13241	2.924×10^{-3}	7.622×10^{-6}	2.098×10^{-3}	0.7174
500	0.00288	3.4572	0.15620	4.363×10^{-3}	8.967×10^{-6}	3.117×10^{-3}	0.7146
1000	0.00189	3.5127	0.20989	8.776×10^{-3}	1.201×10^{-5}	6.354×10^{-3}	0.7241
1500	0.00141	3.6317	0.26381	1.432×10^{-2}	1.477×10^{-5}	1.048×10^{-2}	0.7323
2000	0.00112	3.7656	0.31923	2.098×10^{-2}	1.734×10^{-5}	1.544×10^{-2}	0.7362

(Continued)

TABLE A–10E

Properties of gases at 1 atm pressure (*Continued*)

Temp. T, °F	Density ρ, lbm/ft^3	Specific Heat c_p, Btu/lbm·R	Thermal Conductivity k, Btu/h·ft·R	Thermal Diffusivity α, ft^2/s	Dynamic Viscosity μ, lbm/ft·s	Kinematic Viscosity ν, ft^2/s	Prandtl Number Pr
				Nitrogen, N$_2$			
−50	0.09364	0.2320	0.01176	1.504×10^{-4}	9.500×10^{-6}	1.014×10^{-4}	0.6746
0	0.08346	0.2441	0.01300	1.773×10^{-4}	1.043×10^{-5}	1.251×10^{-4}	0.7056
50	0.07527	0.2480	0.01420	2.113×10^{-4}	1.134×10^{-5}	1.507×10^{-4}	0.7133
100	0.06854	0.2489	0.01537	2.502×10^{-4}	1.221×10^{-5}	1.783×10^{-4}	0.7126
200	0.05815	0.2487	0.01760	3.379×10^{-4}	1.388×10^{-5}	2.387×10^{-4}	0.7062
300	0.05050	0.2492	0.01970	4.349×10^{-4}	1.543×10^{-5}	3.055×10^{-4}	0.7025
500	0.03997	0.2535	0.02359	6.466×10^{-4}	1.823×10^{-5}	4.559×10^{-4}	0.7051
1000	0.02628	0.2697	0.03204	1.255×10^{-3}	2.387×10^{-5}	9.083×10^{-4}	0.7232
1500	0.01958	0.2831	0.04002	2.006×10^{-3}	2.829×10^{-5}	1.445×10^{-3}	0.7202
2000	0.01560	0.2927	0.04918	2.992×10^{-3}	3.212×10^{-5}	2.059×10^{-3}	0.6882
				Oxygen, O$_2$			
−50	0.10697	0.2331	0.01216	1.355×10^{-4}	1.104×10^{-5}	1.032×10^{-4}	0.7622
0	0.09533	0.2245	0.01346	1.747×10^{-4}	1.218×10^{-5}	1.277×10^{-4}	0.7312
50	0.08598	0.2209	0.01475	2.157×10^{-4}	1.326×10^{-5}	1.543×10^{-4}	0.7152
100	0.07830	0.2200	0.01601	2.582×10^{-4}	1.429×10^{-5}	1.826×10^{-4}	0.7072
200	0.06643	0.2221	0.01851	3.484×10^{-4}	1.625×10^{-5}	2.446×10^{-4}	0.7020
300	0.05768	0.2262	0.02096	4.463×10^{-4}	1.806×10^{-5}	3.132×10^{-4}	0.7018
500	0.04566	0.2352	0.02577	6.665×10^{-4}	2.139×10^{-5}	4.685×10^{-4}	0.7029
1000	0.03002	0.2520	0.03698	1.357×10^{-3}	2.855×10^{-5}	9.509×10^{-4}	0.7005
1500	0.02236	0.2626	0.04701	2.224×10^{-3}	3.474×10^{-5}	1.553×10^{-3}	0.6985
2000	0.01782	0.2701	0.05614	3.241×10^{-3}	4.035×10^{-5}	2.265×10^{-3}	0.6988
				Water Vapor, H$_2$O			
−50	0.06022	0.4512	0.00797	8.153×10^{-5}	4.933×10^{-6}	8.192×10^{-5}	1.0050
0	0.05367	0.4484	0.00898	1.036×10^{-4}	5.592×10^{-6}	1.041×10^{-4}	1.0049
50	0.04841	0.4472	0.01006	1.291×10^{-4}	6.261×10^{-6}	1.293×10^{-4}	1.0018
100	0.04408	0.4473	0.01121	1.579×10^{-4}	6.942×10^{-6}	1.574×10^{-4}	0.9969
200	0.03740	0.4503	0.01372	2.263×10^{-4}	8.333×10^{-6}	2.228×10^{-4}	0.9845
300	0.03248	0.4557	0.01648	3.093×10^{-4}	9.756×10^{-6}	3.004×10^{-4}	0.9713
500	0.02571	0.4707	0.02267	5.204×10^{-4}	1.267×10^{-5}	4.931×10^{-4}	0.9475
1000	0.01690	0.5167	0.04134	1.314×10^{-3}	2.014×10^{-5}	1.191×10^{-3}	0.9063
1500	0.01259	0.5625	0.06315	2.477×10^{-3}	2.742×10^{-5}	2.178×10^{-3}	0.8793
2000	0.01003	0.6034	0.08681	3.984×10^{-3}	3.422×10^{-5}	3.411×10^{-3}	0.8563

Note: For ideal gases, the properties c_p, k, μ, and Pr are independent of pressure. The properties ρ, ν, and α at a pressure P (in atm) other than 1 atm are determined by multiplying the values of ρ at the given temperature by P and by dividing ν and α by P.

Source: Data generated from the EES software developed by S. A. Klein and F. L. Alvarado. Originally based on various sources.

TABLE A–11E

Properties of the atmosphere at high altitude

Altitude, ft	Temperature, °F	Pressure, psia	Gravity, g, ft/s^2	Speed of Sound, ft/s	Density, lbm/ft^3	Viscosity μ, lbm/ft·s	Thermal Conductivity, Btu/h·ft·R
0	59.00	14.7	32.174	1116	0.07647	1.202×10^{-5}	0.0146
500	57.22	14.4	32.173	1115	0.07536	1.199×10^{-5}	0.0146
1000	55.43	14.2	32.171	1113	0.07426	1.196×10^{-5}	0.0146
1500	53.65	13.9	32.169	1111	0.07317	1.193×10^{-5}	0.0145
2000	51.87	13.7	32.168	1109	0.07210	1.190×10^{-5}	0.0145
2500	50.09	13.4	32.166	1107	0.07104	1.186×10^{-5}	0.0144
3000	48.30	13.2	32.165	1105	0.06998	1.183×10^{-5}	0.0144
3500	46.52	12.9	32.163	1103	0.06985	1.180×10^{-5}	0.0143
4000	44.74	12.7	32.162	1101	0.06792	1.177×10^{-5}	0.0143
4500	42.96	12.5	32.160	1099	0.06690	1.173×10^{-5}	0.0142
5000	41.17	12.2	32.159	1097	0.06590	1.170×10^{-5}	0.0142
5500	39.39	12.0	32.157	1095	0.06491	1.167×10^{-5}	0.0141
6000	37.61	11.8	32.156	1093	0.06393	1.164×10^{-5}	0.0141
6500	35.83	11.6	32.154	1091	0.06296	1.160×10^{-5}	0.0141
7000	34.05	11.3	32.152	1089	0.06200	1.157×10^{-5}	0.0140
7500	32.26	11.1	32.151	1087	0.06105	1.154×10^{-5}	0.0140
8000	30.48	10.9	32.149	1085	0.06012	1.150×10^{-5}	0.0139
8500	28.70	10.7	32.148	1083	0.05919	1.147×10^{-5}	0.0139
9000	26.92	10.5	32.146	1081	0.05828	1.144×10^{-5}	0.0138
9500	25.14	10.3	32.145	1079	0.05738	1.140×10^{-5}	0.0138
10,000	23.36	10.1	32.145	1077	0.05648	1.137×10^{-5}	0.0137
11,000	19.79	9.72	32.140	1073	0.05473	1.130×10^{-5}	0.0136
12,000	16.23	9.34	32.137	1069	0.05302	1.124×10^{-5}	0.0136
13,000	12.67	8.99	32.134	1065	0.05135	1.117×10^{-5}	0.0135
14,000	9.12	8.63	32.131	1061	0.04973	1.110×10^{-5}	0.0134
15,000	5.55	8.29	32.128	1057	0.04814	1.104×10^{-5}	0.0133
16,000	+1.99	7.97	32.125	1053	0.04659	1.097×10^{-5}	0.0132
17,000	−1.58	7.65	32.122	1049	0.04508	1.090×10^{-5}	0.0132
18,000	−5.14	7.34	32.119	1045	0.04361	1.083×10^{-5}	0.0130
19,000	−8.70	7.05	32.115	1041	0.04217	1.076×10^{-5}	0.0129
20,000	−12.2	6.76	32.112	1037	0.04077	1.070×10^{-5}	0.0128
22,000	−19.4	6.21	32.106	1029	0.03808	1.056×10^{-5}	0.0126
24,000	−26.5	5.70	32.100	1020	0.03553	1.042×10^{-5}	0.0124
26,000	−33.6	5.22	32.094	1012	0.03311	1.028×10^{-5}	0.0122
28,000	−40.7	4.78	32.088	1003	0.03082	1.014×10^{-5}	0.0121
30,000	−47.8	4.37	32.082	995	0.02866	1.000×10^{-5}	0.0119
32,000	−54.9	3.99	32.08	987	0.02661	0.986×10^{-5}	0.0117
34,000	−62.0	3.63	32.07	978	0.02468	0.971×10^{-5}	0.0115
36,000	−69.2	3.30	32.06	969	0.02285	0.956×10^{-5}	0.0113
38,000	−69.7	3.05	32.06	968	0.02079	0.955×10^{-5}	0.0113
40,000	−69.7	2.73	32.05	968	0.01890	0.955×10^{-5}	0.0113
45,000	−69.7	2.148	32.04	968	0.01487	0.955×10^{-5}	0.0113
50,000	−69.7	1.691	32.02	968	0.01171	0.955×10^{-5}	0.0113
55,000	−69.7	1.332	32.00	968	0.00922	0.955×10^{-5}	0.0113
60,000	−69.7	1.048	31.99	968	0.00726	0.955×10^{-5}	0.0113

Source: U.S. Standard Atmosphere Supplements, U.S. Government Printing Office, 1966. Based on year-round mean conditions at 45° latitude and varies with the time of the year and the weather patterns. The conditions at sea level (z = D) are taken to be P = 14.696 psia, T = 59°F, ρ = 0.076474 lbm/ft^3, g = 32.1741 ft^2/s.

GLOSSARY

Guest Author: James G. Brasseur, The Pennsylvania State University

Note: **Boldface color** glossary terms correspond to **boldface color** terms in the text. *Italics* indicates a term defined elsewhere in the glossary.

Boldface terms without page numbers are concepts that are not defined in the text but are defined or cross-referenced in the glossary for students to review.

absolute pressure: See *stress, pressure stress.* Contrast with *gage pressure.*

absolute viscosity: See *viscosity.*

acceleration field: See *field.*

adiabatic process: A process with no heat transfer.

advective acceleration: In order to reduce confusion of terminology in flows where *buoyancy forces* generate convective fluid motions, the term "convective acceleration" is often replaced with the term "advective acceleration."

aerodynamics: The application of *fluid dynamics* to air, land, and water-going vehicles. Often the term is specifically applied to the flow surrounding, and forces and moments on, flight vehicles in air, as opposed to vehicles in water or other liquids (*hydrodynamics*).

angle of attack: The angle between an airfoil or wing and the free-stream flow velocity vector.

average: An area/volume/time average of a fluid property is the integral of the property over an area/volume/time period divided by the corresponding area/volume/time period. Also called *mean.*

axisymmetric flow: A flow that when specified appropriately using cylindrical coordinates (r, θ, x) does not vary in the azimuthal (θ) direction. Thus, all partial derivatives in θ are zero. The flow is therefore either one-dimensional or two-dimensional (see also *dimensionality* and *planar flow*).

barometer: A device that measures atmospheric pressure.

basic dimensions: See *dimensions.*

Bernoulli equation: A useful reduction of *conservation of momentum* (and *conservation of energy*) that describes a balance between pressure (*flow work*), velocity (*kinetic energy*), and position of *fluid particles* relative to the gravity vector (potential energy) in regions of a fluid flow where frictional force on fluid particles is negligible compared to pressure force in that region of the flow (see *inviscid flow*).

Note: This glossary covers **boldface color terms** found in Chapters 1 to 11.

There are multiple forms of the Bernoulli equation for incompressible vs. compressible, steady vs. nonsteady, and derivations through *Newton's law* vs. the *first law of thermodynamics.* The most commonly used forms are for steady incompressible fluid flow derived through conservation of momentum.

bluff (or **blunt**) **body:** A moving object with a blunt rear portion. Bluff bodies have *wakes* resulting from massive *flow separation* over the rear of the body.

boundary condition: In solving for flow field variables (velocity, temperature) from governing equations, it is necessary to mathematically specify a function of the variable at the surface. These mathematical statements are called boundary conditions. The no-slip condition that the flow velocity must equal the surface velocity at the surface is an example of a boundary condition that is used with the Navier–Stokes equation to solve for the velocity field.

boundary layer: At high *Reynolds numbers* relatively thin "boundary layers" exist in the flow adjacent to surfaces where the flow is brought to rest (see *no-slip condition*). Boundary layers are characterized by high *shear* with the highest velocities away from the surface. *Frictional force, viscous stress,* and *vorticity* are significant in boundary layers. The approximate form of the two components of the Navier– Stokes equation, simplified by neglecting the terms that are small within the boundary layer, are called the *boundary layer equations.* The associated approximation based on the existence of thin boundary layers surrounded by *irrotational* or *inviscid* flow is called the *boundary layer approximation.*

boundary layer approximation: See *boundary layer.*

boundary layer equations: See *boundary layer.*

boundary layer thickness measures: Different measures of the thickness of a boundary layer as a function of downstream distance are used in fluid flow analyses. These are:

boundary layer thickness: The full thickness of the viscous layer that defines the boundary layer, from the surface to the edge. Defining the edge is difficult to do precisely, so the "edge" of the boundary layer is often defined as the point where the boundary layer velocity is a large fraction of the free-stream velocity (e.g., δ_{99} is the distance from the surface to the point where the streamwise velocity component is 99 percent of the free-stream velocity).

displacement thickness: A boundary layer thickness measure that quantifies the deflection of fluid streamlines in the direction away from the surface as a result of friction-induced reduction in mass flow adjacent to the surface. Displacement thickness ($\delta*$) is a measure of the thickness of this mass flow rate deficit layer. In all boundary layers, $\delta* < \delta$.

momentum thickness: A measure of the layer of highest deficit in momentum flow rate adjacent to the surface as a result of frictional resisting force (shear stress). Because Newton's second law states that force equals time rate of momentum change, momentum thickness θ is proportional to surface shear stress. In all boundary layers, $\theta < \delta*$.

Buckingham Pi theorem: A mathematical theorem used in *dimensional analysis* that predicts the number of nondimensional groups that must be functionally related from a set of dimensional parameters that are thought to be functionally related.

buffer layer: The part of a turbulent boundary layer, close to the wall, lying between the *viscous* and *inertial sublayers*. This thin layer is a transition from the friction-dominated layer adjacent to the wall where *viscous stresses* are large, to the inertial layer where *turbulent stresses* are large compared to viscous stresses.

bulk modulus of elasticity: See *compressibility*.

buoyant force: The net upward hydrostatic pressure force acting on an object submerged, or partially submerged, in a fluid.

cavitation: The formation of vapor bubbles in a liquid as a result of pressure going below the *vapor pressure*.

center of pressure: The effective point of application of pressure distributed over a surface. This is the point where a counteracting force (equal to integrated pressure) must be placed for the net moment from pressure about that point to be zero.

centripetal acceleration: Acceleration associated with the change in the direction of the velocity (vector) of a material particle.

closed system: See *system*.

coefficient of compressibility: See *compressibility*.

compressibility: The extent to which a *fluid particle* changes volume when subjected to either a change in pressure or a change in temperature.

bulk modulus of elasticity: Synonymous with *coefficient of compressibility*.

coefficient of compressibility: The ratio of pressure change to relative change in volume of a *fluid particle*. This coefficient quantifies compressibility in response to pressure change, an important effect in high Mach number flows.

coefficient of volume expansion: The ratio of relative density change to change in temperature of a *fluid particle*. This coefficient quantifies compressibility in response to temperature change.

computational fluid dynamics (CFD): The application of the conservation laws with boundary and initial conditions in mathematical discretized form to estimate field variables quantitatively on a discretized grid (or mesh) spanning part of the flow field.

conservation laws: The fundamental principles upon which all engineering analysis is based, whereby the material properties of mass, momentum, energy, and entropy can change only in balance with other physical properties involving forces, work, and *heat transfer*. These laws are predictive when written in mathematical form and appropriately combined with boundary conditions, initial conditions, and constitutive relationships.

conservation of energy principle: This is the *first law of thermodynamics*, a fundamental law of physics stating that the time rate of change of total *energy* of a fixed mass (*system*) is balanced by the net rate at which *work* is done on the mass and *heat energy* is transferred to the mass.

Note: To mathematically convert the time derivative of mass, momentum, and energy of fluid mass in a system to that in a *control volume*, one applies the *Reynolds transport theorem*.

conservation of mass principle: A fundamental law of physics stating that a volume always containing the same atoms and molecules (*system*) must always contain the same mass. Thus the time rate of change of mass of a system is zero. This law of physics must be revised when matter moves at speeds approaching the speed of light so that mass and energy can be exchanged as per Einstein's laws of relativity.

conservation of momentum: This is *Newton's second law* of motion, a fundamental law of physics stating that the time rate of change of momentum of a fixed mass (*system*) is balanced by the net sum of all forces applied to the mass.

constitutive equations: An empirical relationship between a physical variable in a *conservation law of physics* and other physical variables in the equation that are to be predicted. For example, the energy equation written for temperature includes the *heat* flux vector. It is known from experiments that heat flux for most common materials is accurately approximated as proportional to the gradient in temperature (this is called Fourier's law). In *Newton's law* written for a *fluid particle*, the *viscous stress tensor* (see *stress*) must be written as a function of velocity to solve the equation. The most common constitutive relationship for viscous stress is that for a *Newtonian fluid*. See also *rheology*.

continuity equation: Mathematical form of *conservation of mass* applied to a *fluid particle* in a flow.

continuum: Treatment of matter as a continuous (without holes) distribution of finite mass *differential volume* elements. Each volume element must contain huge numbers of molecules so that the macroscopic effect of the molecules can be modeled without considering individual molecules.

contour plot: Also called an *isocontour plot,* this is a way of plotting data as lines of constant variable through a flow *field. Streamlines,* for example, may be identified as lines of constant *stream function* in *two-dimensional* incompressible steady flows.

control mass: See *system.*

control volume: A volume specified for analysis where flow enters and/or exits through some portion(s) of the volume surface. Also called an *open system* (see *system*).

convective acceleration: Synonymous with *advective acceleration,* this term must be added to the partial time derivative of velocity to properly quantify the acceleration of a *fluid particle* within an *Eulerian* frame of reference. For example, a fluid particle moving through a contraction in a *steady flow* speeds up as it moves, yet the time derivative is zero. The additional convective acceleration term required to quantify fluid acceleration (e.g., in *Newton's second law*) is called the *convective derivative.* See also *Eulerian description, Lagrangian description, material derivative,* and *steady flow.*

convective derivative: See *material derivative* and *convective acceleration.*

creeping flow: Fluid flow in which frictional forces dominate fluid accelerations to the point that the flow can be well modeled with the acceleration term in Newton's second law set to zero. Such flows are characterized by Reynolds numbers that are small compared to 1 (Re ≪ 1). Since Reynolds number typically can be written as characteristic velocity times characteristic length divided by kinematic viscosity (VL/ν), creeping flows are often slow-moving flows around very small objects (e.g., sedimentation of dust particles in air or motion of spermatozoa in water), or with very viscous fluids (e.g., glacier and tar flows). Also called Stokes flow.

deformation rate: See *strain rate.*

derived dimensions: See *dimensions.*

deviatoric stress tensor: Another term for *viscous stress tensor.* See *stress.*

differential analysis: Analysis at a point in the flow (as opposed to over a *control volume*).

differential volume/area/length: A small volume δV, area δA, or length δx in the limit of the volume/area/length shrinking to a point. Derivatives are often produced in this limit. (Note that δ is sometimes written as Δ or d.)

dimensional analysis: A process of analysis based solely on the variables of relevance to the flow system under study, the dimensions of the variables, and dimensional homogeneity. After determining the other variables on which a variable of interest depends (e.g., drag on a car depends on the speed and size of the car, fluid viscosity, fluid density, and surface roughness), one applies the principle of dimensional homogeneity with the *Buckingham Pi theorem* to relate an appropriately nondimensionalized variable of interest (e.g., drag) with

the other variables appropriately nondimensionalized (e.g., Reynolds numbers, roughness ratio, and Mach number).

dimensional homogeneity: The requirement that summed terms must have the same *dimensions* (e.g., ρV^2, pressure P, and shear stress τ_{xy} are dimensionally homogeneous while *power,* specific enthalpy h, and $P\dot{m}$ are not). Dimensional homogeneity is the basis of *dimensional analysis.*

dimensionality: The number of spatial coordinates in whose direction velocity components and/or other variables vary for a specified coordinate system. For example, *fully developed* flow in a tube is one-dimensional (1-D) in the radial direction r since the only nonzero velocity component (the axial, or x-, component) is constant in the x- and θ-directions, but varies in the r-direction. *Planar flows* are two-dimensional (2-D). Flows over *bluff bodies* such as cars, airplanes, and buildings are three-dimensional (3-D). Spatial derivatives are nonzero only in the directions of dimensionality.

dimensions: The required specification of a physical quantity beyond its numerical value. See also *units.*

   derived (or **secondary**) **dimensions:** Combinations of fundamental dimensions. Examples of derived dimensions are: velocity (L/t), stress or pressure (F/L^2 = m/(Lt^2)), energy or work (mL^2/t^2 = FL), density (m/L^3), specific weight (F/L^3), and specific gravity (unitless).

   fundamental (primary, basic) dimensions: Mass (m), length (L), time (t), temperature (T), electrical current (I), amount of light (C), and amount of matter (N) without reference to a specific system of units. Note that the force dimension is obtained through Newton's law as F = mL/t^2 (thus, the mass dimension can be replaced with a force dimension by replacing m with Ft^2/L).

drag coefficient: Nondimensional drag given by the *drag force* on an object nondimensionalized by *dynamic pressure* of the free-stream flow times frontal area of the object:

$$C_D \equiv \frac{F_D}{\frac{1}{2}\rho V^2 A}$$

Note that at high Reynolds numbers (Re ≫ 1), C_D is a normalized variable, whereas at Re ≪ 1, C_D is nondimensional but is not normalized (see *normalization*). See also *lift coefficient.*

drag force: The force on an object opposing the motion of the object. In a frame of reference moving with the object, this is the force on the object in the direction of flow. There are multiple components to drag force:

   friction drag: The part of the drag on an object resulting from integrated surface *shear stress* in the direction of flow relative to the object.

   induced drag: The component of the drag force on a finite-span wing that is "induced" by lift and associated with the *tip vortices* that form at the tips of the wing and "downwash" behind the wing.

pressure (or **form**) **drag:** The part of the drag on an object resulting from integrated surface *pressure* in the direction of flow relative to the object. Larger pressure on the front of a moving *bluff body* (such as a car) relative to the rear results from massive *flow separation* and *wake* formation at the rear.

dynamic pressure: When the *Bernoulli equation* in *incompressible steady flow* and/or the *conservation of energy* equation along a streamline are written in forms where each term in the equations has the *dimensions* force/area, dynamic pressure is the *kinetic energy* (per unit volume) term (i.e., $\frac{1}{2}\rho V^2$).

dynamic similarity: See *similarity*.

dynamic viscosity: See *viscosity*.

dynamics: When contrasted with *statics* the term refers to the application of Newton's second law of motion to moving matter. When contrasted with *kinematics* the term refers to forces or accelerations through Newton's law force balances.

eddy viscosity: See *turbulence models*.

efficiency: A ratio that describes levels of losses of useful power obtained from a device. Efficiency of 1 implies no losses in the particular function of the device for which a particular definition of efficiency is designed. For example, mechanical efficiency of a pump is defined as the ratio of useful mechanical power transferred to the flow by the pump to the mechanical energy, or shaft work, required to drive the pump. Pump-motor efficiency of a pump is defined as the ratio of useful mechanical power transferred to the flow over the electrical power required to drive the pump. Pump-motor efficiency, therefore, includes additional losses and is thus lower than mechanical pump efficiency.

energy: A state of matter described by the first law of thermodynamics that can be altered at the macroscopic level by work, and at the microscopic level through adjustments in thermal energy.

flow energy: Synonymous with *flow work*. The work associated with *pressure* acting on a flowing *fluid*.

heat (transfer): The term "heat" is generally used synonymously with *thermal energy*. Heat transfer is the transfer of thermal energy from one physical location to another.

internal energy: Forms of energy arising from the microscopic motions of molecules and atoms, and from the structure and motions of the subatomic particles comprising the atoms and molecules, within matter.

kinetic energy: Macroscopic (or mechanical) form of energy arising from the speed of matter relative to an inertial frame of reference.

mechanical energy: The nonthermal components of energy; examples include kinetic and potential energy.

potential energy: A mechanical form of energy that changes as a result of macroscopic displacement of matter relative to the gravitational vector.

thermal energy: Internal energy associated with microscopic motions of molecules and atoms. For single-phase systems, it is the energy represented by temperature.

total energy: Sum of all forms of energy. Total energy is the sum of kinetic, potential, and internal energies. Equivalently, total energy is the sum of mechanical and thermal energies.

work energy: The integral of force over the distance in which a mass is moved by the force. Work is energy associated with the movement of matter by a force.

energy grade line: See *grade lines*.

English system: See *units*.

entry length: The entry flow region in a pipe or duct flow where the wall boundary layers are thickening toward the center with axial distance x of the duct, so that axial derivatives are nonzero. As with the *fully developed* region, the *hydrodynamic entry length* involves growth of a velocity boundary layer, and the *thermal entry length* involves growth of a temperature boundary layer.

Eulerian derivative: See *material derivative*.

Eulerian description: In contrast with a *Lagrangian description*, an Eulerian analysis of fluid flow is developed from a frame of reference through which the *fluid particles* move. In this frame the acceleration of fluid particles is not simply the time derivative of fluid velocity, and must include another term, called *convective acceleration*, to describe the change in velocity of fluid particles as they move through a *velocity field*. Note that velocity fields are always defined in an Eulerian frame of reference.

extensional strain rate: See *strain rate*.

extensive property: A fluid property that depends on total volume or total mass (e.g., total internal energy). See *intensive property*.

field: The representation of a flow variable as a function of Eulerian coordinates (x, y, z). For example, the *velocity* and *acceleration fields* are the fluid velocity and acceleration vectors ($\vec{V}$, $\vec{a}$) as functions of position (x, y, z) in the *Eulerian description* at a specified time t.

flow field: The field of flow variables. Generally, this term refers to the velocity field, but it may also mean all field variables in a fluid flow.

first law of thermodynamics: See *conservation laws, conservation of energy*.

flow separation: A phenomenon where a *boundary layer* adjacent to a surface is forced to leave, or "separate" from, the surface due to "adverse" pressure forces (i.e., increasing pressure) in the flow direction. Flow separation occurs in regions of high surface curvature, for example, at the rear of an automobile and other bluff bodies.

flow work: The work term in *first law of thermodynamics* applied to fluid flow associated with pressure forces on the flow. See *energy, flow energy*.

fluid: A material that when sheared deforms continuously in time during the period that shear forces are applied. By contrast, shear forces applied to a *solid* cause the material either to deform to a fixed static position (after which deformation stops), or cause the material to fracture. Consequently, whereas solid deformations are generally analyzed using strain and shear, fluid flows are analyzed using rates of strain and shear (see *strain rate*).

fluid mechanics/dynamics: The study and analysis of fluids through the macroscopic conservation laws of physics, i.e., conservation of mass, momentum (*Newton's second law*), and energy (first law of thermodynamics), and the second law of thermodynamics.

fluid particle/element: A *differential* particle, or element, embedded in a fluid flow containing always the same atoms and molecules. Thus a fluid particle has fixed mass δm and moves with the flow with local flow velocity $\vec{V}$, acceleration $\vec{a}_{particle} = D\vec{V}/Dt$ and trajectory ($x_{particle}(t)$, $y_{particle}(t)$, $t_{particle}(t)$). See also *material derivative, material particle, material position vector,* and *pathline.*

forced flow: Flow resulting from an externally applied force. Examples include liquid flow through tubes driven by a pump and fan-driven airflow for cooling computer components. *Natural flows,* in contrast, result from internal buoyancy forces driven by temperature (i.e., density) variations within a fluid in the presence of a gravitational field. Examples include buoyant plumes around a human body or in the atmosphere.

friction/frictional: See *Newtonian fluid, viscosity,* and *viscous force.*

friction factor: It can be shown from *dimensional analysis* and *conservation of momentum* applied to a *steady fully developed* pipe flow that the frictional contribution to the pressure drop along the pipe, nondimensionalized by flow *dynamic pressure* ($\frac{1}{2}\rho V_{avg}^2$), is proportional to the length-to-diameter ratio (L/D) of the pipe. The proportionality factor f is called the friction factor. The friction factor is quantified from experiment (turbulent flow) and theory (laminar flow) in empirical relationships, and in the *Moody chart,* as a function of the Reynolds number and nondimensional roughness. Conservation of momentum shows that the friction factor is proportional to the nondimensional wall shear stress (i.e., the *skin friction*).

frictionless flow: Mathematical treatments of fluid flows sometimes use conservation of momentum and energy equations without the frictional terms. Such mathematical treatments "assume" that the flow is "frictionless," implying no *viscous force* (*Newton's second law*), nor *viscous dissipation* (*first law of thermodynamics*). However, no real fluid flow of engineering interest can exist without viscous forces, dissipation, and/or head losses in regions of practical importance. The engineer should always identify the flow

regions where frictional effects are concentrated. When developing models for prediction, the engineer should consider the role of these viscous regions in the prediction of variables of interest and should estimate levels of error in simplified treatments of the viscous regions. In high *Reynolds number* flows, frictional regions include boundary layers, *wakes, jets, shear layers,* and flow regions surrounding *vortices.*

Froude number: An order-of-magnitude estimate of the ratio of the inertial term in Newton's law of motion to the gravity force term. The Froude number is an important nondimensional group in free-surface flows, as is generally the case in channels, rivers, surface flows, etc.

fully developed: Used by itself, the term is generally understood to imply hydrodynamically fully developed, a flow region where the velocity field is constant along a specified direction in the flow. In the fully developed region of pipe or duct flow, the velocity field is constant in the axial direction, x (i.e., it is independent of x), so that x-derivatives of velocity are zero in the fully developed region. There also exists the concept of "thermally fully developed" for the temperature field; however, unlike hydrodynamically fully developed regions where both the magnitude and shape of the velocity profile are constant in x, in thermally fully developed regions only the shape of the temperature profile is constant in x. See also *entry length.*

fundamental dimensions: See *dimensions.*

gage pressure: *Pressure* (P) relative to atmospheric pressure (P_{atm}). That is, $P_{gage} = P - P_{atm}$. See also *stress, pressure stress.* Thus $P_{gage} > 0$ or $P_{gage} < 0$ is simply the pressure above or below atmospheric pressure.

gas dynamics: The study and analysis of gases and vapors through the macroscopic conservation laws of physics (see *fluid mechanics/dynamics*).

geometric similarity: See *similarity.*

grade lines: Lines of *head* summations.

  energy grade line: Line describing the sum of *pressure head, velocity head,* and *elevation head.* See *head.*

  hydraulic grade line: Line describing the sum of *pressure head* and *elevation head.* See *head.*

Hagen–Poiseuille flow: See *Poiseuille flow.*

head: A quantity (pressure, kinetic energy, etc.) expressed as an equivalent column height of a fluid. *Conservation of energy* for *steady flow* written for a *control volume* surrounding a central *streamline* with one inlet and one outlet, or shrunk to a streamline, can be written such that each term has the *dimensions* of length. Each of these terms is called a head term:

  elevation head: The term in the head form of *conservation of energy* (see *head*) involving distance in the direction opposite to the gravitational vector relative to a predefined datum (z).

head loss: The term in the head form of *conservation of energy* (see *head*) that contains frictional losses and other irreversibilities. Without this term, the energy equation for streamlines becomes the *Bernoulli equation* in head form.

pressure head: The term in the head form of *conservation of energy* (see *head*) involving pressure ($P/\rho g$).

velocity head: The (*kinetic energy*) term in the head form of *conservation of energy* (see *head*) involving velocity ($V^2/2g$).

heat: See *energy*.

hot-film anemometer: Similar to a *hot-wire anemometer* except using a metallic film rather than a wire; used primarily for liquid flows. The measurement portion of a hot-film probe is generally larger and more rugged than that of a hot-wire probe.

hot-wire anemometer: A device used to measure a velocity component locally in a gas flow based on the relationship between the flow around a thin heated wire (the hot wire), temperature of the wire, and heating of the wire resulting from a current. See also *hot-film anemometer*.

hydraulic grade line: See *grade lines*.

hydraulics: The *hydrodynamics* of liquid and vapor flow in pipes, ducts, and open channels. Examples include water piping systems and ventilation systems.

hydrodynamic entry length: See *entry length*.

hydrodynamically fully developed: See *fully developed*.

hydrodynamics: The study and analysis of liquids through the macroscopic conservation laws of physics (see *fluid mechanics/dynamics*). The term is sometimes applied to *incompressible* vapor and gas flows, but when the fluid is air, the term *aerodynamics* is generally used instead.

hydrostatic pressure: The component of *pressure* variation in a fluid flow that would exist in the absence of flow as a result of gravitational body force. This term appears in the hydrostatic equation and in the *Bernoulli equation*. See also *dynamic* and *static pressure*.

hypersonic: An order of magnitude or more above the speed of sound (Mach number $\gg 1$).

ideal fluid: See *perfect fluid*.

ideal gas: A gas at low enough density and/or high enough temperature that (a) density, pressure, and temperature are related by the ideal-gas equation of state, $P = \rho RT$, and (b) specific internal energy and enthalpy are functions only of temperature.

incompressible flow: A fluid flow where variations in density are sufficiently small to be negligible. Flows are generally incompressible either because the fluid is incompressible (liquids) or because the Mach number is low (roughly < 0.3).

induced drag: See *drag force*.

inertia/inertial: The acceleration term in Newton's second law, or effects related to this term. Thus, a flow with higher inertia requires larger deceleration to be brought to rest.

inertial sublayer: A highly turbulent part of a turbulent boundary layer, close to the wall but just outside the *viscous sublayer* and *buffer layer*, where *turbulent stresses* are large compared to *viscous stresses*.

intensive property: A fluid property that is independent of total volume or total mass (i.e., an *extensive property* per unit mass or sometimes per unit volume).

internal energy: See *energy*.

inviscid (region of) flow: Region of a fluid flow where viscous forces are sufficiently small relative to other forces (typically, pressure force) on *fluid particles* in that region of the flow to be neglected in *Newton's second law* of motion to a good level of approximation (compare with *viscous flow*). See also *frictionless flow*. An inviscid region of flow is not necessarily *irrotational*.

irrotational (region of) flow: A region of a flow with negligible *vorticity* (i.e., *fluid particle* rotation). Also called *potential flow*. An irrotational region of flow is also <u>inviscid</u>.

isocontour plot: See *contour plot*.

jet: A friction-dominated region issuing from a tube or orifice and formed by surface boundary layers that have been swept behind by the mean velocity. Jets are characterized by high *shear* with the highest velocities in the center of the jet and lowest velocities at the edges. *Frictional force, viscous stress,* and *vorticity* are significant in jets.

Kármán vortex street: The *two-dimensional* alternating unsteady pattern of *vortices* that is commonly observed behind circular cylinders in a flow (e.g., the vortex street behind wires in the wind is responsible for the distinct tone sometimes heard).

kinematic similarity: See *similarity*.

kinematic viscosity: Fluid *viscosity* divided by density.

kinematics: In contrast with *dynamics,* the kinematic aspects of a fluid flow are those that do not directly involve Newton's second law force balance. Kinematics refers to descriptions and mathematical derivations based only on conservation of mass (continuity) and definitions related to flow and deformation.

kinetic energy: See *energy*.

kinetic energy correction factor: *Control volume* analysis of the *conservation of energy* equation applied to tubes contains area integrals of kinetic energy flux. The integrals are often approximated as proportional to kinetic energy formed with area-averaged velocity, V_{avg}. The inaccuracy in this approximation can be significant, so a kinetic energy correction factor, α, multiplies the term to improve the approximation. The correction α depends on the shape of the *velocity profile,* is largest for *laminar* profiles (*Poiseuille flow*), and is closest to 1 in *turbulent* pipe flows at very high *Reynolds numbers.*

Lagrangian derivative: See *material derivative.*

Lagrangian description: In contrast with the *Eulerian description,* a Lagrangian analysis is developed from a frame of reference attached to moving material particles. For example, solid particle acceleration in the standard Newton's second law form, $\vec{F} = m\vec{a}$, is in a coordinate system that moves with the particle so that acceleration $\vec{a}$ is given by the time derivative of particle velocity. This is the typical analytical approach used for analysis of the motion of solid objects.

laminar flow: A stable well-ordered state of fluid flow in which all pairs of adjacent *fluid particles* move alongside one another forming laminates. A flow that is not laminar is either *turbulent* or *transitional* to turbulence, which occurs above a critical *Reynolds number.*

laser Doppler velocimetry (LDV): Also called laser Doppler anemometry (LDA). A technique for measuring a velocity component locally in a flow based on the Doppler shift associated with the passage of small particles in the flow through the small target volume formed by the crossing of two laser beams. Unlike *hot-wire* and *hot-film anemometry* and like *particle image velocimetry,* there is no interference to the flow.

lift coefficient: Nondimensional lift given by the lift force on a lifting object (such as an airfoil or wing) nondimensionalized by dynamic pressure of the free-stream flow times planform area of the object:

$$C_L \equiv \frac{F_L}{\frac{1}{2}\rho V^2 A}$$

Note that at high Reynolds numbers (Re $\gg$ 1), C_L is a normalized variable, whereas at Re $\ll$ 1, C_L is nondimensional but is not normalized (see *normalization*). See also *drag coefficient.*

lift force: The net aerodynamic force on an object perpendicular to the motion of the object.

linear strain rate: Synonymous with *extensional strain rate.* See *strain rate.*

losses: Frictional *head losses* in pipe flows are separated into those losses in the fully developed pipe flow regions of a piping network, the *major losses,* plus head losses in other flow regions of the network, the *minor losses.* Minor loss regions include *entry lengths,* pipe couplings, bends, valves, etc. It is not unusual for minor losses to be larger than major losses.

Mach number: *Nondimensional* ratio of the characteristic speed of the flow to the speed of sound. Mach number

characterizes the level of *compressibility* in response to pressure variations in the flow.

major losses: See *losses.*

manometer: A device that measures pressure based on hydrostatic pressure principles in liquids.

material acceleration: The acceleration of a *fluid particle* at the point (x, y, z) in a flow at time t. This is given by the *material derivative* of fluid velocity: $D\vec{V}(x, y, z, t)/Dt$.

material derivative: Synonymous terms are *total derivative, substantial derivative,* and *particle derivative.* These terms mean the time rate of change of fluid variables (temperature, velocity, etc.) moving with a *fluid particle.* Thus, the material derivative of temperature at a point (x, y, z) at time t is the time derivative of temperature attached to a moving *fluid particle* at the point (x, y, z) in the flow at the time t. In a *Lagrangian* frame of reference (i.e., a frame attached to the moving particle), particle temperature $T_{particle}$ depends only on time, so a time derivative is a total derivative $dT_{particle}(t)/dt$. In an Eulerian frame, the temperature *field* $T(x, y, z, t)$ depends on both position (x, y, z) and time t, so the *material derivative* must include both a partial derivative in time and a *convective derivative:* $dT_{particle}(t)/dt \equiv DT(x, y, z, t)/Dt = \partial T/\partial t + \vec{V}\cdot\vec{\nabla}T$. See also *field.*

material particle: A *differential* particle, or element, that contains always the same atoms and molecules. Thus a material particle has fixed mass δm. In a fluid flow, this is the same as a *fluid particle.*

material position vector: A vector $[x_{particle}(t), y_{particle}(t), z_{particle}(t)]$ that defines the location of a *material particle* as a function of time. Thus the material position vector in a fluid flow defines the trajectory of a *fluid particle* in time.

mean: Synonymous with *average.*

mechanical energy: See *energy.*

mechanics: The study and analysis of matter through the macroscopic conservation laws of physics (mass, momentum, energy, second law).

minor losses: See *losses.*

mixing length: See *turbulence models.*

momentum: The momentum of a *material particle* (or *fluid particle*) is the mass of the material particle times its velocity. The momentum of a macroscopic volume of material particles is the integrated momentum per unit volume over the volume, where momentum per unit volume is the density of the material particle times its velocity. Note that momentum is a vector.

momentum flux correction factor: A correction factor added to correct for approximations made in the simplification of the area integrals for the momentum flux terms in the control volume form of *conservation of momentum.*

Moody chart: A commonly used plot of the *friction factor* as a function of the Reynolds number and roughness parameter for fully developed pipe flow. The chart is a combination of flow theory for laminar flow with a graphical representation of an empirical formula by Colebrook to a large set of experimental data for turbulent pipe flow of various values of "sandpaper" roughness.

natural flow: Contrast with *forced flow.*

Navier–Stokes equation: *Newton's second law* of fluid motion (or *conservation of momentum*) written for a fluid particle (the *differential* form) with the *viscous stress tensor* replaced by the *constitutive relationship* between *stress* and *strain rate* for Newtonian fluids. Thus the Navier–Stokes equation is simply Newton's law written for Newtonian fluids.

Newtonian fluid: When a fluid is subjected to a *shear stress,* the fluid continuously changes shape (deformation). If the fluid is Newtonian, the rate of deformation (i.e., strain rate) is proportional to the applied shear stress and the constant of proportionality is called *viscosity.* In general flows, the rate of deformation of a *fluid particle* is described mathematically by a *strain rate* tensor and the *stress* by a *stress tensor.* In flows of Newtonian fluids, the stress tensor is proportional to the strain rate tensor, and the constant of proportionality is called *viscosity.* Most common fluids (water, oil, gasoline, air, most gases and vapors) without particles or large molecules in suspension are Newtonian.

Newton's second law: See *conservation of momentum.*

nondimensionalization: The process of making a dimensional variable dimensionless by dividing the variable by a *scaling parameter* (a single variable or a combination of variables) that has the same dimensions. For example, the surface pressure on a moving ball might be nondimensionalized by dividing it by ρV^2, where ρ is fluid density and V is free-stream velocity. See also *normalization.*

non-Newtonian fluid: A non-Newtonian fluid is one that deforms at a rate that is not linearly proportional to the stress causing the deformation. Depending on the manner in which *viscosity* varies with *strain rate,* non-Newtonian fluids can be labeled shear thinning (viscosity decreases with increasing strain rate), shear thickening (viscosity increases with increasing strain rate), and viscoelastic (when the shearing forces are removed, the fluid particles return partially to an earlier shape). Suspensions and liquids with long-chain molecules are generally non-Newtonian. See also *Newtonian fluid* and *viscosity.*

normal stress: See *stress.*

normalization: A particular *nondimensionalization* where the *scaling parameter* is chosen so that the nondimensionalized variable attains a maximum value that is of order 1 (say, within roughly 0.5 to 2). Normalization is more restrictive (and more difficult to do properly) than nondimensionalization. For example, $P/(\rho V^2)$ discussed under *nondimensionalization* is also normalized pressure on a flying baseball (where Reynolds number Re $\gg$ 1), but is simply nondimensionalization

of surface pressure on a small glass bead dropping slowly through honey (where Re $\ll$ 1).

no-slip condition: The requirement that at the interface between a fluid and a solid surface, the fluid velocity and surface velocity are equal. Thus if the surface is fixed, the fluid must obey the *boundary condition* that fluid velocity $= 0$ at the surface.

one-dimensional: See *dimensionality.*

open system: Same as *control volume.*

particle derivative: See *material derivative.*

particle image velocimetry (PIV): A technique for measuring a velocity component locally in a flow based on tracking the movement of small particles in the flow over a short time using pulsed lasers. Unlike *hot-wire* and *hot-film anemometry* and like laser Doppler velocimetry, there is no interference to the flow.

pathline: A curve mapping the trajectory of a *fluid particle* as it travels through a flow over a period of time. Mathematically, this is the curve through the points mapped out by the *material position vector* [$x_{particle}(t)$, $y_{particle}(t)$, $z_{particle}(t)$] over a defined period of time. Thus, pathlines are formed over time, and each fluid particle has its own pathline. In a steady flow, fluid particles move along streamlines, so pathlines and streamlines coincide. In a nonsteady flow, however, pathlines and streamlines are generally very different. Contrast with *streamline.*

perfect fluid: Also called an *ideal fluid,* the concept of a fictitious fluid that can flow in the absence of all frictional effects. There is no such thing as a perfect fluid, even as an approximation, so the engineer need not consider the concept further.

periodic: An unsteady flow in which the flow oscillates about a steady mean.

Pitot-static probe: A device used to measure fluid velocity through the application of the Bernoulli equation with simultaneous measurement of *static* and *stagnation pressures.* Also called a Pitot-Darcy probe.

planar flow: A *two-dimensional* flow with two nonzero components of velocity in Cartesian coordinates that vary only in the two coordinate directions of the flow. Thus, all partial derivatives perpendicular to the plane of the flow are zero. See also *axisymmetric flow* and *dimensionality.*

Poiseuille flow: *Fully developed laminar flow* in a pipe or duct. Also called *Hagen–Poiseuille flow.* The mathematical model relationships for Poiseuille flow relating the flow rate and/or velocity profile to the pressure drop along the pipe/duct, fluid viscosity and geometry are sometimes referred to as *Poiseuille's law* (although strictly not a "law" of mechanics). The velocity profile of all Poiseuille flows is parabolic, and the rate of axial pressure drop is constant.

Poiseuille's law: See *Poiseuille flow.*

potential energy: See *energy.*

potential flow: Synonymous with *irrotational flow.* This is a region of a flow with negligible *vorticity* (i.e., *fluid particle* rotation). In such regions, a velocity *potential function* exists (thus the name).

potential function: If a region of a flow has zero *vorticity* (*fluid particle* spin), the velocity vector in that region can be written as the gradient of a scalar function called the velocity potential function, or simply the potential function. In practice, potential functions are often used to model flow regions where vorticity levels are small but not necessarily zero.

power: *Work* per unit time; time rate at which work is done.

pressure: See *stress.*

pressure force: As applicable to Newton's second law, this is the force acting on a *fluid particle* that arises from spatial gradients in pressure within the flow. See also *stress, pressure stress.*

pressure work: See *flow work.*

primary dimension: See *dimensions.*

profile plot: A graphical representation of the spatial variation of a fluid property (temperature, pressure, strain rate, etc.) through a region of a fluid flow. A profile plot defines property variations in part of a *field* (e.g., a temperature profile might define the variation of temperature along a line within the temperature field).

   velocity profile: The spatial variation in a velocity component or vector through a region of a fluid flow. For example, in a pipe flow the velocity profile generally defines the variation in axial velocity with radius across the pipe cross section, while a *boundary layer* velocity profile generally defines variation in axial velocity normal to the surface. The velocity profile is part of a velocity *field.*

quasi-steady flow: See *steady flow.*

Reynolds number: An order-of-magnitude estimate of the ratio of the following two terms in Newton's second law of motion over a region of the flow: the *inertial* (or acceleration) term over the viscous force term. Most but not all Reynolds numbers can be written as an appropriate characteristic velocity V times a characteristic length scale L consistent with the velocity V, divided by the kinematic viscosity ν of the fluid: $Re = VL/\nu$. The Reynolds number is arguably the most important nondimensional *similarity* parameter in fluid flow analysis since it gives a rough estimate of the importance of frictional force in the overall flow.

Reynolds stress: Velocity components (and other variables) in turbulent flows are separated into mean plus fluctuating components. When the equation for mean streamwise velocity component is derived from the *Navier–Stokes equation,* six new terms appear given by fluid density times the averaged product of two velocity components. Because these terms have the same units as *stress* (force/area), they are called turbulent stresses or Reynolds stresses (in memory of Osborne Reynolds who first quantified turbulent variables as mean + fluctuation). Just as *viscous stresses* can be written as a tensor (or matrix), we define a Reynolds stress tensor with Reynolds normal stress components and Reynolds shear stress components. Although Reynolds stresses are not true stresses, they have qualitatively similar effects as do viscous stresses, but as a result of the large chaotic *vortical* motions of turbulence rather than the microscopic molecular motions that underlie viscous stresses.

Reynolds transport theorem: The mathematical relationship between the time rate of change of a fluid property in a *system* (volume of fixed mass moving with the flow) and the time rate of change of a fluid property in a *control volume* (volume, usually fixed in space, with fluid mass moving across its surface). This finite volume expression is closely related to the *material (time) derivative* of a fluid property attached to a moving *fluid particle.* See also *conservation laws.*

rheology: The study and mathematical representation of the deformation of different fluids in response to surface forces, or *stress.* The mathematical relationships between stress and deformation rate (or strain rate) are called *constitutive equations.* The Newtonian relationship between *stress* and *strain rate* is the simplest example of a rheological constitutive equation. See also *Newtonian* and *non-Newtonian fluid.*

rotation rate: The angular velocity, or rate of spin, of a *fluid particle* (a vector, with units rad/s, given by 1/2 the curl of the velocity vector). See also *vorticity.*

rotational flow: Synonymous with *vortical flow,* this term describes a flow field, or a region of a flow field, with significant levels of *vorticity.*

saturation pressure: The pressure at which the phase of a simple compressible substance changes between liquid and vapor at fixed temperature.

saturation temperature: The temperature at which the phase of a simple compressible substance changes between liquid and vapor at fixed pressure.

scaling parameter: A single variable, or a combination of variables, that is chosen to nondimensionalize a variable of interest. See also *nondimensionalization* and *normalization.*

schlieren technique: An experimental technique to visualize flows based on the refraction of light from varying fluid density. The illuminance level in a schlieren image responds to the first spatial derivative of density.

secondary dimensions: See *dimensions.*

shadowgraph technique: An experimental technique to visualize flows based on the refraction of light from varying fluid density. The illuminance level in a shadowgraph image responds to the second spatial derivative of density.

shear: Refers to gradients (derivatives) in velocity components in directions normal to the velocity component.

> **shear force:** See *stress, shear stress.*

> **shear layer:** A quasi two-dimensional flow region with a high gradient in streamwise velocity component in the transverse flow direction. Shear layers are inherently *viscous* and *vortical* in nature.

> **shear rate:** The gradient in streamwise velocity in the direction perpendicular to the velocity. Thus, if streamwise (*x*) velocity *u* varies in *y*, the shear rate is *du/dy*. The term is applied to *shear flows,* where shear rate is twice the *shear strain rate.* See also *strain rate.*

> **shear strain:** See *strain rate.*

> **shear stress:** See *stress, shear stress.*

shear thickening fluid: See *non-Newtonian fluid.*

shear thinning fluid: See *non-Newtonian fluid.*

SI system: See *units.*

similarity: The principle that allows one to quantitatively relate one flow to another when certain conditions are met. *Geometric similarity,* for example, must be true before one can hope for *kinematic* or *dynamic similarity.* The quantitative relationship that relates one flow to another is developed using a combination of dimensional analysis and data (generally, experimental, but also numerical or theoretical).

> **dynamic similarity:** If two objects are *geometrically* and *kinematically similar,* then if the ratios of all forces (pressure, viscous stress, gravity force, etc.) between a point in the flow surrounding one object, and the same point scaled appropriately in the flow surrounding the other object, are all the same at all corresponding pairs of points, the flow is *dynamically similar.*

> **geometric similarity:** Two objects of different size are geometrically similar if they have the same geometrical shape (i.e., if all dimensions of one are a constant multiple of the corresponding dimensions of the other).

> **kinematic similarity:** If two objects are *geometrically similar,* then if the ratios of all velocity components between a point in the flow surrounding one object, and the same point scaled appropriately in the flow surrounding the other object, are all the same at all corresponding pairs of points, the flow is *kinematically similar.*

skin friction: Surface shear stress τ_w nondimensionalized by an appropriate *dynamic pressure* $\frac{1}{2}\rho V^2$. Also called the skin friction coefficient, C_f.

solid: A material that when sheared either deforms to a fixed static position (after which deformation stops) or fractures. See also *fluid.*

sonic: At the speed of sound (*Mach number* = 1).

specific gravity: Fluid density nondimensionalized by the density of liquid water at 4°C and atmospheric pressure (1 g/cm^3 or 1000 kg/m^3). Thus, specific gravity, SG = ρ/ρ_{water}.

specific weight: The weight of a fluid per unit volume, i.e., fluid density times acceleration due to gravity (specific weight, $\gamma \equiv \rho g$).

spin: See *rotation rate* and *vorticity.*

stability: A general term that refers to the tendency of a material particle or object (fluid or solid) to move away from or return when displaced slightly from its original position.

> **neutrally stable:** See *stability.* When displaced slightly, the particle or object will remain in its displaced position.

> **stable:** See *stability.* When displaced slightly, the particle or object will return to its original position.

> **unstable:** See *stability.* When displaced slightly, the particle or object will continue to move from its original position.

stagnation point: A point in a fluid flow where the velocity goes to zero. For example, the point on the *streamline* that intersects the nose of a moving projectile is a stagnation point.

stall: The phenomenon of massive *flow separation* from the surface of a wing when *angle of attack* exceeds a critical value, and consequent dramatic loss of lift and increase in drag. A plane in stall drops rapidly and must have its nose brought down to reestablish attached boundary layer flow and regenerate lift and reduce drag.

static pressure: Another term for *pressure,* used in context with the *Bernoulli equation* to distinguish it from *dynamic pressure.*

statics: The mechanical study and analysis of material that is fully at rest in a specific frame of reference.

steady flow: A flow in which all fluid variables (velocity, pressure, density, temperature, etc.) at all fixed points in the flow are constant in time (but generally vary from place to place). Thus, in steady flows all partial derivatives in time are zero. Flows that are not precisely steady but that change sufficiently slowly in time to neglect time derivative terms with relatively little error are called *quasi-steady.*

Stokes flow: See *creeping flow.*

strain: See *strain rate.*

strain rate: Strain rate can also be called *deformation rate.* This is the rate at which a *fluid particle* deforms (i.e., changes shape) at a given position and time in a fluid flow. To fully quantify all possible changes in shape of a *three-dimensional* fluid particle require six numbers. Mathematically, these are the six independent components of a second-rank symmetric strain rate tensor, generally written as a symmetric 3 × 3 matrix. Strain is time-integrated strain rate and describes deformation of a fluid particle after a period of time. See *stress.*

extensional strain rate: The components of strain rate that describe elongation or compression of a *fluid particle* in one of the three coordinate directions. These are the three diagonal elements of the strain rate tensor. The definition of extensional strain depends on one's choice of coordinate axes. Also called *linear strain rate*.

shear strain rate: The components of strain rate that describe deformation of a *fluid particle* in response to shear changing an angle between planes mutually perpendicular to the three coordinate axes. These are the off-diagonal elements of the strain rate tensor. The definition of shear strain depends on one's choice of coordinate axes.

volumetric strain rate: Rate of change of volume of a *fluid particle* per unit volume. Also called bulk strain rate and rate of volumetric dilatation.

streakline: Used in flow visualization of fluid flows, this is a curve defined over time by the release of a marker (dye or smoke) from a fixed point in the flow. Contrast with *pathline* and *streamline*. In a steady flow, streamlines, *pathlines,* and *streaklines* all coincide. In a nonsteady flow, however, these sets of curves are each different from one another.

stream function: The two velocity components in a *two-dimensional* steady incompressible flow can be defined in terms of a single two-dimensional function ψ that automatically satisfies conservation of mass (the continuity equation), reducing the solution of the two-component velocity field to the solution of this single stream function. This is done by writing the two velocity components as spatial derivatives of the stream function. A wonderful property of the stream function is that *(iso)contours* of constant ψ define *streamlines* in the flow.

streamline: A curve that is everywhere tangent to a velocity vector in a fluid velocity *field* at a fixed instant in time. Thus, the streamlines indicate the direction of the fluid motions at each point. In a *steady flow,* streamlines are constant in time and *fluid particles* move along streamlines. In a *nonsteady flow* the streamlines change with time and fluid particles do not move along streamlines. Contrast with *pathline*.

streamtube: A bundle of streamlines. A streamtube is usually envisioned as a surface formed by an infinite number of streamlines initiated within the flow on a circular circuit and tending to form a tubelike surface in some region of the flow.

stress: A component of a force distributed over an area is written as the integral of a stress over that area. Thus, stress is the force component dF_i on a differential area element divided by the area of the element dA_j (in the limit $dA_j \rightarrow 0$), where i and j indicate a coordinate direction x, y, or z. Stress $\sigma_{ij} = dF_i/dA_j$ is therefore a force component per unit area in the i-direction on surface j. To obtain the surface force from stress, one integrates stress over the corresponding surface area. Mathematically, there are six independent components of a second-rank symmetric *stress tensor,* generally written as a symmetric 3 × 3 matrix.

normal stress: A stress (force component per unit area) that acts perpendicular to the area. Therefore σ_{xx}, σ_{yy}, and σ_{zz} are normal stresses. The normal force over a surface is the net force from shear stress, given by integrating the shear stress over the surface area. The normal stresses are the diagonal elements of the *stress tensor*.

pressure stress: In a fluid at rest all stresses are normal stresses and all stresses act inward on a surface. At a fixed point, the three normal stresses are equal and the magnitude of these equal normal stresses is called pressure. Thus, in a static fluid $\sigma_{xx} = \sigma_{yy} = \sigma_{zz} = -P$, where P is pressure. In a moving fluid, stresses in addition to pressure are *viscous stresses*. A pressure force on a surface is the pressure stress integrated over the surface. The pressure force per unit volume on a *fluid particle* for Newton's second law, however, is the negative of the gradient (spatial derivatives) of pressure at that point.

Reynolds stress: See *Reynolds stress*.

shear stress: A stress (force component per unit area) that acts tangent to the area. Therefore, σ_{xy}, σ_{yx}, σ_{xz}, σ_{zx}, σ_{yz}, and σ_{zy} are shear stresses. The shear force over a surface is the net force from shear stress, given by integrating the shear stress over the surface area. The shear stresses are the off-diagonal elements of the *stress tensor*.

turbulent stress: See *Reynolds stress*.

viscous stress: Flow creates stresses in the fluid that are in addition to hydrostatic pressure stresses. These additional stresses are viscous since they arise from friction-induced fluid deformations within the flow. For example, $\sigma_{xx} = -P + \tau_{xx}$, $\sigma_{yy} = -P + \tau_{yy}$, and $\sigma_{zz} = -P + \tau_{zz}$, where τ_{xx}, τ_{yy}, and τ_{zz} are viscous normal stresses. All shear stresses result from friction in a flow and are therefore viscous stresses. A viscous force on a surface is a viscous stress integrated over the surface. The viscous force per unit volume on a *fluid particle* for Newton's second law, however, is the divergence (spatial derivatives) of the viscous stress tensor at that point.

stress tensor: See *stress*.

subsonic: Below the speed of sound (*Mach number* < 1).

substantial derivative: See *material derivative*.

supersonic: Above the speed of sound (*Mach number* > 1).

surface tension: The force per unit length at a liquid– vapor or liquid–liquid interface resulting from the imbalance in attractive forces among like liquid molecules at the interface.

system: Usually when the word *system* is used by itself, *closed system* is implied, in contrast with a *control volume* or *open system*.

closed system: A volume specified for analysis that encloses always the same *fluid particles*. Therefore, no flow crosses any part of the volume's surface and a closed system must move with the flow. Note that Newton's law analysis of solid particles is generally a *closed system* analysis, sometimes referred to as a free body.

open system: A volume specified for analysis where flow crosses at least part of the volume's surface. Also called a *control volume*.

thermal energy: See *energy*.

three-dimensional: See *dimensionality*.

timeline: Used for visualization of fluid flows, this is a curve defined at some instant in time by the release of a marker from a line in the flow at some earlier instant in time. The timeline, often used to approximate a *velocity profile* in a laboratory flow, is very different from *streaklines, pathlines,* and *streamlines*.

tip vortex: *Vortex* formed off each tip of an airplane wing as a byproduct of lift. Synonymous with *trailing vortex*. See also *induced drag*.

total derivative: See *material derivative*.

total energy: See *energy*.

trailing vortex: See *tip vortex*.

trajectory: See *pathline*.

transient period: A time-dependent period of flow evolution leading to a new equilibrium period that is generally, but not necessarily, steady. An example is the start-up period after a jet engine is switched on, leading to a steady (equilibrium) jet flow.

transitional flow: An unstable *vortical* fluid flow at a Reynolds number higher than a critical value that is large relative to 1, but is not sufficiently high that the flow has reached a fully *turbulent flow* state. Transitional flows often oscillate randomly between *laminar* and turbulent states.

turbulence models: Constitutive model relationships between *Reynolds stresses* and the mean velocity field in turbulent flows. Such model equations are necessary to solve the equation for mean velocity. A simple and widely used modeled form for the Reynolds stresses is to write them like the Newtonian relationship for viscous stresses, as proportional to the mean strain rate, with the proportionality being a *turbulent viscosity* or *eddy viscosity*. However, unlike Newtonian fluids, the eddy viscosity is a strong function of the flow itself, and the different ways in which eddy viscosity is modeled as a function of other calculated flow field variables constitute different eddy viscosity models. One traditional approach to modeling eddy viscosity is in terms of a *mixing length,* which is made proportional to a length set by the flow.

turbulent flow: An unstable disordered state of *vortical* fluid flow that is inherently *unsteady* and that contains eddying motions over a wide range of sizes (or scales). Turbulent flows are always at *Reynolds numbers* above a critical value that is large relative to 1. Mixing is hugely enhanced, surface shear stresses are much higher, and head loss is greatly increased in turbulent flows as compared to corresponding *laminar flows*.

turbulent stress: See *Reynolds stress*.

turbulent viscosity: See *turbulence models*.

two-dimensional: See *dimensionality*.

units: A specific system to quantify numerically the dimensions of a physical quantity. The most common systems of units are SI (kg, N, m, s), English (lbm, lbf, ft, s), BGS (slug, lb, ft, s), and cgs (g, dyne, cm, s). See also *dimensions*.

unsteady flow: A flow in which at least one variable at a fixed point in the flow changes with time. Thus, in unsteady flows a partial derivative in time is nonzero for at least one point in the flow.

vapor pressure: The pressure below which a fluid, at a given temperature, will exist in the vapor state. See also *cavitation* and *saturation pressure*.

velocity: A vector that quantifies the rate of change in position and the direction of motion of a material particle.

velocity field: See *field*.

velocity profile: See *profile plots*.

viscoelastic fluid: See *non-Newtonian fluid*.

viscosity: See *Newtonian fluid*. Viscosity is a property of a fluid that quantifies the ratio of shear stress to rate of deformation (strain rate) of a *fluid particle*. (Therefore viscosity has the dimensions of stress/strain rate, or $Ft/L^2 = m/Lt$.) Qualitatively, viscosity quantifies the level by which a particular fluid resists deformation when subjected to shear stress (frictional resistance or *friction*). Viscosity is a measured property of a fluid and is a function of temperature. For Newtonian fluids, viscosity is independent of the rate of applied stress and strain rate. The viscous nature of *non-Newtonian fluids* is more difficult to quantify in part because viscosity varies with strain rate. The terms *absolute viscosity, dynamic viscosity,* and *viscosity* are synonymous. See also *kinematic viscosity*.

viscous (regions of) flow: Regions of a fluid flow where *viscous forces* are significant relative to other forces (typically, pressure force) on *fluid particles* in that region of the flow, and therefore cannot be neglected in *Newton's second law* of motion (compare with *inviscid flow*).

viscous (or frictional) force: As applicable to Newton's second law, this is the force acting on a *fluid particle* that arises from spatial gradients in viscous (or frictional) stresses within the flow. The viscous force on a surface is the viscous stress integrated over the surface. See also *stress, viscous stress*.

viscous stress tensor: See *stress*. Also called the *deviatoric stress tensor*.

viscous sublayer: The part of a turbulent boundary layer adjacent to the surface that contains the highest *viscous stresses*. The velocity gradient in this layer adjacent to the wall is exceptionally high. See also *inertial layer* and *buffer layer.*

vortex: A local structure in a fluid flow characterized by a concentration of vorticity (i.e., *fluid particle* spin or rotation) in a tubular core with circular streamlines around the core axis. A tornado, hurricane, and bathtub vortex are common examples of vortices. Turbulent flow is filled with small vortices of various sizes, strengths, and orientations.

vortical flow: Synonymous with *rotational flow,* this term describes a flow field, or a region of a flow field, with significant levels of *vorticity.*

vorticity: Twice the angular velocity, or rate of spin, of a *fluid particle* (a vector, with units rad/s, given by the curl of the velocity vector). See also *rotation rate.*

wake: The friction-dominated region behind a body formed by surface boundary layers that are swept to the rear by the free-stream velocity. Wakes are characterized by high *shear* with the lowest velocities in the center of the wake and highest velocities at the edges. *Frictional force, viscous stress,* and *vorticity* are significant in wakes.

work: See *energy.*

INDEX

Abscissa, 149
Absolute (dynamic) viscosity, 52, 54
Absolute pressure, 76–77, 90
Absolute temperature (T), 46–47
Absolute velocity, 164, 191, 245–246
Acceleration (a), 106–107, 135, 136–140,
    199–200, 265, 561
  advective (convective), 137–138
  Bernoulli equation and, 199–200
  centripetal, 265, 561
  convective (advective), 137–138
  Euclidean flow description, 135, 136–139
  field (vector), 135, 136–139
  first-order difference approximation,139
  fluid (material) flow, 135–140, 199–200
  fluid rigid-body motion, 106–107
  gradient (del) operator, 137
  Lagrangian description, 136–139
  local, 137
  material (derivative of), 139–140
  material (fluid) particles, 136–138
  material position vectors, 136–137
  Newton's second law for, 136
  normal (a_n), 199
  partial derivative operator (δ), 137
  particle streamlines (paths) of, 199
  point function, as a, 139
  residence time, 138–139
  rotation and, 265
  straight path, 106–107
  streamwise (a_s), 199
  total derivative operator (d), 137
Accuracy error, 28–30
Adhesive forces, 58
Adiabatic duct flow, 702–711.
    See also Fanno flow
Adiabatic process, 215
Advective (convective) acceleration, 137–138
Aerodynamic drag, 302–303, 548–550
Aerodynamic drag coefficient, 322–323
Aerodynamic shoulder, 549–550
Aerodynamics, study of, 2
Affinity laws, 829–830
Air, properties of at 1 atm pressure, 948, 966
Air flow, 14, 40–41. *See also* Ideal gases
Aircraft, 578–583, 612–613, 616–617, 634–643,
    662–663, 674–688, 688–693, 816–819
  airfoils, 634–638
  angle of attack (α), 612, 617
  back pressure (P_b) effects, 674–678
  blade twist for, 816–818
  boundary layer approximation for, 578–583
  compression of air in, 662–663
  converging–diverging nozzles, 675–678
  drag force on, 612–613
  efficiency of, 637–638
  finite-span wings, 638–639
  flaps, 636–637
  flow fields, 637
  flow separation, 578–583, 616–617
  induced drag and, 638–639
  lift force, 612–613, 634–643
  lift-to-drag ratio, 635–636

National Advisory Committee for Aeronautics
    (NACA) standards for, 637–638
  normal shockwaves, 675–676
  oblique shockwaves, 676
  open axial-flow fans, 816, 816–819
  pitch angle (θ), 816–819
  Prandtl–Meyer expansion waves, 688–693
  pressure forces acting on, 612–613, 634–635
  pressure gradient effects, 578–583
  propellers (rotor), 816–819
  propulsion, 674–678
  reverse thrust of, 819
  rotor airflow swirl, 817–818
  separation bubble, 579–580
  separation point, 580–581
  shock (wave) angle (δ), 684–685
  stall conditions, 580, 616, 637
  starting vortex, 635
  takeoff and landing speeds, 636–637
  turning (deflective) angle (θ), 684–685
  variable pitch, 819
  viscous forces acting on, 612–613, 634–635
  vortex shedding, 617
  wings, 612–613, 617, 634–639
  wingspan (span), 634
  Wright Brothers' impact on, 643
Airfoils, 634–638. *See also* Aircraft
Alternate depth, 734
Ammonia saturation properties, 944, 962
Ampere (A), unit of, 16
Analytical problem approaches, 21–22
Anemometers, 402–404. *See also* Flow rate
Angle of attack (α), 315, 612, 617, 634–637
Angle of deformation (α), 2
Angle valves, 375, 380
Angular displacement (α), 2
Angular momentum, 244–245, 263–273
  conservation of, 245
  equations, 244–245, 264–267
  external forces and, 265–268
  Euler's turbine formula for, 269–270
  fixed control volume (CV), 267
  moment, 266
  moment forces (F) of, 264–265
  momentum analysis and, 244–245, 263–265
  Newton's second law and, 244–245
  no external moments, 268
  radial-flow devices, 269–270
  Reynolds transport theorem (RTM) and, 266
  rotation (ω) and, 244–245, 263–265
  steady flow, 267–268
Angular velocity (rate of rotation), 151–152,
    264–265
Apparent viscosity, 52
Approximate solutions, 515–616. *See also*
    Navier–Stokes equation
Archimedes number (Ar), 309
Area, moments of, 90–91
Aspect ratio (AR), 309, 639
Atmosphere properties at high altitudes, 951, 962
Atmospheric pressure (P_{atm}), 81–83, 89–90, 249
Available wind power ($W_{available}$), 850–851
Average velocity (V_{avg}), 188, 348–349

Axial-flow turbine, 839
Axial pumps, 806, 816–824
Axisymmetric flow, 14, 457, 534, 536–537,
    610, 687–688

Back pressure (P_b), 669–678
Backward-inclined blades, 807–808
Ball valve, 380
Barometers, 81–84
Barometric pressure (P_{atm}), 81–83
Beam splitter, 405
Bends, pipe flow losses at, 377–380
Bernoulli equation, 199–214, 221, 294,
    392–393, 526–527, 531–534, 812
  acceleration of fluid particles and, 199–200
  applications of, 207–214
  approximation solutions using, 526–527,
    531–534
  boundary layers and wake regions for, 199
  compressible flow, 200–202
  dimensional homogeneity of, 294
  energy grade line (EGL), 205–207
  fluid particle acceleration and path, 199–200
  frictionless flow and, 199, 204
  hydraulic grade line (HGL), 205–207
  impeller rotation, 812
  incompressible flow, 201, 205, 221
  internal flow rate from, 392–393
  inviscid flow regions and, 199, 526–527
  irrotational flow regions, 531–534
  law of thermodynamics for, 202, 221
  limitations on use of, 204–205
  linear momentum for, 199–201
  mechanical energy balance, 201–202, 207
  Navier–Stokes equation and, 526–527,
    531–534
  negligible viscous effects and, 204
  Newton's second law for, 200–202
  negligible heat transfer and, 205
  no shaft work and, 204–205
  pressure representation, 202–204
  rotating reference frame, 812
  stagnation pressure, 203–204
  steady flow, 199–202, 204
  streamlines and, 199–200, 202–205
  unsteady, compressible flow, 202
  vector identity for, 526–527
  velocity measurement using, 392–393
Bernoulli head, 790
Best efficiency point (BEP), 791
Betz limit, 853–854
Bias error, 28
Bingham plastic fluids, 466
Biofluid mechanics, 408–416, 493–497
  blood flow studies, 410–415
  cardiovascular system, 408–410
  differential analysis of flow, 493–497
  flow measurement and, 408–416
  particle image velocimetry (PIV) for, 410, 416
  Poiseuille flow comparisons, 493–496
  pulsatile pediatric ventricular device (PVAD)
    for, 410–411

Biological systems, drag coefficient (C_D), 618–621
Biot number (Bi), 309
Blades, 806–809, 815–823, 907–915.
    See also Runners
Blasius similarity variable (η), 565–567
Blockage, 321
Blood flow studies, 410–415
Blower, 788
Bluff (blunt) bodies, 610
Body cavitation, 62
Body forces, 104–105, 246–249
Boiling properties, 941, 959
Bond number, 309
Boundary conditions, 14–15, 161, 214–215,
    438, 475–477, 888–893
    axis, 892
    closed system, 14–15
    computational fluid dynamics (CFD) use of,
        888–893
    continuity equations, 475–477
    control surface, 161
    control volume (CV), 15, 161
    differential analysis use of, 438, 475–477
    energy transfer and, 214–215
    exact solution for equations using, 475–477
    fan, 892
    flow passage (fan), 891
    fluid flow systems and, 14–15
    free-surface, 477
    inflow/outflow, 889–890
    initial, 477
    inlet, 477
    interface, 476
    interior, 892–893
    Navier–Stokes equations, 475–477
    no-slip, 476
    open systems, 15
    outlet, 477
    periodic, 891
    pressure far field, 891
    pressure inlet/outlet, 889–890
    reverse flow, 890
    rotational, 891
    symmetry, 477, 891–892
    translational, 891
    velocity inlet, 889
    wall functions, 889
    water–air interface, 477
Boundary layers, 7, 9, 199, 351–352, 364–367,
    525–526, 554–591, 625–627, 712
    approximation, 525–526, 554–591
    arbitrary shapes, 578–583
    Bernoulli equation for, 199,
    Blasius similarity variable (η) for, 565–567
    buffer, 364–365, 626
    computational fluid dynamics (CFD)
        calculations for, 580–583, 588–591
    coordinate system, 559
    curvature and, 561–562
    displacement thickness (δ^*), 568–571, 574
    equations, 555, 559–563
    external parallel flow, 625–627
    flat plates, 556–558, 572–578, 583–591
    flow region and, 9, 526–527
    free shear layers and, 557
    historical significance of, 7
    inertial sublayer, 364
    internal flow, 351–352
    inviscid regions of flow, 525–526, 554–555
    irrotational (core) regions of flow, 351, 625
    laminar flow, 557–572

logarithmic (log) law for, 366, 576
momentum integral technique for, 583–591
momentum thickness (θ), 571–572, 574
Navier–Stokes equation for, 525–526,
    560–563
no slip condition and, 525–526, 555
one-seventh-power law for, 366, 573–574
outer (turbulent), 364–366, 626
overlap (transition), 364–366, 626
pressure gradient for, 561–562, 564–565,
    578–583
procedure for approximation, 564–568
profile comparisons, 573–578
regions, 351–352, 554–591, 625–627
Reynolds number (Re) for, 557–559
shock wave interactions, 712
Spaulding's law of the wall for, 365–366,
    576, 578
thickness (δ), 556, 562–564, 574, 625
transitional flow, 557–559
turbulent flow, 364–367, 557–558,
    572–578
velocity profile for, 351–352, 364–367
viscous sublayer, 364–367, 626
wall-wake law for, 576–577
zero pressure gradient, 561, 564–565
Bourdon tube, 88
Bow wave, 686–687, 928
Brake horsepower (bhp), 791
Bridge scour, 771
British thermal unit (btu), unit of, 18, 43
Broad-crested weir, 766–767
Buckets, 790
Buckingham Pi theorem, 303–319
Buffer layer, 364–365, 626
Bulk modulus of elasticity (κ), 44–46
Bumps, open-channel flow, 764–765, 929
Buoyancy, 32, 47, 98–103

Calorie (cal), unit of, 18, 43
Candela (cd), unit of, 16
Capacity (volume flow rate), 790, 805–806
Capillary effect, 58–60
Cardiovascular system, 408–410
Cartesian coordinates, 13–14, 157, 247–249,
    440–442, 445, 450–456, 468, 470–472
    continuity equation in, 440–442, 445, 468
    control volume (CV) and, 247–249
    differential analysis applications, 470–472
    fluid flow dimensions, 13–14
    gravitational forces in, 247
    Navier–Stokes equation in, 468
    rotational flow, 157
    stream functions in, 450–456
    vorticity in, 157
Cauchy's equation, 459–464
Cavitation, 41–43, 62, 207, 797–800
    avoidance of, 797–800
    bubbles, 42, 797–798
    net positive suction head (NPSH) for,
        798–800
    pumps and, 797–800
    saturation pressure (P_{sat}) and, 41–43, 797–798
    saturation temperature (T_{sat}) and, 41–43
    sonar dome study of, 63
    sonoluminescence, 63
    vapor pressure (P_V) and, 41–43, 797
    vaporous (gaseous), 63
Cavitation number (Ca), 309
Centrifugal pumps, 806–815

Centripetal acceleration and force, 265, 561
Centroid, 91–92
Channels, 727–729, 737–759, 771
    hydraulic cross sections for, 728–729
    hydraulic diameter (D_h), 728
    hydraulic radius (R_h) for, 728–729
    open-channel flow and, 727–729, 737–759,
        771
    rectangular, 745
    trapezoidal, 745–746
Characteristic (performance) curves, 383–384,
    791–797
Chezy coefficient (C), 738–739
Choked flow, 670, 700, 708–711, 765
Circular fluid flows, comparison of, 159–160
Circular pipe flow, 348, 353–357, 952
Circulatory flow loss, 815
Closed system, 14–15
Closed volume, 805–806
Cohesive forces, 58
Colebrook equation, 367–368
Compressible flow, 10–11, 44–50, 200–202,
    439–445, 610, 659–723, 953–956
    adiabatic, 702–711
    aircraft and, 662–663, 674–688
    Bernoulli equation and, 200–202
    bulk modulus of elasticity (κ), 44–46
    CFD calculations for, 922–928
    choked, 670, 700, 708–711
    compressing flow, 678–688
    computational fluid mechanics (CFD) for,
        922–928
    continuity equations for, 439–445
    converging nozzles, 665, 670–674
    density and, 10–11, 46–47, 667–669
    external, 610
    Fanno flow, 702–711, 956
    friction and, 702–711
    heat transfer (Q) and, 693–702
    ideal gases, 45–46, 663, 667–669, 693–702,
        953–956
    isentropic, 663–678, 953
    Mach number (Ma) for, 11, 50, 663–669,
        670–672
    nozzles for, 665–678, 923–927
    one-dimensional, 663–669, 953–954
    Prandtl–Meyer expansion waves, 688–693
    property tables for, 953–956
    Rayleigh flow, 693–702, 955
    shock waves, 675–688, 712, 927–928
    sonic, 11, 50, 666–667
    speed of sound (c) and, 11, 48–50, 663–665
    stagnation properties of, 660–663, 704–705
    steady, 200–201
    subsonic, 11, 50, 666–667, 687
    supersonic, 11, 50, 666–667, 687
    three-dimensional, 610
    transonic, 50
    viscous-inviscid interactions, 712
Compressible stream function (ψ_p), 458–459
Compressing flow, 678–688. *See also* Shock
    waves
Compressive force, 77–78
Compressors, 788–789
Computational fluid dynamics (CFD), 27, 32,
    141, 149–151, 318–319, 406, 472–473,
    580–583, 588–591, 879–938
    boundary conditions for, 888–893
    boundary layer approximation, 580–583,
        588–591
    bumps, calculations for flow over, 929

cells, 881–882, 887–888
compressible flow calculations, 922–928
computational domain for, 881
continuity equation solution, 472–473
contour plots, 150–151, 916–917, 924–928,
converging–diverging nozzle flow
    calculations, 923–927
creeping flow approximation, 588–591
cross-flow heat exchanger, 915–917
cylinders, calculations of flow around, 897–
    902, 905–907
differential analysis using, 472–473
direct numerical simulations (DHS), 903
engineering software use, 27, 32
equations of motion solutions, 880–883
Euler equation solution, 580–581
flow separation, 580–583, 921–922
flow visualization and, 141, 582
grids, 883–888
heat transfer calculations, 915–922
hydraulic jump calculations, 930
integrated circuit (IC) chips, 917–922
laminar flow calculations, 893–902
magnetic resonance image (MRI)
    simulations, 931
multigridding for, 883
oblique shock wave calculations, 927–928
open-channel flow calculations, 928–930
particle image velocimetry (PIV) for, 406
pipe flow calculations, 893–896
plots for flow data, 149–151, 916–917,
    924–928
postprocessors for, 882
pressure correction algorithms for, 473
pressure gradients, 580–583
pressure contour plots, 924–928
residual of terms, 882
stator blade (vane) design, 907–915
turbulence models, 903–905
turbulent flow calculations, 902–915
vector plots, 149–150
velocity overshoot and, 590–591
Conservation of energy, see Energy (E)
Conservation of mass, see Continuity
    equations; Mass (m)
Constant error, 28
Constitutive equations, 464–465. See also
    Navier–Stokes equation
Contact angle (ϕ), 58
Continuity equation, 186, 438–450, 468–469,
    475–493, 517–518, 529–530, 694, 703,
    880
    boundary conditions for, 475–477
    Cartesian coordinates, 440–442, 468
    compressible flow, 439–445
    computational fluid dynamics (CFD)
        solution to, 472–473, 880
    conservation of mass and, 186, 438–450
    cylindrical coordinates, 442–443, 444–445, 469
    differential analysis using, 470–493
    divergence (Gauss') theorem for, 439–440,
        443–444
    exact solutions of, 475–493
    Fanno flow
    incompressible flow, 443–445
    infinitesimal control volume and, 440–443
    irrotational flow, 529–530
    Laplacian operator (∇) for, 530
    continuity equation for, 529–530
    material element (derivative) for, 443–444,
        463–464

Navier–Stokes equation coupled with,
    475–493
Rayleigh flow, 694
steady, compressible flow, 444–445
Taylor series expansion for, 440–443
velocity potential function (ϕ) and, 529–530
Continuum, 38–39, 134
Contour plots, 150–151, 916–917, 924–928
Contracted/inner (dot) product, 248
Control mass, 15
Control points, 754
Control volume (CV), 15, 32, 134–134, 160–168,
    186–187, 189–191, 217–219, 245–251,
    267, 440–443, 460–463, 583–591, 809–
    810, 851–854
    analysis, 32
    atmospheric pressure acting on, 249
    body forces, 246–248
    boundary (fixed position) of, 15, 161
    Cauchy's equation derived using, 460–463
    closed system relationship to, 160–162
    conservation of energy and, 186–187, 217–219
    conservation of mass (m_{CV}) and, 186,
        189–191, 440–443
    continuity equation derived using, 440–443
    deforming, 191, 245–246
    energy transfer and, 217–219
    extensive property of, 162
    fixed, 163, 217–219, 245, 250, 267
    forces acting on, 246–249
    gravity acting on, 247
    infinitesimal, 440–443, 460–463
    Leibniz theorem for, 165–167
    material derivative and, 167–168
    momentum analysis of, 245–249, 267
    relative velocity of, 163–165, 191, 245–246
    Reynolds transport theorem (RTT) for,
        160–168
    selection of, 245–246
    steady flow, 251
    stress tensor for, 247–249
    surface forces, 246–248
    Taylor series expansion for, 440–441, 462
    unit outer normal of, 162–163
    velocity (moving) of, 245–246
    vortex structures, 32
    wind turbine power, 851–854
Convective (advective) acceleration, 137–138
Converging–diverging nozzles, 665, 674–678,
    923–927
    back pressure (P_b) effects, 675–678
    isentropic flow though, 674–678
    supersonic flow and, 674, 678–681
Converging nozzles, 665, 670–673
    back pressure (P_b) effects, 670–674
    choked flow, 670
    isentropic flow though, 670–674
    Mach number (Ma) and, 665, 670–672
Couette flow, 477–484
Counter-rotating axial-flow fans, 819–820
Coupled equations, 438, 470–475
Creeping flow, 520–525, 588–591
    approximation, 520–525, 588–591
    CFD calculations for, 588–591
    drag force on a sphere in, 523–525
    Navier–Stokes equation for, 520–525
    Reynolds number (Re) for, 520–522
    terminal velocity of, 523–524
Critical depth (y_c), 730, 734
Critical energy, 724
Critical flow, 729–732, 739–737

987
INDEX

Critical property values (*), 668–669
Critical ratios, 668
Critical Reynolds number (Re_{cr}), 350, 557–559
Critical uniform flow, 736–740
Critical velocity (V_c), 734
Cross-flow heat exchanger, 915–917
Curvature of boundary layer, 561–562
Curved surfaces, hydrostatic forces on, 95–97
Cut-in and cut-out speeds, 850
Cylinders, 629–633, 897–902, 905–907
    CFD calculations for flow around, 897–902,
        905–907
    drag force on, 629–633
    external diameter (D), 629
    external flow over, 629–633
    flow separation, 631
    Kármán vortex street formation, 899–902
    laminar flow around, 897–902
    Reynolds number (Re) for, 629–633
    stagnation points, 629–631
    surface roughness effects, 632–633
    turbulent flow around, 905–907
Cylindrical coordinates, 13–14, 107–110, 158–159,
    442–445, 457–458, 469, 473–475
    continuity equation in, 442–445, 469
    differential analysis in, 442–445, 457–458,
        469, 473–475
    fluid flow dimensions, 13–14
    Navier–Stokes equation in, 469
    stream functions in, 457–458
    vorticity in, 158–159

d'Alembert's paradox, 548–549
Darcy friction factor (f), 309, 317, 355–356,
    358, 367–369, 952
    dimensional analysis use of, 317
    laminar flow analysis, 355–356
    Moody charts for, 367–369, 952
    pipe cross sections and, 358
    pipe flow and, 952
    ratio of significance, 309
    relative roughness (ε/D) and, 367–369
    turbulent flow analysis, 367–369
Deadweight testers, 88–89
Deformation, 2, 151–156
    angle of (α), 2
    fluid kinematic properties of, 151–156
    linear strain rate, 151–153
    rates of flow, 2, 151–156
    rotation, rate of, 151–152
    shear strain rate, 153–154
    volumetric (bulk) strain rate, 153
Density (ρ), 10–11, 32, 39–41, 46–47,
    667–669, 850–851
    compressible flow and, 10–11, 46–47,
        667–669
    critical, 667–669
    fluid properties of, 39–41
    ideal gases, 40–41, 667–669
    isentropic flow, 667–669
    Mach number (Ma) and, 11, 668
    specific gravity (SG) and, 39, 41
    volume expansion (β) and, 46–47
    vortex structure and, 32
    wind power, 850–851
Dependent Π (Pi), 300–301
Depth, pressure variation with, 78–81
Derivatives, study of, 22
Derived (secondary) dimensions, 15
Detached oblique shock wave, 686–687, 928

Deviatoric (viscous) stress tensor, 465–467
Differential analysis, 32, 437–514, 515–606,
       705–708
    applications of, 470–497
    approximate solutions of, 515–606
    biofluid mechanics flows, 493–497
    boundary conditions for, 438, 475–477
    Cartesian coordinates, 440–442, 450–456,
       468, 470–472
    Cauchy's equation for, 459–464
    compressible flow, 439–445
    computational fluid dynamics (CFD) for,
       472–473
    conservation of mass and, 438–450
    constitutive equations for, 464–465
    continuity equation for, 438–450, 468–469,
       475–493
    Couette flow, 477–484
    coupled equations, 438, 470–475
    cylindrical coordinates, 442–445, 457–458,
       469, 473–475
    differential linear momentum equation,
       459–464
    divergence (Gauss') theorem for, 439–440,
       443–444, 459–460
    error function (erf), 491–492
    exact solutions for, 475–493
    Fanno flow, friction effects and, 705–708
    flow domain, 438
    incompressible flow, 443–445, 466–468
    infinitesimal control volume for, 440–443,
       460–463
    Navier–Stokes equation for, 464–469,
       475–493, 515–606
    Poiseuille flow, 484–490, 493–496
    pressure field calculation, 470–475
    similarity solution for, 491
    steady flow, 470–490
    stream functions (ψ) for, 450–459
    stress tensors (σ) for, 459–460, 465
    Taylor series expansion for, 440–441
    unsteady flow, 490–493
    viscosity (μ) of fluids, 480–481
Differential equations, use of, 22
Differential manometer, 85–86
Diffusers for minor loses, 379
Dilatant (shear thickening) fluids, 52, 466
Dimensional analysis, 291–345, 824–827,
       855–857
    Froude number (Fr), 296–299, 323–325
    geometric similarity, 299–300
    incomplete similarity, 320–323
    insect flight and, 326
    inspectional analysis, 294–299
    kinematic similarity, 299–300
    method of repeating variables, 303–319
    models and prototypes, 299–303, 320–325
    nondimensional equations for, 294–299
    nondimensional parameters, 294–319
    scaling laws, 824–827, 855–857
    pumps, 824–827
    repeating variables, method of, 303–319
    Reynolds number (Re), 320–326
    similarity of models and prototypes, 299–303
    turbines, 855–857
    units and, 292
Dimensional homogeneity, 19–20, 293–299
    Bernoulli equation example of, 294
    Froude number (Fr) for, 296–299
    inspectional analysis and, 294–299

    law of, 293
    nondimensionalization of equations, 294–299
    nondimensional parameters, 294–296
    normalized equations, 294
    pure constants, 295
    scaling parameters, 295–296
    units and, 19–20
Dimensional variables, 294
Dimensionless (nondimensional) variables, 295
Dimensions, 15–16, 292
Direct numerical simulations (DNS), 903
Discharge coefficient (C_d), 393–394, 762–764
Disk area (A), 850–851
Displacement thickness (δ^*), 568–571
Distorted flow models, 323–325
Divergence (Gauss') theorem, 439–440,
       443–444, 459–460
Dividing streamline, 552, 581–582
Doppler-effect ultrasonic flowmeters, 399–401
Double-regulated turbine, 839
Doublet, 544–545, 546–547
Downwash, 816
Draft tube, 840
Drafting for drag reduction, 622–623
Drag coefficient (C_D), 309, 320–323, 525,
       612–625, 630–631
    aerodynamic, 322–323
    average, 612–613
    biological systems and, 618–621
    creeping flow, 525
    drafting, 622–623
    drag reduction and, 615–616
    external flow and, 612–614, 617–625,
       630–631
    frontal area, 612
    planform area, 612
    dynamic pressure, 612
    ratio of significance, 309
    Reynolds number for, 320–323, 617–618
    Stokes law, 618
    streamlining effects on, 615–616
    superposition of, 623–625
    surface roughness effects, 612, 614–615,
       626–628, 632–633
    three-dimensional bodies, 611, 620–621
    total, 617–618
    two-dimensional bodies, 611–612, 619
    vehicles, 621–623
Drag force (F_D), 51, 302–303, 523–525,
       548–550, 587–588, 610–617, 629–633,
       638–639, 645
    aerodynamic, 302–303, 548–550
    angle of attack (α), 612, 617
    balance and, 302–303
    creeping flow approximation and, 523–525
    cylinders with external flow over, 629–633
    d'Alembert's paradox, 548–549
    differential analysis of, 302–303
    external flow and, 610–617, 638–639
    flat plates with external flow over, 612,
       625–629
    flow separation, 616–617
    friction and, 51, 612, 614–617
    incompressible flow and, 610–617, 638–639
    induced, 638–639
    irrotational flow and, 548–550
    lift force and, 610–613, 638–639
    lift-to-drag ratio, 638–639
    pressure, 612–613, 614–617
    skin friction (wall shear), 567, 612, 614

    spheres and, 523–525, 629–633
    streamlining, 615–616
    surface roughness effects, 612, 614–615,
       626–628, 632–633
    vehicles, 621–623
    wing design and, 612–613
Droplet-on Demand (DoD), 593
Drum gate, 762
Ducted axial-flow fan, 816–817, 819–824
Ducted pumps, 790
Ducts, 348, 358, 663–665, 693–711. See
       Converging-diverging nozzles; Pipe flow
Dynamic machines, 790, 806–824, 834–855
Dynamic pressure, 202–203, 612
Dynamic similarity, 300–301, 315–316, 318–319
Dynamic temperature, 667
Dynamic (absolute) viscosity, 52, 54
Dynamics, study of, 2

Eckert number (Ec), 309
Eddies, 361–364, 902–903
Eddy (turbulent) viscosity, 363–364
Edge of computational domain (2-D flow), 881
Efficiency (η), 195–197, 381–390, 637–638,
       791–797, 815, 841–842
    aircraft, 637–638
    best efficiency point (BEP), 791
    centrifugal pumps, 815
    characteristic (performance) curves for,
       383–384
    circulatory flow loss, 815
    combined (overall), 196–197
    generator, 196
    impeller blades and, 815
    mechanical, 195–197
    motor, 196
    operating point, 384, 792–796
    passage losses, 815
    performance (characteristic) curves,
       383–384, 791–797
    piping systems and pump selection,
       381–390, 791–797
    pump, 196, 383–390, 791–797
    pump–motor, 383–384
    reaction turbines, 841–842
    turbine, 196, 383, 815, 841–842
Elbows, pipe flow losses at, 377–380
Electromagnetic flowmeters, 401–402
Elevation head, 205
Enclosed pumps and turbines, 790
Energy (E), 38, 43–44, 186–187, 194–199,
       214–228, 403–404, 693–702, 703,
       733–737. See also Heat transfer; Power
    conservation of, 186–187, 198–199,
       214–228, 736–737
    control volume (CV) and, 186, 217–219
    enthalpy (h), 43–44
    Fanno flow, 703
    first law of thermodynamics, 214
    flow work (P/ρ), 43–44, 194–195, 218–219
    fluid properties of, 43–44
    friction slope (S_f) and, 737
    heat (Q), transfer by, 215, 693–702
    incompressible flow, analysis of, 221–222
    internal (U), 43
    loss, 216, 219–221
    kinetic (ke), 43, 195
    kinetic energy correction factor (α), 221–222
    macroscopic, 43

mechanical (E_{mech}), 194–199, 215–228
microscopic, 43
open-channel flow, 733–737
potential (pe), 43, 195
Rayleigh flow, 694–702
single-stream devices, 219
specific, 733–736
specific heats and, 43–44
specific total (e), 38, 44
steady flows, analysis of, 219–221
thermal, 43
total, 43–44
units of, 43
work, transfer by, 215–219
Energy absorbing devices, 788. *See also* Pumps
Energy dissipation ratio, 759
Energy grade line (EGL), 205–207, 790, 840–841
Energy pattern factor (K_e), 851
Energy producing devices, 788. *See also* Turbines
Engineering Equation Solver (EES), 26–27
Engineering software packages, 25–27
English system of units, 16–19, 292, 957–969
Enthalpy (h), 43–44, 660–662
compressed flow and, 660–662
energy and, 43–44
ideal gas, 661
stagnation and, 660–662
total, 660–662
Entrance region, 351–352, 893–896
Entropy change, 694–695, 703
Entry length, 352–353
Equations, 40, 105–106, 108, 185–242,
  244–245, 249–250, 264–267, 269–270,
  293–299, 437–514, 517–520, 555, 559–
  563, 880–883
angular momentum, 244–245, 264–267
approximate solutions for, 517–520
Bernoulli, 199–214
boundary layer, 555, 559–563
Cauchy's, 459–464
computational fluid dynamic (CFD)
  solution, 880–883
conservation of momentum principle, 186
continuity, 438–450, 468–469, 475–493, 880
coupled, 438, 470–475
differential analysis, 437–514
dimensional homogeneity and, 293–299
efficiency, 194–199
energy, 186–187, 194–199, 214–228
Euler's turbine formula, 269–270
ideal gases, 40
inspectional analysis of, 294–299
linear momentum, 186, 244–245, 249–250
mass, 186, 187–194
motion, 105–106, 108, 517–520, 880–883
Navier–Stokes, 464–469, 475–493, 880
nondimensionalization of, 294–299, 517–520
normalized, 294
of state, 40
Taylor series expansion, 440–441
Equipotential lines, 535–536
Equivalent length (L_{equiv}), 375
Equivalent roughness (ε), 368
Errors, 28–31, 88–89, 221–222, 251–253, 491–492
Euler equation, 525–526, 531, 580–581
Euler number (Eu), 309, 519
Eulerian description, 134–140, 167–168
acceleration field (vector), 135, 136–139
field variables, 134–136
flow domain (control volume), 134

flow field, 135
gradient (del) operator, 137
material (substantial) derivative, 139–140,
  167–168
material position vectors, 136–137
partial derivative operator (δ), 137
pressure field (scalar), 134
Reynolds transport theorem (RTM) and,
  167–168
total derivative operator (d), 137
vector variables, 134–135
velocity field (vector), 134
Euler's turbine formula, 269–270
Euler's turbomachine equation, 810
Expanding flow, 688–693
Expansion fan, 688–689
Experimental problem approaches, 21–22
Extensional (linear) strain, 151
Extensive property of fluid flow, 38, 162
External diameter (D), 629
External flow, 10, 607–657, 678–693
aircraft, 612–613, 616–617, 634–643
angle of attack (α), 612, 617
axisymmetric, 610
bluff (blunt) bodies, 610
compressible, 610, 678–693
cylinders, over, 629–633
drag coefficient (C_D), 612–614, 617–625,
  630–631
drag force (F_D), 610–617, 638–639
drag reduction, 610–611, 615–616, 645
flat plates, over, 612, 625–629
flow fields for, 608–610
flow separation, 616–617, 630–631
free-stream velocity, 608
friction and, 612, 614–617, 625–629
incompressible, 610–657
internal flow compared to, 10
lift force, 610–613, 634–643
parallel, 625–629
Prandtl–Meyer expansion waves, 688–690
resultant forces of, 607
Reynolds number (Re) for, 612, 617–618,
  629–631
shock waves, 678–693, 712
spheres, over, 629–633
stagnation points, 629–631, 635, 639
streamlined bodies, 610, 614–616
surface roughness effects, 612, 614–615,
  626–628, 632–633
three-dimensional bodies, 610, 611, 620–621
two-dimensional bodies, 608, 610,
  611–612, 619
External forces, 249–250, 254–255, 265–268

Face of computational domain (3-D flow), 881
Fanning friction factor (C_f), 309, 317, 355–356
Fanno flow, 702–711, 956
choked, 708–711
continuity equation for, 703, 705
differential analysis of, 705–708
energy equation for, 703, 705
entropy change of, 703
equation of state for, 704
friction effects on, 704–705
ideal-gases, 702–711, 956
Mach number (Ma) for, 704–705
maximum duct (sonic or critical) length,
  707–708

momentum equation for, 703–705
property functions, 956
property relations of, 705–708
Reynolds number (Re) for, 707–708
T-s diagrams for, 704–705
Fanno line (curve), 679–682
Fans, 788, 816–817, 819–824, 907–915
axial pumps, 788, 816, 819–824
blade row (cascade), 821
CFD model for, 907–915
counter-rotating, 819–820
ducted-axial flow, 816–817, 819–824
open-axial flow, 816
stator blades (vanes), 819–822, 907–915
tube-axial, 819–824
vane-axial, 819–820, 907–915
Field variables, 134–136
Finite-span wings, 638–639
First-order difference approximation, 139
Flaps, lift effects from, 636–637
Flat plate analysis, 89–92, 98, 556–558,
  572–578, 583–591, 612, 625–629
boundary layer approximation of, 556–558,
  572–578, 583–591
boundary layer regions, 625–627
buffer layer, 626
buoyant force on, 98
drag force and, 612, 625–629
external flow over, 612, 625–629
friction coefficient for, 627–629
hydrostatic forces on, 89–92
irrotational flow region, 625
Kármán integral equation, 585–588
laminar flow and, 556–558
momentum integral technique for, 583–591
overlap layer, 626
parallel flow over, 625–629
Reynolds number (Re) for, 626–628
skin friction and, 612
surface, 89–92
surface roughness and, 628
turbulent flow and, 572–578
turbulent layer, 626
velocity boundary layer, 625
viscous sublayer, 626
Flow depth, 726–727, 751
Flow domain (control volume), 134, 438
Flow field regions, 9–10, 156–157, 199, 273–274,
  350–352, 517, 525–591, 608–610
Flow rate, 187–189, 191, 367–370, 391–416,
  670–674, 805–806, 808–809
discharge coefficient (C_d) for, 393–394
Doppler-effect ultrasonic flowmeters,
  399–401
electromagnetic flowmeters, 401–402
flowmeters, 392–402
internal flow, 367–370, 391–408
isentropic nozzle flow, 670–674
King's law for energy balance, 403–404
laser Doppler velocimetry (LDV),
  404–406
mass ($\dot{m}$), 187–188, 191, 670–674
obstruction flowmeters, 392–396
paddlewheel flowmeters, 397–398
particle image velocimetry (PIV), 406–408,
  410, 416
pipe length and, 369–370
Pitot formula for, 392
Pitot probes, 391–392
positive displacement flowmeters, 396

Flow rate (*continued*)
  pump capacity, 805–806, 808–809
  Strouhal number (St) for, 402
  thermal anemometers, 402–404
  turbine (propeller) flowmeters, 397
  ultrasonic flowmeters, 399–401
  variable-area flowmeters (rotameters), 398
  velocimetry, 404–408
  velocity measurement, 391–408
  volume, 188–189, 391, 805–806, 808–809
  vortex flowmeters, 402
Flow separation, 9, 578–583, 616–617,
    630–631, 921–922
  angle of attack (α), 612, 617
  boundary layer approximation for, 578–583
  CFD calculations for, 921–922
  friction drag and, 616–617
  pressure drag and, 616–617
  pressure gradient effects, 578–583
Flow visualization, 32, 141–148, 582, 631
Flow work (P/ρ), 43–44, 194–195, 218–219
Flowmeters, 392–402. *See also* Flow rate
Fluctuating components of turbulent flow,
    361–362
Fluid, defined, 2
Fluid flow, 1–35, 52–54, 133–184, 191–195,
    199–214, 250–273, 299–303, 312–325,
    347–436, 437–514, 663–669, 726,
    880, 882
  acceleration of, 135–140, 199–200
  accuracy of measurements, 28–30
  analysis, 1–35
  Bernoulli equation applications of, 199–214
  biofluid mechanics, 408–416
  boundary layers, 8–9, 351–352, 364–367
  circular, 159–160
  classification of, 9–14
  compressible, 10–11, 200–202
  control volume, 15, 160–168
  differential analysis of, 32, 437–514
  engineering software packages for, 25–27
  Eulerian description, 134–140
  external, 10
  field variables, 134–136
  forced, 11
  forces (*F*) and, 2–3
  frictionless, 199, 204
  fully developed (one-dimensional), 13–14
  global properties of, CFD, 880, 882
  incompressible, 10–11, 192–194, 201, 205, 221
  internal, 10, 347–436
  inviscid, 10, 199
  irrotational, 156–157
  Lagrangian description, 134–140
  laminar, 11, 349–361
  modeling (mathematical), 21–23
  molecular structure and, 3–4
  natural (unforced), 11
  no-slip condition, 8–9
  no-temperature-jump condition, 9
  one-dimensional (fully developed), 13–14,
    663–669
  precision of measurements, 28–29
  problem-solving technique, 23–25
  Reynolds transport theorem (RTM) for,
    160–168
  rotational (ω), 151–152, 156–159
  significant digits and, 28–31
  steady, 12–13, 191, 199–202, 251, 253–254
  strain rate and, 2–3

streaklines, 144–146
streamlines, 141–142
systems, 14–15
three-dimensional, 13–14
turbulent, 11, 349–353, 361–374
two-dimensional, 13–14
units of measurement, 15–21
unsteady, 12–13, 202, 726
viscous, 10
Fluid kinematics, 133–184
  acceleration field (vector), 135, 136–139
  angular velocity (rate of rotation), 151–152
  circular flows, comparison of, 159–160
  deformation of flow, 151–156
  Eulerian description, 134–140, 167–168
  Lagrangian description, 134–140, 167–168
  linear strain rate (ε), 151–153
  material derivative, 139–140, 167–168
  motion of flow, 151–160
  pathline, 142–144
  plots, 148–151
  refractive flow visualization, 147–148
  Reynolds transport theorem (RTM), 160–168
  rotation, 151–152, 156–160
  streaklines, 144–146
  streamlines, 141–142
  surface flow visualization, 148
  timelines, 146–147
  translation, 151
  velocity vector (rate of translation), 134, 151
  vorticity, 156–160
Fluid machines, *see* Turbomachines
Fluid mechanics, 2, 4–5, 6–8, 14–15, 21–31,
    185–242, 243–289, 291–345
  applications of, 4–5
  Bernoulli equation, 199–214, 221
  categories of, 2
  dimensional analysis, 291–345
  efficiency (η), 195–197
  energy (*E*), conservation of, 186–187, 198–
    199, 214–228
  energy grade line, 205–207
  engineering and, 21–31
  equations for, 185–242
  flow systems, 14–15, 243–289, 293
  history of, 6–8
  hydraulic grade line, 205–207
  linear momentum, 186
  mass (*M*) conservation of, 186, 187–194
  mechanical energy, 194–199, 201–202, 207,
    215–228
  momentum analysis, 243–289
Fluid properties, 37–73
  capillary effect, 58–60
  cavitation, 41–43, 62
  compressibility (κ), 44–50
  continuum, 38–39
  density, 39–41
  energy, 43–44
  equations of state, 40
  ideal gases, 40–41
  saturation and, 41–43
  specific gravity (SG), 39, 41
  specific heats, 43–44
  speed of sound, 48–50
  state postulate for, 38
  surface tension, 55–60
  vapor pressure, 41–43
  viscosity, 50–55
  volume expansion, 46–48

Fluid statics, 75–131. *See also* Pressure
  buoyancy, 98–103
Force (*F*), 2–4, 10, 17–19, 50–52, 58–59,
    89–103, 104–105, 216–219, 244–250,
    254–255, 264–267, 607–617, 634–643
  adhesive, 58
  angular momentum and, 244–245, 264–266
  body, 104–105, 246–248
  buoyant, 98–103
  capillary effect and, 58–59
  centripetal, 265
  cohesive, 58
  compressive, 77–78
  control volume, acting on, 246–250
  drag, 51, 610–617, 638–639
  external, 249–250, 254–255, 265–267
  flow and, 2–4, 249–250, 254–255
  friction, 10, 50–51
  gravity as, 18, 247
  hydrostatic, 89–97
  intermolecular (pressure), 3–4
  lift, 51, 610–613, 634–643
  linear momentum and, 249–250, 254–255
  moment of, 265–266
  momentum analysis and, 244–250,
    254–255, 264–267
  net, 244–245
  pressure and, 76–81, 216–219
  resultant, 90–91, 95–96, 607
  rigid-body motion, 104–105
  shear, 52
  stresses and, 2–3
  surface, 104–105, 246–248
  thrust, 254–255
  units of, 17–19
  viscosity and, 10, 50–52
  weight as, 17–18
  work as, 18–19, 216–219
Forced flow, 11
Forward-inclined blades, 808
Fourier number (Fo), 309
Fractional factorial test matrix, 319
Francis turbine, 838–864
Free delivery, 384, 791
Free shear layers, approximation of, 557
Free-stream velocity, 608
Free-surface boundary conditions, 477
Free-surface flow, 323–325
Freezing point properties, 941, 959
Friction, 10, 50–51, 199, 204, 348–349,
    355–356, 358, 365, 367–369, 567, 612,
    614–617, 625–629, 702–711, 952
  absence of (frictionless) in flow, 199, 204
  coefficient, 355–356, 614–615, 627–629
  Darcy friction factor (f), 355–356, 367–369
  drag, 612, 614–617, 625–629
  equivalent roughness and, 368
  external flow, effects on, 612, 614–617,
    625–629
  factor (*f*), 355–356, 358, 614, 952
  Fanning friction factor, 309, 317, 355–356
  Fanno flow, 702–711
  flow separation, 616–617
  force, 10, 50–51, 199, 204
  internal flow and, 348–349
  laminar flow, 355–356, 358
  local coefficient, 567
  Moody chart for pipe flow, 367–369, 952
  parallel flow on flat plates, 625–629
  pipe flow and, 348–349, 355–356, 358

relative roughness (ε/D) and, 367–369
skin (wall shear), 567, 612, 614
streamlining, 615–616
surface roughness and, 612, 614–615, 628
turbulent flow and, 365, 367–368, 576
velocity, 365, 576
viscosity (μ) and, 10, 50–51
Friction lines, 148
Friction slope (S_f), 737, 751
Frictionless flow, 199, 204
Fringe lines, 405
Frontal area, 612
Froude number (Fr), 296–299, 309, 323–325, 519, 729–730, 750–751, 929–930
CFD calculations using, 929–930
critical depth of, 730
critical flow, 729–730
dimensional analysis using, 323–325
free-surface flow, 323–325
Navier–Stokes nondimensionalization using, 519
nondimensionalization using parameter of, 296–299
open-channel flow, 729–730, 929–930
ratio of significance, 309
subcritical (tranquil) flow and, 729–730
supercritical (rapid) flow and, 729–730
surface profiles, 750–751
wave speed (c_0) and, 729–730
Full dynamic pressure, 548
Full factorial test matrix, 319
Fully developed flow, 13–14. *See also* One-dimensional flow
Fundamental (primary) dimensions, 15, 292

Gage pressure (P_{gage}), 76–77, 82–83
Gas dynamics, 2
Gas turbines, 833, 847
Gaseous cavitation, 62
Gases, 3–4, 53–54, 99, 940, 948, 958, 967. *See also* Air; Ideal gases
Gate valves, 374, 380
Gates, 761–770, 930. *See also* Flow control
Gaukler-Manning equations, 738–739
Gauss' (divergence) theorem, 439–440, 443–444, 459–460
General Conference of Weights and Measures (CGMP), 16
Generator efficiency ($\eta_{generator}$), 196
Geometric similarity, 299–300
Geometric similarity, 32
Global properties of flow, 880, 882
Globe valve, 380
Gradient (del) operator, 137
Gradually varied flow (GVF), 727, 747–756
Grashof number (Gr), 309
Gravity, 18, 39, 41, 102–103, 247, 357–358, 461, 519, 526–527
buoyancy and, 102–103
Cauchy's equation and, 461
center of, 102
control volume, acting on, 247
force of, 18, 247
hydrostatic pressure and, 519
inviscid flow regions and, 526–527
laminar flow, effects of on, 357–358
metacentric height, 102–103
specific (SG), 39, 41
stability and, 102–103

Grids, 881–883, 883–888
equiangle skewness, 885
face (3-D flow), 881
generation, 883–888
hexahedral cells, 887–888
hybrid, 886–887
independence, 888
intervals, 884
multiblock analysis using, 886–887
nodes, 884
polyhedral meshes, 887–888
prism cells, 887–888
skewness, 885
structured, 884, 886
tetrahedral cells, 887–888
three-dimensional (3-D), 881, 886–887
two-dimensional (2-D), 881, 885–887
unstructured, 884–885
Gross head (H_{gross}), 839

Harmonic functions, 535–537
Head gate, 840
Head loss (h_L), 220–221, 348, 356–357, 369–370, 375–376, 736–737, 758–759
head loss (h_L), 736–737, 758–759
friction slope (S_f) for, 737
internal flow, 348, 356–357, 369–370, 375–376
irreversible, 220–221
laminar flow, 356–357
minor losses, 375–376
open-channel flow, 736–737, 758–759
steady flow analysis and, 220–221
total, 375–376
turbulent flow, 369–370
Heads (h), 79, 205–207, 221, 383–384, 790, 791–800, 839–842
Bernoulli, 790
cavitation and, 797–800
elevation, 205
energy grade line and, 205–207, 790, 840–841
gross for, 839, 841
hydraulic grade line, 205–207
net (total), 205, 790–795, 841–842
net positive suction, 798–800
piping system efficiency and, 383–384, 791–797
pressure, 79, 205
pump, 221, 384, 792–793
pump performance and, 790–797
required net, 792–793
shutoff, 384, 791–792
turbine, 221, 383–384
turbine performance and, 839–842
useful pump, 792
velocity, 205
Heat transfer (Q), 43, 205, 215, 693–702, 915–922
CFD calculations for, 915–922
compressible duct flow, 693–702
cross-flow heat exchanger, 915–917
energy (E) transfer and, 43, 215
integrated circuit (IC) chips, cooling of, 917–922
negligible effects of, 205
printed circuit boards (PCB) and, 917–922
rate, 215, 919
Rayleigh flow, 693–702

reattachment line, 921–922
separation bubbles, 921–922
temperature contour plots, 916–917
thermal energy compared to, 43, 215
Homologous operating points, 826
Horizontal axis wind turbines (HAWTs), 847–850
Horsepower (hp), unit of, 18–19
Hot film anemometers, 403
Hot wire anemometers, 403–404
Hydraulic cross sections, 728–729
Hydraulic depth (y_h), 732–733
Hydraulic diameter (D_h), 350, 728
Hydraulic grade line (HGL), 205–207
Hydraulic jump, 731–733, 757–761, 930
Hydraulic radius (R_h), 728–729
Hydraulic turbines, 833
Hydraulics, study of, 2
Hydrodynamic entrance region, 351–352
Hydrodynamic entry length (L_h), 351–353
Hydrodynamics, 2
Hydrogen bubble wire, 147
Hydrostatics, 83–84, 89–97, 464, 479, 519–520
absolute pressure, 76–77, 90
area, moments of, 90–91
atmospheric pressure, 89–90
center of pressure (point of application), 89–91
centroid, 91–92
curved surfaces, 95–97
inertia, moments of, 91–92
forces, 89–97
gravity effects on, 519
magnitude, 90–91, 96
modified pressure and, 520
multilayered fluid and, 96
nondimensionalized equations and, 519–520
rectangular plates, 92–94
plane surfaces, 89–92
pressure distribution, 479, 519
pressure, 83–84, 464
pressure prism, 91–92
resultant forces (F_R), 90–91, 95–96
submerged surfaces, 89–97
Hypersonic flow, 11, 50

Ideal gases, 40–41, 44–46, 661, 663–665, 667–669, 693–711, 940, 953–956, 958
compressibility of (κ), 45–46
compressible flow of, 45–46, 661, 663–665, 667–669, 693–702, 953–956
converging-diverging duct flow of, 663–665
critical property values (*), 668–669
density of, 40–41
duct flow, 693–702
energy of, 44
energy equation for, 694, 703
equation of state, 40, 695, 704
enthalpy of, 661
Fanno flow, 702–711, 956
fluid properties of, 44–46
isentropic flow, 663–665, 667–669, 953
Mach number (Ma) for, 663–665, 667–669
property relations, 699–700, 705–708
property tables for, 940, 953–956, 958
flow property variations, 667–669
Rayleigh flow, 693–702, 955
shock functions for, 954
specific heat of, 44, 667–669, 940, 958
speed of sound (c) for, 663
volume expansion (β) of, 46

Image sink, 551
Immersed bodies, buoyancy of, 101–103
Impact pressure, 548
Impellers (rotor), 806–808, 811–812, 819–822
  axial pumps, 819–822
  backward-inclined blades, 807–808
  centrifugal pumps, 806–807, 811–812
  forward-inclined blades, 808
  radial (straight) blades, 808
  shroud, 807
  stator blades (vanes) for, 819–822
Impulse turbines, 835–837
Impurities, 57
In series pipes, 381–382
Inclined manometer, 85
Inclined pipe flow, 357–358
Incomplete similarity, 320–323
Incompressible flow, 10–11, 44, 192–194, 201,
    205, 221–222, 443–445, 466–468,
    610–613, 634–643. See also External flow
  Bernoulli equation for, 201–202, 205, 221
  compressible flow compared to, 10–11
  conservation of mass and, 192–194, 443–445
  continuity equations for, 443–445
  determination of, 44
  differential analysis of, 443–445, 466–468
  drag force and, 610–617, 638–639
  energy analysis of, 221–222
  external, 610–657
  isothermal, 466–468
  kinetic energy correction factor (α), 221–222
  lift force and, 610–613, 634–643
  mass balance for, 192–194
  material elements and, 443–444
  Navier–Stokes equation for, 466–468
  separation, 630–631
  specific heat of, 44
  steady, 201–202, 221
  streamlined bodies, 610, 614–616
  surface roughness effects, 626–628, 632–633
  three-dimensional, 610, 611
  two-dimensional, 608, 610, 611–612
Independent Π (Pi), 300–301
Induced drag, 638–639
Inertia, moments of, 91–92
Inertial sublayer, 364
Infinitesimal control volume, 440–443, 460–463
Initial boundary conditions, 477
Inlet and outlet flow, 164–165, 251, 254
Inlet boundary conditions, 477
Inner flow region, 555. See also Boundary layers
Insect flight, dimensional analysis and, 326
Insertion electromagnetic flowmeter, 401
Inspectional analysis, 294–299
Integral properties of flow, 882
Integrated circuit chips (ICs), 917–922
Intensive property of fluid flow, 38, 162
Interface boundary conditions, 476
Interferometry, 147
Intermolecular bonds (pressure), 3–4
Internal energy (U), 43
Internal flow, 10, 347–436
  Bernoulli equation for, 392–393
  biofluid mechanics, 408–416
  Colebrook equation for, 367–374
  ducts, 348, 358
  entrance region, 351–352
  entry length, 352–353
  external flow compared to, 10
  flow rate, 369–370
  friction effects on, 348–349

gravity effects on, 357–358
head loss (h_L), 348, 356–357, 369–370,
    375–376
laminar, 349–361
measurement of velocity, 391–408
minor losses, 374–381
Moody chart for, 367–374
pipes, 348–349, 351–390
  Hagen-Poiseuille flow, 356
  pressure drop (ΔP), 348, 355–357, 369
  pumps, 381–390
  Reynolds number (Re) for, 350–351
  transitional, 349–351
  tubes, 348
  turbulent, 349–353, 361–374
  valves and, 364–365, 367
  velocity of, 348–349, 357–358, 364–367,
      391–408
International System (SI) of units, 16–19, 292,
    939–965
Interrogation regions, PIV, 407
Inviscid flow, 10, 199, 525–529
  approximation, 525–529
  Bernoulli equation for, 199, 526–527
  boundary layers and, 525–526
  Euler equation for, 525–526
  gravity effects on, 526–527
  Navier–Stokes equation and, 525–529
Irrotational flow, 156–157, 351, 529–554, 625
  aerodynamic drag and, 548–550
  approximation, 529–554
  axisymmetric regions, 534, 536–537
  Bernoulli equation for regions of,
      531–534
  boundary layers and, 351, 625
  circular, 156–157
  circulation (vortex strength) of, 542–543,
      545–546
  continuity equation for, 529–530
  core region, 351
  d'Alembert's paradox, 548–549
  doublet of, 544–545, 546–547
  doublet strength of, 544–545
  equipotential lines for, 535–536
  Euler equation for, 531
  external parallel flow, 625
  harmonic functions for, 535–537
  Laplace equation for, 530, 535–537
  Laplacian operator, 530
  line sink of, 540–542, 545
  line source of, 540–542
  line vortex of, 542–543, 545
  momentum equation for, 531
  Navier–Stokes equations for, 529–554
  planar regions, 534–537, 538–554
  regions of potential flow, 351, 529–530
  stream functions for, 535–537, 545–547
  superposition of, 538, 545–554
  two-dimensional regions of, 534–537
  uniform stream of, 539, 546–549
  velocity components for, 537
  velocity potential function for, 529–530,
      535–537
  vortex strength (circulation) of, 542–543,
      545–546
  vorticity (ζ) of, 156–157
  zero pressure point of, 549–550
Isentropic flow, 663–678, 953
Isobars, 106
Isolated system, 15
Isothermal compressibility (α), 46

Jakob number (Ja), 309
Joule (J), unit of, 18, 43

Kaplan turbine, 838–839
Kármán integral equation, 585–588
Kármán vortex street, 145, 900–902
Kelvin (°K) temperature scale, 16, 40
Kilogram (kg), unit of, 16–17
Kilojoule (kj), unit of, 43
Kilopascal (kPa), unit of, 76
Kilowatt-hour (kWh), unit of, 19
Kinematic eddy (turbulent) viscosity, 363–364
Kinematic similarity, 32, 299–300
Kinematic viscosity (ν), 53
Kinematics, laws of, 32. See also Fluid kinematics
Kinetic energy (ke), 43, 194–195, 265
Kinetic energy correction factor (α), 221–222,
    357
Knudsen number (Kn), 309

Lagrangian description, 134–140, 167–168
Laminar flow, 11, 349–361, 556–572, 578–583,
    727–729, 893–902
  Blasius similarity variable (η) for, 565–567
  boundary layer approximation, 556–572,
      578–583
  circular pipe, 353–357
  computational fluid dynamic (CFD)
      calculations, 580–583, 893–902
  cylinders, CFD calculations for flow around,
      897–902
  Darcy friction factor (f), 355–356
  dividing streamline, 581–582
  entrance region, 351–352, 893–896
  entry length, 352–353
  flat plate boundary layer, 556–558
  flow separation, 580–583
  fluid behavior (flow regime) of, 349–351
  friction factor (f), 355–356, 358
  fully developed, 335–355
  head loss (h_L), 356–357
  hydraulic diameter and radius for, 728–729
  hydrodynamic entry length (L_h), 352–353
  inclined pipes, 357–358
  internal, 349–361
  kinetic energy correction factor (α) for, 357
  Navier–Stokes equation for, 559–563
  noncircular pipes, 358
  open-channel flow, 727–729
  pipe flow, CFD calculations for, 893–896
  Pouseuille's law for, 356
  pressure drop (ΔP), 355–357
  pressure gradient effects, 578–583
  pressure loss (ΔP_L), 355
  Reynolds number (Re) for, 350–351, 556,
      727–729
  skin (local) friction coefficient for, 567
  transition to turbulent, 349–351, 557–558
  turbulent flow compared to, 557–559,
      580–583
  velocity profile for, 351–352, 353–355
  zero pressure gradient, 561, 564–565
Laplace equation for irrotational flow, 530,
    535–537
Laplacian operator (∇), 530
Large eddy simulation (LES), 903
Laser Doppler anemometry (LDA), 404
Laser Doppler velocimetry (LDV), 404–406, 410
Laval nozzles, 665, 678

Law of the wall, 365–366, 576, 578
Leading edge angle, 809, 843
Leading edge flaps, 636
Leibniz theorem, 165–167
Lewis number (Le), 309
Lift coefficient, 309, 314–315, 612–613,
    634–637
Lift force, 313–316, 610–613, 634–643
    airfoils, 634–639
    angle of attack, 315, 612, 634–637
    aspect ratio (AR) for, 639
    dimensional analysis of, 313–316
    drag and, 610–613, 638–639
    finite-span wings, 638–639
    flaps, effects from, 636–637
    frontal area, 612
    incompressible flow and, 610–613, 634–643
    induced drag and, 638–639
    lift-to-drag ratio, 635–636
    Mach number (Ma) and, 315–316
    Magnus effect, 639–643
    planform area, 313, 612, 639
    pressure effects on, 634–635
    spinning, generation of by, 639–643
    stagnation points, 635, 639
    stall conditions, 637
    starting vortex, 635
    takeoff and landing speeds, 636–637
    viscous effects on, 634–635
    wing prototype for, 313–316
Lift-to-drag ratio, 635–636
Line of action, 91
Line (contour) plot, 150–151
Line sink, 160, 540–542, 545
Line source, 540–542
Line source strength, 540–541
Line vortex, 542–543, 545
Linear momentum, 186, 199–201, 244,
    249–263, 459–464
    Cauchy's equation for, 459–464
    conservation of, 199, 244
    control volume (CV), 250–251, 459–460
    differential analysis and, 459–464
    divergence theorem for, 459–460
    energy balance and, 199–200
    equation, 186, 244, 249–250
    external forces and, 249–250, 254–255
    momentum analysis and, 244, 249–263
    momentum-flux correction factor (b), 251–253
    Newton's second law as, 200, 244, 249–250
    no external forces and, 254–255
    Reynolds transport theorem for, 250
    steady flow, 251, 253–254
    thrust, 254–255
    uniform flow, 251
Linear strain rate (ε), 151–153
Liquids, 3–4, 53–60, 946, 964
    properties of, 946, 964
    viscosity of, 53–54
Local acceleration, 137
Local friction coefficient, 567
Logarithmic (log) law, 366, 576
Loss (resistance) coefficient (K_L), 374–379
Loss of energy in transfer, 216, 219–221

Mach angle (μ), 688
Mach number (Ma), 11, 50, 310, 315–316,
    663–669, 670–672, 679–681, 685–688,
    688–689, 691, 695–696, 704
    back pressure (P_b) and, 669–678

compressible flow, 663–669, 670–672, 679–
    681, 685–688, 688–689
converging nozzles, 665, 670–672
converging–diverging nozzles, 665
critical property values (*), 668–669
dimensional analysis use of, 315–316
estimation of, 691
Fanno flow, 704
hypersonic flow, 50
ideal gases, 663–665, 667–669, 695–696, 704
isentropic flow, 663–669, 672
lift force analysis and, 315–316
nozzle shapes and, 665–669
Rayleigh flow, 695–696
shock waves and, 679–681, 685–688
sonic flow, 50, 666–667
subsonic flow, 50, 666–667, 687
supersonic flow, 50, 666–667, 687
transonic flow, 50
unity, 665–665, 668
Mach wave, 688
Macroscopic energy, 43
Magnetic resonance image (MRI) simulations, 931
Magnitude, 90–91, 96
Magnus effect, 639–643
Manning coefficient (n), 739
Mass (m), 21, 165, 186, 187–194, 349, 382, 393,
    438–450, 670–674, 736, 790, 940, 958
    absolute velocity for, 191
    average velocity for, 188, 349
    balance, 189, 191–194, 393
    conservation of, 186, 187–194, 349, 382,
        438–450, 736
    continuity equations for, 186, 438–450
    control volume, 186, 189–191, 440–443
    differential analysis and, 438–450
    divergence (Gauss') theorem for, 439–440
    incompressible flow, 192–194
    molar (M), 940, 958
    open-channel flow, 736
    pipe flow analysis and, 382
    principle of conservation, 189–191
    pumps, 790
    relative velocity for, 191
    steady-flow processes, 191
    Taylor series expansion for, 440–441
    time rate of change, 187
    volume flow rate and, 188–189, 790
    weight and, 21
Material acceleration, 139–140
Material derivative, 139–140, 167–168,
    443–444, 463–464
    acceleration, 139–140
    Cauchy's equation and, 463–464
    continuity equation and, 443–444
    control volume and, 167–168
    differential analysis using, 443–444, 463–464
    fluid particle flow and, 139–140
    pressure, 139–140
    Reynolds transport theorem (RTM) and,
        167–168
Material element, 443–444, 463–464
Material (fluid) particles, 134, 136–138
Material position vectors, 136–137
Material volume, 165–167
Mean pressure (P_m), 465
Mechanical energy (E_{mech}), 194–199, 201–202,
    207, 215–228
    conversion of, 195–197
    efficiency, 195–197
    energy grade line (EGL), 207

flow energy, 195
flow work, 194–195, 218–129
head loss, 220–221
hydraulic grade line (HGL), 207
irreversible losses, 220–221
kinetic energy and, 194–195
loss, 216, 219–221
potential energy and, 194–195
power transfer from, 215–217
pressure forces ($W_{pressure}$) and, 216–219
shaft work (W_{shaft}) as, 195–196, 199, 216
steady flows, analysis of, 219–228
transfer of, 195–197, 205–228
Mechanical pressure (P_m), 465
Mechanics, study of, 2
Megapascal (MPa), unit of, 76
Meniscus, 58
Mesh, *see* Grids
Metacentric height (GM), 102–103
Metals, properties of liquid, 947, 965
Meter (m), unit of, 16–17
Microelectrokinetic actuator arrays (MEKA), 645
Micromechanical systems (MEMS), 645
Microscopic energy, 43
Minor losses, 374–381
Mixed-flow pumps, 790
Mixed-flow turbines, 838–840
Mixing length (l_m), 364
Models, 299–303. *See also* Dimensional
    analysis; Similarity
Modified pressure (P'), 520
Molar mass (M), 940, 958
Mole (mol), unit of, 16
Moment forces (F), 264–265
Moments, 90–92, 102
    area, 90–91
    inertia, 91–92
    restoring, 102
Momentum, 186, 244, 531, 571–572, 583–591,
    694, 703–705
    approximation solution using, 531
    conservation of, 186, 244
    equation, 531, 694, 703–705
    Fanno flow, 703–705
    integral technique, 583–591
    irrotational flow, 531
    Rayleigh flow, 694
    thickness (θ), 571–572
Momentum analysis, 243–289
    angular momentum and, 244–245, 263–273
    atmospheric pressure (P_{atm}) and, 249
    body forces, 246–248
    conservation of momentum, 244–245
    control volume (CV), forces acting on,
        246–249
    gravity and, 247
    linear momentum and, 244, 249–263
    net forces and, 244–245
    Newton's laws and, 244–245
    Reynolds transport theorem (RTM) for, 250
    right-hand rule for, 266
    rotational motion and, 244–245, 247–248,
        263–265
    surface forces, 246–248
    tensors for, 247–248
    thrust, 254–255
    torque, 263–264
    vortex shedding, 273–274
Momentum-flux correction factor (β),
    251–253
Moody chart, 367–374, 952

Motion, 103–110, 134–140, 151–168, 199–200, 243–289, 464–465, 880–883. *See also* Continuity equation; Navier–Stokes equation
acceleration (*a*), 106–107, 135, 136–139, 199–200, 265
angular momentum, 244–245, 263–273
angular velocity, 151–152
Bernoulli's equation and, 199–200
circular, 159–160
computational fluid dynamics (CFD) solutions, 880–883
constitutive equations for, 464–465
deformation rates of flow, 151–156
equations of, 105–106, 108
Eulerian description, 134–140, 167–168
field variables, 134–136
fluid flow, 149–160
Lagrangian description, 134–140, 167–168
linear momentum, 199–200, 244, 249–263
material (substantial) derivative, 139–140
momentum analysis, 243–289
Newton's laws of, 103, 136, 244–245
Reynolds transport theorem (RTM), 160–168
rigid bodies, fluids in, 103–110
rotation (*ω*), 107–110, 151–152, 156–159, 244–245, 247–248, 263–265
thrust, 254–255
torque, 263–264
translation (velocity vector), 134, 151
Motor efficiency (η_{motor}), 196
Moving belt, 302
Multilayered fluid, hydrostatics and, 96
Multistage axial-flow turbomachines, 822–823, 847
Mutual orthogonality, 535–536

National Advisory Committee for Aeronautics (NACA) standards, 637–638
Natural (unforced) flow, 11
Navier–Stokes equation, 464–469, 475–493, 515–606, 903
approximate solutions of, 515–606
boundary conditions for, 475–477
boundary layer approximation, 525–526, 560–563
Cartesian coordinates of, 468
exact and approximate solution comparison, 516–517
computational fluid dynamics (CFD) solution to, 472–473, 580–583
constitutive equations for, 464–465
continuity equation coupled with, 475–493
creeping flow, 520–525, 588–591
cylindrical coordinates of, 469
differential analysis using, 475–493
Euler equation for, 525–526, 531, 580–581
Euler number for, 519
exact solutions of, 475–493
Froude number for, 519
incompressible, isothermal flow, 466–468
inviscid flow regions, 525–529
irrotational flow, 529–554
Laplacian operator for, 530
modified pressure and, 520
Newtonian versus non-Newtonian fluids, 465–466
nondimensionalization of equations for, 517–520

Reynolds-averaged Navier–Stokes (RANS) equation, 903
Reynolds number (Re) for, 519, 520–522, 524
scaling parameters for, 517–518
Strouhal number (St) for, 519
velocity components for, 467–468
viscous (deviatoric) stress tensor for, 465–467
Net (total) head (*H*), 205, 790–795, 841–842
Net positive suction head (NPSH), 798–800
Neutrally stable, 101–102
Newton (N), unit of, 16–17
Newtonian fluids, 52, 465–466
Newton's laws of motion, 17, 78, 103, 136, 200–202, 244–245, 249–250, 463–464
Non-Newtonian fluids, 465–466
Noncircular pipe flow, 348
Nondimensional equations, 294–299, 517–520
approximate solution using, 517–520
Buckingham Pi theorem, 303–319
computational fluid dynamics (CFD) prediction for, 318–319
continuity, 517–518
Euler number (Eu) for, 519
Froude number (Fr), 296–299, 519
generation of, 303–319
independent Π (Pi), 300–301
motion, 517–520
Navier–Stokes, 519–520
parameters for, 294–319, 518–519
persons honored by, 311
Pi (Π) grouping parameters, 300–319, 516–517
process for (nondimensionalization), 294–299, 517–520
ratios of significance, 309–310
repeating parameters, 304–306, 312, 314
Reynolds number (Re), 301, 315–318
scaling parameters, 295–296, 517–518
similarity of model and prototype using, 300–303
Strouhal number (St) for, 519
Nonfixed control volume, 164
Normal acceleration (a_n), 199
Normal depth (y_n), 727, 738
Normal shock waves, 675–676, 678–684
Normal stress (σ), 3, 247
Normalized equations, 294
No-slip boundary condition, 8–9, 51, 476, 498–500, 525–526, 555
No-temperature-jump condition, 9
Nozzle meter, 393–394
Nozzles, 379, 665–678, 837–838, 923–927
Nusselt number (Nu), 310
Nutating discs, 834
Nutating disk flowmeters, 396

Oblique shock waves, 676, 684–688, 691–692, 927–928
back pressure (P_b) and, 676
bow wave, 686–687, 928
calculations for, 691–692
CFD calculations for, 927–928
detached, 686–687, 928
isentropic flow and, 676, 688
Mach angle, 688
Mach number for, 685–688
Mach wave, 688
strong, 687
turning (deflective) angle, 684–685
wedges, over a, 927–928

Obstruction flowmeters, 392–396
Ohnesorge number (Oh), 593
One-dimensional (fully developed) flow, 13–14, 351–352, 353–355, 663–669, 953–954
compressible, 663–669, 953–954
entrance region for, 13, 351–352
flow functions, 953
fluid velocity variation, 665–667
hydrodynamically fully developed region, 351–352
ideal gases, 663–665, 667–669, 953–954
isentropic, 663–669, 953
laminar, 353–355
property functions for, 953–954
radial direction of, 13–14
shock functions, 954
velocity profile for, 13–14, 351–352, 353–355
One-dimensional variables, 726
One-seventh-power law velocity profile, 366, 573–574
Open pumps and turbines, 790
Open system, *see* Control volume (CV)
Open-channel flow, 10, 725–785, 928–930
bridge scour, 771
bumps, 764–765, 929
CFD calculations for, 928–930
channels, 727–729, 737–747, 771
choked, 765
classification of, 726–729
conservation of mass, 736
critical depth (y_c) of, 730, 734
critical, 729–732, 436–740
discharge coefficient (C_d) for, 762–764
energy equations for, 736–737
flow control and measurement, 761–770
flow depth, 726–727
friction slope (S_f) of, 737
Froude number (F_r) for, 729–730, 929–930
gates, 761–770, 930
gradually varied flow (GVF), 727, 747–756
head loss (h_L), 736–737, 758–759
hydraulic cross sections, 743–747
hydraulic depth (y_h), 732–733
hydraulic diameter (D_h), 728
hydraulic jump, 731–733, 757–761, 930
hydraulic radius (R_h) for, 728–729
laminar, 727–728
one-dimensional variables for, 726
rapidly varied flow (RVF), 727, 757–761
Reynolds number (Re) for, 727–728
slope (*S*) of, 737, 748–752
specific energy (E_s) of, 733–736
steady/unsteady determination, 726
subcritical (tranquil), 729–732
supercritical (rapid), 729–732
surface profiles, 749–756
surface waves, 731–733
turbulent, 727–728
uniform (UF), 726–727, 737–743
varied (nonuniform), 726–727, 740–743
wave speed (c_0), 729–733
weirs, 761, 766–770
Operating (duty) point, 384, 792–793, 826
Ordinary differential equations (ODE), 485
Ordinate, 149
Orifice meter, 393–394
Outer (turbulent) layer, 364–366, 626
Outer flow region, 555. *See also* Inviscid flow; Irrotational flow

Outlet boundary conditions, 477
Overflow gates, 764–770
Overlap (transition) layer, 364–366, 626

Paddlewheel flowmeters, 397–398
Paraboloids of revolution, 108–109
Parabolic velocity profile, 351–352, 354–355
Parallel axis theorem, 91
Parallel pipe flow, 382
Parameters, *see* Nondimensional equations
Partial derivative operator (δ), 137
Partial differential equations (PDE), 485
Partial pressure, 41
Particle image velocimetry (PIV), 143,
    406–408, 410, 416
Pascal (Pa), unit of, 52, 76
Pascal's law, 80–81, 85–86
Passage losses, 815
Path functions, 187–188
Pathlines, 142–144
Peclet number (Pe), 310
Pelton wheel, 835–837
Penstock, 840
Performance (characteristic) curves, 383–384,
    791–797
Pi (Π) grouping parameters, 300–319,
    518–519
Piezoelectric transducers, 88
Piezometer (tube), 203
Pipe flow, 348–349, 351–390, 791–797,
    893–896, 952
    analysis of networks for, 381–382
    bend sections, 377–380
    CFD calculations for, 893–896
    circular, 348, 353–357, 952
    conservation of mass for, 382
    Darcy friction factor for, 355–356,
      367–368
    diameter (D) problems, 369–370
    efficiency (η) of, 381–390, 791–797
    elbow sections, 377–380
    entrance region, 351–352, 893–896
    entry length, 352–353
    equivalent length, 375
    equivalent roughness and, 368
    friction effects on, 348–349, 355–356,
      367–369
    friction factor for, 355–356, 358, 952
    in series, 381–382
    inclined, 357–358
    internal, 348–349, 351–390
    kinetic energy correction factor for, 357
    laminar, 351–361, 893–896
    loss (resistance) coefficient (K_L), 374–379
    minor losses, 374–381
    Moody chart for, 367–374, 952
    noncircular shapes, 348
    parallel, 382
    pump efficiency and, 383–390, 791–797
    relative roughness of, 367–369
    sizes of pipe for, 369
    sudden expansion, 379
    systems (networks), 381–390, 791–797
    turbulent, 361–374
    valves for, 374–375, 378, 380
    velocity in, 384–385
    velocity profile for, 351–352, 353–355,
      364–367
    vena contracta region, 376

Pitch angle (θ), 816–819
Pitching moment, 611
Pitot formula for flow rate, 392
Pitot probes, 391–392
Pitot tube, 203
Pitot-static probes, 391–392
Planar flow, 457, 484, 534–537, 538–554
    doublet of, 544–545, 546–547
    harmonic functions for, 535–537
    irrotational regions of, 534–537, 538–554
    line sink of, 540–542, 545
    line source of, 540–542
    line vortex of, 542–543, 545
    Poiseuille flow, 484
    singular point (singularity) of, 540–541
    stream functions for, 457, 535–537, 545–547
    superposition of, 538, 545–554
    two-dimensional, 534–537
    uniform stream of, 539, 546–549
    velocity potential function for, 535–537
Plane surfaces, 89–94, 98
Planform area, 313, 612, 639
Plastic fluids, 466
Plots, 148–151, 916–917, 924–928
    abscissa, 149
    computational fluid dynamics (CFD) use of,
      149–151, 916–917, 924–928
    contour, 150–151, 916–917, 924–928
    fluid flow data using, 148–151
    ordinate, 149
    pressure contour, CFD, 924–928
    profile, 149
    temperature, 916–917
    vector, 149–150
Poise, unit of, 52
Poiseuille flow, 484–490, 493–496
Poisson's equation, 473
Polar coordinates, *see* Cylindrical coordinates
Position vector, 134
Positive-displacement machines, 396, 789,
    803–806, 834, 834
    closed volume for, 805–806
    flowmeters, 396
    fluid flow in, 803–806
    nutating discs, 834
    peristaltic pumps, 803–804
    pumps, 789, 803–806
    rotary pumps, 804–805
    self-priming pumps, 805
    turbines, 789, 834
    water meters, 834
Potential energy, 43, 194–195
Potential flow, *see* Irrotational flow
Potential function (ϕ), 529–530, 535–537
Pound-force (lbf), unit of, 17–18, 76
Pound-mass (lbm), unit of, 17–18
Pouseuille's law, 356
Power, 18–19, 215–217, 847–855
    available wind power ($W_{available}$) for, 850–851
    Betz limit, 853–854
    coefficient, 851, 853–854
    energy pattern factor, 851
    wind power density, 850–851
    wind turbines, 847–855
Power-law velocity profile, 366–367, 496–497
Power number (N_P), 310
Prandtl equation, 356
Prandtl number (Pr), 310
Prandtl–Meyer expansion waves, 688–693.
    See also Expanding flow

Prandtl–Meyer function, 689
Precision error, 28–29
Pressure (P), 3–4, 41–44, 75–131, 139–140,
    202–204, 216–219, 249, 464–465, 470–
    475, 519–520, 547–550, 593, 612–617,
    634–635, 661–662, 667–678, 705,
    924–928
    absolute, 76–77, 90
    aerodynamic drag, 548–550
    aerostatics, 89
    aircraft wing design and, 612–613
    atmospheric, 81–83, 89–90, 249
    back, 669–678
    barometric, 81–83
    Bernoulli equation representation, 202–204
    buoyant force, 98–103
    capillary, 593
    calibration for, 88–89
    center of pressure (point of application),
      89–91
    choked flow from, 670
    compressible flow, 661–662, 669–678, 705,
      924–928
    compressive force as, 77–78
    constitutive equations for, 464–465
    contour plots, CFD, 924–928
    critical, 667–669
    depth, variation of with, 78–81
    drag affected by, 612–617
    differential analysis of fields, 470–475
    dynamic, 202–203, 612
    external flow and, 612–617, 634–634
    flow separation and, 616–617
    flow work (P/ρ), 43–44, 218–219
    force as, 76–81, 216–219
    fluid flow and, 3–4, 202–204
    fluid statics and, 75–131
    full dynamic, 548
    gage, 76–77, 82–83
    hydrostatic, 83–84, 464, 519–520
    hydrostatic forces and, 89–97
    ideal gases, 644–646
    irrotational flow, 547–550
    isentropic nozzle flow and, 669–678
    lift affected by, 612–613, 634–635
    Mach number (Ma) and, 669–678
    material derivative of, 139–140
    measurement of, 4, 81–89, 203–204
    mechanical, 465
    modified, 520
    Navier–Stokes equation and, 464–465,
      470–475
    nondimensionalized equations and, 519–520
    normal force as, 76–77
    partial, 41
    Pascal's law for, 80–81, 85–86
    power from work of, 216–217
    prism, 91–92
    rigid-body motion of fluids and, 103–110
    saturation, 41–43
    scalar quantity of, 77–78
    shock waves from, 675–678
    stability and, 98–103
    stagnation (P_0), 203–204, 661–662, 705
    static, 202–203
    streamlining, 615–616
    superposition and, 547–550
    thermodynamic (hydrostatic), 464
    total, 203
    units of, 76–77

Pressure (*continued*)
  vacuum, 76–77
  vapor, 41–43
  work by, 216–219
Pressure coefficient (C_P), 310, 523, 547–549
Pressure correction algorithms, CFD, 473
Pressure drop (ΔP), 348, 355–357, 369–370, 382
Pressure field (scalar), 134
Pressure gradient, 104, 561–562, 564–565, 578–583
  boundary layer approximation with, 561–562, 564–565, 578–583
  computational fluid dynamics (CFD) calculations for, 580–583
  curvature and, 561–562
  dividing streamline, 581–582
  favorable, 579
  flow separation, 578–583
  reverse flow and, 579–580, 582
  rigid-body motion, 104
  separation bubble, 579–580
  stall condition, 580
  unfavorable (adverse), 579
  velocity profile of, 580
  zero, 561, 564–565
Pressure head, 79, 205
Pressure loss (ΔP_L), 355
Pressure transducers, 88
Preswirl, 820
Primary (fundamental) dimensions, 15–16, 292
Problem-solving technique, 23–25
Profile plots, 149
Propane saturation properties, 945, 963
Propellers (rotor), 816–819, 839–840
Property, defined, 38
Prototypes, 299–303. *See also* Dimensional analysis; Similarity
Pseudoplastic (shear thinning) fluids, 52, 466
Pulsatile pediatric ventricular device (PVAD), 410–411
Pump–motor efficiency, 383–384
Pump–turbines, 858
Pumps, 196, 221, 381–390, 788–790, 790–833
  affinity laws, 829–830
  axial, 806, 816–824
  best efficiency point (BEP), 791
  blower, 788
  brake horsepower (bhp), 791
  capacity (volume flow rate) for, 790, 805–806
  cavitation and, 797–800
  centrifugal, 806–815
  compressors, 788–789
  dimensional analysis for, 824–827
  ducted, 790
  dynamic, 790, 806–824
  efficiency, 196, 383–390, 791–797
  enclosed, 790
  energy absorbing devices, as, 788
  energy grade line (EGL), 790
  fans, 788, 816, 819–824
  flow analysis for selection of, 381–382
  flow rate curves for, 383–384
  free delivery, 384, 791
  head, 221, 383–384, 792–793
  impellers, 790, 806–808, 811–812, 819–822
  in-series systems, 381–382, 800–803
  internal flow and, 381–390
  mass flow rate for, 790
  net head for, 790–791
  net positive suction head (NPSH), 798–800
  open, 790

operating (duty) point, 384, 792–793, 826
parallel systems, 382, 800–803
performance (characteristic) curves, 383–384, 791–797
piping networks and, 383–390791–797
positive-displacement, 789, 803–806
Reynolds number (Re) for, 825–826
rotary (rotodynamic), 806
scaling laws, 824–833
selection of, 381–390, 791–797
shutoff head, 384, 791–792
specific speed, 827–829
system (demand) curve, 383–384
volume flow rate (capacity), 790
Pure constants, 295

Quasi-steady flow, 519

Radial (straight) blades, 808
Radial-flow devices, 269–270, 806–815, 838–839. *See also* Centrifugal pumps
Rake of streaklines, 145
Rankine temperature scale, 40
Rapidly varied flow, 727, 757–761
Rarified gas flow theory, 39
Rate of rotation (angular velocity), 151–152
Rated speed, 850
Rayleigh flow, 693–702, 955
  choked, 700
  continuity equation for, 694
  control volume energy equations for, 694–695
  effects of heating and cooling on, 696
  energy balance equations, 694
  entropy change of, 694–695
  equation of state for, 695
  heat transfer and, 693–702
  heating and cooling effects on, 696
  ideal gases, 693–702
  Mach number (Ma) and properties of, 695–696
  momentum equation for, 694
  property functions for, 955
  property relations for, 699–700
Rayleigh line (curve), 679–680, 695
Rayleigh number, 310
Reaction turbines, 837–846
  axial-flow, 839
  double-regulated, 839
  draft tube, 840
  efficiency of, 841–842
  Francis, 838–864
  efficiency of, 841–842
  energy grade line (EGL), 840–841
  gross head for, 839
  head gate, 840
  Kaplan, 838–839
  mixed-flow, 838–840
  net head, 841–842
  propeller, 839–840
  radial-flow, 838–839
  runner blades (rotor), 837–839, 843–846
  single-regulated, 839
  stay vanes, 837–838
  tailrace, 840
  volute, 838
  wicket gates, 837–838
Reattachment line, 921–922
Rectangular hydraulic cross sections, 745
Refractive flow visualization, 147–148
Refrigerant-134a saturation properties, 943, 961

Regression analysis, 320
Relative density, 39
Relative roughness (ε/D), 367–369
Relative velocity, 163–165, 191, 245–246
Repeating variables, method of, 303–319
Required net head, ($H_{required}$), 792–793
Residence time, 138–139
Resistance (loss) coefficient (K_L), 374–379
Restoring moment, 102
Resultant forces (F_R), 90–91, 95–96, 607
Reverse flow, 579–580, 582
Reverse swirl, 843–844
Reverse thrust, 819
Reversible adiabatic flow, *see* Isentropic flow
Revolution, *see* Rotation
Revolutions per minute, 264–165
Reynolds-averaged Navier-Stokes (RANS) equation, 903
Reynolds number (Re), 11, 301, 310, 315–318, 320–326, 350–351, 519, 520–522, 524, 557–559, 617–618, 626–633, 707–708, 727–729
  aerodynamic drag coefficient of, 322–323
  boundary layer approximation, 557–559
  creeping flow approximation and, 520–522, 524
  critical, 350, 557–559
  cylinders, 629–633
  dimensional analysis use of, 301, 315–318, 320–326
  drag coefficient (C_D) and, 320–303, 617–618
  external flow, 612, 617–618, 629–631
  Fanno flow, 707–708
  flat plate analysis, 626–628
  free-surface flow, 323–325
  hydraulic diameter for, 350, 728
  hydraulic radius for, 728–729
  incomplete similarity, 320–321
  independent flow, 321–322
  insect flight and, 326
  internal flow, 350–351
  Mach number (Ma) and, 315–316
  Navier–Stokes nondimensionalization using, 519
  open-channel flow, 727–729
  ratio of significance, 310
  spheres, 629–633
  Stokes law, 618
  surface roughness and, 626–628, 632–633
  transition, 557–559
  transitional flow and, 350–351, 557–559
Reynolds (turbulent) stresses, 363
Reynolds transport theorem (RTM), 160–168, 250, 266
  angular momentum, 266
  closed/open system relationships, 160–162
  control volume (CV) approach, 160–168
  extensive flow properties, 162
  fixed control volume, 163
  inlet and outlet crossings, 164–165
  intensive flow properties, 162
  Leibniz theorem and, 165–167
  linear momentum, 250
  mass flow rate and, 165
  material derivative and, 167–168
  material volume applications, 165–167
  momentum analysis and, 250, 266
  relative velocity for, 163–165
  streamlines of flow and, 161–162
  system-to-control volume transformation, 163
  nonfixed control volume, 164
  unit outer normal for, 162–163

Richardson number (Ri), 310
Right-hand rule, 266
Rigid-body motion, 103–110
  acceleration (a) on a straight path, 106–107
  body forces for, 104–105
  equations of motion for, 105–106, 108
  fluids at rest, 105
  fluids in, 103–110
  forced vortex motion, 107–108
  free-fall of a fluid body, 105
  isobars, 106
  Newton's second law of motion for, 103
  parabloids of revolution, 108–109
  pressure gradient, 104
  rotation in cylindrical containers, 107–110
  surface forces for, 104–105
Rocket propulsion, 675–688. See also Aircraft;
    Shock waves
Rolling moment, 611
Rotameters (variable-area flowmeters), 398
Rotary fuel atomizers, 861
Rotary (rotodynamic) pumps, 806. See also
    Dynamic machines
Rotation (ω), 107–110, 151–152, 156–159,
    244–245, 247–248, 263–265
  angular momentum and, 244–245, 263–265
  angular velocity, 151–152, 264–265
  Cartesian coordinates for, 157, 247–248
  centripetal acceleration and force, 265
  control volume (CV) surface forces, 247–248
  cylindrical containers, fluid in, 107–110
  cylindrical coordinates for, 158–159
  forced vortex motion, 107–108
  kinetic energy and, 265
  momentum analysis and, 244–245, 247–248
  parabloids of revolution, 108–109
  rate of, 151–152
  rigid-body motion of fluids, 107–110
  shaft power and, 264–265
  tensor notation for, 247–248
  torque, 263–264
  vorticity (ζ) and, 156–159
Rotation swirl, 843–844
Rotational stability, 102
Rotational viscometer, 480–481
Runners, 790, 833, 837–839, 843–844, 847
  band, 838–839
  blades, 790, 837–839, 843–846
  buckets, 790, 847
  gas and steam turbines, 847
  leading edge (angle), 843
  reaction turbines, 837–839, 843–844
  reverse swirl, 843–844
  rotation swirl, 843–844
  trailing edge (angle), 843

Saturation, 41–43, 797–798, 942–945, 960–963
  ammonia properties, 944, 962
  cavitation and, 41–43, 797–798
  fluid properties and, 41–43
  pressure, 41–43, 797–798
  propane properties, 945, 963
  refrigerant-134a properties, 943, 961
  temperature, 41–42
  vapor pressure and, 41–43
  water properties, 942, 960
Scaling laws, 824–833, 855–860
  affinity laws, 829–830
  dimensional analysis for, 824–827, 855–857
  homologous operating points, 826

performance (characteristic) curves for, 826
pumps, 824–833
  Reynolds number (Re) and, 825–826
  similarity ratios and, 825–826, 829–830
  specific speed (N_{Sp} or N_{St}), 827–829, 857–860
  turbines, 855–860
  turbomachinery, 824–833, 855–860
Scaling parameters, 295–296, 517–518
Schlieren images, 678, 684
Schlieren technique, 147–148
Schmidt number (Sc), 310
Scroll (diffuser), 806–807
Second (s), unit of, 16–17
Secondary (derived) dimensions, 15
Seeding particles (seeds), 4016
Separation bubble, 579–580, 921–922
Shadowgraph technique, 147–148
Shaft work, 195–196, 199, 204–205, 216,
    264–265. See also Mechanical energy
  efficiency of, 195–196
  energy transfer (W_{shaft}) by, 216
  mechanical energy as, 195–196, 199
Shape factor (H), 585
Sharp-crested weir, 767–770
Shear force (F), 52
Shear strain (ε), 2, 151, 153–154
Shear stress (τ), 2–3, 52, 247, 363–364
Shear thinning and thickening fluids, 466
Sherwood number (Sh), 310
Shock (wave) angle (δ), 684–685
Shock waves, 32, 445, 675–688, 712,
    927–928, 954
  aircraft and rocket propulsion, 675–688
  axisymmetric flow, 687–688
  back pressure (P_b) and, 675–678
  boundary layer interactions, 712
  bow wave, 686–687, 928
  CFD calculations for, 927–928
  compressible flow and, 675–678, 712,
    927–928
  continuity equation for, 445
  converging–diverging nozzles and, 675–678
  detached oblique, 686–687, 928
  Fanno line (curve), 679–682
  ideal gas, 954
  incompressible flow and, 445
  isentropic flow, 675–678
  Mach angle (μ), 688
  Mach number (Ma) for, 679–681, 685–688
  normal, 675–676, 678–684
  oblique, 676, 684–688, 927–928
  property functions for, 954
  Rayleigh line (curve), 679–680
  Schlieren images, 678, 684
  shock (wave) angle (δ), 684–685
  turning (deflective) angle, 684–685
  viscous-inviscid interactions, 712
  vorticity, 32
Shockless entry condition, 809
Shutoff head, 384, 791–792
Significant digits, 28–31
Similarity, 299–303, 315–316, 320–323, 491,
    565–567, 825–826, 829–830
  characteristics of, 299–303
  dimensional analysis for, 299–303,
    315–316, 318–319
  dynamic, 300–301, 315–316, 318–319
  geometric, 299–300
  incomplete, 320–323
  kinematic, 299–300
  models and prototypes, 299–303, 320–323

pumps, 825–826, 829–830
scaling laws, 825–826, 829–830
ratios and, 825–826, 829–830
solution, 491
variable (η), 565–567
Single-regulated turbines, 839
Single-stream devices, 219
Single-stream systems, 254
Singular point (singularity), 540–541
Skin friction coefficient, 567, 585
Skin friction (surface) drag, 9
Slope (S), 498, 727, 737–740, 748–754
  adverse, 750–751, 753
  classification of, 750–751, 756
  critical (S_c), 739–740
  friction (S_f), 737, 751
  gradually varied flow (GVF), 748–749
  length, 498
  mild, 750–751, 753
  open-channel flow, 727, 737–740, 748–752
  steep, 750–751, 753
  surface profiles and, 749–752
  transition connections, 752–754
  uniform flow, 727, 738–739, 753
Sluice gate, 762, 930
Smoke wire, 144–145
Software packages for engineering, 25–27
Solid phase, 3–4
Sonar dome, 62
Sonic flow, 11, 50, 666–667
Sonic speed, see Speed of sound
Sonoluminescence, 62
Sound navigation and ranging (sonar), 62
Specific energy (E), 733–735
Specific gravity (SG), 39, 41
Specific heat, 43–44, 310, 667–669, 940, 958
  energy and, 43–44
  ideal gases, 940, 958
  isentropic gas flow and, 667–669
  Mach number (Ma) relationships, 667–669
Specific properties, 38
Specific Reynolds stress tensor, 903–904
Specific speed (N_{Sp} or N_{St}), 827–829, 857–860
Specific total energy (e), 38, 44
Specific volume (v), 39
Specific weight (γ_s), 17, 39
Speed of sound (c), 11, 48–50, 315–316, 663–665
  analysis of, 48–50
  compressible flow and, 11, 48–50, 663–665
  flow direction variations, 665
  fluid flow regimes, 11, 48–50
  ideal gas, 663
  isentropic flow, 663–665
  lift on a wing and, 315–316
  Mach number (Ma) for, 11, 50, 663–665
Speeds for takeoff and landing, 636–637
Spheres, 523–525, 629–633
  creeping flow approximation, 525–525
  drag force on, 523–525, 629–633
  external diameter (D), 629
  external flow over, 629–633
  flow separation, 631
  Reynolds number (Re) for, 524, 629–633
  stagnation points, 629–631
  surface roughness effects, 632–633
Spinning, generation of by lift, 639–643
Stability, 101–103
  buoyancy and, 101–103
  center of gravity and, 102
  metacentric height, 102–103
  rotational, 102

Stagnation, 203–204, 629–631, 635, 639, 660–663, 667–669, 704–705
  enthalpy, 660–662
  external flow and, 629–631, 635, 639
  friction effects on, 704–705
  isentropic ideal gas flow properties, 667–669
  isentropic state, 661
  lift force and, 635, 639
  Mach number (Ma) relationships, 667–669
  points, 204, 629–631, 635, 639
  pressure, 203–204, 661–662, 705
  properties of, 660–663, 705
  specific heat, 667–669
  spheres, flow over and, 629–631
  streamline, 204
  temperature, 661–662, 705
Stall conditions, 580, 617, 637
Stanton number (St), 310
Starting vortex, 635
State postulate, 38
Static enthalpy (h), 660
Static pressure, 202–203
Static pressure tap, 203, 547
Statics, study of, 2
Stator blades (vanes), 819–822, 907–915
Stay vanes, 837–838
Steady flow, 12–13, 191, 199–202, 204, 219–221, 251, 253–254, 267–268, 444–445, 470–490, 726
  angular momentum of, 267–268
  Bernoulli equation for, 199–202, 204
  compressible, 200–202, 219–221, 444–445
  continuity equation for, 444–445
  devices, 12
  differential analysis of, 470–490
  energy analysis of, 219–221
  energy losses in, 219–221
  ideal (no mechanical loss), 219
  incompressible, 201–202, 221
  inlet/outlet crossings, 251, 254
  irreversible head loss, 220–221
  linear momentum of, 251, 253–254
  mass balance for, 191
  mechanical energy balance, 220–221
  open-channel flow, 726
  real (mechanical loss), 219
  single-stream devices, 219
  unsteady flow compared to, 12–13
Steam turbines, 833, 847
Stoke, unit of, 53
Stokes flow, 520. See also Creeping flow
Stokes law, 618
Stokes number (Stk or St), 310
Straight path, fluid acceleration on, 106–107
Strain (ε), 2, 151–154
Strain-gage pressure transducers, 88
Streaklines, 144–146
Stream functions (ψ), 450–459, 535–537, 545–547
  axisymmetric flow, 457, 536–537
  Cartesian coordinates, 450–456
  compressible (ψ_ρ), 458–459
  cylindrical coordinates, 457–458
  differential analysis and, 450–459
  irrotational flow, 535–537, 545–547
  mutual orthogonality of, 535–536
  planar flow, 457, 535–537, 545–547
  streamlines and, 451–456
  velocity potential function and, 535–536
Streamlined bodies, 610, 614–616

Streamlines, 141–142, 161–162, 199–200, 202–205, 451–456, 552, 615–616
  Bernoulli's equation and, 199–200, 202–205
  dividing, 552
  drag reduction by, 615–616
  flow visualization using, 141–142
  force balance across, 202
  nondimensional, 552
  particle acceleration and, 199–200
  Reynolds transport theorem (RTT) and, 161–162
  stagnation, 204
  stream functions and, 451–456
  superposition of irrotational flow and, 552
Streamtubes, 142
Streamwise acceleration (a_s), 199
Stress (σ), 2–3, 52, 247–249, 363–364, 459–460, 465–467
  control volume (CV) and, 247–249
  differential analysis and, 459–460, 465–467
  flow and, 2–3
  normal (σ), 3, 247
  Reynolds (turbulent), 363–364
  shear (τ), 2–3, 52, 247, 363–364
  tensor (σ), 247–249, 459–460, 465
  turbulent shear (τ_{turb}), 363–364
  viscous (deviatoric) tensor, 465–467
  yield, 466
Strouhal number (St or Sr), 273, 310, 402, 519, 902
  Kármán vortex street, 902
  Navier–Stokes nondimensionalization using, 519
  ratio of significance, 310
  vortex flowmeters use of, 402
  vortex shedding use of, 273
Subcritical (tranquil) flow, 729–732
Submerged bodies, 89–101
  buoyancy of, 98–101
  curved surfaces, 95–97
  hydrostatics of, 89–97
  plane surfaces, 89–94, 98
Subsonic flow, 11, 50, 666–667, 687
  downstream, 687
  Mach number (Ma) for, 11, 50, 666–667
Substantial derivative, see Material derivative
Sudden expansion, 379
Supercritical (rapid) flow, 729–732
Superposition, 538, 545–554, 623–625, 740
  aerodynamic drag and, 548–550
  approximation solutions using, 538, 545–554
  d'Alembert's paradox, 548–549
    dividing streamline for, 552
  drag coefficient (C_D), 623–625
  flow over a circular cylinder, 546–554
  irrotational flow, 538, 545–554
  line sink and line vortex, 545
  method of images for, 551
  open-channel flow, 740
  planar flow, 538, 545–554
  pressure coefficient (C_p) for, 547–549
  two-dimensional flow, 538
  uniform flow with nonuniform parameters, 740
  uniform stream and doublet, 546–554
  velocity of composite flow field from, 538
Supersonic flow, 11, 50, 666–667, 674, 678–681, 687

downstream, 687
  Mach number (Ma) for, 11, 50, 666–667
  shock waves from, 674, 678–681
Surface (skin friction) drag, 9, 567, 612, 614
Surface forces, 104–105, 246–248
Surface oil visualization, 148
Surface profiles, 749–756
Surface roughness, 612, 614–615, 626–628, 632–633
Surface tension (σ_s), 55–60
  capillary effect and, 58–60
  contact angle (ϕ), 58
  fluid properties of, 55–58
  impurities and, 57
  meniscus, 58
  work by expansion ($W_{expansion}$), 57–58
  visualization of, 56
Surface waves, open-channel flow, 731–733
Surfactants, 57
Surge tower, 45
Swirl, 817–818, 820, 843–844
Symmetry boundary conditions, 477, 891–892
System (demand) curve 383–384
Systematic error, 28
Systems, 14–21, 160–168, 243–289, 293.
    See also Pipe flow
  boundary conditions for, 14–15
  closed, 14–15
  closed and open relationships, 160–162
  control mass, 15
  control volume (CV), 15, 160–165
  dimensions of, 15, 19–20
  isolated, 15
  Leibniz theorem for, 165–167
  material derivative and, 167–168
  material volume, 166–168
  momentum analysis of, 243–289
  open, 15
  relative velocity of, 163–165
  Reynolds transport theorem (RTT) for, 160–168
  single-stream, 254
  surroundings of, 14
  total energy of, 293
  units of, 15–21
System-to-control volume transformation, 163

T-s diagrams, 695–696, 704–705
Tailrace, 840
Taylor series expansion, 440–441
Temperature (T), 16, 41–42, 46–47, 667, 661–662, 668–669, 705, 915–922
  absolute, 46–47
  cavitation and, 41–43
  computational fluid mechanics (CFD) for, 915–922
  contour plots, 916–917
  cooling integrated circuit chips (ICs), 917–922
  critical (T^*), 668–669
  dynamic, 667
  ideal gases, 661–662
  Kelvin (°K) scale, 16, 40
  Mach number (Ma) and, 668–669
  Rankine (R) scale, 40
  rise in cross-flow heat exchanger, 915–917
  saturation, 41–42
  stagnation, 661–662, 705
  volume expansion (β) and, 46–47

999
INDEX

Tensors, 154, 247–249
  contracted (inner) product of, 248
  control volume and, 247–249
  fluid element deformation and, 154
  notation, 247
  strain rate, 154
  stress, 247–249
Terminal velocity, 523–524, 613
Thermal anemometers, 402–404
Thermal energy, 43, 215
Thermodynamics, first law of, 202, 221
Thickness (δ), boundary layers, 556, 562–564, 574, 625
Three-dimensional bodies, drag coefficients (C_D) for, 611, 620–621
Three-dimensional (3-D) CFD grids, 881, 886–887
Three-dimensional flow, 13–14
Throat size of nozzles, 665, 668, 672
Thrust (force), 254–255, 819
Time rate of change, 187
Timelines, 146–147
Tip vortex, 638
Torque, 263–264. See also Angular momentum
Total derivative operator (d), 137
Total energy (E), 43–44
Total enthalpy, (h_0), 660–662
Total head (H) of flow, 205–206
Total head loss (h_L), 375–376
Total pressure, 203
Tracer particles, 143–144
Trailing edge angle, 809, 843
Trailing edge flaps, 636
Trailing vortices, 638–639
Transducers, 88
Transient flow, 12
Transit-time ultrasonic flowmeter, 399
Transition connections, 752–754
Transition Reynolds number, 557–559
Transitional flow, 11, 349–351, 557–559
Translation, rate of, 151
Transonic flow, 50
Trapezoidal hydraulic cross sections, 745–746
Trip wires, 558
Tube-axial fans, 819–824
Tubes, 348. See also Pipe flow
Tufts, 148
Turbine (propeller) flowmeters, 397
Turbines, 196, 221, 380, 788–790, 833–860
  dimensional analysis for, 855–857
  dynamic, 834–855
  efficiency ($\eta_{turbine}$), 196, 380, 815, 841–842
  enclosed, 790
  energy grade line (EGL), 840–841
  energy producing devices, as, 788
  gas, 833, 847
  gross head for, 839
  head ($h_{turbine}$), 221, 383–384
  hydraulic, 833
  impulse, 835–837
  net head for, 841–842
  open, 790
  positive-displacement, 790, 834
  power coefficient, 851, 853–854
  reaction, 837–846
  runners, 790, 833, 837–839
  scaling laws, 855–860
  specific speed, 857–860
  steam, 833, 847
  turbomachinery, as, 788–790, 843–855
  wind, 833, 847–855

Turbofan engine, 822–823
Turbomachinery, 265, 787–877
  angular momentum equation for, 265
  classifications of, 788–790
  dynamic machines, 790, 806–824, 834–855
  energy grade line (EGL), 790, 840–841
  multistage, 822–823, 847
  positive-displacement machines, 789, 803–806, 834
  pump–turbines, 858
  pumps, 788–790, 790–833
  rotary fuel atomizers, 861
  scaling laws, 824–833, 855–860
  turbines, 788–790, 833–860
Turbulence dissipation rate (ε), 904
Turbulence intensity (I), 904
Turbulence kinetic energy (k), 904
Turbulent flow, 11, 253, 349–353, 361–374, 557–558, 572–583, 727–728, 902–915
  boundary layer approximation, 557–558, 572–583
  buffer layer, 364–365
  Colebrook equation for, 367–368
  computational fluid dynamics (CFD) calculations, 580–583, 902–915
  cylinders, CFD calculations for flow around, 905–907
  Darcy friction factor for, 367–368
  direct numerical simulations (DNS) for, 903
  eddies, 361–364, 902–903
  eddy viscosity of, 363–364
  entrance region, 351–352
  entry length, 352–353
  flat-plate boundary layer, 572–578
  flow separation, 580–583
  flow visualization, 582
  fluctuating components of, 361–362
  fluid behavior (flow regime) of, 349–351
  free shear layers, 557
  friction velocity of, 365
  head loss (h_L), 369–370
  hydraulic diameter and radius for, 728–729
  hydrodynamic entry length (L_h), 353
  internal, 349–353, 361–374
  kinematic eddy viscosity of, 363–364
  laminar flow compared to, 557–559, 580–583
  large eddy simulation (LES) for, 903
  law of the wall, 365–366, 576, 578
  logarithmic (log) law, 366, 576
  mixing length, 364
  momentum-flux correction factor for, 253
  Moody chart for, 367–374
  Navier–Stokes equation for, 557–558, 572–578
  one-seventh-power law for, 366, 573–574
  open-channel flow, 727–728
  outer (turbulent) layer, 364–366
  overlap (transition) layer, 364–366
  pipes, 361–374
  power-law velocity profile for, 366–367
  Prandtl equation for, 368
  pressure drop (ΔP), 369
  pressure gradient effects, 578–583
  Reynolds number (Re) for, 350–351, 727–728
  Reynolds (turbulent) stresses, 363–364
  shear stress (τ_{turb}) of, 363–364
  Spaulding's law of the wall, 576, 578
  stator design, CFD model for, 907–915
  transitional flow to, 349–351, 557–558
  trip wires for transition, 558
  turbulence models for, 903–905

  vane-axial flow fan, CFD model for, 907–915
  velocity defect law, 366
  velocity profile for, 351–352, 364–367
  viscous length, 365
  viscous sublayer, 364–367
  von Kármán equation for, 368–369
Turbulent length scale (l), 904
Turning (deflective) angle (θ), 684–685
Twist on propeller blades, 816–818
Two-dimensional bodies, drag coefficients (C_D) for, 612–613, 619
Two-dimensional (2-D) CFD grids, 881, 885–887
Two-dimensional flow, 13–14, 534–537, 608, 610–613

Ultrasonic flowmeters, 399–401
Underflow gates, 761–764
Uniform flow, 12, 251, 539, 546–549, 726–727, 737–743
  Chezy coefficient for, 738–739
  critical, 736–740
  Gaukler-Manning equations for, 738–739
  irrotational flow, 539, 546–549
  linear momentum of, 251
  Manning coefficient for, 739
  normal depth for, 727, 738
  open-channel flow, 726–727, 737–743
  stream of, 539, 546–549
  superposition for nonuniform parameters of, 740
  varied (nonuniform) flow compared to, 726–727
  velocity (V_0), 738
Unit outer normal, 162–163
United States Customary System (USCS) of units, 16–19
Units of measurement, 15–21, 43, 52, 53, 76–77, 292, 827–828, 859–860
  dimensional homogeneity and, 19–20
  dimensions and, 15, 292
  energy, 43
  English system, 16–19, 292
  General Conference of Weights and Measures (CGMP), 16
  importance of, 15–19
  International System (SI), 16–19, 292
  pressure, 76–77
  specific speed, 827–828, 859–860
  United States Customary System (USCS), 16–19
  unity conversion ratios, 20–21
  viscosity (μ), 52, 53
Unity conversion ratios, 20–21
Universal (turbulent) velocity profile, 366
Universal gas constant (R_u), 40
Unstable situations, 101–102
Unsteady flow, 12–13, 202, 490–493, 726
Useful pump head, ($h_{pump, u}$), 792

Vacuum pressure, 76–77
Valves for pipe flow, 374–375, 378, 380
Vane-axial fan, 819–820, 907–915
Vanes, 819–822, 837–838
Vapor phase, 4
Vapor pressure (P_v), 41–43, 797
Vaporous cavitation, 62
Variable-area flowmeters (rotameters), 398

Variable pitch, 819
Varied (nonuniform) flow, 726–727, 740–743
Vector identity, 526–527
Vector plots, 149–150
Vector variables, 134–135
Vehicles, drag coefficient (C_D) for, 621–623
Velocimetry, 404–408. *See also* Flow rate
Velocity (V), 51–52, 151–152, 163–165, 188,
   245–246, 264–265, 348–349, 357–358,
   364–367, 391–408, 523–524, 537, 576,
   608, 610, 612–613, 665–667, 734
   absolute, 164, 191, 245–246
   angular (rate of rotation), 151–152, 264–265
   average, 188, 348–349
   conservation of mass and, 188, 191, 349
   control volume, 163–165, 245–246
   critical, 734
   external flow, 608, 610, 612–613
   flow rate and, 391–408
   fluid motion and, 151–152
   free-stream, 608
   friction, 365, 576
   internal flow, 348–349, 357–358, 364–367,
      391–408
   isentropic flow, variations in, 665–667
   Mach number (Ma) and, 665–667
   measurement of, 391–408
   nozzle shapes and, 665–667
   relative, 163–165, 191, 245–246
   Reynolds transport theorem (RTT), 163–165
   rotation and, 151–152, 264–265
   terminal, 523–524, 613
   two-dimensional irrotational flow,
      components for, 537
   viscosity and, 51–52
Velocity boundary layer, 351
Velocity defect law, 366
Velocity field (vector), 134
Velocity gradient, 51–52
Velocity head, 205
Velocity overshoot, 590–591
Velocity potential function (ϕ), 529–530,
   535–537
Velocity profile, 13–14, 51, 351–352, 353–355,
   364–367, 573–578, 738
   boundary layer approximation, 573–578
   boundary layer development, 351–352,
      364–367
   boundary layer profile comparisons, 573–578
   entrance region and, 351–352
   flat (full), 351–352, 365
   internal (pipe) flow, 351–352, 353–355,
      364–367
   laminar flow, 351–352, 353–355
   logarithmic (log) law, 366, 576
   one-dimensional (fully developed) flow,
      13–14, 351–352
   one-seventh-power law, 366, 573–574
   parabolic, 351–352, 354
   power-law velocity profile for, 366–367
   Spaulding's law of the wall, 365–366, 576, 578
   turbulent flow, 351–352, 364–367, 573–578
   uniform-flow, 738
   universal, 366
   velocity defect law, 366
   viscosity and, 51, 364–367

viscous length, 365
viscous sublayer, 364–367
   wall-wake law, 576–577
Velocity vector (rate of translation), 134, 151
Vena contracta region, 376
Venturi meter, 393–394
Venturi nozzles, 665
Vertical axis wind turbines (VAWTs), 847, 849
Viscoelastic fluid, 465
Viscometer, 55
Viscosity (μ), 9, 11, 50–55, 204, 363–364,
   480–481, 612
   apparent, 52
   coefficient of, 52–54
   differential analysis for, 480–481
   dilatant (shear thickening) fluids, 52
   drag force and, 51, 612
   dynamic (absolute), 52, 54
   eddy, 363–364
   fluid flow and, 9, 11, 52–54
   fluid properties of, 50–55
   friction force and, 10, 50–51
   gases, 53–54
   kinematic, 53
   kinematic eddy, 363–364
   liquids, 53–54
   Newtonian fluids and, 52
   negligible effects of, 204
   no-slip condition and, 9, 51
   pseudoplastic (shear thinning) fluids, 52
   Reynolds number (Re) and, 11
   rotational viscometer for, 480–481
   shear force and, 52
   turbulent pipe flow and, 363–364
   units for, 52, 53
   velocity gradient of, 51–52
   velocity profile for, 51
Viscous flow, 10, 612–613, 634–635
Viscous-inviscid interactions, 712
Viscous length, 365
Viscous (deviatoric) stress tensor, 465–467
Viscous sublayer, 364–367, 626
Volume (V), 39, 46–47, 165–167, 186,
   188–189, 790, 805–806. *See also*
      Control volume (CV)
   capacity, 790, 805–806
   closed, 805–806
   conservation of mass and, 188–189
   expansion (β), coefficient of, 46–47
   flow rate, 188–189, 790, 805–806, 808–809
   Leibniz theorem for, 165–167
   linear momentum and, 186
   material, 166–167
   pumps, 790, 805–806, 808–809
   specific volume, 39
Volumetric (bulk) strain rate, 153
Volute, 838
von Kármán equation, 368–369
Vortex flood formation and damages, 771
Vortex flowmeters, 402
Vortex motion, forced, 107–108
Vortex shedding, 273–274, 617
Vortex strength (circulation) (Γ) of, 542–543,
   545–546
Vortices, 32, 144–145, 900–902
Vorticity (ζ), 32, 156–159

Wake, flow region of, 199, 273–274
Wake function, 576–577
Wall-wake law, 576–577
Water–air interface boundary conditions,
   477
Water hammer, 45
Water meters, 834
Water saturation properties, 942, 960
Watt (W), unit of, 18–19
Wave (shock) angle (δ), 684–685
Wave speed (c_0), 729–733
Weber number (We), 310, 313, 593
Wedges, oblique shock waves over,
   927–928
Weight (W), 17–18, 21, 39
Weirs, 761, 766–770. *See also* Flow control
Wicket gates, 837–838
Wind power density, 850–851
Wind tunnels, 313–316, 320–323, 823–824
   lift determination using, 313–316
   testing, 320–323
   vane-axial flow fan design for, 823–824
Wind turbines, 833, 847–855
   available wind power for, 850–851
   Betz limit, 853–854
   disk area for, 850–851
   energy pattern factor, 851
   horizontal axis, 847–850
   power coefficient, 851, 853–854
   rated speed, 850
   vertical axis, 847, 849
   wind power density, 850–851
Wings, 612–613, 617, 634–639
   endplates, 639–640
   external flow efficiency, 612–613
   finite-span, 638–639
   flaps, 636–637
   flow fields, 635
   ice and snow loads, 617
   lift force, 612–613, 634–639
   loading, 634
   shape, 612–613
   tip vortex, 638
   trailing vortices, 638–639
   winglets, 639–640
   wingspan (span), 634
Wingspan (span), 634
Work (W), 18–19, 43–44, 57–58, 194–195,
   215–219
   by expansion, 57–58
   energy transfer by, 215–129
   flow, 43–44, 194–195, 218–219
   mechanical energy and, 194–195
   power as, 18–19, 215–216
   pressure force, 216–219
   shaft, 216
   surface tension and, 57–58
   units of, 18–19

Yawing moment, 611
Yield stress, 466

Zero pressure gradient, 561, 564–565
Zero pressure point, 549–550

Credits

Full Table of Contents: *Instructor submitted material* 1

Introduction and Basic Concepts: *Chapter 1 from Fluid Mechanics: Fundamentals and Applications, Third Edition by Çengel, Cimbala, 2014* 25

Properties of Fluids: *Chapter 2 from Fluid Mechanics: Fundamentals and Applications, Third Edition by Çengel, Cimbala, 2014* 61

Pressure and Fluid Statics: *Chapter 3 from Fluid Mechanics: Fundamentals and Applications, Third Edition by Çengel, Cimbala, 2014* 99

Fluid Kinematics: *Chapter 4 from Fluid Mechanics: Fundamentals and Applications, Third Edition by Çengel, Cimbala, 2014* 157

Bernoulli and Energy Equations: *Chapter 5 from Fluid Mechanics: Fundamentals and Applications, Third Edition by Çengel, Cimbala, 2014* 209

Momentum Analysis of Flow Systems: *Chapter 6 from Fluid Mechanics: Fundamentals and Applications, Third Edition by Çengel, Cimbala, 2014* 267

Dimensional Analysis and Modeling: *Chapter 7 from Fluid Mechanics: Fundamentals and Applications, Third Edition by Çengel, Cimbala, 2014* 315

Internal Flow: *Chapter 8 from Fluid Mechanics: Fundamentals and Applications, Third Edition by Çengel, Cimbala, 2014* 371

Property Tables and Charts (SI Units): *Chapter Appendix 1 from Fluid Mechanics: Fundamentals and Applications, Third Edition by Çengel, Cimbala, 2014* 461

Property Tables and Charts (English Units): *Chapter Appendix 2 from Fluid Mechanics: Fundamentals and Applications, Third Edition by Çengel, Cimbala, 2014* 479

Glossary: *Chapter from Fluid Mechanics: Fundamentals and Applications, Third Edition by Çengel, Cimbala, 2014* 493

Index: *Chapter from Fluid Mechanics: Fundamentals and Applications, Third Edition by Çengel, Cimbala, 2014* 507